D0203934

Dedicated to our parents Bob and Sarah Kincaid,
Carleton and Elliott (in memoriam) Cheney.

SECOND EDITION

Numerical Analysis

Mathematics of Scientific Computing

David Kincaid
Ward Cheney

The University of Texas at Austin

Brooks/Cole Publishing Company

I(T)P® An International Thomson Publishing Company

Pacific Grove • Albany • Bonn • Boston • Cincinnati • Detroit • London • Madrid • Melbourne
Mexico City • New York • Paris • San Francisco • Singapore • Tokyo • Toronto • Washington

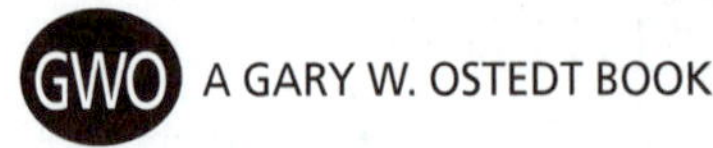

Sponsoring Editor: *Gary W. Ostedt*
Marketing Team: *Patrick Farrant, Romy Fineroff*
Marketing Representative: *Ragu Raghavan*
Editorial Assistant: *Carol Ann Benedict*
Production Coordinator: *Marlene Thom*
Production: *Integre Technical Pub. Co., Inc.*
Manuscript Editor: *Carol Reitz*
Cover Design: *Roy Neuhaus*
Interior Illustration: *Integre Technical Pub. Co., Inc.*
Typesetting: *Integre Technical Pub. Co., Inc.*
Cover Printing: *Color Dot Litho, Inc.*
Printing and Binding: *Quebecor/Fairfield*

For more information, contact:

BROOKS/COLE PUBLISHING COMPANY
511 Forest Lodge Road
Pacific Grove, CA 93950
USA

International Thomson Editores
Campos Eliseos 385, Piso 7
Col. Polanco
11560 México D.F., México

International Thomson Publishing Europe
Berkshire House 168–173
High Holborn
London WC1V 7AA
England

International Thomson Publishing GmbH
Königswinterer Strasse 418
53227 Bonn
Germany

Thomas Nelson Australia
102 Dodds Street
South Melbourne, 3205
Victoria, Australia

International Thomson Publishing Asia
221 Henderson Road
#05–10 Henderson Building
Singapore 0315

Nelson Canada
1120 Birchmount Road
Scarborough, Ontario
Canada M1K 5G4

International Thomson Publishing Japan
Hirakawacho Kyowa Building, 3F
2-2-1 Hirakawacho
Chiyoda-ku, Tokyo 102
Japan

Printed in the United States of America

10 9 8 7 6 5 4 3 2 1

Library of Congress Cataloging in Publication Data
Kincaid, David (David Ronald)
Numerical analysis : mathematics of scientific computing / David Kincaid, Ward Cheney. — 2nd ed.
p. cm.
Includes bibliographical references and index.
ISBN 0-534-33892-5
1. Numerical analysis. I. Cheney, E. W. (Elliott Ward), [date]
. II. Title.
QA297.K563 1996
519.4—dc20 95-53860
CIP

The picture on the cover shows fractals corresponding to the basins of attraction for the roots of $z^4 + 4 = 0$. Additional information about basins of attraction is found at the end of Section 3.5.

Contents

PREFACE

This book has evolved over many years from lecture notes that accompany certain upper-division and graduate courses in mathematics and computer sciences at our university. These courses introduce students to the algorithms and methods that are commonly needed in scientific computing. The mathematical underpinnings of these methods are emphasized as much as their algorithmic aspects. The students have been diverse: mathematics, engineering, science, and computer science undergraduates, as well as graduate students from various disciplines. Portions of the book also have been used to lay the groundwork in several graduate courses devoted to special topics in numerical analysis, such as the numerical solution of differential equations, numerical linear algebra, and approximation theory. Our approach has always been to treat the subject from a mathematical point of view, with attention given to its rich offering of theorems, proofs, and interesting ideas. From these arise many computational procedures and intriguing questions of computer science. Of course, our motivation comes from the practical world of scientific computing, which dictates the choice of topics and the manner of treating each. For example, with some topics it is more instructive to discuss the theoretical foundations of the subject and not attempt to analyze algorithms in detail. In other cases, the reverse is true, and the students learn much from programming simple algorithms themselves and experimenting with them—although we offer a blanket admonition to use well-tested software, such as from program libraries, on problems that arise from applications.

There is some overlap between this book and our more elementary text, *Numerical Mathematics and Computing, Third Edition* (Brooks/Cole). That book is addressed to students having more modest mathematical preparation (and sometimes less enthusiasm for the theoretical side of the subject). In that text, there is a different menu of topics, and no topic is pursued to any great depth. The present book, on the other hand, is intended for a course that offers a more scholarly treatment of the subject; many topics are dealt with at length. Occasionally we broach topics that heretofore have not found their way into standard textbooks at this level. In this category are the multigrid method, procedures for multivariate interpolation, homotopy (or continuation) methods, and delay differential equations.

The algorithms in the book are presented in a *pseudocode* that contains additional details beyond the mathematical formulas. The reader can easily write computer routines based on the pseudocode in any standard computer language. We believe that students learn and understand numerical methods best by seeing how algorithms are developed from the mathematical theory and then writing and testing computer implementations of them. Of course, such computer programs will not contain all the complicated procedures and sophisticated checks found in robust routines available in scientific libraries. Examples of general-purpose mathematical libraries are found in the appendix on mathematical software. For most applications, such libraries are strongly preferred to code written yourself.

An important constituent of the book (and essential to its pedagogic purpose) is the abundance of problems for the student. These are of two types: analytic problems and computer problems. The computer problems are, in turn, of two types: those in which students write their own code, and those in which they employ existing software. We believe that both kinds of programming practice are necessary. Using someone else's software is not always a trivial exercise, even when it is as well documented as the large libraries mentioned above. On the other hand, students usually acquire much more insight into an algorithm after coding and testing it themselves, rather than simply using a software package. In most cases the computer problems require access to a computer that has at least a 32-bit word length.

Software, errata, and teaching aids are available via the Internet as discussed in the appendix on mathematical software. Also, the publisher has made available a Solution Manual for instructors who adopt the book for their classes.

The second edition differs from the first as follows: **(i)** The entire book has been reformated for improved appearance. **(ii)** Corrections and improvements have been made throughout. **(iii)** Many sections have been rewritten and enlarged. **(iv)** Problems have been separated into Problems and Computer Problems, and new problems have been added of both types. **(v)** The bibliography has been updated. **(vi)** An appendix with pointers to mathematical software has been added.

A standard course of one semester can be based on selected sections from Chapters 1–4 and 6–8. A two-semester course could cover selected sections in Chapters 1–9 plus other topics of interest. Chapters 4 and 5 could be taught independently from the previous chapters as a short course on numerical linear algebra. Because of the ambitious scope of this book, some sections make greater demands on the preparation of the reader. These sections usually occur late in any given chapter so that the reader will not be unduly challenged at the start. Such sections are marked with an asterisk and they may be skipped at the reader's discretion.

Acknowledgments

We are glad to express our indebtedness to many persons who have assisted us in the writing of this book.

First Edition

Administrative support was provided by Sheri Brice, Margaret Combs, Jan Duffy, Katherine Mueller, and Jenny Tsao of The University of Texas at Austin. Foremost among these is Margaret Combs of the Mathematics Department, who rendered into TeX innumerable versions of each section, patiently preparing new ones as they were needed for classroom notes in successive years. No technical problem of typesetting was too difficult for her as she mastered the arcane art of dissecting and reassembling TeX macros to serve unusual needs. It is appropriate at this point that we also express our public thanks to Donald Knuth for his magnificent contribution to the scientific community embodied in the TeX typesetting system. We appreciate the suggestions made by the following astute reviewers of preliminary versions of the manuscript: Thomas A. Atchison, Frederick J. Carter, Philip Crooke, Jim D'Archangelo, R. S. Falk, J. R. Hubbard, Patrick Lang, Giles Wilson Maloof, A. K. Rigler, F. Schumann, A. J. Worsey, and Charles Votaw. In addition, thanks are due the following persons for technical help and critical reading of the manuscript: Victoria Hunter, Carole Kincaid, Tad Liszka, Rio Hirowati Shariffudin, and Laurette Tuckerman. David Young was always generous with suggestions and advice. A number of advanced students who served as teaching assistants in our classes also helped; in particular, we thank David Bruce, Nai-ching Chen, Ashok Hattangady, Ru Huang, Wayne Joubert, Irina Mukherjee, Bill Nanry, Tom Oppe, Marcos Raydan, Malathi Ramdas, John Respess, Phien Tran, Linda Tweedy, and Baba Vemuri. The editors and technical staff at Brooks/Cole Publishing Company have been most cooperative and supportive during this project. In particular, we are pleased to thank Jeremy Hayhurst and Marlene Thom for their assistance. Stacey Sawyer of Sawyer and Williams was responsible for a careful copyediting of the manuscript, and Ralph Youngen of the American Mathematical Society provided technical assistance and supervision in turning our TeX files into final printed copy.

Second Edition

We would like to acknowledge the reviewers for their work: Dan Boley, University of Minnesota; Min Chen, Pennsylvania State University; John Harper, University of Rochester; Ramon Moore, Ohio State University; Yves Nievergelt, Eastern Washington University; and Elinor Velasquez, University of California-Berkeley. We especially thank Ron Boisvert for clarifying our understanding of the different categories of mathematical software and for the examples given in the appendix. Also, we want to thank those who took the trouble to contact us with suggestions and corrections in the first edition. Some of these are Victor M. Afram, Roger Alexander, A. Awwal, Carl de Boor, T. P. Brown, James Caveny, George J. Davis, Hakan

Ekblom, Mariano Gasca, Bill Gearhart, Patrick Goetz, Gary L. Gray, Bob Gregorac, Katherine Hua Guo, Cecilia Jea, Liz Jessup, Grant Keady, Baker Kearfott, Junjiang Lei, Teck C. Lim, Julio Lopez, C. Lu, Taketomo Mitsui, Irina Mukherjee, Teresa Perez, Robert Piche, Sherman Riemenschneider, Maria Teresa Rodriquez, Ulf Roennow, Larry Schumaker, Wei-Chang Shann, Christopher J. van Wyk, Kang Zhao, and Mark Zhou.

We are grateful to the Center for Numerical Analysis, the Mathematics Department, the Computer Sciences Department, and the Computation Center of The University of Texas at Austin for technical support and for furnishing excellent computing facilities.

Finally, the editors and technical staff at Brooks/Cole Publishing Company have been most helpful and supportive during the revision of this book.

We would appreciate any comments, suggestions, questions, criticisms, or corrections that readers may take the trouble of communicating to us.

David Kincaid
Ward Cheney

NUMERICAL ANALYSIS: WHAT IS IT?

Numerical analysis involves the *study*, *development*, and *analysis* of algorithms for obtaining numerical solutions to various mathematical problems. Frequently, numerical analysis is called the *mathematics of scientific computing*.

The algorithms that we study invariably are destined for use on high-speed computers, and therefore another crucial step intervenes before the solution to a problem can be obtained: a computer *program* or *code* must be written to communicate the algorithm to the computer. This is, of course, a nontrivial matter, but there are so many choices of computers and computer languages that it is a topic best left out of the science of numerical analysis per se.

There are certainly many other purposes to which computers can be put besides the numerical solution of mathematical problems: providing basic communications, keeping large data bases, playing games, "net surfing," writing novels, accounting, and so on. Solving *mathematical* problems numerically on the computer is *scientific computing*. The development of the associated algorithms (procedures) and the study of their behavior are the mathematics of scientific computing.

Often the development of an algorithm is stimulated by a *constructive* proof in mathematics. In classical analysis, nonconstructive methods are frequently used, but generally they do not lead to algorithms. For example, existence and uniqueness theorems might be established by assuming that they are *not* true and then following the trail of a logical argument until arriving at a contradiction. Not every constructive proof will lead to a successful algorithm, however. A difficulty that may arise is that an *analytical* solution to a given problem may be several steps away from a *numerical* solution. Or it might be completely impractical because of slow convergence or the need for lengthy computation.

As an example of the gap between an existence theorem and a numerical solution of a problem, consider the ubiquitous matrix equation $Ax = b$. We know that it has a unique solution whenever A is nonsingular. But this fact may be of little solace when we are faced with a large linear system containing empirical data and we wish to compute an *approximate* numerical solution.

In general, in this book, we will begin each topic with a basic mathematical problem that arises frequently in practical applications. Then a certain amount of analysis will be presented in order to arrive at an algorithm for solving the problem. Algorithms are usually given in the form of a pseudocode. Finally, additional analysis of the algorithm may be given to help in understanding its behavior, such as its convergence or its resistance to corruption by roundoff error. Such analysis may take the form of either *forward* or *backward* error analysis.

Behind each basic mathematical problem to be considered there are always physical applications. Let us illustrate all this with a heat–flow problem. The temperature in a solid piece of metal with various boundary conditions is governed by mathematical equations that must be satisfied at every point and at every instant of time. The principal equation here might be the *heat equation*

$$\frac{\partial^2 u}{\partial x^2} = \frac{\partial u}{\partial t}$$

It is a parabolic, linear, second–order, partial differential equation. It models the heat flow inside a rod under certain assumptions on the actual physical problem. The variable x is the space coordinate, and t is the time. The temperature is $u = u(x, t)$. To solve the model problem on the computer, the space-time region is *discretized* by a mesh of grid points, and the numerical solution is sought at each of these points. The partial derivatives in the heat equation can be approximated by finite differences such as

$$\begin{aligned}
\frac{\partial v(x,t)}{\partial t} &\approx \frac{1}{k}[v(x, t+k) - v(x,t)] \\
\frac{\partial^2 v(x,t)}{\partial x^2} &\approx \frac{1}{h^2}[v(x+h, t) - 2v(x,t) + v(x-h, t)]
\end{aligned}$$

Here, k and h are the mesh spacings in the t-direction and the x-direction, respectively. Also, we have changed to the variable v to emphasize that we are solving an approximation to the model problem rather than the original problem. Replacing the partial derivatives by these approximations and simplifying, we arrive at a linear equation at each grid point (x_i, t_j). Using the abbreviation v_{ij} for $v(x_i, t_j)$, we obtain

$$v_{i,j+1} = s v_{i-1,j} + (1 - 2s)v_{ij} + s v_{i+1,j}$$

where $s = k/h^2$. The numerical solution can be advanced step by step in the t-direction using the preceding equation. This procedure is called an *explicit* method because the new values $v_{i,j+1}$ are explicitly determined one at a time from the previous values $v_{i-1,j}, v_{i,j}, v_{i+1,j}$. This is all very elegant, and one would not anticipate any difficulties. But the *analysis* as well as *numerical experience* indicate that the method is seriously flawed! We turn then to an *implicit* method. In it, all of the new values are determined at the same time by solving a linear system of the special form

$$V_{j+1} = AV_j$$

Here A is a certain tridiagonal matrix and $V_j = [v_{1j}, v_{2j}, \ldots, v_{nj}]^T$. Each of these methods requires a stability analysis to determine the permissible range of values for the mesh sizes h and k and the associated convergence behavior. It is here that the explicit method competes poorly. Complete details can be found in Chapter 9.

CHAPTER ONE

Mathematical Preliminaries

Introduction

This chapter starts with a review of some important topics in calculus that will be required in the subsequent chapters. We encourage readers to skip boldly over matters that are already familiar to them. In fact, some may wish to begin with Chapter 2.

1.1 Basic Concepts and Taylor's Theorem

We begin with a review of some basic concepts from calculus. At this point, one might ask: *Why do we need to discuss such topics if we are primarily interested in numerical algorithms*? A solid background in basic mathematical concepts is essential to understanding the derivation of most numerical algorithms. Taylor's Theorem in various forms is fundamental to many numerical procedures and is an excellent starting point for the study of scientific computing since no advanced mathematical concepts are required.

Limit, Continuity, and Derivative

If f is a real-valued function of a real variable, then the **limit** of the function f at c (if it exists) is defined as follows: The equation

$$\lim_{x \to c} f(x) = L$$

means that to each positive ε there corresponds a positive δ such that the distance between $f(x)$ and L is less than ε whenever the positive distance between x and c is less than δ; that is, $|f(x) - L| < \varepsilon$ whenever $0 < |x - c| < \delta$. If there is no number L with this property, the limit of f at c does not exist.

For example, the equation

$$\lim_{x \to 2} x^2 = 4$$

is true because $|x^2 - 4| < \varepsilon$ whenever $0 < |x - 2| < \varepsilon(5 + \varepsilon)^{-1}$. (See Problem 1.)

As another example, note that the equation

$$\lim_{x \to 0} \frac{|x|}{x} = L$$

is not true for *any* number L. To see this, let $\varepsilon = 1$, and suppose that $\big||x|/x - L\big| < 1$ whenever $0 < |x| < \delta$. We have $|x|/x = 1$ when $x = \frac{1}{2}\delta$ and $|x|/x = -1$ when $x = -\frac{1}{2}\delta$. There is no number L satisfying $|1 - L| < 1$ and $|-1 - L| < 1$, and we say that the limit does not exist. See Figure 1.1.

If f is defined only on a specified subset X of the real line, the definition of limit is modified so that $|f(x) - L| < \varepsilon$ whenever $x \in X$ and $0 < |x - c| < \delta$.

The function f is said to be **continuous** at c if

$$\lim_{x \to c} f(x) = f(c)$$

Thus, the function $f(x) = x^2$ is continuous at the point 2, whereas the function $|x|/x$ is not continuous at 0, no matter how it is defined at 0. These assertions follow from remarks made previously.

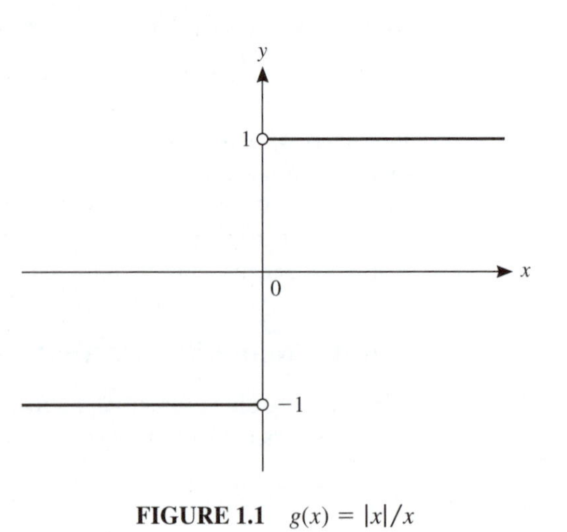

FIGURE 1.1 $g(x) = |x|/x$

The **derivative** of f at c (if it exists) is defined by the equation

$$f'(c) = \lim_{x \to c} \frac{f(x) - f(c)}{x - c}$$

Since this limit need not exist for a particular function and a particular c, it is possible for the derivative not to exist for such a function. If f is a function for which $f'(c)$ exists, we say that f is **differentiable** at c. If f is differentiable at c, then f must be continuous at c. (See Problem 18.) But the converse is not true!

For example, if

$$f(x) = |x|$$

then $f'(0)$ does not exist. See Figure 1.2.

The set of all functions that are continuous on the entire real line $\mathbb{R}$ is denoted by $C(\mathbb{R})$. The set of functions for which f' is continuous everywhere is denoted by $C^1(\mathbb{R})$. If $f \in C^1(\mathbb{R})$, then f' is continuous at all points in $\mathbb{R}$ and, thereby, differentiable throughout the real line. Since the differentiability of a function at a point implies its continuity at that point, we have $C^1(\mathbb{R}) \subset C(\mathbb{R})$. The set $C^1(\mathbb{R})$ is a *proper* subset of $C(\mathbb{R})$ because there are (many) continuous functions whose derivatives do not exist. The function $f(x) = |x|$ is such an example.

We denote by $C^2(\mathbb{R})$ the set of all functions for which f'' is continuous everywhere. By reasoning similar to that given above, $C^2(\mathbb{R}) \subset C^1(\mathbb{R}) \subset C(\mathbb{R})$. Again, these are proper inclusions since there are functions that are once differentiable but not twice, such as $f(x) = x^2 \sin(1/x)$.

Similarly, we define $C^n(\mathbb{R})$, for each natural number n, to be the set of all functions for which $f^{(n)}(x)$ is continuous. Finally, $C^\infty(\mathbb{R})$ is the set of functions each of whose derivatives is continuous. We have now

$$C^\infty(\mathbb{R}) \subset \cdots \subset C^2(\mathbb{R}) \subset C^1(\mathbb{R}) \subset C(\mathbb{R})$$

A familiar function in $C^\infty(\mathbb{R})$ is $f(x) = e^x$.

In the same way, we define $C^n[a, b]$ to be the set of functions f for which $f^{(n)}$ exists and is continuous on the interval $[a, b]$.

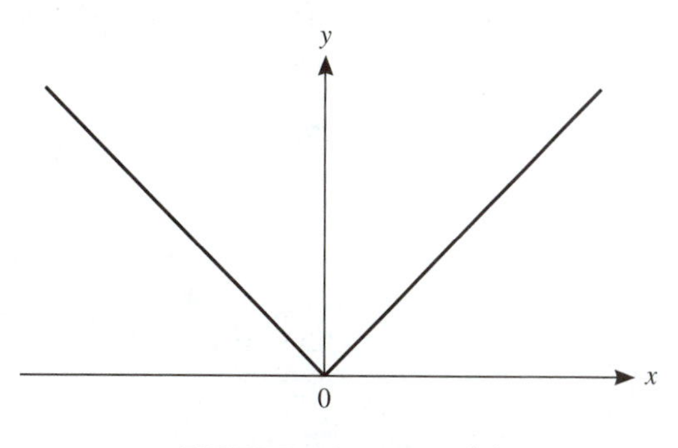

FIGURE 1.2 $f(x) = |x|$

Taylor's Theorem

An important theorem concerning functions in $C^n[a, b]$ is Taylor's Theorem, which arises throughout the study of numerical analysis.

THEOREM 1 **Taylor's Theorem with Lagrange Remainder** *If $f \in C^n[a, b]$ and if $f^{(n+1)}$ exists on (a, b), then for any points c and x in $[a, b]$,*

$$f(x) = \sum_{k=0}^{n} \frac{1}{k!} f^{(k)}(c)(x - c)^k + E_n(x) \tag{1}$$

where, for some point ξ between c and x,

$$E_n(x) = \frac{1}{(n + 1)!} f^{(n+1)}(\xi)(x - c)^{n+1}$$

A special case is when $c = 0$. With an analysis of $E_n(x)$ as $n \to \infty$, Equation (1) becomes the **Maclaurin series** for $f(x)$. (See Section 6.7.)

Example 1 Illustrate Taylor's Theorem with the function

$$f(x) = \ln x$$

taking $a = 1, b = 2$, and $c = 1$.

Solution The formula requires various derivatives of f, which are $f'(x) = x^{-1}$, $f''(x) = -x^{-2}$, $f'''(x) = 2x^{-3}$, $f^{(4)}(x) = -6x^{-4}$, and so on. Next, we obtain the general term*

$$f^{(k)}(x) = (-1)^{k-1}(k - 1)!x^{-k} \qquad (k \geqq 1)$$

and thereby at $x = 1$, we have

$$f^{(k)}(1) = (-1)^{k-1}(k - 1)! \qquad (k \geqq 1)$$

Of course, $f^{(0)}(1) = \ln 1 = 0$. Putting all of this into Taylor's formula (1) gives us

$$\ln x = \sum_{k=1}^{n} (-1)^{k-1} \frac{1}{k}(x - 1)^k + E_n(x) \qquad (1 \leqq x \leqq 2)$$

where

$$E_n(x) = (-1)^n \frac{1}{n + 1} \xi^{-(n+1)}(x - 1)^{n+1} \qquad (1 < \xi < x)$$

■

*Throughout when we use inequalities involving integer values such as $1 \leqq i \leqq m$, we mean $i = 1, 2, \ldots, m$, and similarly, $n \geqq N$ means $n = N, N + 1, \ldots$.

In the equation given for $\ln x$, the summation $\sum$ on the right of the equal sign produces a *polynomial* function in x. It can be regarded as a simple function that approximates the more complicated function $\ln x$. The last term on the right, $E_n(x)$, can be regarded as an error term. It tells us how the polynomial approximation differs from $\ln x$. Note that this term is *not* a polynomial because ξ depends on x in a nonpolynomial way.

We can use Taylor's formula for computing approximations of functions at particular points. For example, writing out the formula for $\ln x$, we have

$$\ln x = (x-1) - \frac{1}{2}(x-1)^2 + \frac{1}{3}(x-1)^3 - \cdots + (-1)^{n-1}\frac{1}{n}(x-1)^n + E_n(x)$$

where

$$|E_n(x)| = \frac{1}{n+1}\xi^{-(n+1)}(x-1)^{n+1} < \frac{1}{n+1}(x-1)^{n+1}$$

In this estimate, we simply note that $1 < \xi$ and $\xi^{-(n+1)} < 1$.

Now suppose that the series is to be used to compute $\ln 2$ with accuracy of one part in 10^8. We have then

$$\begin{aligned} \ln 2 &= 1 - \frac{1}{2} + \frac{1}{3} - \frac{1}{4} + \cdots + (-1)^{n-1}\frac{1}{n} + E_n(2) \\ &= \sum_{k=1}^{n} (-1)^{k-1}\frac{1}{k} + E_n(2) \end{aligned}$$

with $|E_n(2)| < 1/(n+1)$. The term $E_n(2)$ is the numerical error. So to compute $\ln 2$ with the desired accuracy would require n to be chosen so that $E_n(2) \leqq 10^{-8}$. This means $1/(n+1) \leqq 10^{-8}$ or $n+1 \geqq 10^8$. Thus, at least *100 million* terms in the polynomial would be required to compute $\ln 2$ to the desired accuracy. A similar calculation shows that $\ln 1.5$ can be computed to the same accuracy with only 22 terms! (See Problem 19.)

The special case $n = 0$ of Taylor's Theorem is often used in mathematical arguments. It is known as the **Mean-Value Theorem**.

THEOREM 2 **Mean-Value Theorem** *If f is in $C[a, b]$ and if f' exists on (a, b), then for x and c in $[a, b]$,*

$$f(x) = f(c) + f'(\xi)(x-c)$$

where ξ is between c and x.

Taking $x = b$ and $c = a$ and rearranging, we have the important equation

$$f(b) - f(a) = f'(\xi)(b-a) \quad \text{where} \quad a < \xi < b$$

A special case of the Mean-Value Theorem is **Rolle's Theorem**.

THEOREM 3 **Rolle's Theorem** *If f is continuous on $[a, b]$, if f' exists on (a, b), and if $f(a) = f(b) = 0$, then $f'(\xi) = 0$ for some ξ in (a, b).*

This is an immediate consequence of the preceding equation. (Actually, in a formal development, Rolle's Theorem is proved first, and from it, Taylor's Theorem is derived.) In both Rolle's Theorem and the Mean-Value Theorem, there may be more than one point ξ in the interval $[a, b]$ that satisfies the given equation.

In Section 7.6, we shall need another form of Taylor's Theorem as given in the following theorem. It has several nice features—the proof is straightforward and Theorem 1 can be obtained directly from Theorem 4.

THEOREM 4 **Taylor's Theorem with Integral Remainder** *If $f \in C^{n+1}[a, b]$, then for any points x and c in $[a, b]$,*

$$f(x) = \sum_{k=0}^{n} \frac{1}{k!} f^{(k)}(c)(x-c)^k + R_n(x) \tag{2}$$

where

$$R_n(x) = \frac{1}{n!} \int_c^x f^{(n+1)}(t)(x-t)^n \, dt$$

Proof Recall the formula for integration by parts

$$\int u \, dv = uv - \int v \, du$$

and apply it to the integral R_n with

$$u = \frac{(x-t)^n}{n!} \qquad dv = f^{(n+1)}(t) \, dt$$

The result is

$$\begin{aligned} R_n &= \frac{1}{n!} \left[f^{(n)}(t)(x-t)^n \Big|_{t=c}^{t=x} + n \int_c^x f^{(n)}(t)(x-t)^{n-1} \, dt \right] \\ &= -\frac{1}{n!} f^{(n)}(c)(x-c)^n + R_{n-1} \end{aligned}$$

If this process of integrating by parts is repeated, we eventually obtain

$$R_n = -\sum_{k=1}^{n} \frac{1}{k!} f^{(k)}(c)(x-c)^k + R_0$$

Since

$$R_0 = \int_c^x f'(t) \, dt = f(x) - f(c)$$

we have

$$f(x) = f(c) + \sum_{k=1}^{n} \frac{1}{k!} f^{(k)}(c)(x-c)^k + R_n$$

and this completes the proof. ■

Other Forms of Taylor's Formula

Yet another form of the series and the error term in Taylor's formula can be obtained by replacing x by $x + h$ and c by x in Taylor's Theorem with Lagrange Remainder. The result follows.

THEOREM 5 **Alternative Form of Taylor's Theorem** *If $f \in C^{n+1}[a, b]$, then for any points x and $x + h$ in $[a, b]$,*

$$f(x+h) = \sum_{k=0}^{n} \frac{1}{k!} f^{(k)}(x)h^k + E_n(h) \tag{3}$$

where

$$E_n(h) = \frac{1}{(n+1)!} f^{(n+1)}(\xi)h^{n+1}$$

in which the point ξ lies between x and $x + h$.

This is an important form for many applications.

Example 2 Determine Taylor's formula for C^{x+h} and approximate $10^{1.0001}$.

Solution Letting $f(x) = C^x$, we find that $f^{(n)}(x) = C^x(\ln C)^n$. Using Equation (3), we have

$$C^{x+h} = C^x \left(1 + \sum_{k=1}^{n} \frac{h^k}{k!}(\ln C)^k\right) + E_n(h)$$

Next let $C = 10$, $x = 1$, and $h = 10^{-4}$. Thus, we obtain

$$\begin{aligned} 10^{1.0001} &= 10(1 + 10^{-4}(\ln 10) + \frac{1}{2}10^{-8}(\ln 10)^2 + \cdots) \\ &\approx 10(1 + 2.30259 \times 10^{-4} + 2.65095 \times 10^{-8}) \\ &\approx 10.00230\,00265\,095 \end{aligned}$$

■

For vector-valued functions of vectors, there also exist Taylor series and Taylor formulas. Thus, if f is a mapping of $\mathbb{R}^n$ to $\mathbb{R}^m$, formulas are available for expressing $f(x + h)$ in terms of $f(x)$, $f'(x)$, $f''(x)$, and so on. Of course, the main difficulty lies in defining the appropriate derivatives. These matters are taken up in many textbooks, such as Bartle [1976], K.T. Smith [1971], and Dieudonné [1960]. Here we shall mention some special cases needed in later chapters.

For a function $f : \mathbb{R}^2 \to \mathbb{R}$, the simplest expression of Taylor's formula is a symbolic one:

THEOREM 6 **Taylor's Theorem in Two Variables** *If $f \in C^{n+1}([a, b] \times [c, d])$, then for any points $x + h$ and $y + k$ in $[a, b] \times [c, d] \subseteqq \mathbb{R}^2$,*

$$f(x+h, y+k) = \sum_{i=0}^{n} \frac{1}{i!}\left(h\frac{\partial}{\partial x} + k\frac{\partial}{\partial y}\right)^i f(x, y) + E_n(h, k) \tag{4}$$

where

$$E_n(h, k) = \frac{1}{(n+1)!}\left(h\frac{\partial}{\partial x} + k\frac{\partial}{\partial y}\right)^{n+1} f(x + \theta h, y + \theta k)$$

in which θ lies between 0 and 1.

The meaning of the mysterious terms in this theorem is as follows:

$$\left(h\frac{\partial}{\partial x} + k\frac{\partial}{\partial y}\right)^0 f(x, y) = f(x, y)$$

$$\left(h\frac{\partial}{\partial x} + k\frac{\partial}{\partial y}\right)^1 f(x, y) = \left(h\frac{\partial f}{\partial x} + k\frac{\partial f}{\partial y}\right)(x, y)$$

$$\left(h\frac{\partial}{\partial x} + k\frac{\partial}{\partial y}\right)^2 f(x, y) = \left(h^2\frac{\partial^2 f}{\partial x^2} + 2hk\frac{\partial^2 f}{\partial x\,\partial y} + k^2\frac{\partial^2 f}{\partial y^2}\right)(x, y)$$

and so on.

Example 3 What are the first few terms in the Taylor formula for $f(x, y) = \cos(xy)$?

Solution For the given function, we find that

$$\frac{\partial f}{\partial x} = -y\sin(xy) \qquad \frac{\partial f}{\partial y} = -x\sin(xy)$$

$$\frac{\partial^2 f}{\partial x^2} = -y^2\cos(xy) \qquad \frac{\partial^2 f}{\partial x\,\partial y} = -xy\cos(xy) - \sin(xy) \qquad \frac{\partial^2 f}{\partial y^2} = -x^2\cos(xy)$$

Thus, if we let $n = 1$ in Taylor's formula (4), the result is

$$\cos[(x+h)(y+k)] = \cos(xy) - hy\sin(xy) - kx\sin(xy) + E_1(h, k)$$

The remainder E_1 is the sum of three terms—namely,

$$-\frac{1}{2}h^2(y+\theta k)^2\cos[(x+\theta h)(y+\theta k)]$$
$$-hk\{(x+\theta h)(y+\theta k)\cos[(x+\theta h)(y+\theta k)]+\sin[(x+\theta h)(y+\theta k)]\}$$
$$-\frac{1}{2}k^2(x+\theta h)^2\cos[(x+\theta h)(y+\theta k)]$$

■

PROBLEMS 1.1

1. Show that $|x^2-4|<\varepsilon$ when $0<|x-2|<\varepsilon(5+\varepsilon)^{-1}$ and prove $\lim_{x\to 2} x^2 = 4$.
2. Show that $\lim_{x\to 1}(4x+2)=6$ by means of an ε-δ proof.
3. Show that $\lim_{x\to 2}(1/x)=\frac{1}{2}$ by means of an ε-δ proof.
4. For the function $f(x)=3-2x+x^2$ and the interval $[a,b]=[1,3]$, find the number ξ that occurs in the Mean-Value Theorem.
5. Find the Taylor series for $f(x)=\cosh x$ about the point $c=0$.
6. If the series for $\ln x$ is truncated after the term involving $(x-1)^{1000}$ and is then used to compute $\ln 2$, what bound on the error can be given?
7. Find the Taylor series for $f(x)=e^x$ about the point $c=3$. Then simplify the series and show how it could have been obtained directly from the series for f about $c=0$.
8. Show that the function $f(x)=x\sin(1/x)$, with $f(0)=0$, is continuous at 0 but not differentiable at 0.
9. Determine whether the following function is continuous, once differentiable, or twice differentiable:

$$f(x)=\begin{cases} x^3+x-1 & x \leqq 0 \\ x^3-x-1 & x>0 \end{cases}$$

10. Repeat Problem 9 for the function

$$f(x)=\begin{cases} x & x \leqq 1 \\ x^2 & x>1 \end{cases}$$

11. Repeat Problem 4 with the function $f(x)=x^6+x^4-1$ and the interval $[0,1]$.
12. Let $f(x)=x^{-3}(x-\sin x)$ for $x\neq 0$. How should $f(0)$ be defined in order that f be continuous? Will it also be differentiable?
13. Let k be a positive integer and let $0<\alpha<1$. To what classes $C^n(\mathbb{R})$ does the function $x^{k+\alpha}$ belong?
14. Prove: If $f\in C^n(\mathbb{R})$, then $f'\in C^{n-1}(\mathbb{R})$ and $\int_a^x f(t)\,dt$ belongs to $C^{n+1}(\mathbb{R})$.
15. Prove Rolle's Theorem directly (not as a special case of the Mean-Value Theorem).

16. Prove: If $f \in C^n(\mathbb{R})$ and $f(x_0) = f(x_1) = \cdots = f(x_n) = 0$ for $x_0 < x_1 < \cdots < x_n$, then $f^{(n)}(\xi) = 0$ for some $\xi \in (x_0, x_n)$. *Hint*: Use Rolle's Theorem n times.

17. Prove that the function $f(x) = x^2$ is continuous everywhere.

18. Show that a function is necessarily continuous at any point where it is differentiable.

19. **(a)** Derive the Taylor series at 0 for the function $f(x) = \ln(x + 1)$. Write this series in summation notation. Give two expressions for the remainder when the series is truncated.

(b) Determine the smallest number of terms that must be taken in the series to yield $\ln 1.5$ with an error less than 10^{-8}.

(c) Determine the number of terms necessary to compute $\ln 1.6$ with error 10^{-10} at most.

20. For small values of x, the approximation $\sin x \approx x$ is often used. Estimate the error in using this formula with the aid of Taylor's Theorem. For what range of values of x will this approximation give results correct to six decimal places?

21. For small values of x, how good is the approximation $\cos x \approx 1 - \frac{1}{2}x^2$? For what range of values will this approximation give correct results rounded to three decimal places?

22. Criticize this reasoning: The function f defined by

$$f(x) = \begin{cases} x^3 + x & x \leqq 0 \\ x^3 - x & x \geqq 0 \end{cases}$$

has the properties

$$\lim_{x \to 0^+} f''(x) = \lim_{x \to 0^+} 6x = 0$$

$$\lim_{x \to 0^-} f''(x) = \lim_{x \to 0^-} 6x = 0$$

Therefore, f'' is continuous.

23. Use Taylor's Theorem with $n = 2$ to prove that the inequality $1 + x < e^x$ is valid for all real numbers except $x = 0$.

24. Derive the Taylor series with remainder term for $\ln(1 + x)$ about 1. Derive an inequality that gives the number of terms that must be taken to yield $\ln 4$ with error less than 2^{-m}.

25. What is the third term in the Taylor expansion of $x^2 + x - 2$ about the point 3?

26. Using the series for e^x, how many terms are needed to compute e^2 correctly to four decimal places (rounded)?

27. Prove that if f is differentiable at x, then

$$\lim_{h \to 0} \frac{f(x + h) - f(x - h)}{2h} = f'(x)$$

Show that for some functions that are not differentiable at x, the preceding limit exists. (See Eggermont [1988] or Problem 36.)

28. Develop the Taylor series for $f(x) = \ln x$ about e, writing the results in summation notation and giving the remainder term. Suppose $|x - e| < 1$ and accuracy $\frac{1}{2} \times 10^{-1}$ is desired. What is the minimum number of terms in the series required to achieve this accuracy?

29. Determine the first two terms of the Taylor series for x^x about 1 and the remainder term E_1.

30. Determine the Taylor polynomial of degree 2 for

$$f(x) = e^{(\cos x)}$$

expanded about the point π.

31. First develop the function $\sqrt{x}$ in a series of powers of $(x - 1)$ and then use it to approximate $\sqrt{0.99999\,99995}$ to ten decimal places.

32. Assume that $|x| < \frac{1}{2}$ and determine by Taylor's Theorem the best upper bound.

(a) $|\cos x - (1 - x^2/2)|$

(b) $|\sin x - x(1 - x^2/6)|$

33. Determine a function that can be termed the **linearization** of $x^3 - 2x$ at 2.

34. How many terms are required in the series

$$e = \sum_{k=0}^{\infty} \frac{1}{k!}$$

to give e with an error of at most $6/10$ unit in the 20th decimal place?

35. Find the first two terms in the Taylor expansion of $x^{1/5}$ about the point $x = 32$. Approximate the fifth root of 31.999999 using these two terms in the series. How accurate is your answer?

36. Prove or disprove this assertion: If f is differentiable at x, then for $\alpha \neq 1$,

$$f'(x) = \lim_{h \to 0} \frac{f(x + h) - f(x + \alpha h)}{h - \alpha h}$$

37. Find the Taylor polynomial of degree 2 for the function $f(x) = e^{2x} \sin x$ expanded about the point $\pi/2$.

38. Determine the Lagrange form of the remainder when Taylor's Theorem is applied to the function $f(x) = \cos x$, with $n = 2$ and $c = \pi/2$. How small must we make $|x - \pi/2|$ if this remainder term is not to exceed $\frac{1}{2} \times 10^{-4}$ in absolute value?

39. An error term of the form $(-1)^n(n + 1)^{-1}\xi^{-n-1}(x - 1)^{n+1}$ was obtained in the example illustrating Taylor's formula. Compare this to the error term that arises from the integral form of the remainder.

40. Use Taylor's Theorem with Integral Remainder and the Mean-Value Theorem for Integrals (in Section 1.2) to deduce Taylor's Theorem with Lagrange Remainder.

1.2 Orders of Convergence and Additional Basic Concepts

In numerical calculations, especially on high-performance computers, it often happens that the answer to a problem is not produced all at once. Rather, a sequence of approximate answers is produced, usually exhibiting progressively higher accuracy. Convergence of sequences is an important subject that will be taken up again later, such as in Chapter 3. Here we present just a few introductory concepts.

Convergent Sequences

Let us consider an idealized situation in which a single real number is sought as the answer to a problem. It might be, for example, a zero of a complicated equation or the numerical value of an intractable definite integral. In such a case, a computer program may produce a sequence of real numbers $x_1, x_2, x_3, \ldots$ that are *approaching* the correct answer.

We write

$$\lim_{n\to\infty} x_n = L$$

if there corresponds to each positive ε a real number r such that $|x_n - L| < \varepsilon$ whenever $n > r$. (Here n is an integer.)

For example,

$$\lim_{n\to\infty} \frac{n+1}{n} = 1$$

because

$$\left|\frac{n+1}{n} - 1\right| < \varepsilon$$

whenever $n > \varepsilon^{-1}$.

For another example, recall the equation

$$e = \lim_{n\to\infty}\left(1 + \frac{1}{n}\right)^n$$

by which the important irrational number e can be defined. If we compute the sequence $x_n = \left(1 + 1/n\right)^n$, some of the elements are

$$
\begin{aligned}
x_1 &= 2.000000\,0 \\
x_{10} &= 2.593742\,2 \\
x_{30} &= 2.674319\,9 \\
x_{50} &= 2.691588\,8 \\
x_{1000} &= 2.716924\,4
\end{aligned}
$$

This is an example of a sequence that is converging rather slowly, since the limit is $e = 2.71828\,18\ldots$ and in the 1000th term there is still an error of $0.00135\,8$. Using double-precision computations, we find numerical evidence that

$$\frac{|x_{n+1} - e|}{|x_n - e|} \longrightarrow 1$$

This property is worse than **linear convergence** (to be defined precisely later).

An example of a sequence that converges to 0 at a slightly faster rate is

$$x_{n+1} = x_n - \frac{x_n^2}{x_n^2 + x_{n-1}^2}$$

Selecting two initial values, we have

$$
\begin{aligned}
x_0 &= 20.00 \\
x_1 &= 15.00 \\
x_2 &= 14.64 \\
x_3 &= 14.15 \\
x_{33} &= 0.54 \\
x_{34} &= 0.27
\end{aligned}
$$

While faster than the previous example, the convergence is still quite slow. Using double-precision computations, we find numerical evidence that

$$\frac{|x_{n+1}|}{|x_n|} \longrightarrow 0$$

which is called **superlinear** convergence.

As an example of a rapidly convergent sequence, consider the one defined recursively by putting

$$
\begin{cases}
x_1 = 2 \\
x_{n+1} = \dfrac{1}{2}x_n + \dfrac{1}{x_n} & (n \geqq 1)
\end{cases}
$$

The elements of this sequence are

$$\begin{aligned} x_1 &= 2.000000 \\ x_2 &= 1.500000 \\ x_3 &= 1.416667 \\ x_4 &= 1.414216 \end{aligned}$$

The limit is $\sqrt{2} = 1.41421\,3562\ldots$ and the sequence is converging to its limit with great rapidity. Using double-precision computations, we find evidence that

$$\frac{|x_{n+1} - \sqrt{2}|}{|x_n - \sqrt{2}|^2} \leqq 0.36$$

Such a condition corresponds to **quadratic convergence** as we shall shortly see.

Orders of Convergence

Some special terminology is used to describe the rapidity with which a sequence converges. Let $[x_n]$ be a sequence of real numbers tending to a limit x^*. We say that the rate of convergence is at least **linear** if there are a constant $c < 1$ and an integer N such that

$$|x_{n+1} - x^*| \leqq c|x_n - x^*| \qquad (n \geqq N)$$

We say that the rate of convergence is at least **superlinear** if there exist a sequence ε_n tending to 0 and an integer N such that

$$|x_{n+1} - x^*| \leqq \varepsilon_n|x_n - x^*| \qquad (n \geqq N)$$

The convergence is at least **quadratic** if there are a constant C (not necessarily less than 1) and an integer N such that

$$|x_{n+1} - x^*| \leqq C|x_n - x^*|^2 \qquad (n \geqq N)$$

In general, if there are positive constants C and α and an integer N such that

$$|x_{n+1} - x^*| \leqq C|x_n - x^*|^\alpha \qquad (n \geqq N)$$

we say that the rate of convergence is of **order** α at least.

Big $\mathcal{O}$ and Little o Notation

We shall now consider several standard ways of comparing two sequences or two functions. We begin with sequences.

Let $[x_n]$ and $[\alpha_n]$ be two different sequences. We write

$$x_n = \mathcal{O}(\alpha_n)$$

if there are constants C and n_0 such that $|x_n| \leqq C|\alpha_n|$ when $n \geqq n_0$. Here we say that x_n is "Big oh" of α_n.

As an example, suppose that $f(n)$ and $g(n)$ are nonnegative functions of n. We write

$$f(n) = \mathcal{O}(g(n))$$

if there exist positive constants C and n_0 such that $0 \leqq f(n) \leqq Cg(n)$ for all $n \geqq n_0$. The Big $\mathcal{O}$-notation gives an upper bound for the function $f(n)$ to within a constant factor; that is, for all values of n greater than some initial value, n_0, the function value $f(n)$ always lies on or below $Cg(n)$. See Figure 1.3. If $g(n) \neq 0$ for all n, this means that the ratio $|f(n)/g(n)|$ remains bounded (by C) as $n \to \infty$.

The equation

$$x_n = o(\alpha_n)$$

means that $\lim_{n\to\infty} (x_n/\alpha_n) = 0$. Here we say that x_n is "little oh" of α_n.

As an example, suppose that $f(n)$ and $g(n)$ are nonnegative functions of n. We write

$$f(n) = o(g(n))$$

if for each $\varepsilon > 0$ there is an n_0 such that $0 \leqq f(n) \leqq \varepsilon g(n)$ when $n \geqq n_0$. (In $\mathcal{O}$-notation, this inequality holds for *some* constant $C > 0$.) In o-notation, we have

$$\lim_{n\to\infty} \frac{f(n)}{g(n)} = 0$$

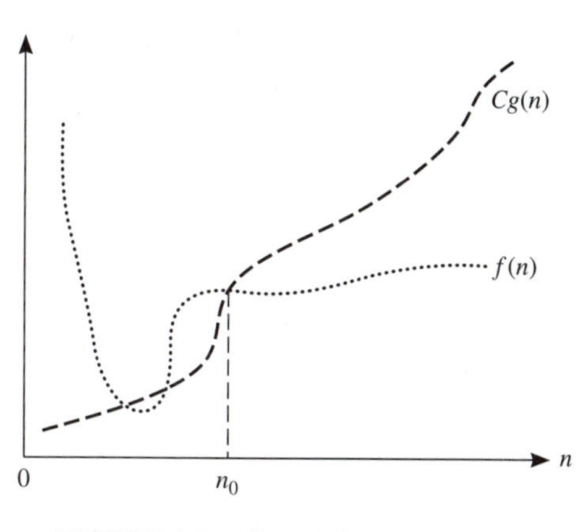

FIGURE 1.3 $f(n) = \mathcal{O}(g(n))$ as $n \to \infty$

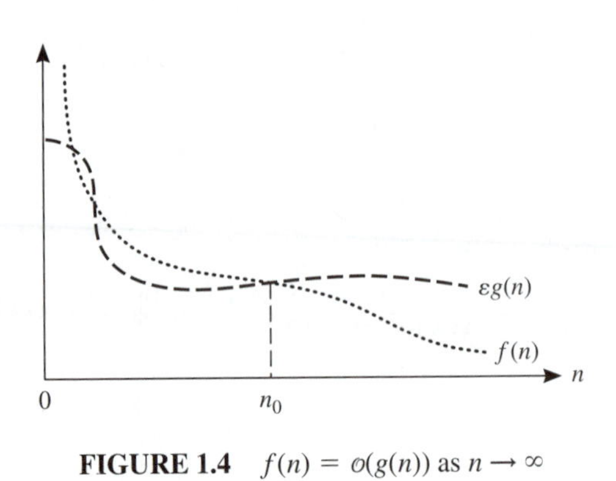

FIGURE 1.4 $f(n) = o(g(n))$ as $n \to \infty$

Intuitively, the function $f(n)$ becomes insignificant relative to $g(n)$ as n approaches infinity. See Figure 1.4.

These two notions give a coarse method of comparing two sequences. They are frequently used when both sequences converge to 0. If $x_n \to 0$, $\alpha_n \to 0$, and $x_n = \mathcal{O}(\alpha_n)$, then x_n converges to 0 at least as rapidly as α_n. If $x_n = o(\alpha_n)$, then x_n converges to 0 more rapidly than α_n does.

Here are some examples:

$$\frac{n+1}{n^2} = \mathcal{O}\left(\frac{1}{n}\right) \tag{1}$$

$$\frac{1}{n \ln n} = o\left(\frac{1}{n}\right) \tag{2}$$

$$\frac{1}{n} = o\left(\frac{1}{\ln n}\right) \tag{3}$$

$$\frac{5}{n} + e^{-n} = \mathcal{O}\left(\frac{1}{n}\right) \tag{4}$$

$$e^{-n} = o\left(\frac{1}{n^2}\right) \tag{5}$$

Taking an example from Section 1.1, we can write

$$\ln 2 - \sum_{k=1}^{n-1} (-1)^{k-1} \frac{1}{k} = \mathcal{O}\left(\frac{1}{n}\right) \tag{6}$$

This is an illustration of very **slow convergence**. On the other hand,

$$e^x - \sum_{k=0}^{n-1} \frac{1}{k!} x^k = \mathcal{O}\left(\frac{1}{n!}\right) \qquad (|x| \leqq 1) \tag{7}$$

and this illustrates very **rapid convergence**.

The notation just introduced is used also for functions other than sequences. For example, we can write

$$\sin x = x - \frac{x^3}{6} + \mathcal{O}(x^5) \qquad (x \to 0) \tag{8}$$

This means that there exists a neighborhood of 0 and a constant C such that on that neighborhood,

$$\left|\sin x - x + \frac{x^3}{6}\right| \leqq C\left|x^5\right|$$

The correctness of this assertion can be verified by using Taylor's Theorem with $n = 4$ and $f(x) = \sin x$.

An equation of the form

$$f(x) = \mathcal{O}(g(x)) \qquad (x \to \infty)$$

means that there exist constants r and C so that $|f(x)| \leqq C|g(x)|$ whenever $x \geqq r$. Thus, for example,

$$\sqrt{x^2 + 1} = \mathcal{O}(x) \qquad (x \to \infty)$$

since $\sqrt{x^2 + 1} \leqq 2x$ when $x \geqq 1$.

In general, we write

$$f(x) = \mathcal{O}(g(x)) \qquad (x \to x^*)$$

when there is a constant C and a neighborhood of x^* such that $|f(x)| \leqq C|g(x)|$ in that neighborhood. Similarly,

$$f(x) = o(g(x)) \qquad (x \to x^*)$$

means that $\lim_{x \to x^*} [f(x)/g(x)] = 0$.

Mean-Value Theorem for Integrals

Another mean-value theorem often needed in numerical analysis follows.

THEOREM 1 **Mean-Value Theorem for Integrals** *Let u and v be continuous real-valued functions on an interval $[a, b]$, and suppose that $v \geqq 0$. Then there exists a point ξ in $[a, b]$ such that*

$$\int_a^b u(x)v(x)\,dx = u(\xi)\int_a^b v(x)\,dx$$

Proof Let α and β denote the least and greatest values of $u(x)$ on $[a, b]$, respectively. Then

$$\alpha \leqq u(x) \leqq \beta \qquad (a \leqq x \leqq b)$$

Since $v(x) \geqq 0$, we have

$$\alpha v(x) \leqq u(x)v(x) \leqq \beta v(x)$$

Now integrate throughout this inequality and put $I = \int_a^b v(x)\,dx$. The result is

$$\alpha I \leqq \int_a^b u(x)v(x)\,dx \leqq \beta I$$

If $I = 0$, we can conclude that $v(x) \equiv 0$, and the result we wish to prove is trivial. If I is not 0, we have

$$\alpha \leqq I^{-1} \int_a^b u(x)v(x)\,dx \leqq \beta$$

By the Intermediate-Value Theorem for Continuous Functions, there exists a point ξ in $[a, b]$ for which

$$u(\xi) = I^{-1} \int_a^b u(x)v(x)\,dx$$

The Intermediate-Value Theorem states that if f is continuous on $[a, b]$ and if r is any real number between the greatest and least values of f on $[a, b]$, then there exists a point c in $[a, b]$ where $f(c) = r$. ■

Nested Multiplication

An elementary concept that is needed in various places throughout the text is nested multiplication of polynomials. A polynomial can be rewritten in a nested form so that it requires little more than the minimum number of multiplications when evaluating it. The polynomial

$$p(x) = a_n x^n + a_{n-1} x^{n-1} + \cdots + a_2 x^2 + a_1 x + a_0$$

can be written using standard mathematical notation involving the sum $\sum$ and product $\prod$ as follows:

$$p(x) = \sum_{k=0}^{n} a_k x^k = \sum_{k=0}^{n} \left(a_k \prod_{j=1}^{k} x \right)$$

Recall that if $n \leqq m$, then

$$\sum_{k=n}^{m} s_k = s_n + s_{n+1} + \cdots + s_m \qquad \text{and} \qquad \prod_{k=n}^{m} y_k = y_n y_{n+1} \cdots y_m$$

and, thereby,

$$\sum_{k=n}^{m} r = (m - n + 1)r \qquad \text{and} \qquad \prod_{k=n}^{m} x = x^{(m-n+1)}$$

By convention, if $m < n$, then

$$\sum_{k=n}^{m} s_k = 0 \qquad \text{and} \qquad \prod_{k=n}^{m} y_k = 1$$

To evaluate the polynomial efficiently, we can group the terms using **nested multiplication**:

$$p(x) = a_0 + x(a_1 + x(a_2 + \cdots + x(a_{n-1} + x(a_n))\cdots))$$

This corresponds to the following simple algorithm involving only n multiplications and n additions:

$p \leftarrow a_n$
for $k = n - 1$ **to** 0 **step** -1 **do**
    $p \leftarrow xp + a_k$
end do

This procedure is also known as **Horner's method** or **synthetic division**.

Upper and Lower Bounds

Two important concepts that arise frequently in numerical analysis are **supremum** (**least upper bound**) and **infimum** (**greatest lower bound**). Let S be a nonempty, bounded set of real numbers. **Boundedness** is the property that for appropriate real numbers a and b,

$$a \leqq x \leqq b \qquad (x \in S)$$

In this situation, we call b an upper bound for S and a a lower bound for S. Of course, S has many upper bounds: If $c \geqq b$, then c is also an upper bound for S. Obviously, there is no *largest* upper bound, but there is a *smallest* upper bound for the set S.

AXIOM 1 **Least Upper Bound Axiom** *Any nonempty set of real numbers that possesses an upper bound has a least upper bound.*

The property expressed in the axiom is one of the deeper characteristics of the real number system. (The rational number system does not have this property, for example.) The **least upper bound** of a set S (if it has one) is denoted by lub S. The word **supremum** is synonymous and is abbreviated by $\sup S$. Thus, we can write $v = \sup S$ if **(i)** v is an upper bound for S and **(ii)** no real number smaller than v is an upper bound for S.

Example 1 What is $\sup S$ if $S = \{x : x^2 < 2\}$?

Solution Since $S = \{x : -\sqrt{2} < x < \sqrt{2}\}$, the least upper bound is $\sqrt{2}$. ■

If F is a function, the notation $\sup_{x \in A} f(x)$ means $\sup\{f(x) : x \in A\}$. For example,

$$\sup_{0<x<\frac{\pi}{6}} \sin x = \frac{1}{2}$$

The corresponding concept of **greatest lower bound** or **infimum** is defined as follows: We write $u = \text{glb}\, S$ or $u = \inf S$ if **(i)** u is a lower bound for S and **(ii)** no real number greater than u is a lower bound for S. Here the basic result is that if a nonempty set of real numbers has a lower bound, then it has a greatest lower bound. This follows from the axiom when we apply it to the set $-S = \{-x : x \in S\}$.

Explicit and Implicit Functions

A function is usually defined via an explicit formula, from which a value of the function can be computed for each argument. For example, we can define a function f by the formula

$$f(x) = \sqrt{7x^3 - 2x}$$

After enlarging our repertoire of familiar functions, we can define more complicated functions such as

$$f(x) = \ln(\arctan x) + \cos(e^x)$$

There are, however, many other methods of defining functions, such as by means of a differential equation or an integral. Thus, a function $y = f(x)$ is well defined by the following differential equation with an initial condition:

$$\begin{cases} y' = 1 + \sin y \\ y(0) = 0 \end{cases}$$

An important function, called the **error function** and denoted by $\text{erf}(x)$, is defined by an integral:

$$\text{erf}(x) = \frac{2}{\sqrt{\pi}} \int_0^x e^{-t^2}\, dt$$

In this section, we consider functions defined **implicitly**. This means that a function G of *two* variables is given and, from the equation $G(x, y) = 0$, we hope to recover y as a function of x. In some cases, we can unravel the equation $G(x, y) = 0$

to obtain $y = f(x)$. For example, in the equation

$$y^2 + 3xy - 7 = 0$$

we can solve for y and get

$$y = \frac{1}{2}\left(-3x \pm \sqrt{9x^2 + 28}\right)$$

thereby recovering *two* explicit functions. Likewise, from the equation

$$\sin(y + 7) = x^3 - 2$$

we can recover an explicit function by solving for y:

$$y = \sin^{-1}(x^3 - 2) - 7$$

Here too there are multiple functions because various **branches** of the inverse sine function can be chosen.

In general, if G is a prescribed function, and if (x_0, y_0) is a particular point where $G(x_0, y_0) = 0$, then we expect that there will be additional nearby points that satisfy the equation $G(x, y) = 0$. Thus, when x is altered, we expect that a suitable change in y will be necessary to keep the equation $G(x, y) = 0$ true. Hence, y should be a function of x in some neighborhood of x_0. Figure 1.5 shows a typical case.

Implicit Functions

The following important theorem governs this situation.

THEOREM 2 **Implicit Function Theorem** *Let G be a function of two real variables defined and continuously differentiable in a neighborhood of (x_0, y_0). If $G(x_0, y_0) = 0$ and*

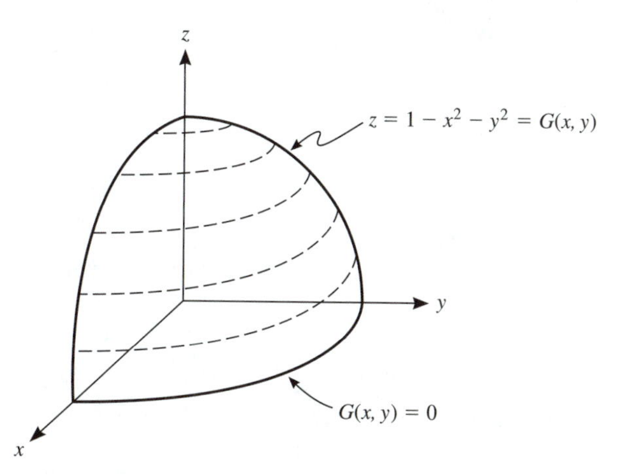

FIGURE 1.5 Surface G intersects xy-plane showing curve $G(x, y) = 0$

$\partial G/\partial y \neq 0$ at (x_0, y_0), then there is a positive δ and a continuously differentiable function f defined for $|x - x_0| < \delta$ such that $f(x_0) = y_0$ and $G(x, f(x)) = 0$ for $|x - x_0| < \delta$.

Example 2 Does the equation

$$x^7 + 2y^8 - y^3 = 0$$

define y as a continuously differentiable function of x in some neighborhood of $x = -1$?

Solution To answer this, let $(x_0, y_0) = (-1, 1)$ and put

$$G(x, y) = x^7 + 2y^8 - y^3$$

Then $G(x_0, y_0) = 0$ and

$$\frac{\partial G}{\partial y} = 16y^7 - 3y^2$$

Hence, $\partial G(x_0, y_0)/\partial y \neq 0$. By the Implicit Function Theorem, y is a continuously differentiable function of x near $x_0 = -1$. ■

If f is a function defined implicitly by $G(x, y) = 0$, then for x in some interval, $G(x, f(x)) = 0$. If we desire to compute $f'(x)$, a familiar technique from calculus can be used. We differentiate in the equation $G(x, y) = 0$ with respect to x, remembering that y is interpreted as a function of x. The result is

$$\frac{\partial G}{\partial x} + \frac{\partial G}{\partial y}\frac{dy}{dx} = 0$$

from which we obtain

$$\frac{dy}{dx} = -\frac{\partial G}{\partial x}\Big/\frac{\partial G}{\partial y}$$

or

$$f'(x) = -\frac{\partial G}{\partial x}\Big/\frac{\partial G}{\partial y}$$

The derivative thus obtained is likely to be a function of x and y.

Example 3 What is the value of dy/dx at the point (2, 1), if $y(x)$ is defined implicitly by the equation $x^3 - y^7 + 4x^2 + y^4 - 24 = 0$?

Solution Total differentiation with respect to x produces

$$3x^2 - 7y^6\frac{dy}{dx} + 8x + 4y^3\frac{dy}{dx} = 0$$

Substituting $x = 2$ and $y = 1$, we get

$$12 - 7\frac{dy}{dx} + 16 + 4\frac{dy}{dx} = 0$$

from which $dy/dx = 28/3$. ■

Some numerical problems involving implicit functions are discussed in Section 3.2.

PROBLEMS 1.2

1. When the sequence $x_n = (1 + 1/n)^n$ is computed, it appears to be monotone increasing. Prove that this is so. *Hints*: First, if $\ln f(x)$ is increasing, then so is $f(x)$. Second, if $f'(x) > 0$, then f is increasing. Finally, $\ln x$ is defined to be $\int_1^x t^{-1}\,dt$.
2. Show that the elements of the sequence in Problem 1 remain less than 3.
3. Prove that $x_n = x + o(1)$ if and only if $\lim_{n\to\infty} x_n = x$.
4. Verify the correctness of each expression in Equation (1) in the text.
5. For fixed n, show that $\sum_{k=0}^{n} x^k = 1/(1-x) + o(x^n)$ as $x \to 0$.
6. Show that if $x_n = \mathcal{O}(\alpha_n)$, then $cx_n = \mathcal{O}(\alpha_n)$.
7. Show that if $x_n = \mathcal{O}(\alpha_n)$, then $x_n/\ln n = o(\alpha_n)$.
8. Find the best integer value of k in the assertion $\cos x - 1 + x^2/2 = \mathcal{O}(x^k)$ as $x \to 0$.
9. Prove: If $x_n = o(\alpha_n)$, then $x_n = \mathcal{O}(\alpha_n)$. Show that the converse is not true.
10. Prove: If $x_n = \mathcal{O}(\alpha_n)$ and $y_n = \mathcal{O}(\alpha_n)$, then $x_n + y_n = \mathcal{O}(\alpha_n)$.
11. Prove: If $x_n = o(\alpha_n)$ and $y_n = o(\alpha_n)$, then $x_n + y_n = o(\alpha_n)$.
12. Show that for any $r > 0$, $x^r = \mathcal{O}(e^x)$ as $x \to \infty$.
13. Show that for any $r > 0$, $\ln x = \mathcal{O}(x^r)$ as $x \to \infty$.
14. Prove: If $\alpha_n \to 0$, $x_n = \mathcal{O}(\alpha_n)$, and $y_n = \mathcal{O}(\alpha_n)$, then $x_n y_n = o(\alpha_n)$.
15. Prove: If $x_n = \mathcal{O}(\alpha_n)$, then $\alpha_n^{-1} = \mathcal{O}(x_n^{-1})$. Show that the same is true for the o-relationship.
16. Determine the best integer value of k in the equation $\arctan x = x + \mathcal{O}(x^k)$, as $x \to 0$.
17. Let a sequence x_n be defined inductively by $x_{n+1} = F(x_n)$. Suppose that $x_n \to x$ as $n \to \infty$ and $F'(x) = 0$. Show that $x_{n+2} - x_{n+1} = o(x_{n+1} - x_n)$. *Hint*: Use the Mean-Value Theorem and assume that F is a continuously differentiable function.
18. Prove that every sufficiently smooth function can be approximated on an interval of length h by a polynomial of degree n with an error that is $\mathcal{O}(h^{n+1})$ as $h \to 0$.

19. **(a)** Consider the series

$$e^{\tan x} = 1 + x + \frac{x^2}{2!} + \frac{3x^3}{3!} + \frac{9x^4}{4!} + \cdots \qquad (|x| \leqq \pi/2)$$

Retaining three terms in the series, estimate the remaining series using o-notation with the best integer value possible, as $x \to 0$.

(b) Repeat the problem using $\mathcal{O}$-notation and the series

$$\ln \tan x = \ln x + \frac{x^2}{3} + \frac{7x^4}{90} + \frac{62x^6}{2835} + \cdots \qquad (0 < |x| < \pi/2)$$

20. Establish the range of integer values of γ and δ for which the $+\cdots$ in the series

$$\ln(1+x) = \sum_{k=1}^{n-1} (-1)^{k-1} \frac{x^k}{k} + \cdots$$

can be replaced with either $\mathcal{O}(x^\gamma)$ or $o(x^\delta)$ as $x \to 0$.

21. For the pair (x_n, α_n), is it true that $x_n = \mathcal{O}(\alpha_n)$ as $n \to \infty$?

(a) $x_n = 5n^2 + 9n^3 + 1, \quad \alpha_n = n^2$

(b) $x_n = 5n^2 + 9n^3 + 1, \quad \alpha_n = 1$

(c) $x_n = \sqrt{n+3}, \quad \alpha_n = 1$

(d) $x_n = 5n^2 + 9n^3 + 1, \quad \alpha_n = n^3$

(e) $x_n = \sqrt{n+3}, \quad \alpha_n = 1/n$

22. Choose the correct assertions (in each, $n \to \infty$).

(a) $(n+1)/n^2 = o(1/n)$

(b) $(n+1)/\sqrt{n} = o(1)$

(c) $1/\ln n = \mathcal{O}(1/n)$

(d) $1/(n \ln n) = o(1/n)$

(e) $e^n/n^5 = \mathcal{O}(1/n)$

23. The expressions e^h, $(1-h^4)^{-1}$, $\cos(h)$, and $1 + \sin(h^3)$ all have the same limit as $h \to 0$. Express each in the following form with the best integer values of α and β.

$$f(h) = c + \mathcal{O}(h^\alpha) = c + o(h^\beta)$$

24. What are the limit and the rate of convergence of the following expression as $h \to 0$?

$$\frac{(1+h) - e^h}{h^2}$$

Express the limit in the form given in Problem 23.

25. Determine the best integer value for β such that for fixed n,

$$\frac{1}{1-x} = 1 + x + x^2 + \cdots + x^n + \mathcal{O}(x^\beta) \qquad (0 < x < 1)$$

as $x \to 0$. Repeat for $o(x^\beta)$. Is there a best integer value in this case? Explain.

26. Show that these assertions are not true.

(a) $e^x - 1 = \mathcal{O}(x^2) \quad \text{as} \quad x \to 0$

(b) $x^{-2} = \mathcal{O}(\cot x) \quad \text{as} \quad x \to 0$

(c) $\cot x = o(x^{-1}) \quad \text{as} \quad x \to 0$

27. Let $[a_n] \to 0$ and let $\lambda > 1$. Show that $\sum_{k=0}^{n} a_k \lambda^k = o(\lambda^n)$ as $n \to \infty$.

28. Explain why the least upper bound axiom does not apply to the empty set.

29. Find two functions in explicit form that are defined implicitly by the equation

$$(x^3 - 1)y + e^x y^2 + \cos x - 1 = 0$$

30. In solving differential equations, one often obtains solutions in implicit form. Show that the equation

$$2x^3y^2 + x^2y + e^x = c$$

defines a solution of the differential equation

$$\frac{dy}{dx} = -(6x^2y^2 + 2xy + e^x)/(4x^3y + x^2)$$

31. Kepler's equation in astronomy is $x - y + \varepsilon \sin y = 0$, where ε is a parameter in the range $0 \leqq \varepsilon \leqq 1$. Show that for each real x there is a real y that makes the equation true. Show that if $0 \leqq \varepsilon < 1$, then dy/dx is continuous everywhere. *Hint*: Write $x = y - \varepsilon \sin y$ and consider the behavior of $y - \varepsilon \sin y$ as $y \to +\infty$ and as $y \to -\infty$. Use the Implicit Function Theorem for the second part of this problem.

32. Find the points x at which the equation

$$y - \ln(x + y) = 0$$

defines y implicitly as a function of x. Compute dy/dx.

33. Give an example to show why the least upper bound axiom does not apply to the set of all rational numbers.

34. Does the least upper bound axiom apply to the set of all integers?

35. What are the values of the following?

(a) $\sup_{x \in \mathbb{R}} \arctan x$

(b) $\sup_{x \geqq 0} e^{-x}$

(c) $\inf_{x\in\mathbb{R}} e^{-x}$

(d) $\sup_{x\in\mathbb{R}}(x^2+1)^{-1}$

36. Use the Mean-Value Theorem for Integrals to prove that

$$\int_0^{\pi/2} e^x \cos x\,dx = e^y$$

for some y in $(0, \pi/2)$.

37. Give an example to show why the continuity of u cannot be dropped from the hypotheses in Theorem 1.

38. Prove that if $0 < \theta < 1$, then $(1 + a\theta^n)/(1 + a\theta^{n-1})$ converges to 1 linearly.

39. Are the following equivalent?

(i) $|f(x)| = \mathcal{O}(|x|^{-n-\varepsilon})$ for some $\varepsilon > 0$ as $|x| \to \infty$

(ii) $|f(x)| = o(|x|^{-n})$

COMPUTER PROBLEMS 1.2

1. Consider the recursive relation

$$\begin{cases} x_0 = 1 \qquad x_1 = c \\ x_{n+1} = x_n + x_{n-1} \qquad (n \geqq 1) \end{cases}$$

(a) It can be shown that when $c = \left(1 + \sqrt{5}\right)/2$, a closed-form formula is given by

$$x_n = \left(\frac{1+\sqrt{5}}{2}\right)^n$$

(b) Similarly, when $c = \left(1 - \sqrt{5}\right)/2$,

$$x_n = \left(\frac{1-\sqrt{5}}{2}\right)^n$$

(c) If $c = 1$, then

$$x_n = \frac{1}{\sqrt{5}}\left(\frac{1+\sqrt{5}}{2}\right)^{n+1} - \frac{1}{\sqrt{5}}\left(\frac{1-\sqrt{5}}{2}\right)^{n+1}$$

For all values of n in the range $1 \leqq n \leqq 30$, compute x_n by both the recursive relation and the formula for each case. Explain the results. This recursion defines the famous **Fibonacci sequence**.

2. Using a symbolic manipulation program, find the Taylor series about 0, up to and including the term x^{10}, for the function $(\tan x)^2$. Express the missing terms in $\mathcal{O}$-notation.

*1.3 Difference Equations

Algorithms for numerical computation are often designed to produce *sequences* of numbers. It is useful, therefore, to develop a little of the theory of linear sequence spaces. This theory will be needed in Chapter 8 for the analysis of linear multistep methods. (We present it here because it requires little mathematical background to understand and thus lends itself to an introductory section that stands alone.)

Basic Concepts

In the following discussion, V will stand for the set of all infinite sequences of complex numbers, such as

$$x = [x_1, x_2, x_3, \ldots]$$
$$y = [y_1, y_2, y_3, \ldots]$$

A sequence is, formally, a complex-valued function defined on the set of positive integers $\mathbb{N} = \{1, 2, 3, \ldots\}$. It is simply a matter of convenience that we write x_n in place of $x(n)$ for the value of the function x at the argument n.

In the set V, we define two operations:

$$x + y = [x_1 + y_1, x_2 + y_2, x_3 + y_3, \ldots]$$
$$\lambda x = [\lambda x_1, \lambda x_2, \lambda x_3, \ldots]$$

A more compact way of writing these equations is

$$(x + y)_n = x_n + y_n$$
$$(\lambda x)_n = \lambda x_n$$

There is a 0 element in V—namely, $0 = [0, 0, 0, \ldots]$. With the adoption of these definitions, V becomes a vector space. The verification of the vector-space postulates is tedious and not very instructive. The vector space V is infinite dimensional; indeed, the following set of vectors is linearly independent:

$$v^{(1)} = [1, 0, 0, 0, \ldots]$$
$$v^{(2)} = [0, 1, 0, 0, \ldots]$$
$$v^{(3)} = [0, 0, 1, 0, \ldots]$$
$$v^{(4)} = [0, 0, 0, 1, \ldots]$$
$$\vdots$$

*Throughout this text, some sections are marked with an asterisk. These sections may be skipped at the instructor's discretion.

We shall be interested in linear operators $L : V \to V$. One of the most important of these is the **displacement operator**, denoted by E and defined by the equation

$$Ex = [x_2, x_3, x_4, \ldots]$$

where

$$x = [x_1, x_2, x_3, \ldots]$$

Thus,

$$(Ex)_n = x_{n+1}$$

It is clear that E can be applied several times in succession to produce, for example,

$$(EEx)_n = x_{n+2}$$

or

$$(E^k x)_n = x_{n+k}$$

In the remainder of this section, our attention will be confined to linear operators that can be expressed as linear combinations of powers of E. Such an operator is termed a **linear difference operator** (with constant coefficients and finite rank). The general form of such an operator is

$$L = \sum_{i=0}^{m} c_i E^i \tag{1}$$

Of course, E^0 is defined as the identity operator,

$$(E^0 x)_n = (Ix)_n = x_n$$

From Equation (1), we see that the linear difference operators of the form (1) make up a linear subspace in the set of all linear operators from V into V. The powers of E provide a basis for this subspace.

Notice that L in Equation (1) is a *polynomial* in E; in other words, it is a linear combination of powers of E. Thus, we could write

$$L = p(E)$$

where p is a polynomial called the **characteristic polynomial** of L and defined by

$$p(\lambda) = \sum_{i=0}^{m} c_i \lambda^i$$

The problem to be studied here is the determination of all the solutions to an equation of the form $Lx = 0$, where L is an operator of the type (1). From the linearity of L, it follows at once that the set $\{x : Lx = 0\}$ is a linear subspace of V; it is called the **null space** of L. We can regard the equation $Lx = 0$ as being solved if a basis has been found for the null space of L.

To see what to expect in general, let us consider a concrete example of L, say by taking $c_0 = 2$, $c_1 = -3$, $c_2 = 1$, and all other $c_i = 0$. The resulting equation, which is known as a **difference equation**, can be written in three forms:

$$\begin{aligned} \left(E^2 - 3E^1 + 2E^0\right) x &= 0 \\ x_{n+2} - 3x_{n+1} + 2x_n &= 0 \qquad (n \geqq 1) \\ p(E)x &= 0 \qquad p(\lambda) = \lambda^2 - 3\lambda + 2 \end{aligned} \tag{2}$$

It is very easy to generate sequences that solve (2). Indeed, we can choose x_1 and x_2 arbitrarily and then determine $x_3, x_4, \ldots$ by use of (2). For instance, we can obtain in this way the various solutions

$$\begin{aligned} &[1, 0, -2, -6, -14, -30, \ldots] \\ &[1, 1, 1, 1, \ldots] \\ &[2, 4, 8, 16, \ldots] \end{aligned}$$

The first solution is more mysterious than the second two because it is not clear at first glance what its general term is. The second two solutions are obviously of the form $x_n = \lambda^n$, with $\lambda = 1$ or 2. It is natural to inquire whether any other such solutions exist. Putting $x_n = \lambda^n$ in (2) yields

$$\begin{aligned} \lambda^{n+2} - 3\lambda^{n+1} + 2\lambda^n &= 0 \\ \lambda^n(\lambda^2 - 3\lambda + 2) &= 0 \\ \lambda^n(\lambda - 1)(\lambda - 2) &= 0 \end{aligned}$$

This simple analysis shows that there is just one other solution of the type sought—namely, $[0, 0, 0, \ldots]$. This we call the **trivial solution**. Now, it happens that the solutions u defined by $u_n = 1$ and v defined by $v_n = 2^n$ form a basis for the solution space of (2). To prove this, let x be **any solution** of (2). We seek constants α and β such that $x = \alpha u + \beta v$. This equation means that $x_n = \alpha u_n + \beta v_n$ for all n. In particular, for $n = 1$ and 2, we have

$$\begin{cases} x_1 = \alpha + 2\beta \\ x_2 = \alpha + 4\beta \end{cases} \tag{3}$$

Equation (3) uniquely determines α and β because the determinant of the matrix

$$\begin{bmatrix} 1 & 2 \\ 1 & 4 \end{bmatrix}$$

is not 0. (It is the determinant of a **Vandermonde matrix** as defined in Section 6.1.) But now we can prove by induction that $x_n = \alpha u_n + \beta v_n$ for *all n*. Indeed, if this equation is true for indices less than n, then it is true for n because

$$\begin{aligned} x_n &= 3x_{n-1} - 2x_{n-2} \\ &= 3(\alpha u_{n-1} + \beta v_{n-1}) - 2(\alpha u_{n-2} + \beta v_{n-2}) \\ &= \alpha(3u_{n-1} - 2u_{n-2}) + \beta(3v_{n-1} - 2v_{n-2}) \\ &= \alpha u_n + \beta v_n \end{aligned}$$

This example illustrates the case of simple zeros of the characteristic polynomial.

Simple Zeros

THEOREM 1 *If p is a polynomial and λ is a zero of p, then one solution of the difference equation $p(E)x = 0$ is $[\lambda, \lambda^2, \lambda^3, \dots]$. If all the zeros of p are simple and nonzero, then each solution of the difference equation is a linear combination of such special solutions.*

Proof First, if λ is any complex number and $u = [\lambda, \lambda^2, \lambda^3, \dots]$, then $Eu = \lambda u$ because

$$(Eu)_n = u_{n+1} = \lambda^{n+1} = \lambda(\lambda^n) = \lambda u_n$$

By reapplying the operator E, one obtains in general $E^i u = \lambda^i u$. Since E^0 has been defined as the identity map, we have $E^0 u = \lambda^0 u$. Thus, if p is a polynomial defined by $p(\lambda) = \sum_{i=0}^{m} c_i \lambda^i$, then

$$p(E)u = \left(\sum_{i=0}^{m} c_i E^i\right) u = \sum_{i=0}^{m} c_i (E^i u) = \sum_{i=0}^{m} c_i \lambda^i u = p(\lambda)u$$

If $p(\lambda) = 0$, then $p(E)u = 0$, as asserted.

Let p be a polynomial all of whose zeros, $\lambda_1, \lambda_2, \dots, \lambda_m$, are simple and nonzero. Corresponding to any zero λ_k there is a solution of the difference equation $p(E)x = 0$; namely, we have the solution $u^{(k)} = [\lambda_k, \lambda_k^2, \lambda_k^3, \dots]$. Let x denote an arbitrary solution of the difference equation. We seek to express x in the form $x = \sum_{k=1}^{m} a_k u^{(k)}$. Taking the first m components of the sequences in this equation, we obtain

$$x_i = \sum_{k=1}^{m} a_k \lambda_k^i \qquad (1 \leqq i \leqq m) \tag{4}$$

The $m \times m$ matrix having elements λ_k^i is nonsingular because its singularity would imply a nontrivial equation

$$\sum_{i=1}^{m} b_i \lambda_k^i = 0 \qquad \text{or} \qquad \sum_{i=1}^{m} b_i \lambda_k^{i-1} = 0$$

(This last equation exhibits a polynomial of degree $m-1$ having m zeros.) Equation (4) thus determines $a_1, a_2, \ldots, a_m$ uniquely. It remains to be proven that Equation (4) is valid for all values of i. Put $z = x - \sum_{k=1}^{m} a_k u^{(k)}$. Then $p(E)z = 0$ or equivalently $\sum_{i=0}^{m} c_i z_{n+i} = 0$ for all n. In other words,

$$z_{n+m} = -c_m^{-1}(c_0 z_n + c_1 z_{n+1} + \cdots + c_{m-1} z_{n+m-1}) \qquad (n \geqq 1) \tag{5}$$

Note that $c_m \neq 0$ because the polynomial p has m distinct zeros and is therefore of degree m. Since $z_i = 0$ for $i = 1, 2, \ldots, m$, Equation (5) used repeatedly shows that

$$z_{m+1} = z_{m+2} = \cdots = 0$$

■

Multiple Zeros

There remains the problem of solving a difference equation $p(E)x = 0$ when p has multiple zeros. Define $x(\lambda) = [\lambda, \lambda^2, \lambda^3, \ldots]$. If p is any polynomial, we have seen that

$$p(E)x(\lambda) = p(\lambda)x(\lambda) \tag{6}$$

A differentiation with respect to λ yields

$$p(E)x'(\lambda) = p'(\lambda)x(\lambda) + p(\lambda)x'(\lambda) \tag{7}$$

If λ is a multiple zero of p, then $p(\lambda) = p'(\lambda) = 0$, and Equations (6) and (7) show that $x(\lambda)$ and $x'(\lambda)$ are solutions of the difference equation. Thus, a solution is the sequence $x'(\lambda) = [1, 2\lambda, 3\lambda^2, \ldots]$. If $\lambda \neq 0$, it is independent of the solution $x(\lambda)$ because

$$\det \begin{bmatrix} \lambda & \lambda^2 \\ 1 & 2\lambda \end{bmatrix} \neq 0$$

and thus, if the sequences are truncated at the second term, the resulting pair of vectors in $\mathbb{R}^2$ is linearly independent.

By extending this reasoning, one can prove that if λ is a zero of p having multiplicity k, then the following sequences are solutions of the difference equation $p(E)x = 0$:

$$\begin{aligned} x(\lambda) &= [\lambda, \lambda^2, \lambda^3, \ldots] \\ x'(\lambda) &= [1, 2\lambda, 3\lambda^2, \ldots] \\ x''(\lambda) &= [0, 2, 6\lambda, \ldots] \\ &\vdots \\ x^{(k-1)}(\lambda) &= \frac{d^{(k-1)}}{d\lambda^{k-1}} [\lambda, \lambda^2, \lambda^3, \ldots] \end{aligned}$$

THEOREM 2 *Let p be a polynomial satisfying $p(0) \neq 0$. Then a basis for the null space of $p(E)$ is obtained as follows: With each zero λ of p having multiplicity k, associate the k basic solutions $x(\lambda), x'(\lambda), \ldots, x^{(k-1)}(\lambda)$, where $x(\lambda) = [\lambda, \lambda^2, \lambda^3, \ldots]$.*

Example 1 Determine the general solution of this difference equation:

$$4x_n + 7x_{n-1} + 2x_{n-2} - x_{n-3} = 0$$

Solution The given equation is of the form $p(E)x = 0$, where $p(\lambda) = 4\lambda^3 + 7\lambda^2 + 2\lambda - 1$. The factors of p are $(\lambda + 1)^2$ and $(4\lambda - 1)$. Thus, p has a double zero at -1 and a simple zero at $1/4$. The basic solutions are

$$\begin{aligned} x(-1) &= [-1, 1, -1, 1, \ldots] \\ x'(-1) &= [1, -2, 3, -4, \ldots] \\ x(1/4) &= [1/4, 1/16, 1/64, \ldots] \end{aligned}$$

The general solution is

$$x = \alpha x(-1) + \beta x'(-1) + \gamma x(1/4)$$

or

$$x_n = \alpha(-1)^n + \beta n(-1)^{n-1} + \gamma(1/4)^n \qquad \blacksquare$$

Stable Difference Equations

An element $x = [x_1, x_2, \ldots]$ of V is said to be **bounded** if there is a constant c such that $|x_n| \leqq c$ for all n. In other words, $\sup_n |x_n| < \infty$. A difference equation of the form $p(E)x = 0$ is said to be **stable** if all of its solutions are bounded. The difference equation (2) is not stable since one of its solutions is given by $x_n = 2^n$. (Conditioning and stability, discussed elsewhere, are not related to the concept of a stable difference equation.)

We now ask whether there is an easy method of identifying a stable difference equation.

THEOREM 3 *For a polynomial p satisfying $p(0) \neq 0$, these properties are equivalent:*

(i) *The difference equation $p(E)x = 0$ is stable.*
(ii) *All zeros of p satisfy $|z| \leqq 1$, and all multiple zeros satisfy $|z| < 1$.*

Proof Assume that **(ii)** is true of p. Let λ be a zero of p. Then one solution of the corresponding difference equation is $x(\lambda) = [\lambda, \lambda^2, \lambda^3, \ldots]$. Since $|\lambda| \leqq 1$, this sequence is bounded. If λ is a multiple zero, then one or more of $x'(\lambda), x''(\lambda), \ldots$ will also be a solution of the difference equation. In this case, $|\lambda| < 1$, by **(ii)**. Since, by elementary calculus (L'Hôpital's Rule),

$$\lim_{n\to\infty} n^k \lambda^n = 0 \qquad (k \geqq 0)$$

we see that each sequence $x'(\lambda), x''(\lambda), \ldots$ is bounded. (See Problems 22 and 23.)

For the converse, suppose that **(ii)** is false. If p has a zero λ satisfying $|\lambda| > 1$, then the sequence $x(\lambda)$ is unbounded. If p has a multiple zero λ satisfying $|\lambda| \geqq 1$, then $x'(\lambda)$ is unbounded since its general term satisfies the inequality

$$|x_n| = |n\lambda^{n-1}| = n|\lambda|^{n-1} \geqq n$$

■

Example 2 Determine whether this difference equation is stable:

$$4x_n + 7x_{n-1} + 2x_{n-2} - x_{n-3} = 0$$

Solution The given equation is of the form $p(E)x = 0$, where $p(\lambda) = 4\lambda^3 + 7\lambda^2 + 2\lambda - 1$. By the preceding example, p has a double zero at -1 and a simple zero at $1/4$. The equation is therefore **unstable**. ■

An example of a difference equation that has *nonconstant* coefficients arises in the theory of Bessel functions. The **Bessel functions** J_n are defined by the formula

$$J_n(x) = \frac{1}{\pi}\int_0^{\pi} \cos(x\sin\theta - n\theta)\,d\theta$$

It is obvious from the definition that $|J_n(x)| \leqq 1$. Not so obvious, but true, is the recurrence formula

$$J_n(x) = 2(n-1)x^{-1}J_{n-1}(x) - J_{n-2}(x)$$

If (for a certain x) we know $J_0(x)$ and $J_1(x)$, then the recurrence relation can be used to compute $J_2(x), J_3(x), \ldots, J_n(x)$. This procedure becomes unstable and useless when $2n > |x|$ because roundoff errors that inevitably occur will be multiplied by the factor $2nx^{-1}$. This factor eventually becomes very large. (See Computer Problem 2.)

For further information on computing functions by means of recurrence relations, see Abramowitz and Stegun [1964, p. xiii], Cash [1979], Gautschi [1961, 1967, 1975], and Wimp [1984].

PROBLEMS 1.3

1. For the sequences following Equation (2), express the first as a linear combination of the second and third.
2. Let p be a polynomial of degree m. Is the solution space of $p(E)x = 0$ necessarily of dimension m?
3. Let p be a polynomial of degree m, with $p(0) \neq 0$. If a sequence x contains m consecutive zeros and $p(E)x = 0$, then $x = 0$.

4. Is the operator E **injective (one to one)**? Does it have a right or left inverse? Is it **surjective (onto)**? Define an operator F by $(Fx)_n = x_{n-1}$, with $(Fx)_1 = 0$, and answer the same questions for F. Explore the relationship between E and F. Suppose that V were redefined as the space of all functions on the set $\{\dots, -3, -2, -1, 0, 1, 2, 3, \dots\}$, and suppose that F were defined simply by $(Fx)_n = x_{n-1}$. How are the answers to the previous questions affected?

5. What are the eigenvalues and eigenvectors of the operator E?

6. Consider the infinite series $\sum_{n=1}^{\infty} x_n v^{(n)}$. What can you say about the question of convergence? Prove that $x = \sum_{n=1}^{\infty} x_n v^{(n)}$ in the pointwise sense.

7. If $\{v^{(1)}, v^{(2)}, \dots\}$ is adopted as a basis for V, show that $\sum_{i=0}^{m} c_i E^i$ can be represented by an infinite matrix.

8. Prove that any two operators of the form described in Problem 7 commute with each other.

9. Prove that if L_1 and L_2 are linear combinations of powers of E, and if $L_1 x = 0$, then $L_1 L_2 x = 0$.

10. Develop a complete theory for the difference equation $E^r x = 0$.

11. Give bases consisting of real sequences for each solution space.

(a) $(4E^0 - 3E^2 + E^3)x = 0$

(b) $(3E^0 - 2E + E^2)x = 0$

(c) $(2E^6 - 9E^5 + 12E^4 - 4E^3)x = 0$

(d) $(\pi E^2 - \sqrt{2}E + \log 2 \cdot E^0)x = 0$

12. Prove that if p is a polynomial with real coefficients, and if $z \equiv [z_1, z_2, \dots]$ is a (complex) solution of $p(E)z = 0$, then the conjugate of z, the real part of z, and the imaginary part of z are also solutions.

13. Solve.

(a) $x_{n+1} - nx_n = 0$

(b) $x_{n+1} - x_n = n$

(c) $x_{n+1} - x_n = 2$

14. Define an operator Δ by putting

$$\Delta x = [x_2 - x_1, x_3 - x_2, x_4 - x_3, \dots]$$

Show that $E = I + \Delta$. Show that if p is a polynomial, then

$$p(E) = p(I) + p'(I)\Delta + \frac{1}{2}p''(I)\Delta^2 + \cdots + \frac{1}{m!}p^{(m)}(I)\Delta^m$$

15. (Continuation) Prove that if $x = [\lambda, \lambda^2, \lambda^3, \dots]$ and p is a polynomial, then $p(\Delta)x = p(\lambda - 1)x$. Describe how to solve a difference equation written in the form $p(\Delta)x = 0$.

16. (Continuation) Show that

$$\Delta^n = (-1)^n \left[E^0 - nE + \frac{1}{2}n(n-1)E^2 - \frac{1}{6}n(n-1)(n-2)E^3 + \cdots + (-1)^n E^n \right]$$

17. Give a complete proof of Theorem 2.

18. Let p be a polynomial such that $p(0) = 0$. Describe the null space of $p(E)$.

19. For $\lambda \in \mathbb{C}$, define $x(\lambda) = [\lambda, \lambda^2, \lambda^2, \ldots]$. Prove that if $\lambda_1, \lambda_2, \ldots, \lambda_m$ are distinct nonzero complex numbers, then $\{x(\lambda_1), x(\lambda_2), \ldots, x(\lambda_m)\}$ is a linearly independent set in V.

20. If λ is a nonzero root of p having multiplicity k, then the equation $p(E)x = 0$ has solutions $u^{(1)}, u^{(2)}, \ldots, u^{(k)}$ in which $u_n^{(j)} = n^{j-1}\lambda^n$.

21. Prove that if $\mu \in (0, \infty)$ and $|\lambda| < 1$, then $\lim_{n \to \infty} n^{\mu}\lambda^n = 0$.

22. Prove in detail that a convergent sequence is bounded.

23. Prove Theorem 3 without the hypothesis that $p(0) \neq 0$.

24. Define a sequence inductively by the equation $x_{n+1} = x_n + x_n^{-1}$, where $x_0 > 0$. Determine the behavior of x_n as $n \to \infty$.

25. Consider the recurrence relation $x_n = 2(x_{n-1} + x_{n-2})$. Show that the general solution is $z_n = \alpha\left(1 + \sqrt{3}\right)^n + \beta\left(1 - \sqrt{3}\right)^n$. Show that the solution with starting values $x_1 = 1$ and $x_2 = 1 - \sqrt{3}$ corresponds to $\alpha = 0$ and $\beta = \left(1 - \sqrt{3}\right)^{-1}$.

26. Determine whether the difference equation $x_n = x_{n-1} + x_{n-2}$ is stable.

27. If x is a solution of the difference equation $p(E)x = 0$, then so is Ex.

COMPUTER PROBLEMS 1.3

1. Consider the difference equation $x_{n+2} - 2x_{n+1} - 2x_n = 0$. One of its solutions is $x_n = \left(1 - \sqrt{3}\right)^{n-1}$. This solution oscillates in sign and converges to 0. Compute and print out the first 100 terms of this sequence by use of the equation $x_{n+2} = 2(x_{n+1} + x_n)$ starting with $x_1 = 1$ and $x_2 = 1 - \sqrt{3}$. Explain the curious phenomenon that occurs.

2. Starting with $J_0(1) = 0.76519\,76866$ and $J_1(1) = 0.44005\,05857$, use the recurrence formula in the text to compute $J_2(1), J_3(1), \ldots, J_{20}(1)$. What evidence is there that the values are in gross error?

3. Consider the difference equation $4x_{n+2} - 8x_{n+1} + 3x_n = 0$. Determine whether it is stable. Find the general solution. Assuming that $x_0 = 0$ and $x_1 = -2$, compute x_{100} using the most efficient method.

4. Compute the particular solution in Problem 25 in the following three ways. Compute these solutions for $1 \leqq n \leqq 100$ and compare them.

(a) x_n directly from the recurrence relation

(b) $y_n = \beta(1 - \sqrt{3})^n$

(c) $z_n = \alpha(1 + \sqrt{3})^n + \beta(1 - \sqrt{3})^n$, where α is chosen as the unit roundoff error for your computer

5. Solve the difference equation $x_{n+2} - (\pi + \pi^{-1})\, x_{n+1} + x_n = 0$ numerically with $x_0 = 1$ and $x_1 = \pi$. Compute 49 terms and determine the relative error in x_{50}. Then do the same with x_1 changed to π^{-1} and explain the difference in the relative errors of the two cases.

Chapter Two
Computer Arithmetic

2.1 Floating-Point Numbers and Roundoff Errors

Most computers deal with real numbers in the binary system, in contrast to the decimal system that humans prefer to use. The binary system uses 2 as the base in the same way that the decimal system uses 10. To make this comparison, recall first how our familiar number representation works. When a real number such as 427.325 is written out in more detail, we have

$$\begin{aligned} 427.325 = {} & 4 \times 10^2 + 2 \times 10^1 + 7 \times 10^0 \\ & + 3 \times 10^{-1} + 2 \times 10^{-2} + 5 \times 10^{-3} \end{aligned}$$

The expression on the right contains powers of 10 together with the digits 0, 1, 2, 3, 4, 5, 6, 7, 8, 9. If we admit the possibility of having an infinite number of digits to the right of the decimal point, then *any* real number can be expressed in the manner just illustrated, with a sign (+ or −) affixed to it. Thus, for example, $-\pi$ is

$$-\pi = -3.14159\,26535\,89793\,23846\,26433\,8\ldots$$

The last 8 written here stands for 8×10^{-26}.

In the binary system, only the two digits 0 and 1 are used. A typical number in the binary system can also be written out in detail; for example,

$$\begin{aligned} (1001.11101)_2 = {} & 1 \times 2^3 + 0 \times 2^2 + 0 \times 2^1 + 1 \times 2^0 \\ & + 1 \times 2^{-1} + 1 \times 2^{-2} + 1 \times 2^{-3} + 0 \times 2^{-4} + 1 \times 2^{-5} \end{aligned}$$

This is the same real number as 9.90625 in decimal notation. (Verify.)

In general, any integer $\beta > 1$ can be used as the base for a number system. Numbers represented in base β will contain digits $0, 1, 2, 3, 4, \ldots, (\beta - 1)$. If the context does not make it clear what base is being used for the number N, the notation $(N)_\beta$ can be employed. Thus, from above we have

$$(1001.11101)_2 = (9.90625)_{10}$$

Since the typical computer communicates with its human users in the **decimal system** but works internally in the **binary system**, conversion procedures must be executed by the computer. These come into use at input and output time. Ordinarily the user need not be concerned with these conversions, but they do involve small roundoff errors.

Computers are not able to operate using real numbers expressed with more than a fixed number of digits. The word length of the computer places a restriction on the precision with which real numbers can be represented. Even a simple number like $1/10$ cannot be stored exactly in the `Marc-32` computer (or in any binary machine). It requires an infinite binary expression:

$$\frac{1}{10} = (0.0001\ 1001\ 1001\ 1001\ \ldots)_2 \tag{1}$$

For example, if we read 0.1 into a 32-bit computer workstation and then print it out to 40 decimal places, we obtain the following result:

0.10000 00014 90116 11938 47656 25000 00000 00000

Usually, we won't notice this conversion error since printing using the default format would show us 0.1.

Also in using computers, we should be aware that there are *two* conversion techniques involved—from and to decimal because we prefer to do our calculations in decimal while the computer prefers binary. Errors can be involved in each conversion.

Rounding

Why do we discuss rounding now? Later in this chapter we present a detailed discussion of rounding and relate it to machine computations. Here we mention rounding only as it is related to computations done by hand or by using a pocket calculator. The reason for this is that in scientific computation the number of digits in the intermediate results may become larger and larger, while the number of significant digits remains fixed or decreases. For example, the product of two numbers that have eight digits to the right of the decimal point will be a number that has 16 digits to the right of the decimal point.

Rounding is an important concept in scientific computing. Consider a positive decimal number x of the form $0.\square\square\square \cdots \square\square\square$ with m digits to the right of the decimal point. One **rounds** x to n decimal places ($n < m$) in a manner that depends on the value of the $(n+1)$st digit. If this digit is a 0, 1, 2, 3, or 4, then the nth digit is not changed and all following digits are discarded. If it is a 5, 6, 7, 8, or 9, then the nth digit is increased by one unit and the remaining digits are discarded. (The situation with 5 as the $(n+1)$st digit can be handled in a variety of ways. For example, some choose to **round up** only when the previous digit is *even*, assuming that this will happen about half of the time. For simplicity, we always choose to round up in this situation.)

Here are some examples of seven-digit numbers being correctly rounded to four digits:

$$
\begin{aligned}
0.1735499 &\longrightarrow 0.1735 \\
0.9999500 &\longrightarrow 1.000 \\
0.4321609 &\longrightarrow 0.4322
\end{aligned}
$$

If x is *rounded* so that $\widetilde{x}$ is the n-digit approximation to it, then

$$|x - \widetilde{x}| \leqq \frac{1}{2} \times 10^{-n} \tag{2}$$

To see why this is true, we reason as follows: If the $(n+1)$st digit of x is 0, 1, 2, 3, or 4, then $x = \widetilde{x} + \varepsilon$ with $\varepsilon < \frac{1}{2} \times 10^{-n}$ and (2) follows. If it is 5, 6, 7, 8, or 9, then $\widetilde{x} = \widehat{x} + 10^{-n}$ where $\widehat{x}$ is a number with the same n digits as x and all digits beyond the nth are 0. Now $x = \widehat{x} + \delta \times 10^{-n}$ with $\delta \geqq \frac{1}{2}$ and $\widetilde{x} - x = (1 - \delta) \times 10^{-n}$. Since $1 - \delta \leqq \frac{1}{2}$, Inequality (2) follows.

If x is a decimal number, the **chopped** or **truncated** n-digit approximation to it is the number $\widehat{x}$ obtained by simply discarding all digits beyond the nth. For it, we have

$$|x - \widehat{x}| < 10^{-n} \tag{3}$$

The relationship between x and $\widehat{x}$ is such that $x - \widehat{x}$ has 0 in the first n places and $x = \widehat{x} + \delta \times 10^{-n}$ with $0 \leqq \delta < 1$. Hence, we have $|x - \widehat{x}| = |\delta| \times 10^{-n} < 10^{-n}$ and Inequality (3) follows.

Normalized Scientific Notation

In the decimal system, any real number can be expressed in **normalized scientific notation**. This means that the decimal point is shifted and appropriate powers of 10 are supplied so that all the digits are to the right of the decimal point and the first digit displayed is not 0. Examples are

$$
\begin{aligned}
732.5051 &= 0.7325051 \times 10^{3} \\
-0.005612 &= -0.5612 \times 10^{-2}
\end{aligned}
$$

In general, a nonzero real number x can be represented in the form

$$x = \pm r \times 10^{n}$$

where r is a number in the range $\frac{1}{10} \leqq r < 1$ and n is an integer (positive, negative, or zero). Of course, if $x = 0$, then $r = 0$; in all other cases, we can adjust n so that r lies in the given range.

In exactly the same way, we can use scientific notation in the *binary* system. Now we have

$$x = \pm q \times 2^m \tag{4}$$

where $\frac{1}{2} \leqq q < 1$ (if $x \neq 0$) and m is an integer. The number q is called the **mantissa** and the integer m the **exponent**. Both q and m are base 2 numbers.

A slight modification of scientific notation for binary numbers is to require that the leading binary digit 1 appears just to the left of the binary point. In this case, the representation would be the same as in Equation (4) except $q = (1.f)_2$ and $1 \leqq q < 2$. This form is useful when storing binary numbers in a computer word because one bit of space can be saved by *not* actually storing the leading bit 1 but assuming that it is there. This should become clear in the next subsection.

Hypothetical Computer `Marc-32`

Within a typical computer, numbers are represented in the way just described, but with certain restrictions placed on q and m that are imposed by the available word length. To illustrate this, we consider a hypothetical computer that we call the `Marc-32`. It has a word length of 32 bits (binary digits) and, therefore, is similar to many personal and workstation computers.

The floating-point representation for a single-precision real number in the hypothetical 32-bit computer `Marc-32` is divided into three fields as shown in Figure 2.1.

The bits composing a word in the `Marc-32` are allocated in the following way when representing a nonzero real number $x = \pm q \times 2^m$:

sign of the real number x	1 bit
biased exponent (integer e)	8 bits
mantissa part (real number f)	23 bits

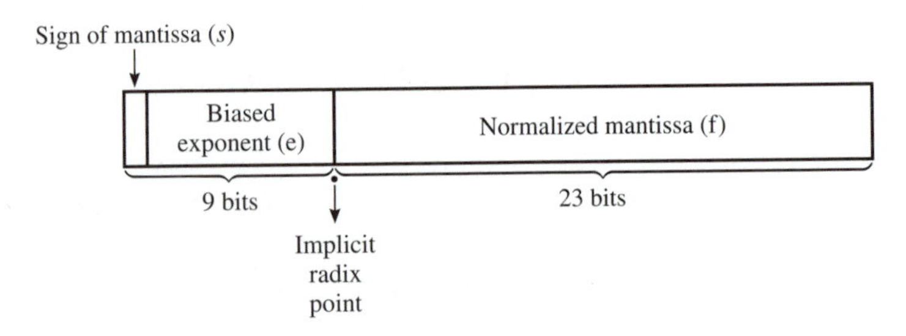

FIGURE 2.1 `Marc-32` single-precision field

A nonzero real number $x = \pm q \times 2^m$ can be written as a *normalized* binary number such that the first nonzero bit in the mantissa is just before the binary point, that is, $q = (1.\text{f})_2$. This bit can be assumed to be 1 and does not require storage. The mantissa is in the range $1 \leqq q < 2$. The 23 bits reserved in the word for the mantissa can be used to store 23 bits from f. So in effect, the machine has a 24-bit mantissa for its floating-point numbers.

Hence, nonzero normalized machine numbers are bit strings whose values are decoded as follows:

$$x = (-1)^s q \times 2^m \tag{5}$$

where

$$q = (1.\text{f})_2 \quad \text{and} \quad m = \text{e} - 127$$

Here $1 \leqq q < 2$ and the most significant bit in q is 1 and is not explicitly stored. Also, here s is the bit representing the sign of x (positive: bit 0, negative: bit 1), $m = \text{e} - 127$ is the 8-bit biased exponent, and f is the 23-bit fractional part of the real number x that, together with an implicit leading bit 1, yields the significant digit field $(1.\square\square\square \cdots \square\square\square)_2$.

A real number expressed as in Equation (5) is said to be in **normalized floating-point form**. If it can then be represented with $|m|$ occupying 8 bits and q occupying 23 bits, it is a **machine number** in the `Marc-32`. That is, it can be precisely represented within this particular computer. Most real numbers are not precisely representable within the `Marc-32`. When such a number occurs as an input datum or as the result of a computation, an inevitable error will arise in representing it as accurately as possible by a *machine* number.

The restriction that $|m|$ require no more than 8 bits means that

$$0 < \text{e} < (11\,111\,111)_2 = 2^8 - 1 = 255$$

and the values $\text{e} = 0$ and $\text{e} = 255$ are reserved for special cases such as ± 0, $\pm\infty$, and `NaN` (Not a Number). Since $m = \text{e} - 127$, we take $-126 \leqq m \leqq 127$ and the `Marc-32` can handle numbers as small as $2^{-126} \approx 1.2 \times 10^{-38}$ and as large as $(2-2^{-23})2^{127} \approx 3.4 \times 10^{38}$. This is not a sufficiently generous range of magnitudes for some scientific calculations, and for this reason and others, we occasionally must write a program in **double-precision** or **extended-precision** arithmetic. A floating-point number in double precision is represented in two computer words, and the mantissa usually has at least twice as many bits. Hence, there are roughly twice the number of decimal places of accuracy in double precision as in single precision. In double precision, calculations are much slower than in single precision, often by a factor of 2 or more. This is because double-precision arithmetic is usually done using software, whereas single-precision is done by the hardware.

The restriction that the mantissa part require no more than 23 bits means that our machine numbers have a limited precision of roughly six decimal places, since the least significant bit in the mantissa represents units of 2^{-23} (or approximately 1.2×10^{-7}). Thus, numbers expressed with more than six decimal digits will be *approximated* when given as input to the computer. Also, some *simple* decimal numbers such as $1/100$ are not machine numbers on a binary computer!

Floating-point numbers in a binary computer are distributed rather unevenly, more of them being concentrated near 0. (In fact, there are lots of gaps or holes in the real number line. One of these is often called the "hole at zero.") There are only a finite number of floating-point numbers in the computer, and between adjacent powers of 2 there are always the same number of machine numbers. Since gaps between powers of 2 are smaller near 0 and larger away from 0, this produces a nonuniform distribution of floating-point numbers, with higher density near the origin.

An *integer* can use all of the computer word in its representation except for a single bit that must be reserved for the sign. Hence in the `Marc-32`, integers range from $-(2^{31}-1)$ to $2^{31}-1 = 21474\,83647$. In scientific computations, purely integer calculations are not common.

Zero, Infinity, NaN

In IEEE standard arithmetic, there are two forms of 0: $+0$ and -0, in single precision, are represented in the computer by the words $[00000000]_{16}$ and $[80000000]_{16}$, respectively. Most arithmetic operations that result in a 0 value are given the value $+0$. A very tiny negative number that is 0 to machine precision is given the value -0.

Similarly, there are two forms of infinity: $+\infty$ and $-\infty$, in single precision, are represented by the computer words $[7F800000]_{16}$ and $[FF800000]_{16}$, respectively. Usually, infinity is treated like a very large number whenever it makes sense to do so. For example, suppose that x is a floating-point number in the range $0 < x < \infty$; then each of the computations $x + \infty$, $x * \infty$, and ∞/x is given the value $+\infty$, whereas x/∞ becomes $+0$. Here ∞ is understood to be $+\infty$. Similar results hold for $-\infty$.

`NaN` means **Not a Number** and results from an indeterminate operation such as $0/0$, $\infty - \infty$, $x + \text{NaN}$, and so on. `NaN`'s are represented by computer words with $\text{e} = 255$ and $\text{f} \neq 0$.

Machine Rounding

In addition to rounding input data, rounding is needed after most arithmetic operations. The result of an arithmetic operation resides in a long 80-bit computer register and must be rounded to single-precision before being placed in memory. An analogous situation occurs for double-precision operations.

The usual (default) rounding mode is **round to nearest**: The closer of the two machine numbers on the left and right of the real number is selected. In the case of a tie, **round to even** is used: If the real number is exactly halfway between the machine numbers to its left and right, then the even machine number is chosen. With this default rounding scheme (round to nearest plus round to even), the maximum error is half a unit in the least significant place.

Other modes of rounding are **directed rounding** such as **round toward** 0 (also known as **truncation**), **round toward** $+\infty$, and **round toward** $-\infty$.

TABLE 2.1 Results from some Fortran 90 Intrinsic Procedures: Sun 4

X	single precision	double precision
`DIGITS(X)`	24	53
`EPSILON(X)`	$1.19 \times 10^{-7} \approx 2^{-23}$	$2.22 \times 10^{-16} \approx 2^{-52}$
`HUGE(X)`	$3.40 \times 10^{38} \approx (2 - 2^{-23})2^{127}$	$1.80 \times 10^{308} \approx (2 - 2^{-23})2^{1023}$
`PRECISION(X)`	6	15
`TINY(X)`	$1.18 \times 10^{-38} \approx 2^{-126}$	$2.23 \times 10^{-308} \approx 2^{-1022}$

Fortran 90

In Fortran 90, a large number of generic intrinsic procedures are available for determining the numerical environment in the computer being used. In general, they return a number of the same **type** (real, integer, complex, etc.) and **kind** (single precision, double precision, etc.) as the argument. For example, some of these procedures related to floating-point numbers are: `DIGITS`—number of significant (binary) digits, `EPSILON`—a positive number that is almost negligible compared to unity (smallest floating-point number ε such that $1 + \varepsilon \neq 1$), `HUGE`—largest number, `PRECISION`—decimal precision, and `TINY`—smallest positive number. Here we assume the machine is a binary computer.

Table 2.1 is constructed from the results of calling these procedures on a 32-bit-word workstation with IEEE standard arithmetic (Sun 4). For `X` integer, `DIGITS(X)` is 31 and `HUGE(X)` is $2147483647 \approx 2^{31} - 1$.

Table 2.2 is constructed from the results of calling these procedures on a 64-bit-word supercomputer without IEEE standard arithmetic (Cray Y-MP). For `X` integer, `DIGITS(X)` is 46 and `HUGE(X)` is $70368744177663 \approx 2^{46} - 1$.

IEEE Standard Floating-Point Arithmetic

The `Marc-32` representation for real numbers is patterned after the usual floating-point representation in 32-bit machines, which is the **IEEE standard floating-point representation**. We have chosen to give only a brief description here. For example, computers that implement floating-point arithmetic according to the current *official* standard use 80 bits for internal calculations. There are many additional concepts—**guard bit**, **round bit**, **sticky bit**, **denormalized numbers**, **unnormalized numbers**, **double rounding**, and others—that enter into any detailed discussion of this subject. We choose not to define or discuss them here because of space considerations, but we invite the interested reader to consult the following

TABLE 2.2 Results from some Fortran 90 Intrinsic Procedures: Cray Y-MP

X	single precision	double precision
`DIGITS(X)`	47	95
`EPSILON(X)`	$1.42 \times 10^{-14} \approx 2^{-46}$	$5.05 \times 10^{-29} \approx 2^{-94}$
`HUGE(X)`	$1.36 \times 10^{2465} \approx (2 - 2^{-46})2^{8188}$	$1.36 \times 10^{2465} \approx (2 - 2^{-46})2^{8188}$
`PRECISION(X)`	13	28
`TINY(X)`	$7.33 \times 10^{-2466} \approx 2^{-8189}$	$7.33 \times 10^{-2466} \approx 2^{-8189}$

references for additional information: *Standard for Binary Floating-Point Arithmetic* ANSI/IEEE [1985] and *A Radix-Independent Standard for Floating-Point Arithmetic* ANSI/IEEE [1987]. Among the articles of interest on this topic are Cody [1988], Cody et al. [1984], Coonen [1981], Fosdick [1993], Hough [1981], Raimi [1969], and Scott [1985].

Nearby Machine Numbers

We now want to assess the error involved in approximating a given positive real number x by a nearby machine number in the `Marc-32`. We assume that

$$x = q \times 2^m \qquad 1 \leqq q < 2 \qquad -126 \leqq m \leqq 127$$

We ask, *What is the machine number closest to x?* First, we write

$$x = (1.a_1 a_2 \dots a_{23} a_{24} a_{25} \dots)_2 \times 2^m$$

in which each a_i is either 0 or 1. One nearby machine number is obtained by simply discarding the excess bits $a_{24}a_{25}\dots$. This procedure is usually called **chopping**. The resulting number is

$$x_- = (1.a_1 a_2 \dots a_{23})_2 \times 2^m$$

Observe that x_- lies to the left of x on the real line. Another nearby machine number lies to the right of x. It is obtained by **rounding up**; that is, we drop the excess bits as before but increase the last remaining bit a_{23} by one unit. This number is

$$x_+ = \left((1.a_1 a_2 \dots a_{23})_2 + 2^{-23}\right) \times 2^m$$

There are two situations, as shown in Figure 2.2. The closer of x_- and x_+ is chosen to represent x in the computer.

If x is represented better by x_-, then we have

$$|x - x_-| \leqq \frac{1}{2}|x_+ - x_-| = \frac{1}{2} \times 2^{m-23} = 2^{m-24}$$

In this case, the *relative* error is bounded as follows:

$$\left|\frac{x - x_-}{x}\right| \leqq \frac{2^{m-24}}{q \times 2^m} = \frac{1}{q} \times 2^{-24} \leqq 2^{-24}$$

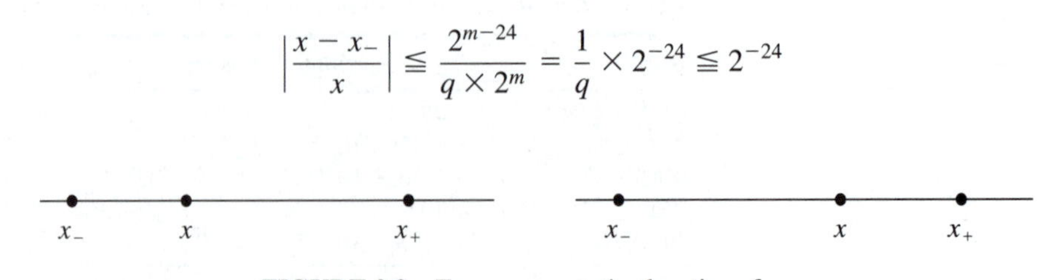

FIGURE 2.2 Two representative locations for x

In the second case, x is closer to x_+ than to x_- and we have

$$|x - x_+| \leqq \frac{1}{2}|x_+ - x_-| = 2^{m-24}$$

The same analysis then shows the relative error to be no greater than 2^{-24}. (A different bound applies to the *chopping* procedure; see Problem 28.)

Occasionally, in the course of a computation, a number is produced of the form $\pm q \times 2^m$ where m is *outside* the range permitted by the computer. If m is too large, we say that an **overflow** has occurred. If m is too small, we say that an **underflow** has occurred. In the `Marc-32`, this means $m > 127$ or $m < -126$, respectively. In the first case (overflow), a fatal error condition exists. In many computers, the execution of the program will be halted automatically when the variable containing the `NaN` or machine infinity is used in a nonsensical calculation. In the second case (underflow), the variable is simply set to 0 in many computers and the computations are allowed to proceed. (A message is issued to alert the user that an underflow has occurred.)

What has just been said about the `Marc-32` can be summarized: If x is a nonzero real number within the range of the machine, then the machine number x^* closest to x satisfies the inequality

$$\left|\frac{x - x^*}{x}\right| \leqq 2^{-24}$$

By letting $\delta = (x^* - x)/x$, we can write this inequality in the form

$$\text{fl}(x) = x(1 + \delta) \qquad |\delta| \leqq 2^{-24} \tag{6}$$

The notation $\text{fl}(x)$ is used to denote the floating-point machine number x^* closest to x. The number 2^{-24} that occurs in the preceding inequalities is called the **unit roundoff error** for the `Marc-32`. Our analysis shows that the number of bits allocated to the mantissa is directly related to the unit roundoff error of the machine and determines the accuracy of the computer arithmetic. (For the `Marc-32`, the machine epsilon is twice as big as the unit roundoff error.)

The results we have described for the `Marc-32` hold true, with suitable modifications, for other machines. If a machine operates with base β and carries n places in the mantissa of its floating-point numbers, then

$$\text{fl}(x) = x(1 + \delta) \qquad |\delta| \leqq \varepsilon$$

where $\varepsilon = \frac{1}{2}\beta^{1-n}$ in the case of correct rounding and $\varepsilon = \beta^{1-n}$ in the case of chopping. The number ε is the unit roundoff error and is a characteristic of the computing machine, its operating system, and the mode of computation (whether single or multiple precision).

The value of the unit roundoff error (ε) varies widely on modern computers because of differences in the word lengths of the machines, differences in the base used for the arithmetic, the type of rounding employed, and so on. The word lengths vary from 64 and 60 for large scientific computers to 32 and 36 for mid-size machines and down to 16 bits for some personal computers. Programmable calculators may have even less precision. Many computers use binary arithmetic, but hexadecimal and octal are also used. Different types of rounding used are **perfect rounding**, **pseudo-rounding**, and **chopping**; some compilers allow the user to set a switch to select the type of rounding used in the calculations.

Suppose that $x = q \times 2^m$, a positive nonzero *machine* number. By changing the last bit in q, we obtain the next (larger) machine number on the right

$$x_r = (q + 2^{-23}) \times 2^m$$

and the previous (smaller) machine number on the left

$$x_\ell = (q - 2^{-23}) \times 2^m$$

Clearly, we have

$$x_r - x = x - x_\ell = 2^{m-23}$$

Hence,

$$\frac{x_r - x}{x} = \frac{x - x_\ell}{x} = \frac{1}{q} \times 2^{-23}$$

Since $1 \leqq q < 2$, we have

$$2^{-24} < \frac{x_r - x}{x} = \frac{x - x_\ell}{x} \leqq 2^{-23}$$

Hence, the relative spacing between any machine number x and the two machine numbers on either side of it, x_ℓ and x_r, is approximately a constant value, namely, 2^{-23}. This value is precisely the precision of the machine representation.

Example 1 What is the binary form of the number $x = 2/3$? What are the two nearby machine numbers x_- and x_+ in the `Marc-32`? Which of these is taken to be $\text{fl}(x)$? What are the absolute roundoff error and the relative roundoff error in representing x by $\text{fl}(x)$?

Solution To determine the binary representation, we write

$$\frac{2}{3} = (0.a_1 a_2 a_3 \ldots)_2$$

We multiply by 2 to obtain

$$\frac{4}{3} = (a_1.a_2 a_3 \ldots)_2$$

Therefore, we get $a_1 = 1$ by taking the integer part of both sides. Subtracting 1 from both sides, we have

$$\frac{1}{3} = (0.a_2a_3a_4\ldots)_2$$

Repeating the previous steps, we eventually arrive at

$$x = \frac{2}{3} = (0.1010\ldots)_2 = (1.010101\ldots)_2 \times 2^{-1}$$

The two nearby machine numbers are

$$\begin{aligned} x_- &= (1.0101\ldots010)_2 \times 2^{-1} \\ x_+ &= (1.0101\ldots011)_2 \times 2^{-1} \end{aligned}$$

Here x_- is obtained by chopping and x_+ by rounding up. There are 23 bits to the right of the binary points.

To determine which is closer to x, we compute $x - x_-$ and $x_+ - x$ and, thereby, decide which of x_- or x_+ should be taken to be $\text{fl}(x)$:

$$\begin{aligned} x - x_- &= (0.1010\ldots)_2 \times 2^{-24} = \frac{2}{3} \times 2^{-24} \\ x_+ - x &= (x_+ - x_-) - (x - x_-) = 2^{-24} - \frac{2}{3} \times 2^{-24} = \frac{1}{3} \times 2^{-24} \end{aligned}$$

Hence, we set $\text{fl}(x) = x_+$, and the absolute roundoff error is

$$|\text{fl}(x) - x| = \frac{1}{3} \times 2^{-24}$$

The relative roundoff error is then

$$\frac{|\text{fl}(x) - x|}{|x|} = \frac{\frac{1}{3} \times 2^{-24}}{2/3} = 2^{-25}$$

It is of interest to determine the internal representation of the machine numbers x_- and x_+. We find that $\text{e} = (126)_{10} = (176)_8 = (001\ 111\ 110)_2$ since the exponent is -1. Thus, the internal representation is

$$\begin{aligned} x_- &= [0011\ 1111\ 0010\ 1010\ 1010\ 1010\ 1010\ 1010]_2 = [\text{3F2AAAAB}]_{16} \\ x_+ &= [0011\ 1111\ 0010\ 1010\ 1010\ 1010\ 1010\ 1011]_2 = [\text{3F2AAAAC}]_{16} \end{aligned}$$

When printed out, the resulting decimal numbers are

$$\begin{aligned} x_- &= 0.66666\ 66865\ 34881\ 59179\ 68750\ 000 \\ x_+ &= 0.66666\ 67461\ 39526\ 36718\ 75000\ 000 \end{aligned}$$

Here $0.00000\,00596\,04644\,77539\,06250\,000 = 2^{-24}$ is the absolute spacing between them. ■

Floating-Point Error Analysis

In continuing our study of the errors in machine computation that are attributable directly to the limited word length, we shall use the `Marc-32` as a model. We assume that the design of this machine is such that whenever two machine numbers are to be combined arithmetically, the combination is first correctly formed, then normalized, rounded off, and finally stored in memory in a machine word. To make this clearer, let the symbol $\odot$ stand for any one of the four basic arithmetic operations $+$, $-$, $*$, or $\div$. If x and y are machine numbers, and if $x \odot y$ is to be computed and stored, then the closest that we can come to actually fitting $x \odot y$ in a single machine word is to round it off to $\text{fl}(x \odot y)$ and store that number.

Example 2 To illustrate this process, let us use a decimal machine operating with five decimal digits in its floating-point number system, and determine the relative errors in adding, subtracting, multiplying, and dividing the two machine numbers

$$x = 0.31426 \times 10^3 \qquad y = 0.92577 \times 10^5$$

Solution Using a double-length accumulator for the intermediate results, we have

$$\begin{aligned} x + y &= 0.92891\,26000 \times 10^5 \\ x - y &= -0.92262\,74000 \times 10^5 \\ x * y &= 0.29093\,24802 \times 10^8 \\ x \div y &\approx 0.33945\,79647 \times 10^{-2} \end{aligned}$$

The computer with five decimal digits will store these in rounded form as

$$\begin{aligned} \text{fl}(x + y) &= 0.92891 \times 10^5 \\ \text{fl}(x - y) &= -0.92263 \times 10^5 \\ \text{fl}(x * y) &= 0.29093 \times 10^8 \\ \text{fl}(x \div y) &= 0.33946 \times 10^{-2} \end{aligned}$$

The relative errors in these results are 2.3×10^{-6}, 2.8×10^{-6}, 8.5×10^{-6}, and 6.0×10^{-6}, respectively—all less than 10^{-5}. ■

In any computer, it is desirable to know that the four arithmetic operations satisfy equations like

$$\text{fl}(x \odot y) = [x \odot y](1 + \delta) \qquad |\delta| \leqq \varepsilon$$

We shall assume, in dealing with general computers, that this equation is valid and that ε can be taken to be the **unit roundoff** for that machine. In any well-designed computer, this assumption will be either literally true or so close to being true that the discrepancy can be neglected in roundoff analysis.

By Equation (6), we previously established that for the `Marc-32`,

$$\text{fl}(x) = x(1 + \delta) \qquad |\delta| \leqq 2^{-24}$$

for any real number x within the range of the `Marc-32`. Hence, we have

$$\text{fl}(x \odot y) = (x \odot y)(1 + \delta) \qquad |\delta| \leqq 2^{-24}$$

if x and y are machine numbers. Thus, the unit roundoff, 2^{-24}, gives a bound for the *relative error* in any single basic arithmetic operation. An examination of the preceding numerical examples will show why roundoff error must be expected in every arithmetic operation. If x and y are not necessarily machine numbers, the corresponding result is

$$\text{fl}\big(\text{fl}(x) \odot \text{fl}(y)\big) = \big(x(1 + \delta_1) \odot y(1 + \delta_2)\big)(1 + \delta_3) \qquad |\delta_i| \leqq 2^{-24}$$

Using the basic result just given, we can analyze compound arithmetic operations. To illustrate, suppose that x, y, and z are machine numbers in the `Marc-32`, and we want to compute $x(y + z)$. We have

$$\begin{aligned}
\text{fl}[x(y + z)] &= [x\,\text{fl}(y + z)](1 + \delta_1) && |\delta_1| \leqq 2^{-24}\\
&= [x(y + z)(1 + \delta_2)](1 + \delta_1) && |\delta_2| \leqq 2^{-24}\\
&= x(y + z)(1 + \delta_2 + \delta_1 + \delta_2\delta_1)\\
&\approx x(y + z)(1 + \delta_1 + \delta_2)\\
&= x(y + z)(1 + \delta_3) && |\delta_3| \leqq 2^{-23}
\end{aligned}$$

since $\delta_2\delta_1$ is negligible in comparison with 2^{-23}.

Since the `Marc-32` is a hypothetical computer, we may make whatever assumptions we wish about how it computes and stores floating-point numbers. For a real computer, the type of assumption we have made about $\text{fl}(x \odot y)$ is very close to the truth and can be used for reliable error estimates. However, a moment's reflection will show that a computer will not be able to form all combinations $x \odot y$ with complete accuracy before rounding them to obtain machine numbers. We saw in Example 1 that $2/3$ cannot be computed precisely in floating point. In actual practice, many computers will carry out arithmetic operations in special registers that have more bits than the usual machine numbers. These extra bits are called **guard bits** and allow for numbers to exist temporarily with extra precision. Then, of course, a rounding procedure is applied to a number in this special register to create a machine number. The number of guard bits and other details differ from machine to machine, and it is sometimes difficult to learn just how a specific machine handles these matters. This subject can be studied further in Sternbenz [1974] and in numerous papers in the computing literature. See also Feldstein and Turner [1986], Gregory [1980], Rall [1965], Scott [1985], and Waser and Flynn [1982].

Relative Error Analysis

We present next a theorem that illustrates how the relative roundoff error in long calculations can be analyzed. The theorem states roughly that when $n + 1$ *positive* machine numbers are added, the relative roundoff error should not exceed $n\varepsilon$, where ε is the unit roundoff error of the machine being used. (This result should be easy to remember because n is the number of additions involved.)

THEOREM 1 *Let $x_0, x_1, \ldots, x_n$ be positive machine numbers in a computer whose unit roundoff error is ε. Then the relative roundoff error in computing $\sum_{i=0}^{n} x_i$ in the usual way is at most $(1 + \varepsilon)^n - 1$. This quantity is approximately $n\varepsilon$.*

Proof Let $S_k = x_0 + x_1 + \cdots + x_k$, and let S_k^* be what the computer calculates instead of S_k. The recursive formulas for these quantities are

$$\begin{cases} S_0 = x_0 \\ S_{k+1} = S_k + x_{k+1} \end{cases} \qquad \begin{cases} S_0^* = x_0 \\ S_{k+1}^* = \text{fl}(S_k^* + x_{k+1}) \end{cases}$$

For the analysis, we define

$$\rho_k = \frac{S_k^* - S_k}{S_k} \qquad \delta_k = \frac{S_{k+1}^* - (S_k^* + x_{k+1})}{S_k^* + x_{k+1}}$$

Thus, $|\rho_k|$ is the relative error in approximating the kth partial sum S_k by the computed partial sum S_k^*, and $|\delta_k|$ is the relative error in approximating $S_k^* + x_{k+1}$ by the quantity $\text{fl}(S_k^* + x_{k+1})$. Using the equations that define ρ_k and δ_k, we have

$$\begin{aligned} \rho_{k+1} &= \left(S_{k+1}^* - S_{k+1}\right)/S_{k+1} \\ &= \left[(S_k^* + x_{k+1})(1 + \delta_k) - (S_k + x_{k+1})\right]/S_{k+1} \\ &= \left\{\left[S_k(1 + \rho_k) + x_{k+1}\right](1 + \delta_k) - (S_k + x_{k+1})\right\}/S_{k+1} \\ &= \delta_k + \rho_k\left(S_k/S_{k+1}\right)(1 + \delta_k) \end{aligned}$$

Since $S_k < S_{k+1}$ and $|\delta_k| \leqq \varepsilon$, we can conclude that

$$|\rho_{k+1}| \leqq \varepsilon + |\rho_k|(1 + \varepsilon) = \varepsilon + \theta|\rho_k|$$

where we have set $\theta = 1 + \varepsilon$. Thus, we have the successive inequalities

$$\begin{aligned} &|\rho_0| = 0 \\ &|\rho_1| \leqq \varepsilon \\ &|\rho_2| \leqq \varepsilon + \theta\varepsilon \\ &|\rho_3| \leqq \varepsilon + \theta(\varepsilon + \theta\varepsilon) = \varepsilon + \theta\varepsilon + \theta^2\varepsilon \\ &\quad\vdots \end{aligned}$$

The general result is

$$\begin{aligned}
|\rho_n| &\leqq \varepsilon + \theta\varepsilon + \theta^2\varepsilon + \theta^3\varepsilon + \cdots + \theta^{n-1}\varepsilon \\
&= \varepsilon(1 + \theta + \cdots + \theta^{n-1}) \\
&= \varepsilon\left[(\theta^n - 1)/(\theta - 1)\right] \\
&= \varepsilon\left\{\left[(1+\varepsilon)^n - 1\right]/\varepsilon\right\} \\
&= (1+\varepsilon)^n - 1
\end{aligned}$$

By the Binomial Theorem, we have

$$(1+\varepsilon)^n - 1 = 1 + \binom{n}{1}\varepsilon + \binom{n}{2}\varepsilon^2 + \binom{n}{3}\varepsilon^3 + \cdots - 1 \approx n\varepsilon \qquad ■$$

Example 3 Suppose that a number x is defined by an infinite series

$$x = \sum_{k=1}^{\infty} s_k$$

where s_k are given real numbers. We intend to approximate x by a two-stage process. First we compute the partial sum

$$x_n = \sum_{k=1}^{n} s_k$$

for a large value of n, and then we round off x_n, retaining a certain number of digits after the decimal point. Let us say that m digits are retained. Can we be sure that the last digit (the mth one after the decimal point) is correct?

Solution Let $\widetilde{x}_n$ be the rounded value of x_n. We want

$$|\widetilde{x}_n - x| \leqq \frac{1}{2} \times 10^{-m}$$

If x_n is correctly rounded to obtain $\widetilde{x}_n$, then

$$|\widetilde{x}_n - x_n| \leqq \frac{1}{2} \times 10^{-m}$$

But

$$|\widetilde{x}_n - x| \leqq |\widetilde{x}_n - x_n| + |x_n - x| \leqq \frac{1}{2} \times 10^{-m} + |x_n - x|$$

This inequality cannot be improved, and it shows that unless $x_n = x$, we cannot achieve $|\widetilde{x}_n - x| \leqq \frac{1}{2} \times 10^{-m}$.

A more realistic demand is that $|\tilde{x}_n - x| < \frac{6}{10} \times 10^{-m}$. Then we are led to the following requirement on x_n:

$$\frac{1}{2} \times 10^{-m} + |x_n - x| < \frac{6}{10} \times 10^{-m} \tag{7}$$

or $|x_n - x| < 10^{-m-1}$. From this requirement, the number n can be obtained. The inequality is equivalent to

$$\left| \sum_{k=n+1}^{\infty} s_k \right| < 10^{-m-1}$$ ■

PROBLEMS 2.1

1. If $1/10$ is correctly rounded to the normalized binary number $(1.a_1a_2 \ldots a_{23})_2 \times 2^m$, what is the roundoff error? What is the relative roundoff error?
2. **(a)** If $3/5$ is correctly rounded to the binary number $(1.a_1a_2 \ldots a_{24})_2$, what is the roundoff error?

   (b) Answer part **(a)** for the number $2/7$.
3. Is $\frac{2}{3}(1 - 2^{-24})$ a machine number in the Marc-32? Explain.
4. Let $x_1, x_2, \ldots, x_n$ be positive machine numbers in the Marc-32. Let S_n denote the sum $x_1 + x_2 + \cdots + x_n$, and let S_n^* be the corresponding sum in the computer. (Assume that the addition is carried out in the order given.) Prove the following: If $x_{i+1} \geqq 2^{-24} S_i$ for each i, then

$$|S_n^* - S_n|/S_n \leqq (n-1)2^{-24}$$

5. Prove this slight improvement of Inequality (6)

$$\left| \frac{x - x^*}{x} \right| \leqq \frac{1}{1 + 2^{24}}$$

   for the representation of numbers in the Marc-32.
6. How many normalized machine numbers are there in the Marc-32? (Do not count 0.)
7. Does each machine number in the Marc-32 have a unique normalized representation?
8. Let $x = (1.11 \ldots 111000 \ldots)_2 \times 2^{16}$, in which the fractional part has 26 1's followed by 0's. For the Marc-32, determine x_-, x_+, $\text{fl}(x)$, $x - x_-$, $x_+ - x$, $x_+ - x_-$, and $|x - \text{fl}(x)|/x$.
9. Let $x = 2^3 + 2^{-19} + 2^{-22}$. Find the machine numbers on the Marc-32 that are just to the right and just to the left of x. Determine $\text{fl}(x)$, the absolute error $|x - \text{fl}(x)|$, and the relative error $|x - \text{fl}(x)|/|x|$. Verify that the relative error in this case does not exceed 2^{-24}.

10. Find the machine number just to the right of $1/9$ in a binary computer with a 43-bit normalized mantissa.

11. What is the exact value of $x^* - x$, if $x = \sum_{n=1}^{26} 2^{-n}$ and x^* is the nearest machine number on the `Marc-32`?

12. Let $S_n = x_1 + x_2 + \cdots + x_n$, where each x_i is a machine number. Let S_n^* be what the machine computes. Then $S_n^* = \mathrm{fl}(S_{n-1}^* + x_n)$. Prove that on the `Marc-32`,

$$S_n^* \approx S_n + S_2\delta_2 + \cdots + S_n\delta_n \qquad |\delta_k| \leqq 2^{-24}$$

13. Which of these is not necessarily true on the `Marc-32`? (Here x, y, and z are machine numbers and $|\delta| \leqq 2^{-24}$.)

(a) $\mathrm{fl}(xy) = xy(1 + \delta)$

(b) $\mathrm{fl}(x + y) = (x + y)(1 + \delta)$

(c) $\mathrm{fl}(xy) = xy/(1 + \delta)$

(d) $|\,\mathrm{fl}(xy) - xy| \leqq |xy|2^{-24}$

(e) $\mathrm{fl}(x + y + z) = (x + y + z)(1 + \delta)$

14. Use the `Marc-32` for this problem. Determine a bound on the relative error in computing $(a \cdot b)/(c \cdot d)$ for machine numbers a, b, c, d.

15. Are these machine numbers in the `Marc-32`?

(a) 10^{40}

(b) $2^{-1} + 2^{-26}$

(c) $1/5$

(d) $1/3$

(e) $1/256$

16. Let $x = 2^{16} + 2^{-8} + 2^{-9} + 2^{-10}$. Let x^* be the machine number closest to x in the `Marc-32`. What is $|x - x^*|$?

17. Criticize the following argument: When two machine numbers are combined arithmetically in the `Marc-32`, the relative roundoff error cannot exceed 2^{-24}. Therefore, when n such numbers are combined, the relative roundoff error cannot exceed $(n - 1)2^{-24}$.

18. Let $x = 2^{12} + 2^{-12}$.

(a) Find the machine numbers x_- and x_+ in the `Marc-32` that are just to the left and right of x, respectively.

(b) For this number show that the relative error between x and $\mathrm{fl}(x)$ is no greater than the unit roundoff error in the `Marc-32`.

19. What relative roundoff error is possible in computing the product of n machine numbers in the `Marc-32`? How is your answer changed if the n numbers are not necessarily machine numbers (but are within the range of the machine)?

20. Give examples of real numbers x and y for which $\mathrm{fl}(x \odot y) \neq \mathrm{fl}(\mathrm{fl}(x) \odot \mathrm{fl}(y))$. Illustrate all four arithmetic operations, using a machine with five decimal digits.

21. When we write $\prod_{i=1}^{n}(1+\delta_i) = 1+\varepsilon$, where $|\delta_i| \leqq 2^{-24}$, what is the range of possible values for ε? Is $|\varepsilon| \leqq n2^{-24}$ a realistic bound?

22. Suppose that numbers $z_1, z_2, \ldots$ are computed from data $x, a_1, a_2, \ldots$ by means of the algorithm

$$\begin{cases} z_1 = a_1 \\ z_n = xz_{n-1} + a_n \qquad (n \geqq 2) \end{cases}$$

(This is **Horner's algorithm**.) Assume that the data are machine numbers. Show that the z_n produced in the computer are the numbers that would result from applying exact arithmetic to perturbed data. Bound the perturbation in terms of the machine's unit roundoff error.

23. The quantity $(1+\varepsilon)^n - 1$ occurs in the theorem of this section. Prove that if $n\varepsilon < 0.01$, then $(1+\varepsilon)^n - 1 < 0.01006$.

24. Establish Equation (7) from the hypotheses given in the text.

25. How many floating-point numbers are there between successive powers of 2 in the `Marc-32`?

26. What numbers other than positive integers can be used as a base for a number system? For example, can we use π? (See, for example, Rousseau [1995].)

27. What is the unit roundoff error for a binary machine carrying 48-bit mantissas?

28. If the `Marc-32` did not round off numbers correctly but simply dropped excess bits, what would the unit roundoff be?

29. What is the unit roundoff error for a *decimal* machine that allocates 12 decimal places to the mantissa? Such a machine stores numbers in the form $x = \pm r \times 10^n$ with $\frac{1}{10} \leqq r < 1$.

30. Prove that 4/5 is not representable exactly on the `Marc-32`. What is the closest machine number? What is the relative roundoff error involved in storing this number on the `Marc-32`?

31. What numbers are representable with a finite expression in the binary system but are not finitely representable in the decimal system?

32. What can be said of the relative roundoff error in adding n machine numbers? (Make no assumption about the numbers being positive because this case is covered by a theorem in the text.)

33. Find a real number x in the range of the `Marc-32` such that $\mathrm{fl}(x) = x(1+\delta)$ with $|\delta|$ as large as possible. Can the bound 2^{-24} be attained by $|\delta|$?

34. Show that, under the assumptions made about the `Marc-32`, we shall have $\mathrm{fl}(x) = x/(1+\delta)$ with $|\delta| \leqq 2^{-24}$.

35. Show that $\mathrm{fl}(x^k) = x^k(1+\delta)^{k-1}$, where $|\delta| \leqq \varepsilon$, if x is a floating-point machine number in a computer that has unit roundoff ε.

36. Show by examples that often $\mathrm{fl}[\mathrm{fl}(xy)z] \neq \mathrm{fl}[x\,\mathrm{fl}(yz)]$ for machine numbers x, y, and z. This phenomenon is often described informally by saying *machine multiplication is not associative.*

37. Prove that if x and y are machine numbers in the `Marc-32`, and if $|y| \leqq |x|2^{-25}$, then $\mathrm{fl}(x+y) = x$.

38. Suppose that x is a machine number in the range $-\infty < x < 0$. In IEEE standard arithmetic, what values will the computer return for the computations $-\infty + x$, $-\infty * x$, $x/-\infty$, and $-\infty/x$.

39. Evaluate the repeating decimals.

(a) 0.181818... **(b)** 2.70270 27027... **(c)** 98.19819 81981...

40. Fix an integer N, and call a real number x *representable* if $x = q2^n$, where $1/2 \leqq q < 1$ and $|n| \leqq N$. Prove that if $x_1, x_2, \ldots, x_k$ are representable and if their product is representable, then so is uv, where $u = \max(x_i)$ and $v = \min(x_i)$.

41. Consider machine epsilon, defined in the text to be $\varepsilon = \frac{1}{2}\beta^{1-n}$, for correct rounding in a machine operating with base β and carrying n places in its mantissa. Prove that ε is the smallest machine number that satisfies the inequality $fl(1 + \varepsilon) > 1$.

COMPUTER PROBLEMS 2.1

1. Without using the Fortran 90 intrinsic procedures, write a program to *compute* the value of the machine precision ε on your computer in both single and double precision. Is this the exact value or an approximation to it? *Hint:* Determine the smallest positive machine number ε of the form 2^{-k} such that $1.0 + \varepsilon \neq 1.0$.

2. (Continuation) Repeat for the largest and smallest machine numbers.

3. Students are sometimes confused about the difference between `TINY` and `EPSILON`. Explain the difference. Then design and execute a numerical computer experiment to demonstrate the difference.

4. By repeated division by 2 and printing the results, you may observe that it seems that real numbers smaller than the smallest machine number `TINY` are obtained. Explain why this is not so and what is happening.

5. Obviously, one can determine the sign of a real number by examining a single bit in the computer representation of it. Similarly, one can determine whether an integer is even or odd from a single bit. Explain why this is so. Design a computer experiment to illustrate this situation.

6. Set `X = 1./3.` and print out both the internal machine representation and the decimal number stored in the computer that corresponds to it. Use a large format field for the decimal number. Explain and discuss the results.

7. Read in 97.6 and 12.9 and echo print them. Next subtract the smaller from the larger number and print the result. Initially, use the default printing format and then a large format field. Discuss the results.

8. **(a)** Verify that the IEEE double-precision floating-point numbers are uniformly spaced in the interval $[1, 2]$ with gaps $\varepsilon = 2^{-52}$. In other words, double-precision numbers between 1 and 2 may be represented as $x = 1 + k\varepsilon$, where $k = 1, 2, \ldots, 2^{52} - 1$ and $\varepsilon = 2^{-52}$.

(b) Show that the interval $[\frac{1}{2}, 1]$ contains as many floating-point numbers with gaps $\varepsilon/2$.

9. **(a)** Find any IEEE double-precision floating-point number x in the range $1 < x < 2$ such that $x * (1/x) \neq 1$; that is, $\mathrm{fl}(x\,\mathrm{fl}(1/x))$ is not exactly 1.

(b) Using a brute force search, find the smallest such number.

(Edelman [1994] shows how to prune the search substantially using mathematical analysis.)

2.2 Absolute and Relative Errors: Loss of Significance

When a real number x is approximated by another number x^*, the **error** is $x - x^*$. The **absolute error** is

$$|x - x^*|$$

and the **relative error** is

$$\left|\frac{x - x^*}{x}\right|$$

In scientific measurements, it is almost always the relative error that is significant. Information about the absolute error is usually of little use in the absence of knowledge about the magnitude of the quantity being measured. (An error of only 1 meter in determining the distance from Jupiter to Earth would be quite remarkable, but you would not want a surgeon to make such an error in an incision!)

We have already considered relative error in our investigation of roundoff errors. The inequality

$$\left|\frac{x - \mathrm{fl}(x)}{x}\right| \leqq \varepsilon$$

is a statement about the relative error involved in representing a real number x by a nearby floating-point machine number.

Loss of Significance

Although roundoff errors are inevitable and difficult to control, other types of errors in computation are under our control. The subject of numerical analysis is largely preoccupied with understanding and controlling errors of various kinds. Here we shall take up one type of error that is often the result of careless programming.

To see the sort of situation in which a large relative error can occur, we consider a subtraction of two numbers that are close to each other. For example,

$$\begin{aligned} x &= 0.37214\,78693 \\ y &= 0.37202\,30572 \\ x - y &= 0.00012\,48121 \end{aligned}$$

If this calculation were to be performed in a decimal computer having a five-digit mantissa, we would see

$$\begin{aligned} \text{fl}(x) &= 0.37215 \\ \text{fl}(y) &= 0.37202 \\ \text{fl}(x) - \text{fl}(y) &= 0.00013 \end{aligned}$$

The relative error is then very large:

$$\left| \frac{x - y - [\text{fl}(x) - \text{fl}(y)]}{x - y} \right| = \left| \frac{0.00012\,48121 - 0.00013}{0.00012\,48121} \right| \approx 4\%$$

Whenever the computer must shift the digits in the mantissa to achieve a *normalized* floating-point number, additional 0's are supplied on the right. These 0's are spurious and do not represent additional accuracy. Thus, $\text{fl}(x) - \text{fl}(y)$ is represented in the computer as 0.13000×10^{-3}, but the 0's in the mantissa serve only as place-holders.

Subtraction of Nearly Equal Quantities

As a rule, we should avoid situations where accuracy can be jeopardized by a subtraction of nearly equal quantities. A careful programmer will be alert to this situation. To illustrate what can be done by reprogramming, we consider an example.

Example 1 The assignment statement

$$y \leftarrow \sqrt{x^2 + 1} - 1$$

involves subtractive cancellation and loss of significance for small values of x. How can we avoid this trouble?

Solution Rewrite the function in this way

$$y = \left(\sqrt{x^2 + 1} - 1\right)\left(\frac{\sqrt{x^2 + 1} + 1}{\sqrt{x^2 + 1} + 1}\right) = \frac{x^2}{\sqrt{x^2 + 1} + 1}$$

Thus, the difficulty is avoided by reprogramming with a different assignment statement

$$y \leftarrow x^2/\left(\sqrt{x^2 + 1} + 1\right)$$

■

Loss of Precision

An interesting question is: *Exactly how many significant binary bits are lost in the subtraction $x - y$ when x is close to y?* The precise answer depends on the particular values of x and y. However, we can obtain bounds in terms of the quantity $|1 - y/x|$, which is a convenient measure for the closeness of x and y. The following theorem contains useful upper and lower bounds. (The theorem is machine-independent.)

THEOREM 1 **Loss of Precision Theorem** *If x and y are positive normalized floating-point binary machine numbers such that $x > y$ and*

$$2^{-q} \leqq 1 - \frac{y}{x} \leqq 2^{-p}$$

then at most q and at least p significant binary bits are lost in the subtraction $x - y$.

Proof We shall prove the lower bound and leave the upper bound as an exercise. The normalized binary floating-point forms for x and y are

$$x = r \times 2^n \qquad \left(\tfrac{1}{2} \leqq r < 1\right)$$
$$y = s \times 2^m \qquad \left(\tfrac{1}{2} \leqq s < 1\right)$$

Since x is larger than y, the computer may have to *shift* y so that y has the same exponent as x before performing the subtraction of y from x. Hence, we must write y as

$$y = (s \times 2^{m-n}) \times 2^n$$

and then we have

$$x - y = (r - s \times 2^{m-n}) \times 2^n$$

The mantissa of this number satisfies

$$r - s \times 2^{m-n} = r\left(1 - \frac{s \times 2^m}{r \times 2^n}\right) = r\left(1 - \frac{y}{x}\right) < 2^{-p}$$

To normalize the computer representation of $x - y$, a shift of at least p bits to the left is required. Then at least p spurious 0's are attached to the right end of the mantissa, which means that at least p bits of precision have been lost. ■

Example 2 Consider the assignment statement

$$y \leftarrow x - \sin x$$

Since $\sin x \approx x$ for small values of x, this calculation will involve a loss of significance. How can this be avoided?

Solution Let us find an alternative form for the function $y = x - \sin x$. The Taylor series for $\sin x$ is helpful here. Thus,

$$\begin{aligned} y &= x - \sin x \\ &= x - \left(x - \frac{x^3}{3!} + \frac{x^5}{5!} - \frac{x^7}{7!} + \cdots\right) \\ &= \frac{x^3}{3!} - \frac{x^5}{5!} + \frac{x^7}{7!} - \cdots \end{aligned}$$

If x is near 0, then a truncated series can be used, as in this assignment statement

$$y \leftarrow (x^3/6)(1 - (x^2/20)(1 - (x^2/42)(1 - x^2/72)))$$

If values of y are needed for a wide range of x-values in this function, it would be best to use both assignment statements, each in its proper range.

In this example, further analysis is needed to determine the appropriate range of x-values for each of these assignment statements. Using the Theorem on Loss of Precision, we see that the loss of bits in the subtraction of the first assignment statement can be limited to at most *one* bit by restricting x so that

$$1 - \frac{\sin x}{x} \geqq \frac{1}{2}$$

(Here we are considering only the case when $\sin x > 0$.) With a calculator, it is easy to determine that x must be at least 1.9. Thus for $|x| \geqq 1.9$, we should use the first assignment statement involving $x - \sin x$, and for $|x| < 1.9$ we should use a truncated series. We can verify that for the worst case ($x = 1.9$), seven terms in the series give y with an error of at most 10^{-9}. (See Problem 26.)

To construct a subprogram for $y = x - \sin x$ using a certain number of terms in the preceding series, notice that each term can be obtained from the previous term by using the formulas

$$\begin{cases} t_1 &= x^3/6 \\ t_{n+1} &= -t_n x^2/[(2n+2)(2n+3)] \qquad (n \geqq 1) \end{cases}$$

The partial sums

$$s_n = \sum_{k=1}^{n} t_k$$

can be obtained inductively by

$$\begin{cases} s_1 &= t_1 \\ s_{n+1} &= s_n + t_{n+1} \qquad (n \geqq 1) \end{cases}$$

■

In some calculations, a loss of significance can be avoided or ameliorated by the use of *double precision*. In this mode of computation, each real number is allotted two words of storage. This at least doubles the number of bits in the mantissa. Certain crucial parts of a computation can be performed in double precision while the remaining parts are carried out in single precision. This is more economical than executing an entire problem in double precision because the latter mode increases computing time (and thereby the cost) by a factor of between 2 and 4. The reason for this is that double-precision arithmetic is usually implemented by software, whereas single-precision arithmetic is performed by the hardware.

Evaluation of Functions

There is another situation in which a drastic loss of significant digits will occur. This is in the evaluation of certain functions for very large arguments. Let us illustrate with the cosine function, which has the periodicity property

$$\cos(x + 2n\pi) = \cos x \qquad (n \text{ an integer})$$

By the use of this property, the evaluation of $\cos x$ for any argument can be effected by evaluating at a **reduced argument** in the interval $[0, 2\pi]$. The library subroutines available on computers exploit this property in a process called **range reduction**. Other properties may also be used, such as

$$\cos(-x) = \cos x = -\cos(\pi - x)$$

For example, the evaluation of $\cos x$ at $x = 33278.21$ proceeds by finding the *reduced* argument

$$y = 33278.21 - 5296 \times 2\pi = 2.46$$

Here we retain only two decimal digits because only two decimal places of accuracy are present in the original argument. The reduced argument has three significant figures, although the original argument may have had seven significant figures. The cosine will then have at most three significant figures. We must not be misled into thinking that the infinite precision available in $5296 \times 2\pi$ is conveyed to the reduced argument y. Also, one should not be deceived by the apparent precision in the printed output from a subroutine. If the cosine subroutine is given an argument y with three significant digits, the value $\cos y$ will have no more than three significant figures, even though it may be displayed as

$$\cos(2.46) = -0.77657\,02835$$

(The reason for this is that the subroutine treats the argument as being accurate to full machine precision, which of course it is not.)

Interval Arithmetic

A method of controlling computations in order to know the extent of roundoff error is *interval arithmetic*. In this manner of computing, each calculated number is accompanied by an interval that is guaranteed to contain the correct value. Ideally, of course, these intervals are very small, and final answers can be given with only small uncertainties. However, the cost of carrying intervals (instead of simple machine numbers) throughout a lengthy computation may make the procedure cumbersome. Consequently, it is used only when great reliance must be placed on the computations. Also, it may be difficult to keep the intervals from growing much larger than is realistic. Recently, a number of software packages have been developed that have contributed significantly to the use of interval arithmetic in computations. Interval arithmetic has its own literature, including a journal devoted to research. Some books concerning it are Alefeld and Grigorieff [1980], Alefeld and Herzberger [1983], Kulisch and Miranker [1981], and Moore [1966, 1979]. Recent research developments on interval arithmetic can be found on a *homepage* on the Internet. (See the appendix on mathematical software.)

PROBLEMS 2.2

1. How many bits of precision are lost in a computer when we carry out the subtraction $x - \sin x$ for $x = \frac{1}{2}$?
2. How many bits of precision are lost in the subtraction $1 - \cos x$ when $x = \frac{1}{4}$?
3. Find a suitable trigonometric identity so that $1 - \cos x$ can be accurately computed for small x with calls to the system functions for $\sin x$ or $\cos x$. (There are two good answers.)
4. For the function in Problem 2, find a suitable Taylor series by which it can be accurately computed.
5. Find a way of computing $\sqrt{x^4 + 4} - 2$ without undue loss of significance.
6. Using the definition $\sinh x \equiv \frac{1}{2}(e^x - e^{-x})$, discuss the problem of computing $\sinh x$.
7. In solving the quadratic equation $ax^2 + bx + c = 0$ by use of the formula

$$x = \left(-b \pm \sqrt{b^2 - 4ac}\right)/(2a)$$

there will be loss of significance when $4ac$ is small relative to b^2 because then

$$\sqrt{b^2 - 4ac} \approx |b|$$

Suggest a method to circumvent this difficulty.

8. Suggest ways to avoid loss of significance in these calculations.
   (a) $\sqrt{x^2 + 1} - x$
   (b) $\log x - \log y$
   (c) $x^{-3}(\sin x - x)$
   (d) $\sqrt{x + 2} - \sqrt{x}$
   (e) $e^x - e$
   (f) $\log x - 1$

(g) $(\cos x - e^{-x})/\sin x$

(h) $\sin x - \tan x$

(i) $\sinh x - \tanh x$

(j) $\ln\left(x + \sqrt{x^2 + 1}\right)$ *Hint:* This is the function $\sinh^{-1} x$.

9. For any $x_0 > -1$, the sequence defined recursively by

$$x_{n+1} = 2^{n+1}\left[\sqrt{1 + 2^{-n}x_n} - 1\right]$$

converges to $\ln(x_0 + 1)$. (See Henrici [1962, p. 243].) Arrange this formula in a way that avoids loss of significance.

10. Arrange the following formulas in order of merit for computing $\tan x - \sin x$ when x is near 0.

(a) $\sin x[(1/\cos x) - 1]$

(b) $\frac{1}{2}x^3$

(c) $(\sin x)/(\cos x) - \sin x$

(d) $(x^2/2)(1 - x^2/12)\tan x$

(e) $\frac{1}{2}x^2 \tan x$

(f) $\tan x \sin^2 x/(\cos x + 1)$

11. Find ways to compute these functions without serious loss of significance.

(a) $(1 - x)/(1 + x) - 1/(3x + 1)$

(b) $\sqrt{x + (1/x)} - \sqrt{x - (1/x)}$

(c) $e^x - \cos x - \sin x$

12. Discuss the calculation of e^{-x} for $x > 0$ from the series

$$e^{-x} = 1 - x + \frac{x^2}{2!} - \frac{x^3}{3!} + \cdots$$

Suggest a better way, assuming that the system function for e^x is *not* available.

13. There will be subtractive cancellation in computing $f(x) = 1 + \cos x$ for some values of x. What are these values of x and how can the loss of precision be averted?

14. Consider the function $f(x) = x^{-1}(1 - \cos x)$.

(a) What is the *correct* definition of $f(0)$—that is, the value that will make f continuous?

(b) Near what points will there be loss of significance if the given formula is used?

(c) How can we circumvent the difficulty in part **(b)**? Find a method that does not use the Taylor series.

(d) If the new formula that you gave in part **(c)** involves subtractive cancellation at some other point, describe how to avoid that difficulty.

15. Let $f(x) = -e^{-2x} + e^x$. Which of these formulas x, $3x$, $3x(1 - x/2)$, $2 - 3x$, or $e^x(1 - e^{-3x})$ for f is most accurate for small values of x?

16. If at most 2 bits of precision are to be lost in the computation $y = \sqrt{x^2 + 1} - 1$, what restriction must be placed on x?

17. For what range of values of θ will the approximation $\sin\theta \approx \theta$ give results correct to three (rounded) decimal places?

18. For small values of x, how good is the approximation $\cos x \approx 1$? How small must x be to have $\frac{1}{2} \times 10^{-8}$ accuracy?

19. The series $\sum_{k=1}^{\infty} k^{-1}$ is called the **harmonic series**. It diverges. The partial sums, $S_n = \sum_{k=1}^{n} k^{-1}$, can be computed recursively by setting $S_1 = 1$ and using $S_n = S_{n-1} + n^{-1}$. If this computation were carried out on your computer, what is the largest S_n that would be obtained? (Do *not* do this experimentally on the computer; it is too expensive.) See Schechter [1984].

20. Find a way of accurately computing $f(x) = x + e^x - e^{2x}$ for small values of x.

21. **(a)** Find a way to calculate accurate values near 0 for the function

$$f(x) = (e^{\tan x} - e^x)/x^3$$

(b) Determine $\lim_{x \to 0} f(x)$. *Hint:* See Problem 19, Section 1.2.

22. Explain why loss of significance due to subtraction is not serious in using the approximation

$$x - \sin x \approx (x^3/6)(1 - (x^2/20)(1 - x^2/42))$$

23. In computing the sum of an infinite series $\sum_{n=1}^{\infty} x_n$, suppose that the answer is desired with an absolute error less than ε. Is it safe to stop the addition of terms when their magnitude falls below ε? Illustrate with the series $\sum_{n=1}^{\infty} (0.99)^n$.

24. Repeat Problem 23 under the additional assumptions that the terms x_n are alternately positive and negative and that $|x_n|$ converges monotonically downward to 0. (Use a theorem in calculus about alternating series.)

25. Show that if x is a machine number on the `Marc-32` and if $x > 2^{25}\pi$, then $\cos x$ will be computed with *no* significant digits.

26. By using the error term in Taylor's Theorem, show that at least seven terms are required in the series of Example 2, if the error is not to exceed 10^{-9}.

COMPUTER PROBLEMS 2.2

1. Write and execute a program to compute

$$f(x) = \sqrt{x^2 + 1} - 1$$
$$g(x) = x^2/\left(\sqrt{x^2 + 1} + 1\right)$$

for a succession of values of x such as $8^{-1}, 8^{-2}, 8^{-3}, \ldots$. Although $f = g$, the computer will produce different results. Which results are reliable and which are not?

2. Write and test a subroutine that accepts a machine number x and returns the value $y = x - \sin x$ with nearly full machine precision.

3. Using your computer, print the values of the functions

$$f(x) = x^8 - 8x^7 + 28x^6 - 56x^5 + 70x^4 - 56x^3 + 28x^2 - 8x + 1$$
$$g(x) = (((((((x - 8)x + 28)x - 56)x + 70)x - 56)x + 28)x - 8)x + 1$$
$$h(x) = (x - 1)^8$$

at 101 equally spaced points covering the interval [0.99, 1.01]. Calculate each function in a straightforward way without rearranging or factoring. Observe that the three functions are identical. Account for the fact that the printed values are not all positive as they should be. If a plotter is available, plot these functions near 1.0 using a magnified scale for the function values in order to see the variations involved. (See Rice [1992, p. 43].)

4. Write and test a code to supply accurate values of $1 - \cos x$ for $-\pi \leqq x \leqq \pi$. Use a Taylor series near 0 and the subprogram for cosine otherwise. Determine carefully the range where each method should be used in order to lose at most one bit.

5. Write and test a function subprogram for $f(x) = x^{-2}(1 - \cos x)$. Avoid loss of significance in subtraction for all arguments x and (of course) take care of the difficulty at $x = 0$.

6. An interesting numerical experiment is to compute the dot product of the following two vectors:

$$x = [2.71828\,1828,\ -3.14159\,2654,\ 1.41421\,3562,\ 0.57721\,56649,\ 0.30102\,99957]$$
$$y = [1486.2497,\ 8\,78366.9879,\ -22.37492,\ 47\,73714.647,\ 0.00018\,5049]$$

Compute the summation in four ways:

(a) Forward order $\sum_{i=1}^{n} x_i y_i$

(b) Reverse order $\sum_{i=n}^{1} x_i y_i$

(c) Largest-to-smallest order (add positive numbers in order from largest to smallest, then add negative numbers in order from smallest to largest, and then add the two partial sums)

(d) Smallest-to-largest order (reverse the order of adding in the previous method)

(e) Use both single and double precision for a total of eight answers. Compare the results with the *correct* value to seven decimal places, 1.006571×10^{-9}. Explain your results.

7. (Continuation) Repeat Problem 6 but drop the final 9 from x_4 and the final 7 from x_5. What effect does this small change have on the results?

8. On a vector computer such as a Cray supercomputer, run a Fortran 90 program similar to the following one and explain the results.

```
integer, parameter :: n = 1000000
real a(n), x
   a(1 : n) = epsilon(x)/4.0
!DIR$ NOVECTOR
   s = 1.0
   do i = 1, n
      s = s + a(i)
   end do
   print "(F20.15)", s
!DIR$ VECTOR
   s = 1.0
   do i = 1, n
      s = s + a(i)
   end do
   print "(F20.15)", s
end
```

Here epsilon(x) returns the value of the machine precision. The statement !DIR$ NOVECTOR is a control directive that turns vectorization off, and the corresponding statement later in the code turns it back on.

9. In 1994, a flaw was found in the Intel Pentium computer chip related to the division of certain large integers. For example, it was discovered that

(a) 55 05001 divided by 2 94911 gave 18.66600 09290 9

(b) 4.99999 9 divided by 14.99999 9 gave 0.33332 9

(c) 41 95835 divided by 31 45727 gave 1.33382

Analyze these results and give the absolute and relative errors involved.

10. Define the function $f(x, y) = 9x^4 - y^4 + 2y^2$. Our objective is to compute $f(40545, 70226)$. Carry out this calculation as follows:

(a) Use each of integer, single-precision, and double-precision computations.

(b) Using elementary algebra, show that $f(x, y) = (3x^2 - y^2 + 1)(3x^2 + y^2 - 1) + 1$. Use this formula and repeat part **(a)**.

(c) Use the original formulation of the function and a symbolic manipulation program with increasing precision. Compute first with 6 decimal digits of accuracy and then 7, 8, . . . , 25.

(d) Use the formula from part **(b)** and repeat the instructions in part **(c)**.

What does this exercise teach you?

11. Some rational approximations to π are

$$\frac{22}{7}, \frac{333}{106}, \frac{355}{113}, \frac{104348}{33215}, \frac{1148183}{365478}, \frac{1252531}{398693}, \frac{2400714}{764171},$$
$$\frac{18057529}{5747890}, \frac{56573301}{18007841}, \frac{208235675}{66283474}, \frac{681280326}{216858263}$$

Explore the absolute and relative errors involved in them. Use a symbolic manipulation program to compute these rational approximations for π.

12. Kahan [1993] reports that many automated algebra systems have trouble simplifying constants such as

$$\sqrt{(e-\pi)^2}$$

They sometimes give $e - \pi$, which has the wrong sign because $\pi > e$. Also, he says that they have difficulties determining the sign of transcendental constants such as

$$\sqrt{\sqrt{10}+3}\sqrt{\sqrt{5}+2} - \sqrt{\sqrt{10}-3}\sqrt{\sqrt{5}-2} - \sqrt{10\sqrt{2}+10}$$

Using a symbolic manipulation program, test these two situations.

13. Using a symbolic manipulation program, find the first prime number greater than 27448.

2.3 Stable and Unstable Computations: Conditioning

In this section, we introduce another theme that occurs repeatedly in numerical analysis: the distinction between numerical processes that are **stable** and those that are not. Closely related are the concepts of **well conditioned** problems and **badly conditioned** problems.

Numerical Instability

Speaking informally, we say that a numerical process is **unstable** if small errors made at one stage of the process are magnified in subsequent stages and seriously degrade the accuracy of the overall calculation.

An example will help to explain this concept. Consider the sequence of real numbers defined inductively by

$$\begin{cases} x_0 = 1 \qquad x_1 = \dfrac{1}{3} \\ x_{n+1} = \dfrac{13}{3}x_n - \dfrac{4}{3}x_{n-1} \qquad (n \geqq 1) \end{cases} \tag{1}$$

It is easily seen that this recurrence relation generates the sequence

$$x_n = \left(\frac{1}{3}\right)^n \tag{2}$$

Indeed, Equation (2) is obviously true for $n = 0$ and $n = 1$. If its validity is granted for $n \leqq m$, then its validity for $n = m + 1$ follows from

$$\begin{aligned} x_{m+1} &= \frac{13}{3}x_m - \frac{4}{3}x_{m-1} = \frac{13}{3}\left(\frac{1}{3}\right)^m - \frac{4}{3}\left(\frac{1}{3}\right)^{m-1} \\ &= \left(\frac{1}{3}\right)^{m-1}\left[\frac{13}{9} - \frac{4}{3}\right] = \left(\frac{1}{3}\right)^{m+1} \end{aligned}$$

If the *inductive* definition (1) is used to generate the sequence numerically, say on the `Marc-32`, then some of the computed terms will be grossly inaccurate. Here are a few of them, calculated on a 32-bit computer similar to the `Marc-32`:

$x_0 = 1.00000\,00$

$x_1 = 0.33333\,33$ (7 correctly rounded significant digits)

$x_2 = 0.11111\,12$ (6 correctly rounded significant digits)

$x_3 = 0.03703\,73$ (5 correctly rounded significant digits)

$x_4 = 0.01234\,66$ (4 correctly rounded significant digits)

$x_5 = 0.00411\,87$ (3 correctly rounded significant digits)

$x_6 = 0.00138\,57$ (2 correctly rounded significant digits)

$x_7 = 0.00051\,31$ (1 correctly rounded significant digit)

$x_8 = 0.00037\,57$ (0 correctly rounded significant digits)

$x_9 = 0.00094\,37$

$x_{10} = 0.00358\,87$

$x_{11} = 0.01429\,27$

$x_{12} = 0.05715\,02$

$x_{13} = 0.22859\,39$

$x_{14} = 0.91437\,35$

$x_{15} = 3.65749\,3$ (incorrect with relative error of 10^8)

This algorithm is therefore *unstable*. Any error present in x_n is multiplied by $13/3$ in computing x_{n+1}. Hence there is a possibility that the error in x_1 will be propagated into x_{15} with a factor of $(13/3)^{14}$. Since the absolute error in x_1 is around 10^{-8} and since $(13/3)^{14}$ is roughly 10^9, the error in x_{15} due solely to the error in x_1 could be as much as 10. In fact, additional roundoff errors occur in computing each of $x_2, x_3, \ldots$, and these errors may also be propagated into x_{15} with various factors of the form $(13/3)^k$.

Another way of explaining this example is to observe that Equation (1) is a *difference* equation, of which the general solution is

$$x_n = A\left(\frac{1}{3}\right)^n + B(4)^n$$

Here A and B are constants determined by the initial values of x_0 and x_1. (The reader should consult Section 1.3 for the theory of linear difference equations.) Although we want to compute the pure solution (2) corresponding to $A = 1$ and $B = 0$, it is impossible to avoid contamination by the unwanted component 4^n. The latter will therefore eventually dominate the desired solution.

Whether a process is numerically stable or unstable should be decided on the basis of *relative* errors. Thus, if there are large errors in a computation, that situation may be quite acceptable, if the answers are large. In the preceding example, let us start with initial values $x_0 = 1$ and $x_1 = 4$. The recurrence relation (1) is unchanged, and therefore errors will still be propagated and magnified as before. But the correct solution is now $x_n = 4^n$, and the results of computation are correct to seven significant figures. Here are three of them:

$$\begin{aligned} x_1 &= 4.000006 \\ x_{10} &= 1.048576 \times 10^6 \\ x_{20} &= 1.099512 \times 10^{12} \end{aligned}$$

In this case, the correct values are large enough to overwhelm the errors. The absolute errors are undoubtedly large (as before), but they are *relatively* negligible.

Another example of numerical instability is provided by a calculation of the numbers

$$y_n = \int_0^1 x^n e^x \, dx \qquad (n \geqq 0) \tag{3}$$

If we apply integration by parts to the integral defining y_{n+1}, the result is this recurrence relation:

$$y_{n+1} = e - (n+1)y_n \tag{4}$$

From this and the obvious fact that $y_0 = e - 1$, we obtain y_1:

$$y_1 = e - y_0 = e - (e - 1) = 1$$

Starting with $y_1 = 1$, let us generate $y_2, y_3, \ldots, y_{15}$ in a computer like the `Marc-32`, using relation (4). Three of the results are

$$\begin{aligned} y_2 &= 0.7182817 \\ y_{11} &= 1.422453 \\ y_{15} &= 39711.43 \end{aligned}$$

These *cannot* be correct. Indeed, it is obvious from Equation (3) that the y sequence satisfies $y_1 > y_2 > \cdots > 0$ and that $\lim_{n\to\infty} y_n = 0$. (Indeed, for $0 < x < 1$, the expression x^n decreases monotonically to 0.) Once we know this, we see from Equation (4) that $\lim_{n\to\infty}(n + 1)y_n = e$.

In this example, an error of δ units in y_2 is multiplied by 3 in computing y_3. This error of 3δ is multiplied by 4 in computing y_4. The resulting error of 12δ is multiplied by 5 in computing y_5. This process continues, so that in computing y_{10}, the error may be as much as $\frac{1}{2}10!\,\delta \approx 2 \times 10^6\delta$. For y_{20}, the corresponding figure is $10^{18}\delta$. Since $\delta \approx 2^{-23}$ on the Marc-32, $10^{18}\delta \approx 10^{10}$. Thus, the errors *completely overwhelm* the correct values of y_n; the correct values are tending rapidly to 0.

Conditioning

The words **condition** and **conditioning** are used informally to indicate how sensitive the solution of a problem may be to small relative changes in the input data. A problem is **ill conditioned** if small changes in the data can produce large changes in the answers. For certain types of problems, a **condition number** can be defined. If that number is large, it indicates an ill-conditioned problem. Examples of this will occur later, but here we shall discuss some elementary illustrations.

While we cannot discuss conditioning in any substantial way at this point, our elementary examples illustrate an important numerical concept. It has been shown that a condition number is intimately associated with the numerical behavior in solving a given problem independent of the particular method of solution. Basically, if the condition number is large, then be prepared for trouble!

Suppose our problem is simply to evaluate a function f at a point x. We ask: *If x is perturbed slightly, what is the effect on $f(x)$?* If this question refers to *absolute* errors, we can invoke the Mean-Value Theorem and write

$$f(x + h) - f(x) = f'(\xi)h \approx hf'(x)$$

Thus, if $f'(x)$ is not too large, the effect of the perturbation on $f(x)$ is small. Usually, however, it is the *relative* error that is of significance in such questions. In perturbing x by the amount h, we have h/x as the *relative* size of the perturbation. Likewise, when $f(x)$ is perturbed to $f(x + h)$, the relative size of that perturbation is

$$\frac{f(x + h) - f(x)}{f(x)} \approx \frac{hf'(x)}{f(x)} = \left[\frac{xf'(x)}{f(x)}\right]\left(\frac{h}{x}\right)$$

Thus, the factor $xf'(x)/f(x)$ serves as a **condition number** for this problem.

Example 1 What is the condition number for the evaluation of the inverse sine function?

Solution Let $f(x) = \arcsin x$. Then

$$\frac{xf'(x)}{f(x)} = \frac{x}{\sqrt{1 - x^2}\arcsin x}$$

For x near 1, $\arcsin x \approx \pi/2$ and the condition number becomes infinite as x approaches 1 since it is approximated by $2x/\left(\pi\sqrt{1-x^2}\right)$. Hence, small relative errors in x may lead to large relative errors in $\arcsin x$ near $x = 1$. ■

We now consider the problem of locating a zero (or *root*) of a function f. (This problem is studied from the algorithmic standpoint in Chapter 3.) Let f and g be two functions that belong to class C^2 in a neighborhood of r, where r is a root of f.

We assume that r is a **simple root**, so that $f'(r) \neq 0$. If we perturb the function f to $F \equiv f + \varepsilon g$, where is the new root? Let us suppose that the new root is $r + h$; we shall derive an approximate formula for h. The perturbation h satisfies the equation $F(r + h) = 0$ or

$$f(r + h) + \varepsilon g(r + h) = 0$$

Since f and g belong to C^2, we can use Taylor's Theorem to express $F(r + h)$:

$$\left[f(r) + hf'(r) + \frac{1}{2}h^2 f''(\xi)\right] + \varepsilon\left[g(r) + hg'(r) + \frac{1}{2}h^2 g''(\eta)\right] = 0$$

Discarding terms in h^2 and using the fact that $f(r) = 0$, we obtain

$$h \approx -\varepsilon\frac{g(r)}{f'(r) + \varepsilon g'(r)} \approx -\varepsilon\frac{g(r)}{f'(r)}$$

Example 2 As an illustration of this analysis, we consider a classic example given by Wilkinson. Let

$$\begin{aligned} f(x) &= \textstyle\prod_{k=1}^{20}(x - k) = (x - 1)(x - 2)\cdots(x - 20) \\ g(x) &= x^{20} \end{aligned}$$

The roots of f are obviously the integers $1, 2, \ldots, 20$. How is the root $r = 20$ affected by perturbing f to $f + \varepsilon g$?

Solution The answer is

$$h \approx -\varepsilon\frac{g(20)}{f'(20)} = -\varepsilon\frac{20^{20}}{19!} \approx -\varepsilon 10^9$$

Thus, a change of ε in the coefficient of x^{20} in $f(x)$ may cause a perturbation of the root 20 by an amount $10^9\varepsilon$. The roots of this polynomial are therefore extremely sensitive to perturbations in the coefficients. (See Problem 6.) ■

Yet another type of condition number is associated with solving linear systems $Ax = b$; it will be discussed in detail in Section 4.4. Briefly, the **condition number** of the matrix A is denoted by $\kappa(A)$ and is defined as the product of the magnitudes of A and its inverse; that is,

$$\kappa(A) = \|A\| \cdot \|A^{-1}\|$$

where $\| \cdot \|$ is a matrix norm. If the solution of $Ax = b$ is rather insensitive to small changes in the right-hand side b, then small perturbations in b result in only small perturbations in the computed solution x. In this case, A is said to be **well conditioned**. This situation corresponds to the condition number $\kappa(A)$ being of only modest size. On the other hand, if the condition number is large, then A is **ill conditioned** and any numerical solution of $Ax = b$ must be accepted with a great deal of skepticism.

The nth-order **Hilbert matrix** $H_n = (h_{ij})$ is defined by $h_{ij} = 1/(i + j - 1)$, where $1 \leqq i \leqq n$ and $1 \leqq j \leqq n$. The Hilbert matrix has been the subject of intense study over many years. Some of its history can be read in Chapter IX of *Mathematicians Learning to Use Computers* by Hestenes and Todd [1991]. This matrix is often used for test purposes because of its ill-conditioned nature. In fact, $\kappa(H_n)$ increases rapidly with n, as can be seen from the formula $\kappa(H_n) = ce^{3.5n}$. This condition number has been computed by using the norm $\|A\| = \max_{ij} |a_{ij}|$. An interesting test problem is to solve the linear system $H_n x = b$ on a computer with increasing values for n and with $b_i = \sum_{j=1}^{n} h_{ij}$. The solution should be $x = (1, 1, \dots, 1)^T$. For small values of n, the computed solution is quite accurate, but as n increases, the precision quickly degenerates. In fact, when n is approximately equal to the number of decimal places carried by the computer, the numerical solution will probably contain no significant figures. These matters will be discussed in more detail in Chapter 4.

The Hilbert matrix arises in least-squares approximation when we attempt to minimize the expression

$$\int_0^1 \left[\sum_{j=0}^{n} a_j x^j - f(x) \right]^2 dx$$

Upon differentiating this expression with respect to a_i and setting the result equal to 0, we obtain the **normal equations**

$$\sum_{j=0}^{n} a_j \int_0^1 x^i x^j \, dx = \int_0^1 x^i f(x) \, dx \qquad (0 \leqq i \leqq n)$$

Since the integral on the left is

$$\left. \frac{x^{i+j+1}}{i+j+1} \right|_0^1 = \frac{1}{i+j+1}$$

the coefficient matrix in the normal equations is the Hilbert matrix of order $n+1$. The functions $x \longrightarrow x^i$ form a very badly conditioned basis for the space of polynomials of degree n. For this problem, a good basis can be provided by an orthogonal set of polynomials. This will be discussed in Section 6.8.

PROBLEMS 2.3

1. Find analytically the solution of this difference equation with the given initial values:

$$\begin{cases} x_0 = 1 \qquad x_1 = 0.9 \\ x_{n+1} = -0.2x_n + 0.99x_{n-1} \end{cases}$$

Without computing the solution recursively, predict whether such a computation would be stable or not.

2. The **exponential integrals** are the functions E_n defined by

$$E_n(x) = \int_1^\infty (e^{xt}t^n)^{-1}dt \qquad (n \geqq 0,\ x > 0)$$

These functions satisfy the equation

$$nE_{n+1}(x) = e^{-x} - xE_n(x)$$

If $E_1(x)$ is known, can this equation be used to compute $E_2(x), E_3(x), \ldots$ accurately? *Hint:* Determine whether $E_n(x)$ is increasing or decreasing as a function of n.

3. The condition number of the function $f(x) = x^\alpha$ is independent of x. What is it?

4. What are the condition numbers of the following functions? Where are they large?

(a) $(x-1)^\alpha$

(b) $\ln x$

(c) $\sin x$

(d) e^x

(e) $x^{-1}e^x$

(f) $\cos^{-1} x$

5. Consider the example in the text for which $y_{n+1} = e - (n+1)y_n$. How many decimal places of accuracy should be used in computing $y_1, y_2, \ldots, y_{20}$ if y_{20} is to be accurate to five decimal places?

6. There is a function f of the form

$$f(x) = \alpha x^{12} + \beta x^{13}$$

for which $f(0.1) = 6.06 \times 10^{-13}$ and $f(0.9) = 0.03577$. Determine α and β, and assess the sensitivity of these parameters to slight changes in the values of f.

7. Show that the recurrence relation

$$x_n = 2x_{n-1} + x_{n-2}$$

has a general solution of the form

$$x_n = A\lambda^n + B\mu^n$$

Is the recurrence relation a good way to compute x_n from arbitrary initial values x_0 and x_1?

8. The **Fibonacci sequence** is generated by the formulas

$$\begin{cases} r_0 = 1 & r_1 = 1 \\ r_{n+1} = r_n + r_{n-1} \end{cases}$$

The sequence therefore starts out 1, 1, 2, 3, 5, 8, 13, 21, 34, ... Prove that the sequence $[2r_n/r_{n-1}]$ converges to $1+\sqrt{5}$. Is the convergence linear, superlinear, quadratic?

9. (Continuation) If the recurrence relation in Problem 8 is used with starting values $r_0 = 1$ and $r_1 = (1-\sqrt{5})/2$, what is the theoretically correct value of r_n $(n \geqq 2)$? Will the recurrence relation provide a stable means for computing r_n in this case?

COMPUTER PROBLEMS 2.3

1. Let sequences $[A_n]$ and $[B_n]$ be generated as follows:

$$\begin{cases} A_0 = 0 & A_1 = 1 \\ A_n = nA_{n-1} + A_{n-2} \end{cases} \qquad \begin{cases} B_0 = 1 & B_1 = 1 \\ B_n = nB_{n-1} + B_{n-2} \end{cases}$$

What is $\lim_{n\to\infty}(A_n/B_n)$?

2. The **Bessel functions** Y_n satisfy the same recurrence formula that the functions J_n satisfy. (See Section 1.3.) However, they use *different starting values*. For $x = 1$, they are

$$Y_0(1) = 0.08825\,69642 \qquad Y_1(1) = -0.78121\,28213$$

Compute $Y_2(1), Y_3(1), \ldots, Y_{20}(1)$ using the recurrence formula. Try to decide whether the results are reliable or not. *Hint:* The numbers $|Y_n(1)|$ should grow rapidly. Perhaps you can prove an inequality such as $|Y_n(1)/Y_{n-1}(1)| > n$.

3. Define

$$x_n = \int_0^1 t^n(t+5)^{-1}\,dt$$

Show that $x_0 = \ln 1.2$ and that $x_n = n^{-1} - 5x_{n-1}$ for $n \geqq 1$. Compute $x_0, x_1, \ldots, x_{10}$ using this recurrence formula and estimate the accuracy of x_{10}.

4. (Continuation) Find a way of computing x_{20} accurately, perhaps by replacing the integrand by a truncated Taylor series. After computing x_{20} to full machine precision, use the recurrence backward to get $x_{19}, x_{18}, \ldots, x_0$. Is x_0 correct? What about the other x_n? Does the recurrence relation behave differently when used backward, and if so, why?

5. The **Bessel functions** $J_0(x), J_1(x), \ldots$ can be computed by the recurrence relation in Section 1.3 if it is used with *descending* n. Thus, we start with $n = N$ and proceed downward by the formula

$$J_{n-1}(x) = \frac{2n}{x} J_n(x) - J_{n+1}(x) \qquad (N \leqq n \leqq 1)$$

To get started, we set $J_{N+1}(x) = 0$ and $J_N(x) = 1$. These are tentative values. After computing $J_{N-1}(x), J_{N-2}(x), \ldots, J_0(x)$, we scale them (that is, replace J_n with λJ_n) by using the identity

$$J_0(x)^2 + 2\sum_{n=1}^{\infty} J_n(x)^2 = 1$$

Obviously N must be large enough so that the fiction $J_{N+1}(x) = 0$ is reasonable for the precision expected. This technique is due to J. C. P. Miller. Taking $N = 51$ in the process just described, compute $J_0(1), J_1(1), \ldots, J_{50}(1)$.

6. ("The Perfidious Polynomial," Wilkinson [1984])

(a) Using a root-finding subroutine from the program library of your computer center, compute the 20 zeros of the polynomial P where

$$\begin{aligned} P(x) = {} & x^{20} - 210x^{19} + 20615x^{18} - 12\,56850x^{17} + 533\,27946x^{16} \\ & - 16722\,80820x^{15} + 4\,01717\,71630x^{14} - 75\,61111\,84500x^{13} \\ & + 1131\,02769\,95381x^{12} - 13558\,51828\,99530x^{11} \\ & + 1\,30753\,50105\,40395x^{10} - 10\,14229\,98655\,11450x^{9} \\ & + 63\,03081\,20992\,94896x^{8} - 311\,33364\,31613\,90640x^{7} \\ & + 1206\,64780\,37803\,73360x^{6} - 3599\,97951\,79476\,07200x^{5} \\ & + 8037\,81182\,26450\,51776x^{4} - 12870\,93124\,51509\,88800x^{3} \\ & + 13803\,75975\,36407\,04000x^{2} - 8752\,94803\,67616\,00000x \\ & + 2432\,90200\,81766\,40000 \end{aligned}$$

Use a routine that is capable of computing complex roots in double precision. The formula given for P is the expanded form of Wilkinson's polynomial discussed in the text:

$$\begin{aligned} p(x) = {} & (x-20)(x-19)(x-18)(x-17)(x-16)(x-15)(x-14) \\ & \times (x-13)(x-12)(x-11)(x-10)(x-9)(x-8)(x-7) \\ & \times (x-6)(x-5)(x-4)(x-3)(x-2)(x-1) \end{aligned}$$

Check the computed roots z_k, for $1 \leqq k \leqq 20$, by computing $|P(z_k)|$, $|p(z_k)|$, and $|z_k - k|$. Explain.

(b) Repeat part **(a)** eight times after changing the coefficient of x^{20} of $P(x)$ to $1 + \varepsilon$, where $\varepsilon = 10^{-2k}$ for $k = 8, 7, \ldots, 1$. Do your computer results agree with the analysis presented in the text? Explain.

(c) Wilkinson showed that by changing the coefficient -210 to $-210 - 2^{-23}$, the roots 16 and 17 changed to the complex pair $16.73\ldots \pm (2.812\ldots)i$. Reproduce this numerical experiment.

Cultural note: In the late 1940s, Jim Wilkinson was a young scientist working with Alan Turing and others on a new machine called the *Pilot ACE computer* at the National Physical Laboratory in England. As a routine test for this machine, he wrote a program using Newton's method to calculate the roots of this polynomial. Anticipating no difficulties, he began the Newton iteration with $x_0 = 21$ and expected immediate convergence to the largest zero 20. When the numerical results did not turn out this way, he investigated further. Eventually, he was led by this and other numerical experiments to the development of an entirely new area of numerical mathematics—**backward error analysis**. For additional details on the work of James Hardy Wilkinson, see the *Biographical Memoirs of Fellows of the Royal Society* by L. Fox [1987].

CHAPTER THREE

Solution of Nonlinear Equations

Introduction

This chapter is devoted to the problem of locating **roots** of equations (or **zeros** of functions). The problem occurs frequently in scientific work. In this chapter, we are interested in *solving* a **nonlinear** equation or a system of nonlinear equations—namely, finding x such that $f(x) = 0$ or finding $X = (x_1, x_2, \dots, x_n)^T$ so that $F(X) = 0$. In these equations, one or more of the variables appear in any number of nonlinear ways. In contrast to this, we discuss finding the solution of linear systems of the form $Ax = b$ in Chapter 4.

The general problem, posed in the simplest case of a real-valued function of a real variable, is this: Given a function $f\colon \mathbb{R} \to \mathbb{R}$, find the values of x for which $f(x) = 0$. We shall consider several standard procedures for solving this problem.

Examples of nonlinear equations can be found in many applications. In the theory of diffraction of light, we need the roots of the equation

$$x - \tan x = 0$$

In the calculation of planetary orbits, we need the roots of **Kepler's equation**

$$x - a\sin x = b$$

for various values of a and b.

Since locating the zeros of functions has been an active area of study for several hundred years, numerous methods have been developed. In this chapter, we begin with three simple methods that are quite useful—the bisection method, Newton's method, and the secant method. Then we explore the general theory of fixed-point methods and continuation methods. Also, we discuss special methods for computing the zeros of polynomials.

When a computer is used to find an *approximate zero* of a function, there may be many approximate solutions even though the exact solution is unique. This is nicely illustrated by considering the polynomial

$$p_4(x) = x^4 - 4x^3 + 6x^2 - 4x + 1$$

If we happen to notice that this polynomial can be factored so that $p_4(x) = (x-1)^4$, then it is obvious that 1 is the only zero (of multiplicity four). Suppose that we are given the expanded polynomial and do not see that it has such a zero. If we write a single-precision program in a computer such as the `Marc-32` and evaluate the expanded polynomial at points 0.001 distance apart in the interval [0.975, 1.035], we find a large number of sign changes indicating the presence of apparent zeros. Plotting these points connected by straight lines, we obtain the graph shown in Figure 3.1. Instead of a nice clean curve for the polynomial, we obtain a *fuzzy* one. Any value in the interval [0.981, 1.026] could be taken as an approximation to the true solution. The reasons for this are related to the way the polynomial was evaluated, the limited precision of the computer arithmetic being used, and the associated roundoff errors. This example follows a similar one found in Conte and de Boor [1980, p. 73], and it illustrates one of the dangers involved in finding roots.

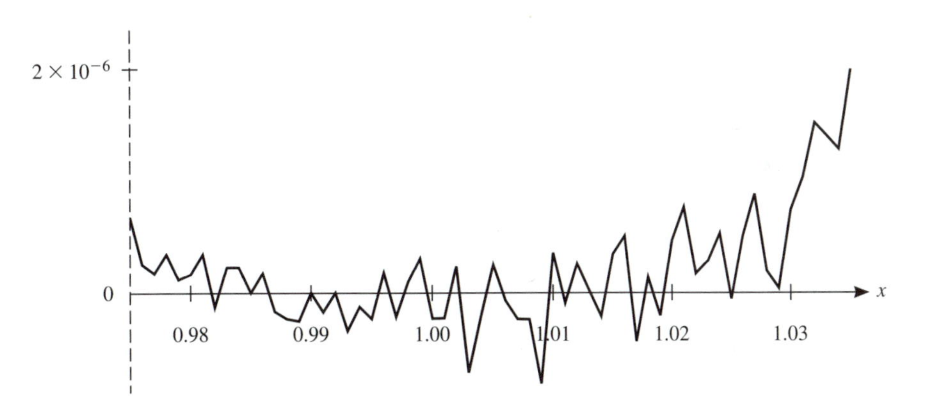

FIGURE 3.1 Plot of expanded polynomial

3.1 Bisection (Interval Halving) Method

If f is a continuous function on the interval $[a, b]$ and if $f(a)f(b) < 0$, then f must have a zero in (a, b). Since $f(a)f(b) < 0$, the function f changes sign on the interval $[a, b]$ and, therefore, it has at least one zero in the interval. This is a consequence of the following fundamental theorem.

THEOREM 1 **Intermediate-Value Theorem for Continuous Functions** *If f is continuous on $[a, b]$, and if $f(a) < y < f(b)$, then $f(x) = y$ for some $x \in (a, b)$.*

The bisection method exploits this idea in the following way. If $f(a)f(b) < 0$, then we compute $c = \frac{1}{2}(a + b)$ and test whether $f(a)f(c) < 0$. If this is true, then f has a zero in $[a, c]$. So we rename c as b and start again with the new interval $[a, b]$, which is half as large as the original interval. If $f(a)f(c) > 0$, then $f(c)f(b) < 0$, and in this case we rename c as a. In either case, a new interval containing a zero of f has been produced, and the process can be repeated. Figures 3.2(a) and (b) show the two cases, assuming $f(a) > 0 > f(b)$. These figures makes it clear why the bisection method finds one zero but not all the zeros in the interval $[a, b]$. Of

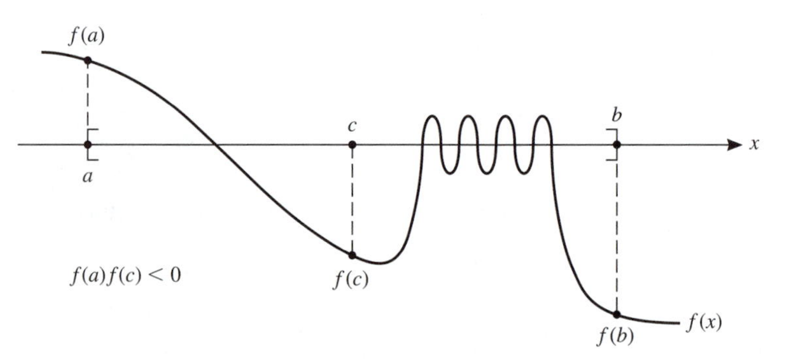

FIGURE 3.2(a) Bisection method selects left subinterval

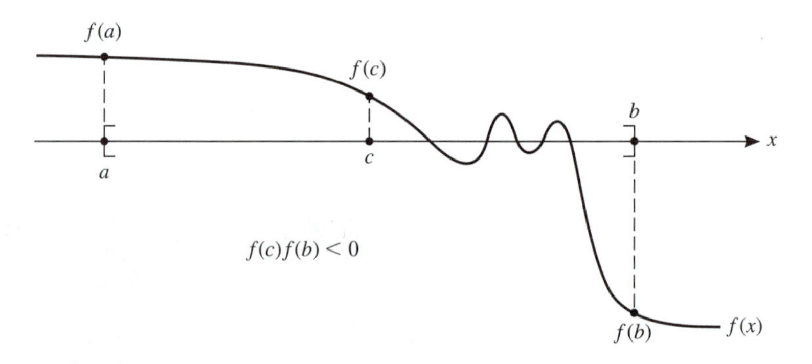

FIGURE 3.2(b) Bisection method selects right subinterval

course, if $f(a)f(c) = 0$, then $f(c) = 0$ and a zero has been found. However, it is quite unlikely that $f(c)$ will be exactly 0 in the computer because of roundoff errors. Thus, the **stopping criterion** should *not* be whether $f(c) = 0$. A reasonable tolerance must be allowed, such as $|f(c)| < 10^{-5}$ on the `Marc-32`. (See Section 2.1 for a description of the hypothetical `Marc-32` computer.) The bisection method is also known as the **method of interval halving**.

Example 1 Use the bisection method to find the root of the equation $e^x = \sin x$ closest to 0.

Solution If the graphs of e^x and $\sin x$ are roughly plotted, it becomes clear that there are no positive roots of $f(x) = e^x - \sin x$ and that the first root to the left of 0 is in the interval $[-4, -3]$. When the bisection algorithm was run on a machine similar to the `Marc-32`, the following output was produced, starting with the interval $[-4, -3]$:

k	c	$f(c)$
1	−3.5000	−0.321
2	−3.2500	-0.694×10^{-1}
3	−3.1250	0.605×10^{-1}
4	−3.1875	0.625×10^{-1}
⋮	⋮	⋮
13	−3.1829	0.122×10^{-3}
14	−3.1830	0.193×10^{-4}
15	−3.1831	-0.124×10^{-4}
16	−3.1831	0.345×10^{-5}

■

Bisection Algorithm

Some parts of the bisection pseudocode (given on page 83) need some additional explanation. First, the midpoint c is computed as $c \leftarrow a + (b - a)/2$ rather than as $c \leftarrow (a + b)/2$ to adhere to the general stratagem that in numerical calculations it is best to compute a quantity by adding a small correction term to a previous approximation. Forsythe, Malcolm, and Moler [1977, p. 162] present an example in which the midpoint computed by $(a + b)/2$ moves outside of the interval $[a, b]$ on a machine with limited precision! Second, it is better to determine whether the function changes sign over the interval using $\text{sgn}(w) \neq \text{sgn}(u)$, rather than $wv < 0$, since the latter requires an unnecessary multiplication and could cause an underflow or overflow. Next, e is the computation of the error bound that is established in the theorem to follow. Finally, notice that *three* stopping criteria are present in the algorithm. First, M gives the maximum number of steps that the user will permit. Such a safeguard should *always* be present to reduce the possibility of the computation going into an infinite loop. Next, the calculation can be stopped when either the error is small enough or the value of $f(c)$ is small enough. The parameters δ and ε control this. Examples can be easily given in which one of the latter two stopping criteria is satisfied but the other is not. For example, consider the two sketches in Figures 3.3(a) and (b). In Figure 3.3(a), the graph is flat near the zero, which corresponds to a multiple root, and the bisection method may have difficulty

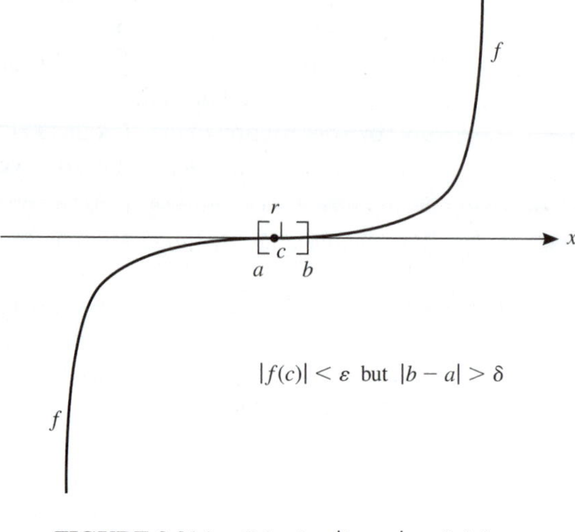

FIGURE 3.3(a) Criterion $|b - a| < \delta$ failure

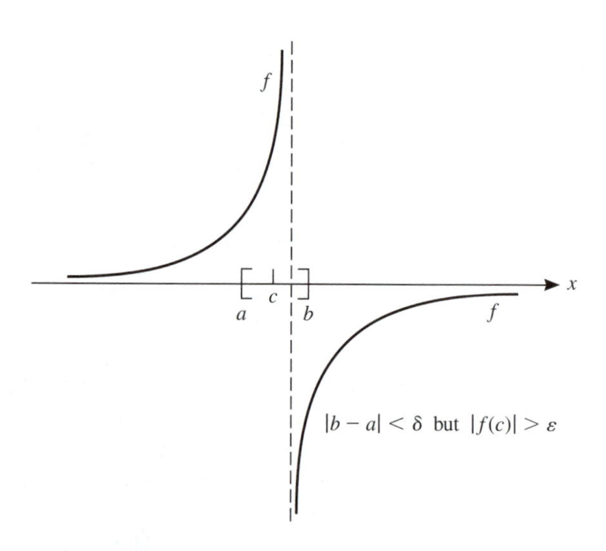

FIGURE 3.3(b) Criterion $|f(c)| < \varepsilon$ failure

determining this zero to high precision. The curve in Figure 3.3(b) is, of course, not continuous, but the continuity of a function may be hard to verify beforehand. Although these are pathological examples, we choose to stop the algorithm when any one of the three stopping criteria is satisfied, in the interest of having a robust code.

The bisection algorithm can be written as follows:

```
input a, b, M, δ, ε
u ← f(a)
v ← f(b)
e ← b − a
output a, b, u, v
if sgn(u) = sgn(v) then stop
for k = 1 to M do
    e ← e/2
    c ← a + e
    w ← f(c)
    output k, c, w, e
    if |e| < δ or |w| < ε then stop
    if sgn(w) ≠ sgn(u) then
        b ← c
        v ← w
    else
        a ← c
        u ← w
    end if
end do
```

Error Analysis

To analyze the bisection method, let us denote the successive intervals that arise in the process by $[a_0, b_0]$, $[a_1, b_1]$, and so on. Here are some observations about these numbers:

$$a_0 \leqq a_1 \leqq a_2 \leqq \cdots \leqq b_0 \tag{1}$$

$$b_0 \geqq b_1 \geqq b_2 \geqq \cdots \geqq a_0 \tag{2}$$

$$b_{n+1} - a_{n+1} = \frac{1}{2}(b_n - a_n) \qquad (n \geqq 0) \tag{3}$$

Since the sequence $[a_n]$ is increasing and bounded above, it converges. Likewise, $[b_n]$ converges. If we apply Equation (3) repeatedly, we find that

$$b_n - a_n = 2^{-n}(b_0 - a_0)$$

Thus,

$$\lim_{n\to\infty} b_n - \lim_{n\to\infty} a_n = \lim_{n\to\infty} 2^{-n}(b_0 - a_0) = 0$$

If we put

$$r = \lim_{n\to\infty} a_n = \lim_{n\to\infty} b_n$$

then, by taking a limit in the inequality $0 \geqq f(a_n)f(b_n)$, we obtain $0 \geqq [f(r)]^2$, whence $f(r) = 0$.

Suppose that, at a certain stage in the process, the interval $[a_n, b_n]$ has just been defined. If the process is now stopped, the root is certain to lie in this interval. The best estimate of the root at this stage is not a_n or b_n but the *midpoint* of the interval:

$$c_n = (a_n + b_n)/2$$

The error is then bounded as follows:

$$|r - c_n| \leqq \frac{1}{2}(b_n - a_n) = 2^{-(n+1)}(b_0 - a_0)$$

Summarizing this discussion, we have the following theorem on the bisection method.

THEOREM 2 **Bisection Method Theorem** *If $[a_0, b_0], [a_1, b_1], \ldots, [a_n, b_n], \ldots$ denote the intervals in the bisection method, then the limits $\lim_{n\to\infty} a_n$ and $\lim_{n\to\infty} b_n$ exist, are equal, and represent a zero of f. If $r = \lim_{n\to\infty} c_n$ and $c_n = \frac{1}{2}(a_n + b_n)$, then*

$$|r - c_n| \leqq 2^{-(n+1)}(b_0 - a_0)$$

Example 2 Suppose that the bisection method is started with the interval [50, 63]. How many steps should be taken to compute a root with relative accuracy of one part in 10^{-12}?

Solution The stated requirement on relative precision means that

$$|r - c_n|/|r| \leqq 10^{-12}$$

We know that $r \geqq 50$, and thus it will suffice to secure the inequality

$$|r - c_n|/50 \leqq 10^{-12}$$

By means of Theorem 2, we infer that the following condition will be sufficient:

$$2^{-(n+1)} \times (13/50) \leqq 10^{-12}$$

Solving this for n, we conclude that $n \geqq 37$. ■

PROBLEMS 3.1

1. Let $c_n = \frac{1}{2}(a_n + b_n)$, $r = \lim_{n\to\infty} c_n$, and $e_n = r - c_n$. Here $[a_n, b_n]$, with $n \geqq 0$, denotes the successive intervals that arise in the bisection method when it is applied to a continuous function f.
   (a) Show that $|e_n| \leqq 2^{-n}(b_1 - a_1)$.
   (b) Show that $e_n = \mathcal{O}(2^{-n})$ as $n \to \infty$.

(c) Is it true that $|e_0| \geqq |e_1| \geqq \cdots$? Explain.

(d) Show that $|c_n - c_{n+1}| = 2^{-n-2}(b_0 - a_0)$.

(e) Show that for all n and m, $a_m \leqq b_n$.

(f) Show that r is the unique element in $\bigcap_{n=0}^{\infty}[a_n, b_n]$.

(g) Show that for all n, $[a_n, b_n] \supseteqq [a_{n+1}, b_{n+1}]$.

2. In the bisection method, an interval $[a_{n-1}, b_{n-1}]$ is divided in half, and one of these halves is chosen for the next interval. Define $d_n = 0$ if $[a_n, b_n]$ is the left half of the interval $[a_{n-1}, b_{n-1}]$, and let $d_n = 1$ otherwise. Express the root determined by the algorithm in terms of the sequence $d_1, d_2, \ldots$. *Hint:* Consider the case $[a_0, b_0] = [0, 1]$ first, and think about the binary representation of the root.

3. Using the notation of the previous two problems, find the formulas that relate a_n, b_n, and c_n to d_n.

4. Give an example (or prove that none exists) in which $a_0 < a_1 < a_2 < \cdots$.

5. Give an example in which $a_0 = a_1 < a_2 = a_3 < a_4 = a_5 < a_6 = \cdots$.

6. In the bisection method, does $\lim_{n\to\infty} |r - c_{n+1}|/|r - c_n|$ exist? Explain.

7. Give a formula involving $b_0 - a_0$ and ε for the number of steps that must be taken in the bisection method to guarantee that $|r - c_n| < \varepsilon$.

8. Find a positive root of

$$x^2 - 4x \sin x + (2 \sin x)^2 = 0$$

accurate to two significant figures. Use a hand calculator.

9. Consider the bisection method starting with the interval $[1.5, 3.5]$.

(a) What is the width of the interval at the nth step of this method?

(b) What is the maximum distance possible between the root r and the midpoint of this interval?

10. Let the bisection method be applied to a continuous function, resulting in intervals $[a_0, b_0]$, $[a_1, b_1]$, and so on. Let $r = \lim_{n\to\infty} a_n$. Which of these statements can be false?

(a) $a_0 \leqq a_1 \leqq a_2 \leqq \cdots$

(b) $|r - 2^{-1}(a_n + b_n)| \leqq 2^{-n}(b_0 - a_0) \qquad (n \geqq 0)$

(c) $|r - 2^{-1}(a_{n+1} + b_{n+1})| \leqq |r - 2^{-1}(a_n + b_n)| \qquad (n \geqq 0)$

(d) $[a_{n+1}, b_{n+1}] \subseteqq [a_n, b_n] \qquad (n \geqq 0)$

(e) $|r - a_n| = \mathcal{O}(2^{-n})$ as $n \to \infty$

(f) $|r - c_n| < |r - c_{n-1}| \qquad (n \geqq 1)$

11. Give a formula involving a_0, b_0, and ε for the number of steps that should be taken in the bisection algorithm to ensure that the root is determined with *relative* precision $\leqq \varepsilon$. Assume $a_0 > 0$.

12. What happens in Problem 11 if $a_0 < 0 < b_0$?

13. If the bisection method is used (in single precision) on the `Marc-32` starting with the interval [128,129], can we compute the root with absolute precision $< 10^{-6}$? Answer the same question for the *relative* precision.

14. Prove that the point c computed in the bisection method is the point where the line through $(a, \operatorname{sgn}(f(a)))$ and $(b, \operatorname{sgn}(f(b)))$ intersects the x-axis.

15. Suppose that $|a_n - b_n| \leqq \lambda_n |a_{n-1} - b_{n-1}|$ for all n with $\lambda_n < 1$. Find an upper bound on $|a_n - b_n|$ in terms of $|a_0 - b_0|$ and $\lambda = \max_{1 \leqq i \leqq n}\{\lambda_i\}$.

COMPUTER PROBLEMS 3.1

1. Write and test a subprogram or procedure to implement the bisection algorithm. Test the program on these functions and intervals:

(a) $x^{-1} - \tan x$ on $[0, \pi/2]$

(b) $x^{-1} - 2^x$ on $[0, 1]$

(c) $2^{-x} + e^x + 2\cos x - 6$ on $[1, 3]$

(d) $(x^3 + 4x^2 + 3x + 5)/(2x^3 - 9x^2 + 18x - 2)$ on $[0, 4]$

2. Find numbers $a < b$ such that the two mathematically equivalent computations $c \leftarrow (a + b)/2$ and $c \leftarrow a + 0.5(b - a)$ produce different results on your computer. Do not choose an example in which there is underflow or overflow.

3. Find a root of $f(x) = x - \tan x$ in the interval $[1, 2]$.

4. Find a root of

$$x^8 - 36x^7 + 546x^6 - 4536x^5 + 22449x^4 - 67284x^3 + 118124x^2 - 109584x + 40320 = 0$$

in the interval $[5.5, 6.5]$. Change -36 to -36.001 and repeat.

5. Write and test a recursive version of the bisection algorithm.

3.2 Newton's Method

Newton's method is a general procedure that can be applied in many diverse situations. When specialized to the problem of locating a zero of a real-valued function of a real variable, it is often called the **Newton-Raphson iteration**. In general, Newton's method is faster than the bisection and the secant methods since its convergence is quadratic rather than linear or superlinear. Once the quadratic convergence becomes effective, that is, the values of Newton's method sequence are sufficiently close to the root, the convergence is so rapid that only a few more values are needed. Unfortunately, the method is not guaranteed always to converge. Often Newton's method is combined with other slower methods in a hybrid method that is numerically globally convergent.

As in Section 3.1, we have a function f whose zeros are to be determined numerically. Let r be a zero of f and let x be an approximation to r. If f'' exists and is continuous, then by Taylor's Theorem,

$$0 = f(r) = f(x+h) = f(x) + hf'(x) + \mathcal{O}(h^2)$$

where $h = r - x$. If h is small (that is, x is near r), then it is reasonable to ignore the $\mathcal{O}(h^2)$-term and solve the remaining equation for h. If we do this, the result is $h = -f(x)/f'(x)$. If x is an approximation to r, then $x - f(x)/f'(x)$ should be a better approximation to r. Newton's method begins with an estimate x_0 of r and then defines inductively

$$x_{n+1} = x_n - \frac{f(x_n)}{f'(x_n)} \qquad (n \geqq 0) \tag{1}$$

Newton's Algorithm

A simple algorithm that applies M steps of Newton's method, starting with an initial value for x, is:

input x, M
$y \leftarrow f(x)$
output $0, x, y$
for $k = 1$ **to** M **do**
  $x \leftarrow x - y/f'(x)$
  $y \leftarrow f(x)$
  output k, x, y
end do

A more detailed pseudocode including stopping criteria follows:

input $x_0, M, \delta, \varepsilon$
$v \leftarrow f(x_0)$
output $0, x_0, v$
if $|v| < \varepsilon$ **then stop**
for $k = 1$ **to** M **do**
  $x_1 \leftarrow x_0 - v/f'(x_0)$
  $v \leftarrow f(x_1)$
  output k, x_1, v
  if $|x_1 - x_0| < \delta$ **or** $|v| < \varepsilon$ **then stop**
  $x_0 \leftarrow x_1$
end do

A computer program based on either of these two algorithms would need subprograms or procedures for $f(x)$ and $f'(x)$.

Example 1 Use Newton's method, with double-precision computation, to find the negative zero of the function $f(x) = e^x - 1.5 - \tan^{-1} x$.

Solution The preceding algorithm was executed with double precision on a machine with 48-bit mantissas in single precision. (Double-precision machine numbers have 96 bits, corresponding to about 28 decimal places.) The function $f'(x) = e^x - (1 + x^2)^{-1}$ as well as f had to be programmed. A starting point of $x_0 = -7$ was chosen. The output from the computer program is shown here.

k	x	$f(x)$
0	− 7.00000 00000 00000 00000 00000 0	-0.702×10^{-1}
1	−10.67709 61766 40013 99296 98438 6	-0.226×10^{-1}
2	−13.27916 73756 32712 90859 78631 9	-0.437×10^{-2}
3	−14.05365 58542 69238 73474 83175 3	-0.239×10^{-3}
4	−14.10110 99568 66413 47616 31270 6	-0.800×10^{-6}
5	−14.10126 97709 39415 94621 57950 6	-0.901×10^{-11}
6	−14.10126 97727 39968 42508 30031 4	-0.114×10^{-20}
7	−14.10126 97727 39968 42531 15512 2	0.000
8	−14.10126 97727 39968 42531 15512 2	0.000

The output shows rapid convergence of the iterates; in fact, in each step the number of correct digits in the approximation seems to *double*. Our analysis will reveal why this is true. ■

Graphical Interpretation

Before examining the theoretical basis for Newton's method, let's give a graphical interpretation of it. From the description already given, we can say that Newton's method involves **linearizing the function**. That is, f was replaced by a linear function. The usual way of doing this is to replace f by the first two terms in its Taylor series. Thus, if

$$f(x) = f(c) + f'(c)(x - c) + \frac{1}{2!} f''(c)(x - c)^2 + \cdots$$

then the linearization (at c) produces the linear function

$$\ell(x) = f(c) + f'(c)(x - c)$$

Notice that ℓ is a good approximation to f in the vicinity of c, and in fact we have $\ell(c) = f(c)$ and $\ell'(c) = f'(c)$. Thus, the linear function has the same value and the same slope as f at the point c. So in Newton's method we are constructing the tangent to the f-curve at a point near r, and finding where the tangent line intersects the x-axis. (See Figure 3.4.)

Keeping in mind this graphical interpretation, we can easily imagine functions and starting points for which the Newton iteration will *fail*. Such a function is shown in Figure 3.5. In this example, the shape of the curve is such that for certain starting

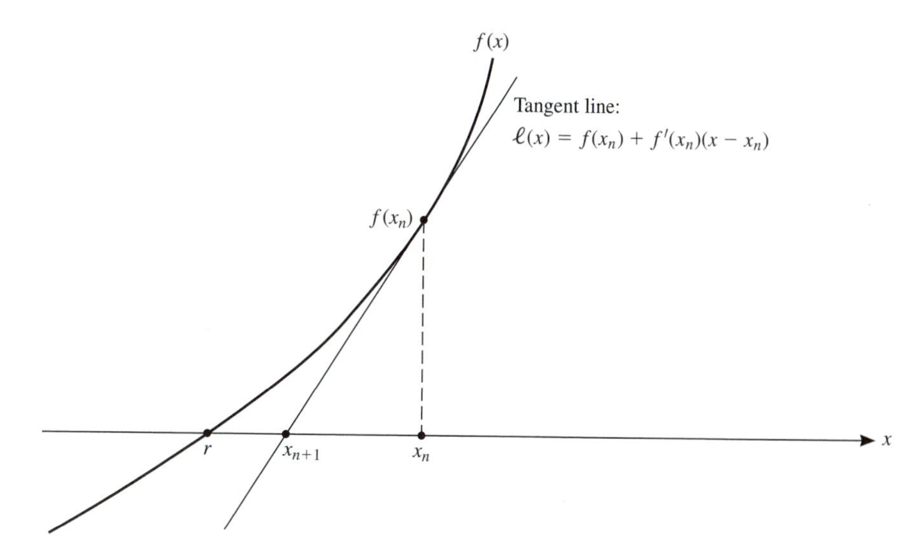

FIGURE 3.4 Geometric interpretation of Newton's method

values, the sequence $[x_n]$ will diverge. Thus, any formal statement about Newton's method must involve an assumption that x_0 is **sufficiently close to a zero**, or that the graph of f has a prescribed shape.

Error Analysis

Now we shall analyze the errors in Newton's method. By **errors**, we mean the quantities

$$e_n = x_n - r$$

(We are *not* considering roundoff errors.) Let us assume that f'' is continuous and r is a **simple zero** of f, so that $f(r) = 0 \neq f'(r)$. From the definition of the Newton

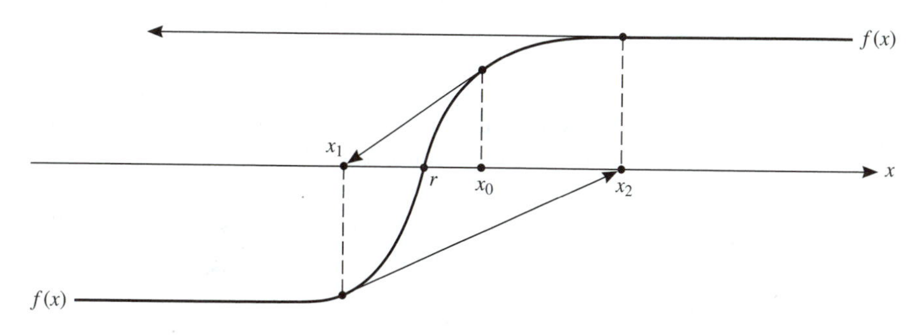

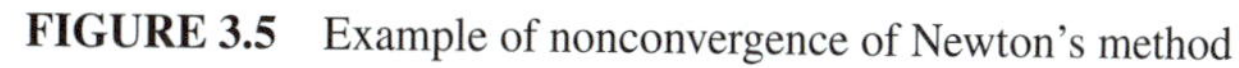

FIGURE 3.5 Example of nonconvergence of Newton's method

iteration, we have

$$\begin{aligned} e_{n+1} &= x_{n+1} - r = x_n - \frac{f(x_n)}{f'(x_n)} - r \\ &= e_n - \frac{f(x_n)}{f'(x_n)} = \frac{e_n f'(x_n) - f(x_n)}{f'(x_n)} \end{aligned} \tag{2}$$

By Taylor's Theorem, we have

$$0 = f(r) = f(x_n - e_n) = f(x_n) - e_n f'(x_n) + \frac{1}{2} e_n^2 f''(\xi_n)$$

where ξ_n is a number between x_n and r. A rearrangement of this equation yields

$$e_n f'(x_n) - f(x_n) = \frac{1}{2} f''(\xi_n) e_n^2$$

Putting this in Equation (2) leads to

$$e_{n+1} = \frac{1}{2} \frac{f''(\xi_n)}{f'(x_n)} e_n^2 \approx \frac{1}{2} \frac{f''(r)}{f'(r)} e_n^2 = C e_n^2 \tag{3}$$

Suppose that $C \approx 1$ and $e_n \approx 10^{-4}$. Then by Equation (3), we have $e_{n+1} \approx 10^{-8}$ and $e_{n+2} \approx 10^{-16}$. We are impressed that only a few additional iterations are needed to obtain more than machine precision!

This equation tells us that e_{n+1} is roughly a constant times e_n^2. This desirable state of affairs is called **quadratic convergence**. It accounts for the apparent doubling of precision with each iteration of Newton's method in many applications.

We have yet to establish the convergence of the method. By Equation (3), the idea of the proof is simple: If e_n is small and if the factor $\frac{1}{2} f''(\xi_n)/f'(x_n)$ is not too large, then e_{n+1} will be smaller than e_n. Define a quantity $c(\delta)$ dependent on δ by

$$c(\delta) = \frac{1}{2} \max_{|x-r| \leqq \delta} \left| f''(x) \right| \Big/ \min_{|x-r| \leqq \delta} \left| f'(x) \right| \qquad (\delta > 0) \tag{4}$$

We select δ small enough to be sure that the denominator in Equation (4) is positive, and then if necessary we decrease δ so that $\delta c(\delta) < 1$. This is possible because as δ converges to 0, $c(\delta)$ converges to $\frac{1}{2}|f''(r)|/|f'(r)|$, and so $\delta c(\delta)$ converges to 0. Having fixed δ, set $\rho = \delta c(\delta)$. Suppose that we start the Newton iteration with a point x_0 satisfying $|x_0 - r| \leqq \delta$. Then $|e_0| \leqq \delta$ and $|\xi_0 - r| \leqq \delta$. Hence, by the definition of $c(\delta)$, we have

$$\frac{1}{2} \left| f''(\xi_0)/f'(x_0) \right| \leqq c(\delta)$$

Therefore, Equation (3) yields

$$|x_1 - r| = |e_1| \leqq e_0^2 c(\delta) = |e_0||e_0|c(\delta) \leqq |e_0|\delta c(\delta) = |e_0|\rho < |e_0| \leqq \delta$$

This shows that the next point, x_1, also lies within δ units of r. Hence, the argument can be repeated, with the results

$$\begin{aligned} |e_1| &\leqq \rho|e_0| \\ |e_2| &\leqq \rho|e_1| \leqq \rho^2|e_0| \\ |e_3| &\leqq \rho|e_2| \leqq \rho^3|e_0| \\ &\vdots \end{aligned}$$

In general, we have

$$|e_n| \leqq \rho^n|e_0|$$

Since $0 \leqq \rho < 1$, we have $\lim_{n\to\infty} \rho^n = 0$ and so $\lim_{n\to\infty} e_n = 0$. Summarizing, we obtain the following theorem on Newton's method.

THEOREM 1 **Newton's Method Theorem** *Let f'' be continuous and let r be a simple zero of f. Then there is a neighborhood of r and a constant C such that if Newton's method is started in that neighborhood, the successive points become steadily closer to r and satisfy*

$$|x_{n+1} - r| \leqq C(x_n - r)^2 \qquad (n \geqq 0)$$

In some situations Newton's iteration can be guaranteed to converge from an *arbitrary* starting point. We give one such theorem as a sample.

THEOREM 2 *If f belongs to $C^2(\mathbb{R})$, is increasing, is convex, and has a zero, then the zero is unique, and the Newton iteration will converge to it from any starting point.*

Proof Recall that a function f is **convex** if $f''(x) > 0$ for all x. Since f is increasing, $f' > 0$ on $\mathbb{R}$. By Equation (3), $e_{n+1} > 0$. Thus, $x_n > r$ for $n \geqq 1$. Since f is increasing, $f(x_n) > f(r) = 0$. Hence, by Equation (2), $e_{n+1} < e_n$. Thus, the sequences $[e_n]$ and $[x_n]$ are decreasing and bounded below (by 0 and r, respectively). Therefore, the limits $e^* = \lim_{n\to\infty} e_n$ and $x^* = \lim_{n\to\infty} x_n$ exist. From Equation (2), we have $e^* = e^* - f(x^*)/f'(x^*)$, whence $f(x^*) = 0$ and $x^* = r$. ■

Example 2 Find an efficient method for computing square roots based on the use of Newton's method.

Solution Let $R > 0$ and set $x = \sqrt{R}$. Then x is a root of the equation $x^2 - R = 0$. If we use Newton's method (1) on the function $f(x) = x^2 - R$, the iteration formula can be written as

$$x_{n+1} = \frac{1}{2}\left(x_n + \frac{R}{x_n}\right)$$

This formula, combined with range reduction, is often used in the subroutine that computes square roots. (This formula is very ancient and is credited to Heron, a Greek engineer and architect who lived sometime between 100 B.C. and A.D. 100.) If, for example, we wish to compute $\sqrt{17}$ and begin with $x_0 = 4$, the successive approximations are as follows (given in rounded form to exhibit only correct figures):

$$\begin{aligned}
x_1 &= 4.12\\
x_2 &= 4.123106\\
x_3 &= 4.12310\,56256\,177\\
x_4 &= 4.12310\,56256\,17660\,54982\,14098\,56
\end{aligned}$$

The value given by x_4 is correct to 28 figures, and we observe the expected doubling of significant digits in the results. ■

Implicit Functions

One interesting application of Newton's method occurs in the evaluation of functions defined implicitly. In Section 1.2, the Implicit Function Theorem was discussed; it states that under rather general conditions, an equation of the form $G(x, y) = 0$ will define y as a function of x. If x is prescribed, then the equation $G(x, y) = 0$ can be solved for y by using Newton's method. From a suitable starting point y_0, we define $y_1, y_2, \ldots$ by

$$y_{k+1} = y_k - G(x, y_k) \Big/ \frac{\partial G}{\partial y}(x, y_k)$$

This method can be used to construct a table of the function $y(x)$. If the table contains an entry (x_n, y_n), and if we desire to compute a nearby entry (x_{n+1}, y_{n+1}), we *start* the Newton iteration with (x_{n+1}, y_n). The *result* of the iteration will be the correct value y_{n+1} that makes the equation $G(x_{n+1}, y_{n+1}) = 0$ true. Since $G(x_n, y_n) = 0$ and x_{n+1} is close to x_n, we can expect that $G(x_{n+1}, y_n)$ is small and that a few steps of the Newton method will produce the correction to y_n necessary to achieve the exact equality $G(x_{n+1}, y_{n+1}) = 0$.

Example 3 Produce a table of x *versus* y, where y is defined implicitly as a function of x. Use $G(x, y) = 3x^7 + 2y^5 - x^3 + y^3 - 3$ and start at $x = 0$, proceeding in steps of 0.1 to $x = 10$.

Solution Let $x = 0$ and $y = 1$. We assume that four steps of Newton's method will suffice to give full machine precision. In the algorithm, we have M steps in the x-variable, and at each step there are N iterations in Newton's method. The program based on the algorithm would need two subprograms or procedures, the first to compute $G(x, y)$ and the second to compute $\partial G/\partial y$. (The latter happens to be independent of x in this example.) These functions are

$$G(x, y) = 3x^7 + 2y^5 - x^3 + y^3 - 3$$
$$\frac{\partial G}{\partial y}(x, y) = 10y^4 + 3y^2$$

Here is a suitable algorithm:

input $x \leftarrow 0;\ y \leftarrow 1;\ h \leftarrow 0.1;\ M \leftarrow 100;\ N \leftarrow 4$
output $0, x, y, G(x, y)$
for $i = 1$ **to** M **do**
  $x \leftarrow x + h$
  for $j = 1$ **to** N **do**
    $y \leftarrow y - G(x, y) \Big/ \frac{\partial G}{\partial y}(x, y)$
  end do
  output $i, x, y, G(x, y)$
end do

Some of the values of the implicit function, computed by the foregoing algorithm, are shown here:

i	x	y	$G(x, y)$
0	0.0	1.0000000	0.00
1	0.1	1.0000077	0.00
2	0.2	1.0000612	0.89×10^{-15}
⋮	⋮	⋮	⋮
20	2.0	− 2.8106339	-0.82×10^{-10}
⋮	⋮	⋮	⋮
80	8.0	−19.92635	0.56×10^{-9}
⋮	⋮	⋮	⋮
99	9.9	−26.85618	0.12×10^{-7}
100	10.0	−27.23685	-0.15×10^{-8}

■

Systems of Nonlinear Equations

Newton's method for systems of nonlinear equations follows the same strategy that was used for a single equation. Thus, we *linearize* and *solve*, repeating the steps as often as necessary. Let us illustrate with a pair of equations involving two variables:

$$\begin{cases} f_1(x_1, x_2) = 0 \\ f_2(x_1, x_2) = 0 \end{cases} \tag{5}$$

Supposing that (x_1, x_2) is an approximate solution of (5), let us compute corrections h_1 and h_2 so that $(x_1 + h_1, x_2 + h_2)$ will be a better approximate solution. Using only linear terms in the Taylor expansion in two variables (see Section 1.1), we have

$$\begin{cases} 0 = f_1(x_1 + h_1, x_2 + h_2) \approx f_1(x_1, x_2) + h_1 \dfrac{\partial f_1}{\partial x_1} + h_2 \dfrac{\partial f_1}{\partial x_2} \\ 0 = f_2(x_1 + h_1, x_2 + h_2) \approx f_2(x_1, x_2) + h_1 \dfrac{\partial f_2}{\partial x_1} + h_2 \dfrac{\partial f_2}{\partial x_2} \end{cases} \tag{6}$$

The partial derivatives appearing in (6) are to be evaluated at (x_1, x_2). Equation (6) constitutes a pair of *linear* equations for determining h_1 and h_2. The coefficient matrix is the **Jacobian matrix** of f_1 and f_2:

$$J = \begin{bmatrix} \partial f_1/\partial x_1 & \partial f_1/\partial x_2 \\ \partial f_2/\partial x_1 & \partial f_2/\partial x_2 \end{bmatrix}$$

To solve (6), we require J to be nonsingular. If this condition is met, the solution is

$$\begin{bmatrix} h_1 \\ h_2 \end{bmatrix} = -J^{-1} \begin{bmatrix} f_1(x_1, x_2) \\ f_2(x_1, x_2) \end{bmatrix}$$

Hence, Newton's method for two nonlinear equations in two variables is

$$\begin{bmatrix} x_1^{(k+1)} \\ x_2^{(k+1)} \end{bmatrix} = \begin{bmatrix} x_1^{(k)} \\ x_2^{(k)} \end{bmatrix} + \begin{bmatrix} h_1^{(k)} \\ h_2^{(k)} \end{bmatrix}$$

where the Jacobian linear system

$$J \begin{bmatrix} h_1^{(k)} \\ h_2^{(k)} \end{bmatrix} = - \begin{bmatrix} f_1(x_1^{(k)}, x_2^{(k)}) \\ f_2(x_1^{(k)}, x_2^{(k)}) \end{bmatrix}$$

is solved using Gaussian elimination—a subject discussed in Chapter 4. Solving the Jacobian system may be difficult if J is nearly singular.

To discuss larger systems involving many more variables, no new ideas are needed. However, a matrix-vector formalism is very convenient in this situation. The system of equations

$$f_i(x_1, x_2, \dots, x_n) = 0 \qquad (1 \leqq i \leqq n) \tag{7}$$

can be expressed simply as

$$F(X) = 0 \tag{8}$$

by letting $X = (x_1, x_2, \ldots, x_n)^T$ and $F = (f_1, f_2, \ldots, f_n)^T$. The analogue of Equation (6) is

$$0 = F(X + H) \approx F(X) + F'(X)H \tag{9}$$

where $H = (h_1, h_2, \ldots, h_n)^T$ and $F'(X)$ is the $n \times n$ Jacobian matrix $J(X)$ with elements $\partial f_i / \partial x_j$; namely,

$$F'(X) = \begin{bmatrix} \partial f_1/\partial x_1 & \partial f_1/\partial x_2 & \cdots & \partial f_1/\partial x_n \\ \partial f_2/\partial x_1 & \partial f_2/\partial x_2 & \cdots & \partial f_2/\partial x_n \\ \vdots & \vdots & \ddots & \vdots \\ \partial f_n/\partial x_1 & \partial f_n/\partial x_2 & \cdots & \partial f_n/\partial x_n \end{bmatrix}$$

The correction vector H is obtained by solving the linear system of equations in (9). Theoretically, this means

$$H = -F'(X)^{-1}F(X) \tag{10}$$

but in practice H would usually be determined by Gaussian elimination from (9), thus bypassing the more costly computation of the inverse in (10). Hence, Newton's method for n nonlinear equations in n variables is given by

$$X^{(k+1)} = X^{(k)} + H^{(k)} \tag{11}$$

where the Jacobian system is

$$F'(X^{(k)})H^{(k)} = -F(X^{(k)}) \tag{12}$$

Example 4 Starting with $(1, 1, 1)^T$, carry out six iterations of Newton's method for finding a root of the nonlinear system

$$\begin{cases} xy = z^2 + 1 \\ xyz + y^2 = x^2 + 2 \\ e^x + z = e^y + 3 \end{cases}$$

Solution Let

$$F(X) = \begin{bmatrix} f_1(x_1, x_2, x_3) \\ f_2(x_1, x_2, x_3) \\ f_3(x_x, x_2, x_3) \end{bmatrix} = \begin{bmatrix} x_1x_2 - x_3^2 - 1 \\ x_1x_2x_3 - x_1^2 + x_2^2 - 2 \\ e^{x_1} - e^{x_2} + x_3 - 3 \end{bmatrix}$$

Taking partial derivatives, we get the Jacobian matrix

$$F'(X) = \begin{bmatrix} x_2 & x_1 & -2x_3 \\ x_2x_3 - 2x_1 & x_1x_3 + 2x_2 & x_1x_2 \\ e^{x_1} & -e^{-x_2} & 1 \end{bmatrix}$$

Using the starting value $X^{(0)} = (1, 1, 1)^T$, we carry out the nonlinear Newton's method given in Equations (11) and (12) and obtain the following:

n	x_1	x_2	x_3
0	1.00000 00	1.00000 00	1.00000 00
1	1.75479 11	1.53639 88	1.14559 50
2	1.85029 07	1.42072 00	1.28849 21
3	1.77808 28	1.42349 40	1.23846 41
4	1.77739 18	1.42394 65	1.23730 05
5	1.77766 36	1.42396 02	1.23746 60
6	1.77767 17	1.42396 06	1.23747 10

■

The solution of systems such as (8) is often challenging. The standard reference on the subject is Ortega and Rheinboldt [1970]. See also Rheinboldt [1974], Ostrowski [1966], Byrne and Hall [1973], Schnabel and Frank [1984], Eaves, Gould, Peitgen, and Todd [1983], and Allgower, Glasshoff, and Peitgen [1981]. For a discussion of the convergence of Newton's method in higher dimensions, refer to Goldstein [1966] or Ortega and Rheinboldt [1970].

PROBLEMS 3.2

1. Find the smallest positive starting point for which Newton's method diverges when it is applied to $f(x) = \tan^{-1} x$.
2. Let Newton's method be used on $f(x) = x^2 - q$ (where $q > 0$). Show that if x_n has k correct digits after the decimal point, then x_{n+1} will have at least $2k - 1$ correct digits after the decimal point, provided that $q > 0.006$ and $k \geqq 1$.
3. Prove that if Newton's method is used on a function f for which f'' is continuous and $f(r) = 0 \neq f'(r)$, then $\lim_{n\to\infty} e_{n+1}e_n^{-2}$ exists and equals $f''(r)/[2f'(r)]$. How can this fact be used in a program to test whether convergence is quadratic?
4. (**Steffensen's method**) Consider the iteration formula

$$x_{n+1} = x_n - f(x_n)/g(x_n)$$

where

$$g(x) = [f(x + f(x)) - f(x)]/f(x)$$

Show that this is quadratically convergent, under suitable hypotheses.

5. What is the purpose of the following iteration formula?

$$x_{n+1} = 2x_n - x_n^2 y$$

Identify it as the Newton iteration for a certain function.

6. To compute reciprocals without division, we can solve $x = 1/R$ by finding a zero of the function $f(x) = x^{-1} - R$. Write a short algorithm to find $1/R$ by

Newton's method applied to f. Do not use division or exponentiation in your algorithm. For positive R, what starting points are suitable?

7. Define $x_0 = 0$ and $x_{n+1} = x_n - [(\tan x_n - 1)/\sec^2 x_n]$. What is $\lim_{n\to\infty} x_n$ in this example? Relate this to Newton's method.

8. Perform four iterations of Newton's method for the polynomial

$$p(x) = 4x^3 - 2x^2 + 3$$

starting with $x_0 = -1$. Use a hand calculator.

9. If Newton's method is used on $f(x) = x^3 - 2$ starting with $x_0 = 1$, what is x_2?

10. Devise a Newton iteration formula for computing $\sqrt[3]{R}$ where $R > 0$. Perform a graphical analysis of your function $f(x)$ to determine the starting values for which the iteration will converge.

11. Devise a Newton algorithm for computing the fifth root of any positive real number.

12. The function $f(x) = x^2 + 1$ has zeros in the complex plane at $x = \pm i$. Is there a *real* starting point for complex Newton's method such that the iterates converge to either of these zeros? What complex starting points will work?

13. If Newton's method is used with $f(x) = x^2 - 1$ and $x_0 = 10^{10}$, how many steps are required to obtain the root with accuracy 10^{-8}? (Solve analytically, not experimentally.)

14. Suppose that r is a double zero of the function f. Thus, $f(r) = f'(r) = 0 \neq f''(r)$. Show that if f'' is continuous, then in Newton's method we shall have $e_{n+1} \approx \frac{1}{2}e_n$ (linear convergence).

15. Consider a variation of Newton's method in which only one derivative is needed; that is,

$$x_{n+1} = x_n - \frac{f(x_n)}{f'(x_0)}$$

Find C and s such that

$$e_{n+1} = Ce_n^s$$

16. Prove that Newton's iteration will diverge for these functions, no matter what (real) starting point is selected.

(a) $f(x) = x^2 + 1$

(b) $f(x) = 7x^4 + 3x^2 + \pi$

17. Which of these sequences converges quadratically?

(a) $1/n^2$ **(b)** $1/2^{2^n}$ **(c)** $1/\sqrt{n}$ **(d)** $1/e^n$ **(e)** $1/n^n$

18. Find the conditions on α to ensure that the iteration

$$x_{n+1} = x_n - \alpha f(x_n)$$

will converge linearly to a zero of f if started near the zero.

19. Prove that if r is a zero of multiplicity k of the function f, then quadratic convergence in Newton's iteration will be restored by making this modification:

$$x_{n+1} = x_n - kf(x_n)/f'(x_n)$$

20. (Continuation) In the course of using Newton's method, how can a multiple zero be detected by examining the behavior of the points $(x_n, f(x_n))$?

21. **Halley's method** for solving the equation $f(x) = 0$ uses the iteration formula

$$x_{n+1} = x_n - \frac{f_n f_n'}{(f_n')^2 - (f_n f_n'')/2}$$

where $f_n = f(x_n)$ and so on. Show that this formula results when Newton's iteration is applied to the function $f/\sqrt{f'}$.

22. Starting with (0, 0, 1), carry out an iteration of Newton's method for nonlinear systems on

$$\begin{cases} xy - z^2 = 1 \\ xyz - x^2 + y^2 = 2 \\ e^x - e^y + z = 3 \end{cases}$$

Explain the results.

23. Perform two iterations of Newton's method on these systems.

(a) Starting with (0, 1)

$$\begin{cases} 4x_1^2 - x_2^2 = 0 \\ 4x_1x_2^2 - x_1 = 1 \end{cases}$$

(b) Starting with (1, 1)

$$\begin{cases} xy^2 + x^2y + x^4 = 3 \\ x^3y^5 - 2x^5y - x^2 = -2 \end{cases}$$

COMPUTER PROBLEMS 3.2

1. Write a computer program to solve the equation $x = \tan x$ by means of Newton's method. Find the roots nearest 4.5 and 7.7.

2. Find the positive minimum point of the function $f(x) = x^{-2}\tan x$ by computing the zeros of f' using Newton's method.

3. Write a brief computer program to solve the equation $x^3 + 3x = 5x^2 + 7$ by Newton's method. Take ten steps starting at $x_0 = 5$.

4. The equation $2x^4 + 24x^3 + 61x^2 - 16x + 1 = 0$ has two roots near 0.1. Determine them by means of Newton's method.

5. In the first example of this section, investigate the sensitivity of the root to perturbations in the constant term.

6. For the function $f(z) = z^4 - 1$, carry out the complex Newton's method using as starting values all grid points on the mesh with uniform spacing 0.1 in the circle $|z| < 2$ in the complex plane. Assign the same color to all grid points that engender sequences convergent to the same zero. Display the resulting four-color mesh on a color graphics computer terminal or on a color plotter.

7. Program the Newton algorithm in complex arithmetic, and test it on these functions with the given starting points.

(a) $f(z) = z^2 + 1\,, \quad z = 3 + i$

(b) $f(z) = z + \sin z - 3\,, \quad z = 2 - i$

(c) $f(z) = z^4 + z^2 + 2 + 3i\,, \quad z = 1$

8. Write and test a program to compute the first ten roots of the equation $\tan x = x$. (This is much more difficult than Computer Problem 1.) *Cultural note:* If $\lambda_1, \lambda_2, \ldots$ are *all* the positive roots of this equation, then $\sum_{i=1}^{\infty} \lambda_i^{-2} = 1/10$. (See *Amer. Math. Monthly*, Oct. 1986, p. 660.)

9. Program and test Steffensen's method (Problem 4) using the test functions in Computer Problems 1 and 2.

10. The polynomial $p(x) = x^3 + 94x^2 - 389x + 294$ has zeros 1, 3, and -98. The point $x_0 = 2$ should therefore be a good starting point for computing either of the small zeros by the Newton iteration. Carry out the calculation and explain what happens.

11. Carry out five iterations of the complex Newton's method applied to the complex-valued function $f(z) = 1 + z^2 + e^z$, using the starting value $z_0 = -1 + 4i$.

12. Find the first four zeros of the function $f(z) = 1 + z^2 + e^z$ ordered according to increasing complex absolute values. How do you know these are the *first* four zeros and that you have not missed some?

13. Carry out five iterations of Newton's method (for two nonlinear functions in two variables) on the following system:

$$\begin{cases} f_1(x, y) = 1 + x^2 - y^2 + e^x \cos y \\ f_2(x, y) = 2xy + e^x \sin y \end{cases}$$

Use starting values $x_0 = -1$ and $y_0 = 4$. Is this problem related to Computer Problem 11 and do they have similar numerical behavior? Explain.

14. Using Newton's method, find a roots of the nonlinear systems.

(a) $$\begin{cases} 4y^2 + 4y + 52x = 19 \\ 169x^2 + 3y^2 + 111x - 10y = 10 \end{cases}$$

(b) $$\begin{cases} x + e^{-1x} + y^3 = 0 \\ x^2 + 2xy - y^2 + \tan(x) = 0 \end{cases}$$

3.3 Secant Method

Recall that the Newton iteration is defined by the equation

$$x_{n+1} = x_n - \frac{f(x_n)}{f'(x_n)} \tag{1}$$

One of the drawbacks of Newton's method is that it involves the derivative of the function whose zero is sought. To overcome this disadvantage, a number of methods have been proposed. For example, Steffensen's iteration (Problem 4 in Section 3.2)

$$x_{n+1} = x_n - \frac{[f(x_n)]^2}{f(x_n + f(x_n)) - f(x_n)}$$

gives one approach to this problem. Another is to replace $f'(x)$ in Equation (1) by a **difference quotient** such as

$$f'(x_n) \approx \frac{f(x_n) - f(x_{n-1})}{x_n - x_{n-1}} \tag{2}$$

The approximation given in Equation (2) comes directly from the definition of f' as a limit; namely,

$$f'(x) = \lim_{u \to x} \frac{f(x) - f(u)}{x - u}$$

When this replacement is made, the resulting algorithm is called the **secant method**. Its formula is

$$x_{n+1} = x_n - f(x_n)\left[\frac{x_n - x_{n-1}}{f(x_n) - f(x_{n-1})}\right] \qquad (n \geqq 1) \tag{3}$$

Since the calculation of x_{n+1} requires x_n and x_{n-1}, two initial points must be prescribed at the beginning. However, each new x_{n+1} requires only one new evaluation of f. (Steffensen's algorithm requires two evaluations of f for each new x_{n+1}.)

The graphical interpretation of the secant method is similar to that of Newton's method. The tangent line to the curve is replaced by a secant line. (See Figure 3.6.)

Example 1 Use the secant method to find a zero of the function

$$f(x) = x^3 - \sinh x + 4x^2 + 6x + 9$$

Solution A rough plot suggests that there is a zero between 7 and 8. We take these points as x_0 and x_1 in the algorithm. When this code was run on a machine similar to the

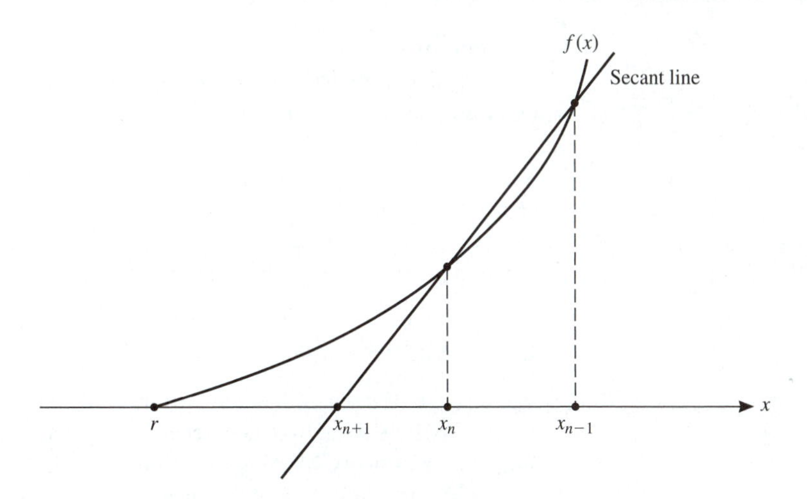

FIGURE 3.6 Geometric interpretation of secant method

`Marc-32`, the following results were obtained:

n	x_n	$f(x_n)$
0	8.00000	-0.665×10^3
1	7.00000	0.417×10^2
2	7.05895	0.208×10^2
3	7.11764	-0.183×10^1
4	7.11289	0.710×10^{-1}
5	7.11306	0.244×10^{-3}
6	7.11306	0.191×10^{-4}

■

Secant Algorithm

An algorithm for the secant method can be written as follows. We modify the standard secant method slightly in order to obtain nonincreasing function values.

```
input a, b, M, δ, ε
fa ← f(a); fb ← f(b)
output 0, a, fa
output 1, b, fb
for k = 2 to M do
    if |fa| > |fb| then
        a ↔ b; fa ↔ fb
    end if
    s ← (b − a)/(fb − fa)
    b ← a
    fb ← fa
    a ← a − fa * s
    fa ← f(a)
```

```
    output k, a, fa
    if |fa| < ε or |b − a| < δ then stop
end do
```

Notice that in the pseudocode the endpoints of $[a, b]$ may be interchanged to keep $|f(a)| \leqq |f(b)|$. Thus, the pair $\{x_n, x_{n-1}\}$ has $|f(x_n)| \leqq |f(x_{n-1})|$, and the next pair $\{x_{n+1}, x_n\}$ has $|f(x_{n+1})| \leqq |f(x_n)|$. This ensures that the absolute value of the function is nonincreasing beginning in step 2.

Error Analysis

The errors in the secant method will now be analyzed. The pseudocode includes a possible switch of the two most recent root guesses. However, we consider only the simple case of no exchanging of endpoints in the following analysis.

From the definition of the secant method (3), we have, with $e_n = x_n - r$,

$$\begin{aligned} e_{n+1} = x_{n+1} - r &= [f(x_n)x_{n-1} - f(x_{n-1})x_n]/[f(x_n) - f(x_{n-1})] - r \\ &= [f(x_n)e_{n-1} - f(x_{n-1})e_n]/[f(x_n) - f(x_{n-1})] \end{aligned}$$

Factoring out $e_n e_{n-1}$ and inserting $(x_n - x_{n-1})/(x_n - x_{n-1})$, we obtain

$$e_{n+1} = \left[\frac{x_n - x_{n-1}}{f(x_n) - f(x_{n-1})}\right]\left[\frac{f(x_n)/e_n - f(x_{n-1})/e_{n-1}}{x_n - x_{n-1}}\right] e_n e_{n-1} \tag{4}$$

By Taylor's Theorem,

$$f(x_n) = f(r + e_n) = f(r) + e_n f'(r) + \frac{1}{2}e_n^2 f''(r) + \mathcal{O}(e_n^3)$$

Since $f(r) = 0$, this gives us

$$f(x_n)/e_n = f'(r) + \frac{1}{2}e_n f''(r) + \mathcal{O}(e_n^2)$$

Changing the index to $n - 1$ yields

$$f(x_{n-1})/e_{n-1} = f'(r) + \frac{1}{2}e_{n-1} f''(r) + \mathcal{O}(e_{n-1}^2)$$

By subtraction of these equations, we get

$$f(x_n)/e_n - f(x_{n-1})/e_{n-1} = \frac{1}{2}(e_n - e_{n-1})f''(r) + \mathcal{O}(e_{n-1}^2)$$

Since $x_n - x_{n-1} = e_n - e_{n-1}$, we arrive at the equation

$$\frac{f(x_n)/e_n - f(x_{n-1})/e_{n-1}}{x_n - x_{n-1}} \approx \frac{1}{2}f''(r)$$

The first bracketed expression in Equation (4) can be written as

$$\frac{x_n - x_{n-1}}{f(x_n) - f(x_{n-1})} \approx \frac{1}{f'(r)}$$

Hence, we have shown that

$$e_{n+1} \approx \frac{1}{2}\frac{f''(r)}{f'(r)} e_n e_{n-1} = C\, e_n e_{n-1} \tag{5}$$

This equation is similar to the one encountered in the analysis of Newton's method—namely, Equation (3) in Section 3.2. To discover the **order of convergence** of the secant method, we postulate the following asymptotic relationship:

$$|e_{n+1}| \sim A|e_n|^\alpha \tag{6}$$

where A is a positive constant. This means that the ratio $|e_{n+1}|/(A|e_n|^\alpha)$ tends to 1 as $n \longrightarrow \infty$ and implies α-order convergence. Hence,

$$|e_n| \sim A|e_{n-1}|^\alpha \quad \text{and} \quad |e_{n-1}| \sim (A^{-1}|e_n|)^{1/\alpha} \tag{7}$$

In Equation (5), we substitute the asymptotic values of $|e_{n+1}|$ and $|e_{n-1}|$ from relations (6) and (7). The result is

$$A|e_n|^\alpha \sim |C|\,|e_n|A^{-1/\alpha}|e_n|^{1/\alpha}$$

This can be written as

$$A^{1+1/\alpha}\,|C|^{-1} \sim |e_n|^{1-\alpha+1/\alpha} \tag{8}$$

Since the left side of this relation is a nonzero constant while $e_n \longrightarrow 0$, we conclude that $1 - \alpha + 1/\alpha = 0$, or $\alpha = (1+\sqrt{5})/2 \approx 1.62$ taking the positive root. Hence, the secant method's rate of convergence is **superlinear** (that is, better than linear). Now we can determine A since the right side of relation (8) is 1. Thus, using the equation $1 + 1/\alpha = \alpha$, we have

$$A = |C|^{1/(1+1/\alpha)} = |C|^{1/\alpha} = |C|^{\alpha-1} = |C|^{0.62} = \left|\frac{f''(r)}{2f'(r)}\right|^{0.62}$$

With A as just given, we finally have for the secant method

$$|e_{n+1}| \approx A\,|e_n|^{(1+\sqrt{5})/2}$$

Since $(1+\sqrt{5})/2 \approx 1.62 < 2$, the rapidity of convergence of the secant method is not as good as Newton's method but is better than the bisection method. However, each step of the secant method requires only *one* new function evaluation, whereas each step of the Newton algorithm requires *two* function evaluations—namely, $f(x)$ and $f'(x)$. Since function evaluations constitute the principal computational burden in these algorithms, a *pair* of steps in the secant method is comparable to one step in the Newton method. For two steps of the secant method, we have

$$|e_{n+2}| \sim A\,|e_{n+1}|^{\alpha} \sim A^{1+\alpha}|e_n|^{\alpha^2} = A^{1+\alpha}|e_n|^{(3+\sqrt{5})/2}$$

This is considerably *better* than the quadratic convergence of the Newton method since $(3+\sqrt{5})/2 \approx 2.62$. Of course, two steps of the secant method would require more work per iteration.

The three methods we have discussed (bisection, Newton, and secant) illustrate a common phenomenon in numerical analysis and particularly scientific computing: the trade-off between speed and reliability. Speed is directly related to computing cost. For some computationally intensive problems (such as the numerical solution of partial differential equations), speed is the overriding consideration. For general-purpose software that will be used by many different customers, reliability and robustness are overriding concerns. An algorithm or subprogram is **robust** if it is able to cope with a wide variety of different numerical situations without the intervention of the user. Herculean efforts have been expended over many years to produce general-purpose mathematical libraries that possess both of these attributes. Two fine examples of such libraries are the IMSL [1995] and NAG [1995] collections.

In all scientific computing tasks, established software packages are recommended in preference to programs written by oneself, except for special applications. For the root-finding problem, the best software is rather complicated because it must incorporate several procedures in order to secure both global convergence and locally rapid convergence. These goals conflict somewhat with each other, as mentioned previously.

Unfortunately, there are just too many methods and variations of them to try to include them all in this book. Examples of algorithms of interest that we do not discuss are ones developed by Brent [1973], Dekker [1969], and Le [1985]. They combine the good features of the bisection method and the secant method and assume nothing about the function whose root is sought other than $f(a)f(b) \leqq 0$. The idea behind Brent's method is to combine the bisection method with the secant method and include an inverse quadratic interpolation in order to get a more robust procedure. Le's algorithm combines the bisection method with second- and third-order methods that use derivative estimates from the objective function values. The interested reader is encouraged to find out more about these algorithms by referring to the papers cited above for a complete description of them. Since the resulting codes are rather long and complicated, we recommend obtaining the software for them via the Internet. Our general advice, to use established and proven software whenever possible, holds true here. Usually such software is freely available and has been carefully written and tested. For example, by using a browser to the World Wide Web, we can connect to the address `http://gams.nist.gov`, which is a Guide to Available Mathematical Software. A problem decision tree is found there, and we go down it from *F. Solution of Nonlinear Equations* to *F1. Single Equation* to *F1b. Nonpolynomial.* There we find that on `Netlib` a version of Brent's algorithm in the C-language (for finding a minimum or a zero of a univariate function within a given range) is available. (See also the appendix on mathematical software.)

PROBLEMS 3.3

1. Establish Equation (4).

2. In the secant method, prove that if $x_n \to q$ as $n \to \infty$ and if $f'(q) \neq 0$, then q is a zero of f.

3. Using Taylor expansions for $f(x+h)$ and $f(x+k)$, derive the following approximation to $f'(x)$:

$$f'(x) \approx \frac{k^2 f(x+h) - h^2 f(x+k) + (h^2 - k^2) f(x)}{(k-h)kh}$$

4. If the secant method is applied to the function $f(x) = x^2 - 2$ with $x_0 = 0$ and $x_1 = 1$, what is x_2?

5. What is x_2 if $x_0 = 1$, $x_1 = 2$, $f(x_0) = 2$, and $f(x_1) = 1.5$ in an application of the secant method?

6. The relation of asymptotic equality between two sequences is written $x_n \sim y_n$ and signifies that $\lim_{n\to\infty}(x_n/y_n) = 1$. Prove that if $x_n \sim y_n$, $u_n \sim v_n$, and $c \neq 0$, then

   (a) $cx_n \sim cy_n$

   (b) $x_n^c \sim y_n^c$

   (c) $x_n u_n \sim y_n v_n$

   (d) If $y_n \sim u_n$, then $x_n \sim v_n$.

   (e) $y_n \sim x_n$

7. Prove that the formula for the secant method can be written in the form

$$x_{n+1} = \frac{f(x_n)x_{n-1} - x_n f(x_{n-1})}{f(x_n) - f(x_{n-1})}$$

   Explain why this formula is inferior to Equation (3) in practice.

8. (A polynomial of very high degree) An **annuity** is a fund of money to which payments (not necessarily equal in value) are made at regular intervals of time. In one form of annuity, the fund is invested at a constant rate of interest, r, per interval. The interest is compounded at the end of each interval. Let the payments be in amounts $a_1, a_2, \ldots$. Let V_i denote the accumulated value of the annuity, computed just after payment a_i has been made. Then $V_1 = a_1$ and

$$V_i = V_{i-1}(1+r) + a_i \qquad (i = 2, 3, \ldots)$$

The factor $(1+r)$ accounts for the interest rV_{i-1} earned on V_{i-1} during one time interval. Prove that $V_n = \sum_{i=1}^{n} a_i x^{n-i}$, where $x = 1 + r$. (Computer Problem 6 is related.)

9. Two men follow two different savings strategies over a period of 44 years. The first saves \$1000 per year for six years and leaves the account to earn interest for the remaining 38 years. The second invests nothing in the first six years, but thereafter saves \$1000 per year. At the end of 44 years, the values of the two accounts are the same. Assume the earnings in both accounts are compounded annually at the same rate of interest. What is the rate and what is the value of each account?

10. How does the interchange in the secant method pseudocode affect the error analysis? Write out the details for a modified error analysis.

COMPUTER PROBLEMS 3.3

1. Write a subprogram to carry out the secant method on a function f, assuming that two starting points are given. Test the routine on these functions.

(a) $\sin(x/2) - 1$

(b) $e^x - \tan x$

(c) $x^3 - 12x^2 + 3x + 1$

2. Program and test a refinement of the secant method that uses the approximate value of $f'(x)$ given in Problem 3. That is, use this approximation for $f'(x)$ in the formula for the Newton method. Three starting points are now needed. The first two can be arbitrary, and the third can be obtained by the secant method.

3. Write subprograms for carrying out the bisection method, Newton's method, and the secant method. They should apply to an arbitrary function F. In each case, the calling sequence should include a parameter M for the maximum number of steps the user will allow. The user should be able to specify also the accuracy desired (ε and δ as in the pseudocode in this section). These codes should be in single precision.

(a) Test your subprograms on the function

$$f(x) = \tan^{-1} x - \frac{2x}{1+x^2}$$

Try to obtain the positive zero with full machine precision.

(b) Use the zero found in part **(a)** as the starting point for Newton's method on the function

$$g(x) = \tan^{-1} x$$

(c) Combine two of the programs that you have written to create a hybrid method that has good global and local characteristics.

4. Select a routine from the program library available on your computer for solving an equation $f(x) = 0$ without using derivatives. Test the code on these functions on the intervals given.
   (a) $x^{20} - 1$ on $[0, 10]$
   (b) $\tan x - 30x$ on $[1, 1.57]$
   (c) $x^2 - (1 - x)^{10}$ on $[0, 1]$
   (d) $x^3 + 10^{-4}$ on $[-0.75, 0.5]$
   (e) $x^{19} + 10^{-4}$ on $[-0.75, 0.5]$
   (f) x^5 on $[-1, 10]$
   (g) x^9 on $[-1, 10]$
   (h) xe^{-x^2} on $[-1, 4]$
   (See Nerinckx and Haegemans [1976].)
5. Program and test the secant algorithm using the same example as in the text. Then repeat the computation using 3 and 10 as the initial points. Explain what happened.
6. (Refer to Problem 8.) A succession of 60 monthly payments is made into an annuity. In the first to fifth years the payments are, respectively, \$200, \$275, \$312, \$380, and \$400. Just after the last payment has been made, the accumulated value of the annuity is \$24,738. What was the monthly rate of interest? Use the secant method to find the zero of the polynomial that arises.

*3.4 Fixed Points and Functional Iteration

Newton's method and Steffensen's method are examples of procedures whereby a sequence of points is computed by a formula of the form

$$x_{n+1} = F(x_n) \qquad (n \geqq 0) \tag{1}$$

The algorithm defined by such an equation is called **functional iteration**. In Newton's method, the function F is given by

$$F(x) = x - \frac{f(x)}{f'(x)}$$

whereas in Steffensen's method, we have

$$F(x) = x - \frac{[f(x)]^2}{f(x + f(x)) - f(x)}$$

Many other instances of the general procedure of functional iteration could be cited, and we will study the general theory briefly.

Formula (1) can be used to generate sequences that do not converge—for example, the sequence $1, 3, 9, 27, \ldots,$ which arises if $x_0 = 1$ and $F(x) = 3x$. However, our interest lies mainly in those cases where $\lim_{n\to\infty} x_n$ exists. Suppose, then, that

$$\lim_{n\to\infty} x_n = s$$

What is the relation between s and F? If F is continuous, then

$$F(s) = F\left(\lim_{n\to\infty} x_n\right) = \lim_{n\to\infty} F(x_n) = \lim_{n\to\infty} x_{n+1} = s$$

Thus, $F(s) = s$, and we call s a **fixed point** of the function F. We could think of a fixed point as a value that the function "locks onto" in the iterative process.

Often a mathematical problem can be reduced to the problem of finding a fixed point of a function. Very interesting applications occur in differential equations, optimization theory, and other areas. Usually the function F whose fixed points are sought will be a mapping from one vector space into another. We intend to analyze the simplest case when F maps some closed set $C \subseteqq \mathbb{R}$ into itself. The theorem to be proved concerns **contractive mappings**. A mapping (or function) F is said to be **contractive** if there exists a number λ *less than* 1 such that

$$|F(x) - F(y)| \leqq \lambda |x - y| \tag{2}$$

for all points x and y in the domain of F. As illustrated in Figure 3.7, the distance between x and y is mapped into a smaller distance between $F(x)$ and $F(y)$ by the contractive function F.

THEOREM 1 **Contractive Mapping Theorem** *Let C be a closed subset of the real line. If F is a contractive mapping of C into C, then F has a unique fixed point. Moreover, this fixed point is the limit of every sequence obtained from Equation (1) with a starting point $x_0 \in C$.*

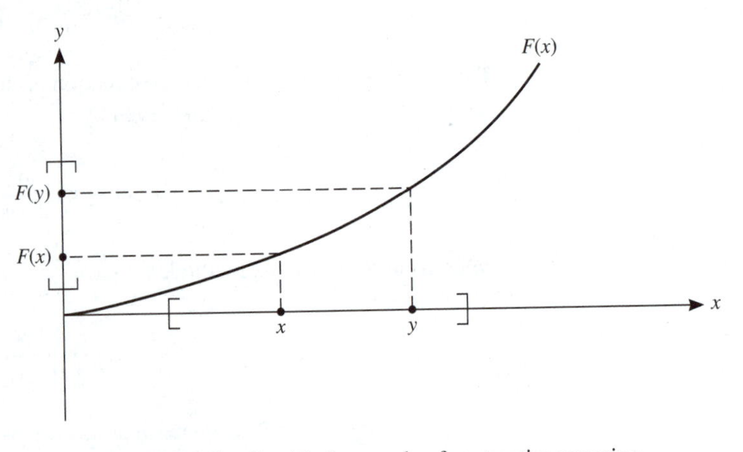

FIGURE 3.7 Graphical example of contractive mapping

Proof We use the contractive property (2) together with Equation (1) to write

$$|x_n - x_{n-1}| = |F(x_{n-1}) - F(x_{n-2})| \leqq \lambda |x_{n-1} - x_{n-2}| \tag{3}$$

This argument can be repeated to get

$$|x_n - x_{n-1}| \leqq \lambda |x_{n-1} - x_{n-2}| \leqq \lambda^2 |x_{n-2} - x_{n-3}| \leqq \cdots \leqq \lambda^{n-1} |x_1 - x_0|$$

Since x_n can be written in the form

$$x_n = x_0 + (x_1 - x_0) + (x_2 - x_1) + \cdots + (x_n - x_{n-1})$$

we see that the sequence $[x_n]$ converges if and only if the series

$$\sum_{n=1}^{\infty} (x_n - x_{n-1})$$

converges. To prove that this series converges, it suffices to prove that the series

$$\sum_{n=1}^{\infty} |x_n - x_{n-1}|$$

converges. This is easy because we can use the comparison test and the previous work:

$$\sum_{n=1}^{\infty} |x_n - x_{n-1}| \leqq \sum_{n=1}^{\infty} \lambda^{n-1} |x_1 - x_0| = \frac{1}{1-\lambda} |x_1 - x_0|$$

Since the sequence converges, let $s = \lim_{n\to\infty} x_n$. Then $F(s) = s$, as noted previously. (Observe that the contractive property implies continuity of F.) As for the unicity of the fixed point, if x and y are fixed points, then

$$|x - y| = |F(x) - F(y)| \leqq \lambda |x - y|$$

Since $\lambda < 1$, $|x - y| = 0$. Finally note that the point s that is obtained belongs to C since s is the limit of a sequence lying in C. ■

Theorem 1 can be proved for a contractive map of any complete metric space into itself.

Example 1 Show that the sequence $[x_n]$ in the Contractive Mapping Theorem satisfies the **Cauchy criterion** for convergence.

Solution Recall the Cauchy criterion for the sequence $[x_n]$: Given any ε, there exists an integer N such that $|x_n - x_m| < \varepsilon$ whenever $n, m \geqq N$. If $n \geqq m \geqq N$, then by the triangle inequality and the equation following (3),

$$\begin{aligned}
|x_n - x_m| &\leqq |x_n - x_{n-1}| + |x_{n-1} - x_{n-2}| + \cdots + |x_{m+1} - x_m| \\
&\leqq \lambda^{n-1}|x_1 - x_0| + \lambda^{n-2}|x_1 - x_0| + \cdots + \lambda^m|x_1 - x_0| \\
&= \lambda^m|x_1 - x_0| \left(1 + \lambda + \lambda^2 + \cdots + \lambda^{n-1-m}\right) \\
&\leqq \lambda^N|x_1 - x_0| \left(1 + \lambda + \lambda^2 + \cdots\right) \\
&= \lambda^N|x_1 - x_0|(1 - \lambda)^{-1}
\end{aligned}$$

For any $\varepsilon > 0$, there exists an N such that $|x_n - x_m| < \varepsilon$ whenever $n, m \geqq N$. We just increase N until $\lambda^N|x_1 - x_0|(1 - \lambda)^{-1} < \varepsilon$. ■

Example 2 Prove that the sequence $[x_n]$ defined recursively as follows is convergent.

$$\begin{cases} x_0 = -15 \\ x_{n+1} = 3 - \frac{1}{2}|x_n| \qquad (n \geqq 0) \end{cases}$$

Solution The function $F(x) = 3 - \frac{1}{2}|x|$ is a contraction because

$$|F(x) - F(y)| = \left|3 - \frac{1}{2}|x| - 3 + \frac{1}{2}|y|\right| = \frac{1}{2}\left||y| - |x|\right| \leqq \frac{1}{2}|y - x|$$

by the triangle inequality. By the Contractive Mapping Theorem, the sequence described must converge to the unique fixed point of F, which is readily seen to be 2. In the theorem, we can take C to be $\mathbb{R}$. ■

Example 3 Use the Contractive Mapping Theorem to compute a fixed point of the function

$$F(x) = 4 + \frac{1}{3}\sin 2x$$

Solution By the Mean-Value Theorem, we have

$$|F(x) - F(y)| = \frac{1}{3}|\sin 2x - \sin 2y| = \frac{2}{3}|\cos 2\zeta|\,|x - y| \leqq \frac{2}{3}|x - y|$$

for some ζ between x and y. This shows that F is contractive, with $\lambda = 2/3$. By the Contractive Mapping Theorem, F has a fixed point. A computer program to compute this fixed point can be based on the following algorithm, which carries out 20 iterations starting with an initial value of 4:

input $x \leftarrow 4$; $M \leftarrow 20$
for $k = 1$ **to** M **do**
  $x \leftarrow 4 + \frac{1}{3}\sin 2x$
  output k, x
end do

The program produces 20 lines of output; several are shown here:

k	x
1	4.32978 61
2	4.23089 51
3	4.27363 38
⋮	⋮
14	4.26148 30
15	4.26148 40
16	4.26148 36
⋮	⋮
20	4.26148 37

The fixed point is given to seven decimal digits of accuracy in the last line. ■

Error Analysis

Now let us analyze the errors in the process of functional iteration. We suppose that F has a fixed point, s, and that a sequence $[x_n]$ has been defined by the formula $x_{n+1} = F(x_n)$. Let

$$e_n = x_n - s$$

If F' exists and is continuous, then by the Mean-Value Theorem,

$$x_{n+1} - s = F(x_n) - F(s) = F'(\zeta_n)(x_n - s)$$

or

$$e_{n+1} = F'(\zeta_n)e_n$$

where ζ_n is a point between x_n and s. The condition $|F'(x)| < 1$ for all x ensures that the errors decrease in magnitude. If e_n is small, then ζ_n is near s, and $F'(\zeta_n) \approx F'(s)$. One would expect rapid convergence if $F'(s)$ is small. An ideal situation would be $F'(s) = 0$. In that case, an additional term in the Taylor series will be useful. To handle several cases at once, let us suppose that q is an integer such that

$$F^{(k)}(s) = 0 \quad \text{for} \quad 1 \leqq k < q \quad \text{but} \quad F^{(q)}(s) \neq 0$$

By the Taylor series for $F(x_n)$ expanded about s, we have

$$\begin{aligned} e_{n+1} &= x_{n+1} - s = F(x_n) - F(s) \\ &= F(s + e_n) - F(s) \end{aligned}$$

$$= \left[F(s) + e_n F'(s) + \frac{1}{2}e_n^2 F''(s) + \cdots\right] - F(s)$$

$$= e_n F'(s) + \frac{1}{2}e_n^2 F''(s) + \cdots + \frac{1}{(q-1)!}e_n^{q-1}F^{(q-1)}(s) + \frac{1}{q!}e_n^q F^{(q)}(\zeta_n)$$

and therefore

$$e_{n+1} = \frac{1}{q!}e_n^q F^{(q)}(\zeta_n) \tag{4}$$

If we know that $\lim_{n\to\infty} x_n = s$, then Equation (4) implies that

$$\lim_{n\to\infty} \frac{|e_{n+1}|}{|e_n|^q} = \frac{1}{q!}F^{(q)}(s) \tag{5}$$

Recall from Section 1.2 that for any sequence $[x_n]$ converging to a point s (whether it arises from a functional iteration or not), we define the **order of convergence** to be the largest real number q such that the limit

$$\lim_{n\to\infty} \frac{|x_{n+1} - s|}{|x_n - s|^q}$$

exists and is nonzero. Such a q need not exist, and when it does, it need not be an integer.

If, for example, $F'(s) = 0$ and $F''(s) \neq 0$, then $q = 2$ and we have

$$e_{n+1} = \frac{1}{2}e_n^2 F''(\zeta_n)$$

This is reminiscent of Newton's method, Equation (3) in Section 3.2. In fact, Newton's method uses

$$F(x) = x - \frac{f(x)}{f'(x)}$$

and for it we have

$$F'(x) = 1 - \frac{f'(x)f'(x) - f(x)f''(x)}{[f'(x)]^2} = \frac{f(x)f''(x)}{[f'(x)]^2}$$

Since a fixed point of F is a zero of f, we have $F'(s) = 0$. Moreover,

$$F''(x) = \frac{[f'(x)]^2[f(x)f'''(x) + f''(x)f'(x)] - [f(x)f''(x)][2f'(x)f''(x)]}{[f'(x)]^4}$$

and therefore we usually have

$$F''(s) = \frac{f''(s)}{f'(s)} \neq 0$$

If F is represented near a fixed point s by a Taylor series

$$\sum_{k=0}^{\infty} \frac{1}{k!} F^{(k)}(s)(x-s)^k$$

and if $x_{n+1} = F(x_n)$, then the order of convergence is the first integer q such that $F^{(q)}(s) \neq 0$. Problem 16 asks for a proof of this assertion.

PROBLEMS 3.4

1. If F is contractive from $[a, b]$ to $[a, b]$ and $x_{n+1} = F(x_n)$, with $x_0 \in [a, b]$, then $|x_n - s| \leqq C\lambda^n$ for an appropriate C. Prove this and give an upper bound for C. Here s is the fixed point of F.

2. Prove that if $F\colon [a, b] \to \mathbb{R}$, if F' is continuous, and if $|F'(x)| < 1$ on $[a, b]$, then F is a contraction. Does F necessarily have a fixed point?

3. Prove that if F is a continuous map of $[a, b]$ into $[a, b]$, then F must have a fixed point. Then determine whether this assertion is true for functions from $\mathbb{R}$ to $\mathbb{R}$.

4. Show that these functions are contractive on the indicated intervals. Determine the best values of λ in Equation (2).

   (a) $(1 + x^2)^{-1}$ on an arbitrary interval

   (b) $\frac{1}{2}x$ on $1 \leqq x \leqq 5$

   (c) $\tan^{-1} x$ on an arbitrary closed interval excluding 0

   (d) $|x|^{\frac{3}{2}}$ on $|x| \leqq \frac{1}{3}$

5. Kepler's equation in astronomy reads: $x = y - \varepsilon \sin y$, with $0 < \varepsilon < 1$. Show that for each $x \in [0, \pi]$, there is a y satisfying the equation. Interpret this as a fixed-point problem.

6. Consider an iteration function of the form $F(x) = x + f(x)g(x)$, where $f(r) = 0$ and $f'(r) \neq 0$. Find the precise conditions on the function g so that the method of functional iteration will converge cubically to r if started near r.

7. If you enter a number into a handheld calculator and then repeatedly press the cosine button, what number will eventually appear? Provide a proof.

8. Prove that the sequence generated by the iteration $x_{n+1} = F(x_n)$ will converge if $|F'(x)| \leqq \lambda < 1$ on the interval $[x_0 - \rho,\ x_0 + \rho]$, where $\rho = |F(x_0) - x_0|/(1 - \lambda)$.

9. What special properties must a function f have if Newton's method applied to f converges cubically to a zero of f?

10. If we attempt to find a fixed point of F by using Newton's method on the equation $F(x) - x = 0$, what iteration formula results?

11. If f' is continuous and positive on $[a, b]$ and if $f(a)f(b) < 0$, then f has *exactly* one zero in (a, b). Prove this, and show that with a suitable parameter λ, the zero can be obtained by applying the method of functional iteration, with $F(x) = x + \lambda f(x)$.

12. Let p be a positive number. What is the value of the following expression?

$$x = \sqrt{p + \sqrt{p + \sqrt{p + \cdots}}}$$

Note that this can be interpreted as meaning $x = \lim_{n\to\infty} x_n$, where $x_1 = \sqrt{p}$, $x_2 = \sqrt{p + \sqrt{p}}$, and so forth. *Hint:* Observe that $x_{n+1} = \sqrt{p + x_n}$.

13. Let $p > 1$. What is the value of the following continued fraction?

$$x = \cfrac{1}{p + \cfrac{1}{p + \cfrac{1}{p + \cdots}}}$$

Use the ideas of Problem 12 to solve this one. Prove that the sequence of values converges by using the Contractive Mapping Theorem.

14. (Continuation) Solve for the roots of the quadratic equation $x^2 + px + q = 0$ by developing an iterative method.

15. Let F be a contractive mapping of an interval $[a, b]$ into itself, and let s be the fixed point of F. If $a \leqq x \leqq b$ and $|F(x) - x| < \varepsilon$, does it follow that $|x - s| < \varepsilon$? Prove that $|x - s| < \varepsilon(1 - \lambda)^{-1}$, where λ is the constant in Equation (2).

16. Prove the statement at the end of this section in the text concerning the order of convergence in the method of functional iteration.

17. Most iterative processes are not as simple as the one expressed by $x_{n+1} = F(x_n)$, with $F\colon \mathbb{R} \to \mathbb{R}$. For example, we might have a map $F\colon \mathbb{R}^2 \to \mathbb{R}^2$. Show that the bisection method and the secant method are of this type. In each case, define F explicitly.

18. Prove that if F' is continuous and if $|F'(x)| < 1$ on the interval $[a, b]$, then F is contractive on $[a, b]$. Show that this is not necessarily true for an open interval.

19. If the method of functional iteration is applied to the function $F(x) = 2 + (x - 2)^4$, starting at $x = 2.5$, what order of convergence results? Find the range of starting values for which this functional iteration converges. Note that 2 is a fixed point.

20. Show that the following functions are contractive on the given domains, yet they have no fixed point on these domains. Why does this not contradict the Contractive Mapping Theorem?

(a) $F(x) = 3 - x^2$ on $[-\frac{1}{4}, \frac{1}{4}]$

(b) $F(x) = -x/2$ on $[-2, -1] \cup [1, 2]$

21. Prove that if f is continuous on $[a, b]$ and satisfies $a \leqq f(a)$ and $f(b) \leqq b$, then f has a fixed point in the interval $[a, b]$. Note that we do not assume $a \leqq f(x) \leqq b$ for all x in $[a, b]$.

22. What is the weakest condition that can be put on the interval $[c, d]$ in order that each continuous map of $[a, b]$ into $[c, d]$ shall have a fixed point?

23. Find the order of convergence of these sequences.

(a) $x_n = (1/n)^{\frac{1}{2}}$

(b) $x_n = \sqrt[n]{n}$

(c) $x_n = (1 + 1/n)^{\frac{1}{2}}$

(d) $x_{n+1} = \tan^{-1} x_n$

24. To find a zero of the function f, we can look for a fixed point of the function $F(x) = x - f(x)/f'(x)$. To find a fixed point of F, we can solve $F(x) - x = 0$ by Newton's method. When this is done, what is the formula for generating the sequence x_n?

25. Prove that the function F defined by $F(x) = 4x(1 - x)$ maps the interval $[0, 1]$ into itself and is not a contraction. Prove that it has a fixed point. Why does this not contradict the Contractive Mapping Theorem?

26. If the method of functional iteration is used on the function $f(x) = \frac{1}{2}(1 + x^2)^{-1}$ starting at $x_0 = 7$, will the resulting sequence converge? If so, what is the limit? Establish your answers rigorously.

27. Prove or disprove: If $F: \mathbb{R} \longrightarrow [a, b]$ and if F is contractive on $[a, b]$, then F has a unique fixed point, which can be obtained by the method of functional iteration, starting at any real value.

28. Give examples of functions that *do not* have fixed points but *do* have these characteristics:

(a) $f: [0, 1] \longrightarrow [0, 1]$

(b) $f: (0, 1) \longrightarrow (0, 1)$ and is continuous

(c) $f: A \longrightarrow A$ and is continuous, with $A = [0, 1] \bigcup [2, 3]$

(d) $f: \mathbb{R} \longrightarrow \mathbb{R}$ and is continuous

29. Prove that the function $f(x) = 2 + x - \tan^{-1} x$ has the property $|f'(x)| < 1$. Prove that f does not have a fixed point. Explain why this does not contradict the Contractive Mapping Theorem.

30. This problem concerns the function $F(x) = 10 - 2x$. Prove that F has a fixed point. Let x_0 be arbitrary, and define $x_{n+1} = F(x_n)$ for $n \geqq 0$. Find a nonrecursive formula for x_n. Prove that the method of functional iteration does not produce a convergent sequence unless x_0 is given a particular value. What is this special value for x_0? Why does this not contradict the Contractive Mapping Theorem?

31. Let F be continuously differentiable in an open interval, and suppose that F has a fixed point s in this open interval. Prove that if $|F'(s)| < 1$, then the sequence defined by functional iteration will converge to s if started sufficiently close to s. *Hint:* Select λ so that $|F'(r)| < \lambda < 1$, and consider an interval centered at r in which $|F'(x)| < \lambda$.

32. Let $\frac{1}{2} \leqq q \leqq 1$, and define $F(x) = 2x - qx^2$. On what interval can it be guaranteed that the method of iteration using F will converge to a fixed point? (This problem is related to Problem 5 in Section 3.2.)

33. Write down two different fixed-point procedures for finding a zero of the function $f(x) = 2x^2 + 6e^{-x} - 4$.

34. On which of these intervals $[1/2, \infty]$, $[1/8, 1]$, $[1/4, 2]$, $[0, 1]$, $[1/5, 3/2]$ is the function $f(x) = \sqrt{x}$ contractive?

35. A function F is called an **iterated contraction** if

$$|F(F(x)) - F(x)| \leqq \lambda |F(x) - x| \qquad (\lambda < 1)$$

Show that every contraction is an iterated contraction. Show that an iterated contraction need not be a contraction nor continuous.

36. If the method of functional iteration is used on $F(x) = x^2 + x - 2$ and produces a convergent sequence of positive numbers, what is the limit of that sequence and what was the starting point?

37. Consider a function of the form $F(x) = x - f(x)f'(x)$, where $f(r) = 0$ and $f'(r) \neq 0$. Find the precise conditions on the function f so that the method of functional iteration will converge at least *cubically* to r if started near r.

38. Analyze Steffensen's method (Problem 4, Section 3.2) as an example of functional iteration. Determine its order of convergence.

39. A student incorrectly recalls Newton's method and writes $x_{n+1} = f(x_n)/f'(x_n)$. Will this method find a zero of f? What is the order of convergence?

40. Show that the following method has third-order convergence for computing $\sqrt{R}$:

$$x_{n+1} = \frac{x_n(x_n^2 + 3R)}{3x_n^2 + R}$$

41. Consider an iterative method of the form $x_{n+1} = x_n - f(x_n)/g(x_n)$. Assume that it converges to a point ξ that is a simple zero of the function f but not a zero of the function g. Establish the relationship between f and g so that the order of convergence of the method is 3 or greater.

*3.5 Computing Zeros of Polynomials

Methods developed in the preceding sections—in particular, Newton's method—can certainly be applied to polynomials. But in finding the zeros of a polynomial, we should exploit the special structure of such a function, if possible. Moreover, the polynomial problem is frequently complicated by our wish to compute the complex zeros (even if the polynomial has real coefficients) or all the zeros of a given polynomial. Thus, the topic of finding the zeros of a polynomial deserves special treatment, which indeed it has received for almost 400 years!

We begin with some important theoretical results, most of which are probably familiar to the reader. We write a polynomial in the form

$$p(z) = a_n z^n + a_{n-1} z^{n-1} + \cdots + a_2 z^2 + a_1 z + a_0 \tag{1}$$

in which the coefficients a_k and the variable z may be complex numbers. If $a_n \neq 0$, then p has **degree n**. We are interested in finding the *zeros* of p. Are there any? The following theorem (first proved by Gauss in 1799) answers this question. The proof given here is by G. H. Hardy [1960, Appendix I] and Fefferman [1967].

THEOREM 1 **Fundamental Theorem of Algebra** *Every nonconstant polynomial has at least one zero in the complex field.*

Proof Let p be a nonconstant polynomial. We want to show that $p(z_0) = 0$ for some $z \in \mathbb{C}$. Since p is not constant, $|p(z)| \to \infty$ when $|z| \to \infty$. Let D be a disk centered at 0 outside of which $|p(z)| \geqq |p(0)|$. Let z_0 be a point where $\inf_{z \in D} |p(z)|$ is attained. Since $0 \in D$, $|p(z_0)| \leqq |p(0)|$. Thus, $|p(z_0)| \leqq |p(z)|$ for all $z \in \mathbb{C}$. Put $q(z) = p(z + z_0)$. We want to prove that $q(0) = 0$, so that $p(z_0) = 0$. Write $q(z) = c_0 + c_j z^j + \cdots + c_n z^n = c_0 + c_j z^j + z^{j+1} r(z)$, in which $c_j \neq 0$ and r is a polynomial (possibly 0). Now we want to prove that $c_0 = 0$. Suppose $c_0 \neq 0$. Let w be any complex number such that $c_j w^j = -c_0$. Define $N = \sup_{0<\varepsilon<1} |r(\varepsilon w)|$. Select ε in (0, 1) so small that $\varepsilon |w|^{j+1} N < |c_0|$. Then we obtain a contradiction as follows:

$$\begin{aligned} |q(\varepsilon w)| &\leqq |c_0 + c_j \varepsilon^j w^j| + \varepsilon^{j+1} |w^{j+1}| |r(\varepsilon w)| \\ &= |c_0 - c_0 \varepsilon^j| + \varepsilon^j \varepsilon |w|^{j+1} N \\ &< |c_0|(1 - \varepsilon^j) + \varepsilon^j |c_0| = |c_0| = |q(0)| \\ &= |p(z_0)| \leqq |p(z_0 + \varepsilon w)| = |q(\varepsilon w)| \end{aligned}$$

■

The history of this theorem and of attempts to prove it can be read in Kline [1972, pp. 597–606]. The usual *modern* proof is as a corollary to Liouville's Theorem. See, for example, Bak and Newman [1982], Henrici [1974], Ahlfors [1966], or other texts on complex analysis.

The Fundamental Theorem does *not* assert the existence of real zeros, and the simplest examples, such as $z^2 + 1$, show that a polynomial need not have real zeros, even if its coefficients are all real.

If the polynomial p, having degree n at least 1, is divided by a linear factor $z - c$, the result is a quotient q and a remainder r. The latter is a complex number, and the former is a polynomial of degree $n - 1$. We can represent the process by the equation

$$p(z) = (z - c)q(z) + r$$

From this we see (by letting $z = c$) that $p(c) = r$. This fact is known as the **Remainder Theorem**. If c is a zero of p, then $r = 0$, and we have

$$p(z) = (z - c)q(z)$$

Thus, $z - c$ is a factor of $p(z)$. This implication is known as the **Factor Theorem**.

Let us write $p(z) = (z - r_1)q_1(z)$, where r_1 is any zero of p. By the Fundamental Theorem, q_1 has a zero, say r_2 (unless q_1 is of degree 0). Hence, $q_1(z)$ has a factor $z - r_2$, and we can write $p(z) = (z - r_1)(z - r_2)q_2(z)$. Continuing in this manner, we eventually arrive at a stopping point because the degrees of the successive q_k decrease by 1 at each step. Hence, q_n will be a constant, and the last equation is

$$p(z) = (z - r_1)(z - r_2) \cdots (z - r_n)q_n$$

This proves that a polynomial of degree n has a factorization into a product of n linear factors, each corresponding to a zero of p. It is clear that p can have no other zeros because if z is any complex number different from $r_1, r_2, \ldots, r_n$, then the product $\prod_{k=1}^{n}(z - r_k)$ cannot be 0. Since some of the zeros r_k may be equal to each other, we see that a polynomial of degree n can have at most n zeros. The **multiplicity** of a zero is the number of times that the factor corresponding to it appears in the factorization of p. If we combine the preceding remarks with the Fundamental Theorem, we obtain the following theorem:

THEOREM 2 *A polynomial of degree n has exactly n zeros in the complex plane, it being agreed that each zero shall be counted a number of times equal to its multiplicity.*

It is often desirable to know roughly where the zeros of a polynomial are situated in the complex plane. Many results—such as Descartes' rule of signs—are available in the literature for this problem of **localizing the zeros**. See, for example, Section 5.5 of Young and Gregory [1972], Section 5.5 of Stoer and Bulirsch [1980], the treatise of Marden [1966], or Volume 1 of the three-volume treatise of Henrici [1974]. The following theorem adapted from the textbook of Conte and de Boor [1980] provides an easily computed upper bound for the moduli of the zeros.

THEOREM 3 *All zeros of the polynomial in Equation (1) lie in the open disk whose center is at the origin of the complex plane and whose radius is*

$$\rho = 1 + |a_n|^{-1} \max_{0 \leqq k < n} |a_k|$$

Proof Put $c = \max_{0 \leqq k < n} |a_k|$ so that $c|a_n|^{-1} = \rho - 1$. If $c = 0$, our result is trivially true. Hence, assume $c > 0$. Then $\rho > 1$. If $|z| \geqq \rho$, then (because $\rho > 1$)

$$
\begin{aligned}
|p(z)| &\geqq |a_n z^n| - |a_{n-1} z^{n-1} + \cdots + a_0| \\
&\geqq |a_n z^n| - c \sum_{k=0}^{n-1} |z|^k \\
&> |a_n z^n| - c|z|^n (|z| - 1)^{-1} \\
&= |a_n z^n| \{1 - c|a_n|^{-1} (|z| - 1)^{-1}\} \\
&\geqq |a_n z^n| \{1 - c|a_n|^{-1} (\rho - 1)^{-1}\} = 0
\end{aligned}
$$

■

Example 1 Find a disk centered at the origin that contains all the zeros of the polynomial

$$p(z) = z^4 - 4z^3 + 7z^2 - 5z - 2$$

Solution By Theorem 3, one such disk has radius

$$\rho = 1 + |a_4|^{-1} \max_{0 \leqq k < 4} |a_k| = 8$$

■

Another useful idea available for the analysis of polynomials is this: Take the polynomial p of Equation (1), and consider the function $s(z) = z^n p(1/z)$. Then

$$
\begin{aligned}
s(z) &= z^n \left[a_n \left(\frac{1}{z} \right)^n + a_{n-1} \left(\frac{1}{z} \right)^{n-1} + \cdots + a_0 \right] \\
&= a_n + a_{n-1} z + a_{n-2} z^2 + \cdots + a_0 z^n
\end{aligned}
$$

This shows that s is a polynomial of degree at most n. Its coefficients can be written down at once, since they are the same as the coefficients of p except in the reverse order. Now it is obvious that for a nonzero complex number z_0, the condition $p(z_0) = 0$ is equivalent to the condition $s(1/z_0) = 0$. Thus, we have the following result.

THEOREM 4 *If all the zeros of s are in the disk $\{z : |z| \leqq \rho\}$, then all the nonzero zeros of p are outside the disk $\{z : |z| < \rho^{-1}\}$.*

Example 2 Find a disk centered at the origin in which there are no zeros of p, where p is the polynomial of Example 1.

Solution The polynomial s referred to in Theorem 4 is

$$s(z) = -2z^4 - 5z^3 + 7z^2 - 4z + 1$$

So by Theorem 2, all zeros of s lie in a disk centered at the origin having radius $\rho = 1 + |a_4|^{-1} \max_{0 \leqq k < 4} |a_k| = 9/2$. Therefore, by Theorem 4, the zeros of p lie outside the disk of radius $2/9$. Hence, all zeros of p lie in the ring $2/9 < |z| < 8$ in the complex plane.

■

Horner's Algorithm

For the efficient computation of values of a polynomial, **Horner's algorithm** is needed. This algorithm is also known as **nested multiplication** and as **synthetic division**. We shall find that it is useful for other purposes as well. If a polynomial p and a complex number z_0 are given, Horner's algorithm will produce the number $p(z_0)$ and the polynomial

$$q(z) = \frac{p(z) - p(z_0)}{z - z_0} \tag{2}$$

The polynomial q is of degree 1 less than the degree of p. From this equation we have

$$p(z) = (z - z_0)q(z) + p(z_0) \tag{3}$$

Let the unknown polynomial q be represented by

$$q(z) = b_0 + b_1 z + \cdots + b_{n-1} z^{n-1}$$

When this form for $q(z)$ and the analogous form for $p(z)$ are substituted in Equation (3), the coefficients of like powers of z on the two sides of the equation can be set equal to each other. These equations arise from doing so:

$$\begin{aligned}
b_{n-1} &= a_n \\
b_{n-2} &= a_{n-1} + z_0 b_{n-1} \\
&\ \vdots \\
b_0 &= a_1 + z_0 b_1 \\
p(z_0) &= a_0 + z_0 b_0
\end{aligned}$$

In compact form, Horner's algorithm can be written like this:

input $n, (a_i : 0 \leqq i \leqq n), z_0$
$b_{n-1} \leftarrow a_n$
for $k = n - 1$ **to** 0 **step** -1 **do**
    $b_{k-1} \leftarrow a_k + z_0 b_k$
end do
output $(b_i : -1 \leqq i \leqq n - 1)$

Notice that $b_{-1} = p(z_0)$ in this pseudocode. If the calculation in Horner's algorithm is to be carried out with pencil and paper, the following arrangement is often used:

$$\begin{array}{c|ccccc}
 & a_n & a_{n-1} & a_{n-2} & \cdots & a_0 \\
z_0 & & z_0 b_{n-1} & z_0 b_{n-2} & \cdots & z_0 b_0 \\
\hline
 & b_{n-1} & b_{n-2} & b_{n-3} & \cdots & b_{-1}
\end{array}$$

Example 3 Use Horner's algorithm to evaluate $p(3)$, where p is the polynomial

$$p(z) = z^4 - 4z^3 + 7z^2 - 5z - 2$$

Solution We arrange the calculation as suggested above.

$$\begin{array}{c|rrrrr} & 1 & -4 & 7 & -5 & -2 \\ 3 & & 3 & -3 & 12 & 21 \\ \hline & 1 & -1 & 4 & 7 & 19 \end{array}$$

Thus $p(3) = 19$, and we can write

$$p(z) = (z - 3)(z^3 - z^2 + 4z + 7) + 19$$

■

Horner's algorithm is used also for **deflation**. This is the process of removing a linear factor from a polynomial. If z_0 is a zero of the polynomial p, then $z - z_0$ is a factor of p, and conversely. The remaining zeros of p are the $n - 1$ zeros of $p(z)/(z - z_0)$.

Example 4 Deflate the polynomial p of Example 3 using the fact that 2 is one of its zeros.

Solution We use the same arrangement of computations as explained previously:

$$\begin{array}{c|rrrrr} & 1 & -4 & 7 & -5 & -2 \\ 2 & & 2 & -4 & 6 & 2 \\ \hline & 1 & -2 & 3 & 1 & 0 \end{array}$$

Thus,

$$z^4 - 4z^3 + 7z^2 - 5z - 2 = (z - 2)(z^3 - 2z^2 + 3z + 1)$$

■

A third application of Horner's algorithm is in finding the Taylor expansion of a polynomial about any point. Let $p(z)$ be as in Equation (1), and suppose that we desire the coefficients c_k in the equation

$$\begin{aligned} p(z) &= a_n z^n + a_{n-1} z^{n-1} + \cdots + a_0 \\ &= c_n(z - z_0)^n + c_{n-1}(z - z_0)^{n-1} + \cdots + c_0 \end{aligned}$$

Of course, Taylor's Theorem asserts that $c_k = p^{(k)}(z_0)/k!$, but we seek a more efficient algorithm. Notice that $p(z_0) = c_0$, so that this coefficient is obtained by applying Horner's algorithm to the polynomial p with the point z_0. The algorithm also yields the polynomial

$$q(z) = \frac{p(z) - p(z_0)}{z - z_0} = c_n(z - z_0)^{n-1} + c_{n-1}(z - z_0)^{n-2} + \cdots + c_1$$

This shows that the second coefficient, c_1, can be obtained by applying Horner's algorithm to the polynomial q with point z_0 because $c_1 = q(z_0)$. (Notice that the first application of Horner's algorithm does not yield q in the form shown but rather as a sum of powers of z.) This process is repeated until all coefficients c_k are found.

Example 5 Find the Taylor expansion of the polynomial in the preceding examples about the point $z_0 = 3$.

Solution The work can be arranged as follows:

$$
\begin{array}{c|ccccc}
 & 1 & -4 & 7 & -5 & -2 \\
3 & & 3 & -3 & 12 & 21 \\
\hline
 & 1 & -1 & 4 & 7 & 19 \\
3 & & 3 & 6 & 30 & \\
\hline
 & 1 & 2 & 10 & 37 & \\
3 & & 3 & 15 & & \\
\hline
 & 1 & 5 & 25 & & \\
3 & & 3 & & & \\
\hline
 & 1 & 8 & & &
\end{array}
$$

The calculation shows that

$$p(z) = (z-3)^4 + 8(z-3)^3 + 25(z-3)^2 + 37(z-3) + 19$$ ■

We call the algorithm just described the **complete Horner's algorithm.** The pseudocode for executing it is arranged so that the coefficients c_k *overwrite* the input coefficients a_k.

```
input n, (a_i : 0 ≦ i ≦ n), z_0
for k = 0 to n − 1 do
    for j = n − 1 to k step − 1 do
        a_j ← a_j + z_0 a_{j+1}
    end do
end do
output (a_i : 0 ≦ i ≦ n)
```

We have everything necessary for showing how Newton's iteration can be carried out on a polynomial. Recall that this iteration is defined by the equation

$$z_{k+1} = z_k - \frac{f(z_k)}{f'(z_k)}$$

If this is applied to a polynomial p, an efficient method incorporating Horner's algorithm can be used to compute $p(z)$ and $p'(z)$. We saw previously that if only two steps inside the complete Horner's algorithm are used, we obtain $c_0 = p(z_0)$ and $c_1 = p'(z_0)$. These two steps can be combined in the pseudocode. Also, we desist from overwriting the input coefficients because they will be required in subsequent steps of the iteration. Here is a pseudocode that produces $\alpha = p(z_0)$ and $\beta = p'(z_0)$, assuming that p is given as in Equation (1) and that z_0 is given:

```
input n, (a_i : 0 ≦ i ≦ n), z_0
α ← a_n
β ← 0
for k = n − 1 to 0 step − 1 do
    β ← α + z_0 β
    α ← a_k + z_0 α
end do
output α, β
```

If we refer to this pseudocode as `horner` $(n, (a_i : 0 \leqq i \leqq n), z_0, \alpha, \beta)$, then a pseudocode for taking M steps in Newton's method on the given polynomial, starting at z_0, would look like this:

```
input n, (a_i : 0 ≦ i ≦ n), z_0, M, ε
for j = 1 to M do
    call horner (n, (a_i : 0 ≦ i ≦ n), z_0, α, β)
    z_1 ← z_0 − α/β
    output α, β, z_1
    if |z_1 − z_0| < ε then stop
    z_0 ← z_1
end do
```

Example 6 Carry out Newton's iteration on the polynomial used in all the previous examples, starting at $z_0 = 0$.

Solution In the first step, we use $z_0 = 0$, and compute the values $p(0) = -2$ and $p'(0) = -5$ by the algorithms previously explained. The new value of z is

$$z_1 = z_0 - \frac{p(z_0)}{p'(z_0)} = 0 - \frac{-2}{-5} = -0.4$$

Executing the algorithm on a computer with the precision of the `Marc-32` produces the following results:

k	$p(z_k)$	$p'(z_k)$	z_k
1	−2.00000	−5.00000	−0.40000
2	1.40160	−12.77600	−0.29029
3	1.46322	−10.17322	−0.27591
4	0.00226	−9.86030	−0.27568
5	0.00000	−9.85537	−0.27568

Observe that z_k converges rapidly to the zero -0.27568. ■

A formal proof that Horner's algorithm is correct follows.

THEOREM 5 *Let $p(x) = a_n x^n + \cdots + a_1 x + a_0$. Define pairs (α_j, β_j) for $j = n, n-1, \ldots, 0$ by the algorithm*

$$\begin{cases} (\alpha_n, \beta_n) = (a_n, 0) & \\ (\alpha_j, \beta_j) = (a_j + x\alpha_{j+1}, \alpha_{j+1} + x\beta_{j+1}) & (n - 1 \geqq j \geqq 0) \end{cases}$$

Then $\alpha_0 = p(x)$ and $\beta_0 = p'(x)$.

Proof If $n = 0$, then $p(x) = a_0$ and

$$(\alpha_0, \beta_0) = (a_0, 0)$$

So obviously $\alpha_0 = p(x)$ and $\beta_0 = p'(x)$. Thus, the theorem is true for $n = 0$. If $n = 1$, then $p(x) = a_1 x + a_0$ and

$$(\alpha_0, \beta_0) = (a_0 + x\alpha_1, \alpha_1 + x\beta_1) = (a_0 + xa_1, a_1) = (p(x), p'(x))$$

Thus, the theorem is true for $n = 1$. Suppose the theorem is true for all indices n less than m. Let

$$p(x) = c_m x^m + \cdots + c_1 x + c_0 = c_0 + x(c_m x^{m-1} + \cdots + c_2 x + c_1) = c_0 + xq(x)$$

Consequently,

$$p'(x) = xq'(x) + q(x)$$

Set $n = m - 1$, $a_n = c_m$, $a_{n-1} = c_{m-1}, \ldots, a_2 = c_3$, $a_1 = c_2$, and $a_0 = c_1$. Since the theorem is true for q, we apply the algorithm to $q(x) = a_n x^n + \cdots + a_1 x + a_0$. We get $\alpha_0 = q(x)$ and $\beta_0 = q'(x)$ by the induction hypothesis. If the algorithm is applied to p, we get the same set of pairs $(\alpha_n, \beta_n), \ldots, (\alpha_1, \beta_1), (\alpha_0, \beta_0)$ (because $(c_m, \ldots, c_2, c_1) = (a_n, \ldots, a_1, a_0)$), but there is one more pair to be computed at the end for p; namely,

$$\begin{aligned} (\alpha_{-1}, \beta_{-1}) &= (a_{-1} + x\alpha_0, \alpha_0 + x\beta_0) \\ &= (c_0 + xq(x), q(x) + xq'(x)) \\ &= (p(x), p'(x)) \end{aligned}$$

■

Note that we could define our sequence by

$$\begin{cases} (\alpha_{n-1}, \beta_{n-1}) = (a_{n-1} + xa_n, a_n) & \\ (\alpha_j, \beta_j) = (a_j + x\alpha_{j+1}, \alpha_{j+1} + x\beta_{j+1}) & (n - 2 \geqq j \geqq 0) \end{cases}$$

or by

$$\begin{cases}(\alpha_n, \beta_n) = (a_{n-1} + x a_n, a_n) \\ (\alpha_j, \beta_j) = (a_{j-1} + x\alpha_{j+1}, \alpha_{j+1} + x\beta_{j+1}) \qquad (n-1 \geqq j \geqq 1)\end{cases}$$

These are *inferior* because they do not work when $n = 0$.

THEOREM 6 *Let x_k and x_{k+1} be two successive iterates when Newton's method is applied to a polynomial p of degree n. Then there is a zero of p within distance $n|x_k - x_{k+1}|$ of x_k in the complex plane.*

Proof Let $r_1, r_2, \ldots, r_n$ be the zeros of p. Then $p(z) = c\prod_{j=1}^{n}(z - r_j)$. The correction term in the Newton iteration is $-p(z)/p'(z)$. The derivative of p is

$$p'(z) = c\sum_{k=1}^{n}\prod_{\substack{i=1\\ i\neq k}}^{n}(z - r_i) = \sum_{k=1}^{n} p(z)/(z - r_k) = p(z)\sum_{k=1}^{n}(z - r_k)^{-1}$$

We want to show that for any z (playing the role of x_k) there is an index j for which $|z - r_j| \leqq n|p(z)/p'(z)|$. If no index j satisfies the desired inequality, then for all j, $|z - r_j| > n|p(z)/p'(z)|$. From this it would follow that

$$|z - r_j|^{-1} < \frac{1}{n}|p'(z)/p(z)| = \frac{1}{n}\left|\sum_{k=1}^{n}(z - r_k)^{-1}\right| \leqq \frac{1}{n}\sum_{k=1}^{n}|z - r_k|^{-1}$$

But this is not possible because the average of n numbers cannot be greater than each of them. ■

This theorem is due to Bodewig [1946].

Bairstow's Method

If a polynomial has only real coefficients, its zeros may nevertheless be complex. In this case, it is possible to compute the complex zeros two at a time using only real arithmetic. The procedure for doing this is called **Bairstow's method**; it will be considered next.

For any complex number $z = x + iy$, the **conjugate** number is $\bar{z} = x - iy$. A basic result that will be exploited follows.

THEOREM 7 *If p is a polynomial whose coefficients are all real, and if w is a nonreal zero of p, then $\overline{w}$ is also a zero, and $(z - w)(z - \overline{w})$ is a real quadratic factor of p.*

Proof Let $p(z) = a_n z^n + a_{n-1}z^{n-1} + \cdots + a_1 z + a_0$, with all a_k being real. Since w is a zero of p, we have

$$0 = a_n w^n + a_{n-1}w^{n-1} + \cdots + a_1 w + a_0$$

Take the conjugate of both sides, using repeatedly the rules $\overline{z_1 + z_2} = \overline{z}_1 + \overline{z}_2$ and $\overline{z_1 z_2} = \overline{z}_1 \overline{z}_2$. Since the a_k are real, the result is

$$0 = a_n \overline{w}^n + a_{n-1} \overline{w}^{n-1} + \cdots + a_1 \overline{w} + a_0$$

Thus, $\overline{w}$ is a zero of p. Since w is not real, w and $\overline{w}$ are distinct zeros of p. Consequently, p contains the quadratic factor

$$(z - w)(z - \overline{w}) = z^2 - (w + \overline{w})z + w\overline{w}$$

This factor is real because $w + \overline{w}$ and $w\overline{w}$ are real. ■

The nonreal zeros of the real polynomial p occur in conjugate pairs, and these pairs induce quadratic real factors of p. Hence, it is reasonable to search for quadratic factors; this can be done with the aid of Newton's iteration.

THEOREM 8 *If the polynomial $p(z) = a_n z^n + a_{n-1} z^{n-1} + \cdots + a_0$ is divided by the quadratic polynomial $z^2 - uz - v$, then the quotient and remainder*

$$\begin{aligned} q(z) &= b_n z^{n-2} + b_{n-1} z^{n-3} + \cdots + b_3 z + b_2 \\ r(z) &= b_1 (z - u) + b_0 \end{aligned}$$

can be computed recursively by setting $b_{n+1} = b_{n+2} = 0$ and then using

$$b_k = a_k + u b_{k+1} + v b_{k+2} \qquad (n \geqq k \geqq 0)$$

Proof The relationship between p, q, and r is expressed by

$$p(z) = q(z)(z^2 - uz - v) + r(z)$$

In detail, this equation reads:

$$\sum_{k=0}^{n} a_k z^k = \left(\sum_{k=2}^{n} b_k z^{k-2} \right) (z^2 - uz - v) + b_1 (z - u) + b_0$$

If we equate the coefficients of z^k on the two sides of this equation, the result is

$$\begin{aligned} a_k &= b_k - u b_{k+1} - v b_{k+2} \qquad (0 \leqq k \leqq n - 2) \\ a_{n-1} &= b_{n-1} - u b_n \\ a_n &= b_n \end{aligned}$$

The first of these three equations will include the last two if we define b_{n+1} and b_{n+2} to be 0. ■

Let us specialize the analysis to the case where all the coefficients a_k in p are real. We shall seek a real quadratic factor of the type mentioned in the theorem;

thus, u and v will be real. In the division process above, b_0 and b_1 are functions of u and v, and we write $b_0 = b_0(u, v)$ and $b_1 = b_1(u, v)$. In order that q be a *factor* of p, the remainder r should vanish; this leads to the two equations

$$\begin{aligned} b_0(u, v) &= 0 \\ b_1(u, v) &= 0 \end{aligned}$$

In Bairstow's method, this pair of simultaneous nonlinear equations is solved by Newton's method. We require the partial derivatives

$$c_k = \frac{\partial b_k}{\partial u} \qquad d_k = \frac{\partial b_{k-1}}{\partial v} \qquad (0 \leqq k \leqq n)$$

These are obtained by differentiating the recurrence relation already established for b_k in Theorem 7. The result is the following pair of additional recurrences:

$$\begin{aligned} c_k &= b_{k+1} + uc_{k+1} + vc_{k+2} \qquad (c_{n+1} = c_n = 0) \\ d_k &= b_{k+1} + ud_{k+1} + vd_{k+2} \qquad (d_{n+1} = d_n = 0) \end{aligned} \tag{4}$$

Since these recurrence relations obviously generate the same two sequences, we need only the first of them. An outline of the process is as follows: Starting values are assigned to u and v. We seek corrections—denoted by δu and δv—so that the equations

$$b_0(u + \delta u, v + \delta v) = b_1(u + \delta u, v + \delta v) = 0$$

are true. As in Section 3.2, we linearize these equations by writing

$$\begin{aligned} b_0(u, v) + \frac{\partial b_0}{\partial u}\delta u + \frac{\partial b_0}{\partial v}\delta v &= 0 \\ b_1(u, v) + \frac{\partial b_1}{\partial u}\delta u + \frac{\partial b_1}{\partial v}\delta v &= 0 \end{aligned}$$

In view of the preceding remarks, this system becomes

$$\begin{bmatrix} c_0 & c_1 \\ c_1 & c_2 \end{bmatrix} \begin{bmatrix} \delta u \\ \delta v \end{bmatrix} = - \begin{bmatrix} b_0(u, v) \\ b_1(u, v) \end{bmatrix}$$

The solution of this system follows:

$$\begin{aligned} \delta u &= (c_1 b_1 - c_2 b_0)/J \\ \delta v &= (c_1 b_0 - c_0 b_1)/J \\ J &= c_0 c_2 - c_1^2 \end{aligned}$$

Notice that J is the Jacobian determinant for the pair of nonlinear functions $b_0(u, v)$ and $b_1(u, v)$.

A pseudocode to execute M steps of Bairstow's method, starting at a prescribed point (u, v), is given next. It follows the preceding description closely.

```
input n, (a_i : 0 ≦ i ≦ n), u, v, M
b_n ← a_n
c_n ← 0
c_{n-1} ← a_n
for j = 1 to M do
    b_{n-1} ← a_{n-1} + u b_n
    for k = n - 2 to 0 step - 1 do
        b_k ← a_k + u b_{k+1} + v b_{k+2}
        c_k ← b_{k+1} + u c_{k+1} + v c_{k+2}
    end do
    J ← c_0 c_2 - c_1^2
    u ← u + (c_1 b_1 - c_2 b_0)/J
    v ← v + (c_1 b_0 - c_0 b_1)/J
    output j, u, v, b_0, b_1
end do
```

Example 7 Find a real quadratic factor of the polynomial in the previous examples, using Bairstow's method and starting with the point $(u, v) = (3, -4)$.

Solution A computer program based on the pseudocode above was written in double precision, and at the seventh iteration, we found

$$\begin{aligned} u &= 2.27568\,22036\,510 \\ v &= -3.62736\,50847\,118 \\ b_0 &= -0.2 \times 10^{-14} \\ b_1 &= 0.0 \end{aligned}$$

Since b_0 and b_1 are practically 0, we can accept u and v as the approximate values of the coefficients in the quadratic factor $z^2 - uz - v$. By combining the results of this and previous examples, we can exhibit the factorization of the polynomial p. (We display only three-digit precision in the equation.)

$$z^4 - 4z^3 + 7z^2 - 5z - 2 = (z - 2)(z + 0.276)(z^2 - 2.28z + 3.63)$$

The values of the coefficients have been given previously with higher accuracy. Two complex zeros of p can be computed from the quadratic factor (using its more accurate coefficients); they are

$$1.13784\,11018\,255 \pm (1.52731\,22508\,866)i \qquad \blacksquare$$

To complete the analysis of the Bairstow method, we need to establish (under reasonable hypotheses) that the Jacobian J is not 0 at the solution point.

THEOREM 9 *Let (u_0, v_0) be a point such that the zeros of $z^2 - u_0 z - v_0$ are simple zeros of p. Then the Jacobian in Bairstow's method is not 0 at (u_0, v_0).*

Proof At every step in the process, we have

$$p(z) = (z^2 - uz - v)q(z) + b_1(z - u) + b_0$$

Partial differentiation with respect to u and v gives us these two equations:

$$0 = -zq(z) + (z^2 - uz - v)\frac{\partial q}{\partial u} - b_1 + \frac{\partial b_1}{\partial u}(z - u) + \frac{\partial b_0}{\partial u}$$
$$0 = -q(z) + (z^2 - uz - v)\frac{\partial q}{\partial v} + \frac{\partial b_1}{\partial v}(z - u) + \frac{\partial b_0}{\partial v}$$

Assuming the hypotheses, we know that $z^2 - u_0 z - v_0$ has two zeros, say z_1 and z_2. It follows that $p(z_1) = 0$, $p(z_2) = 0$, $b_1 = 0$, and $b_0 = 0$. In the preceding equations, let $u = u_0$, $v = v_0$, and $z = z_1$ or $z = z_2$. Four equations result from this:

$$0 = -z_j q(z_j) + c_1(z_j - u_0) + c_0 \qquad (j = 1, 2)$$
$$0 = -q(z_j) + c_2(z_j - u_0) + c_1 \qquad (j = 1, 2)$$

Here we use the notation and analysis that immediately follow the proof of Theorem 7. In matrix form, these equations look like this:

$$\begin{bmatrix} c_0 & c_1 \\ c_1 & c_2 \end{bmatrix} \begin{bmatrix} 1 & 1 \\ z_1 - u_0 & z_2 - u_0 \end{bmatrix} = \begin{bmatrix} z_1 q(z_1) & z_2 q(z_2) \\ q(z_1) & q(z_2) \end{bmatrix}$$

To see that the Jacobian matrix is not singular, it suffices to prove that the matrix on the right of the preceding equation is not singular. The determinant of this matrix is $(z_1 - z_2)q(z_1)q(z_2)$. Since z_1 and z_2 are simple zeros of p, they cannot be zeros of q. Also, $z_1 \neq z_2$, and therefore this determinant is not 0. ■

The discussion of Bairstow's method follows the exposition of Henrici [1964].

Laguerre Iteration

We turn next to Laguerre's method for finding the zeros of a polynomial p. This method is used in several modern software packages because it has very favorable convergence properties and is rather robust. It exhibits third-order convergence in the vicinity of each simple zero. The algorithm is iterative and proceeds from one approximate zero z to a new one by calculating

$$\begin{aligned} A &= -p'(z)/p(z) \\ B &= A^2 - p''(z)/p(z) \\ C &= n^{-1}\left[A \pm \sqrt{(n-1)(nB - A^2)}\,\right] \\ z_{\text{new}} &= z + 1/C \end{aligned}$$

Here p is a polynomial of degree n. In the definition of C, the sign is chosen to make $|C|$ as large as possible.

For this algorithm, the following theorem of Kahan [1967] is the analogue of Theorem 6.

THEOREM 10 *If p is a polynomial of degree n, if z is any complex number, and if C is computed as in Laguerre's algorithm, then p has a zero in the complex plane within distance $\sqrt{n}/C$ of z.*

Proof As in Theorem 5, we denote the zeros of our polynomial by $r_1, r_2, \ldots, r_n$. We can assume that $p(x) = \prod_{j=1}^{n}(x - r_j)$, with leading coefficient unity, since the formulas for A, B, and C are not affected by any scalar factor attached to p. Differentiating p and putting $u_j = (r_j - z)^{-1}$ yield

$$p'(z) = \sum_{j=1}^{n} \prod_{\substack{k=1 \\ k \neq j}}^{n} (z - r_k) = \sum_{j=1}^{n} p(z)/(z - r_j) = -p(z) \sum_{j=1}^{n} u_j$$

Hence,

$$A = -\frac{p'(z)}{p(z)} = \sum_{j=1}^{n} u_j$$

Differentiating in this formula and using the definition of B lead to

$$B = \frac{-p(z)p''(z) + [p'(z)]^2}{[p(z)]^2} = \sum_{j=1}^{n} (r_j - z)^{-2} = \sum_{j=1}^{n} u_j^2$$

From the definition of C, we have

$$(nC - A)^2 = (n-1)(nB - A^2) \tag{5}$$

Define $D = (A - C)/(n - 1)$, so that

$$A = (n-1)D + C \tag{6}$$

From Equation (5), it follows that

$$\begin{aligned} n^2C^2 - 2nCA + A^2 &= (n-1)nB - nA^2 + A^2 \\ (n-1)B &= nC^2 - 2CA + A^2 \end{aligned} \tag{7}$$

In Equation (7), replace A by $(n - 1)D + C$ to get

$$\begin{aligned} (n-1)B &= nC^2 - 2C\left[(n-1)D + C\right] + \left[(n-1)D + C\right]^2 \\ (n-1)B &= nC^2 - 2(n-1)CD - 2C^2 + (n-1)^2D^2 + 2(n-1)CD + C^2 \\ (n-1)B &= (n-1)C^2 + (n-1)^2D^2 \\ B &= C^2 + (n-1)D^2 \end{aligned} \tag{8}$$

Starting with (7) and then using (6) and (8) yield

$$\begin{aligned} nB - A^2 &= nC^2 - 2CA + B \\ &= nC^2 - 2C\left[(n-1)D + C\right] + C^2 + (n-1)D^2 \\ &= (n-1)C^2 - (n-1)2CD + (n-1)D^2 \\ &= (n-1)(C-D)^2 \end{aligned} \tag{9}$$

Starting with Equation (9) and using formulas for A and B, we obtain

$$\begin{aligned} (n-1)|C-D|^2 &= |nB - A^2| \\ &= n^{-1}|n^2B - 2nA^2 + nA^2| \\ &= n^{-1}\left|\sum_{j=1}^{n}(n^2u_j^2 - 2nAu_j + A^2)\right| \\ &= n^{-1}\left|\sum_{j=1}^{n}(nu_j - A)^2\right| \\ &\leqq n^{-1}\sum_{j=1}^{n}|nu_j - A|^2 \\ &= n^{-1}\sum_{j=1}^{n}\{n^2|u_j|^2 - 2n\,\mathrm{Re}(\overline{A}u_j) + |A|^2\} \\ &= n\sum_{j=1}^{n}|u_j|^2 - 2\overline{A}A + |A|^2 \\ &\leqq n^2\max_j|u_j|^2 - |A|^2 \\ &= n^2\max_j|u_j|^2 - |(n-1)D + C|^2 \\ &= n^2\max_j|u_j|^2 - |(C-D) + nD|^2 \\ &= n^2\max_j|u_j|^2 - |C-D|^2 - 2n\,\mathrm{Re}\left[\overline{D}(C-D)\right] - n^2|D|^2 \end{aligned}$$

Hence,

$$\begin{aligned} n|C-D|^2 &\leqq n^2\max_j|u_j|^2 - n^2|D|^2 - 2n\,\mathrm{Re}(\overline{D}C) + 2n|D|^2 \\ |C-D|^2 &\leqq n\max_j|u_j|^2 - n|D|^2 - 2\,\mathrm{Re}(\overline{D}C) + 2|D|^2 \\ |C|^2 - 2\,\mathrm{Re}(C\overline{D}) + |D|^2 &\leqq n\max_j|u_j|^2 - n|D|^2 - 2\,\mathrm{Re}(\overline{D}C) + 2|D|^2 \\ |C|^2 + (n-1)|D|^2 &\leqq n\max_j|u_j|^2 = n/\min_j|z - r_j|^2 \end{aligned}$$

This gives

$$\min_j |z - r_j|^2 \leqq n/\{|C|^2 + (n-1)|D|^2\}$$

and

$$\min_j |z - r_j| \leqq \sqrt{n} \Big/ \sqrt{|C|^2 + (n-1)|D|^2} \tag{10}$$

This inequality is stronger than the one in the statement of the theorem. The proof is from Kahan [1967]. ■

A simple pseudocode to execute Laguerre's method on a given polynomial with a prescribed starting point z_0 is offered next:

```
input n, (a_i : 0 ≦ i ≦ n), z_0, M, ε
for k = 1 to M do
    α ← a_n
    β ← 0
    γ ← 0
    for j = n − 1 to 0 step − 1 do
        γ ← z_0 γ + β
        β ← z_0 β + α
        α ← z_0 α + a_j
    end do
    A ← −β/α
    B ← A² − 2γ/α
    C ← [A ± √((n − 1)(nB − A²))]/n
    z_1 ← z_0 + 1/C
    output k, z
    if |z_1 − z_0| < ε then stop
    z_0 ← z_1
end do
```

In this algorithm, α is $p(z_0)$, β is $p'(z_0)$, and γ is $\frac{1}{2}p''(z_0)$.

LEMMA 1 *Let $v_1, v_2, \ldots, v_n$ be any real numbers. Put $\alpha = \sum_{i=1}^n v_i$ and $\beta = \sum_{i=1}^n v_i^2$. Then the numbers v_i lie in the closed interval whose endpoints are*

$$n^{-1}\left[\alpha \pm \sqrt{(n-1)(n\beta - \alpha^2)}\,\right] \tag{11}$$

Proof It suffices to prove that v_1 lies in the interval described. Recall the Cauchy-Schwarz Inequality:

$$\left(\sum_{i=1}^m x_i y_i\right)^2 \leqq \left(\sum_{i=1}^m x_i^2\right)\left(\sum_{j=1}^m y_j^2\right)$$

Applying this, we have

$$\begin{aligned}\alpha^2 - 2\alpha v_1 + v_1^2 &= (\alpha - v_1)^2 = (v_2 + v_3 + \cdots + v_n)^2 \\ &\leqq (1^2 + 1^2 + \cdots + 1^2)(v_2^2 + v_3^2 + \cdots + v_n^2) \\ &= (n-1)(v_2^2 + v_3^2 + \cdots + v_n^2) \\ &= (n-1)(\beta - v_1^2) - (n-1)\beta - nv_1^2 + v_1^2\end{aligned}$$

Rearranging this inequality gives us

$$nv_1^2 - 2\alpha v_1 + \alpha^2 - (n-1)\beta \leqq 0$$

This shows that the quadratic function $q(x) = nx^2 - 2\alpha x + \alpha^2 - (n-1)\beta$ has the property $q(v_1) \leqq 0$. For large $|x|$, obviously $q(x) > 0$. Hence, v_1 lies between the two zeros of q, and they are the endpoints in Formula (11). ■

LEMMA 2 *Let p be a real polynomial of degree n whose zeros, $r_1, r_2, \dots, r_n$, are real. For any real x different from all the r_j, the numbers $(x - r_j)^{-1}$ lie in the interval whose endpoints are*

$$\left[np(x)\right]^{-1}\left\{p'(x) \pm \sqrt{\left[(n-1)p'(x)\right]^2 - n(n-1)p(x)p''(x)}\right\} \qquad \textbf{(12)}$$

Proof The polynomial p is of the form $p(x) = c\prod_{j=1}^{n}(x - r_j)$. Since the assertion to be proved is independent of c, we can assume that $c = 1$. As in the proof of Theorem 9, we have, with $v_j = (x - r_j)^{-1}$, $\alpha = \sum_{j=1}^{n} v_j$, and $\beta = \sum_{j=1}^{n} v_j^2$,

$$p'(x)/p(x) = \sum_{j=1}^{n} v_j = \alpha \qquad \textbf{(13)}$$

$$\left[(p'(x))^2 - p(x)p''(x)\right]/(p(x))^2 = \sum_{j=1}^{n} v_j^2 = \beta \qquad \textbf{(14)}$$

By Lemma 1, all the numbers v_j lie in an interval with endpoints given in Formula (11). When the values of α and β in Equations (13) and (14) are substituted in this formula, the result is Formula (12). ■

THEOREM 11 *Let p be a real polynomial whose roots are all real. The sequence produced by the Laguerre algorithm with an arbitrary starting point converges monotonically to a zero of p.*

Proof Let the zeros of p be $r_1 \leqq r_2 \leqq \cdots \leqq r_n$. Suppose that x is not a zero. Then by Lemma 2, all the numbers $(x - r_i)^{-1}$ lie in an interval whose endpoints are

$$u(x) = \left[p'(x) + w(x)\right]/np(x) \quad \text{and} \quad v(x) = \left[p'(x) - w(x)\right]/np(x)$$

where

$$w(x) = \sqrt{[(n-1)p'(x)]^2 - n(n-1)p''(x)p(x)}$$

First, let us consider the case when $r_j < x < r_{j+1}$ for some j. Suppose that $p(y) > 0$ on (r_j, r_{j+1}); the other case is similar. Then we have

$$v(x) \leqq (x - r_{j+1})^{-1} < 0 < (x - r_j)^{-1} \leqq u(x)$$

From this it follows that

$$r_j \leqq x - \frac{1}{u(x)} < x < x - \frac{1}{v(x)} \leqq r_{j+1}$$

Thus, if we start with x in (r_j, r_{j+1}) and compute $x - 1/u(x)$ and $x - 1/v(x)$, as in the Laguerre iteration, then these two new points will lie in $[r_j, r_{j+1}]$. Of course, if either of these two new points is an endpoint of this interval, then a zero of p has been computed. In the contrary case, the two new points are in (r_j, r_{j+1}), and the process can be repeated. Proceeding formally, we define two sequences $[y_k]$ and $[z_k]$ by putting

$$\begin{cases} y_0 = x & y_{k+1} = y_k - 1/u(y_k) \qquad (k \geqq 1) \\ z_0 = x & z_{k+1} = z_k - 1/v(z_k) \qquad (k \geqq 1) \end{cases}$$

Our analysis above shows that

$$r_j \leqq y_{k+1} < y_k < \cdots < x < z_1 < \cdots < z_k < z_{k+1} \leqq r_{j+1}$$

Since the sequence $[y_k]$ decreases and is bounded below by r_j, it must converge. Say $y_k \downarrow y$. From the iteration formulas

$$1/u(y_k) = y_k - y_{k+1} \to 0$$

Hence, $u(y_k) \to \infty$. The formula for u now indicates that $p(y_k) \to 0$, so $p(y) = 0$ and $y = r_j$. Similarly, we conclude that $z_k \uparrow r_{j+1}$.

In the remaining case, the starting point is not between two zeros of p. Say $x > r_n$. Suppose also that $p(y) > 0$ on (r_n, ∞). Proceeding as before, we have

$$0 < (x - r_n)^{-1} \leqq u(x)$$

From this it follows that $r_n \leqq x - 1/u(x) < x$. The sequence $[y_n]$ defined as before converges downward to r_n by the same reasoning. ■

Discussions of Laguerre's method can be found in papers and books by Bodewig [1946], van der Corput [1946], Durand [1960], Foster [1981], Galeone [1977], Householder [1970], Kahan [1967], Ostrowski [1966], Parlett [1964], Redish [1974], and Wilkinson [1965].

Complex Newton's Method

For a polynomial having complex coefficients, Newton's method should be programmed in complex arithmetic. After one root has been found, the deflation process (also programmed in complex arithmetic) should be used. Newton's method can then be applied to the reduced polynomial. This process can be repeated until all zeros have been determined. Further analysis and experience indicate that the procedure will be satisfactory in general, provided that two precautions are taken:

(i) The zeros should be computed in order of increasing magnitude.

(ii) Any zero obtained by using Newton's method on a reduced polynomial should be immediately refined by applying Newton's method to the original polynomial with the best estimate of the zero as the starting value. Only after this has been done should the next step of deflation be carried out.

We recommend that the reader consult Wilkinson [1984] and Peters and Wilkinson [1971] for further guidance on this overall strategy. Many other methods could be discussed for the polynomial root-finding problem. The method of Laguerre is especially attractive because of its favorable global convergence behavior. A robust method suitable for general software has been developed by Jenkins and Traub [1970a]. Other references not already mentioned are Allgower, Glasshoff, and Peitgen [1981], Gautschi [1979, 1984], Henrici [1974], Householder [1970], Ostrowski [1966], Jenkins and Traub [1970b], Marden [1966], Smale [1981], Stoer and Bulirsch [1980], Ralston and Rabinowitz [1978], and Traub [1964].

Here is a brief explanation of how the picture on the cover of this book was obtained. It involves Newton's method in the complex plane.

Let p be a polynomial of degree at least 2, and let ξ be one of its zeros. If Newton's method is started at a point z in the complex plane, it will produce a sequence defined by the equations

$$\begin{cases} z_0 = z \\ z_{n+1} = z_n - p(z_n)/p'(z_n) \qquad (n \geqq 0) \end{cases}$$

If $\lim_{n\to\infty} z_n = \xi$, we say that z (the starting point) is **attracted** to ξ. The set of all points z that are attracted to ξ is called the **basin of attraction** corresponding to ξ. Each zero of p has a basin of attraction, and these sets are mutually disjoint because a sequence that converges to one zero of p cannot converge to another zero. Some complex numbers belong to no basin of attraction; these are starting points for which the Newton iteration does not converge. These exceptional points constitute the **Julia set** of p, so named in honor of G. Julia, a French mathematician who published an important memoir on this subject in 1918. If all zeros of p are simple, then the basins of attraction are open sets in the complex plane, and the Julia set will be the boundary of each basin of attraction.

The basins of attraction for the polynomial $p(z) = z^4 + 4$ are shown on the cover of this book. The four roots of p are:

$$\omega_1 = 1 + i \qquad \omega_2 = 1 - i \qquad \omega_3 = -1 + i \qquad \omega_4 = -1 - i$$

We took a square region in the complex plane and generated a large number of lattice points in that square. For each lattice point, we made a rough test to determine to which basin of attraction it belonged. This test consisted of monitoring the first 20 Newton iterates and checking whether any of them was within distance 0.25 of a root. If so, then subsequent iteration would produce quadratic convergence to that root. This fact was established by use of the standard convergence theory of Newton's method in the complex plane. In this way, a list of lattice points belonging to each basin of attraction was generated. The basin of attraction for root ω_1 was assigned one color, and the basins for roots ω_2, ω_3, and ω_4 were assigned three other colors. These four sets (actually just the lattice points in them) were displayed on the color screen of a workstation, and then a color plot was produced from it. The four color sets fit together in an incredible way that displays a **fractal** appearance. That is, on magnifying a portion of the plane where two sets meet, we see the same general patterns repeated. This will persist on repeated magnification. Furthermore, each boundary point of any one of these four sets is also a boundary point of the other three sets!

Recently, a number of articles and books have been published on fractals and chaos. Additional information can be found, for example, in Barnsley [1988], Curry, Garnett, and Sullivan [1983], Dewdney [1988], Glieck [1987], Mandelbrot [1982], Peitgen and Richter [1986], Peitgen, Saupe, and Haeseler [1984], Pickover [1988], and Sander [1987].

PROBLEMS 3.5

1. Use Horner's algorithm to find $p(4)$, where

$$p(z) = 3z^5 - 7z^4 - 5z^3 + z^2 - 8z + 2$$

2. For the polynomial of Problem 1, find its expansion in a Taylor series about the point $z_0 = 4$.

3. For the polynomial of Problem 1, start Newton's method at the point $z_0 = 4$. What is z_1?

4. For the polynomial of Problem 1, apply Bairstow's method with the initial point $(u, v) = (3, 1)$. Compute the corrections δu and δv.

5. For the polynomial of Problem 1, find a disk centered at the origin that contains all the zeros.

6. For the polynomial of Problem 1, find a disk centered at the origin that contains none of the zeros.

7. Does Theorem 3 sometimes give the radius of the smallest circle centered at the origin that contains all the zeros of a given polynomial?

8. Prove that every polynomial having real coefficients can be factored into a product of linear and quadratic factors having real coefficients.

9. Verify the recurrence relations and starting values given for c_k and d_k in the discussion of Bairstow's method.

10. For the polynomial $p(z) = 9z^4 - 7z^3 + z^2 - 2z + 5$, compute $p(6)$, $p'(6)$, and the next point in the Newton iteration starting at $z = 6$.

11. Using the definition of multiplicity adopted in the text, prove that if z is a root of multiplicity m of a polynomial p, then $p(z) = p'(z) = \cdots = p^{(m-1)}(z) = 0$ and $p^{(m)}(z) \neq 0$.

12. Prove the converse of the assertion in Problem 11.

13. Does Bairstow's method yield a quadratically convergent sequence (u_k, v_k)?

14. Write an algorithm that deflates $p(z)$ when a root z_0 is known, but computes the coefficients of the reduced polynomial in ascending order—that is, constant term first.

15. Derive Equation (4) in the discussion of Bairstow's method.

16. Concerning the polynomial $p(x) = a_0 + a_1 x + \cdots + a_n x^n$, prove the following result. For a given x, we set $(\alpha_n, \beta_n, \gamma_n) = (a_n, 0, 0)$ and define inductively

$$(\alpha_j, \beta_j, \gamma_j) = (a_j + x\alpha_{j+1}, \alpha_{j+1} + x\beta_{j+1}, \beta_{j+1} + x\gamma_{j+1})$$

for $j = n-1, n-2, \ldots, 0$. Then $p(x) = \alpha_0$, $p'(x) = \beta_0$, and $p''(x) = 2\gamma_0$.

17. In the analysis of Laguerre's method, prove that

$$C^2 + (n-1)D^2 = \sum_{j=1}^{n} u_j^2$$

18. In the description of Laguerre's method, the quantities A and B are functions of z and depend on the given polynomial p. Let r be a zero of p. Show that the corresponding functions A and B for $p(z)/(z-r)$ are, respectively,

$$A + (z-r)^{-1} \quad \text{and} \quad B - (z-r)^2$$

19. Prove that if p is a polynomial of degree n having real coefficients, then

$$(n-1)\left[p'(x)\right]^2 \geqq np(x)p''(x)$$

Assume that the roots are real.

COMPUTER PROBLEMS 3.5

1. Write a program that takes as input the coefficients of a polynomial p and a specific point z_0, and produces as output the values $p(z_0)$, $p'(z_0)$, and $p''(z_0)$. Write the pseudocode with only one loop. Test on the polynomial in Problem 1, taking $z_0 = 4$.

2. Write a complex Newton method for a polynomial having complex coefficients, with a given starting point in the complex plane and a given number of iterations. Test your program on the polynomial of Problem 1, starting with $z_0 = 3 - 2i$.

3. Experiment with Laguerre's method coded in complex arithmetic. Find all four roots of the polynomial used in the examples of this section.

4. Using Newton's method and the polynomial $p(z) = z^3 - 1$, find three nearby starting points (within 0.01 of each other) so that the resulting sequences converge to different roots. Using a plotter, show the paths of these sequences of points within a square containing the roots by connecting successive points with line segments.

5. Write a computer routine that uses Newton's method and deflation for finding all roots of a polynomial. Test the routine using Wilkinson's perfidious polynomial given in Computer Problem 6 of Section 2.3.

6. Write and test a program to compute all the zeros of a complex polynomial. Use Laguerre's method, and compute the zeros in order of ascending magnitude. Use *deflation*, and refine any zero obtained from a deflated polynomial by Laguerre's iteration with the original polynomial. To test your program, compute all roots of the polynomial

$$x^8 - 36x^7 + 546x^6 - 4536x^5 + 22449x^4 - 67284x^3 + 118124x^2 - 109584x + 40320 = 0$$

The correct roots are the integers $1, 2, \ldots, 8$. Next, solve the same equation when the coefficient of x^7 is changed to -37. Observe how a minor perturbation in the coefficients can cause massive changes in the roots. Thus, the roots are unstable functions of the coefficients.

7. Modify the Laguerre algorithm so that the coefficients in the quotient polynomials are saved in the application of the Horner algorithm. A separate calculation of the deflated polynomial will then be unnecessary after a zero has been computed. Test your program on

$$z^4 - (10 + 26i)z^3 - (216 - 190i)z^2 + (1140 + 636i)z - (72 - 68i) = 0$$

8. Write a computer program using the complex Newton's method to compute the basins of attraction of the roots of $f(z) = z^4 + 4$ as discussed in this section. Display the results on a graphics computer terminal or workstation.

*3.6 Homotopy and Continuation Methods

In this section, we consider the problem of finding the roots of an equation

$$f(x) = 0 \tag{1}$$

Here f is a mapping from one linear space to another, say $f : X \to Y$. This problem is sufficiently general to include systems of algebraic equations, integral equations, differential equations, and so on. The methods to be considered here involve a strategy somewhat different from that in the preceding sections, and this new approach requires the numerical solution of systems of ordinary differential equations—a topic not taken up in this book until Chapter 8. Also, one of our illustrations of the method involves the linear programming problem, which is also taken up in a later chapter.

Basic Concepts

The basic idea of the continuation method is to embed the given problem in a one-parameter family of problems, using a parameter t that runs over the interval $[0, 1]$. The original problem is made to correspond to $t = 1$, and another problem with known solution is made to correspond to $t = 0$. For example, we can define

$$h(t, x) = tf(x) + (1 - t)g(x) \tag{2}$$

The equation $g(x) = 0$ should have a known solution. The next step is to select points $t_0, t_1, \ldots, t_m$ so that

$$0 = t_0 < t_1 < t_2 < \cdots < t_m = 1$$

We then attempt to solve each equation $h(t_i, x) = 0$, $0 \leqq i \leqq m$. Assuming that some iterative method will be used (such as Newton's method), it is prudent to use the solution at the ith step as the starting point in computing a solution at the $(i+1)$st step.

This whole procedure can be regarded as a cure for the difficulty that besets Newton's method: the need for a good starting point.

The relationship (2), which embeds the original problem (1) in a family of problems, gives an example of a **homotopy** that connects the two functions f and g. In general, a homotopy can be *any* continuous connection between f and g. Formally, a homotopy between two functions $f, g : X \to Y$ is a continuous map

$$h : [0, 1] \times X \to Y \tag{3}$$

such that $h(0, x) = g(x)$ and $h(1, x) = f(x)$. If such a map exists, we say that f is *homotopic* to g. This is an equivalence relation among the continuous maps from X to Y, where X and Y can be any two topological spaces.

A simple homotopy that is often used in the continuation method is

$$\begin{aligned} h(t, x) &= tf(x) + (1 - t)\big[f(x) - f(x_0)\big] \\ &= f(x) + (t - 1)f(x_0) \end{aligned} \tag{4}$$

Here x_0 can be any point in X, and it is clear that x_0 will be a solution of the problem when $t = 0$.

If the equation $h(t, x) = 0$ has a unique root for each $t \in [0, 1]$, then that root is a function of t, and we can write $x(t)$ as the unique member of X that makes the equation $h(t, x(t)) = 0$ true. The set

$$\{x(t) : 0 \leqq t \leqq 1\} \tag{5}$$

can be interpreted as an arc or curve in X, parametrized by t. This arc leads from the known point $x(0)$ to the solution of our problem, $x(1)$. The continuation method attempts to determine this curve by computing points on it, $x(t_0), x(t_1), \ldots, x(t_m)$.

If the function $t \longmapsto x(t)$ is differentiable and if h is differentiable, then the Implicit Function Theorem enables us to compute $x'(t)$. By pursuing this idea, we can describe the curve in (5) by a differential equation. Assuming an arbitrary homotopy, we have

$$0 = h\big(t, x(t)\big) \tag{6}$$

On differentiating with respect to t, we obtain

$$0 = h_t\big(t, x(t)\big) + h_x\big(t, x(t)\big)x'(t) \tag{7}$$

in which subscripts denote partial derivatives. Thus,

$$x'(t) = -\left[h_x\big(t, x(t)\big)\right]^{-1} h_t\big(t, x(t)\big) \tag{8}$$

This is a differential equation for x. It has a known initial value because $x(0)$ is supposed known. On integrating this differential equation (usually by numerical procedures), we shall have the value $x(1)$, which is the solution.

Example 1 We let $X = Y = \mathbb{R}^2$, and define

$$f(x) = \begin{bmatrix} \xi_1^2 - 3\xi_2^2 + 3 \\ \xi_1 \xi_2 + 6 \end{bmatrix} \qquad x = (\xi_1, \xi_2) \in \mathbb{R}^2$$

Solution A convenient homotopy is given by Equation (4), and we select $x_0 = (1, 1)$. The derivatives on the right side of Equation (8) are computed to be

$$h_x = f'(x) = \begin{bmatrix} \partial f_1/\partial \xi_1 & \partial f_1/\partial \xi_2 \\ \partial f_2/\xi_1 & \partial f_2/\xi_2 \end{bmatrix} = \begin{bmatrix} 2\xi_1 & -6\xi_2 \\ \xi_2 & \xi_1 \end{bmatrix}$$

$$h_t = f(x_0) = \begin{bmatrix} f_1(x_0) \\ f_2(x_0) \end{bmatrix} = \begin{bmatrix} 1 \\ 7 \end{bmatrix}$$

The inverse of $f'(x)$ is

$$h_x^{-1} = \left[f'(x)\right]^{-1} = \frac{1}{\Delta} \begin{bmatrix} \xi_1 & 6\xi_2 \\ -\xi_2 & 2\xi_1 \end{bmatrix} \qquad \Delta = 2\xi_1^2 + 6\xi_2^2$$

The differential equation that governs the path leading away from the point x_0 is Equation (8). In this concrete case, it is a pair of ordinary differential equations:

$$\begin{bmatrix} \xi_1' \\ \xi_2' \end{bmatrix} = -\frac{1}{\Delta}\begin{bmatrix} \xi_1 & 6\xi_2 \\ -\xi_2 & 2\xi_1 \end{bmatrix}\begin{bmatrix} 1 \\ 7 \end{bmatrix} = -\frac{1}{\Delta}\begin{bmatrix} \xi_1 + 42\xi_2 \\ -\xi_2 + 14\xi_1 \end{bmatrix}$$

This system was integrated numerically (using one of the methods to be presented in Chapter 8) over the interval $0 \leqq t \leqq 1$, and at $t = 1$ the solution was $(-2.961,\ 1.978)$. Notice that f has a root $(-3,\ 2)$.

To complete this problem and find its numerical solution, we can use Newton's iteration starting at the point produced by the homotopy method. The Newton iteration replaces any approximate root x by $x - \delta$, where the correction δ is given by

$$\delta = [f'(x)]^{-1} f(x)$$

In this example, the vector δ is

$$\begin{bmatrix} \delta_1 \\ \delta_2 \end{bmatrix} = \frac{1}{\Delta}\begin{bmatrix} \xi_1 & 6\xi_2 \\ -\xi_2 & 2\xi_1 \end{bmatrix}\begin{bmatrix} \xi_1^2 - 3\xi_2^2 + 3 \\ \xi_1\xi_2 + 6 \end{bmatrix}$$

Three steps of the Newton iteration produced these results:

k	ξ_1	ξ_2
0	−2.96100 00000 00	1.97800 00000 00
1	−3.00025 32813 14	2.00032 02744 78
2	−3.00000 00057 80	2.00000 00378 24
3	−3.00000 00000 00	2.00000 00000 00

■

The following formal result, due to Ortega and Rheinboldt [1970], gives some conditions under which the homotopy method will succeed.

THEOREM 1 *If $f: \mathbb{R}^n \to \mathbb{R}^n$ is continuously differentiable and if $\|[f'(x)]^{-1}\| \leqq M$ on $\mathbb{R}^n$, then for any $x_0 \in \mathbb{R}^n$ there is a unique curve $\{x(t) : 0 \leqq t \leqq 1\}$ in $\mathbb{R}^n$ such that $f(x(t)) + (t-1)f(x_0) = 0$, with $0 \leqq t \leqq 1$. The function $t \mapsto x(t)$ is a continuously differentiable solution of the initial-value problem $x' = -[f'(x)]^{-1} f(x_0)$, where $x(0) = x_0$.*

Tracing the Path

Another way of tracing the path $x(t)$ has been described by Garcia and Zangwill [1981]. We start with the equation $h(t, x) = 0$, supposing that $x \in \mathbb{R}^n$ and $t \in [0, 1]$. A vector $y \in \mathbb{R}^{n+1}$ is defined by

$$y = (t, \xi_1, \xi_2, \ldots, \xi_n)$$

where $\xi_1, \xi_2, \ldots, \xi_n$ are the components of x. Hence, our equation is simply $h(y) = 0$. Each component of y, including t, is now allowed to be a function of an independent variable s, and we write $h(y(s)) = 0$. Differentiating with respect to s, we obtain the basic differential equation

$$h'(y(s))y'(s) = 0 \tag{9}$$

The variable s starts at 0, as does t. The initial value of x is $x(0) = x_0$. Thus, suitable starting values are available for the differential equation (9).

Since f and g are maps of $\mathbb{R}^n$ into $\mathbb{R}^n$, h is a map of $\mathbb{R}^{n+1}$ into $\mathbb{R}^n$. The derivative $h'(y)$ is therefore represented by an $n \times (n+1)$ matrix, A. The vector $y(s)$ has $n+1$ components, which we denote by $\eta_1, \eta_2, \ldots, \eta_{n+1}$. By appealing to the lemma below, we can obtain another form for Equation (9); namely,

$$\eta_j' = (-1)^{j+1} \det(A_j) \qquad (1 \leqq j \leqq n+1) \tag{10}$$

where A_j is the $n \times n$ matrix that results from A by deleting the jth column. Let us illustrate this formalism with the same problem as in Example 1.

Example 2 Taking f and x_0 as in Example 1, we have

$$h(t, x) = \begin{bmatrix} \xi_1^2 - 3\xi_2^2 + 2 + t \\ \xi_1\xi_2 - 1 + 7t \end{bmatrix}$$

Solution The differential equation (9) is given by

$$\begin{bmatrix} \partial h_1/\partial t & \partial h_1/\partial \xi_1 & \partial h_1/\partial \xi_2 \\ \partial h_2/\partial t & \partial h_2/\partial \xi_1 & \partial h_2/\partial \xi_2 \end{bmatrix} \begin{bmatrix} t' \\ \xi_1' \\ \xi_2' \end{bmatrix} = \begin{bmatrix} 0 \\ 0 \end{bmatrix}$$

$$\begin{bmatrix} 1 & 2\xi_1 & -6\xi_2 \\ 7 & \xi_2 & \xi_1 \end{bmatrix} \begin{bmatrix} t' \\ \xi_1' \\ \xi_2' \end{bmatrix} = \begin{bmatrix} 0 \\ 0 \end{bmatrix} \tag{11}$$

It is preferable to use Equation (10), however, and to write the differential equations in the form

$$\begin{cases} t' = 2\xi_1^2 + 6\xi_2^2 & t(0) = 0 \\ \xi_1' = -\xi_1 - 42\xi_2 & \xi_1(0) = 1 \\ \xi_2' = \xi_2 - 14\xi_1 & \xi_2(0) = 1 \end{cases} \tag{12}$$

The derivatives in this system are with respect to s. Performing a numerical integration, we arrive at these two points:

$$s = 0.087 \quad t = 0.969 \quad \xi_1 = -2.94 \quad \xi_2 = 1.97$$
$$s = 0.088 \quad t = 1.010 \quad \xi_1 = -3.02 \quad \xi_2 = 2.01$$

Either of these can be used to start a Newton iteration, as was done in Example 1. The path generated by this homotopy is shown in Figure 3.8. ■

A drawback to the method used in Example 2 is that we have no *a priori* knowledge of the value of s corresponding to $t = 1$. In practice, this may necessitate several computer runs.

LEMMA 1 *Let A be an $n \times (n+1)$ matrix. A solution of the homogeneous equation $Ax = 0$ is given by $x_j = (-1)^j \det(A_j)$, where A_j is A without column j.*

Proof Select any row (for example, the ith row) in A and adjoin a copy of it as a new row at the top of A. This creates an $(n+1) \times (n+1)$ matrix B that is obviously singular because row i of A occurs twice in B. In expanding the determinant of B by the elements in its top row, we obtain

$$0 = \det B = \sum_{j=1}^{n+1} (-1)^{j+1} a_{ij} \det(A_j) = -\sum_{j=1}^{n+1} a_{ij} x_j$$

Since this is true for $i = 1, 2, \ldots, n$, we have $Ax = 0$. ■

Relation to Newton's Method

The connection between the homotopy methods and Newton's method is deeper than may be seen at first glance. Let us start with the homotopy

$$h(t, x) = f(x) - e^{-t} f(x_0) \tag{13}$$

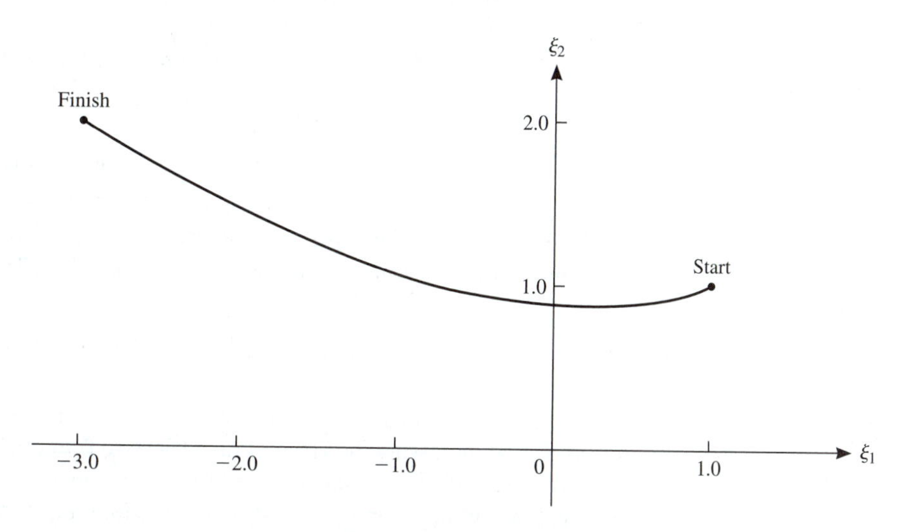

FIGURE 3.8 The path generated in Example 2

In this equation, t will run from 0 to ∞. We seek a curve or path, $x = x(t)$, on which

$$0 = h\big(t, x(t)\big) = f\big(x(t)\big) - e^{-t} f(x_0)$$

As usual, differentiation with respect to t will lead to a differential equation describing the path:

$$\begin{aligned} 0 &= f'\big(x(t)\big)x'(t) + e^{-t} f(x_0) \\ &= f'\big(x(t)\big)x'(t) + f\big(x(t)\big) \\ x'(t) &= -\left[f'\big(x(t)\big)\right]^{-1} f\big(x(t)\big) \end{aligned} \tag{14}$$

If this differential equation is integrated with Euler's method (described in Section 8.2), with step size 1, the result is the formula

$$x_{n+1} = x_n - [f'(x_n)]^{-1} f(x_n)$$

This is, of course, the formula for Newton's method. It is clear that one can expect to obtain better results by integrating Equation (14) with a more accurate numerical method and a variable step size. (For these matters, consult Chapter 8.)

Linear Programming

The homotopy method can be used to solve linear programming problems. (Such problems are discussed in Chapter 10.) This approach leads naturally to the algorithm proposed by Karmarkar [1984]. In explaining the homotopy method in this context, we follow closely the description given by Brophy and Smith [1988]. The reader who wishes to explore these ideas further can profitably consult other references.

Consider the standard linear programming problem

$$\begin{cases} \text{maximize } c^T x \\ \text{subject to } Ax = b \text{ and } x \geqq 0 \end{cases} \tag{15}$$

Here, $c \in \mathbb{R}^n$, $x \in \mathbb{R}^n$, $b \in \mathbb{R}^m$, and A is an $m \times n$ matrix. We start with a **feasible point**—that is, a point x^0 that satisfies the constraints. The **feasible set** is

$$\mathcal{F} = \left\{x \in \mathbb{R}^n : Ax = b \text{ and } x \geqq 0\right\}$$

Our intention is to move from $x^{(0)}$ to a succession of other points, remaining always in $\mathcal{F}$, and increasing the value of the **objective function**, $c^T x$. It is clear that if we move from $x^{(0)}$ to $x^{(1)}$, the difference $x^{(1)} - x^{(0)}$ must lie in the null space of A. We shall try to find a curve $t \mapsto x(t)$ in the feasible set, starting at x^0 and leading to a solution of the extremal problem. Our requirements are

(i) $x(t) \geqq 0$ for $t \geqq 0$
(ii) $Ax(t) = b$ for $t \geqq 0$
(iii) $c^T x(t)$ is increasing for $t \geqq 0$.

The curve will be defined by an initial-value problem:

$$\begin{cases} x' = f(x) \\ x(0) = x^{(0)} \end{cases} \tag{16}$$

The task facing us is to determine a suitable f. To satisfy condition **(i)**, we shall arrange that whenever a component x_i approaches 0, its velocity $x_i'(t)$ will also approach 0. This can be accomplished by putting

$$D(x) = \begin{bmatrix} x_1 & & & 0 \\ & x_2 & & \\ & & \ddots & \\ 0 & & & x_n \end{bmatrix}$$

and assuming that for some bounded function G,

$$f(x) = D(x)G(x) \tag{17}$$

If this is the case, then from Equations (16) and (17) we have

$$x_i' = x_i G_i(x)$$

and clearly $x_i' \longrightarrow 0$ if $x_i \longrightarrow 0$.

To satisfy requirement **(ii)** it suffices to require $Ax' = 0$. Since $x' = f = DG$, we must require $ADG = 0$. This is most conveniently arranged by letting $G = PH$ where H is any function, and P is the orthogonal projection onto the null space of AD.

Finally, to satisfy property **(iii)**, we should select H so that $c^T x(t)$ is increasing. Thus, we want

$$0 < \frac{d}{dt}\left(c^T x(t)\right) = c^T x' = c^T f(x) = c^T DG = c^T DPH$$

A convenient choice for H is Dc because then we have, with $v = Dc$,

$$\begin{aligned} c^T DPH &= c^T DPDc = v^T Pv = \langle v, Pv \rangle \\ &= \langle v - Pv + Pv, Pv \rangle = \langle Pv, Pv \rangle \geqq 0 \end{aligned}$$

Notice that $v - Pv$ is orthogonal to the range of P, and $\langle v - Pv, Pv \rangle = 0$.

The final version of our initial-value problem is

$$x' = D(x)P(x)D(x)c \qquad x(0) = x^{(0)} \tag{18}$$

The theoretical formula for P is

$$P = I - (AD)^T \left[(AD)(AD)^T\right]^{-1} AD \tag{19}$$

The validity of this depends on $B \equiv AD$ having full rank, so that BB^T will be nonsingular. This will, in turn, require $x_i > 0$ for each component. Thus, the points $x(t)$ should remain in the interior of the set

$$\{x : x \geqq 0\}$$

In particular, $x^{(0)}$ should be so chosen. In practice, Pv is computed not by Equation (19) but by solving the equation $BB^T z = Bv$ and noting that

$$Pv = v - B^T z$$

The initial-value problem (18) need not be solved very accurately. A variation of the Euler method can be used. Recall that the Euler method for Equation (16) advances the solution by

$$x(t + \delta) = x(t) + \delta x'(t) = x(t) + \delta f(x)$$

Using this type of formula, we generate a sequence of vectors $x^{(0)}, x^{(1)}, \ldots$ by the equation

$$x^{(k+1)} = x^{(k)} + \delta_k f(x^{(k)})$$

Although it is tempting to take the value of δ_k as large as possible subject to the requirement $x^{(k+1)} \in \mathcal{F}$, that will lead to a point $x^{(k+1)}$ having at least one 0 component. As mentioned previously, that will introduce other difficulties. What seems to work well in practice is to take δ_k approximately $9/10$ of the maximum possible step. The latter is easily computed; it is the maximum λ for which $x^{(k+1)} \geqq 0$. (The constraint $Ax = b$ is maintained automatically.)

PROBLEMS 3.6

1. Solve the system of equations

$$x - 2y + y^2 + y^3 - 4 = -x - y + 2y^2 - 1 = 0$$

by the homotopy method used in Example 2, starting with the point (0,0). (All the calculations can be performed without recourse to numerical methods.)

2. Consider the homotopy $h(t, x) = tf(x) + (1 - t)g(x)$ in which

$$f(x) = x^2 - 5x + 6 \qquad g(x) = x^2 - 1$$

Show that there is no path connecting a zero of g to a zero of f.

3. Let $y = y(s)$ be a differentiable function from $\mathbb{R}$ to $\mathbb{R}^n$ satisfying the differential equation (9). Assume that $h(y(0)) = 0$. Prove that $h(y(s)) = 0$.

4. If the homotopy method of Example 2 is to be used on the system

$$\sin x + \cos y + e^{xy} = \tan^{-1}(x + y) - xy = 0$$

starting at $(0, 0)$, what system of differential equations will govern the path? A computer program to seek the solution will be instructive.

5. Prove that homotopy is an equivalence relation among the continuous maps from one topological space to another.

6. Are the functions $f(x) = \sin x$ and $g(x) = \cos x$ homotopic?

7. Consider these maps of $[0, 1]$ into $[0, 1] \bigcup [2, 3]$:

$$f(x) = 0 \qquad g(x) = 2$$

Are they homotopic?

CHAPTER FOUR

Solving Systems of Linear Equations

Introduction

In this chapter, we shall construct a general-purpose algorithm for solving the problem $Ax = b$. Then we analyze the errors that are associated with the computer solution and study methods for controlling and reducing them. Finally, we introduce the important topic of iterative algorithms for this problem.

The overall objective of this chapter is to discuss the numerical aspects of solving systems of linear equations having the form

$$\begin{cases} a_{11}x_1 + a_{12}x_2 + a_{13}x_3 + \cdots + a_{1n}x_n = b_1 \\ a_{21}x_1 + a_{22}x_2 + a_{23}x_3 + \cdots + a_{2n}x_n = b_2 \\ a_{31}x_1 + a_{32}x_2 + a_{33}x_3 + \cdots + a_{3n}x_n = b_3 \\ \qquad\qquad\qquad\qquad\qquad\qquad\qquad\qquad\vdots \\ a_{n1}x_1 + a_{n2}x_2 + a_{n3}x_3 + \cdots + a_{nn}x_n = b_n \end{cases} \tag{1}$$

This is a system of n equations in the n unknowns $x_1, x_2, \ldots, x_n$. The elements a_{ij} and b_i are assumed to be prescribed real numbers.

Matrices are useful devices for representing systems of equations. Thus, System (1) can be written as

$$\begin{bmatrix} a_{11} & a_{12} & a_{13} & \cdots & a_{1n} \\ a_{21} & a_{22} & a_{23} & \cdots & a_{2n} \\ a_{31} & a_{32} & a_{33} & \cdots & a_{3n} \\ \vdots & \vdots & \vdots & \ddots & \vdots \\ a_{n1} & a_{n2} & a_{n3} & \cdots & a_{nn} \end{bmatrix} \begin{bmatrix} x_1 \\ x_2 \\ x_3 \\ \vdots \\ x_n \end{bmatrix} = \begin{bmatrix} b_1 \\ b_2 \\ b_3 \\ \vdots \\ b_n \end{bmatrix}$$

Then we can denote these matrices by A, x, and b, so that the equation becomes simply

$$Ax = b$$

4.1 Matrix Algebra

This section serves as a review of the basic concepts of matrix theory. Further material on this topic will be introduced and discussed when needed in the subsequent sections. Since this is review material, most readers can skim over or skip it.

A matrix is a rectangular array of numbers such as

$$\begin{bmatrix} 3.0 & 1.1 & -0.12 \\ 6.2 & 0.0 & 0.15 \\ 0.6 & -4.0 & 1.3 \\ 9.3 & 2.1 & 8.2 \end{bmatrix} \qquad \begin{bmatrix} 3 & 6 & \frac{11}{7} & -17 \end{bmatrix} \qquad \begin{bmatrix} 3.2 \\ -4.7 \\ 0.11 \end{bmatrix}$$

These are, respectively, a 4×3 matrix, a 1×4 matrix, and a 3×1 matrix. In describing the dimensions of a matrix, we give the number of **rows** (horizontal lines) first and the number of **columns** (vertical lines) second. A $1 \times n$ matrix is also called a **row vector**. An $m \times 1$ matrix is called a **column vector** or just a **vector**.

If A is a matrix, the notation a_{ij}, $(A)_{ij}$, or $A(i, j)$ is used to denote the element at the intersection of the ith row and jth column. For example, if A denotes the first of the matrices displayed above, then $A(3, 2) = -4.0$. The **transpose** of a matrix is denoted by A^T and is the matrix defined by $(A^T)_{ij} = a_{ji}$. Using the same example to illustrate the transpose, we have

$$A^T = \begin{bmatrix} 3.0 & 6.2 & 0.6 & 9.3 \\ 1.1 & 0.0 & -4.0 & 2.1 \\ -0.12 & 0.15 & 1.3 & 8.2 \end{bmatrix}$$

If a matrix A has the property $A^T = A$, we say that A is **symmetric**.

If A is a matrix and λ is a scalar (that is, a real number in this context), then λA is defined by $(\lambda A)_{ij} = \lambda a_{ij}$. If $A = (a_{ij})$ and $B = (b_{ij})$ are $m \times n$ matrices, then $A + B$ is defined by $(A + B)_{ij} = a_{ij} + b_{ij}$. Of course, $-A$ means $(-1)A$. If A is an $m \times p$ matrix and B is a $p \times n$ matrix, then AB is an $m \times n$ matrix defined by

$$(AB)_{ij} = \sum_{k=1}^{p} a_{ik} b_{kj} \qquad (1 \leqq i \leqq m,\ 1 \leqq j \leqq n)$$

Here are some examples of the algebraic operations:

$$\begin{bmatrix} 1 & 3 \\ 2 & -1 \\ 4 & -4 \end{bmatrix} + \begin{bmatrix} 6 & 0 \\ 3 & -7 \\ 8 & 2 \end{bmatrix} = \begin{bmatrix} 7 & 3 \\ 5 & -8 \\ 12 & -2 \end{bmatrix}$$

$$3 \begin{bmatrix} 1 & 3 \\ 2 & -1 \\ 4 & -4 \end{bmatrix} = \begin{bmatrix} 3 & 9 \\ 6 & -3 \\ 12 & -12 \end{bmatrix}$$

$$\begin{bmatrix} 2 & 1 & 3 \\ 1 & 5 & -6 \\ 2 & 1 & 5 \\ 0 & 1 & -2 \end{bmatrix} \begin{bmatrix} 1 & 0 \\ -5 & 4 \\ 0 & 3 \end{bmatrix} = \begin{bmatrix} -3 & 13 \\ -24 & 2 \\ -3 & 19 \\ -5 & -2 \end{bmatrix}$$

As we deal with systems of linear equations, a concept of equivalence is important. Let two systems be given, each consisting of n equations with n unknowns:

$$Ax = b \qquad Bx = d$$

If the two systems have precisely the same solutions, we call them **equivalent systems**. Thus, to solve a system of equations, we can instead solve any equivalent system; no solutions are lost and no new ones appear. This simple idea is at the heart of our numerical procedures. Given a system of equations to be solved, we transform it by certain elementary operations into a simpler equivalent system, which we then solve instead.

The **elementary operations** alluded to in the previous paragraph are of the following three types. (Here $\mathcal{E}_i$ denotes the ith equation in the system.)

(i) Interchanging two equations in the system: $\mathcal{E}_i \leftrightarrow \mathcal{E}_j$
(ii) Multiplying an equation by a nonzero number: $\lambda \mathcal{E}_i \rightarrow \mathcal{E}_i$
(iii) Adding to an equation a multiple of some other equation: $\mathcal{E}_i + \lambda \mathcal{E}_j \rightarrow \mathcal{E}_i$

THEOREM 1 *If one system of equations is obtained from another by a finite sequence of elementary operations, then the two systems are equivalent.*

Proof It suffices to consider the effect of a single application of each elementary operation. Suppose that an elementary operation transforms the system $Ax = b$ into the system $Bx = d$. If the operation is of type **(i)**, then the two systems consist of precisely the same equations, though written in a different order. Clearly, if x solves the first system, then it solves the second, and vice versa. If the operation is of type **(ii)**, then suppose that the ith equation has been multiplied by a scalar λ, with $\lambda \neq 0$. The ith and jth equations in $Ax = b$ are

$$a_{i1}x_1 + \cdots + a_{in}x_n = b_i \tag{2}$$

and

$$a_{j1}x_1 + \cdots + a_{jn}x_n = b_j \tag{3}$$

and the ith equation in $Bx = d$ is

$$\lambda a_{i1}x_1 + \cdots + \lambda a_{in}x_n = \lambda b_i \tag{4}$$

Any vector x that satisfies Equation (2) satisfies Equation (4), and vice versa, because $\lambda \neq 0$. Finally, suppose that the operation is of type **(iii)**. Assume that λ times the jth equation has been added to the ith. Then the ith equation in $Bx = d$ is

$$(a_{i1} + \lambda a_{j1})x_1 + \cdots + (a_{in} + \lambda a_{jn})x_n = b_i + \lambda b_j \tag{5}$$

Observe particularly that the jth equation in the system $Bx = d$ has *not* been changed. If $Ax = b$, then Equations (2) and (3) are true. Hence, (5) is true. Thus, $Bx = d$. On the other hand, if we suppose that x solves $Bx = d$, then Equations (5) and (3) are true. If λ times Equation (3) is *subtracted* from Equation (5), the result is Equation (2). Hence, $Ax = b$. ■

Matrix Properties

The $n \times n$ matrix

$$I = \begin{bmatrix} 1 & 0 & 0 & \cdots & 0 \\ 0 & 1 & 0 & \cdots & 0 \\ 0 & 0 & 1 & \cdots & 0 \\ \vdots & \vdots & \vdots & \ddots & \vdots \\ 0 & 0 & 0 & \cdots & 1 \end{bmatrix}$$

is called an **identity matrix**. It has the property that $IA = A = AI$ for any matrix A of size $n \times n$.

If A and B are two matrices such that $AB = I$, then we say that B is a **right inverse** of A and that A is a **left inverse** of B. For example,

$$\begin{bmatrix} 1 & 0 & 0 \\ 0 & 1 & 0 \end{bmatrix} \begin{bmatrix} 1 & 0 \\ 0 & 1 \\ \alpha & \beta \end{bmatrix} = \begin{bmatrix} 1 & 0 \\ 0 & 1 \end{bmatrix} \tag{6}$$

We see from this example that if a matrix has a right inverse, then the latter is not necessarily unique. For square matrices the situation is better, as we now show.

THEOREM 2 *A square matrix can possess at most one right inverse.*

Proof Let $AB = I$, where A, B, and I are all $n \times n$ matrices. Denote by $A^{(j)}$ the jth column of A and by $I^{(k)}$ the kth column of I. The equation $AB = I$ means that

$$\sum_{j=1}^{n} b_{jk} A^{(j)} = I^{(k)} \qquad (1 \leqq k \leqq n) \tag{7}$$

Each column of I is therefore a linear combination of the columns of A. Since the columns of I span $\mathbb{R}^n$, the same is true of the columns of A. Hence, the columns of A form a basis for $\mathbb{R}^n$, and, consequently, the coefficients b_{jk} in Equation (7) are *uniquely* determined. ■

THEOREM 3 *If A and B are square matrices such that $AB = I$, then $BA = I$.*

Proof Let $C = BA - I + B$. Then

$$AC = ABA - AI + AB = A - A + I = I$$

Thus, C (as well as B) is a right inverse of A. By Theorem 2, $B = C$; hence, $BA = I$. ■

It follows from Theorems 2 and 3 that if a square matrix A has a right inverse B, then B is unique and $BA = AB = I$. We then call B the **inverse** of A and say that A is **invertible** or **nonsingular**. (Of course, B is therefore invertible and A is *its* inverse.) We write $B = A^{-1}$ and $A = B^{-1}$. Here is an example:

$$\begin{bmatrix} -2 & 1 \\ \frac{3}{2} & -\frac{1}{2} \end{bmatrix} \begin{bmatrix} 1 & 2 \\ 3 & 4 \end{bmatrix} = \begin{bmatrix} 1 & 2 \\ 3 & 4 \end{bmatrix} \begin{bmatrix} -2 & 1 \\ \frac{3}{2} & -\frac{1}{2} \end{bmatrix} = \begin{bmatrix} 1 & 0 \\ 0 & 1 \end{bmatrix}$$

If A is invertible, then the system of equations $Ax = b$ has the solution $x = A^{-1}b$. If A^{-1} is already available, this equation provides a good method of computing x. If A^{-1} is not available, then in general A^{-1} should *not* be computed solely for the purpose of obtaining x. More efficient procedures will be developed in later sections.

The elementary operations discussed earlier can be carried out with matrix multiplications, as we now indicate. An **elementary matrix** is defined to be an $n \times n$ matrix that arises when an elementary operation is applied to the $n \times n$ identity matrix. The elementary operations, expressed in terms of the rows of a matrix A, are:

(i) The interchange of two rows in A: $A_s \leftrightarrow A_t$
(ii) Multiplying one row by a nonzero constant: $\lambda A_s \longrightarrow A_s$
(iii) Adding to one row a multiple of another: $A_s + \lambda A_t \longrightarrow A_s$

In this description, we have denoted the rows of A by subscripts, A_s, A_t, and so on. Each elementary row operation on A can be accomplished by multiplying A

on the left by an elementary matrix. Here are three examples, illustrating the three types of operations:

$$\begin{bmatrix} 1 & 0 & 0 \\ 0 & 0 & 1 \\ 0 & 1 & 0 \end{bmatrix} \begin{bmatrix} a_{11} & a_{12} & a_{13} \\ a_{21} & a_{22} & a_{23} \\ a_{31} & a_{32} & a_{33} \end{bmatrix} = \begin{bmatrix} a_{11} & a_{12} & a_{13} \\ a_{31} & a_{32} & a_{33} \\ a_{21} & a_{22} & a_{23} \end{bmatrix}$$

$$\begin{bmatrix} 1 & 0 & 0 \\ 0 & \lambda & 0 \\ 0 & 0 & 1 \end{bmatrix} \begin{bmatrix} a_{11} & a_{12} & a_{13} \\ a_{21} & a_{22} & a_{23} \\ a_{31} & a_{32} & a_{33} \end{bmatrix} = \begin{bmatrix} a_{11} & a_{12} & a_{13} \\ \lambda a_{21} & \lambda a_{22} & \lambda a_{23} \\ a_{31} & a_{32} & a_{33} \end{bmatrix}$$

$$\begin{bmatrix} 1 & 0 & 0 \\ 0 & 1 & 0 \\ 0 & \lambda & 1 \end{bmatrix} \begin{bmatrix} a_{11} & a_{12} & a_{13} \\ a_{21} & a_{22} & a_{23} \\ a_{31} & a_{32} & a_{33} \end{bmatrix} = \begin{bmatrix} a_{11} & a_{12} & a_{13} \\ a_{21} & a_{22} & a_{23} \\ \lambda a_{21} + a_{31} & \lambda a_{22} + a_{32} & \lambda a_{23} + a_{33} \end{bmatrix}$$

If we wish to apply a succession of elementary row operations to A, we introduce elementary matrices $E_1, E_2, \ldots, E_m$ and then write the transformed matrix as

$$E_m E_{m-1} \cdots E_2 E_1 A$$

If a matrix is invertible, such a sequence of elementary row operations can be applied to A, reducing it to I. Thus,

$$E_m E_{m-1} \cdots E_2 E_1 A = I$$

From this it follows that $A^{-1} = E_m E_{m-1} \cdots E_2 E_1$. Consequently, A^{-1} can be obtained by subjecting I to the same sequence of elementary row operations. Here is an example showing the computation of an inverse:

$$A = \begin{bmatrix} 1 & 2 & 3 \\ 1 & 3 & 3 \\ 2 & 4 & 7 \end{bmatrix} \qquad \begin{bmatrix} 1 & 0 & 0 \\ 0 & 1 & 0 \\ 0 & 0 & 1 \end{bmatrix} = I$$

$$E_1 A = \begin{bmatrix} 1 & 2 & 3 \\ 0 & 1 & 0 \\ 2 & 4 & 7 \end{bmatrix} \qquad \begin{bmatrix} 1 & 0 & 0 \\ -1 & 1 & 0 \\ 0 & 0 & 1 \end{bmatrix} = E_1 I$$

$$E_2 E_1 A = \begin{bmatrix} 1 & 2 & 3 \\ 0 & 1 & 0 \\ 0 & 0 & 1 \end{bmatrix} \qquad \begin{bmatrix} 1 & 0 & 0 \\ -1 & 1 & 0 \\ -2 & 0 & 1 \end{bmatrix} = E_2 E_1 I$$

$$E_3 E_2 E_1 A = \begin{bmatrix} 1 & 0 & 3 \\ 0 & 1 & 0 \\ 0 & 0 & 1 \end{bmatrix} \qquad \begin{bmatrix} 3 & -2 & 0 \\ -1 & 1 & 0 \\ -2 & 0 & 1 \end{bmatrix} = E_3 E_2 E_1 I$$

$$E_4 E_3 E_2 E_1 A = \begin{bmatrix} 1 & 0 & 0 \\ 0 & 1 & 0 \\ 0 & 0 & 1 \end{bmatrix} \qquad \begin{bmatrix} 9 & -2 & -3 \\ -1 & 1 & 0 \\ -2 & 0 & 1 \end{bmatrix} = E_4 E_3 E_2 E_1 I = A^{-1}$$

Here are the elementary matrices:

$$E_1 = \begin{bmatrix} 1 & 0 & 0 \\ -1 & 1 & 0 \\ 0 & 0 & 1 \end{bmatrix} \qquad E_3 = \begin{bmatrix} 1 & -2 & 0 \\ 0 & 1 & 0 \\ 0 & 0 & 1 \end{bmatrix}$$

$$E_2 = \begin{bmatrix} 1 & 0 & 0 \\ 0 & 1 & 0 \\ -2 & 0 & 1 \end{bmatrix} \qquad E_4 = \begin{bmatrix} 1 & 0 & -3 \\ 0 & 1 & 0 \\ 0 & 0 & 1 \end{bmatrix}$$

THEOREM 4 *For an $n \times n$ matrix A, the following properties are equivalent:*

(i) *The inverse of A exists; that is, A is nonsingular.*
(ii) *The determinant of A is nonzero.*
(iii) *The rows of A form a basis for $\mathbb{R}^n$.*
(iv) *The columns of A form a basis for $\mathbb{R}^n$.*
(v) *As a map from $\mathbb{R}^n$ to $\mathbb{R}^n$, A is injective (one to one).*
(vi) *As a map from $\mathbb{R}^n$ to $\mathbb{R}^n$, A is surjective (onto).*
(vii) *The equation $Ax = 0$ implies $x = 0$.*
(viii) *For each $b \in \mathbb{R}^n$, there is exactly one $x \in \mathbb{R}^n$ such that $Ax = b$.*
(ix) *A is a product of elementary matrices.*
(x) *0 is not an eigenvalue of A.*

An important fundamental concept is the **positive definiteness** of a matrix. A matrix A is **positive definite** if $x^T Ax > 0$ for every nonzero vector x. For example, the matrix

$$A = \begin{bmatrix} 2 & 1 \\ 1 & 2 \end{bmatrix}$$

is positive definite since

$$x^T Ax = \begin{bmatrix} x_1 & x_2 \end{bmatrix} \begin{bmatrix} 2 & 1 \\ 1 & 2 \end{bmatrix} \begin{bmatrix} x_1 \\ x_2 \end{bmatrix} = (x_1 + x_2)^2 + x_1^2 + x_2^2 > 0$$

for all x_1 and x_2 except $x_1 = x_2 = 0$. Here $x^T Ax$ is called a **quadratic form**. It follows from Problems 17–19 that when dealing with positive definiteness, we can assume symmetry. Establishing the positive definiteness of a matrix using the definition is usually not an easy task since it involves an *arbitrary* $x \neq 0$. If A is positive definite and symmetric, then its eigenvalues are real and positive.

Partitioned Matrices

It is frequently convenient to partition matrices into submatrices and compute products as if the submatrices were numbers. An example of this procedure follows:

$$\begin{bmatrix} \begin{bmatrix} 1 & 2 \end{bmatrix} & \begin{bmatrix} 1 & -1 & 0 & 1 \end{bmatrix} \\ \begin{bmatrix} -1 & 1 \\ 0 & 1 \\ 1 & -1 \\ 1 & 0 \end{bmatrix} & \begin{bmatrix} 1 & 0 & -1 & 1 \\ -1 & 1 & 0 & 1 \\ 0 & 0 & 1 & 0 \\ 1 & 2 & 1 & 0 \end{bmatrix} \end{bmatrix} \begin{bmatrix} \begin{bmatrix} 1 & 0 & 1 \\ -1 & 1 & 2 \end{bmatrix} & \begin{bmatrix} 2 & 1 \\ 0 & 1 \end{bmatrix} \\ \begin{bmatrix} 1 & 0 & 1 \\ -1 & 1 & 0 \\ 2 & 1 & 0 \\ 0 & 1 & 1 \end{bmatrix} & \begin{bmatrix} 1 & 2 \\ 0 & 1 \\ -2 & 1 \\ -1 & 1 \end{bmatrix} \end{bmatrix}$$

$$= \begin{bmatrix} \begin{bmatrix} 1 & 2 & 7 \end{bmatrix} & \begin{bmatrix} 2 & 5 \end{bmatrix} \\ \begin{bmatrix} -3 & 1 & 3 \\ -3 & 3 & 2 \\ 4 & 0 & -1 \\ 2 & 3 & 2 \end{bmatrix} & \begin{bmatrix} 0 & 2 \\ -2 & 1 \\ 0 & 1 \\ 1 & 6 \end{bmatrix} \end{bmatrix}$$

If these matrices are partitioned as shown, and if the submatrices are denoted by single letters, we have a product of the form

$$\begin{bmatrix} A_{11} & A_{12} \\ A_{21} & A_{22} \end{bmatrix} \begin{bmatrix} B_{11} & B_{12} \\ B_{21} & B_{22} \end{bmatrix} = \begin{bmatrix} C_{11} & C_{12} \\ C_{21} & C_{22} \end{bmatrix}$$

One can verify that $C_{ij} = \sum_{s=1}^{2} A_{is}B_{sj}$. Thus, for example,

$$\begin{bmatrix} 1 & 2 \end{bmatrix} \begin{bmatrix} 1 & 0 & 1 \\ -1 & 1 & 2 \end{bmatrix} + \begin{bmatrix} 1 & -1 & 0 & 1 \end{bmatrix} \begin{bmatrix} 1 & 0 & 1 \\ -1 & 1 & 0 \\ 2 & 1 & 0 \\ 0 & 1 & 1 \end{bmatrix} = \begin{bmatrix} 1 & 2 & 7 \end{bmatrix}$$

and, as shown by the computation, $C_{11} = A_{11}B_{11} + A_{12}B_{21}$.

To establish a general result concerning this technique, let A, B, and C be matrices that have been partitioned into submatrices:

$$A = \begin{bmatrix} A_{11} & A_{12} & \cdots & A_{1n} \\ A_{21} & A_{22} & \cdots & A_{2n} \\ \vdots & \vdots & \ddots & \vdots \\ A_{m1} & A_{m2} & \cdots & A_{mn} \end{bmatrix} \qquad B = \begin{bmatrix} B_{11} & B_{12} & \cdots & B_{1k} \\ B_{21} & B_{22} & \cdots & B_{2k} \\ \vdots & \vdots & \ddots & \vdots \\ B_{n1} & B_{n2} & \cdots & B_{nk} \end{bmatrix}$$

$$C = \begin{bmatrix} C_{11} & C_{12} & \cdots & C_{1k} \\ C_{21} & C_{22} & \cdots & C_{2k} \\ \vdots & \vdots & \ddots & \vdots \\ C_{m1} & C_{m2} & \cdots & C_{mk} \end{bmatrix}$$

THEOREM 5 *If each product $A_{is}B_{sj}$ can be formed and if $C_{ij} = \sum_{s=1}^{n} A_{is}B_{sj}$, then $C = AB$.*

Proof Let the dimensions of A_{ij} be $m_i \times n_j$. Let the dimensions of B_{ij} be $\widehat{m}_i \times \widehat{n}_j$. Since $A_{is}B_{sj}$ exists, we must have $n_s = \widehat{m}_s$ for all s. Then C_{ij} will have dimensions $m_i \times \widehat{n}_j$. Now select an arbitrary element c_{ij} in the matrix C. Suppose that c_{ij} lies in the block

C_{rs} and is in the pth row and qth column of C_{rs}. Then we must have

$$i = m_1 + m_2 + \cdots + m_{r-1} + p \tag{8}$$

$$j = \widehat{n}_1 + \widehat{n}_2 + \cdots + \widehat{n}_{s-1} + q \tag{9}$$

Consequently,

$$\begin{aligned} c_{ij} = (C_{rs})_{pq} &= \left(\sum_{t=1}^{n} A_{rt}B_{ts} \right)_{pq} = \sum_{t=1}^{n} (A_{rt}B_{ts})_{pq} \\ &= \sum_{t=1}^{n} \sum_{\alpha=1}^{n_t} (A_{rt})_{p\alpha}(B_{ts})_{\alpha q} \end{aligned}$$

The elements $(A_{rt})_{p\alpha}$ lie in row i of A by Equation (8). These elements fill out the entire row i of A since $1 \leqq t \leqq n$ and $1 \leqq \alpha \leqq n_t$. Similar reasoning shows that the elements $(B_{ts})_{\alpha q}$ lie in column j of B because of Equation (9). Also the entire column j of B is present and appears in its natural order. Hence,

$$c_{ij} = \sum_{\beta=1}^{n} (A)_{i\beta}(B)_{\beta j} = (AB)_{ij}$$ ■

PROBLEMS 4.1

1. Show that an elementary operation of the first type can be accomplished by four operations of the other types, and that we can restrict λ to be ± 1.
2. Is Theorem 1 true for systems in which the number of equations differs from the number of unknowns?
3. Prove that each of the elementary operations can be undone by an operation of the same type.
4. Consider the system of linear equations $Ax = b$, where A is an $m \times n$ matrix, x is $n \times 1$, and b is $m \times 1$. Denote the column vectors of A by $A_1, A_2, \ldots, A_n$. Prove that the system has a solution if and only if b is in the linear span of $\{A_1, A_2, \ldots, A_n\}$. Prove that if $\{A_1, A_2, \ldots, A_n\}$ is linearly independent, then the system has at most one solution.
5. Let $E(p, q, \lambda)$ be the matrix that results from the $n \times n$ identity matrix when λ times row q is added to row p. (Assume that $p \neq q$.) Prove that the relationship $E(p, q, \lambda)^{-1} = E(p, q, -\lambda)$ holds. Prove that for any $m \times n$ matrix A, the product $AE(p, q, \lambda)$ can be computed by adding λ times column p to column q in A.
6. A **monomial matrix** is a square matrix in which each row and column contains exactly one nonzero entry. Prove that a monomial matrix is nonsingular.
7. Let A have the block form

$$A = \begin{bmatrix} B & C \\ 0 & I \end{bmatrix}$$

in which the blocks are $n \times n$. Prove that if $B - I$ is nonsingular, then, for $k \geqq 1$,

$$A^k = \begin{bmatrix} B^k & (B^k - I)(B - I)^{-1}C \\ 0 & I \end{bmatrix}$$

8. Carry out the multiplication of the following two matrices—first using the block product method and then using the ordinary product.

$$\begin{bmatrix} \begin{bmatrix} 1 & 2 \end{bmatrix} & \begin{bmatrix} 1 & -1 & 0 & 1 \end{bmatrix} \\ \begin{bmatrix} -1 & 1 \\ 0 & 1 \\ 1 & -1 \\ 1 & 0 \end{bmatrix} & \begin{bmatrix} 1 & 0 & -1 & 1 \\ -1 & 1 & 0 & 1 \\ 0 & 0 & 1 & 0 \\ 1 & 2 & 1 & 0 \end{bmatrix} \end{bmatrix} \begin{bmatrix} \begin{bmatrix} 1 & 0 & 1 \\ -1 & 1 & 2 \end{bmatrix} & \begin{bmatrix} 2 & 1 \\ 0 & 1 \end{bmatrix} \\ \begin{bmatrix} 1 & 0 & 1 \\ -1 & 1 & 0 \\ 2 & 1 & 0 \\ 0 & 1 & 1 \end{bmatrix} & \begin{bmatrix} 1 & 2 \\ 0 & 1 \\ -2 & 1 \\ -1 & 1 \end{bmatrix} \end{bmatrix}$$

9. Refer to Problem 7 and find the block structure of A^k when

$$A = \begin{bmatrix} B & 0 \\ C & I \end{bmatrix}$$

Prove your result by mathematical induction.

10. Prove that the set of upper triangular $n \times n$ matrices is a subalgebra of the algebra of all $n \times n$ matrices. In other words, prove that the set is algebraically closed under the operations of addition, multiplication, and multiplication by a scalar.

11. Prove that the inverse of a nonsingular upper triangular matrix is also upper triangular.

12. Let A be an $n \times n$ invertible matrix, and let u and v be two vectors in $\mathbb{R}^n$. Find the necessary and sufficient conditions on u and v in order that the matrix

$$\begin{bmatrix} A & u \\ v^T & 0 \end{bmatrix}$$

be invertible, and give a formula for the inverse when it exists.

13. Let D be a matrix in partitioned form:

$$D = \begin{bmatrix} A & B \\ C & I \end{bmatrix}$$

Prove that if $A - BC$ is nonsingular, then so is D.

14. (Continuation) Prove the stronger result that the dimension of the null space of D is no greater than the dimension of the null space of $A - BC$.

15. Are these matrices positive definite?

(a) $\begin{bmatrix} 1 & -1 \\ 1 & 1 \end{bmatrix}$ **(b)** $\begin{bmatrix} 4 & 2 & 1 \\ 2 & 5 & 2 \\ 1 & 2 & 4 \end{bmatrix}$

16. For what values of a is this matrix positive definite?

$$A = \begin{bmatrix} 1 & a & a \\ a & 1 & a \\ a & a & 1 \end{bmatrix}$$

17. A square matrix A is said to be skew-symmetric if $A^T = -A$. Prove that if A is skew-symmetric, then $x^T A x = 0$ for all x.

18. (Continuation) Prove that the diagonal elements of a skew-symmetric matrix are 0. Also, prove that the determinant is 0 when the matrix is of odd order.

19. (Continuation) Let A be any square matrix, and define $A_0 = \frac{1}{2}(A + A^T)$ and $A_1 = \frac{1}{2}(A - A^T)$. Prove that A_0 is symmetric, that A_1 is skew-symmetric, that $A = A_0 + A_1$, and that for all x, $x^T A x = x^T A_0 x$. This explains why, in discussing quadratic forms, we can confine our attention to symmetric matrices.

20. Give an example of a symmetric matrix A containing all positive elements such that $x^T A x$ is sometimes negative.

21. Can a matrix have a right inverse and a left inverse that are not equal?

COMPUTER PROBLEMS 4.1

1. (Programming project) For readers interested in numerical experimentation, we suggest a programming project that can be carried out in several stages. The project involves writing a number of subroutines to do basic tasks in linear algebra. This set of subroutines will provide a personal software package for solving linear equations, factoring matrices in various ways, and computing eigenvalues and pseudoinverses. The first part of the project is described in Computer Problem 2 in Section 4.2, Computer Problems 1 and 3–7 in Section 4.3, and Computer Problem 1 in Section 4.4.

2. Write and test subroutines or procedures for the following:

(a) `Store` (n, x, y), which replaces the n-vector y by the n-vector x.

(b) `Prod` (m, n, A, x, y), which multiplies the n-vector x by the $m \times n$ matrix A and stores the result in the m-vector y.

(c) `Mult` (k, m, n, A, B, C), which computes $C = AB$, where A is $k \times m$, B is $m \times n$, and C is $k \times n$.

(d) `Dot` (n, x, y, a), which computes (in double-precision arithmetic) the dot product $\sum_{i=1}^{n} x_i y_i$ and stores the answer as a single-precision real number, a. *Note:* x_i, y_i, and a are single-precision numbers.

4.2 The *LU* and Cholesky Factorizations

Let us consider a system of n linear equations in n unknowns $x_1, x_2, \ldots, x_n$. It can be written in the form

$$\begin{bmatrix} a_{11} & a_{12} & a_{13} & \cdots & a_{1n} \\ a_{21} & a_{22} & a_{23} & \cdots & a_{2n} \\ a_{31} & a_{32} & a_{33} & \cdots & a_{3n} \\ \vdots & \vdots & \vdots & \ddots & \vdots \\ a_{n1} & a_{n2} & a_{n3} & \cdots & a_{nn} \end{bmatrix} \begin{bmatrix} x_1 \\ x_2 \\ x_3 \\ \vdots \\ x_n \end{bmatrix} = \begin{bmatrix} b_1 \\ b_2 \\ b_3 \\ \vdots \\ b_n \end{bmatrix}$$

The matrices in this equation are denoted by A, x, and b. Thus, our system is simply

$$Ax = b \tag{1}$$

Easy-to-Solve Systems

We begin by looking for special types of systems that can be *easily* solved. For example, suppose that the $n \times n$ matrix A has a **diagonal structure**. This means that all the nonzero elements of A are on the main diagonal, and System (1) is

$$\begin{bmatrix} a_{11} & 0 & 0 & \cdots & 0 \\ 0 & a_{22} & 0 & \cdots & 0 \\ 0 & 0 & a_{33} & \cdots & 0 \\ \vdots & \vdots & \vdots & \ddots & \vdots \\ 0 & 0 & 0 & \cdots & a_{nn} \end{bmatrix} \begin{bmatrix} x_1 \\ x_2 \\ x_3 \\ \vdots \\ x_n \end{bmatrix} = \begin{bmatrix} b_1 \\ b_2 \\ b_3 \\ \vdots \\ b_n \end{bmatrix}$$

In this case, our system collapses to n simple equations, and the solution is

$$x = \begin{bmatrix} b_1/a_{11} \\ b_2/a_{22} \\ b_3/a_{33} \\ \vdots \\ b_n/a_{nn} \end{bmatrix}$$

If $a_{ii} = 0$ for some index i, and if $b_i = 0$ also, then x_i can be *any* real number. If $a_{ii} = 0$ and $b_i \neq 0$, no solution of the system exists.

Continuing our search for *easy* solutions of System (1), we assume a **lower triangular structure** for A. This means that all the nonzero elements of A are situated on or below the main diagonal, and System (1) is

$$\begin{bmatrix} a_{11} & 0 & 0 & \cdots & 0 \\ a_{21} & a_{22} & 0 & \cdots & 0 \\ a_{31} & a_{32} & a_{33} & \cdots & 0 \\ \vdots & \vdots & \vdots & \ddots & \vdots \\ a_{n1} & a_{n2} & a_{n3} & \cdots & a_{nn} \end{bmatrix} \begin{bmatrix} x_1 \\ x_2 \\ x_3 \\ \vdots \\ x_n \end{bmatrix} = \begin{bmatrix} b_1 \\ b_2 \\ b_3 \\ \vdots \\ b_n \end{bmatrix}$$

To solve this, assume that $a_{ii} \neq 0$ for all i; then obtain x_1 from the first equation. With the known value of x_1 substituted in the second equation, solve the second equation for x_2. We proceed in the same way, obtaining $x_1, x_2, \ldots, x_n$, one at a time *and in this order*. A formal algorithm for the solution in this case is called **forward substitution**:

$$
\begin{array}{l}
\textbf{input } n, (a_{ij}), (b_i) \\
\textbf{for } i = 1 \textbf{ to } n \textbf{ do} \\
\qquad x_i \leftarrow \left(b_i - \sum_{j=1}^{i-1} a_{ij}x_j\right) \Big/ a_{ii} \\
\textbf{end do} \\
\textbf{output } (x_i)
\end{array}
\tag{2}
$$

As is customary, any sum of the type $\sum_{i=\alpha}^{\beta} x_i$ in which $\beta < \alpha$ is interpreted to be 0.

The same ideas can be exploited to solve a system having an **upper triangular structure**. Such a matrix system has the form

$$
\begin{bmatrix}
a_{11} & a_{12} & a_{13} & \cdots & a_{1n} \\
0 & a_{22} & a_{23} & \cdots & a_{2n} \\
0 & 0 & a_{33} & \cdots & a_{3n} \\
\vdots & \vdots & \vdots & \ddots & \vdots \\
0 & 0 & 0 & \cdots & a_{nn}
\end{bmatrix}
\begin{bmatrix} x_1 \\ x_2 \\ x_3 \\ \vdots \\ x_n \end{bmatrix}
=
\begin{bmatrix} b_1 \\ b_2 \\ b_3 \\ \vdots \\ b_n \end{bmatrix}
$$

Again, it must be assumed that $a_{ii} \neq 0$ for $1 \leqq i \leqq n$. The formal algorithm to solve for x is as follows and is called **back substitution**:

$$
\begin{array}{l}
\textbf{input } n, (a_{ij}), (b_i) \\
\textbf{for } i = n \textbf{ to } 1 \textbf{ step } -1 \textbf{ do} \\
\qquad x_i \leftarrow \left(b_i - \sum_{j=i+1}^{n} a_{ij}x_j\right) \Big/ a_{ii} \\
\textbf{end do} \\
\textbf{output } (x_i)
\end{array}
\tag{3}
$$

There is still another simple type of system that can be easily solved using these same ideas—namely, a system obtained by permuting the equations in a triangular system. To illustrate, consider the system

$$
\begin{bmatrix}
a_{11} & a_{12} & 0 \\
a_{21} & a_{22} & a_{23} \\
a_{31} & 0 & 0
\end{bmatrix}
\begin{bmatrix} x_1 \\ x_2 \\ x_3 \end{bmatrix}
=
\begin{bmatrix} b_1 \\ b_2 \\ b_3 \end{bmatrix}
\tag{4}
$$

If we simply reorder these equations, we can get a lower triangular system:

$$
\begin{bmatrix}
a_{31} & 0 & 0 \\
a_{11} & a_{12} & 0 \\
a_{21} & a_{22} & a_{23}
\end{bmatrix}
\begin{bmatrix} x_1 \\ x_2 \\ x_3 \end{bmatrix}
=
\begin{bmatrix} b_3 \\ b_1 \\ b_2 \end{bmatrix}
\tag{5}
$$

This can be solved in the same manner as indicated previously. Putting this another way, we should solve the equations of System (4) *not* in the usual order 1, 2, 3 but rather in the order 3, 1, 2.

Let us try to describe in a formal way the matrix property just considered. We wish to say that one row of A, say row p_1, has zeros in positions $2, 3, \dots, n$. Then another row, say row p_2, has zeros in positions $3, 4, \dots, n$, and so on. If this is the case, we shall use row p_1 to obtain x_1, row p_2 to obtain x_2, and so on. If we were to reorder the rows as $p_1, p_2, \dots, p_n$, the resulting matrix would be lower triangular.

How do we solve $Ax = b$ if A is a permuted lower or upper triangular matrix of the type just considered? Let's assume that the **permutation vector** $(p_1, p_2, \dots, p_n)$ is known or has been determined somehow beforehand. Modifying our previous algorithms, we arrive at **forward substitution** for a **permuted lower triangular system**:

(6)

input $n, (a_{ij}), (b_i), (p_i)$
for $i = 1$ **to** n **do**
  $x_i \leftarrow \left(b_{p_i} - \sum_{j=1}^{i-1} a_{p_i j} x_j\right) \big/ a_{p_i i}$
end do
output (x_i)

Of course, this works only if A has the property $a_{p_i j} = 0$ for $j > i$, and $a_{p_i i} \neq 0$ for all i. Similarly, **back substitution** for a **permuted upper triangular system** is as follows:

(7)

input $n, (a_{ij}), (b_i), (p_i)$
for $i = n$ **to** 1 **step** -1 **do**
  $x_i \leftarrow \left(b_{p_i} - \sum_{j=i+1}^{n} a_{p_i j} x_j\right) \big/ a_{p_i i}$
end do
output (x_i)

This works if $a_{p_i j} = 0$ for $j < i$, and $a_{p_i i} \neq 0$ for all i.

The algorithms (2), (3), (6), and (7) have not been given with sufficient detail for direct translation into most programming languages. For example, algorithm (7) might require this more elaborate set of instructions:

input $n, (a_{ij}), (b_i), (p_i)$
for $i = n$ **to** 1 **step** -1 **do**
  $s \leftarrow b_{p_i}$
  for $j = i + 1$ **to** n **do**
    $s \leftarrow s - a_{p_i j} x_j$
  end do
  $x_i \leftarrow s / a_{p_i i}$
end do
output (x_i)

LU-Factorizations

Suppose that A can be factored into the product of a lower triangular matrix L and an upper triangular matrix U: $A = LU$. Then in order to solve the system of equations $Ax = b$, it is enough to solve this problem in two stages:

$$\begin{aligned} Lz &= b \qquad \text{solve for } z \\ Ux &= z \qquad \text{solve for } x \end{aligned}$$

Our previous analysis indicates that solving these two *triangular* systems is simple.

We shall show how the factorization $A = LU$ can be carried out, provided that in certain steps of the computation 0 divisors are not encountered. Not every matrix has such a factorization, and this difficulty will be investigated presently.

We begin with an $n \times n$ matrix A and search for matrices

$$L = \begin{bmatrix} \ell_{11} & 0 & 0 & \cdots & 0 \\ \ell_{21} & \ell_{22} & 0 & \cdots & 0 \\ \ell_{31} & \ell_{32} & \ell_{33} & \cdots & 0 \\ \vdots & \vdots & \vdots & \ddots & \vdots \\ \ell_{n1} & \ell_{n2} & \ell_{n3} & \cdots & \ell_{nn} \end{bmatrix}$$

$$U = \begin{bmatrix} u_{11} & u_{12} & u_{13} & \cdots & u_{1n} \\ 0 & u_{22} & u_{23} & \cdots & u_{2n} \\ 0 & 0 & u_{33} & \cdots & u_{3n} \\ \vdots & \vdots & \vdots & \ddots & \vdots \\ 0 & 0 & 0 & \cdots & u_{nn} \end{bmatrix}$$

such that

$$A = LU \tag{8}$$

When this is possible, we say that A has an ***LU*-decomposition**. It turns out that L and U are *not* uniquely determined by Equation (8). In fact, for each i, we can *assign* a nonzero value to either ℓ_{ii} or u_{ii} (but not both). For example, one simple choice is to set $\ell_{ii} = 1$ for $i = 1, 2, \ldots, n$, thus making L **unit lower triangular**. Another obvious choice is to make U **unit upper triangular** ($u_{ii} = 1$ for each i). These special cases are of particular importance.

To derive an algorithm for the LU-factorization of A, we start with the formula for matrix multiplication:

$$a_{ij} = \sum_{s=1}^{n} \ell_{is} u_{sj} = \sum_{s=1}^{\min(i,j)} \ell_{is} u_{sj} \tag{9}$$

Here we have used the fact that $\ell_{is} = 0$ for $s > i$ and $u_{sj} = 0$ for $s > j$.

Each step in this process determines one new row of U and one new column in L. At step k, we can assume that rows $1, 2, \ldots, k-1$ have been computed in U and that columns $1, 2, \ldots, k-1$ have been computed in L. (If $k = 1$, this assumption is true vacuously.) Putting $i = j = k$ in Equation (9), we obtain

$$a_{kk} = \sum_{s=1}^{k-1} \ell_{ks} u_{sk} + \ell_{kk} u_{kk} \tag{10}$$

If u_{kk} or ℓ_{kk} has been specified, we use Equation (10) to determine the other. With ℓ_{kk} and u_{kk} now known, we use Equation (9) to write for the kth row ($i = k$) and the kth column ($j = k$), respectively,

$$a_{kj} = \sum_{s=1}^{k-1} \ell_{ks} u_{sj} + \ell_{kk} u_{kj} \qquad (k+1 \leqq j \leqq n) \tag{11}$$

$$a_{ik} = \sum_{s=1}^{k-1} \ell_{is} u_{sk} + \ell_{ik} u_{kk} \qquad (k+1 \leqq i \leqq n) \tag{12}$$

If $\ell_{kk} \neq 0$, Equation (11) can be used to obtain the elements u_{kj}. Similarly, if $u_{kk} \neq 0$, Equation (12) can be used to obtain the elements ℓ_{ik}. It is interesting to note that these two computations can theoretically be carried out in parallel (that is, simultaneously). On some computers this can actually be done, with a considerable savings in execution time. For details, see Kincaid and Oppe [1988]. The computation of the kth row in U and the kth column in L as described completes the kth step in the algorithm. The calculations call for division by ℓ_{kk} and u_{kk}; hence, if these divisors are 0, the computations can usually not be completed. However in some cases, they *can* be completed. (See Problem 43.)

The algorithm based on the preceding analysis is known as **Doolittle's factorization** when L is unit lower triangular ($\ell_{ii} = 1$ for $1 \leqq i \leqq n$) and as **Crout's factorization** when U is unit upper triangular ($u_{ii} = 1$ for $1 \leqq i \leqq n$). When $U = L^T$ so that $\ell_{ii} = u_{ii}$ for $1 \leqq i \leqq n$, the algorithm is called **Cholesky's factorization**. We shall discuss the Cholesky method in more detail later in this section since this factoring requires the matrix A to have several special properties; namely, A should be real, symmetric, and positive definite.

Which of these factorizations is better? Each relates to a different variation of the basic Gaussian elimination procedure. One needs to know about them in order to have a complete understanding of this subject.

The algorithm for the **general LU-factorization** is as follows:

input $n, (a_{ij})$
for $k = 1$ **to** n **do**
  Specify a nonzero value for either
    ℓ_{kk} or u_{kk} and compute the other from

$$\ell_{kk} u_{kk} = a_{kk} - \sum_{s=1}^{k-1} \ell_{ks} u_{sk}$$

for $j = k+1$ **to** n **do**

$$u_{kj} \leftarrow \left(a_{kj} - \sum_{s=1}^{k-1} \ell_{ks} u_{sj} \right) \Big/ \ell_{kk}$$

end do
for $i = k+1$ **to** n **do**

$$\ell_{ik} \leftarrow \left(a_{ik} - \sum_{s=1}^{k-1} \ell_{is} u_{sk} \right) \Big/ u_{kk}$$

end do
end do
output $(\ell_{ij}), (u_{ij})$

Notice the possible parallelism (simultaneity) in computing the kth row of U and the kth column of L. (See Problem 56.)

Algorithms such as the LU and Cholesky factorizations are extensively modified in order to obtain versions of them that are suitable for use on high-performance computers with advanced architectures. Various techniques are used in the redesign process to develop scalable algorithms that exploit the unique features of multiple-instruction multiple-data (MIMD) distributed memory concurrent computers. For example, block-partitioned algorithms are used to reduce the frequency of the movement of data between different levels in the memory hierarchy. Also, special distributed versions of the basic linear algebra subprograms (BLAS) are used as computational and communicational building blocks. Software libraries such as LAPACK and ScaLAPACK have been written that utilize these design features. For a discussion of these issues, see LAPACK Users' Guide by Anderson et al. [1995] and Dongarra and Walker [1995].

In the factorization $A = LU$ with L lower triangular and U upper triangular, n^2 equations are obtained from Equation (9) for the $n^2 + n$ unknown elements of L and U. Obviously, n of them have to be specified. An even more general factorization would allow any n elements from L and U to be specified and solve the resulting system of equations. Unfortunately, this system may be **nonlinear** in ℓ_{ij} and u_{ij}. (See Problem 50.) In the preceding algorithm, we must designate either ℓ_{kk} or u_{kk} and then compute all elements in the kth column of L and in the kth row of U before moving on to the next column and row. Moreover, the order $k = 1, 2, \ldots, n$ is important in these calculations.

The pseudocode for carrying out **Doolittle's factorization** is as follows:

input $n, (a_{ij})$
for $k = 1$ **to** n **do**
$\ell_{kk} \leftarrow 1$
for $j = k$ **to** n **do**

$$u_{kj} \leftarrow a_{kj} - \sum_{s=1}^{k-1} \ell_{ks} u_{sj}$$

end do

for $i = k+1$ **to** n **do**

$$\ell_{ik} \leftarrow \left(a_{ik} - \sum_{s=1}^{k-1} \ell_{is} u_{sk} \right) \Big/ u_{kk}$$

end do

end do

output $(\ell_{ij}), (u_{ij})$

Example 1 Find the Doolittle, Crout, and Cholesky factorizations of the matrix

$$A = \begin{bmatrix} 60 & 30 & 20 \\ 30 & 20 & 15 \\ 20 & 15 & 12 \end{bmatrix}$$

Solution The Doolittle factorization from the algorithm is

$$A = \begin{bmatrix} 1 & 0 & 0 \\ \frac{1}{2} & 1 & 0 \\ \frac{1}{3} & 1 & 1 \end{bmatrix} \begin{bmatrix} 60 & 30 & 20 \\ 0 & 5 & 5 \\ 0 & 0 & \frac{1}{3} \end{bmatrix} \equiv LU$$

Rather than computing the next two factorizations directly, we can obtain them from the Doolittle factorization above. By putting the diagonal elements of U into a diagonal matrix D, we can write

$$A = \begin{bmatrix} 1 & 0 & 0 \\ \frac{1}{2} & 1 & 0 \\ \frac{1}{3} & 1 & 1 \end{bmatrix} \begin{bmatrix} 60 & 0 & 0 \\ 0 & 5 & 0 \\ 0 & 0 & \frac{1}{3} \end{bmatrix} \begin{bmatrix} 1 & \frac{1}{2} & \frac{1}{3} \\ 0 & 1 & 1 \\ 0 & 0 & 1 \end{bmatrix} \equiv LD\widehat{U}$$

By putting $\widehat{L} = LD$, we obtain the Crout factorization

$$A = \begin{bmatrix} 60 & 0 & 0 \\ 30 & 5 & 0 \\ 20 & 5 & \frac{1}{3} \end{bmatrix} \begin{bmatrix} 1 & \frac{1}{2} & \frac{1}{3} \\ 0 & 1 & 1 \\ 0 & 0 & 1 \end{bmatrix} \equiv \widehat{L}\widehat{U}$$

The Cholesky factorization is obtained by splitting D into the form $D^{1/2}D^{1/2}$ in the $LD\widehat{U}$-factorization and associating one factor with L and the other with $\widehat{U}$. Thus,

$$A = \begin{bmatrix} 1 & 0 & 0 \\ \frac{1}{2} & 1 & 0 \\ \frac{1}{3} & 1 & 1 \end{bmatrix} \begin{bmatrix} \sqrt{60} & 0 & 0 \\ 0 & \sqrt{5} & 0 \\ 0 & 0 & \frac{1}{3}\sqrt{3} \end{bmatrix} \begin{bmatrix} \sqrt{60} & 0 & 0 \\ 0 & \sqrt{5} & 0 \\ 0 & 0 & \frac{1}{3}\sqrt{3} \end{bmatrix} \begin{bmatrix} 1 & \frac{1}{2} & \frac{1}{3} \\ 0 & 1 & 1 \\ 0 & 0 & 1 \end{bmatrix}$$

$$= \begin{bmatrix} \sqrt{60} & 0 & 0 \\ \frac{1}{2}\sqrt{60} & \sqrt{5} & 0 \\ \frac{1}{3}\sqrt{60} & \sqrt{5} & \frac{1}{3}\sqrt{3} \end{bmatrix} \begin{bmatrix} \sqrt{60} & \frac{1}{2}\sqrt{60} & \frac{1}{3}\sqrt{60} \\ 0 & \sqrt{5} & \sqrt{5} \\ 0 & 0 & \frac{1}{3}\sqrt{3} \end{bmatrix} \equiv \widetilde{L}\widetilde{L}^T$$

■

The factorization of primary interest is $A = LU$, where L is *unit* lower triangular and U is upper triangular. Henceforth, when we refer to an LU-factorization, we mean one in which L is *unit* lower triangular. Here is a sufficient condition for a square matrix A to have an LU-decomposition.

THEOREM 1 *If all n leading principal minors of the $n \times n$ matrix A are nonsingular, then A has an LU-decomposition.*

Proof Recall that the kth leading **principal minor** of the matrix A is the matrix

$$A_k = \begin{bmatrix} a_{11} & a_{12} & \cdots & a_{1k} \\ a_{21} & a_{22} & \cdots & a_{2k} \\ \vdots & \vdots & \ddots & \vdots \\ a_{k1} & a_{k2} & \cdots & a_{kk} \end{bmatrix}$$

Let A_k, L_k, and U_k denote the kth leading principal minors in the matrices A, L, and U, respectively. Our hypothesis is that $A_1, A_2, \ldots, A_n$ are nonsingular. For the purposes of an inductive proof, suppose that L_{k-1} and U_{k-1} have been obtained. In Equation (9), if i and j are in the range $1, 2, \ldots, k-1$, then so is s. Hence, Equation (9) states that

$$A_{k-1} = L_{k-1}U_{k-1}$$

Since A_{k-1} is nonsingular by hypothesis, L_{k-1} and U_{k-1} are also nonsingular. Since L_{k-1} is nonsingular, we can solve the system

$$\sum_{s=1}^{k-1} \ell_{is}u_{sk} = a_{ik} \qquad (1 \leqq i \leqq k-1)$$

for the quantities u_{sk} with $1 \leqq s \leqq k-1$. These elements lie in the kth column of U. Since U_{k-1} is nonsingular, we can solve the system

$$\sum_{s=1}^{k-1} \ell_{ks}u_{sj} = a_{kj} \qquad (1 \leqq j \leqq k-1)$$

for ℓ_{ks} with $1 \leqq s \leqq k-1$. These elements lie in the kth row of L. From the requirement

$$a_{kk} = \sum_{s=1}^{k} \ell_{ks}u_{sk} = \sum_{s=1}^{k-1} \ell_{ks}u_{sk} + \ell_{kk}u_{kk}$$

we can obtain u_{kk} since ℓ_{kk} has been specified as unity. Thus, all the new elements necessary to form L_k and U_k have been defined. The induction is completed by noting that $\ell_{11}u_{11} = a_{11}$ and, therefore, $\ell_{11} = 1$ and $u_{11} = a_{11}$. ■

Cholesky Factorization

As mentioned earlier in this section, a matrix factorization that is useful in some situations has been given the name of the mathematician André Louis Cholesky, who proved the following result.

THEOREM 2 *If A is a real, symmetric, and positive definite matrix, then it has a unique factorization, $A = LL^T$, in which L is lower triangular with a positive diagonal.*

Proof Recall that a matrix A is **symmetric and positive definite** if $A = A^T$ and $x^TAx > 0$ for every nonzero vector x. It follows at once that A is nonsingular because A obviously cannot map any nonzero vector into 0. Moreover, by considering special vectors of the form $x = (x_1, x_2, \ldots, x_k, 0, 0, \ldots, 0)^T$, we see that the leading principal minors of A are also positive definite. Theorem 1 implies that A has an LU-decomposition. By the symmetry of A, we then have

$$LU = A = A^T = U^TL^T$$

This implies that

$$U(L^T)^{-1} = L^{-1}U^T$$

The left member of this equation is upper triangular, whereas the right member is lower triangular. (See Problem 1.) Consequently, there is a diagonal matrix D such that $U(L^T)^{-1} = D$. Hence, $U = DL^T$ and $A = LDL^T$. By Problem 27, D is positive definite, and thus its elements d_{ii} are positive. Denoting by $D^{1/2}$ the diagonal matrix whose diagonal elements are $\sqrt{d_{ii}}$, we have $A = \widetilde{L}\widetilde{L}^T$ where $\widetilde{L} \equiv LD^{1/2}$, which is the Cholesky factorization. The proof of uniqueness is left as a problem. ■

The algorithm for the Cholesky factorization is a special case of the general LU-factorization algorithm. If A is real, symmetric, and positive definite, then by Theorem 2 it has a unique factorization of the form $A = LL^T$, in which L is lower triangular and has positive diagonal. Thus, in Equation (8), $U = L^T$. In the kth step of the general algorithm, the diagonal entry is computed by

$$\ell_{kk} = \left(a_{kk} - \sum_{s=1}^{k-1} \ell_{ks}^2 \right)^{1/2} \tag{13}$$

The algorithm for the **Cholesky factorization** will then be as follows:

input $n, (a_{ij})$
for $k = 1$ **to** n **do**

$$\ell_{kk} \leftarrow \left(a_{kk} - \sum_{s=1}^{k-1} \ell_{ks}^2 \right)^{1/2}$$

for $i = k+1$ **to** n **do**

$$\ell_{ik} \leftarrow \left(a_{ik} - \sum_{s=1}^{k-1} \ell_{is}\ell_{ks} \right) \Big/ \ell_{kk}$$

end do

end do

output (ℓ_{ij})

Theorem 2 guarantees that $\ell_{kk} > 0$. Observe that Equation (13) gives us the following bound for $j \leqq k$:

$$a_{kk} = \sum_{s=1}^{k} \ell_{ks}^2 \geqq \ell_{kj}^2$$

from which we conclude that

$$|\ell_{kj}| \leqq \sqrt{a_{kk}} \qquad (1 \leqq j \leqq k)$$

Hence, any element of L is bounded by the square root of a corresponding diagonal element in A. This implies that the elements of L do not become large relative to A even without any pivoting. (Pivoting is explained in the next section.)

In both the Cholesky and Doolittle algorithms, the dot products of vectors should be carried out in double precision in order to avoid a buildup of roundoff errors. (See Computer Problem 6 in Section 2.2.)

PROBLEMS 4.2

1. Prove these facts, needed in the proof of Theorem 2.
   (a) If U is upper triangular and invertible, then U^{-1} is upper triangular.
   (b) The inverse of a unit lower triangular matrix is unit lower triangular.
   (c) The product of two upper (lower) triangular matrices is upper (lower) triangular.
2. Prove that if a nonsingular matrix A has an LU-factorization in which L is a *unit* lower triangular matrix, then L and U are unique.
3. Prove that algorithms (2), (3), (6), and (7) always solve $Ax = b$ if A is nonsingular.
4. Prove that an upper or lower triangular matrix is nonsingular if and only if its diagonal elements are all different from 0.
5. Show that if all the principal minors of A are nonsingular and $\ell_{ii} \neq 0$ for each i, then $u_{kk} \neq 0$ for $1 \leqq k \leqq n$.
6. Prove that the matrix $A = \begin{bmatrix} 0 & 1 \\ 1 & 1 \end{bmatrix}$ does not have an LU-factorization. *Caution*: This is *not* a simple consequence of Theorem 1 proved in this section.

7. (a) Write the **row version** of the **Doolittle algorithm** that computes the kth row of L and the kth row of U at the kth step. (Consequently at the kth step, the order of computing is $\ell_{k1}, \ell_{k2}, \ldots, \ell_{k,k-1}, u_{kk}, \ldots, u_{kn}$.)

   (b) Write the **column version** of the **Doolittle algorithm**, which computes the kth column of U and the kth column of L at the kth step. (Consequently, the order of computing is $u_{1k}, u_{2k}, \ldots, u_{kk}, \ell_{k+1,k}, \ldots, \ell_{nk}$ at the kth step.)

8. By use of the equation $UU^{-1} = I$, obtain an algorithm for finding the inverse of an upper triangular matrix. Assume that U^{-1} exists; that is, the diagonal elements of U are all nonzero.

9. Count the number of arithmetic operations involved in algorithms (2), (3), (6), and (7).

10. A matrix $A = (a_{ij})$ in which $a_{ij} = 0$ when $j > i$ or $j < i - 1$ is called a **Stieltjes matrix**. Devise an efficient algorithm for inverting such a matrix.

11. Let A be an $n \times n$ matrix. Let $(p_1, p_2, \ldots, p_n)$ be a permutation of $(1, 2, \ldots, n)$ such that (for $i = 1, 2, \ldots, n$) row i in A contains nonzero elements only in columns $p_1, p_2, \ldots, p_i$. Write an algorithm to solve $Ax = b$.

12. Show that every matrix of the form $A = \begin{bmatrix} 0 & a \\ 0 & b \end{bmatrix}$ has an LU-factorization. Show that even if L is *unit* lower triangular, the factorization is not unique. (This problem, as well as Problems 13 and 14, illustrate Taussky's Maxim: If a conjecture about matrices is false, it can usually be disproved with a 2×2 example.)

13. Show that every matrix of the form $A = \begin{bmatrix} 0 & 0 \\ a & b \end{bmatrix}$ has an LU-factorization. Does it have an LU-factorization in which L is a *unit* lower triangular?

14. Show that every matrix of the following form has an LU-factorization. Does it have an LU-factorization in which L is a *unit* lower triangular?

   (a) $A = \begin{bmatrix} a & 0 \\ b & 0 \end{bmatrix}$

   (b) $A = \begin{bmatrix} a & b \\ 0 & 0 \end{bmatrix}$

15. If A is invertible and has an LU-decomposition, then all principal minors of A are nonsingular.

16. Let the system $Ax = b$ have the following property: There are two permutations of $(1, 2, \ldots, n)$ called $p = (p_1, p_2, \ldots, p_n)$ and $q = (q_1, q_2, \ldots, q_n)$ such that, for each i, equation number p_i contains only the variables $x_{q_1}, x_{q_2}, \ldots, x_{q_i}$. Write an efficient algorithm to solve this system.

17. Count the number of multiplications and/or divisions needed to invert a unit lower triangular matrix.

18. Prove or disprove: If A has an LU-factorization in which L is *unit* lower triangular, then it has an LU-factorization in which U is *unit* upper triangular.

19. Assuming that its LU-factorization is known, give an algorithm for inverting A. (Use Problem 8 and Computer Problem 1.)

20. Develop an algorithm for inverting a matrix A that has the property $a_{ij} = 0$ if $i + j \leqq n$.

21. Use the Cholesky Theorem to prove that these two properties of a symmetric matrix A are equivalent.

(i) A is positive definite.

(ii) There exists a linearly independent set of vectors $x^{(1)}, x^{(2)}, \ldots, x^{(n)}$ in $\mathbb{R}^n$ such that $A_{ij} = (x^{(i)})^T(x^{(j)})$.

22. Establish the correctness of the following algorithm for solving $Ux = b$ in the case that U is upper triangular:

```
for j = n to 1 step − 1 do
    x_j ← b_j/u_jj
    for i = 1 to j − 1 do
        b_i ← b_i − u_ij x_j
    end do
end do
```

23. Prove that if all the leading principal minors of A are nonsingular, then A has a factorization LDU in which L is unit lower triangular, U is unit upper triangular, and D is diagonal.

24. (Continuation) If A is a symmetric matrix whose leading principal minors are nonsingular, then A has a factorization LDL^T in which L is unit lower triangular and D is diagonal.

25. (Continuation) Write an algorithm to compute the LDL^T-factorization of a symmetric matrix A. Your algorithm should do approximately half as much work as the standard Gaussian algorithm. *Note:* This algorithm can fail if some principal minors of A are singular. (This modification of the Cholesky algorithm does not involve square root calculations.)

26. Prove: A is positive definite and B is nonsingular if and only if BAB^T is positive definite.

27. If A is positive definite, does it follow that A^{-1} is also positive definite?

28. Consider

$$A = \begin{bmatrix} 2 & 6 & -4 \\ 6 & 17 & -17 \\ -4 & -17 & -20 \end{bmatrix}$$

Determine *directly* the factorization $A = LDL^T$ where D is diagonal and L is unit lower triangular—that is, do *not* use Gaussian elimination.

29. Develop an algorithm for finding directly the UL-factorization of a matrix A where L is *unit* lower triangular and U is upper triangular. Give an algorithm for solving $ULx = b$.

30. Find the LU-factorization of the matrix

$$A = \begin{bmatrix} 3 & 0 & 1 \\ 0 & -1 & 3 \\ 1 & 3 & 0 \end{bmatrix}$$

in which L is lower triangular and U is unit upper triangular.

31. Factor the matrix $A = \begin{bmatrix} 1 & 2 \\ 2 & 5 \end{bmatrix}$ so that $A = LL^T$, where L is lower triangular.

32. Determine directly the LL^T-factorization, in which L is a lower triangular matrix with positive diagonal elements, for the matrix

$$A = \begin{bmatrix} 4 & \frac{1}{2} & 1 \\ \frac{1}{2} & \frac{17}{16} & \frac{1}{4} \\ 1 & \frac{1}{4} & \frac{33}{64} \end{bmatrix}$$

33. Suppose that the nonsingular matrix A has a Cholesky factorization. What can be said about the determinant of A?

34. Determine the LU-factorization of the matrix $A = \begin{bmatrix} 1 & 5 \\ 3 & 16 \end{bmatrix}$ in which *both* L and U have unit diagonal elements. Repeat with 16 changed to 15.

35. Consider the symmetric tridiagonal positive definite matrix

$$A = \begin{bmatrix} 136.01 & 90.860 & 0.0 & 0.0 \\ 90.860 & 98.810 & -67.590 & 0.0 \\ 0.0 & -67.590 & 132.01 & 46.260 \\ 0.0 & 0.0 & 46.260 & 177.17 \end{bmatrix}$$

Using five significant figures, factor A in the following ways:

(a) $A = LU$, where L is unit lower triangular and U is upper triangular.

(b) $A = LDU$, where L is unit lower triangular, D is diagonal, and U is unit upper triangular.

(c) $A = LU$, where L is lower triangular and U is unit upper triangular.

(d) $A = LL^T$, where L is lower triangular.

36. Determine the LU-factorization of the matrix

$$A = \begin{bmatrix} 6 & 10 & 0 \\ 12 & 26 & 4 \\ 0 & 9 & 12 \end{bmatrix}$$

in which L is a lower triangular matrix with 2's on its main diagonal.

37. Prove or disprove: If a singular matrix has a Doolittle factorization, then that factorization is not unique.

38. Prove the uniqueness of the factorization $A = LL^T$, where L is lower triangular and has positive diagonal.

39. A matrix A that is symmetric and positive definite (SPD) has a square root X that is SPD. Thus $X^2 = A$. Find X if $A = \begin{bmatrix} 13 & 10 \\ 10 & 17 \end{bmatrix}$.

40. Develop algorithms for solving the linear system $Ax = b$ in two special cases:

(a) $a_{ij} = 0$ when $j \leqq n - i$

(b) $a_{ij} = 0$ when $j > n + 1 - i$

41. Using Equations (10), (11), and (12), find all the Doolittle factorizations of the matrix

$$A = \begin{bmatrix} 2 & 1 & -2 \\ 4 & 2 & -1 \\ 6 & 3 & 11 \end{bmatrix}$$

In this example, the algorithm works, although $u_{22} = 0$.

42. Prove that if A is symmetric, then in its LU-factorization the columns of L are multiples of the rows of U.

43. For $A = \begin{bmatrix} 1 & 5 \\ 3 & 17 \end{bmatrix}$, find all LU-factorizations and all UL-factorizations in which L is *unit* lower triangular.

44. Define a P-**matrix** to be one in which $a_{ij} = 0$ if $j \leqq n - i$ and a Q-**matrix** to be a P-matrix in which $a_{i,n-i+1} = 1$ for $i = 1, 2, \ldots, n$. Find the PQ-**factorization** of the matrix $A = \begin{bmatrix} 3 & 15 \\ -1 & -1 \end{bmatrix}$.

45. (Continuation) Devise an algorithm to obtain the PQ-factorization of a given matrix. Similarly, devise an algorithm for solving a system of equations of the form $PQx = b$.

46. Assuming that the LU-factorization of A is available, write an algorithm to solve the equation $x^T A = b^T$.

47. If A has a Doolittle factorization, what is a simple formula for the determinant of A?

48. Let

$$A = \begin{bmatrix} 25 & 0 & 0 & 0 & 1 \\ 0 & 27 & 4 & 3 & 2 \\ 0 & 54 & 58 & 0 & 0 \\ 0 & 108 & 116 & 0 & 0 \\ 100 & 0 & 0 & 0 & 24 \end{bmatrix}$$

Determine the most general LU-factorization of A in which the matrix L is unit lower triangular. Show that the Doolittle algorithm produces one of these LU-factorizations.

49. Let A be a symmetric matrix whose leading principal minors are nonnegative. Does the matrix $A + \varepsilon I$ have the same properties for $\varepsilon > 0$?

50. Consider the LU-factorization of a 2×2 matrix A. Show that if ℓ_{22} and u_{22} are specified, then the equations that determine the remaining elements of L and U are nonlinear.

51. Prove: If A is symmetric and nonnegative definite, then $A = LL^T$ for some lower triangular matrix L. The terminology **nonnegative definite** means that $x^T Ax \geqq 0$ for all x.

52. Find the precise conditions on a, b, and c in order that the matrix $\begin{bmatrix} a & b \\ b & c \end{bmatrix}$ will be nonnegative definite.

53. Prove that if the matrix $\begin{bmatrix} a & b \\ b & c \end{bmatrix}$ is nonnegative definite, then it has a factorization LL^T in which L is lower triangular.

54. Prove or disprove: A symmetric matrix is nonnegative definite if and only if all of its leading principal minors have nonnegative determinant.

55. Find necessary and sufficient conditions on a, b, and c in order that the matrix $\begin{bmatrix} a & b \\ b & c \end{bmatrix}$ has a factorization LL^T in which L is lower triangular.

56. In this problem, use the notation $X_{i:j,k}$ to denote the part of the kth column of the matrix X consisting of entries i to j. Similarly, let $X_{k,i:j}$ be the part of row k in X consisting of entries i to j.

(a) Refer to Equation (11) and show that it can be written as

$$U_{k,k+1:n} = (A_{k,k+1:n} - L_{k,1:k-1}M)/\ell_{kk}$$

in which M is the matrix whose rows are $U_{i,k+1:n}$, for $1 \leqq i \leqq k - 1$.

(b) Carry out the analogous transformation of Equation (12).

The computations discussed in this problem have the form $y \leftarrow y - Mx$. They can be carried out very efficiently on a vector supercomputer. (See Kincaid and Oppe [1988] and Oppe and Kincaid [1988] for further details.)

COMPUTER PROBLEMS 4.2

1. Devise an efficient algorithm for inverting an $n \times n$ lower triangular matrix A. *Suggestion:* Use the fact that A^{-1} is also lower triangular. Code your algorithm and test it on the matrix whose elements are $a_{ij} = (i + j)^2$ when $i \geqq j$. Use $n = 10$. Form the product AA^{-1} as a test of the computed inverse.

2. Solve this system by the Cholesky method:

$$\begin{cases} 0.05x_1 + 0.07x_2 + 0.06x_3 + 0.05x_4 = 0.23 \\ 0.07x_1 + 0.10x_2 + 0.08x_3 + 0.07x_4 = 0.32 \\ 0.06x_1 + 0.08x_2 + 0.10x_3 + 0.09x_4 = 0.33 \\ 0.05x_1 + 0.07x_2 + 0.09x_3 + 0.10x_4 = 0.31 \end{cases}$$

3. Write a subprogram or procedure that implements the general LU-factorization algorithm. The diagonal elements that are prescribed can be stored in an array D. An associated logical array can be used to indicate whether an element of D belongs to the diagonal of L or U. Test the routine on some Hilbert matrices whose elements are $a_{ij} = (i + j - 1)^{-1}$. For each matrix, produce the Doolittle, Crout, and Cholesky factorizations plus one or more others with specified diagonal entries.

4.3 Pivoting and Constructing an Algorithm

In Section 4.2, an abstract version of Gaussian elimination was presented in the guise of the LU-factorization of a matrix. In this section, the traditional form of Gaussian elimination will be described and related to the abstract form. Then we shall take up the modifications of the process necessary to produce a satisfactory computer realization of it. Throughout this discussion, we shall use the words *equation* and *row* of a matrix system interchangeably.

Why do we have Doolittle, Crout, and Cholesky decompositions when Gaussian elimination works well? In the days of desk-top calculators, there were possible advantages in one or more of these procedures over the others. But with the advances in computers and mathematical software, these slight advantages have vanished. Thus, the discussion of Doolittle and Crout is primarily for historical reasons. On the other hand, the Cholesky procedure is particularly well suited for symmetric positive definite systems.

Basic Gaussian Elimination

Here is a simple system of four equations in four unknowns that will be used to illustrate the Gaussian algorithm:

$$\begin{bmatrix} 6 & -2 & 2 & 4 \\ 12 & -8 & 6 & 10 \\ 3 & -13 & 9 & 3 \\ -6 & 4 & 1 & -18 \end{bmatrix} \begin{bmatrix} x_1 \\ x_2 \\ x_3 \\ x_4 \end{bmatrix} = \begin{bmatrix} 12 \\ 34 \\ 27 \\ -38 \end{bmatrix} \tag{1}$$

In the first step of the process, we subtract 2 times the first equation from the second. Then we subtract $\frac{1}{2}$ times the first equation from the third. Finally, we subtract -1 times the first equation from the fourth. The numbers 2, $\frac{1}{2}$, and -1 are called the **multipliers** for the first step in the elimination process. The number 6 used as the divisor in forming each of these multipliers is called the **pivot element** for this step. After the first step has been completed, the system will look like this:

$$\begin{bmatrix} 6 & -2 & 2 & 4 \\ 0 & -4 & 2 & 2 \\ 0 & -12 & 8 & 1 \\ 0 & 2 & 3 & -14 \end{bmatrix} \begin{bmatrix} x_1 \\ x_2 \\ x_3 \\ x_4 \end{bmatrix} = \begin{bmatrix} 12 \\ 10 \\ 21 \\ -26 \end{bmatrix} \tag{2}$$

Although the first row was used in the process, it was not changed. In this first step, we refer to row 1 as the **pivot row**. In the next step of the process, row 2 is used as the pivot row and -4 is the pivot element. We subtract 3 times the second row from the third, and then $-\frac{1}{2}$ times the second row is subtracted from the fourth. The multipliers for this step are 3 and $-\frac{1}{2}$. The result is

$$\begin{bmatrix} 6 & -2 & 2 & 4 \\ 0 & -4 & 2 & 2 \\ 0 & 0 & 2 & -5 \\ 0 & 0 & 4 & -13 \end{bmatrix} \begin{bmatrix} x_1 \\ x_2 \\ x_3 \\ x_4 \end{bmatrix} = \begin{bmatrix} 12 \\ 10 \\ -9 \\ -21 \end{bmatrix} \tag{3}$$

The final step consists of subtracting 2 times the third row from the fourth so that 2 is both the multiplier and the pivot element. The resulting system is

$$\begin{bmatrix} 6 & -2 & 2 & 4 \\ 0 & -4 & 2 & 2 \\ 0 & 0 & 2 & -5 \\ 0 & 0 & 0 & -3 \end{bmatrix} \begin{bmatrix} x_1 \\ x_2 \\ x_3 \\ x_4 \end{bmatrix} = \begin{bmatrix} 12 \\ 10 \\ -9 \\ -3 \end{bmatrix} \tag{4}$$

This system is upper triangular and **equivalent** to the original system in the sense that the solutions of the two systems are the same. The final system is easily solved by starting at the fourth row and working backward up the rows. The solution is

$$x = \begin{bmatrix} 1 \\ -3 \\ -2 \\ 1 \end{bmatrix}$$

The multipliers used in transforming the system can be exhibited in a unit lower triangular matrix $L = (\ell_{ij})$:

$$L = \begin{bmatrix} 1 & 0 & 0 & 0 \\ 2 & 1 & 0 & 0 \\ \frac{1}{2} & 3 & 1 & 0 \\ -1 & -\frac{1}{2} & 2 & 1 \end{bmatrix} \tag{5}$$

Notice that each multiplier is written in the location corresponding to the 0 entry in the matrix it was responsible for creating. The coefficient matrix of the final system is an upper triangular matrix $U = (u_{ij})$:

$$U = \begin{bmatrix} 6 & -2 & 2 & 4 \\ 0 & -4 & 2 & 2 \\ 0 & 0 & 2 & -5 \\ 0 & 0 & 0 & -3 \end{bmatrix} \tag{6}$$

These two matrices give the LU-factorization of A, where A is the coefficient matrix of the original system. Thus,

$$\begin{bmatrix} 6 & -2 & 2 & 4 \\ 12 & -8 & 6 & 10 \\ 3 & -13 & 9 & 3 \\ -6 & 4 & 1 & -18 \end{bmatrix} = \begin{bmatrix} 1 & 0 & 0 & 0 \\ 2 & 1 & 0 & 0 \\ \frac{1}{2} & 3 & 1 & 0 \\ -1 & -\frac{1}{2} & 2 & 1 \end{bmatrix} \begin{bmatrix} 6 & -2 & 2 & 4 \\ 0 & -4 & 2 & 2 \\ 0 & 0 & 2 & -5 \\ 0 & 0 & 0 & -3 \end{bmatrix} \tag{7}$$

It is not hard to see why this must be true. If we know how U was obtained from A, then by reversing the process we can get A from U. If we denote the rows of A by A_1, A_2, A_3, A_4 and the rows of U by U_1, U_2, U_3, U_4, then the elimination process gives us, for example, $U_2 = A_2 - 2A_1$. Hence, $A_2 = 2A_1 + U_2 = 2U_1 + U_2$. The coefficients 2 and 1 occupy the second row of L. Similarly, the row operations leading to row 3 are $U_3 = (A_3 - \frac{1}{2}A_1) - 3U_2$, and finally we have $A_3 = \frac{1}{2}A_1 + 3U_2 + U_3 = \frac{1}{2}U_1 + 3U_2 + U_3$. The coefficients $\frac{1}{2}$, 3, and 1 must therefore occupy the third row of L, and so on.

To describe formally the progress of the Gaussian algorithm, we interpret it as a succession of $n - 1$ major steps resulting in a sequence of matrices as follows:

$$A = A^{(1)} \longrightarrow A^{(2)} \longrightarrow \cdots \longrightarrow A^{(n)}$$

At the conclusion of step $k - 1$, the matrix $A^{(k)}$ will have been constructed; its appearance is shown in the following display in which lines are placed around the kth row and just before the kth column to illustrate the structure produced by the elimination process:

$$\left[\begin{array}{ccc|ccccc} a_{11}^{(k)} & \cdots & a_{1,k-1}^{(k)} & a_{1k}^{(k)} & \cdots & a_{1j}^{(k)} & \cdots & a_{1n}^{(k)} \\ & \ddots & & \vdots & & \vdots & & \vdots \\ & & a_{k-1,k-1}^{(k)} & a_{k-1,k}^{(k)} & \cdots & a_{k-1,j}^{(k)} & \cdots & a_{k-1,n}^{(k)} \\ \hline 0 & \cdots & 0 & a_{kk}^{(k)} & \cdots & a_{kj}^{(k)} & \cdots & a_{kn}^{(k)} \\ \hline 0 & \cdots & 0 & & \cdots & & \cdots & \\ \vdots & & \vdots & \vdots & & \vdots & & \vdots \\ 0 & \cdots & 0 & a_{ik}^{(k)} & \cdots & a_{ij}^{(k)} & \cdots & a_{in}^{(k)} \\ \vdots & & \vdots & \vdots & & \vdots & & \vdots \\ 0 & \cdots & 0 & a_{nk}^{(k)} & \cdots & a_{nj}^{(k)} & \cdots & a_{nn}^{(k)} \end{array}\right]$$

Our task is to describe how $A^{(k+1)}$ is obtained from $A^{(k)}$. To produce 0's in column k below the pivot element $a_{kk}^{(k)}$, we subtract multiples of row k from the rows beneath it. Rows $1, 2, \ldots, k$ are *not* altered. The formula is therefore

$$a_{ij}^{(k+1)} = \begin{cases} a_{ij}^{(k)} & \text{if } i \leqq k \\ a_{ij}^{(k)} - \left(a_{ik}^{(k)} / a_{kk}^{(k)}\right) a_{kj}^{(k)} & \text{if } i \geqq k+1 \text{ and } j \geqq k+1 \\ 0 & \text{if } i \geqq k+1 \text{ and } j \leqq k \end{cases} \tag{8}$$

Then we set $U = A^{(n)}$ and define L by

$$\ell_{ik} = \begin{cases} a_{ik}^{(k)}/a_{kk}^{(k)} & \text{if } i \geqq k+1 \\ 1 & \text{if } i = k \\ 0 & \text{if } i \leqq k-1 \end{cases} \tag{9}$$

Here $A = LU$ is the standard Gaussian factorization of matrix A with L *unit* lower triangular and U upper triangular. It should be clear from Equations (8) and (9), as well as from the previous numerical example, that this entire elimination process will break down if any of the pivot elements are 0. Now we can prove the following theorem.

THEOREM 1 *If all the pivot elements $a_{kk}^{(k)}$ are nonzero in the process just described, then $A = LU$.*

Proof Observe that $a_{ij}^{(k+1)} = a_{ij}^{(k)}$ if $i \leqq k$ or $j \leqq k-1$. Next note that $u_{kj} = a_{kj}^{(n)} = a_{kj}^{(k)}$. Finally, note that $\ell_{ik} = 0$ if $k > i$ and $u_{kj} = 0$ if $k > j$. Now let $i \leqq j$. Using these facts, we have

$$\begin{aligned}
(LU)_{ij} &= \sum_{k=1}^{n} \ell_{ik} u_{kj} = \sum_{k=1}^{i} \ell_{ik} u_{kj}^{(k)} = \sum_{k=1}^{i} \ell_{ik} a_{kj}^{(k)} \\
&= \sum_{k=1}^{i-1} \ell_{ik} a_{kj}^{(k)} + \ell_{ii} a_{ij}^{(i)} \\
&= \sum_{k=1}^{i-1} \left(a_{ik}^{(k)}/a_{kk}^{(k)} \right) a_{kj}^{(k)} + a_{ij}^{(i)} \\
&= \sum_{k=1}^{i-1} (a_{ij}^{(k)} - a_{ij}^{(k+1)}) + a_{ij}^{(i)} \\
&= a_{ij}^{(1)} = a_{ij}
\end{aligned}$$

Similarly, if $i > j$, then

$$\begin{aligned}
(LU)_{ij} &= \sum_{k=1}^{j} \ell_{ik} a_{kj}^{(k)} \\
&= \sum_{k=1}^{j} (a_{ij}^{(k)} - a_{ij}^{(k+1)}) \\
&= a_{ij}^{(1)} - a_{ij}^{(j+1)} \\
&= a_{ij}^{(1)} = a_{ij}
\end{aligned}$$

since $a_{ij}^{(k)} = 0$ if $i \geqq j+1$ and $k \geqq j+1$. ■

An algorithm to carry out the basic Gaussian elimination process just described on the matrix $A = (a_{ij})$ is as follows:

```
input n, (a_ij)
for k = 1 to n − 1 do
    for i = k + 1 to n do
        z ← a_ik / a_kk
        a_ik ← 0
        for j = k + 1 to n do
            a_ij ← a_ij − z a_kj
        end do
    end do
end do
output (a_ij)
```

Here it is assumed that all the pivot elements are nonzero. The multipliers are chosen so that entries below the main diagonal in A are computed to be 0. Rather than carrying out this computation, we simply replace these entries with 0's in the algorithm.

Pivoting

The Gaussian algorithm, in the simple form just described, is not satisfactory since it fails on systems that are in fact easy to solve. To illustrate this remark, we consider three elementary examples. The first is

$$\begin{bmatrix} 0 & 1 \\ 1 & 1 \end{bmatrix} \begin{bmatrix} x_1 \\ x_2 \end{bmatrix} = \begin{bmatrix} 1 \\ 2 \end{bmatrix} \tag{10}$$

The simple version of the algorithm fails because there is no way of adding a multiple of the first equation to the second in order to obtain a 0-coefficient of x_1 in the second equation. (See Problem 6 in Section 4.2.)

The difficulty just encountered will persist for the following system, in which ε is a small number different from 0:

$$\begin{bmatrix} \varepsilon & 1 \\ 1 & 1 \end{bmatrix} \begin{bmatrix} x_1 \\ x_2 \end{bmatrix} = \begin{bmatrix} 1 \\ 2 \end{bmatrix} \tag{11}$$

When applied to (11), the Gaussian algorithm will produce this upper triangular system:

$$\begin{bmatrix} \varepsilon & 1 \\ 0 & 1 - \varepsilon^{-1} \end{bmatrix} \begin{bmatrix} x_1 \\ x_2 \end{bmatrix} = \begin{bmatrix} 1 \\ 2 - \varepsilon^{-1} \end{bmatrix} \tag{12}$$

The solution is

$$\begin{cases} x_2 = (2 - \varepsilon^{-1})/(1 - \varepsilon^{-1}) \approx 1 \\ x_1 = (1 - x_2)\varepsilon^{-1} \approx 0 \end{cases} \tag{13}$$

In the computer, if ε is small enough, $2 - \varepsilon^{-1}$ will be computed to be the same as $-\varepsilon^{-1}$. Likewise, the denominator $1 - \varepsilon^{-1}$ will be computed to be the same as $-\varepsilon^{-1}$. Hence, under these circumstances, x_2 will be computed as 1, and x_1 will be computed as 0. Since the correct solution is

$$\begin{cases} x_1 = 1/(1 - \varepsilon) \approx 1 \\ x_2 = (1 - 2\varepsilon)/(1 - \varepsilon) \approx 1 \end{cases}$$

the computed solution is accurate for x_2 but is extremely inaccurate for x_1!

In the computer, if ε is small enough, why does the computation of $2 - \varepsilon^{-1}$ lead to the same machine number as the computation of $-\varepsilon^{-1}$? The reason is that before the subtraction can take place, the *exponents* in the floating-point form of 2 and ε^{-1} must be made to agree by a shift of the radix point. If this shift is great enough, the mantissa of 2 will be 0. For example, in a seven-place decimal machine similar to the hypothetical `Marc-32`, with $\varepsilon = 10^{-8}$, we have $\varepsilon^{-1} = 0.1000000 \times 10^9$ and $2 = 0.2000000 \times 10^1$. If 2 is rewritten with exponent 9, we have $2 = 0.000000002 \times 10^9$ and $2 - \varepsilon^{-1} = -0.099999998 \times 10^9$, so that $2 - \varepsilon^{-1} = -0.1000000 \times 10^9 = -\varepsilon^{-1}$ in the machine.

The final example will show that it is not actually the smallness of the coefficient a_{11} that is causing the trouble. Rather, it is the smallness of a_{11} *relative* to the other elements in its row. Consider the following system, which is equivalent to System (11):

$$\begin{bmatrix} 1 & \varepsilon^{-1} \\ 1 & 1 \end{bmatrix} \begin{bmatrix} x_1 \\ x_2 \end{bmatrix} = \begin{bmatrix} \varepsilon^{-1} \\ 2 \end{bmatrix} \tag{14}$$

The simple Gaussian algorithm produces

$$\begin{bmatrix} 1 & \varepsilon^{-1} \\ 0 & 1 - \varepsilon^{-1} \end{bmatrix} \begin{bmatrix} x_1 \\ x_2 \end{bmatrix} = \begin{bmatrix} \varepsilon^{-1} \\ 2 - \varepsilon^{-1} \end{bmatrix} \tag{15}$$

The solution of (15) is

$$\begin{cases} x_2 = (2 - \varepsilon^{-1})/(1 - \varepsilon^{-1}) \approx 1 \\ x_1 = \varepsilon^{-1} - \varepsilon^{-1}x_2 \approx 0 \end{cases}$$

Again, for small ε, x_2 will be computed as 1 and x_1 will be computed as 0, which is, as before, wrong!

The difficulties in these examples will disappear if the order of the equations is changed. Thus, interchanging the two equations in System (11) leads to

$$\begin{bmatrix} 1 & 1 \\ \varepsilon & 1 \end{bmatrix} \begin{bmatrix} x_1 \\ x_2 \end{bmatrix} = \begin{bmatrix} 2 \\ 1 \end{bmatrix}$$

Gaussian elimination applied to this system produces

$$\begin{bmatrix} 1 & 1 \\ 0 & 1-\varepsilon \end{bmatrix} \begin{bmatrix} x_1 \\ x_2 \end{bmatrix} = \begin{bmatrix} 2 \\ 1-2\varepsilon \end{bmatrix}$$

The solution is then

$$\begin{cases} x_2 = (1-2\varepsilon)/(1-\varepsilon) \approx 1 \\ x_1 = 2 - x_1 \approx 1 \end{cases}$$

The conclusion to be drawn from these elementary examples is that good algorithms must incorporate the interchanging of equations in a system when the circumstances require it. For reasons of economy in computing, we prefer *not* to move the rows of the matrix around in the computer's memory. Instead, we simply choose the **pivot rows** in a logical manner. Suppose that instead of using the rows in the order $1, 2, \dots, n-1$ as pivot rows, we use rows $p_1, p_2, \dots, p_{n-1}$. Then in the first step, multiples of row p_1 will be subtracted from the other rows. If we introduce entry p_n so that $(p_1, p_2, \dots, p_n)$ is a permutation of $(1, 2, \dots, n)$, then row p_n will not occur as a pivot row, but we can say that multiples of row p_1 will be subtracted from rows $p_2, p_3, \dots, p_n$. In the next step, multiples of row p_2 will be subtracted from rows $p_3, p_4, \dots, p_n$; and so on.

Here is an algorithm to accomplish this. (It is assumed that the permutation array p has been predetermined and consists of the natural numbers $1, 2, \dots, n$ in some order.)

```
input n, (a_ij)
for k = 1 to n − 1 do
    for i = k + 1 to n do
        z ← a_{p_i k} / a_{p_k k}
        a_{p_i k} ← 0
        for j = k + 1 to n do
            a_{p_i j} ← a_{p_i j} − z a_{p_k j}
        end do
    end do
end do
output (a_ij)
```

Comparing this algorithm to the one for the basic Gaussian elimination process, we see that they are identical except for one global change. In the pseudocode above, the first subscript on the elements of the coefficient array A involves the permutation array p. Of course, when the permutation array corresponds to the natural ordering ($p_i = i$), the basic process will be obtained.

Gaussian Elimination with Scaled Row Pivoting

We now describe an algorithm called **Gaussian elimination with scaled row pivoting** for solving an $n \times n$ system

$$Ax = b$$

The algorithm consists of two parts: a **factorization phase** (also called **forward elimination**) and a **solution phase** (involving **updating** and **back substitution**). The factorization phase is applied to A only and is designed to produce the LU-decomposition of PA, where P is a permutation matrix derived from the permutation array p. (PA is obtained from A by permuting its rows.) The permuted linear system is

$$PAx = Pb$$

The factorization $PA = LU$ is obtained from a modified Gaussian elimination algorithm to be explained below. In the solution phase, we consider two equations $Lz = Pb$ and $Ux = z$. First, the right-hand side b is rearranged according to P and the results are stored back in b—that is, $b \leftarrow Pb$. Next, $Lz = b$ is solved for z and the results stored back in b—that is, $b \leftarrow L^{-1}b$. Since L is unit lower triangular, this amounts to a forward substitution process. This procedure is called **updating** b. Then back substitution is used to solve $Ux = b$ for $x_n, x_{n-1}, \dots, x_1$.

In the factorization phase, we begin by computing the **scale** of each row. We put

$$s_i = \max_{1 \leqq j \leqq n} |a_{ij}| = \max \left\{ |a_{i1}|, |a_{i2}|, \dots, |a_{in}| \right\} \qquad (1 \leqq i \leqq n)$$

These are recorded in an array s in the algorithm.

In starting the factorization phase, we do not arbitrarily subtract multiples of row 1 from the other rows. Rather, we select as the **pivot row** the row for which $|a_{i1}|/s_i$ is largest. The index thus chosen is denoted by p_1 and becomes the first entry in the permutation array. Thus, $|a_{p_1 1}|/s_{p_1} \geqq |a_{i1}|/s_i$ for $1 \leqq i \leqq n$. Once p_1 has been determined, we subtract appropriate multiples of row p_1 from the other rows in order to produce 0's in the first column of A. Of course, row p_1 will remain unchanged throughout the remainder of the factorization process.

To keep track of the indices p_i that arise, we begin by setting the permutation vector $(p_1, p_2, \dots, p_n)$ to $(1, 2, \dots, n)$. Then we select an index j for which $|a_{p_j 1}|/s_{p_j}$ is maximal, and interchange p_1 with p_j in the permutation array p. The actual elimination step involves subtracting $(a_{p_i 1}/a_{p_1 1})$ times row p_1 from row p_i for $2 \leqq i \leqq n$.

To describe the general process, suppose that we are ready to create 0's in column k. We scan the numbers $|a_{p_i k}|/s_{p_i}$ for $k \leqq i \leqq n$, looking for a maximal entry. If j is the index of the *first* of the largest of these ratios, we interchange p_k with p_j in the array p, and then subtract $(a_{p_i k}/a_{p_k k})$ times row p_k from row p_i for $k + 1 \leqq i \leqq n$.

Here is how this works on the matrix

$$A = \begin{bmatrix} 2 & 3 & -6 \\ 1 & -6 & 8 \\ 3 & -2 & 1 \end{bmatrix}$$

At the beginning, $p = (1, 2, 3)$ and $s = (6, 8, 3)$. To select the first pivot row, we look at the ratios $\{2/6, 1/8, 3/3\}$. The largest corresponds to $j = 3$, and row 3 is to be the first pivot row. So we interchange p_1 with p_3, obtaining $p = (3, 2, 1)$. Now multiples of row 3 are subtracted from rows 2 and 1 to produce 0's in the first column. The result is

$$\begin{bmatrix} \textcircled{\frac{2}{3}} & \frac{13}{3} & -\frac{20}{3} \\ \textcircled{\frac{1}{3}} & -\frac{16}{3} & \frac{23}{3} \\ 3 & -2 & 1 \end{bmatrix}$$

The entries in locations a_{11} and a_{21} are the multipliers. In the next step, the selection of a pivot row is made on the basis of the numbers $|a_{p_2 2}|/s_{p_2}$ and $|a_{p_3 2}|/s_{p_3}$. The first of these ratios is $(16/3)/8$ and the second is $(13/3)/6$. So $j = 3$, and we interchange p_2 with p_3. Then a multiple of row p_2 is subtracted from row p_3. The result is $p = (3, 1, 2)$ and

$$\begin{bmatrix} \textcircled{\frac{2}{3}} & \frac{13}{3} & -\frac{20}{3} \\ \textcircled{\frac{1}{3}} & \textcircled{-\frac{16}{13}} & -\frac{7}{13} \\ 3 & -2 & 1 \end{bmatrix}$$

The final multiplier is stored in location a_{22}.

If the rows of the original matrix A were interchanged according to the final permutation array p, then we would have an LU-factorization of A. Hence, we have

$$PA = \begin{bmatrix} 1 & 0 & 0 \\ \frac{2}{3} & 1 & 0 \\ \frac{1}{3} & -\frac{16}{13} & 1 \end{bmatrix} \begin{bmatrix} 3 & -2 & 1 \\ 0 & \frac{13}{3} & -\frac{20}{3} \\ 0 & 0 & -\frac{7}{13} \end{bmatrix} = \begin{bmatrix} 3 & -2 & 1 \\ 2 & 3 & -6 \\ 1 & -6 & 8 \end{bmatrix}$$

where

$$P = \begin{bmatrix} 0 & 0 & 1 \\ 1 & 0 & 0 \\ 0 & 1 & 0 \end{bmatrix} \qquad A = \begin{bmatrix} 2 & 3 & -6 \\ 1 & -6 & 8 \\ 3 & -2 & 1 \end{bmatrix}$$

The permutation matrix P is obtained from the permutation array p by setting $(P)_{ij} = \delta_{p_i j}$. In other words, P is formed by permuting the rows of the identity matrix I according to the entries in p.

Here is an algorithm to do the **factorization phase** of Gaussian elimination with scaled row pivoting:

input $n, (a_{ij})$
for $i = 1$ **to** n **do**
  $p_i \leftarrow i$
  $s_i \leftarrow \max_{1 \leqq j \leqq n} |a_{ij}|$
end do
for $k = 1$ **to** $n - 1$ **do**
  select $j \geqq k$ so that
    $|a_{p_j k}|/s_{p_j} \geqq |a_{p_i k}|/s_{p_i}$ for $i = k, k+1, \dots, n$
  $p_k \leftrightarrow p_j$
  for $i = k + 1$ **to** n **do**
    $z \leftarrow a_{p_i k}/a_{p_k k}$; $a_{p_i k} \leftarrow z$
    for $j = k + 1$ **to** n **do**
      $a_{p_i j} \leftarrow a_{p_i j} - z a_{p_k j}$
    end do
  end do
end do
output $(a_{ij}), (p_i)$

Notice that the **multipliers** are being stored in A at the locations where 0's would have been created by the elimination process. Hence, all the necessary data for reconstructing the LU-factorization are stored within this array. Nevertheless, the user should be aware that the algorithm overwrites the values in the original A-array. The matrix A should be saved in a separate array if it will be needed again.

To solve $Ax = b$, once the factorization phase has been carried out on A, we apply the forward phase of the algorithm to b, using the final permutation array p and the multipliers that were determined by the factorization phase of the algorithm. Next we carry out the back substitution—that is, solve for $x_n, x_{n-1}, \dots, x_1$ in that order. Here is the algorithm that carries out the **solution phase** of the algorithm:

input $n, (a_{ij}), (p_i), (b_i)$
for $k = 1$ **to** $n - 1$ **do**
  for $i = k + 1$ **to** n **do**
    $b_{p_i} \leftarrow b_{p_i} - a_{p_i k} b_{p_k}$
  end do
end do
for $i = n$ **to** 1 **step** -1 **do**

$$x_i \leftarrow \left(b_{p_i} - \sum_{j=i+1}^{n} a_{p_i j} x_j \right) \Big/ a_{p_i i}$$

end do
output (x_i)

The first part of this pseudocode *updates* the right-hand side b corresponding to what happened during the factorization phase. The second part recovers x from a lower triangular system—that is, back substitution. Of course, the permutation array p must be used to keep things from being jumbled. Notice that when $i = n$, the j-sum is vacuous, so that $x_n = b_{p_n}/a_{p_n n}$.

Gaussian Elimination with Complete Pivoting

Yet another pivoting strategy is **complete pivoting**. Recall that in partial pivoting, the kth pivot element is determined by examining the $n - k + 1$ elements in the subcolumn below and including the diagonal entry $a_{k,k}^{(k-1)}$. In partial pivoting, the largest of these elements in absolute value designated the pivot row. In scaled partial pivoting, the ratios of these elements with the row scaling constants are formed, and the pivot row is chosen as the one associated with the largest of these ratios in absolute value. In contrast, complete pivoting is similar to partial pivoting except all the $(n - k + 1)^2$ elements in the submatrix below, to the right, and including the diagonal entry $a_{k,k}^{(k-1)}$ are examined. (Scaled complete pivoting can be defined as well.) Generally, it is believed that there is no practical value in selecting complete pivoting over partial pivoting because of additional computational costs. (Selecting the pivot element requires the examination of a much larger number of entries in the matrix.)

Factorizations $PA = LU$

As alluded to earlier, if scaled row pivoting is included in the Gaussian algorithm, then we obtain the LU-factorization of PA, where P is a certain permutation matrix. A formal proof of this result will now be given, modeled on the proof of Theorem 1. The proof is *not* dependent on the strategy used to determine the pivots.

Let $p_1, p_2, \ldots, p_n$ be the indices of the rows in the order in which they become pivot rows. Let $A^{(1)} = A$, and define $A^{(2)}, A^{(3)}, \ldots, A^{(n)}$ recursively by the formula

$$a_{p_i j}^{(k+1)} = \begin{cases} a_{p_i j}^{(k)} & \text{if } i \leqq k \text{ or } i > k > j \\ a_{p_i j}^{(k)} - \left(a_{p_i k}^{(k)}/a_{p_k k}^{(k)}\right) a_{p_k j}^{(k)} & \text{if } i > k \text{ and } j > k \\ a_{p_i k}^{(k)}/a_{p_k k}^{(k)} & \text{if } i > k \text{ and } j = k \end{cases} \qquad \textbf{(16)}$$

THEOREM 2 *Define a permutation matrix P whose elements are $P_{ij} = \delta_{p_i j}$. Define an upper triangular matrix U whose elements are $u_{ij} = a_{p_i j}^{(n)}$ if $j \geqq i$. Define a unit lower triangular matrix L whose elements are $\ell_{ij} = a_{p_i j}^{(n)}$ if $j < i$. Then $PA = LU$.*

Proof From the recursive formula, we have

$$u_{kj} = a_{p_k j}^{(n)} = a_{p_k j}^{(k)} \qquad (j \geqq k)$$

$$\ell_{ik} = a_{p_i k}^{(n)} = a_{p_i k}^{(k+1)} = a_{p_i k}^{(k)}/a_{p_k k}^{(k)} \qquad (i \geqq k)$$

These two equations depend on the fact that row p_k in $A^{(n)}$ became fixed in step k, and column k in $A^{(n)}$ became fixed in step $k+1$. Thus,

$$a^{(n)}_{p_k j} = a^{(k)}_{p_k j} \quad \text{and} \quad a^{(n)}_{p_i k} = a^{(k+1)}_{p_i k}$$

Furthermore, the formula just given for ℓ_{ik} is valid when $i = k$ because the formula produces 1 for ℓ_{kk}. Now suppose that $i \leqq j$. Then

$$\begin{aligned}
(LU)_{ij} &= \sum_{k=1}^{i} \ell_{ik} u_{kj} \\
&= \sum_{k=1}^{i-1} \left(a^{(k)}_{p_i k} / a^{(k)}_{p_k k} \right) a^{(k)}_{p_k j} + \ell_{ii} a^{(i)}_{p_i j} \\
&= \sum_{k=1}^{i-1} \left(a^{(k)}_{p_i j} - a^{(k+1)}_{p_i j} \right) + a^{(i)}_{p_i j} \\
&= a^{(1)}_{p_i j} = a_{p_i j}
\end{aligned}$$

If $i > j$, then

$$\begin{aligned}
(LU)_{ij} &= \sum_{k=1}^{j} \ell_{ik} u_{kj} \\
&= \sum_{k=1}^{j-1} \left(a^{(k)}_{p_i k} / a^{(k)}_{p_k k} \right) a^{(k)}_{p_k j} + \left(a^{(j)}_{p_i j} / a^{(j)}_{p_j j} \right) a^{(j)}_{p_j j} \\
&= \sum_{k=1}^{j-1} \left(a^{(k)}_{p_i j} - a^{(k+1)}_{p_i j} \right) + a^{(j)}_{p_i j} \\
&= a^{(1)}_{p_i j} = a_{p_i j}
\end{aligned}$$

On the other hand,

$$(PA)_{ij} = \sum_{k=1}^{n} P_{ik} a_{kj} = \sum_{k=1}^{n} \delta_{p_i k} a_{kj} = a_{p_i j}$$

Thus we have proved, for all pairs (i, j), that

$$(PA)_{ij} = (LU)_{ij}$$

■

THEOREM 3 *If the factorization $PA = LU$ is produced from the Gaussian algorithm with scaled row pivoting, then the solution of $Ax = b$ is obtained by first solving $Lz = Pb$ and* *then solving $Ux = z$. Similarly, the solution of $y^T A = c^T$ is obtained by solving $U^T z = c$ and then $L^T P y = z$.*

Proof This is left for the reader as Problem 47. ■

The pseudocode for solving $Ax = b$ in terms of L and U looks like this:

input $n, (\ell_{ij}), (u_{ij}), (b_i), (p_i)$
for $i = 1$ **to** n **do**
$$z_i \leftarrow b_{p_i} - \sum_{j=1}^{i-1} \ell_{ij} z_j$$
end do
for $i = n$ **to** 1 **step** -1 **do**
$$x_i \leftarrow \left(z_i - \sum_{j=i+1}^{n} u_{ij} x_j \right) \bigg/ u_{ii}$$
end do
output (x_i)

The same algorithm can be written using entries in the final A-matrix that resulted from the Gaussian process. The result is

input $n, (a_{ij}), (b_i), (p_i)$
for $i = 1$ **to** n **do**
$$z_i \leftarrow b_{p_i} - \sum_{j=1}^{i-1} a_{p_i j} z_j$$
end do
for $i = n$ **to** 1 **step** -1 **do**
$$x_i \leftarrow \left(z_i - \sum_{j=i+1}^{n} a_{p_i j} x_j \right) \bigg/ a_{p_i i}$$
end do
output (x_i)

A pseudocode for solving $y^T A = c^T$ follows. Recall that $P^{-1} = P^T$.

input $n, (a_{ij}), (c_i), (p_i)$
for $j = 1$ **to** n **do**
$$z_j \leftarrow \left(c_j - \sum_{i=1}^{j-1} a_{p_i j} z_i \right) \bigg/ a_{p_j j}$$
end do
for $j = n$ **to** 1 **step** -1 **do**
$$y_{p_j} \leftarrow z_j - \sum_{i=j+1}^{n} a_{p_i j} y_{p_i}$$
end do
output (y_i)

Operation Counts

To estimate the computing effort that goes into the solution of a system of linear equations, we shall count the number of arithmetic operations in the factorization phase and solution phase. Since the operations of multiplication and division are usually comparable in execution time and are much more time consuming than addition and subtraction, it has become traditional to count only multiplications and divisions, lumping them together as **long operations**, or **ops** for short.

Consider the first major step ($k = 1$) in the factorization process, as it operates on an $n \times n$ matrix A. The computation necessary to define p_1 (the index of the pivot row) involves n divisions (n ops). For each of the $n-1$ rows numbered $p_2, p_3, \ldots, p_n$, a multiplier is computed (1 op) and then a multiple of row p_1 is subtracted from row p_i for $2 \leqq i \leqq n$. The 0's created in column 1 are *not* computed. Thus, the multiplier and the elimination process consume n ops per row. Since $n - 1$ rows are processed in this way, we have $n(n - 1)$ ops. To this are added the n ops needed for determining p_1. The total is n^2 ops.

The remainder of the factorization process can be interpreted as a repetition of step 1 on smaller and smaller matrices. Thus, in step 2, the row p_1 is ignored and column 1 is ignored. The entire calculation in step 2 is like step 1 applied to an $(n - 1) \times (n - 1)$ matrix. This observation implies that the factorization requires

$$n^2 + (n-1)^2 + \cdots + 3^2 + 2^2 = \frac{1}{3}n^3 + \frac{1}{2}n^2 + \frac{1}{6}n - 1 \approx \frac{1}{3}n^3 + \frac{1}{2}n^2$$

long operations. Here we use the fact that $\sum_{k=1}^{n} k^2 = \frac{1}{6}n(n+1)(2n+1)$. For large n, the term $\frac{1}{3}n^3$ is the dominant one. Thus, finding the LU-factorization with scaled row pivoting involves roughly $\frac{1}{3}n^3$ long operations for large n.

An examination of the second phase of the algorithm shows that in updating the right-hand side b, there are $n - 1$ steps. In the first of these, there are $n - 1$ long operations. In the second, there are $n - 2$, and so on. The total is therefore

$$(n-1) + (n-2) + \cdots + 1 = \frac{1}{2}n^2 - \frac{1}{2}n$$

Here we use $\sum_{k=1}^{n} k = \frac{1}{2}n(n+1)$. In the back substitution, there is one long operation in the first step (computing x_n). Then there are successively $2, 3, \ldots, n$ long operations. The total is

$$1 + 2 + 3 + \cdots + n = \frac{1}{2}n^2 + \frac{1}{2}n$$

The grand total for this phase of the algorithm is therefore n^2. To summarize these findings, we state the following slightly more general result:

THEOREM 4 *If Gaussian elimination is used with scaled row pivoting, then the solution of the system $Ax = b$ with fixed A and m different vectors b involves approximately*

$$\frac{1}{3}n^3 + \left(\frac{1}{2} + m\right) n^2$$

long operations (multiplications and divisions).

This organization of the Gaussian elimination algorithm permits an efficient treatment of the problem $Ax^{(i)} = b^{(i)}$ for $i = 1, 2, \dots, m$. Here we have m linear systems, each with the same coefficient matrix but with different right-hand sides. One application of the forward elimination phase gives the factorization $PA = LU$. Then m "back solves" using this factorization are needed to obtain the $x^{(i)}$'s. Thus, only $\mathcal{O}(\frac{1}{3}n^3 + (\frac{1}{2} + m)n^2)$ long operations are needed rather than $\mathcal{O}(\frac{1}{3}mn^3)$, as would be the case if the m systems were treated separately. Moreover, we can compute A^{-1} in $\mathcal{O}(\frac{4}{3}n^3)$ ops since it is a special case of the situation previously discussed; that is, $AX = I$ can be solved for each column of A^{-1} by solving $Ax^{(i)} = e_i$ using n values for i. Be advised that A^{-1} should *not* be computed when solving $Ax = b$, but rather x should be solved for directly!

Diagonally Dominant Matrices

Sometimes a system of equations has the property that Gaussian elimination *without* pivoting can be safely used. One class of matrices for which this is true is **diagonally dominant matrices**. This property is expressed by the inequality

$$|a_{ii}| > \sum_{\substack{j=1 \\ j \neq i}}^{n} |a_{ij}| \qquad (1 \leqq i \leqq n) \tag{17}$$

An example of a diagonally dominant matrix is

$$\begin{bmatrix} 4 & -1 & 0 & -1 \\ -1 & 4 & 0 & -1 \\ -1 & 0 & 4 & -1 \\ 0 & -1 & -1 & 4 \end{bmatrix}$$

Such matrices arise naturally in applications involving the discretization by finite differences of partial differential equations. Also, they are seen in the study of splines and many other areas.

If the coefficient matrix has this property, then in the first step of Gaussian elimination we can use row 1 as the pivot row since the pivot element a_{11} is not 0, by Inequality (17). After step 1 has been completed, we would like to know that row 2 can be used as the next pivot row. This situation is governed by the next theorem.

THEOREM 5 *Gaussian elimination without pivoting preserves the diagonal dominance of a matrix.*

Proof It suffices to consider the first step in Gaussian elimination (the step in which 0's are created in column 1) because subsequent steps mimic the first, except for being applied to matrices of smaller size. Thus, let A be an $n \times n$ matrix that is diagonally dominant. Taking account of the 0's created in column 1 as well as the fact that row 1 is unchanged, we have to prove that, for $i = 2, 3, \dots, n$,

$$|a_{ii}^{(2)}| > \sum_{\substack{j=2 \\ j\neq i}}^{n} |a_{ij}^{(2)}|$$

In terms of A, this means

$$|a_{ii} - (a_{i1}/a_{11})a_{1i}| > \sum_{\substack{j=2 \\ j\neq i}}^{n} |a_{ij} - (a_{i1}/a_{11})a_{1j}|$$

It suffices to prove the stronger inequality

$$|a_{ii}| - |(a_{i1}/a_{11})a_{1i}| > \sum_{\substack{j=2 \\ j\neq i}}^{n} \{|a_{ij}| + |(a_{i1}/a_{11})a_{1j}|\}$$

An equivalent inequality is

$$|a_{ii}| - \sum_{\substack{j=2 \\ j\neq i}}^{n} |a_{ij}| > \sum_{j=2}^{n} |(a_{i1}/a_{11})a_{1j}|$$

From the diagonal dominance in the ith row, we know that

$$|a_{ii}| - \sum_{\substack{j=2 \\ j\neq i}}^{n} |a_{ij}| > |a_{i1}|$$

Hence, it suffices to prove that

$$|a_{i1}| \geqq \sum_{j=2}^{n} |(a_{i1}/a_{11})a_{1j}|$$

This is true because of the diagonal dominance in row 1:

$$|a_{11}| > \sum_{j=2}^{n} |a_{1j}| \Rightarrow 1 > \sum_{j=2}^{n} |a_{1j}/a_{11}|$$

■

COROLLARY 1 *Every diagonally dominant matrix is nonsingular and has an LU-factorization.*

Proof Theorem 5 together with Theorem 1 implies that a diagonally dominant matrix A has an LU-decomposition in which L is unit lower triangular. The matrix U, by the preceding theorem, is diagonally dominant. Hence, its diagonal elements are nonzero. Thus, L and U are nonsingular. ■

COROLLARY 2 *If the scaled row pivoting version of Gaussian elimination recomputes the scale array after each major step and is applied to a diagonally dominant matrix, then the pivots will be the natural ones:* $1, 2, \ldots, n$. *Hence, the work of choosing the pivots can be omitted in this case.*

Proof By Theorem 5, it suffices to prove that the first pivot chosen in the algorithm is 1. Thus, it is enough to prove that

$$|a_{11}|/s_1 > |a_{i1}|/s_i \qquad (2 \leqq i \leqq n)$$

By the diagonal dominance, $|a_{ii}| = \max_j |a_{ij}| = s_i$ for all i. Hence, $|a_{11}|/s_1 = 1$. For $i \geqq 2$, we have

$$|a_{i1}| \leqq \sum_{\substack{j=1 \\ j \neq i}}^{n} |a_{ij}| < |a_{ii}| = s_i$$

Thus, $|a_{i1}|/s_i < 1$. ■

In Section 4.2, it was proved (Theorem 2) that a symmetric and positive definite matrix A has a unique Cholesky factorization, $A = LL^T$. It was shown also that the elements of the lower triangle matrix L satisfy the inequality

$$|\ell_{ij}| \leqq \sqrt{a_{ii}} \tag{18}$$

The Cholesky factorization algorithm is a special case of the LU-factorization. Pivoting is not needed in this particular case, because—as Inequality (18) indicates—the elements of L remain modest in magnitude compared to the elements of A.

Tridiagonal System

In applications, systems of equations often arise in which the coefficient matrix has a special structure. It is usually better to solve these systems using tailor-made algorithms that exploit the special structure. We consider one example of this—the tridiagonal system.

A square matrix $A = (a_{ij})$ is said to be **tridiagonal** if $a_{ij} = 0$ for all pairs (i, j) that satisfy $|i - j| > 1$. Thus in the ith row, only $a_{i,i-1}$ and a_{ii} and $a_{i,i+1}$ can be different from 0. Three vectors can be used to store the nonzero elements, and we arrange the notation like this:

$$\begin{bmatrix} d_1 & c_1 & & & & & \\ a_1 & d_2 & c_2 & & & & \\ & a_2 & d_3 & c_3 & & & \\ & & a_3 & d_4 & c_4 & & \\ & & & \ddots & \ddots & \ddots & \\ & & & & a_{n-2} & d_{n-1} & c_{n-1} \\ & & & & & a_{n-1} & d_n \end{bmatrix} \begin{bmatrix} x_1 \\ x_2 \\ x_3 \\ x_4 \\ \vdots \\ x_{n-1} \\ x_n \end{bmatrix} = \begin{bmatrix} b_1 \\ b_2 \\ b_3 \\ b_4 \\ \vdots \\ b_{n-1} \\ b_n \end{bmatrix} \tag{19}$$

In this display, matrix elements not shown are 0's.

We shall assume that the coefficient matrix is such that pivoting is not necessary in solving the system. This is true, for example, if the matrix is symmetric and positive definite. Simple Gaussian elimination will be used, and the right-hand side (vector b) is processed simultaneously in this algorithm. In step 1, a multiple of row 1 is subtracted from row 2 to produce a 0 where a_1 stood originally. Note that d_2 and b_2 are altered but c_2 is not. The appropriate multiple is a_1/d_1. Hence, step 1 consists in these replacements:

$$\begin{aligned} d_2 &\leftarrow d_2 - (a_1/d_1)c_1 \\ b_2 &\leftarrow b_2 - (a_1/d_1)b_1 \end{aligned}$$

All the remaining steps in the forward elimination phase are exactly like the first. In the back substitution phase, the first step is

$$x_n \leftarrow b_n/d_n$$

The next step is

$$x_{n-1} \leftarrow (b_{n-1} - c_{n-1}x_n)/d_{n-1}$$

and all the rest are similar. Here is the algorithm:

input $n, (a_i), (b_i), (c_i), (d_i)$
for $i = 2$ **to** n **do**
  $d_i \leftarrow d_i - (a_{i-1}/d_{i-1})c_{i-1}$
  $b_i \leftarrow b_i - (a_{i-1}/d_{i-1})b_{i-1}$
end do
$x_n \leftarrow b_n/d_n$
for $i = n - 1$ **to** 1 **step** -1 **do**
  $x_i \leftarrow (b_i - c_i x_{i+1})/d_i$
end do
output (x_i)

PROBLEMS 4.3

1. Solve the following linear systems twice. First, use Gaussian elimination and give the factorization $A = LU$. Second, use Gaussian elimination with scaled row pivoting and determine the factorization of the form $PA = LU$.

(a) $$\begin{bmatrix} -1 & 1 & -4 \\ 2 & 2 & 0 \\ 3 & 3 & 2 \end{bmatrix} \begin{bmatrix} x_1 \\ x_2 \\ x_3 \end{bmatrix} = \begin{bmatrix} 0 \\ 1 \\ \frac{1}{2} \end{bmatrix}$$

(b) $$\begin{bmatrix} 1 & 6 & 0 \\ 2 & 1 & 0 \\ 0 & 2 & 1 \end{bmatrix} \begin{bmatrix} x_1 \\ x_2 \\ x_3 \end{bmatrix} = \begin{bmatrix} 3 \\ 1 \\ 1 \end{bmatrix}$$

(c) $$\begin{bmatrix} -1 & 1 & 0 & -3 \\ 1 & 0 & 3 & 1 \\ 0 & 1 & -1 & -1 \\ 3 & 0 & 1 & 2 \end{bmatrix} \begin{bmatrix} x_1 \\ x_2 \\ x_3 \\ x_4 \end{bmatrix} = \begin{bmatrix} 4 \\ 0 \\ 3 \\ 1 \end{bmatrix}$$

(d) $$\begin{bmatrix} 6 & -2 & 2 & 4 \\ 12 & -8 & 4 & 10 \\ 3 & -13 & 3 & 3 \\ -6 & 4 & 2 & -18 \end{bmatrix} \begin{bmatrix} x_1 \\ x_2 \\ x_3 \\ x_4 \end{bmatrix} = \begin{bmatrix} 0 \\ -10 \\ -39 \\ -16 \end{bmatrix}$$

(e) $$\begin{bmatrix} 1 & 0 & 2 & 1 \\ 4 & -9 & 2 & 1 \\ 8 & 16 & 6 & 5 \\ 2 & 3 & 2 & 1 \end{bmatrix} \begin{bmatrix} x_1 \\ x_2 \\ x_3 \\ x_4 \end{bmatrix} = \begin{bmatrix} 2 \\ 14 \\ -3 \\ 0 \end{bmatrix}$$

2. Show that Equation (8) defining the Gaussian elimination algorithm can also be written in the form

$$a_{ij}^{(k+1)} = \begin{cases} a_{ij}^{(k)} & \text{if } i \leqq k \text{ or } j < k \\ a_{ij}^{(k)} - \left(a_{ik}^{(k)}/a_{kk}^{(k)}\right) a_{kj}^{(k)} & \text{if } i > k \text{ and } j \geqq k \end{cases}$$

3. Let $(p_1, p_2, \ldots, p_n)$ be a permutation of $(1, 2, \ldots, n)$ and define the matrix P by $P_{ij} = \delta_{p_i j}$. Let A be an arbitrary $n \times n$ matrix. Describe PA, AP, P^{-1}, and PAP^{-1}.

4. Let A be an $n \times n$ matrix with scale factors $s_i = \max_{1 \leqq j \leqq n} |a_{ij}|$. Assume that all s_i are positive, and let B be the matrix whose elements are (a_{ij}/s_i). Prove that if forward elimination is applied to A and to B, then the two final L-arrays are the same. Find the formulas that relate the final A and B matrices (after processing).

5. It is sometimes advisable to modify a system of equations $Ax = b$ by introducing new variables $y_i = d_i x_i$, where d_i are positive numbers. If the x_i correspond to physical quantities, then this change of variables corresponds to a change in the units by which x_i is measured. Thus, if we decide to change x_1 from centimeters to meters, then $y_1 = 10^{-2} x_1$. In matrix terms, we define a diagonal matrix D with d_i as diagonal entries, and put $y = Dx$. The new system of equations is $AD^{-1}y = b$. If d_j is chosen as $\max_{1 \leqq i \leqq n} |a_{ij}|$, we call this **column**

equilibration. Modify the factorization and solution algorithms to incorporate column equilibration. (The two algorithms together still will solve $Ax = b$.)

6. Show that the multipliers in the Gaussian algorithm with *full* pivoting (both row and column pivoting) lie in the interval $[-1, 1]$. (See Computer Problem 1.)

7. Let the $n \times n$ matrix A be processed by forward elimination, with the resulting matrix called B, and permutation vector $p = (p_1, p_2, \ldots, p_n)$. Let P be the matrix that results from the identity matrix with its rows written in the order $p_1, p_2, \ldots, p_n$. Prove that the LU-decomposition of PA is obtained as follows: Put $C = PB$, $L_{ij} = C_{ij}$ for $j < i$, and $U_{ij} = C_{ij}$ for $i \leqq j$. (Of course, $U_{ij} = 0$ if $i > j$, $L_{ij} = 0$ if $j > i$, and $L_{ii} = 1$.)

8. If the factor U in the LU-decomposition of A is known, what is the algorithm for calculating L?

9. Show how Gaussian elimination with scaled row pivoting works on this example (forward phase only):

$$\begin{bmatrix} 2 & -2 & -4 \\ 1 & 1 & -1 \\ 3 & 7 & 5 \end{bmatrix}$$

Display the scale array (s_1, s_2, s_3) and the *final* permutation array (p_1, p_2, p_3). Show the *final* A-array, with the multipliers stored in the correct locations.

10. Carry out the instructions in Problem 9 on the matrix

$$\begin{bmatrix} 3 & 7 & 3 \\ 1 & \frac{7}{3} & 4 \\ 4 & \frac{4}{3} & 0 \end{bmatrix}$$

11. Assume that A is tridiagonal. Define $c_0 = 0$ and $a_n = 0$. Show that if A is **columnwise diagonally dominant**

$$|d_i| > |a_i| + |c_{i-1}| \qquad (1 \leqq i \leqq n)$$

then the algorithm for tridiagonal systems will, in theory, be successful since no 0 pivot entries will be encountered. Refer to Equation (19) for the notation.

12. Write a special Gaussian elimination algorithm to solve linear equations when A has the property $a_{ij} = 0$ for $i > j + 1$. Do not use pivoting. Include the processing of the right-hand side in the algorithm. Count the operations needed to solve $Ax = b$.

13. Count the operations in the algorithm in the text for tridiagonal systems.

14. Rewrite the algorithm for tridiagonal systems so that the order of processing the equations and variables is reversed.

15. Prove the theorem concerning the number of long operations in Gaussian elimination.

16. Show how Gaussian elimination with scaled row pivoting works on this example:

$$A = \begin{bmatrix} -9 & 1 & 17 \\ 3 & 2 & -1 \\ 6 & 8 & 1 \end{bmatrix}$$

Show the scale array. The final A-array should contain the multipliers stored in the correct positions. Determine P, L, and U, and verify that $PA = LU$.

17. Show how the factorization phase of Gaussian elimination with scaled row pivoting works on the matrix

$$\begin{bmatrix} 1 & -2 & 3 \\ 2 & -4 & 2 \\ 3 & -5 & -1 \end{bmatrix}$$

Show all intermediate steps—that is, multipliers, scale array s, and index array p—and the final array A as it would appear after the algorithm had finished working on it.

18. This problem shows how the solution of a system of equations can be **unstable** relative to perturbations in the data. Solve $Ax = b$ with $b = (100, 1)^T$, and with each of the following matrices. (See Stoer and Bulirsch [1980, p. 185].)

$$A_1 = \begin{bmatrix} 1 & 1 \\ 1 & 0 \end{bmatrix} \qquad A_2 = \begin{bmatrix} 1 & 1 \\ 1 & 0.01 \end{bmatrix}$$

19. Assume that $0 < \varepsilon < 2^{-22}$. If the Gaussian algorithm without pivoting is used to solve the system

$$\begin{cases} \varepsilon x_1 + 2x_2 = 4 \\ x_1 - x_2 = -1 \end{cases}$$

on the `Marc-32`, what will be the solution vector (x_1, x_2)?

20. Solve this system by Gaussian elimination with full pivoting (as described in Computer Problem 1):

$$\begin{bmatrix} -9 & 1 & 17 \\ 3 & 2 & -1 \\ 6 & 8 & 1 \end{bmatrix} \begin{bmatrix} x_1 \\ x_2 \\ x_3 \end{bmatrix} = \begin{bmatrix} 5 \\ 9 \\ -3 \end{bmatrix}$$

21. Solve the system

$$\begin{cases} 0.2641x_1 + 0.1735x_2 + 0.8642x_3 = -0.7521 \\ 0.9411x_1 + 0.0175x_2 + 0.1463x_3 = 0.6310 \\ -0.8641x_1 - 0.4243x_2 + 0.0711x_3 = 0.2501 \end{cases}$$

using Gaussian elimination with:

(a) no pivoting

(b) scaled row pivoting

22. Write an algorithm to solve the system $Ax = b$ under the following conditions: There is a permutation $(p_1, p_2, \ldots, p_n)$ of $(1, 2, \ldots, n)$ such that for each i, equation p_i contains only the variable x_i.

23. Repeat Problem 22 assuming that for each i, equation i contains only the variable x_{p_i}.

24. Repeat Problem 22 assuming that for each i, equation p_i contains only the variable x_{p_i}. In this case, give the *simplest* algorithm.

25. **(a)** Show that if we apply Gaussian elimination without pivoting to a symmetric matrix A, then $\ell_{i1} = a_{1i}/a_{11}$.

(b) From this, show that if the first row and column of $A^{(2)}$ are removed, the remaining $(n-1) \times (n-1)$ matrix is symmetric. Conclude then that elements below the diagonal in this smaller matrix need not be computed. Use induction to infer that this simplification will occur in each succeeding step of the factorization phase.

(c) Show that the computation required is almost halved compared to the nonsymmetric case.

(d) Use this simplification to solve the system

$$\begin{cases} 0.6428x_1 + 0.3475x_2 - 0.8468x_3 = 0.4127 \\ 0.3475x_1 + 1.8423x_2 + 0.4759x_3 = 1.7321 \\ -0.8468x_1 + 0.4759x_2 + 1.2147x_3 = -0.8621 \end{cases}$$

26. Consider the matrix

$$\begin{bmatrix} 0 & 4 & 25 & 79 \\ 9 & 7 & 39 & 89 \\ 0 & 16 & 2 & 99 \\ 0 & 6 & 6 & 49 \end{bmatrix}$$

Circle the entry that will be used as the next pivot element in Gaussian elimination with scaled row pivoting. The scale array is $s = (80, 89, 160, 30)$.

27. Show the resulting matrix after the forward phase of Gaussian elimination with scaled row pivoting is applied to the matrix

$$\begin{bmatrix} 2 & -2 & -4 \\ 1 & 1 & -1 \\ 3 & 7 & 5 \end{bmatrix}$$

In the final matrix, write the multipliers in the appropriate locations.

28. Determine $\det(A)$, where

$$A = \begin{bmatrix} 2 & 1 & 0 \\ 1 & 4 & 1 \\ 0 & 1 & 2 \end{bmatrix}$$

without computing the determinant by expansion by minors.

29. Use Gaussian elimination with scaled row pivoting to find the determinant of

$$A = \begin{bmatrix} 0 & -1 & 0 & 1 \\ 0 & 1 & 0 & 1 \\ 1 & 1 & 2 & 0 \\ 2 & 0 & 1 & 0 \end{bmatrix}$$

30. Consider the system

$$\begin{cases} x_2 + 2x_3 = 1 \\ 2x_1 - x_2 = 2 \\ 2x_2 + x_3 = 3 \end{cases}$$

Determine the factorization $PA = LU$, where P is a permutation matrix. Use this factorization to obtain $\det(A)$.

31. Consider

$$A = \begin{bmatrix} 3 & 2 & -1 \\ 6 & 6 & 2 \\ -1 & 1 & 3 \end{bmatrix}$$

Use Gaussian elimination with scaled row pivoting to obtain the factorization

$$PA = LDU$$

where L is a unit lower triangular matrix, U is a unit upper triangular matrix, D is a diagonal matrix, and P is a permutation matrix.

32. In the next few problems, we fix n and denote by J the set $\{1, 2, \dots, n\}$. A **permutation** of J is a map $p : J \twoheadrightarrow J$, where the double arrow indicates that p is **surjective**. Thus, each element of J is the image, $p(i)$, of some element i in J. The **identity permutation** is defined by $u(i) = i$ for all $i \in J$. Show that if p and q are permutations of J, then so is $p \circ q$, which is defined as usual by the equation $(p \circ q)(i) = p(q(i))$. Prove that $p \circ (q \circ r) = (p \circ q) \circ r$ and that $p \circ u = u \circ p = p$.

33. (Continuation) Prove that each permutation p has an inverse p^{-1} that satisfies $p \circ p^{-1} = u = p^{-1} \circ p$. The set of all permutations of J is called the **symmetric group** on J.

34. (Continuation) Establish an algorithm for determining the inverse of any given permutation. (In a computer, a permutation of J can be represented as a vector $(p(1), p(2), \ldots, p(n))$.)

35. We are given a system of equations, $Ax = b$, in which A is $n \times n$. Let p and q be permutations of $\{1, 2, \ldots, n\}$. Write an algorithm for solving the system under the assumption that for $i = 1, 2, \ldots, n$, the equation numbered p_i contains only the variable x_{q_i}.

36. (Continuation) Repeat Problem 35 under the assumption that for each i, the variables $x_{q_1}, x_{q_2}, \ldots, x_{q_{i-1}}$ do not appear in the equation numbered p_i.

37. (Continuation) Repeat Problem 35 under the assumption that for each i, the variable x_{q_i} occurs only in the equations numbered $p_1, p_2, \ldots, p_i$.

38. (Difficult) Find an algorithm to solve $Ax = b$ under the assumption that all elements a_{ij} are 0 unless $|i - j| \leqq 1$ or $(i, j) = (1, n)$ or $(i, j) = (n, 1)$. Use Gaussian elimination without pivoting.

39. Count the number of long operations involved in the LU-factorization of an $n \times n$ matrix, assuming that no pivoting is done.

40. Let A be an $n \times n$ matrix that is diagonally dominant in its columns. Thus,

$$\sum_{\substack{i=1 \\ i \neq j}}^{n} |a_{ij}| < |a_{jj}| \qquad (1 \leqq j \leqq n)$$

Determine whether Gaussian elimination without pivoting preserves this diagonal dominance.

41. In a diagonally dominant matrix A, define the **excess in row** i by the equation

$$e_i = |a_{ii}| - \sum_{\substack{j=1 \\ j \neq i}}^{n} |a_{ij}|$$

Show that in the proof of Theorem 5 the following is true:

$$|a_{ii} - a_{i1}a_{1i}/a_{11}| \geqq \sum_{\substack{j=2 \\ j \neq i}}^{n} |a_{ij} - a_{i1}a_{1j}/a_{11}| + e_i$$

Thus, the excess in row i is not diminished in Gaussian elimination.

42. Refer to Problems 32–34 if necessary. Let p be a permutation of $\{1, 2, \ldots, n\}$, and let P be the corresponding permutation matrix. (Thus $P_{ij} = \delta_{p_i j}$.) Let q be the inverse of p and Q the permutation matrix corresponding to q. Prove that $P^{-1} = Q$.

43. (Continuation) Prove that if P is a permutation matrix, then $P^{-1} = P^T$.

44. If A is $n \times n$ and B is $n \times m$, how many multiplications and divisions are required to solve $AX = B$ by Gaussian elimination with scaled row pivoting? What if $B = I$?

45. Prove or disprove: If A is tridiagonal and P is a permutation matrix, then PAP^{-1} is tridiagonal.

46. Suppose that the scale array is recomputed in each major step of Gaussian elimination with scaled row pivoting. Prove that for a symmetric and diagonally dominant matrix, Gaussian elimination without pivoting is the same as with scaled row pivoting.

47. Prove Theorem 3.

48. In the scaled row pivoting algorithm for Gaussian elimination, suppose that the scale numbers s_i are redefined by

$$s_i = |a_{i1}| + |a_{i2}| + \cdots + |a_{in}|$$

Prove that if the resulting algorithm is applied to a diagonally dominant matrix, then the natural pivot order $(1, 2, \ldots, n)$ will be chosen. Write and test programs to carry out the ideas in Problem 25. This is Gaussian elimination without pivoting on a symmetric system.

49. Prove or disprove the natural conjecture that if a matrix has the property

$$0 \neq |a_{ii}| \geqq \sum_{\substack{j=1 \\ j \neq i}} |a_{ij}| \qquad (1 \leqq i \leqq n)$$

then the Gaussian elimination without pivoting will preserve this property.

50. **(a)** Prove that computing the determinant of a matrix by expansion by minors involves $(n-1)(n!)$ ops.

(b) Prove that Cramer's rule requires $(n^2 - 1)(n!)$ ops.

(c) Prove that the Gauss-Jordan method involves $\frac{1}{2}n(n+1)^2 \approx \frac{1}{2}n^3$ ops and therefore is 50% more expensive than Gaussian elimination.

COMPUTER PROBLEMS 4.3

1. Gaussian elimination with **full pivoting** treats both rows and columns in an order different from the natural order. Thus, in the first step, the pivot element a_{ij} is chosen so that $|a_{ij}|$ is the largest in the entire matrix. This determines that row i will be the pivot row and column j will be the pivot column. Zeros are created in column j by subtracting multiples of row i from the other rows. Write the algorithm to carry out this process. Two permutation vectors will be required.

2. Refer to Problem 25, and write a program to carry out the factorization phase of Gaussian elimination on a symmetric matrix. Assume that pivoting is not required.

3. Write and test programs to carry out Gaussian elimination with scaled row pivoting. Suitable test cases are in Problems 18, 20, 21, and 25.

4. Write and test programs that include column equilibration (see Problem 5) in the Gaussian algorithms.

5. Write and test programs to solve $Ax = b$ and $y^T A = c^T$ using only one factorization of A (with scaled row pivoting) and two other subprograms to solve for x and y.

6. Write and test programs to solve $Ax = b$ using column equilibration, row equilibration, and full pivoting. (Refer to Problem 5 and Computer Problem 1 for the terminology.)

7. Write a subroutine `Gaussj` (n, A, b, x, p, s, d) that solves an $n \times n$ system $Ax = b$ by the **Gauss-Jordan** method, with column equilibration and scaled row pivoting. In the Gauss-Jordan algorithm (without pivoting) at the kth major step, multiples of row k are subtracted from *all* the other rows so that the coefficient of x_k is 0 in all rows except the kth row. At the end, the matrix will be a diagonal matrix (rather than an upper triangular matrix, as in the Gaussian elimination method). With scaled row pivoting, row p_k is used as the pivot row to produce 0 coefficients of x_k in all rows except row p_k. Column equilibration, as discussed in Problem 6, should be carried out at the beginning. The divisors needed in this process should be stored in array d since they will be needed at the end in obtaining x.

8. Write and test a recursive version of the scaled Gaussian elimination algorithm, in which the scale array is recomputed repeatedly.

4.4 Norms and the Analysis of Errors

To discuss the errors in numerical problems involving vectors, it is useful to employ *norms*. Our vectors are usually in one of the spaces $\mathbb{R}^n$, but a norm can be defined on any vector space.

Vector Norms

On a vector space V, a **norm** is a function $\|\cdot\|$ from V to the set of nonnegative reals that obeys these three postulates:

$$\|x\| > 0 \qquad \text{if } x \neq 0,\ x \in V \tag{1}$$

$$\|\lambda x\| = |\lambda|\,\|x\| \qquad \text{if } \lambda \in \mathbb{R},\ x \in V \tag{2}$$

$$\|x + y\| \leqq \|x\| + \|y\| \qquad \text{if } x, y \in V \quad \text{(triangle inequality)} \tag{3}$$

We can think of $\|x\|$ as the **length** or **magnitude** of the vector x. A norm on a vector space generalizes the notion of absolute value, $|r|$, for a real or complex number. The most familiar norm on $\mathbb{R}^n$ is the **Euclidean ℓ_2-norm** defined by

$$\|x\|_2 = \left(\sum_{i=1}^{n} x_i^2\right)^{1/2} \qquad \text{where} \qquad x = (x_1, x_2, \dots, x_n)^T$$

This is the norm that corresponds to our intuitive concept of length. We use the subscript 2 as an identifier only. In numerical analysis, other norms are also used; the simplest and easiest to compute is called the ℓ_∞-**norm**:

$$\|x\|_\infty = \max_{1 \leqq i \leqq n} |x_i| \tag{4}$$

Again, a subscript is used to distinguish this norm from others. A third important example of a norm on $\mathbb{R}^n$ is called the ℓ_1-**norm**:

$$\|x\|_1 = \sum_{i=1}^{n} |x_i| \tag{5}$$

Example 1 Using the norm $\|\cdot\|_1$, compare the lengths of the following three vectors in $\mathbb{R}^4$. Then repeat the calculation for the norms $\|\cdot\|_2$ and $\|\cdot\|_\infty$.

$$x = (4, 4, -4, 4)^T \qquad v = (0, 5, 5, 5)^T \qquad w = (6, 0, 0, 0)^T$$

Solution The results are displayed here:

	$\|\cdot\|_1$	$\|\cdot\|_2$	$\|\cdot\|_\infty$
x	16.	8.	4.
v	15.	8.66	5.
w	6.	6.	6.

■

To understand these norms better, it is instructive to consider $\mathbb{R}^2$. For the three norms given above, we give sketches in Figure 4.1 of the set

$$\{x : x \in \mathbb{R}^2, \|x\| \leqq 1\}$$

This set is called the **unit cell** or the **unit ball** in two-dimensional vector space.

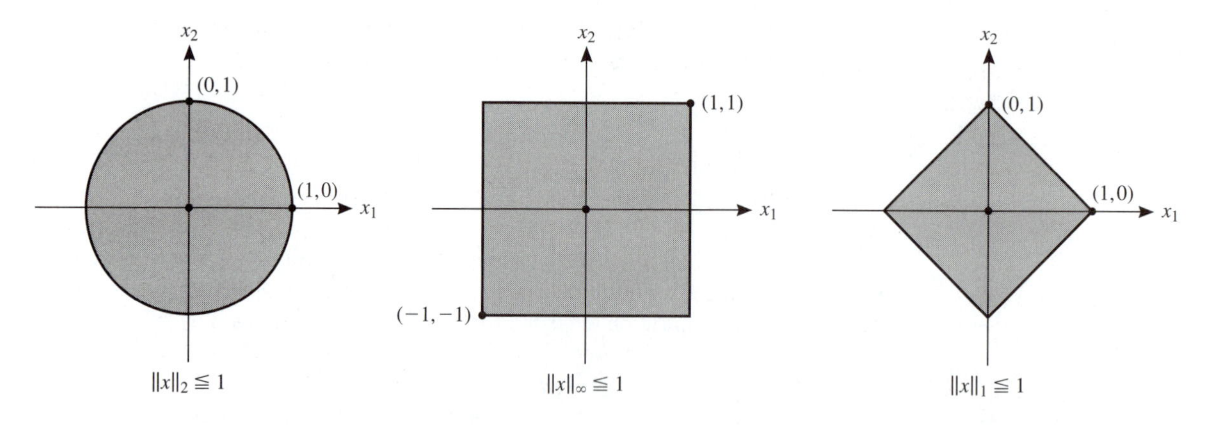

FIGURE 4.1 Unit cells in $\mathbb{R}^2$ for three norms

Matrix Norms

Now we turn to the question of defining norms for matrices. Although we could deal with general matrix norms, subjecting them only to the same requirements (1)–(3), we usually prefer a matrix norm to be intimately related to a vector norm. If a vector norm $\|\cdot\|$ has been specified, the matrix norm **subordinate** to it is defined by

$$\|A\| = \sup\left\{\|Au\| : u \in \mathbb{R}^n, \|u\| = 1\right\} \tag{6}$$

This is also called the matrix norm **associated** with the given vector norm. Here A is understood to be an $n \times n$ matrix.

THEOREM 1 *If $\|\cdot\|$ is any norm on $\mathbb{R}^n$, then Equation (6) defines a norm on the linear space of all $n \times n$ matrices.*

Proof We shall verify the three axioms for a norm. First, if $A \neq 0$, then A has at least one nonzero column, say, $A^{(j)} \neq 0$. Consider the vector x having 1 as its jth component and 0's elsewhere—that is, $x = (0, \dots, 0, 1, 0, \dots, 0)^T$. Obviously, $x \neq 0$ and the vector $v = x/\|x\|$ is of norm 1. Hence, by the definition of $\|A\|$,

$$\|A\| \geqq \|Av\| = \frac{\|Ax\|}{\|x\|} = \frac{\|A^{(j)}\|}{\|x\|} > 0$$

Next, from Property (2) of the vector norm, we have

$$\|\lambda A\| = \sup\{\|\lambda Au\| : \|u\| = 1\} = |\lambda| \sup\{\|Au\| : \|u\| = 1\} = |\lambda|\, \|A\|$$

For the triangle inequality, we use the analogous property of the vector norm and Problem 4 to write

$$\begin{aligned}
\|A + B\| &= \sup\{\|(A + B)u\| : \|u\| = 1\} \\
&\leqq \sup\{\|Au\| + \|Bu\| : \|u\| = 1\} \\
&\leqq \sup\{\|Au\| : \|u\| = 1\} + \sup\{\|Bu\| : \|u\| = 1\} \\
&= \|A\| + \|B\|
\end{aligned}$$

■

An important consequence of Definition (6)—and indeed the motivation for the definition—is that

$$\|Ax\| \leqq \|A\|\, \|x\| \qquad (x \in \mathbb{R}^n) \tag{7}$$

To prove this, observe that it is true for $x = 0$. If $x \neq 0$, then the vector $v = x/\|x\|$ is of norm 1. Hence, from (6),

$$\|A\| \geqq \|Av\| = \frac{\|Ax\|}{\|x\|}$$

As an illustration of this important concept, let us fix as our vector norm the norm $\|x\|_\infty$, defined in Equation (4). What is its subordinate matrix norm? Here is the calculation:

$$\begin{aligned} \|A\|_\infty &= \sup_{\|u\|_\infty=1} \|Au\|_\infty = \sup_{\|u\|_\infty=1} \left\{ \max_{1\leqq i\leqq n} |(Au)_i| \right\} = \max_{1\leqq i\leqq n} \left\{ \sup_{\|u\|_\infty=1} |(Au)_i| \right\} \\ &= \max_{1\leqq i\leqq n} \left\{ \sup_{\|u\|_\infty=1} \left| \sum_{j=1}^{n} a_{ij}u_j \right| \right\} = \max_{1\leqq i\leqq n} \sum_{j=1}^{n} |a_{ij}| \end{aligned} \tag{8}$$

Here we use the fact (Problem 9) that two maximization processes can be interchanged. Also used is the fact that the supremum of $|\sum_{j=1}^{n} a_{ij}u_j|$ for fixed i and $\|u\|_\infty = 1$ is obtained by putting $u_j = +1$ if $a_{ij} \geqq 0$ and $u_j = -1$ if $a_{ij} < 0$.

Thus, we have proved the following theorem.

THEOREM 2 *If the vector norm $\|\cdot\|_\infty$ is defined by*

$$\|x\|_\infty = \max_{1\leqq i\leqq n} |x_i|$$

then its subordinate matrix norm is given by

$$\|A\|_\infty = \max_{1\leqq i\leqq n} \sum_{j=1}^{n} |a_{ij}|$$

A matrix norm subordinate to a vector norm has additional properties besides the basic ones (1)–(3). For example:

$$\|I\| = 1 \tag{9}$$

$$\|AB\| \leqq \|A\|\,\|B\| \tag{10}$$

The proof of (9) is immediate from the definition (6), and Equation (10) follows from Equations (6) and (7).

Another important matrix norm is the **ℓ_2-matrix norm** (also called the **spectral norm**) defined by

$$\|A\|_2 = \sup_{\|x\|_2=1} \|Ax\|_2$$

Here the subordinate vector norm is the Euclidean norm. In Theorem 5 of Section 5.4, we will prove that

$$\|A\|_2 = \max_{1\leqq i\leqq n} |\sigma_i|$$

where σ_i are the **singular values** of A. If $\sigma_1 \geqq \sigma_2 \geqq \cdots \geqq \sigma_n \geqq 0$, then we have $Av_1 = \sigma_1 u_1$ and $A^T u_1 = \sigma_1 v_1$ from the proof of Theorem 1 in Section 5.4.

Hence, $A^T A v_1 = \sigma_1^2 v_1$. So σ_1^2 equals the largest eigenvalue of $A^T A$. Consequently, the 2-matrix norm is frequently defined as

$$||A||_2 = \sqrt{\rho(A^T A)}$$

where $\rho(A^T A)$ is called the **spectral radius** of $A^T A$ and is defined as the largest eigenvalue of $A^T A$.

Condition Number

Now let us put these concepts to work. We consider an equation

$$Ax = b$$

where A is an $n \times n$ matrix. Suppose that A is invertible.

Example 2 If A^{-1} is perturbed to obtain a new matrix B, then the solution $x = A^{-1}b$ is perturbed to become a new vector $\widetilde{x} = Bb$. How large is this latter perturbation in absolute and relative terms?

Solution We have, using any vector norm and its subordinate matrix norm:

$$||x - \widetilde{x}|| = ||x - Bb|| = ||x - BAx|| = ||(I - BA)x|| \leqq ||I - BA||\,||x||$$

This gives the magnitude of the perturbation in x. If the relative perturbation is being measured, we can write

$$\frac{||x - \widetilde{x}||}{||x||} \leqq ||I - BA|| \tag{11}$$

Inequality (11) gives an upper bound on $||x - \widetilde{x}||/||x||$, and this ratio is taken as a measure of the **relative error** between x and $\widetilde{x}$. ■

Example 3 Suppose that the vector b is perturbed to obtain a vector $\widetilde{b}$. If x and $\widetilde{x}$ satisfy $Ax = b$ and $A\widetilde{x} = \widetilde{b}$, by how much do x and $\widetilde{x}$ differ, in absolute and relative terms?

Solution Assuming that A is invertible, we have

$$\begin{aligned} ||x - \widetilde{x}|| &= ||A^{-1}b - A^{-1}\widetilde{b}|| = ||A^{-1}(b - \widetilde{b})|| \\ &\leqq ||A^{-1}||\,||b - \widetilde{b}|| \end{aligned}$$

This gives a measure of the perturbation in x. To estimate the relative perturbation, we write

$$\begin{aligned} ||x - \widetilde{x}|| \leqq ||A^{-1}||\,||b - \widetilde{b}|| &= ||A^{-1}||\,||Ax||\,\frac{||b - \widetilde{b}||}{||b||} \\ &\leqq ||A^{-1}||\,||A||\,||x||\,\frac{||b - \widetilde{b}||}{||b||} \end{aligned}$$

Hence,

$$\frac{\|x - \widetilde{x}\|}{\|x\|} \leqq \kappa(A) \frac{\|b - \widetilde{b}\|}{\|b\|} \tag{12}$$

where

$$\kappa(A) \equiv \|A\| \cdot \|A^{-1}\| \tag{13}$$

The number $\kappa(A)$ is called a **condition number** of the matrix A. ■

Inequality (12) tells us that the relative error in x is no greater than $\kappa(A)$ times the relative error in b. The condition number depends on the vector norm chosen at the beginning of the analysis. From Inequality (12), we see that if the condition number is small, then small perturbations in b lead to small perturbations in x. The inequality $\kappa(A) \geqq 1$ is always true. (See Problem 13.)

For an example to illustrate the condition number, let $\varepsilon > 0$ and

$$A = \begin{bmatrix} 1 & 1+\varepsilon \\ 1-\varepsilon & 1 \end{bmatrix} \qquad A^{-1} = \varepsilon^{-2} \begin{bmatrix} 1 & -1-\varepsilon \\ -1+\varepsilon & 1 \end{bmatrix}$$

If the ∞-norm is used, we have $\|A\|_\infty = 2+\varepsilon$ and $\|A^{-1}\|_\infty = \varepsilon^{-2}(2+\varepsilon)$, by Equation (8). Hence, $\kappa(A) = [(2+\varepsilon)/\varepsilon]^2 > 4/\varepsilon^2$. If $\varepsilon \leqq 0.01$, then $\kappa(A) \geqq 40{,}000$. In such a case, a small relative perturbation in b may induce a relative perturbation *40,000 times greater* in the solution of the system $Ax = b$.

If we solve a system of equations

$$Ax = b$$

numerically, we obtain not the exact solution x but an approximate solution $\widetilde{x}$. One can test $\widetilde{x}$ by forming $A\widetilde{x}$ to see whether it is close to b. Thus, we have the **residual vector**

$$r = b - A\widetilde{x}$$

The difference between the exact solution x and the approximate solution $\widetilde{x}$ is called the **error vector**

$$e = x - \widetilde{x}$$

The following relationship

$$Ae = r \tag{14}$$

between the error vector and the residual vector is of fundamental importance.

Notice that $\widetilde{x}$ is the exact solution of the linear system $A\widetilde{x} = \widetilde{b}$, which has a perturbed right-hand side $\widetilde{b} = b - r$. We now establish relationships between the relative errors in x and b. In other words, we want to relate $\|x - \widetilde{x}\|/\|x\|$ to $\|b - \widetilde{b}\|/\|b\| = \|r\|/\|b\|$. The following theorem shows that the condition number $\kappa(A)$ plays an important role.

THEOREM 3

$$\frac{1}{\kappa(A)} \frac{\|r\|}{\|b\|} \leqq \frac{\|e\|}{\|x\|} \leqq \kappa(A) \frac{\|r\|}{\|b\|}$$

Proof The inequality on the right can be written as

$$\|e\| \, \|b\| \leqq \|A\| \, \|A^{-1}\| \, \|r\| \, \|x\|$$

and this is true since

$$\|e\| \, \|b\| = \|A^{-1}r\| \, \|Ax\| \leqq \|A^{-1}\| \, \|r\| \, \|A\| \, \|x\|$$

(In fact, the inequality on the right in the theorem is Inequality (12).) The inequality on the left can be written as

$$\|r\| \, \|x\| \leqq \|A\| \, \|A^{-1}\| \, \|b\| \, \|e\|$$

and this follows at once from

$$\|r\| \, \|x\| = \|Ae\| \, \|A^{-1}b\| \leqq \|A\| \, \|e\| \, \|A^{-1}\| \, \|b\| \qquad \blacksquare$$

A matrix with a large condition number is said to be **ill conditioned**. For an ill-conditioned matrix A, there will be cases in which the solution of a system $Ax = b$ will be very sensitive to small changes in the vector b. In other words, to attain a certain precision in the determination of x, we shall require significantly higher precision in b. If the condition number of A is of moderate size, the matrix is said to be **well conditioned**.

PROBLEMS 4.4

1. Show that the norms $\|x\|_\infty$, $\|x\|_2$, $\|x\|_1$ satisfy the postulates (1), (2), (3) for norms.
2. Show that $\|x\|_\infty \leqq \|x\|_2 \leqq \|x\|_1$ for all $x \in \mathbb{R}^n$, and that equalities can occur, even for nonzero vectors.
3. Show that $\|x\|_1 \leqq n\|x\|_\infty$ and $\|x\|_2 \leqq \sqrt{n}\,\|x\|_\infty$ for $x \in \mathbb{R}^n$.
4. Show that for any two functions into $\mathbb{R}$,

$$\sup[f(x) + g(x)] \leqq \sup f(x) + \sup g(x)$$

5. For the matrix norm in Theorem 2, prove or disprove: $\|AB\|_\infty = \|A\|_\infty \|B\|_\infty$. What about the special case $\|A^2\|_\infty = \|A\|_\infty^2$?

6. Show that a norm defined on $\mathbb{R}^n$ must involve all components of a vector in some way. Show that the norm $\|x\|_\infty$ is indeed the simplest on $\mathbb{R}^n$. (You will have to define a suitable concept of *simplicity*.)

7. Determine whether these expressions define norms on $\mathbb{R}^n$:

(a) $\max\{|x_2|, |x_3|, \dots, |x_n|\}$

(b) $\sum_{i=1}^{n} |x_i|^3$

(c) $\left\{ \sum_{i=1}^{n} |x_i|^{1/2} \right\}^2$

(d) $\max\{|x_1 - x_2|, |x_1 + x_2|, |x_3|, |x_4|, \dots, |x_n|\}$

(e) $\sum_{i=1}^{n} 2^{-i}|x_i|$

8. Define $\|A\| = \sum_{i=1}^{n} \sum_{j=1}^{n} |a_{ij}|$. Show that this is a matrix norm (that is, a norm on the linear space of all $n \times n$ matrices). Show that it is not subordinate to any vector norm. Does it conform to Equations (9) and (10)?

9. **(a)** Prove that if A and B are arbitrary sets and f is a bounded real function on $A \times B$, then

$$\sup_{a \in A} \sup_{b \in B} f(a, b) = \sup_{b \in B} \sup_{a \in A} f(a, b)$$

(b) Show by example that a supremum and an infimum cannot be interchanged in general.

(c) Show that

$$\sup_{a \in A} \inf_{b \in B} f(a, b) \leqq \inf_{b \in B} \sup_{a \in A} f(a, b)$$

10. Prove that for any vector norm and its subordinate matrix norm, and for any $n \times n$ matrix A, there corresponds a vector $x \neq 0$ such that $\|Ax\| = \|A\| \, \|x\|$.

11. Show that for the vector norm $\|x\|_1$ defined in Equation (5), the subordinate matrix norm is

$$\|A\|_1 = \max_{1 \leqq j \leqq n} \sum_{i=1}^{n} |a_{ij}|$$

12. Do the subordinate matrix norms satisfy $\|AB\| = \|BA\|$? Explain.

13. Prove that the condition number of an invertible matrix must be at least 1.

14. What matrices have condition number equal to 1?

15. Using the one matrix norm (Problem 11), compute the condition number of the matrix $\begin{bmatrix} 1 & 1 + \varepsilon \\ 1 - \varepsilon & 1 \end{bmatrix}$.

16. Using the infinity matrix norm, Equation (8), compute the condition number of the matrix $\begin{bmatrix} 7 & 8 \\ 9 & 10 \end{bmatrix}$.

17. Let A be an $n \times n$ matrix with inverse $C = (c_{ij})$. Show that in the solution of $Ax = b$, a perturbation of amount δ in b_j will cause a perturbation of $c_{ij}\delta$ in x_i.

18. (Continuation) Prove that a perturbation of amount δ in a_{jk} will produce a perturbation of approximately $-c_{ij}x_k\delta$ in x_i.

19. (Continuation) The following quantity is sometimes used for a condition number of an $n \times n$ matrix A:

$$M(A) = n \max_{1 \leqq i, j \leqq n} |a_{ij}| \max_{1 \leqq i, j \leqq n} |c_{ij}|$$

where C is the inverse of A. Prove that if $Ax = b$ and if only one component of b is perturbed (say by an amount ε), then the perturbed x (denoted by $\widetilde{x}$) satisfies

$$\frac{\|x - \widetilde{x}\|_\infty}{\|x\|_\infty} \leqq M(A) \frac{|\varepsilon|}{\|b\|_\infty}$$

20. It is proved in the text that if $Ax = b$ and $A\widetilde{x} = \widetilde{b}$, then

$$\frac{\|x - \widetilde{x}\|}{\|x\|} \leqq \kappa(A) \frac{\|b - \widetilde{b}\|}{\|b\|}$$

Prove that for every nonsingular matrix A, this inequality will become an equality for some vectors b and $\widetilde{b}$. (Of course, we want $b \neq 0$ and $b \neq \widetilde{b}$.) *Hint*: Go through the proof of the above inequality, and find the conditions for equality to occur.

21. Let $n = 3$ and let

$$A = \begin{bmatrix} 4 & -3 & 2 \\ -1 & 0 & 5 \\ 2 & 6 & -2 \end{bmatrix}$$

Among all the vectors x satisfying $\|x\|_\infty \leqq 1$, find one for which $\|Ax\|_\infty$ is as large as possible. Also give the numerical value of $\|A\|_\infty$.

22. A **weighted ℓ_∞-norm** is a norm on $\mathbb{R}^n$ of the form

$$\|x\| = \max_{1 \leqq i \leqq n} w_i |x_i|$$

where $w_1, w_2, \ldots, w_n$ are fixed positive numbers called **weights**. Prove the norm postulates for this norm. What is the subordinate matrix norm?

23. Prove that if $\|\cdot\|$ is a norm on a vector space, and if we define $\|x\|' = \alpha\|x\|$ with a fixed positive α, then $\|\cdot\|'$ is also a norm.

24. If the construction in Problem 23 is applied to a subordinate matrix norm, is the resulting norm also a subordinate matrix norm?

25. Let $\|\cdot\|$ be a norm on a vector space V. For x and y in V, let $d(x, y) = \|x - y\|$. Show that d has these properties:

(a) $d(x, x) = 0$

(b) $d(x, y) = d(y, x)$

(c) $d(x, y) > 0$ if $x \neq y$

(d) $d(x, y) \leqq d(x, z) + d(z, y)$

(A function having these four properties is called a **metric**.)

26. Give an example of a well-conditioned matrix whose determinant is very small.

27. Compute the condition number of the $n \times n$ lower triangular matrix that has $+1$'s on the diagonal and -1's below the diagonal. Use the matrix norm $\|\cdot\|_\infty$.

28. Prove that if A has a nontrivial fixed point (that is, $Ax = x \neq 0$), then $\|A\| \geqq 1$ for any subordinate matrix norm.

29. Give an example of a norm on $\mathbb{R}^2$ such that the norm of $(1, 0)$ is 2 and the norm of $(1,1)$ is 1.

30. Is there a norm on $\mathbb{R}^2$ such that $\|(1, 0)\| = \|(0, 1)\| = \|(\frac{1}{3}, \frac{1}{3})\|$?

31. (See Problems 2 and 3.) Find the precise conditions on x in order that

(a) $\|x\|_\infty = \|x\|_1$

(b) $\|x\|_\infty = \|x\|_2$

(c) $\|x\|_1 = \|x\|_2$

32. Show that $\|A\|$ is the smallest number M such that $\|Ax\| \leqq M\|x\|$ for all x.

33. For any $n \times n$ matrix A, define

$$\|A\|_F = \left(\sum_{i=1}^{n} \sum_{j=1}^{n} a_{ij}^2 \right)^{1/2}$$

(This is called the **Frobenius norm**.) Is this a subordinate matrix norm? Answer the same question for $\|A\| = \max_{1 \leqq i, j \leqq n} |a_{ij}|$. Prove that these equations define norms on the vector space of all $n \times n$ matrices.

34. Is it necessarily true, for any subordinate matrix norm, that the norm of a permutation matrix is unity? Explain.

35. For a vector $x = (x_1, x_2, \ldots, x_n)^T \in \mathbb{R}^n$, we define $|x|$, the absolute value of the vector x, to be the vector $(|x_1|, |x_2|, \ldots, |x_n|)^T$. For vectors x and y, we also define $x \leqq y$ to mean that $x_i \leqq y_i$ for $i = 1, 2, \ldots, n$. Prove that the norms $\|\cdot\|_1$, $\|\cdot\|_2$, and $\|\cdot\|_\infty$ have the property that if $|x| \leqq |y|$, then $\|x\| \leqq \|y\|$.

36. For any real number $p \geqq 1$, the formula

$$\|x\|_p = \left(\sum_{i=1}^{n} |x_i|^p \right)^{1/p}$$

defines a norm. (For the proof, consult Bartle [1976, p. 61].) Prove that for each $x \in \mathbb{R}^n$,

$$\lim_{p \to \infty} \|x\|_p = \|x\|_\infty$$

This explains why the notation $\|\cdot\|_\infty$ is used.

37. Prove these properties of norms:

(a) $\|0\| = 0$

(b) $\|x + y\| \geqq \big|\, \|x\| - \|y\| \,\big|$

(c) $\left\| \sum_{i=1}^{m} x^{(i)} \right\| \leqq \sum_{i=1}^{m} \|x^{(i)}\|$ for vectors $x^{(1)}, x^{(2)}, \ldots, x^{(m)}$

38. Let $\|\cdot\|$ be a norm on $\mathbb{R}^n$. Define

$$\|x\|' = \sup \left\{ u^T x : u \in \mathbb{R}^n ,\ \|u\| = 1 \right\}$$

Prove that this equation defines a norm. Prove that if the process is repeated, then the original norm is obtained; that is, $(\|\cdot\|')' = \|\cdot\|$. Prove that for all $x, y \in \mathbb{R}^n$,

$$|x^T y| \leqq \|x\| \, \|y\|'$$

39. Prove this inequality for condition numbers as defined in Equation (13):

$$\kappa(AB) \leqq \kappa(A)\, \kappa(B)$$

40. Compute condition numbers using norms $\|A\|_1$, $\|A\|_2$, and $\|A\|_\infty$:

(a) $\begin{bmatrix} a+1 & a \\ a & a-1 \end{bmatrix}$

(b) $\begin{bmatrix} 0 & 1 \\ -2 & 0 \end{bmatrix}$

(c) $\begin{bmatrix} \alpha & 1 \\ 1 & 1 \end{bmatrix}$

41. Using Problems 2 and 3, prove that

$$n^{-1}\|A\|_2 \leqq n^{-1/2}\|A\|_\infty \leqq \|A\|_2 \leqq n^{1/2}\|A\|_1 \leqq n\|A\|_2$$

42. In solving the system of equations $Ax = b$ with matrix $A = \begin{bmatrix} 1 & 2 \\ 1 & 2.01 \end{bmatrix}$, predict how slight changes in b will affect the solution x. Test your prediction in the concrete case when $b = (4, 4)$ and $b' = (3, 5)$.

43. Let A be an $m \times n$ matrix. We interpret A as a linear map from $\mathbb{R}^n$ with the norm $\|\cdot\|_1$ to $\mathbb{R}^m$ with the norm $\|\cdot\|_\infty$. What is $\|A\|$ under these circumstances?

What is wanted is a simple formula for

$$\|A\| = \max\left\{\|Ax\|_\infty : \|x\|_1 = 1\right\}$$

44. Try Problem 43 when the norms $\|\cdot\|_1$ and $\|\cdot\|_\infty$ are interchanged.

45. Prove that the condition number $\kappa(A)$ can be expressed by the formula

$$\kappa(A) = \sup_{\|x\|=\|y\|} \|Ax\|/\|Ay\|$$

46. Let $\|\cdot\|$ be a norm on $\mathbb{R}^n$, and let A be an $n \times n$ matrix. Put $\|x\|' = \|Ax\|$. What are the precise conditions on A to ensure that $\|\cdot\|'$ is also a norm?

47. Show that if the Euclidean norm is used, the set

$$H = \left\{x \in \mathbb{R}^n : \|x - a\| = \|x - b\|\right\}$$

will be a hyperplane (that is, a translate of a linear subspace of dimension $n-1$), but for other norms this is not generally true. Illustrate in the case $n = 2$.

48. Prove that the condition number has the property

$$\kappa(\lambda A) = \kappa(A) \qquad (\lambda \neq 0)$$

49. Prove that if a square matrix A satisfies an inequality $\|Ax\| \geqq \theta\|x\|$ for all x, with $\theta > 0$, then A is nonsingular and $\|A^{-1}\| \leqq \theta^{-1}$. This is valid for any vector norm and its subordinate matrix norm.

50. Prove that if A is diagonally dominant, then it will have the property in Problem 49. Give a value for θ when the norm $\|\cdot\|_\infty$ is used.

51. Prove that if A is nonsingular, then there is a $\delta > 0$ with the property that $A + E$ is nonsingular for all matrices E satisfying $\|E\| < \delta$. This can be shown for any norm on the vector space of all $n \times n$ matrices.

52. Show that in Problem 51 we can use

$$\delta = \inf\left\{\|Ax\| : \|x\| = 1\right\}$$

Here a vector norm and its subordinate matrix norm are used.

53. Let A be an $n \times n$ matrix and N its null space (**kernel**). Define

$$\delta = \inf\left\{\|Ax\| : \|x\| = 1 \text{ and } x \perp N\right\}$$

Prove that if $\|E\| < \delta$, then $\operatorname{rank}(A + E) \geqq \operatorname{rank}(A)$.

54. Prove: If A is nonsingular, then there exists a singular matrix B such that $\|B - A\|_2 = \|A^{-1}\|_2$.

55. Establish Equation (14).

56. Prove that

$$\|A\|_2 = \max_{\substack{\|x\|_2=1 \\ \|y\|_2=1}} |y^T Ax|$$

COMPUTER PROBLEM 4.4

1. Write subroutines for computing the norm of a vector and a square matrix. Use the **maximum norm**, $\|x\|_\infty$, and its subordinate matrix norm.

4.5 Neumann Series and Iterative Refinement

An important application of norms occurs in making precise the concept of convergence in a vector space. If a vector space V is assigned a norm $\|\cdot\|$, then the pair $(V, \|\cdot\|)$ is a **normed linear space**. The notion of a convergent sequence of vectors $v^{(1)}, v^{(2)}, \ldots$ is then defined as follows: We say that the given sequence **converges** to a vector v if

$$\lim_{k\to\infty} \|v^{(k)} - v\| = 0$$

This conforms to our intuitive idea that the distances between the vectors $v^{(k)}$ and the limit vector v are approaching 0 as k increases.

For an example in $\mathbb{R}^4$, let

$$v^{(k)} = \begin{bmatrix} 3 - k^{-1} \\ -2 + k^{-\frac{1}{2}} \\ (k+1)k^{-1} \\ e^{-k} \end{bmatrix} \qquad \text{and} \qquad v = \begin{bmatrix} 3 \\ -2 \\ 1 \\ 0 \end{bmatrix}$$

Then

$$v^{(k)} - v = \begin{bmatrix} -k^{-1} \\ k^{-\frac{1}{2}} \\ k^{-1} \\ e^{-k} \end{bmatrix}$$

If we compute $\|v^{(k)} - v\|$, using for example the infinity norm of Section 4.4, we see that $\|v^{(k)} - v\|_\infty \to 0$ as $k \to \infty$. Hence, v is the limit of the sequence $[v^{(k)}]$ in the normed linear space $(\mathbb{R}^4, \|\cdot\|_\infty)$.

At this point it is appropriate to give without proof an important result about normed linear spaces: Any two norms on a finite-dimensional vector space lead to the same concept of convergence. Thus, in the preceding example, having verified that $\|v^{(k)} - v\|_\infty \to 0$, we can conclude without further calculation that $\|v^{(k)} - v\| \to 0$

for *any* norm on $\mathbb{R}^4$. *Caution:* This theorem does not apply in infinite-dimensional normed linear spaces. (See Problem 20 for an example.)

Another important result about finite-dimensional normed linear spaces is that in such a space, each Cauchy sequence converges. Thus, if a sequence $[v^{(k)}]$ in a finite-dimensional normed linear space satisfies the **Cauchy criterion**

$$\lim_{k\to\infty} \sup_{i,j\geqq k} \|v^{(i)} - v^{(j)}\| = 0$$

then there necessarily exists a point $v \in V$ to which the sequence converges.

We shall apply these ideas to vectors in $\mathbb{R}^n$ and to $n \times n$ matrices. In the next theorem, we take $\|\cdot\|$ to be any norm on $\mathbb{R}^n$ and use its subordinate matrix norm $\|\cdot\|$ as defined in Section 4.4.

Neumann Series

THEOREM 1 *If A is an $n \times n$ matrix such that $\|A\| < 1$, then $I - A$ is invertible, and*

$$(I - A)^{-1} = \sum_{k=0}^{\infty} A^k \tag{1}$$

Proof First, we shall show that $I - A$ is invertible. If it is not invertible, then it is singular, and there exists a vector x satisfying $\|x\| = 1$ and $(I - A)x = 0$. From this we have

$$1 = \|x\| = \|Ax\| \leqq \|A\|\,\|x\| = \|A\|$$

which contradicts the hypothesis that $\|A\| < 1$.

We shall show that the partial sums of the Neumann series converge to $(I - A)^{-1}$:

$$\sum_{k=0}^{m} A^k \to (I - A)^{-1} \qquad (\text{as } m \to \infty)$$

It will suffice to prove that

$$(I - A)\sum_{k=0}^{m} A^k \to I \qquad (\text{as } m \to \infty) \tag{2}$$

The left-hand side can be written as

$$(I - A)\sum_{k=0}^{m} A^k = \sum_{k=0}^{m}(A^k - A^{k+1}) = A^0 - A^{m+1} = I - A^{m+1}$$

Since $\|A^{m+1}\| \leqq \|A\|^{m+1} \to 0$ as $m \to \infty$, this establishes (2). (Why?) ■

This theorem occurs in essentially the same form in the theory of continuous linear operators on any Banach space. The theorem has both practical and theoretical consequences of great importance. Observe that from Equation (1) we obtain this estimate:

$$\|(I-A)^{-1}\| \leqq \sum_{k=0}^{\infty} \|A^k\| \leqq \sum_{k=0}^{\infty} \|A\|^k = \frac{1}{1-\|A\|}$$

Example 1 Use the Neumann series to compute the inverse of the matrix

$$B = \begin{bmatrix} 0.9 & -0.2 & -0.3 \\ 0.1 & 1.0 & -0.1 \\ 0.3 & 0.2 & 1.1 \end{bmatrix}$$

Solution Let $B = I - A$, where

$$A = \begin{bmatrix} 0.1 & 0.2 & 0.3 \\ -0.1 & 0.0 & 0.1 \\ -0.3 & -0.2 & -0.1 \end{bmatrix}$$

Since $\|A\|_\infty = 0.6$, the Neumann series $\sum_{k=0}^{\infty} A^k$ will converge to B^{-1}. Using the algorithm in Problem 22, we compute some of the partial sums:

$$\sum_{k=0}^{0} A^k = \begin{bmatrix} 1 & 0 & 0 \\ 0 & 1 & 0 \\ 0 & 0 & 1 \end{bmatrix}$$

$$\sum_{k=0}^{1} A^k = \begin{bmatrix} 1.1 & 0.2 & 0.3 \\ -0.1 & 1.0 & 0.1 \\ -0.3 & -0.2 & 0.9 \end{bmatrix}$$

$$\sum_{k=0}^{2} A^k = \begin{bmatrix} 1.00 & 0.16 & 0.32 \\ -0.14 & 0.96 & 0.06 \\ -0.28 & -0.24 & 0.80 \end{bmatrix}$$

$$\vdots$$

$$\sum_{k=0}^{19} A^k = \begin{bmatrix} 1.00000\,000 & 0.14285\,714 & 0.28571\,429 \\ -0.12500\,000 & 0.96428\,571 & 0.05357\,143 \\ -0.25000\,000 & -0.21428\,571 & 0.82142\,857 \end{bmatrix}$$

The last partial sum shown gives B^{-1} to eight decimal places of accuracy. ■

Here is a variant of Theorem 1.

THEOREM 2 *If A and B are $n \times n$ matrices such that $\|I - AB\| < 1$, then A and B are invertible. Furthermore,*

$$A^{-1} = B\sum_{k=0}^{\infty}(I-AB)^k \quad \text{and} \quad B^{-1} = \sum_{k=0}^{\infty}(I-AB)^k A \tag{3}$$

Proof By the preceding theorem, AB is invertible and its inverse is

$$(AB)^{-1} = \sum_{k=0}^{\infty}(I - AB)^k$$

Hence,

$$A^{-1} = BB^{-1}A^{-1} = B(AB)^{-1} = B\sum_{k=0}^{\infty}(I - AB)^k$$

$$B^{-1} = B^{-1}A^{-1}A = (AB)^{-1}A = \sum_{k=0}^{\infty}(I - AB)^k A$$

■

Iterative Refinement

If $x^{(0)}$ is an approximate solution of the equation

$$Ax = b$$

then the precise solution x is given by

$$x = x^{(0)} + A^{-1}(b - Ax^{(0)}) = x^{(0)} + e^{(0)} \tag{4}$$

where $e^{(0)} = A^{-1}(b - Ax^{(0)})$ and is called the **error vector**. The **residual vector** corresponding to the approximate solution $x^{(0)}$ is $r^{(0)} = b - Ax^{(0)}$. It is computable. Of course, we do not want to compute A^{-1}, but the vector $e^{(0)} = A^{-1}r^{(0)}$ can be obtained by solving the equation

$$Ae^{(0)} = r^{(0)}$$

These remarks lead to a numerical procedure called **iterative improvement** or **iterative refinement**, which we now describe in more detail.

Suppose that the equation $Ax = b$ has been solved by Gaussian elimination as in Section 4.3. Since the result is not expected to be the exact solution (because of roundoff errors), we denote it by $x^{(0)}$ and then compute $r^{(0)}$, $e^{(0)}$, and $x^{(1)}$ by the three equations

$$\begin{cases} r^{(0)} = b - Ax^{(0)} \\ A^{(0)} = r^{(0)} \\ x^{(1)} = x^{(0)} + e^{(0)} \end{cases} \tag{5}$$

To obtain better solutions $x^{(2)}, x^{(3)}, \dots$, this process can be repeated. The success of the method depends on computing the residuals $r^{(i)}$ in *double precision* to avoid the loss of significance expected in the subtraction. Thus, the expression $b_i - \sum_{j=1}^{n} a_{ij}x_j^{(0)}$ is evaluated in double precision. (Remember that, ideally, $r^{(i)}$ should be the zero vector, and thus the subtraction involved in computing $r^{(i)}$ *must* involve nearly equal quantities.)

Example 2 Apply iterative refinement in solving the system

$$\begin{bmatrix} 420 & 210 & 140 & 105 \\ 210 & 140 & 105 & 84 \\ 140 & 105 & 84 & 70 \\ 105 & 84 & 70 & 60 \end{bmatrix} \begin{bmatrix} x_1 \\ x_2 \\ x_3 \\ x_4 \end{bmatrix} = \begin{bmatrix} 875 \\ 539 \\ 399 \\ 319 \end{bmatrix}$$

Solution First, the system is factored by Gaussian elimination with scaled row pivoting, and the solution

$$x^{(0)} = (0.99998\,8,\ 1.00013\,7,\ 0.99967\,0,\ 1.00021\,5)^T$$

is obtained from back substitution. Several steps of iterative improvement give

$$\begin{aligned} x^{(1)} &= (0.99999\,4,\ 1.00006\,9,\ 0.99983\,1,\ 1.00011\,0)^T \\ x^{(2)} &= (0.99999\,6,\ 1.00004\,6,\ 0.99989\,1,\ 1.00007\,0)^T \\ x^{(3)} &= (0.99999\,3,\ 1.00008\,0,\ 0.99981\,2,\ 1.00012\,1)^T \\ x^{(4)} &= (1.00000\,0,\ 1.00000\,6,\ 0.99998\,4,\ 1.00001\,1)^T \end{aligned}$$

The true solution is $x = (1,\ 1,\ 1,\ 1)^T$. Since a computer having approximately the precision of the `Marc-32` was used for these calculations, we consider the final result quite good. ■

To analyze this algorithm theoretically, we adopt the point of view that our solution $x^{(0)}$ is obtained by the formula

$$x^{(0)} = Bb$$

where B is an approximate inverse of A. The iterative process then can be written

$$x^{(k+1)} = x^{(k)} + B(b - Ax^{(k)}) \qquad (k \geqq 0) \tag{6}$$

We shall now show that these vectors are the partial sums of the Neumann series and use this fact to show that the sequence $[x^{(k)}]$ converges to the solution of $Ax = b$.

We interpret the loose expression "B is an approximate inverse of A" to mean that $\|I - AB\| < 1$. By Theorem 2, A^{-1} is given by

$$A^{-1} = B \sum_{k=0}^{\infty} (I - AB)^k \tag{7}$$

Thus, the exact solution of the equation $Ax = b$ is

$$x = B \sum_{k=0}^{\infty} (I - AB)^k b \tag{8}$$

THEOREM 3 *If $\| I - AB \| < 1$, then the method of iterative improvement given by Equation (6) produces the sequence of vectors*

$$x^{(m)} = B \sum_{k=0}^{m} (I - AB)^k b \qquad (m \geqq 0)$$

These are the partial sums in Equation (8) and therefore converge to x.

Proof We use induction. Since $x^{(0)} = Bb$, the case $m = 0$ is true. If the mth case is assumed true, then the $(m + 1)$st case is true since

$$\begin{aligned} x^{(m+1)} &= x^{(m)} + B(b - Ax^{(m)}) = B \sum_{k=0}^{m} (I - AB)^k b + Bb - BAB \sum_{k=0}^{m} (I - AB)^k b \\ &= B \left\{ b + (I - AB) \sum_{k=0}^{m} (I - AB)^k b \right\} = B \sum_{k=0}^{m+1} (I - AB)^k b \end{aligned}$$

■

It is also possible to prove directly that the vectors $x^{(m)}$ converge to x. To do this, use Equation (6) to obtain

$$\begin{aligned} x^{(m+1)} - x &= x^{(m)} - x + B(Ax - Ax^{(m)}) \\ &= (I - BA)(x^{(m)} - x) \end{aligned}$$

It follows that

$$\begin{aligned} \|x^{(m+1)} - x\| &\leqq \|I - BA\| \, \|x^{(m)} - x\| \\ &\leqq \|I - BA\|^2 \, \|x^{(m-1)} - x\| \\ &\vdots \\ &\leqq \|I - BA\|^m \, \|x^{(0)} - x\| \end{aligned}$$

Since we have assumed that $\|I - BA\| < 1$, the errors $\|x^{(m)} - x\|$ converge to 0 as $m \longrightarrow \infty$ for any $x^{(0)}$.

Equilibration

For solving systems of linear equations in extremely critical cases, a number of refinements can be added to the factorization and solution procedures described in Section 4.3. We shall discuss briefly five such techniques:

(i) Preconditioning by row equilibration
(ii) Preconditioning by column equilibration
(iii) Full pivoting
(iv) Preconditioning or scaling within each major step of the elimination procedure
(v) Iterative improvement at the end

Row equilibration is the process of dividing each row of the coefficient matrix by the maximum element in absolute value in that row; that is, multiplying row i by $r_i = 1/\max_{1 \leqq j \leqq n} |a_{i,j}|$ for $1 \leqq i \leqq n$. After this has been done, the new elements, $\widetilde{a}_{ij}$, will satisfy $\max_{1 \leqq j \leqq n} |\widetilde{a}_{ij}| = 1$ for $1 \leqq i \leqq n$. In numerical practice on a binary computer, the factor r_i is taken to be a number of the form 2^m as close as possible to $1/\max_{1 \leqq j \leqq n} |a_{ij}|$. This is done to avoid the introduction of additional roundoff errors. Since the ith equation in our system is

$$\sum_{j=1}^{n} a_{ij}x_j = b_i \qquad \text{or} \qquad \sum_{j=1}^{n} (r_i a_{ij})x_j = r_i b_i$$

we should multiply b_i by r_i also. Hence, the numbers r_i should be stored during the factorization procedure so that they can be used in the solution procedure. In matrix-vector notation, row equilibration is written as

$$(RA)x = (Rb)$$

where $R = \text{diag}\,(r_i)$.

Column equilibration is similar except that we are dealing with the columns: We multiply column j by $c_j = 1/\max_{1 \leqq i \leqq n} |a_{ij}|$ for $1 \leqq j \leqq n$. Again, it is preferable to take c_j to be a number of the form 2^m as close as possible to $1/\max_{1 \leqq i \leqq n} |a_{ij}|$. Our original equations

$$\sum_{j=1}^{n} a_{ij}x_j = b_i \qquad (1 \leqq i \leqq n)$$

now can be written as

$$\sum_{j=1}^{n} (c_j a_{ij}) \left(\frac{x_j}{c_j}\right) = b_i \qquad (1 \leqq i \leqq n)$$

After the solution phase, we shall have computed approximate solutions for the components x_j/c_j. These solutions must be multiplied by c_j to obtain x_j. Thus, $c_1, c_2, \ldots, c_n$ will have to be stored also. Using matrices, we write column equilibration as

$$(AC)(C^{-1}x) = b$$

where $C = \text{diag}(c_j)$.

If row and column equilibrations have been carried out on our system, the full pivoting strategy at the initial step can be safely simplified to a search for the largest element (in magnitude) in the matrix. This element determines both the first pivot row *and* the first column in which 0's will be introduced by the elimination. Thus, we intend to process the columns not in the natural order $1, 2, 3, \ldots, n$ but in an order determined by this more accurate pivoting strategy. Two permutation arrays will be needed, one to list the row numbers of the successive pivot elements and another to list the corresponding column numbers. (Refer to Problem 5 and Computer Problem 1 in Section 4.3.)

The fourth technique in our list of refinements is preconditioning or scaling. It contributes to a more logical organization of the factorization phase because each step in the algorithm will be just like the first except for being applied to smaller matrices.

The fifth technique in our list is iterative improvement, which has been discussed previously.

The value of row and column equilibration as a **preconditioning** procedure is somewhat controversial. We can see that row equilibration followed by Gaussian elimination with *unscaled* row pivoting is virtually the same as Gaussian elimination with *scaled* row pivoting. Hence, one reasonable strategy is to begin with column equilibration followed by scaled row pivoting in Gaussian elimination.

This example shows the difference between row-column equilibration

$$\begin{bmatrix} 1 & 10^8 \\ 2 & 0 \end{bmatrix} \longrightarrow \begin{bmatrix} 10^{-8} & 1 \\ 1 & 0 \end{bmatrix} \longrightarrow \begin{bmatrix} 10^{-8} & 1 \\ 1 & 0 \end{bmatrix}$$

and column-row equilibration

$$\begin{bmatrix} 1 & 10^8 \\ 2 & 0 \end{bmatrix} \longrightarrow \begin{bmatrix} \frac{1}{2} & 1 \\ 1 & 0 \end{bmatrix} \longrightarrow \begin{bmatrix} \frac{1}{2} & 1 \\ 1 & 0 \end{bmatrix}$$

Note that row-column equilibration results in a pre- and post-scaling of the system by diagonal matrices, say R and B, such as

$$((RA)B)(B^{-1}x) = (Rb)$$

whereas column-row equilibration corresponds to

$$(S(AC))(C^{-1}x) = (Sb)$$

for some diagonal matrices C and S.

PROBLEMS 4.5

1. Prove that the set of invertible $n \times n$ matrices is an open set in the set of all $n \times n$ matrices. Thus, if A is invertible, then there is a positive ε such that every matrix B satisfying $\| A - B \| < \varepsilon$ is also invertible.
2. Prove that if A is invertible and $\| B - A \| < \| A^{-1} \|^{-1}$, then B is invertible.
3. Prove that if $\| A \| < 1$, then

$$\| (I - A)^{-1} \| \geqq \frac{1}{1 + \| A \|}$$

4. Prove that if A is invertible and $\| A - B \| < \| A^{-1} \|^{-1}$, then

$$\| A^{-1} - B^{-1} \| \leqq \| A^{-1} \| \frac{\| I - A^{-1}B \|}{1 - \| I - A^{-1}B \|}$$

5. Prove that if $\| AB - I \| < 1$, then (with E denoting $AB - I$) we have

$$A^{-1} = B - BE + BE^2 - BE^3 + \cdots$$

6. Prove that if A is invertible, then for any B,

$$\| B - A^{-1} \| \geqq \frac{\| I - AB \|}{\| A \|}$$

7. Prove or disprove: If $1 = \| A \| > \| B \|$, then $A - B$ is invertible.

8. Prove that if $\| A \| < 1$, then

$$(I + A)^{-1} = I - A + A^2 - A^3 + \cdots$$

9. Prove or disprove: If $\| AB - I \| < 1$, then $\| BA - I \| < 1$.

10. Prove that if $\|A\| < 1$, then $\|(I + A)^{-1}\| \leqq \left(1 - \|A\|\right)^{-1}$.

11. Prove that if A is invertible and $\| B - A \| < \| A^{-1} \|^{-1}$, then

$$B^{-1} = A^{-1} \sum_{k=0}^{\infty} (I - BA^{-1})^k$$

12. For any $n \times n$ matrix A, prove that

$$A^m = I - (I - A) \sum_{k=0}^{m-1} A^k$$

13. Criticize this "proof" that every $n \times n$ matrix is invertible: Given A, select a vector norm so that $\| I - A \| < 1$ when the associated matrix norm is used. Then apply the theorem concerning the Neumann series.

14. Prove that if $\inf_{\lambda \in \mathbb{R}} \| I - \lambda A \| < 1$, then A is invertible.

15. Prove that the invertible $n \times n$ matrices form a dense set in the set of all $n \times n$ matrices. This means that if A is an $n \times n$ matrix and if $\varepsilon > 0$, then there is an invertible matrix B such that $\| A - B \| < \varepsilon$. (See Problem 1.)

16. Show that if $\| AB - I \| = \varepsilon < 1$, then

$$\| A^{-1} - B \| \leqq \| B \| \left(\frac{\varepsilon}{1 - \varepsilon} \right)$$

17. Prove that the operation $x \longmapsto Ax$ is continuous. That is, for fixed A, show that if a sequence $[x^{(k)}]$ converges to x, then $[Ax^{(k)}]$ converges to Ax.

18. Prove that if E is an $n \times n$ matrix for which $\| E \|$ is sufficiently small, then

$$\| (I - E)^{-1} - (I + E) \| \leqq 3 \| E \|^2$$

How small must $\| E \|$ be?

19. Prove that if A is invertible, then

$$\| Ax \| \geqq \|x\| \, \| A^{-1} \|^{-1}$$

20. Consider the vector space V of all continuous functions defined on the interval $[0, 1]$. Two important norms on V are

$$\|x\|_\infty = \max_{0 \leqq t \leqq 1} |x(t)| \qquad \|x\|_1 = \int_0^1 |x(t)| \, dt$$

Show that the sequence of functions $x_n(t) = t^n$ has the properties $\|x_n\|_\infty = 1$ and $\|x_n\|_1 \to 0$ as $n \to \infty$. Thus, these norms lead to different concepts of convergence.

21. Prove that if $\| AB - I \| < 1$, then $2B - BAB$ is a better approximate inverse for A than B, in the sense that $A(2B - BAB)$ is closer to I.

22. Let $B_k = \sum_{j=0}^{k} A^j$. Show that the sequence $[B_k]$ can be computed recursively by the formulas $B_0 = I$ and $B_{k+1} = I + AB_k$.

23. Give a series that represents A^{-1} under the assumption that $\| I - \alpha A \| < 1$ for some known scalar α.

24. In a normed linear space, prove that if a sequence of vectors converges, then it must satisfy the Cauchy criterion.

25. Prove that if A is ill conditioned, then there is a singular matrix within distance $\|A\|_2 / \kappa(A)$ of A.

26. Consider the linear system $\begin{bmatrix} 1 & 2 \\ 1 + \delta & 2 \end{bmatrix} \begin{bmatrix} x_1 \\ x_2 \end{bmatrix} = \begin{bmatrix} 3 \\ 3 + \delta \end{bmatrix}$ for small $\delta > 0$.

(a) Using the approximate solution $\widetilde{x} = (3, 0)^T$, compare the infinity norm of the residual vector and the infinity norm of the error vector. What conclusion can you draw?

(b) Determine the condition number $\kappa_\infty(A)$. What happens if $\delta \to 0$?

(c) Carry out *one* step of iterative improvement based on the approximate solution $\widetilde{x}$.

27. Prove that if $\|I - AB\| < 1$, then BA is invertible. (Here A and B are square matrices. What happens if they are not square?)

28. Prove that if $\|I - cA^n\| < 1$ for some c and some integer $n \geqq 1$, then A is invertible.

29. Prove that if there is a polynomial p without constant term such that

$$\|I - p(A)\| < 1$$

then A is invertible. Generalize so that the coefficients in p may be matrices.

30. Prove that if p is a polynomial with constant term c_0 and if $|c_0| + \|I - p(A)\| < 1$, then A is invertible.

COMPUTER PROBLEMS 4.5

1. The purpose of this project is to write and test procedures that are refinements of the algorithms given in the text. The refinements to be incorporated are these:

(i) Column equilibration

(ii) Row equilibration

(iii) Full pivoting

(iv) Include two steps of iterative refinement at the end. Alternatively, write a separate subprogram to take A, x, and b and improve x twice (or m times). Here, each residual $b_i - \sum_{j=1}^{n} a_{ij}x_j$ should be computed in double precision and then rounded to single precision. Notice that the *original* matrix A and vector b must be saved in order to compute the residuals.

Test your codes by solving $Ax = b$, where $n = 10$ and

$$\begin{cases} a_{ij} = (i/11)^j & (1 \leqq i, j \leqq n) \\ b_i = i[1 - (i/11)^{10}]/(33 - 3i) & (1 \leqq i \leqq n) \end{cases}$$

As a second test of your codes, let $n = 4$, $a_{ij} = 1/(2n - i - j + 1)$, and $b_i = \sum_{j=1}^{n} a_{ij}$. The solution should be $x = (1, 1, 1, 1)^T$. For the third test, repeat the previous test with $n = 10$. Include enough print statements to show lots of detail—in particular, the initial solution and the solutions obtained with iterative improvement.

2. Using a test matrix A of dimensions 3×3 and the method of the preceding problem, compute $B = \sum_{j=0}^{20} A^j$ and see whether $(I - A)B \approx I$. Be sure that $\| A \| < 1$ for some subordinate matrix norm.

3. Solve the following linear system and apply three steps of iterative improvement. Print r, e, and x after each iteration.

$$\begin{bmatrix} 60 & 30 & 20 \\ 30 & 20 & 15 \\ 20 & 15 & 12 \end{bmatrix} \begin{bmatrix} x_1 \\ x_2 \\ x_3 \end{bmatrix} = \begin{bmatrix} 110 \\ 65 \\ 47 \end{bmatrix}$$

---

*4.6 Solution of Equations by Iterative Methods

The Gaussian algorithm and its variants are termed **direct** methods for solving the matrix problem $Ax = b$. They proceed through a finite number of steps and produce a solution x that would be completely accurate were it not for roundoff errors.

An **indirect method**, by contrast, produces a sequence of vectors that ideally *converges* to the solution. The computation is halted when an approximate solution is obtained having some specified accuracy or after a certain number of iterations.

Indirect methods are almost always **iterative** in nature: A simple process is applied repeatedly to generate the sequence referred to previously.

For large linear systems containing thousands of equations, iterative methods often have decisive advantages over direct methods in terms of speed and demands on computer memory. Sometimes, if the accuracy requirements are not stringent, a modest number of iterations will suffice to produce an acceptable solution. For **sparse systems** (in which a large proportion of the elements in A are 0), iterative methods are often very efficient. In sparse problems, the nonzero elements of A are sometimes stored in a sparse-storage format; in other cases, it is not necessary to store A at all! The latter situation is common in the numerical solution of partial differential equations. In this case, each row of A might be generated as needed but not retained after use. Another advantage of iterative methods is that they are usually stable, and they will actually dampen errors (due to roundoff or minor blunders) as the process continues.

To convey the general idea, we describe two fundamental iterative methods.

Example 1 Consider the linear system

$$\begin{bmatrix} 7 & -6 \\ -8 & 9 \end{bmatrix} \begin{bmatrix} x_1 \\ x_2 \end{bmatrix} = \begin{bmatrix} 3 \\ -4 \end{bmatrix}$$

How can it be solved by an iterative process?

Solution A straightforward procedure would be to solve the ith equation for the ith unknown as follows:

$$\begin{aligned} x_1^{(k)} &= \frac{6}{7}x_2^{(k-1)} + \frac{3}{7} \\ x_2^{(k)} &= \frac{8}{9}x_1^{(k-1)} - \frac{4}{9} \end{aligned}$$

This is known as the **Jacobi method** or **iteration**. Initially, we select for $x_1^{(0)}$ and $x_2^{(0)}$ the best available guess for the solution, or simply set them to 0. The equations above then generate what we hope are *improved* values, $x_1^{(1)}$ and $x_2^{(1)}$. The process is repeated a prescribed number of times or until a certain precision appears to have been achieved in the vector $(x_1^{(k)}, x_2^{(k)})^T$. Here are some selected values of the iterates of the Jacobi method for this example:

k	$x_1^{(k)}$	$x_2^{(k)}$
0	0.00000	0.00000
10	0.14865	−0.19820
20	0.18682	−0.24909
30	0.19662	−0.26215
40	0.19913	−0.26551
50	0.19978	−0.26637

It is apparent that this iterative process could be modified so the *newest* value for $x_1^{(k)}$ is used immediately in the second equation. The resulting method is called the **Gauss-Seidel method** or **iteration**. Its equations are

$$
\begin{aligned}
x_1^{(k)} &= \frac{6}{7}x_2^{(k-1)} + \frac{3}{7} \\
x_2^{(k)} &= \frac{8}{9}x_1^{(k)} - \frac{4}{9}
\end{aligned}
$$

Some of the output from the Gauss-Seidel method follows:

k	$x_1^{(k)}$	$x_2^{(k)}$
0	0.00000	0.00000
10	0.21978	−0.24909
20	0.20130	−0.26531
30	0.20009	−0.26659
40	0.20001	−0.26666
50	0.20000	−0.26667

Both the Jacobi and the Gauss-Seidel iterates seem to be converging to the same limit, and the latter is converging faster. Also, notice that, in contrast to a direct method, the precision we obtain in the solution depends on when the iterative process is halted. ■

Basic Concepts

We now consider iterative methods in a more general mathematical setting. A general type of iterative process for solving the system

$$Ax = b \tag{1}$$

can be described as follows: A certain matrix Q—called the **splitting matrix**—is prescribed, and the original problem is rewritten in the equivalent form

$$Qx = (Q - A)x + b \tag{2}$$

Equation (2) suggests an iterative process, defined by writing

$$Qx^{(k)} = (Q - A)x^{(k-1)} + b \qquad (k \geqq 1) \tag{3}$$

The initial vector $x^{(0)}$ can be arbitrary; if a good guess of the solution is available, it should be used for $x^{(0)}$. We shall say that the iterative method in Equation (3) converges if it converges for *any* initial vector $x^{(0)}$. A sequence of vectors $x^{(1)}, x^{(2)}, \ldots$ can be computed from Equation (3), and our objective is to choose Q so that these two conditions are met:

(i) The sequence $[x^{(k)}]$ is easily computed.

(ii) The sequence $[x^{(k)}]$ converges rapidly to a solution.

In this section, we shall see that both of these conditions follow if it is easy to solve $Qx^{(k)} = y$ and if Q^{-1} approximates A^{-1}.

Observe, to begin with, that if the sequence $[x^{(k)}]$ converges, say to a vector x, then x is *automatically* a solution. Indeed, if we simply take the limit in Equation (3) and use continuity of the algebraic operations, the result is

$$Qx = (Q - A)x + b \tag{4}$$

which means that $Ax = b$.

To assure that Equation (1) has a solution for any vector b, we shall assume that A is nonsingular. We shall assume that Q is nonsingular as well, so that Equation (3) can be solved for the unknown vector $x^{(k)}$. Having made these assumptions, we can use the following equation for the *theoretical* analysis:

$$x^{(k)} = (I - Q^{-1}A)x^{(k-1)} + Q^{-1}b \tag{5}$$

It is to be emphasized that Equation (5) is convenient for the analysis, but in numerical work $x^{(k)}$ is almost always obtained by solving Equation (3) without the use of Q^{-1}.

Observe that the actual solution x satisfies the equation

$$x = (I - Q^{-1}A)x + Q^{-1}b \tag{6}$$

Thus, x is a fixed point of the mapping

$$x \longmapsto (I - Q^{-1}A)x + Q^{-1}b \tag{7}$$

By subtracting the terms in Equation (6) from those in Equation (5), we obtain

$$x^{(k)} - x = (I - Q^{-1}A)(x^{(k-1)} - x) \tag{8}$$

Now select any convenient vector norm and its subordinate matrix norm. We obtain from Equation (8)

$$\|x^{(k)} - x\| \leqq \|I - Q^{-1}A\| \, \|x^{(k-1)} - x\| \tag{9}$$

By repeating this step, we arrive eventually at the inequality

$$\|x^{(k)} - x\| \leqq \|I - Q^{-1}A\|^k \, \|x^{(0)} - x\| \tag{10}$$

Thus, if $\|I - Q^{-1}A\| < 1$, we can conclude at once that

$$\lim_{k \to \infty} \|x^{(k)} - x\| = 0 \tag{11}$$

for any $x^{(0)}$. Observe that the hypothesis $\|I - Q^{-1}A\| < 1$ implies (by Theorem 1 in Section 4.5) the invertibility of $Q^{-1}A$ and of A. Hence, we have the next theorem.

THEOREM 1 *If* $\| I - Q^{-1}A \| < 1$ *for some subordinate matrix norm, then the sequence produced by Equation* (3) *converges to the solution of* $Ax = b$ *for any initial vector* $x^{(0)}$.

If the norm $\delta \equiv \|I - Q^{-1}A\|$ is less than 1, then it is safe to halt the iterative process when $\|x^{(k)} - x^{(k-1)}\|$ is small. Indeed, we can prove (Problem 33) that

$$\|x^{(k)} - x\| \leqq \frac{\delta}{1-\delta}\|x^{(k)} - x^{(k-1)}\|$$

Richardson Method

As an illustration of these concepts, we consider the **Richardson method**, in which Q is chosen to be the identity matrix. Equation (3) in this case reads as follows:

$$x^{(k)} = (I - A)x^{(k-1)} + b = x^{(k-1)} + r^{(k-1)} \tag{12}$$

where $r^{(k-1)}$ is the residual vector, defined by $r^{(k-1)} = b - Ax^{(k-1)}$. According to Theorem 1, the Richardson iteration will produce a solution to $Ax = b$ (in the limit) if $\| I - A \| < 1$ for some subordinate matrix norm. (See Problems 2 and 3 for two classes of matrices having the required property.)

An algorithm to carry out the Richardson iteration is as follows:

input $n, (a_{ij}), (b_i), (x_i), M$
for $k = 1$ **to** M **do**
  for $i = 1$ **to** n **do**
    $r_i \leftarrow b_i - \sum_{j=1}^{n} a_{ij}x_j$
  end do
  for $i = 1$ **to** n **do**
    $x_i \leftarrow x_i + r_i$
  end do
end do
output $k, (x_i), (r_i)$

Example 2 Compute 100 iterates on the following problem, using the Richardson method starting with $x = (0, 0, 0)^T$:

$$\begin{bmatrix} 1 & \frac{1}{2} & \frac{1}{3} \\ \frac{1}{3} & 1 & \frac{1}{2} \\ \frac{1}{2} & \frac{1}{3} & 1 \end{bmatrix} \begin{bmatrix} x_1 \\ x_2 \\ x_3 \end{bmatrix} = \begin{bmatrix} \frac{11}{18} \\ \frac{11}{18} \\ \frac{1}{18} \end{bmatrix}$$

Solution A computer program based on the above algorithm was written. A few of the iterates produced by this routine are shown here:

$$x^{(0)} = (\, 0.00000,\;\; 0.00000,\;\; 0.00000 \,)^T$$
$$x^{(1)} = (\, 0.61111,\;\; 0.61111,\;\; 0.61111 \,)^T$$
$$\vdots$$
$$x^{(10)} = (\, 0.27950,\;\; 0.27950,\;\; 0.27950 \,)^T$$
$$\vdots$$
$$x^{(40)} = (\, 0.33311,\;\; 0.33311,\;\; 0.33311 \,)^T$$
$$\vdots$$
$$x^{(80)} = (\, 0.33333,\;\; 0.33333,\;\; 0.33333 \,)^T$$

■

Jacobi Method

Another illustration of our basic theory is provided by the **Jacobi iteration**, in which Q is the diagonal matrix whose diagonal entries are the same as those in the matrix $A = (a_{ij})$. In this case, the generic element of $Q^{-1}A$ is a_{ij}/a_{ii}. The diagonal elements of this matrix are all 1, and hence,

$$\| I - Q^{-1}A \|_\infty = \max_{1 \leqq i \leqq n} \sum_{\substack{j=1 \\ j \neq i}}^{n} |a_{ij}/a_{ii}| \tag{13}$$

THEOREM 2 *If A is diagonally dominant, then the sequence produced by the Jacobi iteration converges to the solution of $Ax = b$ for any starting vector.*

Proof Diagonal dominance means that

$$|a_{ii}| > \sum_{\substack{j=1 \\ j \neq i}}^{n} |a_{ij}| \qquad (1 \leqq i \leqq n)$$

From Equation (13), we then conclude that

$$\| I - Q^{-1}A \|_\infty < 1$$

By Theorem 1, the Jacobi iteration converges. ■

An algorithm for the Jacobi method follows:

input n, (a_{ij}), (b_i), (x_i), M
for $k = 1$ **to** M **do**

```
        for i = 1 to n do
            u_i ← ( b_i − Σ_{j=1, j≠i}^{n} a_ij x_j ) / a_ii
        end do
        for i = 1 to n do
            x_i ← u_i
        end do
        output k, (x_i)
end do
```

This algorithm and others in this section can be made more efficient by performing all divisions before the iteration begins. Thus, we could start the computation with these operations:

```
for i = 1 to n do
    d = 1/a_ii
    b_i ← d b_i
    for j = 1 to n do
        a_ij = d a_ij
    end do
end do
```

Then the replacement statement for u_i becomes simply

$$u_i \leftarrow b_i - \sum_{\substack{j=1 \\ j\neq i}}^{n} a_{ij}x_j$$

Another way to interpret this is that the original system $Ax = b$ has been replaced by $D^{-1}Ax = D^{-1}b$, where $D = \text{diag}(a_{ii})$. Alternatively, divisions can be avoided by preprocessing the system with some other scaling such as the two-sided scaling $(D^{-1/2}AD^{-1/2})(D^{1/2}x) = (D^{-1/2}b)$, where $D^{1/2} = \text{diag}(\sqrt{a_{ii}})$, provided A has positive diagonal entries. Notice that if A is symmetric, then this scaling preserves symmetry. In many iterative methods, some simple preparatory work such as this can pay large dividends in efficiency or speed of convergence.

Analysis

Our next task is to develop some of the theory for *arbitrary* linear iterative processes. We consider that such a process has been defined by an equation of the form

$$x^{(k)} = Gx^{(k-1)} + c \tag{14}$$

in which G is a prescribed $n \times n$ matrix and c is a prescribed vector in $\mathbb{R}^n$. Notice that the iteration defined in Equation (3) will be included in any general theory that we may develop for Equation (14); namely, we can set $G = I - Q^{-1}A$ and

$c = Q^{-1}b$. We want to find a necessary and sufficient condition on G so that the iteration of Equation (14) will converge for any starting vector. Some preliminary matters must be dealt with first.

The **eigenvalues** of a matrix A are the complex numbers λ for which the matrix $A - \lambda I$ is not invertible. These numbers are then the roots of the **characteristic equation** of A:

$$\det(A - \lambda I) = 0$$

(The reader may wish to look ahead at Section 5.1, where these concepts are discussed further.) The **spectral radius** of A is defined by the equation

$$\rho(A) = \max\{|\lambda| : \det(A - \lambda I) = 0\}$$

Thus, $\rho(A)$ is the smallest number such that a circle with that radius centered at 0 in the complex plane will contain all the eigenvalues of A.

A matrix A is said to be **similar** to a matrix B if there is a nonsingular matrix S such that $S^{-1}AS = B$. It follows that similar matrices have the same eigenvalues. Moreover, it is easy to see that the eigenvalues of a triangular matrix are the elements on its diagonal.

THEOREM 3 *Every square matrix is similar to an (possibly complex) upper triangular matrix whose off-diagonal elements are arbitrarily small.*

Proof Let A be an $n \times n$ matrix. We borrow a result known as Schur's Theorem from Section 5.2. This theorem states that A is similar to an upper triangular matrix $T = (t_{ij})$, which may be complex. Now let $0 < \varepsilon < 1$, and let $D = \text{diag}(\varepsilon, \varepsilon^2, \dots, \varepsilon^n)$. The generic element of $D^{-1}TD$ is $t_{ij}\varepsilon^{j-i}$, by an elementary calculation. Since T is upper triangular, the elements below the diagonal in $D^{-1}TD$ are 0, whereas the elements above the diagonal satisfy

$$|t_{ij}\varepsilon^{j-i}| \leqq \varepsilon |t_{ij}|$$

because $j > i$ and $\varepsilon < 1$. This upper bound can be made as small as we wish by decreasing ε. ■

THEOREM 4 *The spectral radius function satisfies the equation*

$$\rho(A) = \inf_{\|\cdot\|} \|A\|$$

in which the infimum is taken over all subordinate matrix norms.

Proof It is easy to prove that $\rho(A) \leqq \inf_{\|\cdot\|} \|A\|$. To do so, let λ be any eigenvalue of A. Select a nonzero eigenvector x corresponding to λ. Then for any vector norm and its subordinate matrix norm, we have

$$|\lambda| \, \|x\| = \|\lambda x\| = \|Ax\| \leqq \|A\| \, \|x\|$$

Hence, $|\lambda| \leqq \| A \|$. It follows that $\rho(A) \leqq \| A \|$. By taking an infimum, we have $\rho(A) \leqq \inf_{\|\cdot\|} \| A \|$.

For the reverse inequality, we use Theorem 3. It asserts that for any $\varepsilon > 0$, there exists a nonsingular matrix S such that $S^{-1}AS = D + T$, where D is diagonal and T is strictly upper triangular, with $\| T \|_\infty \leqq \varepsilon$. Then we have

$$\| S^{-1}AS \|_\infty = \| D + T \|_\infty \leqq \| D \|_\infty + \| T \|_\infty$$

Since D has the eigenvalues of A on its diagonal, it follows that

$$\| D \|_\infty = \max_{1 \leqq i \leqq n} |\lambda_i| = \rho(A)$$

Hence, we have

$$\| S^{-1}AS \|_\infty \leqq \rho(A) + \varepsilon$$

By appealing to Problem 6, we know that the function $\| \cdot \|'_\infty$ defined by

$$\|A\|'_\infty \equiv \| S^{-1}AS \|_\infty$$

is a subordinate matrix norm. Now $\|A\|'_\infty \leqq \rho(A) + \varepsilon$. An infimum over all subordinate matrix norms gives us

$$\inf_{\|\cdot\|} \| A \| \leqq \rho(A) + \varepsilon$$

Since ε was arbitrary, $\inf_{\|\cdot\|} \| A \| \leqq \rho(A)$. ■

Theorem 4 tells us that for any matrix A, its spectral radius lies below its norm value for any subordinate matrix norm and, moreover, there exists a subordinate matrix norm with a value arbitrarily close to the spectral radius.

We now give a necessary and sufficient condition on the iteration matrix G for convergence of the associated iterative method.

THEOREM 5 *For the iteration formula*

$$x^{(k)} = Gx^{(k-1)} + c$$

to produce a sequence converging to $(I - G)^{-1}c$, for any starting vector $x^{(0)}$, it is necessary and sufficient that the spectral radius of G be less than 1.

Proof Suppose that $\rho(G) < 1$. By Theorem 4, there is a subordinate matrix norm such that $\| G \| < 1$. We write

$$\begin{aligned} x^{(1)} &= Gx^{(0)} + c \\ x^{(2)} &= G^2x^{(0)} + Gc + c \\ x^{(3)} &= G^3x^{(0)} + G^2c + Gc + c \end{aligned}$$

The general formula is

$$x^{(k)} = G^k x^{(0)} + \sum_{j=0}^{k-1} G^j c \tag{15}$$

Using the vector norm that engendered our matrix norm, we have

$$\|G^k x^{(0)}\| \leqq \| G^k \| \|x^{(0)}\| \leqq \|G\|^k \|x^{(0)}\| \to 0 \text{ as } k \to \infty$$

By the theorem on the Neumann series (Theorem 1 in Section 4.5), we have

$$\sum_{j=0}^{\infty} G^j c = (I - G)^{-1} c$$

Thus, by letting $k \to \infty$ in Equation (15), we obtain

$$\lim_{k \to \infty} x^{(k)} = (I - G)^{-1} c$$

For the converse, suppose that $\rho(G) \geqq 1$. Select u and λ so that

$$Gu = \lambda u \qquad |\lambda| \geqq 1 \qquad u \neq 0$$

If $|\lambda| = 1$, let $c = u$ and $x^{(0)} = 0$. By Equation (15), we have $x^{(k)} = \sum_{j=0}^{k-1} G^j u = \pm ku$, which diverges as $k \to \infty$. Let $c = u$ and $x^{(0)} = 0$. By Equation (15), $x^{(k)} = \sum_{j=0}^{k-1} G^j u = \sum_{j=0}^{k-1} \lambda^j u$. If $\lambda = 1$, $u^{(k)} = ku$, and this diverges as $k \to \infty$. If $\lambda \neq 1$, then $x^{(k)} = (\lambda^k - 1)(\lambda - 1)^{-1} u$, and this diverges also because $\lim_{k\to\infty} \lambda^k$ does not exist. ■

COROLLARY 1 *The iteration formula (3)—that is, $Qx^{(k)} = (Q - A)x^{(k-1)} + b$—will produce a sequence converging to the solution of $Ax = b$, for any $x^{(0)}$, if $\rho(I - Q^{-1}A) < 1$.*

Gauss-Seidel Method

Let us examine **Gauss-Seidel iteration** in more detail. It is defined by letting Q be the lower triangular part of A, including the diagonal.

THEOREM 6 *If A is diagonally dominant, then the Gauss-Seidel method converges for any starting vector.*

Proof By Corollary 1, it suffices to prove that

$$\rho(I - Q^{-1}A) < 1$$

To this end, let λ be any eigenvalue of $I - Q^{-1}A$. Let x be a corresponding eigenvector. We assume, with no loss of generality, that $\|x\|_\infty = 1$. We have now

$$(I - Q^{-1}A)x = \lambda x \qquad \text{or} \qquad Qx - Ax = \lambda Qx$$

Since Q is the lower triangular part of A, including its diagonal,

$$-\sum_{j=i+1}^{n} a_{ij}x_j = \lambda \sum_{j=1}^{i} a_{ij}x_j \qquad (1 \leqq i \leqq n)$$

By transposing terms in this equation, we obtain

$$\lambda a_{ii}x_i = -\lambda \sum_{j=1}^{i-1} a_{ij}x_j - \sum_{j=i+1}^{n} a_{ij}x_j \qquad (1 \leqq i \leqq n)$$

Select an index i such that $|x_i| = 1 \geqq |x_j|$ for all j. Then

$$|\lambda|\, |a_{ii}| \leqq |\lambda| \sum_{j=1}^{i-1} |a_{ij}| + \sum_{j=i+1}^{n} |a_{ij}|$$

Solving for $|\lambda|$ and using the diagonal dominance of A, we get

$$|\lambda| \leqq \left\{ \sum_{j=i+1}^{n} |a_{ij}| \right\} \left\{ |a_{ii}| - \sum_{j=1}^{i-1} |a_{ij}| \right\}^{-1} < 1$$

■

An algorithm for the **Gauss-Seidel iteration** follows:

input $n, (a_{ij}), (b_i), (x_i), M$
for $k = 1$ **to** M **do**
  for $i = 1$ **to** n **do**

$$x_i \leftarrow \left(b_i - \sum_{\substack{j=1 \\ j \neq i}}^{n} a_{ij}x_j \right) \Big/ a_{ii}$$

  end do
output $k, (x_i)$
end do

Notice that in the Gauss-Seidel algorithm, the updated values of x_i replace the old values immediately, whereas in the Jacobi method, all new components of the x-vector are computed before the replacement takes place. In the Jacobi algorithm, the new components of x (denoted by u_i in the pseudocode) can be computed simultaneously, whereas in the Gauss-Seidel method they must be computed serially since the computation of the new x_i requires all the new values of $x_1, x_2, \ldots, x_{i-1}$. Because of these differences, the Jacobi iteration may be preferable on computers that allow vector or parallel processing. Notice that improved efficiency in the Gauss-Seidel algorithm can be obtained by some preparatory work on the system. In fact, the remarks made about the Jacobi iteration apply here unchanged.

Example 3 Consider the system

$$\begin{bmatrix} 2 & -1 & 0 \\ 1 & 6 & -2 \\ 4 & -3 & 8 \end{bmatrix} \begin{bmatrix} x_1 \\ x_2 \\ x_3 \end{bmatrix} = \begin{bmatrix} 2 \\ -4 \\ 5 \end{bmatrix}$$

Apply Gauss-Seidel iteration starting with $x^{(0)} = (0, 0, 0)^T$.

Solution After the scaling referred to previously, $D^{-1}Ax = D^{-1}b$ where $D = \text{diag}(A)$, the system becomes

$$\begin{bmatrix} 1 & -\frac{1}{2} & 0 \\ \frac{1}{6} & 1 & -\frac{1}{3} \\ \frac{1}{2} & -\frac{3}{8} & 1 \end{bmatrix} \begin{bmatrix} x_1 \\ x_2 \\ x_3 \end{bmatrix} = \begin{bmatrix} 1 \\ -\frac{2}{3} \\ \frac{5}{8} \end{bmatrix}$$

We refer to this system as $Ax = b$. In the Gauss-Seidel method, Q is taken to be the lower triangular part of A, including the diagonal. The formula defining the iteration is

$$Qx^{(k)} = (Q - A)x^{(k-1)} + b$$

or

$$\begin{bmatrix} 1 & 0 & 0 \\ \frac{1}{6} & 1 & 0 \\ \frac{1}{2} & -\frac{3}{8} & 1 \end{bmatrix} \begin{bmatrix} x_1^{(k)} \\ x_2^{(k)} \\ x_3^{(k)} \end{bmatrix} = \begin{bmatrix} 0 & \frac{1}{2} & 0 \\ 0 & 0 & \frac{1}{3} \\ 0 & 0 & 0 \end{bmatrix} \begin{bmatrix} x_1^{(k-1)} \\ x_2^{(k-1)} \\ x_3^{(k-1)} \end{bmatrix} + \begin{bmatrix} 1 \\ -\frac{2}{3} \\ \frac{5}{8} \end{bmatrix}$$

From this we obtain $x^{(k)}$ by solving a lower triangular system. The pertinent formulas in this example are

$$\begin{aligned} x_1^{(k)} &= \frac{1}{2}x_2^{(k-1)} + 1 \\ x_2^{(k)} &= -\frac{1}{6}x_1^{(k)} + \frac{1}{3}x_3^{(k-1)} - \frac{2}{3} \\ x_3^{(k)} &= -\frac{1}{2}x_1^{(k)} + \frac{3}{8}x_2^{(k)} + \frac{5}{8} \end{aligned}$$

The computations yield the following iterates, of which $x^{(13)}$ is correct:

$$x^{(1)} = (1.000000,\ -0.833333,\ -0.187500)^T$$
$$\vdots$$
$$x^{(5)} = (0.622836,\ -0.760042,\ 0.028566)^T$$
$$\vdots$$
$$x^{(10)} = (0.620001,\ -0.760003,\ 0.029998)^T$$
$$\vdots$$
$$x^{(13)} = (0.620000,\ -0.760000,\ 0.030000)^T$$

■

SOR Method

The next example of an important iterative method is known as **successive overrelaxation**, commonly abbreviated as **SOR**. Since we wish to present a general theory for SOR that will apply to matrices and vectors over the field of complex numbers, we first review some of the concepts germane to this setting.

If γ is a complex number, then γ can be written in the form $\gamma = \alpha + i\beta$, where α and β are real and $i^2 = -1$. The **conjugate** of γ is defined to be

$$\overline{\gamma} = \alpha - i\beta$$

The **magnitude** of γ is $|\gamma| = \sqrt{\alpha^2 + \beta^2} = \sqrt{\gamma\overline{\gamma}}$. The space of complex n-vectors is denoted by $\mathbb{C}^n$. In this space, the inner product is defined by the equation

$$\langle x, y\rangle = y^* x = \sum_{i=1}^{n} x_i \overline{y}_i$$

Here y^* is the row vector defined by $y^* = (\overline{y}_1, \overline{y}_2, \ldots, \overline{y}_n)$ and is called the **conjugate transpose** of y. It is easy to see that

$$\langle x, x\rangle > 0 \qquad (\text{if } x \neq 0)$$
$$\langle x, \lambda y\rangle = \overline{\lambda}\langle x, y\rangle$$
$$\langle x, y\rangle = \overline{\langle y, x\rangle}$$

Then it follows that $\langle \alpha x + \beta y, z\rangle = \alpha\langle x, z\rangle + \beta\langle y, z\rangle$, for scalars α, β and vectors x, y, z, and that $\langle x, Ay\rangle = \langle A^* x, y\rangle$. The **Euclidean norm** of x is

$$\|x\|_2 = \sqrt{\langle x, x\rangle} = \sqrt{x^* x} = \left\{\sum_{i=1}^{n} |x_i|^2\right\}^{1/2}$$

A matrix $A = (a_{ij})$ is said to be **Hermitian** if $A^* = A$. Here $A^* = (\overline{a}_{ji})$ is the **conjugate transpose** of A. Finally, a matrix A is said to be **positive definite** if $\langle Ax, x\rangle > 0$ for all $x \neq 0$. Notice that if A is Hermitian, then $\langle Ax, y\rangle = \langle x, Ay\rangle$.

Next, we present a theorem containing conditions for the convergence of the SOR method when A is Hermitian and positive definite.

THEOREM 7 *In the SOR method, suppose that the splitting matrix Q is chosen to be $\alpha D - C$, with α a real parameter, where D is any positive definite Hermitian matrix and C is any matrix satisfying $C + C^* = D - A$. If A is positive definite Hermitian, if Q is nonsingular, and if $\alpha > \frac{1}{2}$, then the SOR iteration converges for any starting vector.*

Proof As in the previous proof, we let $G = I - Q^{-1}A$ and attempt to establish that the spectral radius of G satisfies $\rho(G) < 1$. Let λ be an eigenvalue of G, and let x be an eigenvector corresponding to λ. Put $y = (I - G)x$. The following equations are readily verified:

$$y = x - Gx = x - \lambda x = Q^{-1}Ax \qquad \textbf{(16)}$$

$$Q - A = (\alpha D - C) - (D - C - C^*) = \alpha D - D + C^* \qquad \textbf{(17)}$$

Using (16), we have

$$(\alpha D - C)y = Qy = Ax \qquad \textbf{(18)}$$

Using (17), (18), and (16), we obtain

$$\begin{aligned}(\alpha D - D + C^*)y &= (Q - A)y = Ax - Ay = A(x - y)\\ &= A(x - Q^{-1}Ax) = AGx \qquad \textbf{(19)}\end{aligned}$$

From (18) and (19), we have

$$\alpha\langle Dy, y\rangle - \langle Cy, y\rangle = \langle Ax, y\rangle \qquad \textbf{(20)}$$

$$\alpha\langle y, Dy\rangle - \langle y, Dy\rangle + \langle y, C^*y\rangle = \langle y, AGx\rangle \qquad \textbf{(21)}$$

On adding Equations (20) and (21), we obtain

$$2\alpha\langle Dy, y\rangle - \langle y, Dy\rangle = \langle Ax, y\rangle + \langle y, AGx\rangle \qquad \textbf{(22)}$$

$$(2\alpha - 1)\langle Dy, y\rangle = \langle Ax, y\rangle + \langle y, AGx\rangle \qquad \textbf{(23)}$$

Here we have used $\langle Dy, y\rangle = \langle y, Dy\rangle$, which is true because D is Hermitian. Since $y = (1 - \lambda)x$ and $Gx = \lambda x$, Equation (23) yields

$$\begin{aligned}(2\alpha - 1)|1 - \lambda|^2\langle Dx, x\rangle &= (1 - \overline{\lambda})\langle Ax, x\rangle + \overline{\lambda}(1 - \lambda)\langle x, Ax\rangle\\ &= (1 - |\lambda|^2)\langle Ax, x\rangle \qquad \textbf{(24)}\end{aligned}$$

because A is Hermitian. If $\lambda \neq 1$, the left side of Equation (24) is positive. Hence, the right side must also be positive, and $|\lambda| < 1$. On the other hand, if $\lambda = 1$, then $y = 0$ from $y = (1 - \lambda)x$ and $Ax = 0$ from Equation (18). This contradicts the condition $\langle Ax, x \rangle > 0$ for any $x \neq 0$. Hence, $\rho(G) < 1$ and the SOR method converges. ■

A common choice for D and C in the SOR method is to let D be the diagonal of A and to let C be the lower triangular part of A, excluding the diagonal. However, Theorem 7 does not presuppose this choice. Also, the reader should be alert to the fact that in the literature the parameter α is usually denoted by $1/\omega$. Then $0 < \omega < 2$. The question of choosing ω for the most rapid convergence of the SOR iteration is addressed in Young [1971], Varga [1962], Hageman and Young [1981], Wachspress [1966], Issacson and Keller [1966], and many other books.

Iteration Matrices

Suppose that A is partitioned into

$$A = D - C_L - C_U$$

where $D = \text{diag}(A)$, C_L is the negative of the strictly lower triangular part of A, and C_U is the negative of the strictly upper triangular part of A. Another partitioning is similar but with block components. In the discretization of partial differential equations, the first partitioning corresponds to single grid points, whereas the latter corresponds to groupings such as by lines or blocks of grid points.

For the basic iterative methods presented in this section, we summarize the key matrices and the methods as follows:

Richardson:

$$\begin{cases} Q = I \\ G = I - A \end{cases}$$

$$x^{(k)} = (I - A)x^{(k-1)} + b$$

Jacobi:

$$\begin{cases} Q = D \\ G = D^{-1}(C_L + C_U) \end{cases}$$

$$Dx^{(k)} = (C_L + C_U)x^{(k-1)} + b$$

Gauss-Seidel:

$$\begin{cases} Q = D - C_L \\ G = (D - C_L)^{-1}C_U \end{cases}$$

$$(D - C_L)x^{(k)} = C_U x^{(k-1)} + b$$

SOR:

$$\begin{cases} Q = \omega^{-1}(D - \omega C_L) \\ G = (D - \omega C_L)^{-1}(\omega C_U + (1 - \omega)D) \end{cases}$$

$$(D - \omega C_L)x^{(k)} = \omega(C_U x^{(k-1)} + b) + (1 - \omega)Dx^{(k-1)}$$

SSOR:

$$\begin{cases} Q = (\omega(2 - \omega))^{-1}(D - \omega C_L)D^{-1}(D - \omega C_U) \\ G = (D - \omega C_U)^{-1}(\omega C_L + (1 - \omega)D)(D - \omega C_L)^{-1}(\omega C_U + (1 - \omega)D) \end{cases}$$

$$(D - \omega C_L)x^{(k-1/2)} = \omega(C_U x^{(k-1)} + b) + (1 - \omega)Dx^{(k-1)}$$
$$(D - \omega C_U)x^{(k)} = \omega(C_L x^{(k-1/2)} + b) + (1 - \omega)Dx^{(k-1/2)}$$

Here we have included another basic iterative method—the **symmetric successive overrelaxation (SSOR) method**. Each iteration of the SSOR method consists of first a **forward SOR** iteration that computes the unknowns in a certain order and then a **backward SOR** *sweep* that solves for them in the opposite order. Choosing the *best* relaxation parameters for the SOR and SSOR methods is an intriguing question with rather complicated answers that we shall not discuss here.

Extrapolation

Next, we present a general technique called **extrapolation** that can be used to improve the convergence properties of a linear iterative process. Consider the iteration formula

$$x^{(k)} = Gx^{(k-1)} + c \tag{25}$$

We introduce a parameter $\gamma \neq 0$ and consider the method (25) to be embedded in a one-parameter family of iteration methods given by

$$\begin{aligned} x^{(k)} &= \gamma(Gx^{(k-1)} + c) + (1 - \gamma)x^{(k-1)} \\ &= G_\gamma x^{(k-1)} + \gamma c \end{aligned} \tag{26}$$

where

$$G_\gamma = \gamma G + (1 - \gamma)I$$

Notice that when $\gamma = 1$, we recover the original iteration in Equation (25).

If the iteration in (26) converges, say to x, then by taking a limit, we get

$$x = \gamma(Gx + c) + (1 - \gamma)x$$

or

$$x = Gx + c$$

since $\gamma \neq 0$. Recall that the objective of the iteration (25) is to produce a solution of the equation $x = Gx + c$. If $G = I - Q^{-1}A$ and $c = Q^{-1}b$, then this corresponds to solving $Ax = b$.

Before attempting to determine an optimum value for the parameter γ, we require a result about eigenvalues.

THEOREM 8 *If λ is an eigenvalue of a matrix A and if p is a polynomial, then $p(\lambda)$ is an eigenvalue of $p(A)$.*

Proof Let $Ax = \lambda x$ with $x \neq 0$. Then $A^2x = \lambda Ax = \lambda^2 x$. By induction and a separate verification for $k = 0$, we have

$$A^k x = \lambda^k x \qquad (k \geqq 0)$$

Thus, λ^k is an eigenvalue of A^k. For a polynomial p, write $p(z) = \sum_{k=0}^{m} c_k z^k$. Then

$$p(A)x = \sum_{k=0}^{m} c_k A^k x = \sum_{k=0}^{m} c_k \lambda^k x = p(\lambda)x \qquad \blacksquare$$

By Theorem 5, a criterion that is necessary and sufficient for the convergence of the extrapolated method in Equation (26) is that $\rho(G_\gamma) < 1$. Suppose we do not know the eigenvalues of G precisely, but we know only an interval $[a, b]$ on the real line that contains all of them. By Theorem 8, the eigenvalues of the matrix $G_\gamma \equiv \gamma G + (1 - \gamma)I$ lie in the interval whose endpoints are $\gamma a + 1 - \gamma$ and $\gamma b + 1 - \gamma$. Is it possible to select γ so that $\rho(G_\gamma) < 1$?

Denote by $\Lambda(A)$ the set of eigenvalues of any matrix A. Then

$$\rho(G_\gamma) = \max_{\lambda \in \Lambda(G_\gamma)} |\lambda| = \max_{\lambda \in \Lambda(G)} |\gamma\lambda + 1 - \gamma| \leqq \max_{a \leqq \lambda \leqq b} |\gamma\lambda + 1 - \gamma| \qquad \textbf{(27)}$$

We shall prove that if $1 \notin [a, b]$, then γ can be chosen so that $\rho(G_\gamma) < 1$.

THEOREM 9 *If the only information available about the eigenvalues of G is that they lie in the interval $[a, b]$, and if $1 \notin [a, b]$, then the best choice for γ is $2/(2 - a - b)$. With this value of γ, $\rho(G_\gamma) \leqq 1 - |\gamma| d$, where d is the distance from 1 to $[a, b]$.*

Proof Assume the hypotheses, and let γ be as given. Since $1 \notin [a, b]$, either $a > 1$ or $b < 1$. We give the proof in the second case only, leaving the first to the problems. Since $a \leqq b < 1$, it follows that $\gamma > 0$ and $d = 1 - b$. Hence, any eigenvalue λ of G_γ satisfies the inequality

$$\gamma a + 1 - \gamma \leqq \lambda \leqq \gamma b + 1 - \gamma \qquad \textbf{(28)}$$

as noted in the preceding paragraphs. Thus,

$$\lambda \leqq \gamma b + 1 - \gamma = 1 + \gamma(b - 1) = 1 - \gamma d$$

Furthermore,

$$\lambda \geqq \gamma a + 1 - \gamma = \gamma(a + b - 2) + 1 + \gamma(1 - b) = -1 + \gamma d$$

Thus, we have proved that

$$-1 + \gamma d \leqq \lambda \leqq 1 - \gamma d$$

whence $\rho(G_\gamma) \leqq 1 - \gamma d$. To see that our choice of γ is optimal, note that if γ is increased, then the left-hand endpoint of the interval in Equation (28) moves to the left. If that point happens to be an eigenvalue of G_γ, then $\rho(G_\gamma)$ will *increase*. A similar argument applies if γ is decreased. ■

It should be observed that the extrapolation process just discussed can be applied to methods that are not convergent themselves. All that is required is that the eigenvalues of G be real and lie in an interval that does not contain 1.

If A is a matrix whose eigenvalues $\lambda_1, \lambda_2, \dots, \lambda_n$ are all real, we define

$$m(A) = \min_i \lambda_i \qquad M(A) = \max_i \lambda_i \tag{29}$$

Thus, in Theorem 9, we can let $a = m(G)$ and $b = M(G)$.

Example 4 In the case of Richardson iteration, $Q = I$ and $G = I - A$. If A has only real eigenvalues, then the same is true of G. By Theorem 8, we have

$$M(G) = 1 - m(A) \qquad m(G) = 1 - M(A) \tag{30}$$

If $m(A) > 0$ or $M(A) < 0$, then acceleration is possible in Richardson iteration. The optimal γ, calculated from Theorem 9, is

$$\gamma = 2/\left[m(A) + M(A)\right]$$

The resulting spectral radius, from $d = m(A)$, is

$$\rho(G_\gamma) = \left[M(A) - m(A)\right]/\left[M(A) + m(A)\right]$$

Example 5 Jacobi iteration can be treated as indicated in Problem 9. We let

$$D = \operatorname{diag}(a_{11}, a_{22}, \dots, a_{nn})$$

and apply Richardson iteration to the problem $D^{-1}Ax = D^{-1}b$. From the result of Example 4, we conclude that if $m(D^{-1}A) > 0$ or $M(D^{-1}A) < 0$, then acceleration is possible in Richardson iteration. Furthermore, the spectral radius of the iteration matrix will be

$$\rho(G_\gamma) = \left[M(D^{-1}A) - m(D^{-1}A)\right] / \left[M(D^{-1}A) + m(D^{-1}A)\right]$$

if the optimal $\gamma = 2/[m(D^{-1}A) + M(D^{-1}A)]$ is used.

Chebyshev Acceleration

A more general type of acceleration procedure called **Chebyshev acceleration** can also be applied to a linear iterative algorithm. As before, let us consider a basic iterative method

$$x^{(k)} = Gx^{(k-1)} + c$$

It is assumed that the solution to the problem is a vector x such that $x = Gx + c$. At step k in the process, we shall have computed the vectors $x^{(1)}, x^{(2)}, \dots, x^{(k)}$, and we ask whether some linear combination of these vectors is perhaps a better approximation to the solution than $x^{(k)}$. We assume that $a_0^{(k)} + a_1^{(k)} + \cdots + a_k^{(k)} = 1$, and set

$$u^{(k)} = \sum_{i=0}^{k} a_i^{(k)} x^{(i)}$$

By familiar techniques, we obtain

$$\begin{aligned} u^{(k)} - x &= \sum_{i=0}^{k} a_i^{(k)}(x^{(i)} - x) = \sum_{i=0}^{k} a_i^{(k)} G^i (x^{(0)} - x) \\ &= P(G)(x^{(0)} - x) \end{aligned}$$

where P is the polynomial defined by $P(z) = \sum_{i=0}^{k} a_i^{(k)} z^i$. Taking norms, we get

$$\|u^{(k)} - x\| \leqq \|P(G)\| \, \|x^{(0)} - x\|$$

Here any vector norm and its subordinate matrix norm may be used. Recall from Theorem 4 that the infimum of $\|P(G)\|$ over all subordinate matrix norms is $\rho(P(G))$; this is the quantity that should be made a minimum. If the eigenvalues μ_i of G lie within some bounded set S in the complex plane, then by Theorem 8,

$$\rho\big(P(G)\big) = \max_{1 \leqq i \leqq n} |P(\mu_i)| \leqq \max_{z \in S} |P(z)|$$

The polynomial P should be chosen to minimize this last expression, subject to the constraint $\sum_{i=0}^{k} a_i = 1$, which is $P(1) = 1$. This is a standard problem in approximation theory for which explicit solutions are known in some cases.

For example, if S is an interval $[a, b]$ on the real line, not containing 1, then a scaled and shifted Chebyshev polynomial solves the problem. The classic Chebyshev polynomial T_k ($k \geqq 1$) is the unique polynomial of degree k that minimizes the expression

$$\max_{-1 \leqq z \leqq 1} |T_k(z)|$$

subject to the constraint that the leading coefficient is 2^{k-1}. These polynomials can be generated recursively by the formulas

$$\begin{cases} T_0(z) = 1 \qquad T_1(z) = z \\ T_k(z) = 2zT_{k-1}(z) - T_{k-2}(z) \qquad (k \geqq 2) \end{cases}$$

Now suppose that the eigenvalues of G are contained in an interval $[a, b]$ that does not contain 1, say $b < 1$. We are interested in the following min-max problem:

$$\min_{P_k(1)=1} \left\{ \max_{a \leqq z \leqq b} |P_k(z)| \right\}$$

The solution of this extremum problem is contained in the next four lemmas. These results involve the Chebyshev polynomials, T_k, which are discussed above and in Section 6.1. The set of polynomials having degree at most k is denoted by Π_k.

LEMMA 1 *Let $\beta \in \mathbb{R} \backslash (-1, 1)$. If $p \in \Pi_k$ and $p(\beta) = 1$, then $\|p\| \geqq |\alpha|$, where $\alpha = 1/T_k(\beta)$ and $\|p\| = \max_{-1 \leqq t \leqq 1} |p(t)|$.*

Proof Suppose that p satisfies the hypotheses but not the conclusion. Let $t_i = \cos(i\pi/k)$ for $0 \leqq i \leqq k$. These points are the extrema of T_k. Indeed,

$$T_k(t_i) = \cos(k \cos^{-1} t_i) = \cos i\pi = (-1)^i$$

Let $\sigma = \operatorname{sgn} \alpha$. Then

$$\sigma(-1)^i \left[\alpha T_k(t_i) - p(t_i)\right] \geqq |\alpha| - \|p\| > 0$$

This shows that the polynomial $\alpha T_k - p$ takes alternately positive and negative values at the points $t_0, t_1, \ldots, t_k$. Hence, this polynomial has at least k zeros in the interval $(-1, 1)$. It also vanishes at the point β, giving a total of at least $k + 1$ zeros. Since $\alpha T_k - p$ is of degree at most k, it must be 0, a contradiction. ■

LEMMA 2 *Let $a < b < 1$. If $p \in \Pi_k$ and $p(1) = 1$, then $\|p\| \geqq 1/T_k(w(1))$. Here*

$$\|p\| = \max_{a \leqq t \leqq b} |p(t)| \qquad \text{and} \qquad w(t) = (2t - b - a)/(b - a)$$

The inequality becomes an equality if $p = (T_k \circ w)/T_k(w(1))$.

Proof Put $\beta = w(1)$ and $p = q \circ w$, with $q \in \Pi_k$. Note that $1 = p(1) = q(w(1)) = q(\beta)$ and $\beta > 1$. By Lemma 1, it follows that $\|q\|_{[-1,1]} \geqq 1/T_k(\beta)$. Equivalently, we have $\|p\|_{[a,b]} \geqq 1/T_k(w(1))$. If $p = (T_k \circ w)/T_k(w(1))$, then obviously $p(1) = 1$ and $\|p\|_{[a,b]} = \|T_k\|_{[-1,1]}/T_k(w(1)) = 1/T_k(w(1))$. ■

LEMMA 3 *The polynomials $P_k = (T_k \circ w)/T_k(w(1))$ can be generated by a recurrence relation:*

$$\begin{cases} P_0(t) = 1 \qquad P_1(t) = (2t - b - a)/(2 - b - a) \\ P_k(t) = \rho_k P_1(t) P_{k-1}(t) + (1 - \rho_k) P_{k-2}(t) \qquad (k \geqq 2) \end{cases}$$

in which the coefficients ρ_k are obtained from the equations

$$\rho_1 = 2 \qquad \rho_k = (1 - \alpha\rho_{k-1})^{-1} \qquad \alpha = \left[2w(1)\right]^{-2} \qquad (k \geqq 2)$$

Proof Define $\beta_k = T_k(w(1))$. With the aid of the relations $T_k(t) = 2tT_{k-1}(t) - T_{k-2}(t)$ and $\beta_k P_k(t) = T_k(w(t))$, we obtain

$$\begin{aligned} P_k(t) = \beta_k^{-1} T_k\big(w(t)\big) &= \beta_k^{-1}\Big[2w(t)T_{k-1}\big(w(t)\big) - T_{k-2}\big(w(t)\big)\Big] \\ &= 2\beta_k^{-1}\beta_{k-1} w(t) P_{k-1}(t) - \beta_k^{-1}\beta_{k-2}P_{k-2}(t) \\ &= 2\beta_k^{-1}\beta_{k-1} w(1) P_1(t) P_{k-1}(t) - \beta_k^{-1}\beta_{k-2}P_{k-2}(t) \end{aligned}$$

Here we used the easily verified equation $w(1)P_1(t) = w(t)$. It is convenient to define $\rho_k = 2\beta_k^{-1}\beta_{k-1}w(1) = \alpha^{-1/2}\beta_k^{-1}\beta_{k-1}$. Using again the recurrence relation for Chebyshev polynomials, we obtain

$$\begin{aligned} \beta_k = T_k\big(w(1)\big) &= 2w(1)T_{k-1}\big(w(1)\big) - T_{k-2}\big(w(1)\big) = \alpha^{-1/2}\beta_{k-1} - \beta_{k-2} \\ 1 &= 2w(1)\beta_k^{-1}\beta_{k-1} - \beta_k^{-1}\beta_{k-2} = \rho_k - \beta_k^{-1}\beta_{k-2} \end{aligned}$$

Thus, our recurrence relation for P_k can be written as

$$P_k = \rho_k P_1 P_{k-1} + (1 - \rho_k) P_{k-2}$$

The coefficients ρ_k satisfy the equation

$$\begin{aligned} \rho_k = \alpha^{-1/2}\beta_k^{-1}\beta_{k-1} &= \alpha^{-1/2}\beta_{k-1}\left[\alpha^{-1/2}\beta_{k-1} - \beta_{k-2}\right]^{-1} \\ &= \alpha^{-1/2}\left[\alpha^{-1/2} - \beta_{k-1}^{-1}\beta_{k-2}\right]^{-1} \\ &= \alpha^{-1}\left\{\alpha^{-1} - \alpha^{-1/2}\beta_{k-1}^{-1}\beta_{k-2}\right\}^{-1} \\ &= \alpha^{-1}\left\{\alpha^{-1} - \rho_{k-1}\right\}^{-1} \\ &= (1 - \alpha\rho_{k-1})^{-1} \end{aligned}$$

■

LEMMA 4 *The vectors $u^{(k)}$ in the Chebyshev acceleration method can be computed recursively from an arbitrary starting vector $u^{(0)}$ by the formulas*

$$\begin{cases} u^{(1)} = \gamma[Gu^{(0)} + c] + (1-\gamma)u^{(0)} & \big(\gamma = 2/(2-b-a)\big) \\ u^{(k)} = \rho_k\left[\gamma(Gu^{(k-1)} + c) + (1-\gamma)u^{(k-1)}\right] + (1-\rho_k)u^{(k-2)} & (k \geqq 2) \end{cases}$$

Proof We retain all the notation established previously. In particular,

$$u^{(k)} = \sum_{i=0}^{k} a_i^{(k)} x^{(i)} \qquad P_k(t) = \sum_{i=0}^{k} a_i^{(k)} t^i$$

It follows that $u^{(0)} = x^{(0)}$ and that $u^{(1)}$ is given by the formula in the lemma. This is left to the reader to verify. From equations proved earlier, with x satisfying $x = Gx + c$,

$$\begin{aligned} u^{(k)} - x &= P_k(G)(u^{(0)} - x) \\ &= \left[\rho_k P_1(G)P_{k-1}(G) + (1-\rho_k)P_{k-2}(G)\right](u^{(0)} - x) \\ &= \rho_k P_1(G)(u^{(k-1)} - x) + (1-\rho_k)(u^{(k-2)} - x) \end{aligned}$$

This can be written in the form

$$u^{(k)} = \rho_k P_1(G)u^{(k-1)} + (1-\rho_k)u^{(k-2)} + \rho_k\left[I - P_1(G)\right]x$$

An easy calculation reveals that the last term in this equation equals $\rho_k \gamma c$. Finally, we note that

$$P_1(G) = \gamma G + (1-\gamma)I$$ ■

As in our analysis of the extrapolation method, we can obtain an upper bound on the spectral radius of $P_k(G)$:

$$\begin{aligned} \rho\big(P_k(G)\big) &= \max_{\lambda \in \Lambda(P_k(G))} |\lambda| = \max_{\lambda \in \Lambda(G)} |P_k(\lambda)| \\ &\leqq \max_{a \leqq \lambda \leqq b} |P_k(\lambda)| = 1/T_k\big(w(1)\big) \end{aligned}$$

Here an appeal to Lemma 2 has been made. For computing this bound we can use the result of Problem 36 and arrive at these formulas:

$$\frac{1}{T_k(w(1))} = \frac{2}{b^n + b^{-n}} \qquad b = t + \sqrt{t^2 - 1} \qquad t = w(1)$$

It can be shown that Chebyshev acceleration is an order of magnitude faster than extrapolation. See Hageman and Young [1981] or Kincaid and Young [1979] for additional details.

Example 6 Solve the problem

$$\begin{bmatrix} 4 & -1 & -1 & 0 \\ -1 & 4 & 0 & -1 \\ -1 & 0 & 4 & -1 \\ 0 & -1 & -1 & 4 \end{bmatrix} \begin{bmatrix} x_1 \\ x_2 \\ x_3 \\ x_4 \end{bmatrix} = \begin{bmatrix} -4 \\ 0 \\ 4 \\ -4 \end{bmatrix}$$

using the Chebyshev acceleration of the Jacobi method.

Solution First we scale the system by $D^{-1}Ax = D^{-1}b$, where $D = \text{diag}(A)$, obtaining a Jacobi iteration matrix whose eigenvalues lie in the interval $[-\frac{1}{2}, \frac{1}{2}]$. The convergence of this basic method can be speeded up by applying Chebyshev acceleration. Beginning with the initial vector $u = (0, 0, 0, 0)^T$, we obtain convergence to the approximate solution $u = (-0.999996, -0.500002, 0.500002, -0.999996)^T$ in ten iterations using a computer similar to the `Marc-32`. ■

An algorithm for the **Chebyshev acceleration method** can be written as follows:

input u, a, b, M, δ
$\gamma \leftarrow 2/(2 - b - a)$
$\alpha \leftarrow [\frac{1}{2}(b - a)/(2 - b - a)]^2$
output $0, u$
call `Extrap` (γ, n, G, c, u, v)
output $1, v$
$\rho \leftarrow 1/(1 - 2\alpha)$
call `Cheb` $(\rho, \gamma, n, G, c, u, v)$
output $2, u$
for $k = 3$ **to** M **step** 2 **do**
  $\rho \leftarrow (1 - \rho\alpha)$
  call `Cheb` $(\rho, \gamma, n, G, c, v, u)$
  output k, v
  $\rho \leftarrow 1/(1 - \rho\alpha)$
  call `Cheb` $(\rho, \gamma, n, G, c, u, v)$
  output $k + 1, u$
  if $\|u - v\|_\infty < \delta$ **stop**
end do

Here the first step of this method is just the extrapolation procedure, and each successive iteration involves two acceleration steps. Two subroutines or procedures are needed to carry out the basic iteration steps—namely, `Extrap` and `Cheb`; they are given below. Notice that by appropriate calls to the subroutine or procedure `Cheb`, we obtain an automatic interchange of the desired vectors.

procedure `Extrap` (γ, n, G, c, u, v)
$v \leftarrow \gamma c + (1 - \gamma)u$
$v \leftarrow \gamma G u + v$
return

procedure Cheb$(\rho, \gamma, n, G, c, u, v)$
$u \leftarrow \rho\gamma c + \rho(1 - \gamma)v + (1 - \rho)u$
$u \leftarrow \rho\gamma G v + u$
return

PROBLEMS 4.6

1. Prove that if A is diagonally dominant and if Q is chosen as in the Jacobi method, then

$$\rho(I - Q^{-1}A) < 1$$

2. Prove that if A has this property (**unit row diagonally dominant**):

$$a_{ii} = 1 > \sum_{\substack{j=1 \\ j \neq i}}^{n} |a_{ij}| \qquad (1 \leqq i \leqq n)$$

then Richardson iteration is successful.

3. Repeat Problem 2 with this assumption (**unit column diagonally dominant**):

$$a_{jj} = 1 > \sum_{\substack{i=1 \\ i \neq j}}^{n} |a_{ij}| \qquad (1 \leqq j \leqq n)$$

4. Prove that if A has the property in Problem 2, then the following iteration will solve $Ax = b$ (in the limit):

for $k = 1$ **to** ...
  for $i = 1$ **to** n **do**
    $x_i \leftarrow x_i + b_i - \sum_{j=1}^{n} a_{ij} x_j$
  end do
end do

5. Let $\|\cdot\|$ be a norm on $\mathbb{R}^n$, and let S be an $n \times n$ nonsingular matrix. Define $\|x\|' = \|Sx\|$, and prove that $\|\cdot\|'$ is a norm.
6. (Continuation) Let $\|\cdot\|$ be a subordinate matrix norm, and let S be a nonsingular matrix. Define $\|A\|' = \|SAS^{-1}\|$, and show that $\|\cdot\|'$ is a subordinate matrix norm.
7. Using Q as in the Gauss-Seidel method, prove that if A is diagonally dominant, then $\|I - Q^{-1}A\|_\infty < 1$.
8. Prove that $\rho(A) < 1$ if and only if $\lim_{k\to\infty} A^k x = 0$ for every x.
9. Prove that if the ith equation in the system $Ax = b$ is divided by a_{ii} and if Richardson iteration is then applied, the result is the same as applying Jacobi iteration in the first place.

10. Which of the norm axioms are satisfied by the spectral radius function ρ and which are not? Give proofs and examples, as appropriate.

11. Consider (for fixed n) the set of upper triangular $n \times n$ matrices. Show that this set is a vector space. Prove that the spectral radius function ρ is a **pseudonorm** on the vector space since it satisfies all the norm axioms, except that $\rho(A)$ can be 0 if $A \neq 0$.

12. Explain why, in the proof of Theorem 3, we cannot let $\varepsilon \to 0$ and conclude that A is similar to a diagonal matrix.

13. Let A be invertible, and let f be a function of the form $f(z) = \sum_{j=-m}^{m} c_j z^j$. Show that if λ is an eigenvalue of A, then $f(\lambda)$ is an eigenvalue of $f(A)$.

14. Prove that the eigenvalues of a Hermitian matrix are real. *Hint:* Consider $\langle x, Ax\rangle$ and $\langle Ax, x\rangle$.

15. Let A be diagonally dominant, and let Q be the lower triangular part of A, as in the Gauss-Seidel method. Prove that $\rho(I - Q^{-1}A)$ is no greater than the largest of the ratios

$$r_i = \left\{\sum_{j=i+1}^{n} |a_{ij}|\right\} \Big/ \left\{|a_{ii}| - \sum_{j=1}^{i-1} |a_{ij}|\right\}$$

16. Prove that if A is nonsingular, then AA^* is positive definite.

17. Prove that if A is positive definite, then its eigenvalues are positive.

18. Prove that if A is positive definite, then so are $A^2, A^3, \ldots$ as well as $A^{-1}, A^{-2}, \ldots$.

19. Is there a matrix A such that $\rho(A) < \|A\|$ for all subordinate matrix norms?

20. Prove that if $\rho(A) < 1$, then $I - A$ is invertible and $(I - A)^{-1} = \sum_{k=0}^{\infty} A^k$.

21. Is the inequality $\rho(AB) \leqq \rho(A)\rho(B)$ true for all pairs of $n \times n$ matrices? Is your answer the same when A and B are upper triangular?

22. Show that the basic iteration process given by Equation (3) is equivalent to the following: Given $x^{(k)}$, compute $r^{(k)} = b - Ax^{(k)}$, solve for $z^{(k)}$ in the equation $Qz^{(k)} = r^{(k)}$, and define $x^{(k+1)} = x^{(k)} + z^{(k)}$.

23. Do the Hermitian matrices of order n form a vector space over the complex field?

24. Using the notation of Problem 22, show that

$$r^{(k+1)} = (I - AQ^{-1})r^{(k)}$$
$$z^{(k+1)} = (I - Q^{-1}A)z^{(k)}$$

25. Show that for nonsingular matrices A and B, $\rho(AB) = \rho(BA)$. (Prove a stronger result, if possible.) What is the relevance of this fact to Problem 24?

26. What are the necessary and sufficient conditions on a diagonal matrix D in order that for any A, the positive definiteness of A implies that of DA?

27. Prove that the Gauss-Seidel method is a special case of the SOR method.

28. Show that these matrices

$$\begin{aligned}
\mathcal{R} &= I - A \\
\mathcal{J} &= I - D^{-1}A \\
\mathcal{G} &= I - (D - C_L)^{-1}A \\
\mathcal{L}_\omega &= I - \omega(D - \omega C_L)^{-1}A \\
\mathcal{U}_\omega &= I - \omega(D - \omega C_U)^{-1}A \\
\mathcal{S}_\omega &= I - \omega(2 - \omega)(D - \omega C_U)^{-1}D(D - \omega C_L)^{-1}A
\end{aligned}$$

are the iteration matrices for the Richardson, Jacobi, Gauss-Seidel, forward SOR, backward SOR, and SSOR methods, respectively. Then show that the splitting matrices Q and iteration matrices G given in this section are correct.

29. Find the explicit form for the iteration matrix $I - Q^{-1}A$ in the Gauss-Seidel method when

$$A = \begin{bmatrix}
2 & -1 & & & & \\
-1 & 2 & -1 & & & \\
 & -1 & 2 & -1 & & \\
 & & \ddots & \ddots & \ddots & \\
 & & & -1 & 2 & -1 \\
 & & & & -1 & 2
\end{bmatrix}$$

30. Characterize the family of all $n \times n$ nonsingular matrices A for which *one step* of the Gauss-Seidel algorithm solves $Ax = b$, starting at the vector $x = 0$.

31. Give an example of a matrix A that is not diagonally dominant, yet the Gauss-Seidel method applied to $Ax = b$ converges.

32. How does the Chebyshev acceleration method simplify if the basic method is the Jacobi method?

33. Prove that if the number $\delta = \|I - Q^{-1}A\|$ is less than 1, then

$$\|x^{(k)} - x\| \leqq \frac{\delta}{1 - \delta}\|x^{(k)} - x^{(k-1)}\|$$

34. Prove that

$$T_n(t) = \frac{1}{2}(b^n + b^{-n}) \qquad b = t + \sqrt{t^2 - 1}$$

35. Prove Theorem 9 in the case $a > 1$.

36. Prove that a positive definite matrix A is Hermitian using the definition; namely, A is positive definite if $x^*Ax > 0$ for all nonzero $x \in \mathbb{C}^n$. Thus, in particular, a real positive definite matrix A must be symmetric if this definition is used. However, if the alternative definition $x^TAx > 0$ for all nonzero $x \in \mathbb{R}^n$ is used, then a real positive definite matrix need not be symmetric.

37. The set of all matrices similar to a given fixed $n \times n$ matrix A is an **equivalence class** under the equivalence relation of similarity. Are these equivalence classes closed? Thus, we ask whether the conditions $B^{(k)} \cong A$ and $B^{(k)} \to B$ imply that $B \cong A$. *Hint:* See Problem 12.

38. Show that the converse in Problem 17 is not true.

39. Prove that if A is nonsingular and if $|\lambda| < \|A^{-1}\|^{-1}$, then λ is not an eigenvalue of A. Here the norm can be any subordinate matrix norm.

COMPUTER PROBLEMS 4.6

1. Program the Gauss-Seidel method and test it on these examples:

$$\text{(a)} \quad \begin{cases} 3x + y + z = 5 \\ x + 3y - z = 3 \\ 3x + y - 5z = -1 \end{cases}$$

$$\text{(b)} \quad \begin{cases} 3x + y + z = 5 \\ 3x + y - 5z = -1 \\ x + 3y - z = 3 \end{cases}$$

Analyze what happens when these systems are solved by simple Gaussian elimination without pivoting.

2. Apply Gauss-Seidel iteration to the system in which

$$A = \begin{bmatrix} 0.96326 & 0.81321 \\ 0.81321 & 0.68654 \end{bmatrix} \qquad b = \begin{bmatrix} 0.88824 \\ 0.74988 \end{bmatrix}$$

Use $(0.33116,\ 0.70000)^T$ as the starting point and explain what happens.

*4.7 Steepest Descent and Conjugate Gradient Methods

In this section, some special methods will be developed for solving the system

$$Ax = b$$

for the case when A is a real $n \times n$ **symmetric** and **positive definite matrix**. These hypotheses mean that

$$A^T = A$$

and

$$x^T Ax > 0 \ \text{ for } \ x \neq 0$$

Throughout we use the inner-product notation for real vectors x and y:

$$\langle x, y \rangle = x^T y = \sum_{i=1}^{n} x_i y_i$$

Some immediate properties are:

(i) $\langle x, y \rangle = \langle y, x \rangle$
(ii) $\langle \alpha x, y \rangle = \alpha \langle x, y \rangle$ for any constant α
(iii) $\langle x + y, z \rangle = \langle x, z \rangle + \langle y, z \rangle$
(iv) $\langle x, Ay \rangle = \langle A^T x, y \rangle$

By property **(i)**, the order of the arguments in properties **(ii)**–**(iii)** can be reversed.

We begin by establishing the equivalence of two numerical problems involving A and b.

LEMMA 1 *If A is symmetric and positive definite, then the problem of solving $Ax = b$ is equivalent to the problem of minimizing the quadratic form*

$$q(x) = \langle x, Ax \rangle - 2\langle x, b \rangle$$

Proof First, let us find out how the function q behaves along a **one-dimensional ray**. Here we consider $x + tv$, where x and v are vectors and t is a scalar. Figure 4.2 shows why this can be thought of as a one-dimensional ray. A straightforward calculation reveals that for a scalar t, we have

$$\begin{aligned} q(x + tv) &= \langle x + tv, A(x + tv) \rangle - 2\langle x + tv, b \rangle \\ &= \langle x, Ax \rangle + t\langle x, Av \rangle + t\langle v, Ax \rangle + t^2\langle v, Av \rangle - 2\langle x, b \rangle - 2t\langle v, b \rangle \\ &= q(x) + 2t\langle v, Ax \rangle - 2t\langle v, b \rangle + t^2\langle v, Av \rangle \\ &= q(x) + 2t\langle v, Ax - b \rangle + t^2\langle v, Av \rangle \qquad \textbf{(1)} \end{aligned}$$

since $A^T = A$. Note that the coefficient of t^2 in Equation (1) is positive. Thus, the quadratic function on the ray has a minimum and not a maximum. From Equation

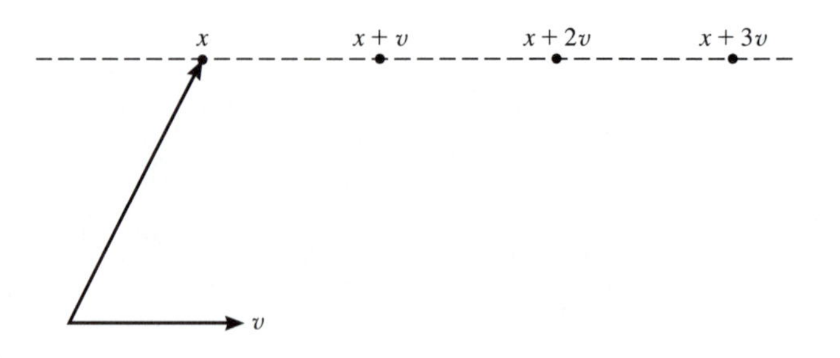

FIGURE 4.2 Example of one-dimensional ray

(1) we compute the derivative with respect to t as

$$\frac{d}{dt}q(x + tv) = 2\langle v, Ax - b\rangle + 2t\langle v, Av\rangle \tag{2}$$

The minimum of q along the ray occurs when the derivative in (2) is 0. The value of t that yields the minimum point is therefore

$$\widehat{t} = \langle v, b - Ax\rangle / \langle v, Av\rangle \tag{3}$$

Using this value, $\widehat{t}$, we compute the minimum of q on the ray:

$$\begin{aligned} q(x + \widehat{t}v) &= q(x) + \widehat{t}\,[2\langle v, Ax - b\rangle + \widehat{t}\,\langle v, Av\rangle] \\ &= q(x) + \widehat{t}\,[2\langle v, Ax - b\rangle + \langle v, b - Ax\rangle] \\ &= q(x) - \widehat{t}\,\langle v, b - Ax\rangle \\ &= q(x) - \langle v, b - Ax\rangle^2 / \langle v, Av\rangle \end{aligned} \tag{4}$$

Our calculation shows that a reduction of the value of q always occurs in passing from x to $x + \widehat{t}v$ unless v is **orthogonal** to the residual—that is, $\langle v, b - Ax\rangle = 0$. If x is not a solution of the system $Ax = b$, then many vectors v exist satisfying $\langle v, b - Ax\rangle \neq 0$. Hence, if $Ax \neq b$, then x does not minimize q. On the other hand, if $Ax = b$, then there is no ray emanating from x on which q takes a lesser value than $q(x)$. Hence, such an x is a minimum point of q. ■

The preceding proof suggests an iterative method for solving $Ax = b$. We proceed by minimizing q along a succession of rays. At the kth step in such an algorithm, $x^{(0)}, x^{(1)}, x^{(2)}, \dots, x^{(k)}$ would be available. Then by use of some rule, a suitable search direction $v^{(k)}$ is chosen. The next point in our sequence is

$$x^{(k+1)} = x^{(k)} + t_k v^{(k)}$$

where

$$t_k = \frac{\langle v^{(k)}, b - Ax^{(k)}\rangle}{\langle v^{(k)}, Av^{(k)}\rangle}$$

Many iterative methods are of the general form

$$x^{(k+1)} = x^{(k)} + t_k v^{(k)}$$

for specific values of the scalar t_k and the vectors $v^{(k)}$. If $\|v^{(k)}\| = 1$, then t_k measures the distance we move from $x^{(k)}$ to obtain $x^{(k+1)}$.

Steepest Descent

The method of **steepest descent** is an algorithm of the type just described. It is defined by stipulating that $v^{(k)}$ should be the negative gradient of q at $x^{(k)}$. It turns out that this negative gradient points in the direction of the residual, $r^{(k)} = b - Ax^{(k)}$. (See Problem 1.) A formal description of steepest descent then goes as follows:

```
input x^(0), A, b, M
output 0, x^(0)
for k = 0 to M − 1 do
    v^(k) ← b − Ax^(k)
    t_k ← ⟨v^(k), v^(k)⟩/⟨v^(k), Av^(k)⟩
    x^(k+1) ← x^(k) + t_k v^(k)
output k + 1, x^(k+1)
end do
```

In the actual programming of this algorithm, the successive vectors $x^{(0)}, x^{(1)}, \ldots$ need not be saved; the *current* x-vector can be overwritten. The same remark holds for the direction vectors $v^{(0)}, v^{(1)}, \ldots$. Thus, we could write alternatively

```
input x, A, b, M
output 0, x
for k = 1 to M do
    v ← b − Ax
    t ← ⟨v, v⟩/⟨v, Av⟩
    x ← x + tv
    output k, x
end do
```

The method of steepest descent is rarely used on this problem because it is too slow. How it might progress on a two-dimensional problem is depicted in Figure 4.3, where we have shown the contours (or **level lines**) of the quadratic form q.

Conjugate Directions

A family of methods called **conjugate direction** methods shares the basic strategy of minimizing the quadratic function along a succession of rays. Usually these search directions are determined within the solution process, one at a time. However, we shall discuss first the case when they are prescribed at the beginning.

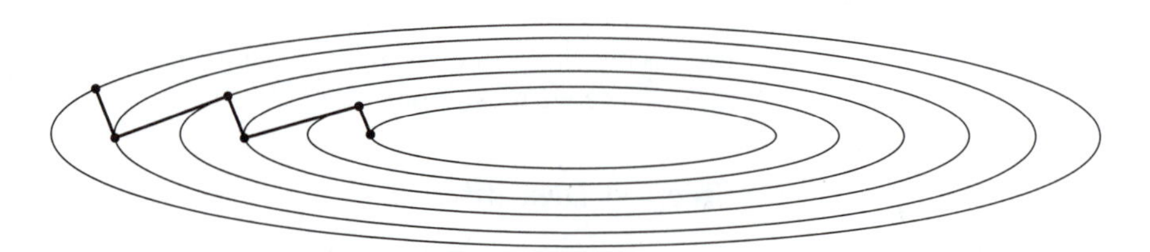

FIGURE 4.3 Geometric interpretation of steepest descent

Assuming that A is an $n \times n$ symmetric and positive definite matrix, suppose that a set of vectors $\{u^{(1)}, u^{(2)}, \dots, u^{(n)}\}$ is provided and has the property

$$\langle u^{(i)}, Au^{(j)} \rangle = \delta_{ij} \qquad (1 \leqq i, j \leqq n)$$

This property is called **A-orthonormality** and is an obvious generalization of ordinary orthonormality. We shall see that if these vectors are used as the search directions in the step-by-step minimization of the quadratic function q, then the solution is obtained by the nth step. Before giving the formal result, we observe that the A-orthonormality condition can be expressed as a matrix equation

$$U^T AU = I$$

where U is the $n \times n$ matrix whose columns are $u^{(1)}, u^{(2)}, \dots, u^{(n)}$. From this it is clear that A and U are nonsingular, and the columns $u^{(1)}, u^{(2)}, \dots, u^{(n)}$ form a basis for $\mathbb{R}^n$.

THEOREM 1 *Let $\{u^{(1)}, u^{(2)}, \dots, u^{(n)}\}$ be an A-orthonormal system. Define*

$$x^{(i)} = x^{(i-1)} + \langle b - Ax^{(i-1)}, u^{(i)} \rangle u^{(i)} \qquad (1 \leqq i \leqq n)$$

in which $x^{(0)}$ is an arbitrary point of $\mathbb{R}^n$. Then $Ax^{(n)} = b$.

Proof Define $t_i = \langle b - Ax^{(i-1)}, u^{(i)} \rangle$ so that the recursive formula becomes simply

$$x^{(i)} = x^{(i-1)} + t_i u^{(i)}$$

From this and with the aid of the relation $\langle Au^{(j)}, u^{(i)} \rangle = \delta_{ij}$, we deduce the following equations:

$$\begin{aligned} Ax^{(i)} &= Ax^{(i-1)} + t_i Au^{(i)} \\ Ax^{(n)} &= Ax^{(0)} + t_1 Au^{(1)} + \cdots + t_n Au^{(n)} \\ \langle Ax^{(n)} - b, u^{(i)} \rangle &= \langle Ax^{(0)} - b, u^{(i)} \rangle + t_i \end{aligned}$$

We can show that the right side of this last equation is 0 as follows:

$$\begin{aligned} t_i &= \langle b - Ax^{(i-1)}, u^{(i)} \rangle \\ &= \langle b - Ax^{(0)}, u^{(i)} \rangle + \langle Ax^{(0)} - Ax^{(1)}, u^{(i)} \rangle + \cdots + \langle Ax^{(i-2)} - Ax^{(i-1)}, u^{(i)} \rangle \\ &= \langle b - Ax^{(0)}, u^{(i)} \rangle + \langle -t_1 Au^{(1)}, u^{(i)} \rangle + \cdots + \langle -t_{i-1} Au^{(i-1)}, u^{(i)} \rangle \\ &= \langle b - Ax^{(0)}, u^{(i)} \rangle \end{aligned}$$

Our analysis shows that $Ax^{(n)} - b$ is orthogonal (in the usual sense) to the vectors $u^{(1)}, u^{(2)}, \dots, u^{(n)}$ and must therefore be 0. ■

We shall now describe a concrete realization of this method. Let A be a symmetric and positive definite $n \times n$ matrix. The equation

$$\langle x, y\rangle_A = \langle x, Ay\rangle = \sum_{i=1}^{n}\sum_{j=1}^{n} a_{ij}x_iy_j$$

defines an inner product, as is easily verified. There is a corresponding quadratic norm, $\|x\|_A^2 = \langle x, x\rangle_A$. If the Gram-Schmidt process is used with this new inner product on the set of standard unit vectors $\{e^{(1)}, e^{(2)}, \dots, e^{(n)}\}$, the result is an A-orthonormal system $\{u^{(1)}, u^{(2)}, \dots, u^{(n)}\}$. The Gram-Schmidt process (discussed more fully in Section 5.3) is described as follows:

$$u^{(i)} = \left(\|v^{(i)}\|_A\right)^{-1} v^{(i)} \qquad \text{where} \qquad v^{(i)} = e^{(i)} - \sum_{j<i}\langle e^{(i)}, u^{(j)}\rangle_A\, u^{(j)}$$

Because of the special nature of the vectors $e^{(i)}$, these formulas reduce to

$$u^{(i)} = \left(\|v^{(i)}\|_A\right)^{-1} v^{(i)} \qquad \text{where} \qquad v^{(i)} = e^{(i)} - \sum_{j<i}(Au^{(j)})_i\, u^{(j)}$$

For example, $u^{(1)} = (1/\sqrt{a_{11}})e^{(1)}$. The formulas indicate that each $u^{(i)}$ is a linear combination of the unit vectors $e^{(1)}, e^{(2)}, \dots, e^{(i)}$. The matrix U whose columns are $u^{(1)}, u^{(2)}, \dots, u^{(n)}$ is therefore upper triangular. The equation $U^TAU = I$ mentioned previously gives us $A = (U^T)^{-1}U^{-1}$, which is an LU-factorization. The solution of $Ax = b$ is thus being computed by a variant of Gaussian elimination, without pivoting.

In numerical work, it is more convenient to use A-orthogonal systems instead of A-orthonormal systems. A set of vectors $v^{(1)}, v^{(2)}, \dots$ is **A-orthogonal** if $\langle v^{(i)}, Av^{(j)}\rangle = 0$ whenever $i \neq j$. From the A-orthogonal system we pass to an A-orthonormal system by the normalization process:

$$u^{(i)} = \left(\|v^{(i)}\|_A\right)^{-1} v^{(i)}$$

If each $v^{(i)}$ is a nonzero vector and if A is a positive definite matrix, then the $u^{(i)}$ will form an A-orthonormal system. The analogue of Theorem 1 follows.

THEOREM 2 *Let $\{v^{(1)}, v^{(2)}, \dots, v^{(n)}\}$ be an A-orthogonal system of nonzero vectors for a symmetric and positive definite $n \times n$ matrix A. Define*

$$x^{(i)} = x^{(i-1)} + \frac{\langle b - Ax^{(i-1)}, v^{(i)}\rangle}{\langle v^{(i)}, Av^{(i)}\rangle}\, v^{(i)} \qquad (1 \leqq i \leqq n)$$

in which $x^{(0)}$ is arbitrary. Then $Ax^{(n)} = b$.

Conjugate Gradient Method

The **conjugate gradient** method of Hestenes and Stiefel [1952] is a particular type of conjugate direction method. It applies to a system $Ax = b$ in which A is symmetric and positive definite. In the conjugate gradient method, the search directions $v^{(i)}$ referred to in Theorem 2 are chosen one by one during the iterative process and form an A-orthogonal system. The distinguishing property, however, is that the residuals, $r^{(i)} = b - Ax^{(i)}$, form an orthogonal system in the ordinary sense; that is, $\langle r^{(i)}, r^{(j)} \rangle = 0$ if $i \neq j$.

The conjugate gradient method can be recommended over simple Gaussian elimination if A is very large and sparse. Theoretically, the conjugate gradient algorithm will yield the solution of the system $Ax = b$ in at most n steps. In practice, however, the algorithm is used as an iterative method to produce a sequence of vectors converging to the solution. In an ill-conditioned problem, roundoff errors often prevent the algorithm from furnishing a sufficiently precise solution at the nth step. When the conjugate gradient method was introduced by Hestenes and Stiefel [1952], there was initially much excitement over it. However, interest quickly waned when it was discovered that the finite-termination property was not obtained in practice. For a **direct method** this was undesirable. Some 20 years later, there was renewed interest in the method when it was viewed as an **iterative method**. Not obtaining the solution to full precision after n steps is the expected behavior of an iterative method. In fact, we hope for a satisfactory answer in many fewer than n steps for extremely large systems. In well-conditioned problems, the number of iterations necessary for satisfactory convergence of the conjugate gradient method can be much less than the order of the system. The history of the method has been written by Golub and O'Leary [1989].

The **formal conjugate gradient algorithm** follows. (This pseudocode is not designed for computer execution. A suitable algorithm for that purpose is given later.)

input $x^{(0)}, M, A, b, \varepsilon$
$r^{(0)} \leftarrow b - Ax^{(0)}$
$v^{(0)} \leftarrow r^{(0)}$
output $0, x^{(0)}, r^{(0)}$
for $k = 0$ **to** $M - 1$ **do**
  if $v^{(k)} = 0$ **then stop**
  $t_k \leftarrow \langle r^{(k)}, r^{(k)} \rangle / \langle v^{(k)}, Av^{(k)} \rangle$
  $x^{(k+1)} \leftarrow x^{(k)} + t_k v^{(k)}$
  $r^{(k+1)} \leftarrow r^{(k)} - t_k A v^{(k)}$
  if $\|r^{(k+1)}\|_2^2 < \varepsilon$ **then stop**
  $s_k \leftarrow \langle r^{(k+1)}, r^{(k+1)} \rangle / \langle r^{(k)}, r^{(k)} \rangle$
  $v^{(k+1)} \leftarrow r^{(k+1)} + s_k v^{(k)}$
  output $k + 1, x^{(k+1)}, r^{(k+1)}$
end do

In the algorithm, if $r^{(k)} = 0$, then $x^{(k)}$ is (theoretically) a solution of the linear system $Ax = b$. The computer realization of the algorithm requires storage for four vectors—namely, $x^{(k)}$, $r^{(k)}$, $v^{(k)}$, and $Av^{(k)}$. Provision must also be made for computing the products $Av^{(k)}$. (This may or may not require the storage of the full matrix A.) The work per iteration is modest, amounting to a single matrix-vector product for $Av^{(k)}$ and just two inner products (after the first iterations) for $\langle v^{(k)}, Av^{(k)}\rangle$ and $\langle r^{(k+1)}, r^{(k+1)}\rangle$. Notice that the stopping test involves $\|r^{(k+1)}\|_2^2 = \langle r^{(k+1)}, r^{(r+1)}\rangle$, which is easy to compute since $r^{(k+1)}$ is already available. A computer code for the **conjugate gradient method** should therefore be based on the following algorithm:

```
input x, A, b, M, ε, δ
r ← b − Ax
v ← r
c ← ⟨r, r⟩
for k = 1 to M do
    if ⟨v, v⟩^{1/2} < δ then exit loop
    z ← Av
    t ← c/⟨v, z⟩
    x ← x + tv
    r ← r − tz
    d ← ⟨r, r⟩
    if d < ε then exit loop
    v ← r + (d/c)v
    c ← d
    output k, x, r
end do
```

THEOREM 3 *In the conjugate gradient algorithm, for any integer $m < n$, if $v^{(0)}, v^{(1)}, \dots, v^{(m)}$ are all nonzero vectors, then $r^{(i)} = b - Ax^{(i)}$ for $0 \leqq i \leqq m$, and $\{r^{(0)}, r^{(1)}, \dots, r^{(m)}\}$ is an orthogonal set of nonzero vectors.*

Proof We shall prove somewhat more—namely, that (under the given hypotheses) each of the following holds:

(a) $\langle r^{(m)}, v^{(i)}\rangle = 0 \qquad (0 \leqq i < m)$

(b) $\langle r^{(i)}, r^{(i)}\rangle = \langle r^{(i)}, v^{(i)}\rangle \qquad (0 \leqq i \leqq m)$

(c) $\langle v^{(m)}, Av^{(i)}\rangle = 0 \qquad (0 \leqq i < m)$

(d) $r^{(i)} = b - Ax^{(i)} \qquad (0 \leqq i \leqq m)$

(e) $\langle r^{(m)}, r^{(i)}\rangle = 0 \qquad (0 \leqq i < m)$

(f) $r^{(i)} \neq 0 \qquad (0 \leqq i \leqq m)$

The proof is by induction on m. For the case $m = 0$, we assume that $v^{(0)} \neq 0$ and we must prove parts **(a)**–**(f)**. Notice that **(a)**, **(c)**, and **(e)** are vacuous in this case. As for **(b)**, **(d)**, and **(f)**, they follow immediately from the definition of the algorithm since

$$r^{(0)} = b - Ax^{(0)} = v^{(0)} \neq 0$$

Now assume that the theorem has been established for a certain m. On this basis, we shall prove it for $m + 1$. To this end, we assume that $v^{(0)}, v^{(1)}, \dots, v^{(m+1)}$ are all nonzero. By the induction hypothesis, parts **(a)**–**(f)** are valid. We want to prove that

(a′) $\langle r^{(m+1)}, v^{(i)} \rangle = 0 \qquad (0 \leqq i \leqq m)$
(b′) $\langle r^{(m+1)}, r^{(m+1)} \rangle = \langle r^{(m+1)}, v^{(m+1)} \rangle$
(c′) $\langle v^{(m+1)}, Av^{(i)} \rangle = 0 \quad (0 \leqq i \leqq m)$
(d′) $r^{(m+1)} = b - Ax^{(m+1)}$
(e′) $\langle r^{(m+1)}, r^{(i)} \rangle = 0 \quad (0 \leqq i \leqq m)$
(f′) $r^{(m+1)} \neq 0$

For **(a′)**, first let $i = m$. We have

$$\begin{aligned}\langle r^{(m+1)}, v^{(m)} \rangle &= \langle r^{(m)} - t_m A v^{(m)}, v^{(m)} \rangle = \langle r^{(m)}, v^{(m)} \rangle - t_m \langle v^{(m)}, Av^{(m)} \rangle \\ &= \langle r^{(m)}, v^{(m)} \rangle - \langle r^{(m)}, r^{(m)} \rangle = 0\end{aligned}$$

using **(b)**. If $0 \leqq i < m$, then by **(a)** and **(c)**,

$$\langle r^{(m+1)}, v^{(i)} \rangle = \langle r^{(m)}, v^{(i)} \rangle - t_m \langle v^{(m)}, Av^{(i)} \rangle = 0$$

For the proof of **(b′)**, we use part **(a′)** to conclude that

$$\langle r^{(m+1)}, v^{(m+1)} \rangle = \langle r^{(m+1)}, r^{(m+1)} + s_m v^{(m)} \rangle = \langle r^{(m+1)}, r^{(m+1)} \rangle$$

In proving part **(c′)** we set $s_{-1} = 0$ and $v^{(-1)} = 0$ whenever these terms appear. If $0 \leqq i \leqq m$, then

$$\begin{aligned}\langle v^{(m+1)}, Av^{(i)} \rangle &= \langle r^{(m+1)} + s_m v^{(m)}, Av^{(i)} \rangle \\ &= \langle r^{(m+1)}, Av^{(i)} \rangle + s_m \langle v^{(m)}, Av^{(i)} \rangle) \\ &= t_i^{-1} \langle r^{(m+1)}, r^{(i)} - r^{(i+1)} \rangle + s_m \langle v^{(m)}, Av^{(i)} \rangle \\ &= t_i^{-1} \langle r^{(m+1)}, v^{(i)} - s_{i-1} v^{(i-1)} - v^{(i+1)} + s_i v^{(i)} \rangle \\ &\quad + s_m \langle v^{(m)}, Av^{(i)} \rangle \\ &= t_i^{-1} [\langle r^{(m+1)}, v^{(i)} \rangle - s_{i-1} \langle r^{(m+1)}, v^{(i-1)} \rangle - \langle r^{(m+1)}, v^{(i+1)} \rangle \\ &\quad + s_i \langle r^{(m+1)}, v^{(i)} \rangle] + s_m \langle v^{(m)}, Av^{(i)} \rangle\end{aligned}$$

If $i < m$, then by part **(a′)**, $r^{(m+1)}$ is orthogonal to $v^{(i)}$, $v^{(i-1)}$, and $v^{(i+1)}$. Also, by part **(c)**, $\langle v^{(m)}, Av^{(i)} \rangle = 0$. Hence in this case, $\langle v^{(m+1)}, Av^{(i)} \rangle = 0$. The case $i = m$ is special. By a previous equation, we have

$$\begin{aligned}\langle v^{(m+1)}, Av^{(m)} \rangle &= t_m^{-1} \langle r^{(m+1)}, v^{(m)} - s_{m-1} v^{(m-1)} - v^{(m+1)} + s_m v^{(m)} \rangle \\ &\quad + s_m \langle v^{(m)}, Av^{(m)} \rangle\end{aligned}$$

By part (**a**′), $r^{(m+1)}$ is orthogonal to $v^{(m)}$ and to $v^{(m-1)}$. Hence,

$$\begin{aligned}
\langle v^{(m+1)}, Av^{(m)}\rangle &= -t_m^{-1}\langle r^{(m+1)}, v^{(m+1)}\rangle + s_m\langle r^{(m)}, Av^{(m)}\rangle \\
&= -\frac{\langle v^{(m)}, Av^{(m)}\rangle}{\langle r^{(m)}, r^{(m)}\rangle}\langle r^{(m+1)}, v^{(m+1)}\rangle + \frac{\langle r^{(m+1)}, r^{(m+1)}\rangle}{\langle r^{(m)}, r^{(m)}\rangle}\langle v^{(m)}, Av^{(m)}\rangle
\end{aligned}$$

We see that this expression is 0 by using (**b**′).

For the proof of (**d**′), we write

$$\begin{aligned}
b - Ax^{(m+1)} &= b - A(x^{(m)} + t_m v^{(m)}) = b - Ax^{(m)} - t_m Av^{(m)} \\
&= r^{(m)} - (r^{(m)} - r^{(m+1)}) = r^{(m+1)}
\end{aligned}$$

For the proof of (**e**′), let $0 \leqq i \leqq m$, and put $s_{-1} = 0$ and $v^{(-1)} = 0$. Then from part (**a**′) we have

$$\begin{aligned}
\langle r^{(m+1)}, r^{(i)}\rangle &= \langle r^{(m+1)}, v^{(i)} - s_{i-1}v^{(i-1)}\rangle \\
&= \langle r^{(m+1)}, v^{(i)}\rangle - s_{i-1}\langle r^{(m+1)}, v^{(i-1)}\rangle = 0
\end{aligned}$$

For part (**f**′), use (**c**′) and the positive definiteness of A as follows:

$$\begin{aligned}
0 < \langle v^{(m+1)}, Av^{(m+1)}\rangle &= \langle r^{(m+1)} + s_m v^{(m)}, Av^{(m+1)}\rangle \\
&= \langle r^{(m+1)}, Av^{(m+1)}\rangle + s_m\langle v^{(m)}, Av^{(m+1)}\rangle \\
&= \langle r^{(m+1)}, Av^{(m+1)}\rangle
\end{aligned}$$

Hence, $r^{(m+1)} \neq 0$. ■

This theorem and its proof are adapted from the book of Stoer and Bulirsch [1980].

Preconditioned Conjugate Gradient

We want to solve the system

$$Ax = b$$

where A is symmetric and positive definite, using a variant of the conjugate gradient method. It would be advantageous to **precondition** this system and obtain a new system that is *better conditioned* than the original system. By this we mean that for some nonsingular matrix S, the preconditioned system

$$\widehat{A}\widehat{x} = \widehat{b}$$

where

$$\begin{cases} \widehat{A} = S^T A S \\ \widehat{x} = S^{-1} x \\ \widehat{b} = S^T b \end{cases}$$

is such that $\kappa(\widehat{A}) < \kappa(A)$. As a by-product, the iterative method used to solve the preconditioned system may converge faster than it would for the original system. Rather than taking an arbitrary matrix S, we suppose that the symmetric and positive definite splitting matrix Q can be factored so that

$$Q^{-1} = SS^T$$

The reason for this will become clear shortly.

The formal conjugate gradient algorithm can be written for the preconditioned system as follows:

$\widehat{r}^{(0)} = \widehat{b} - \widehat{A}\widehat{x}^{(0)}$
$\widehat{v}^{(0)} = \widehat{r}^{(0)}$
for $k = 0$ **to** M **do**
$\quad \widehat{t}_k = \langle \widehat{r}^{(k)}, \widehat{r}^{(k)} \rangle / \langle \widehat{v}^{(k)}, \widehat{A}\widehat{v}^{(k)} \rangle$
$\quad \widehat{x}^{(k+1)} = \widehat{x}^{(k)} + \widehat{t}_k \widehat{v}^{(k)}$
$\quad \widehat{r}^{(k+1)} = \widehat{r}^{(k)} - \widehat{t}_k \widehat{A}\widehat{v}^{(k)}$
$\quad \widehat{s}_k = \langle \widehat{r}^{(k+1)}, \widehat{r}^{(k+1)} \rangle / \langle \widehat{r}^{(k)}, \widehat{r}^{(k)} \rangle$
$\quad \widehat{v}^{(k+1)} = \widehat{r}^{(k+1)} + \widehat{s}_k \widehat{v}^{(k)}$
end do

In the preconditioning of the original system, any sparsity in the matrix A can be destroyed in forming $\widehat{A}$. So rather than explicitly forming the preconditioned system, we would like to use the original system and have the preconditioning done implicitly within the algorithm.

We write

$$\begin{aligned}
\widehat{x}^{(k)} &= S^{-1} x^{(k)} \\
\widehat{v}^{(k)} &= S^{-1} v^{(k)} \\
\widehat{r}^{(k)} &= \widehat{b} - \widehat{A}\widehat{x}^{(k)} = S^T b - (S^T A S)(S^{-1} x^{(0)}) = S^T r^{(k)} \\
\widetilde{r}^{(k)} &= Q^{-1} r^{(k)}
\end{aligned}$$

Then

$$\begin{aligned}
\widehat{t}_k &= \langle \widehat{r}^{(k)}, \widehat{r}^{(k)} \rangle / \langle \widehat{v}^{(k)}, \widehat{A}\widehat{v}^{(k)} \rangle \\
&= \langle S^T r^{(k)}, S^T r^{(k)} \rangle / \langle S^{-1} v^{(k)}, (S^T A S)(S^{-1} v^{(k)}) \rangle \\
&= \langle Q^{-1} r^{(k)}, r^{(k)} \rangle / \langle v^{(k)}, A v^{(k)} \rangle \\
&= \langle \widetilde{r}^{(k)}, r^{(k)} \rangle / \langle v^{(k)}, A v^{(k)} \rangle
\end{aligned}$$

Moreover,

$$\begin{aligned} \widehat{x}^{(k+1)} &= \widehat{x}^{(k)} + \widehat{t}_k \widehat{v}^{(k)} \\ S^{-1}x^{(k+1)} &= S^{-1}x^{(k)} + \widehat{t}_k S^{-1} v^{(k)} \end{aligned}$$

and multiplying by S, we have

$$x^{(k+1)} = x^{(k)} + \widehat{t}_k v^{(k)}$$

Similarly,

$$\begin{aligned} \widehat{r}^{(k+1)} &= \widehat{r}^{(k)} - \widehat{t}_k \widehat{A} \widehat{v}^{(k)} \\ S^T r^{(k+1)} &= S^T r^{(k)} - \widehat{t}_k (S^T A S)(S^{-1} v^{(k)}) \end{aligned}$$

and multiplying by S^{-T}, we have

$$r^{(k+1)} = r^{(k)} - \widehat{t}_k A v^{(k)}$$

Now

$$\begin{aligned} \widehat{s}_k &= \langle \widehat{r}^{(k+1)}, \widehat{r}^{(k+1)} \rangle / \langle \widehat{r}^{(k)}, \widehat{r}^{(k)} \rangle \\ &= \langle S^T r^{(k+1)}, S^T r^{(k+1)} \rangle / \langle S^T r^{(k)}, S^T r^{(k)} \rangle \\ &= \langle Q^{-1} r^{(k+1)}, r^{(k+1)} \rangle / \langle Q^{-1} r^{(k)}, r^{(k)} \rangle \\ &= \langle \widetilde{r}^{(k+1)}, r^{(k+1)} \rangle / \langle \widetilde{r}^{(k)}, r^{(k)} \rangle \end{aligned}$$

Also,

$$\begin{aligned} \widehat{v}^{(k+1)} &= \widehat{r}^{(k+1)} + \widehat{s}_k \widehat{v}^{(k)} \\ S^{-1} v^{(k+1)} &= S^T r^{(k+1)} + \widehat{s}_k S^{-1} v^{(k)} \end{aligned}$$

and multiplying by S, we have

$$\begin{aligned} v^{(k+1)} &= Q^{-1} r^{(k+1)} + \widehat{s}_k v^{(k)} \\ &= \widetilde{r}^{(k+1)} + \widehat{s}_k v^{(k)} \end{aligned}$$

Now we can write the **preconditioned conjugate gradient algorithm** primarily in terms of the original system. The pseudocode given next is not suitable for a computer program. Another more efficient version is given later in this section.

input $x^{(0)}, A, b, M, Q$
$r^{(0)} \leftarrow b - Ax^{(0)}$
solve $Q\widetilde{r}^{(0)} = r^{(0)}$ for $\widetilde{r}^{(0)}$
$v^{(0)} \leftarrow r^{(0)}$
output $0, x^{(0)}$

```
for k = 0 to M − 1 do
    if v^(k) = 0 then exit loop
    t̂_k ← ⟨r̃^(k), r^(k)⟩/⟨v^(k), Av^(k)⟩
    x^(k+1) ← x^(k) + t̂_k v^(k)
    r^(k+1) ← r^(k) − t̂_k Av^(k)
    solve Qr̃^(k+1) = r^(k+1) for r̃^(k+1)
    if ⟨r̃^(k+1), r^(k+1)⟩ < ε then
        if ⟨r^(k+1), r^(k+1)⟩ < ε then exit loop
    end if
    ŝ_k ← ⟨r̃^(k+1), r^(k+1)⟩/⟨r̃^(k), r^(k)⟩
    v^(k+1) ← r̃^(k+1) + ŝ_k v^(k)
    output k + 1, x^(k+1), r^(k+1)
end do
```

The preconditioned conjugate gradient algorithm above reduces to the regular conjugate gradient algorithm when $Q^{-1} = I$, since $\widetilde{r}^{(k)} = r^{(k)}$, $\widehat{t}_k = t_k$, and $\widehat{s}_k = s_k$.

If $Q = A$ and $S = A^{-1/2}$, then the preconditioned system reduces to $\widehat{x} = \widehat{b}$ and would be trivially solved. Unfortunately, this *perfectly* conditioned system ($\kappa(\widehat{A}) = 1$) is of no computational value because determining $\widehat{b} = S^T b$ would be as difficult as solving the original system.

Since we must solve a system of the form $Qx = y$ on each iteration of the preconditioned conjugate gradient algorithm, Q must be selected so that this system is *easy* to solve. If Q is a diagonal matrix, then this condition is met, but more complicated preconditioners may lead to faster convergence. As Q^{-1} becomes a better approximation to A, the preconditioned system becomes better conditioned, and the convergence of the iterative procedure occurs in fewer steps. On the other hand, solving $Qx = y$ becomes more difficult; this illustrates the typical trade-off between iterative methods requiring a few expensive steps and those requiring many inexpensive steps.

What about the convergence test? Since $\|r^{(k+1)}\|_2^2$ is not available in this algorithm, this requires an additional computation. What *is* available is $\langle \widetilde{r}^{(k+1)}, r^{(k+1)} \rangle$. Using it may cause the iterative process to halt either before or after it should for the specified accuracy. Hence, an additional evaluation of $\|r^{(k+1)}\|_2^2$ should be made as a check, once the previous test indicates convergence.

A computer code can be based on the following **preconditioned conjugate gradient algorithm**:

```
input x, A, b, M, Q, δ, ε
r ← b − Ax
solve Qz = r for z
v ← z
c ← ⟨z, r⟩
for k = 1 to M do
    if ⟨v, v⟩^(1/2) < δ then exit loop
    z ← Av
    t ← c/⟨v, z⟩
    x ← x + tv
```

$r \leftarrow r - tz$
solve $Qz = r$ for z
$d \leftarrow \langle z, r \rangle$
if $d < \varepsilon$ **then**
  if $\langle r, r \rangle < \varepsilon$ **then** exit loop
end if
$v \leftarrow z + (d/c)v$
$c \leftarrow d$
output k, x, r
end do

The preconditioned conjugate gradient method is an area of active research. There are a number of choices for Q, such as the symmetric successive over-relaxation matrix. Computer packages are available containing conjugate gradient routines and other iterative methods. Examples of such packages are `ITPACKV 2D` by Kincaid, Oppe, and Young [1989], `NSPCG` by Oppe, Joubert, and Kincaid [1988], `PCGPAK2` [1990], and `PCG` by Joubert et al. [1995]. We have patterned the presentation here for the preconditioned conjugate gradient method after that of Ortega [1988]. Additional details can be found there and in many other books, such as Golub and Van Loan [1989].

PROBLEMS 4.7

1. Prove that if A is symmetric, then the gradient of the function $q(x) = \langle x, Ax \rangle - 2\langle x, b \rangle$ at x is $2(Ax - b)$. Recall that the gradient of a function $g : \mathbb{R}^n \to \mathbb{R}$ is the vector whose components are $\partial g / \partial x_i$, for $i = 1, 2, \ldots, n$.
2. Prove that the minimum value of $q(x)$ is $-\langle b, A^{-1}b \rangle$.
3. Prove that in the method of steepest descent, $v^{(k)} \perp v^{(k+1)}$.
4. Let A be positive definite, and let b be a fixed vector. For any x, the **residual vector** is $r = b - Ax$, and the **error vector** is $e = A^{-1}b - x$. Show that the inner product of the error vector with the residual vector is positive unless $Ax = b$.
5. Let A be symmetric, let $Ax = b$, and let y be any vector. Using q as defined in the text, prove that

$$\langle (x - y), A(x - y) \rangle = \langle b, A^{-1}b \rangle + q(y)$$

   This shows that the minimization of $q(y)$ is equivalent to the minimization of $\langle (x - y), A(x - y) \rangle$.
6. Prove that if $\widehat{t}$ is defined by Equation (3) and if $y = x + \widehat{t}v$, then $v \perp (b - Ay)$; that is, $\langle v, b - Ay \rangle = 0$.
7. Show that in the method of steepest descent,

$$q(x^{(k+1)}) = q(x^{(k)}) - \|r^{(k)}\|^4 / \langle r^{(k)}, Ar^{(k)} \rangle$$

   where $r^{(k)} = b - Ax^{(k)}$.

8. If $\{u^{(1)}, u^{(2)}, \dots, u^{(n)}\}$ is a set of vectors that is orthonormal in the usual sense, and if these are used as the directions of search to minimize q, will the solution be obtained after n steps?

9. Let A be an $n \times n$ matrix, not assumed to be symmetric or positive definite. Assume the existence of an A-orthonormal system $\{u^{(1)}, u^{(2)}, \dots, u^{(n)}\}$ and *prove* that A is symmetric and positive definite.

10. Prove that in the conjugate gradient method,

$$
\begin{aligned}
t_k &= \langle r^{(k)}, v^{(k)} \rangle / \langle v^{(k)}, Av^{(k)} \rangle \\
s_k &= -\langle r^{(k+1)}, Av^{(k)} \rangle / \langle v^{(k)}, Av^{(k)} \rangle
\end{aligned}
$$

11. In the conjugate gradient method, prove that if $v^{(k)} = 0$, then $Ax^{(k)} = b$.

12. For what values of t_k does the method of steepest descent reduce to the Richardson method? What conditions on t_k and A must be fulfilled in order that the method of steepest descent be equivalent to the Jacobi method?

13. Consider the linear system

$$
\begin{bmatrix} 2 & 0 & -1 \\ -2 & -10 & 0 \\ -1 & -1 & 4 \end{bmatrix} \begin{bmatrix} x_1 \\ x_2 \\ x_3 \end{bmatrix} = \begin{bmatrix} 1 \\ -12 \\ 2 \end{bmatrix}
$$

Using the starting vector $x^{(0)} = (0, 0, 0)^T$, carry out two iterations of each of the following methods:

(a) Jacobi **(b)** Gauss-Seidel **(c)** Conjugate gradient

COMPUTER PROBLEMS 4.7

1. Program and test the conjugate gradient method. A good test case is the Hilbert matrix with a simple b-vector:

$$
a_{ij} = (i + j + 1)^{-1} \qquad b_i = \frac{1}{3}\sum_{j=1}^{n} a_{ij} \qquad (1 \leqq i, j \leqq n)
$$

2. Solve the following system using the indicated methods starting with $x^{(0)} = 0$.

(a) Jacobi **(b)** Gauss-Seidel **(c)** Conjugate gradient

$$
\begin{bmatrix} 10 & 1 & 2 & 3 & 4 \\ 1 & 9 & -1 & 2 & -3 \\ 2 & -1 & 7 & 3 & -5 \\ 3 & 2 & 3 & 12 & -1 \\ 4 & -3 & -5 & -1 & 15 \end{bmatrix} \begin{bmatrix} x_1 \\ x_2 \\ x_3 \\ x_4 \\ x_5 \end{bmatrix} = \begin{bmatrix} 12 \\ -27 \\ 14 \\ -17 \\ 12 \end{bmatrix}
$$

Remark: Here the coefficient matrix is symmetric and positive definite but not diagonally dominant.

*4.8 Analysis of Roundoff Error in the Gaussian Algorithm

In this section, we will investigate the roundoff errors that inevitably arise in solving a system of linear equations:

$$Ax = b \qquad (A \in \mathbb{R}^{n\times n},\ x \in \mathbb{R}^n,\ b \in \mathbb{R}^n)$$

An analysis, due originally to Wilkinson, is given for the Gaussian algorithm using unscaled row pivoting. The results are in the form of *a posteriori* bounds on the error. Thus, at the *conclusion* of the computation, an assertion can be made about the magnitude of the error.

In the analysis, we assume that suitable row interchanges have been carried out at the beginning so that the correct pivot elements will always be situated in the desired position. The row pivoting strategy then assures us that in step k of the algorithm, $|a_{kk}^{(k)}| \geqq |a_{ik}^{(k)}|$ for $i \geqq k$. (For the notation, refer to Section 4.3.) The inequality just given implies that the multipliers needed in the elimination process will not exceed unity in magnitude.

The formulas that define the LU-decomposition are as follows:

$$a_{ij}^{(k+1)} = \begin{cases} a_{ij}^{(k)} & i \leqq k \\ a_{ij}^{(k)} - \ell_{ik} a_{kj}^{(k)} & i > k \text{ and } j > k \\ 0 & j \leqq k < i \end{cases}$$

$$\ell_{ik} = \begin{cases} 0 & i < k \\ 1 & i = k \\ a_{ik}^{(k)} / a_{kk}^{(k)} & i > k \end{cases}$$

The process begins with $A^{(1)} = A$ and ends with $A^{(n)} = U$. The matrices L and U are unit lower triangular and upper triangular, respectively.

The numbers that actually occur in the computer will be distinguished notationally with a tilde. Thus, for example, $\widetilde{\ell}_{ik}$ is the number computed and stored in the memory location assigned to ℓ_{ik}. The meaning of the tilde is therefore machine-dependent. We shall assume that no overflow or underflow occurs in the course of the computations.

In describing the actual calculations in the computer, we use the function fl defined in Section 2.1. Recall that if x and y are machine numbers and if $\odot$ denotes an arithmetic operation, then the computer will produce, instead of $x \odot y$, a machine number denoted by $\text{fl}(x \odot y)$. By Equation (8) and by Problem 8 in Section 2.1, the relation between $x \odot y$ and $\text{fl}(x \odot y)$ can be expressed by the equation

$$\text{fl}(x \odot y) = (x \odot y)(1 - \delta) = (x \odot y)/(1 - \delta') \tag{1}$$

in which δ and δ' are numbers whose magnitudes do not exceed the unit roundoff, ε, in the particular computer being considered.

When these matters are taken into account, the computed numbers in the Gaussian algorithm can be described by the equations

$$\widetilde{a}_{ij}^{(k+1)} = \begin{cases} \widetilde{a}_{ij}^{(k)} & i \leqq k \\ \text{fl}\left[\widetilde{a}_{ij}^{(k)} - \text{fl}(\widetilde{\ell}_{ik}\widetilde{a}_{kj}^{(k)})\right] & i > k,\ j > k \end{cases}$$

$$\widetilde{\ell}_{ik} = \begin{cases} 1 & i = k \\ \text{fl}\left(\widetilde{a}_{ik}^{(k)}/\widetilde{a}_{kk}^{(k)}\right) & i > k \end{cases}$$

THEOREM 1 *Let A be an $n \times n$ nonsingular matrix whose elements are machine numbers in a computer with unit roundoff ε. The Gaussian algorithm with row pivoting produces matrices $\widetilde{L}$ and $\widetilde{U}$ such that*

$$\widetilde{L}\widetilde{U} = A + E \qquad \textit{where} \qquad |e_{ij}| \leqq 2n\varepsilon \max_{1 \leqq i,j,k \leqq n} |a_{ij}^{(k)}|$$

Proof Introduce quantities δ_{ij} and δ'_{ij} to play the roles of $\pm\delta$ and $\pm\delta'$ in Equation (1). The computed numbers in the algorithm then obey these equations:

$$\widetilde{a}_{ij}^{(k+1)} = \left[\widetilde{a}_{ij}^{(k)} - (1 - \delta_{ij})\widetilde{\ell}_{ik}\widetilde{a}_{kj}^{(k)}\right]/(1 - \delta'_{ij}) \qquad (i > k,\ j > k)$$

$$\widetilde{\ell}_{ik} = (1 + \delta_{ik})\widetilde{a}_{ik}^{(k)}/\widetilde{a}_{kk}^{(k)} \qquad (i > k)$$

The first of these two equations can also be written in the form

$$\widetilde{a}_{ij}^{(k+1)} = \widetilde{a}_{ij}^{(k)} - \widetilde{\ell}_{ik}\widetilde{a}_{kj}^{(k)} + \delta_{ij}\widetilde{\ell}_{ik}\widetilde{a}_{kj}^{(k)} + \delta'_{ij}\widetilde{a}_{ij}^{(k+1)} \qquad (i > k,\ j > k)$$

Now introduce an $n \times n$ matrix $\widetilde{L}^{(k)}$ that consists entirely of 0's except in column k below the diagonal:

$$\widetilde{L}^{(k)} = \begin{bmatrix} 0 & & & & \\ & \ddots & & & \\ & & 0 & & \\ & & \widetilde{\ell}_{k+1,k} & & \\ & & \vdots & \ddots & \\ & & \widetilde{\ell}_{n,k} & & 0 \end{bmatrix}$$

Then $\widetilde{A}^{(k)} - \widetilde{L}^{(k)}\widetilde{A}^{(k)}$ is the result of applying certain elementary row operations to $\widetilde{A}^{(k)}$; namely, $\widetilde{\ell}_{ik}$ times row k is subtracted from row i, for $k + 1 \leqq i \leqq n$. This is almost the definition of $\widetilde{A}^{(k+1)}$. The difference arises from the roundoff and from the fact that in $\widetilde{A}^{(k+1)}$ we set $\widetilde{a}_{ik}^{(k+1)} = 0$ for $k + 1 \leqq i \leqq n$. It is therefore possible to write

$$\widetilde{A}^{(k+1)} = \widetilde{A}^{(k)} - \widetilde{L}^{(k)}\widetilde{A}^{(k)} + E^{(k)} \tag{2}$$

where $E^{(k)}$ is a matrix that compensates for the errors.

To analyze the matrix $E^{(k)}$, consider first the case $i > k$, $j = k$. From previous equations, we have

$$\begin{aligned} e_{ik}^{(k)} &= \widetilde{a}_{ik}^{(k+1)} - a_{ik}^{(k)} + \widetilde{\ell}_{ik}^{(k)}\widetilde{a}_{kk}^{(k)} \\ &= 0 - \widetilde{a}_{ik}^{(k)} + \widetilde{\ell}_{ik}\widetilde{a}_{kk}^{(k)} \\ &= -\widetilde{a}_{ik}^{(k)} + \left(\widetilde{a}_{ik}^{(k)}/\widetilde{a}_{kk}^{(k)}\right)(1 + \delta_{ik})\widetilde{a}_{kk}^{(k)} \\ &= \delta_{ik}\widetilde{a}_{ik}^{(k)} \qquad\qquad (i > k) \end{aligned}$$

Next, consider the case $i > k$, $j > k$. Here we obtain

$$\begin{aligned} e_{ij}^{(k)} &= \widetilde{a}_{ij}^{(k+1)} - \widetilde{a}_{ij}^{(k)} + \widetilde{\ell}_{ik}^{(k)}\widetilde{a}_{kj}^{(k)} \\ &= \delta_{ij}\widetilde{\ell}_{ik}\widetilde{a}_{kj}^{(k)} + \delta'_{ij}\widetilde{a}_{ij}^{(k+1)} \qquad\qquad (i > k,\ j > k) \end{aligned}$$

All other elements of $E^{(k)}$ are 0. Now add Equations (2) for $k = 1, 2, \ldots, n - 1$. The result can be written

$$\widetilde{L}^{(1)}\widetilde{A}^{(1)} + \cdots + \widetilde{L}^{(n-1)}\widetilde{A}^{(n-1)} + I\widetilde{A}^{(n)} = A^{(1)} + E^{(1)} + \cdots + E^{(n-1)}$$

Notice that $\widetilde{L}^{(k)}\widetilde{A}^{(k)}$ is a matrix whose rows are simply various multiples of the single row vector

$$\left[\widetilde{a}_{k1}^{(k)}, \widetilde{a}_{k2}^{(k)}, \ldots, \widetilde{a}_{kn}^{(k)}\right]$$

Since this row vector is the same as

$$\left[\widetilde{a}_{k1}^{(n)}, \widetilde{a}_{k2}^{(n)}, \ldots, \widetilde{a}_{kn}^{(n)}\right]$$

we conclude that $\widetilde{L}^{(k)}\widetilde{A}^{(k)} = \widetilde{L}^{(k)}\widetilde{A}^{(n)}$. Recall also that $A^{(1)} = A$ and that the computed upper triangular factor U is $\widetilde{U} = \widetilde{A}^{(n)}$. Putting $E = \sum_{k=1}^{n-1} E^{(k)}$, we have

$$\left(\widetilde{L}^{(1)} + \widetilde{L}^{(2)} + \cdots + \widetilde{L}^{(n-1)} + I\right)\widetilde{A}^{(n)} = A^{(1)} + E$$

or

$$\widetilde{L}\widetilde{U} = A + E$$

All that remains now is to produce a bound on $\|E\|$. Let $\rho = \max_{1 \leqq i,j,k \leqq n} |A_{ij}^{(k)}|$. The preliminary row interchanges have ensured that all the multipliers satisfy $|\widetilde{\ell}_{ik}| \leqq 1$.

Thus, from equations given previously for $e_{ij}^{(k)}$, we have

$$|e_{ik}^{(k)}| = |\delta_{ik}|\,|\widetilde{a}_{ik}^{(k)}| \leqq \varepsilon\rho \qquad (i > k)$$

$$|e_{ij}^{(k)}| = |\delta_{ij}\widetilde{\ell}_{ik}\widetilde{a}_{kj}^{(k)} + \delta_{ij}'\widetilde{a}_{ij}^{(k+1)}| \leqq 2\varepsilon\rho \qquad (i > k,\ j > k)$$

From the definition of E it follows that

$$|e_{ij}| = \left|\sum_{k=1}^{n-1} e_{ij}^{(k)}\right| \leqq \sum_{k=1}^{n-1} |e_{ij}^{(k)}| \leqq 2n\varepsilon\rho \qquad \blacksquare$$

To apply this theorem in practice, we must first know the number

$$\rho = \max_{1\leqq i,j\leqq n} |a_{ij}^{(k)}|$$

This can be computed during the progress of the algorithm by adding suitable code.

THEOREM 2 *If $x_1, x_2, \dots, x_n$ and $y_1, y_2, \dots, y_n$ are machine numbers, then the machine value of*

$$\sum_{i=1}^{n} x_i y_i$$

computed in the natural way can be expressed as $\sum_{i=1}^{n} x_i y_i(1 + \delta_i)$, in which the δ_i's satisfy $|\delta_i| \leqq \frac{6}{5}(n + 1)\varepsilon$. (The number ε is the unit roundoff error of the machine, and we assume that $n\varepsilon < \frac{1}{3}$.)

Proof The hypothesis means that the computation will proceed as follows:

$$\begin{cases} z_0 = 0 \\ z_k = \mathrm{fl}\left[z_{k-1} + \mathrm{fl}(x_k y_k)\right] \qquad (1 \leqq k \leqq n) \end{cases}$$

Let us use induction on n to prove that $|\delta_i| \leqq (1 + \varepsilon)^{n+2-i} - 1$. For $n = 1$, we have

$$z_1 = \mathrm{fl}(x_1 y_1) = x_1 y_1(1 + \delta_1) \qquad |\delta_1| \leqq \varepsilon$$

In this case, the assertion being proved is that $|\delta_1| \leqq (1 + \varepsilon)^2 - 1$, and this is true since $\varepsilon \leqq (1 + \varepsilon)^2 - 1$. If the theorem is true for $n = k - 1$, then its proof for $n = k$ goes like this:

$$\begin{aligned} z_k &= \left[z_{k-1} + x_k y_k(1 + \delta')\right](1 + \delta) \\ &= \left[\sum_{i=1}^{k-1} x_i y_i(1 + \delta_i) + x_k y_k(1 + \delta')\right](1 + \delta) \\ &= \sum_{i=1}^{k-1} x_i y_i(1 + \delta_i + \delta + \delta_i\delta) + x_k y_k(1 + \delta + \delta' + \delta\delta') \end{aligned}$$

In the preceding equation, we have these bounds:

$$|\delta| \leqq \varepsilon \qquad |\delta'| \leqq \varepsilon \qquad |\delta_i| \leqq (1+\varepsilon)^{k+1-i} - 1 \qquad (1 \leqq i \leqq k-1)$$

We need to prove that

$$|\delta_i + \delta + \delta_i\delta| \leqq (1+\varepsilon)^{k+2-i} - 1 \qquad |\delta + \delta' + \delta\delta'| \leqq (1+\varepsilon)^2 - 1$$

The first of these inequalities is established by writing

$$\begin{aligned}|\delta_i + \delta + \delta_i\delta| &\leqq |\delta_i| + |\delta|(1 + |\delta_i|)\\ &\leqq (1+\varepsilon)^{k+1-i} - 1 + \varepsilon(1+\varepsilon)^{k+1-i}\\ &= (1+\varepsilon)^{k+2-i} - 1\end{aligned}$$

The second inequality is proved in a similar manner. This completes the induction. The following estimate is now required for $k \leqq n+1$:

$$\begin{aligned}(1+\varepsilon)^k - 1 &= \left[1 + k\varepsilon + \frac{k(k-1)}{2}\varepsilon^2 + \cdots + \varepsilon^k\right] - 1\\ &= k\varepsilon\left[1 + \frac{k-1}{2}\varepsilon + \frac{(k-1)(k-2)}{2\cdot 3}\varepsilon^2 + \cdots\right]\\ &\leqq k\varepsilon\left[1 + \frac{k\varepsilon}{2} + \left(\frac{k\varepsilon}{2}\right)^2 + \cdots\right]\\ &= k\varepsilon\Big/\left(1 - \frac{1}{2}k\varepsilon\right) < \frac{6}{5}k\varepsilon\end{aligned}$$

In the last step, the assumption $k\varepsilon \leqq n\varepsilon < \frac{1}{3}$ was used. Finally,

$$|\delta_i| \leqq (1+\varepsilon)^{n+2-i} - 1 \leqq \frac{6}{5}(n+2-i)\varepsilon \leqq \frac{6}{5}(n+1)\varepsilon \qquad \blacksquare$$

THEOREM 3 *Let L be an $n \times n$ unit lower triangular matrix whose elements are machine numbers. Let b be a vector whose components are machine numbers. The computed solution of $Ly = b$ is a vector $\widetilde{y}$ that is the exact solution of*

$$(L+\Delta)\widetilde{y} = b \qquad \text{with} \qquad |\Delta_{ij}| \leqq \frac{6}{5}(n+1)\varepsilon|\ell_{ij}| \tag{3}$$

Here ε is the machine's unit roundoff error, and it is assumed that $n\varepsilon < \frac{1}{3}$.

Proof The exact solution of the system $Ly = b$ is computed from the formula

$$y_i = b_i - \sum_{j=1}^{i-1} \ell_{ij} y_j \qquad (1 \leqq i \leqq n)$$

The computed values $\widetilde{y}_1, \widetilde{y}_2, \ldots, \widetilde{y}_n$ will satisfy an equation of the form

$$\widetilde{y}_i = \left[b_i - \sum_{j=1}^{i-1} \ell_{ij}\widetilde{y}_j(1 + \delta_{ij}) \right] \Big/ (1 + \delta_{ii}) \tag{4}$$

in which the δ_{ij} are numbers satisfying

$$|\delta_{ij}| \leqq \frac{6}{5}(n + 1)\varepsilon \qquad (1 \leqq j \leqq i \leqq n) \tag{5}$$

This assertion is a consequence of the assumptions made in Section 2.1 and of Theorem 2. Equation (4) can be rearranged like this:

$$(1 + \delta_{ii})\widetilde{y}_i = b_i - \sum_{j=1}^{i-1} \ell_{ij}\widetilde{y}_j(1 + \delta_{ij})$$

and then rewritten like this:

$$\sum_{j=1}^{i} \ell_{ij}\widetilde{y}_j(1 + \delta_{ij}) = b_i$$

We interpret this as a matrix equation

$$(L + \Delta)\widetilde{y} = b \tag{6}$$

with Δ the lower triangular matrix whose elements are $\ell_{ij}\delta_{ij}$ whenever $1 \leqq j \leqq i \leqq n$. Thus,

$$|\Delta_{ij}| = |\ell_{ij}|\,|\delta_{ij}| \leqq \frac{6}{5}(n + 1)\varepsilon|\ell_{ij}| \qquad \blacksquare$$

THEOREM 4 *Let U be an $n \times n$, upper triangular, nonsingular matrix. If the elements of U and c are machine numbers and if $n\varepsilon < \frac{1}{3}$, then the computed solution $\widetilde{y}$ of $Uy = c$ satisfies exactly a perturbed system*

$$(U + \Delta)\widetilde{y} = c \qquad \textit{with} \qquad |\Delta_{ij}| \leqq \frac{6}{5}(n + 1)\varepsilon|u_{ij}|$$

Proof This is left to the reader as Problem 1. ■

THEOREM 5 *Let the elements of A and b be machine numbers. If the Gaussian algorithm with row pivoting is used to solve $Ax = b$, then the computed solution $\widetilde{x}$ is the exact solution of a perturbed system*

$$(A + F)\widetilde{x} = b \qquad \textit{in which} \qquad |f_{ij}| \leqq 10n^2\varepsilon\rho$$

Here n is the order of the matrix A, $\rho = \max_{1 \leqq i,j,k \leqq n} |a_{ij}^{(k)}|$, and ε is the machine's unit roundoff error. It is assumed that $n\varepsilon < \frac{1}{3}$.

Proof By Theorems 1, 3, and 4, since $n\varepsilon < \frac{1}{3}$, we have

$$A + E = \widetilde{L}\widetilde{U} \qquad |e_{ij}| \leqq 2n\varepsilon\rho$$

$$(\widetilde{L} + \Delta)\widetilde{y} = b \qquad |\Delta_{ij}| \leqq \frac{6}{5}(n+1)\varepsilon|\widetilde{\ell}_{ij}|$$

$$(\widetilde{U} + \Delta')\widetilde{x} = \widetilde{y} \qquad |\Delta'_{ij}| \leqq \frac{6}{5}(n+1)\varepsilon|\widetilde{u}_{ij}|$$

Putting these equations together yields

$$\begin{aligned} b &= (\widetilde{L} + \Delta)\widetilde{y} = (\widetilde{L} + \Delta)(\widetilde{U} + \Delta')\widetilde{x} = (\widetilde{L}\widetilde{U} + \Delta\widetilde{U} + \widetilde{L}\Delta' + \Delta\Delta')\widetilde{x} \\ &= (A + E + \Delta\widetilde{U} + \widetilde{L}\Delta' + \Delta\Delta')\widetilde{x} = (A + F)\widetilde{x} \end{aligned}$$

where $F = E + \Delta\widetilde{U} + \widetilde{L}\Delta' + \Delta\Delta'$. To obtain an estimate of F, we use the bounds given above, together with the observations that $|\ell_{ij}| \leqq 1$ and

$$|u_{ij}| = |a_{ij}^{(n)}| \leqq \rho$$

Thus,

$$\begin{aligned} |f_{ij}| &\leqq |e_{ij}| + \sum_{\nu=1}^{n}\{|\Delta_{i\nu}|\,|\widetilde{u}_{\nu j}| + |\widetilde{\ell}_{i\nu}|\,|\Delta'_{\nu j}| + |\Delta_{i\nu}|\,|\Delta'_{\nu j}|\} \\ &\leqq 2n\varepsilon\rho + \frac{6}{5}n(n+1)\varepsilon\rho + \frac{6}{5}n(n+1)\varepsilon\rho + \frac{36}{25}n(n+1)^2\varepsilon^2\rho \\ &= n^2\varepsilon\rho\left\{\frac{2}{n} + \frac{12}{5}\,\frac{n+1}{n} + \frac{36}{25}\varepsilon\left(\frac{n+1}{n}\right)^2\right\} \end{aligned}$$

The quantity within the braces cannot exceed 10. ■

The bound given in Theorem 5 can be improved by paying more attention to details in Theorems 1–5 and by computing $\|F\|_\infty$ instead. The reader should consult Forsythe and Moler [1967], Golub and van Loan [1989], Wilkinson [1965], and Isaacson and Keller [1966].

Growth Factor

As a measure of the stability in the numerical solution of $Ax = b$ by Gaussian elimination with pivoting, the **growth factor** of an $n \times n$ matrix A under Gaussian elimination is defined as

$$g_n(A) = \frac{\max_{i,j,k} |a_{ij}^{(k)}|}{\max_{i,j} |a_{i,j}|}$$

where $a_{ij}^{(k)}$ denotes the (i, j)-element after the kth step of the elimination process. The growth factor measures, in a relative way, how large the numbers become during the elimination process. Gaussian elimination with pivoting is considered numerically stable unless $g_n(A)$ is large.

The growth factor arises in various bounds. Wilkinson [1965] established the bound

$$\frac{\|\tilde{x} - x\|_\infty}{\|x\|_\infty} \leqq 4n^2 g_n(A)\kappa_\infty(A)\varepsilon$$

Here x is the exact solution of $Ax = b$, $\tilde{x}$ is the solution computed in floating-point arithmetic by Gaussian elimination with partial pivoting* on a computer with relative machine precision ε, and $\kappa_\infty(A)$ is the condition number of A using the ∞-norm. Another bound related to the backward error is the following one, which is similar to the bound in Theorem 5:

$$\frac{\|\tilde{A} - A\|_\infty}{\|A\|_\infty} < 8n^3 g_n(A)\varepsilon$$

Here the computed solution $\tilde{x}$ satisfies $\tilde{A}\tilde{x} = b$. (See, for example, Golub and Van Loan [1989].)

For Gaussian elimination with partial pivoting, a sharp upper bound on the growth factor is

$$g_n(A) \leqq 2^{n-1}$$

Examples where this bound is attained can be found in Golub and Van Loan [1989], Higham and Higham [1989], Foster [1994], and Wilkinson [1965]. Trefethen and Schreiber [1990] showed that $g_n(A)$ does not grow exponentially for random matrices. Recently, Foster [1994] and Wright [1993] presented examples where the growth factor does grow exponentially. In applications, the error growth is usually quite small. Consequently, Gaussian elimination with partial pivoting is considered a stable process, and it is widely used with full confidence.

Over 30 years ago, Wilkinson reported that in his experience any substantial increase in the growth factor was extremely uncommon even with partial pivoting. Wilkinson observed that it was difficult to find matrices for which $g_n(A) > n$, and he could find no example arising naturally with a factor larger than 16. Since Wilkinson's observation, his comments have been taken as an intellectual challenge by some researchers and efforts have been made to find practical examples with large growth factors. For example, Dongarra, Bunch, Moler, and Stewart [1979] reported an example where $g_n(A)$ is 23 for Gaussian elimination with partial pivoting.

For Gaussian elimination with complete pivoting, Wilkinson [1961] showed that

$$g_n(A) \leqq n^{1/2}\left[2 \cdot 3^{1/2} \cdot 4^{1/3} \cdots n^{1/(n-1)}\right]^{1/2}$$

While this bound can be quite large, Wilkinson mentioned that it is much smaller than 2^n for large n and that the proof indicates that the true upper bound is much

*This pivoting strategy differs from ours in that it does not involve scaling constants.

smaller still. Moreover, the bound is a slowly growing function of n. Recall that there seems to be little practical value in using complete pivoting over partial pivoting because of the additional computational costs involved.

Cryer [1968] published a statement known as **Wilkinson's conjecture**: If Gaussian elimination with complete pivoting is performed on a matrix A, then $g_n(A) \leqq n$. Gould [1991] and Edelman [1992] found examples where $g_n(A) > n$ for complete pivoting, so finally this conjecture has been settled as false.

In many cases, supercomputers or advanced mathematical software packages were required to find the examples mentioned above since they are not simple ones. On the other hand, some experts speculate that the potential for large error growth in Gaussian elimination has been overlooked. Recently, examples have been discovered that lead to disastrous error growth in Gaussian elimination with partial pivoting. It is certainly a topic worthy of further investigation. (See, for example, papers by Edelman [1992] and Foster [1994].)

PROBLEMS 4.8

1. Prove Theorem 4.

2. In Theorem 1, prove the inequalities

$$\|E^{(k)}\|_\infty \leqq [1 + 2(n-k)]\,\varepsilon\rho$$
$$\|E\|_\infty \leqq n^2\varepsilon\rho$$

3. In Theorem 3, prove that

$$\|\Delta\|_\infty \leqq \frac{3}{5}n(n+1)\varepsilon \max_{1\leqq i,j\leqq n} |\ell_{ij}|$$

CHAPTER FIVE

Selected Topics in Numerical Linear Algebra

Review of Basic Concepts

The study of matrix eigenvalues requires some familiarity with complex numbers, and we will briefly review the basic notions that are needed. The knowledgeable reader may skip over this material.

Basic Concepts

The field $\mathbb{R}$ of real numbers has the deficiency that a polynomial of degree n having real coefficients need not have n real zeros (or *roots*). For example, the polynomial $p(x) = x^2 - 2x + 2$ has no real zeros. This deficiency is overcome in one stroke by enlarging the field so that it contains the (nonreal) element i. The element i is characterized by the equation $i^2 = -1$. The new field thus obtained is denoted by $\mathbb{C}$, and its elements are called **complex numbers**. They have the form

$$\gamma = \alpha + i\beta \qquad (\alpha, \beta \text{ real})$$

The **conjugate** and **modulus** of γ are defined, respectively, by

$$\overline{\gamma} = \alpha - i\beta$$
$$|\gamma| = \sqrt{\alpha^2 + \beta^2}$$

Notice that $\overline{\overline{\gamma}} = \gamma$ and $\gamma\overline{\gamma} = |\gamma|^2$.

The field $\mathbb{C}$ does not have the deficiency that was mentioned as a property of $\mathbb{R}$. Thus, we have the **Fundamental Theorem of Algebra**, which states that every nonconstant polynomial with complex coefficients has at least one zero in the complex plane. It follows that every polynomial of degree n can be expressed as a product of n linear factors.

The vector space $\mathbb{C}^n$ consists of all complex n-tuples such as

$$x = (x_1, x_2, \ldots, x_n)^T$$

where $x_j \in \mathbb{C}$ for $1 \leqq j \leqq n$. If the complex vector x is multiplied by the complex number λ, the result is another complex vector—namely,

$$\lambda x = (\lambda x_1, \lambda x_2, \ldots, \lambda x_n)^T$$

Thus, we can regard $\mathbb{C}^n$ as a vector space over the scalar field $\mathbb{C}$. In the space $\mathbb{C}^n$, the **inner product** and the **Euclidean norm** are defined, respectively, by

$$\langle x, y\rangle = \sum_{j=1}^{n} x_j\overline{y}_j \qquad \text{and} \qquad \|x\|_2 = \sqrt{\langle x, x\rangle}$$

Notice that

$$\langle x, y\rangle = \overline{\langle y, x\rangle} \qquad \langle x, \lambda y\rangle = \overline{\lambda}\langle x, y\rangle$$

and

$$\langle x + y, z\rangle = \langle x, z\rangle + \langle y, z\rangle$$

If A is a matrix having complex elements, A^* will denote its **conjugate transpose**: $(A^*)_{jk} = \overline{A}_{kj}$. In particular, if x is an $n \times 1$ matrix (or column vector), then $x^* = (\overline{x}_j)$ is a $1 \times n$ matrix (or row vector), and

$$y^*x = \langle x, y\rangle = \sum_{j=1}^{n} x_j\overline{y}_j$$

$$x^*x = \langle x, x\rangle = \|x\|_2^2 = \sum_{j=1}^{n} x_j\overline{x}_j = \sum_{j=1}^{n} |x_j|^2$$

Let A be an $n \times n$ matrix (whose elements may be complex numbers). Let λ be a scalar (a complex number). If the equation

$$Ax = \lambda x \tag{1}$$

has a *nontrivial* solution (that is, $x \neq 0$), then λ is an **eigenvalue** of A. A nonzero vector x satisfying Equation (1) is an **eigenvector** of A corresponding to the eigenvalue λ. For example, the equation

$$\begin{bmatrix} 2 & 0 & 1 \\ 5 & -1 & 2 \\ -3 & 2 & -\frac{5}{4} \end{bmatrix} \begin{bmatrix} 1 \\ 3 \\ -4 \end{bmatrix} = -2 \begin{bmatrix} 1 \\ 3 \\ -4 \end{bmatrix}$$

tells us that -2 is an eigenvalue of the given 3×3 matrix and that $(1, 3, -4)^T$ is a corresponding eigenvector. Notice that a nonzero multiple of an eigenvector is another eigenvector corresponding to the same eigenvalue.

The condition that Equation (1) have a nontrivial solution is equivalent to each of these conditions:

$$A - \lambda I \text{ maps some nonzero vector into } 0 \tag{2}$$

$$A - \lambda I \text{ is singular} \tag{3}$$

$$\det(A - \lambda I) = 0 \tag{4}$$

Thus we can, in principle, solve Equation (4) for the unknown values of λ and thereby compute the eigenvalues of A. Equation (4) is known as the **characteristic equation** of the matrix A. If this equation is written out with more detail, we have

$$\det \begin{bmatrix} a_{11} - \lambda & a_{12} & a_{13} & \cdots & a_{1n} \\ a_{21} & a_{22} - \lambda & a_{23} & \cdots & a_{2n} \\ a_{31} & a_{32} & a_{33} - \lambda & \cdots & a_{3n} \\ \vdots & \vdots & \vdots & \ddots & \vdots \\ a_{n1} & a_{n2} & a_{n3} & \cdots & a_{nn} - \lambda \end{bmatrix} = 0$$

By the definition of the determinant function as a sum of terms that are products of elements of the matrix, the left side of Equation (4) has the form of a polynomial of degree n in the variable λ. This polynomial is the **characteristic polynomial** of A. We are led at once to the conclusion that an $n \times n$ matrix has exactly n eigenvalues, provided that these are counted with the multiplicities that they possess as roots of the characteristic equation.

Example 1 To illustrate these concepts, let us determine the eigenvalues of the matrix

$$A = \begin{bmatrix} 1 & 2 & 1 \\ 0 & 1 & 3 \\ 2 & 1 & 1 \end{bmatrix}$$

Solution The characteristic equation is

$$\det\begin{bmatrix} 1-\lambda & 2 & 1 \\ 0 & 1-\lambda & 3 \\ 2 & 1 & 1-\lambda \end{bmatrix} = -\lambda^3+3\lambda^2+2\lambda+8 = -(\lambda-4)(\lambda^2+\lambda+2) = 0$$

The roots of the characteristic equation are

$$\lambda_1 = 4 \qquad \lambda_2 = -\frac{1}{2} + \frac{\sqrt{7}}{2}i \qquad \lambda_3 = -\frac{1}{2} - \frac{\sqrt{7}}{2}i$$

This illustrates that the eigenvalues of a real matrix are not necessarily real numbers. ■

The preceding procedure is the **direct method** for computing eigenvalues. For calculations by hand on small matrices, it may be the best method to use. For large matrices and for automatic computation, it is emphatically *not* recommended. One reason for this proscription is that the roots of a polynomial may be very sensitive functions of the coefficients in the polynomial. Thus, any errors (such as roundoff errors) in the coefficients may lead to gross inaccuracies in the numerically determined roots. A classic example of this, due to Wilkinson, was discussed at the end of Section 2.3.

5.1 Matrix Eigenvalue Problem: Power Method

Power Method

The next numerical method that we shall discuss is the **power method**. This procedure is designed to compute the dominant eigenvalue and an eigenvector corresponding to the dominant eigenvalue. For the theory to proceed smoothly, it is necessary to assume that A has the following two properties:

(i) There is a single eigenvalue of maximum modulus.
(ii) There is a linearly independent set of n eigenvectors.

According to the first assumption, the eigenvalues $\lambda_1, \lambda_2, \dots, \lambda_n$ can be labeled so that

$$|\lambda_1| > |\lambda_2| \geqq |\lambda_3| \geqq \cdots \geqq |\lambda_n|$$

According to the second assumption, there is a basis $\{u^{(1)}, u^{(2)}, \dots, u^{(n)}\}$ for $\mathbb{C}^n$ such that

$$Au^{(j)} = \lambda_j u^{(j)} \qquad (1 \leqq j \leqq n) \tag{1}$$

Let $x^{(0)}$ be any element of $\mathbb{C}^n$ such that when $x^{(0)}$ is expressed as a linear combination of the basis elements $u^{(1)}, u^{(2)}, \ldots, u^{(n)}$, the coefficient of $u^{(1)}$ is not 0. Thus,

$$x^{(0)} = a_1 u^{(1)} + a_2 u^{(2)} + \cdots + a_n u^{(n)} \qquad (a_1 \neq 0) \tag{2}$$

We form then

$$x^{(1)} = Ax^{(0)} \qquad x^{(2)} = Ax^{(1)} \qquad \cdots \qquad x^{(k)} = Ax^{(k-1)}$$

so that

$$x^{(k)} = A^k x^{(0)} \tag{3}$$

In the analysis to follow, there is no loss of generality in absorbing all the coefficients a_j in the vectors $u^{(j)}$ that they multiply. Hence, we may write Equation (2) as

$$x^{(0)} = u^{(1)} + u^{(2)} + \cdots + u^{(n)}$$

By this equation and (3), we have

$$x^{(k)} = A^k u^{(1)} + A^k u^{(2)} + \cdots + A^k u^{(n)}$$

Using Equation (1), we arrive at

$$x^{(k)} = \lambda_1^k u^{(1)} + \lambda_2^k u^{(2)} + \cdots + \lambda_n^k u^{(n)}$$

This last equation can be rewritten in the form

$$x^{(k)} = \lambda_1^k \left[u^{(1)} + \left(\frac{\lambda_2}{\lambda_1}\right)^k u^{(2)} + \cdots + \left(\frac{\lambda_n}{\lambda_1}\right)^k u^{(n)} \right]$$

Since $|\lambda_1| > |\lambda_j|$ for $2 \leqq j \leqq n$, we see that the coefficients $(\lambda_j/\lambda_1)^k$ tend to 0 and the vector within the brackets converges to $u^{(1)}$ as $k \to \infty$.

To simplify the notation, we write $x^{(k)}$ in the form

$$x^{(k)} = \lambda_1^k \left[u^{(1)} + \varepsilon^{(k)} \right]$$

where $\varepsilon^{(k)} \to 0$ as $k \to \infty$. In order to be able to take ratios, let φ be any linear functional on $\mathbb{C}^n$ for which $\varphi(u^{(1)}) \neq 0$. Recall that a linear functional φ satisfies $\varphi(\alpha x + \beta y) = \alpha\varphi(x) + \beta\varphi(y)$, for scalars α and β and vectors x and y. (For example, φ could simply evaluate the jth component of any given vector.) Then

$$\varphi(x^{(k)}) = \lambda_1^k [\varphi(u^{(1)}) + \varphi(\varepsilon^{(k)})] \tag{4}$$

Consequently, the following ratios converge to λ_1 as $k \to \infty$:

$$r_k \equiv \frac{\varphi(x^{(k+1)})}{\varphi(x^{(k)})} = \lambda_1 \left[\frac{\varphi(u^{(1)}) + \varphi(\varepsilon^{(k+1)})}{\varphi(u^{(1)}) + \varphi(\varepsilon^{(k)})} \right] \to \lambda_1$$

This constitutes the **power method** for computing λ_1. Since the direction of the vector $x^{(k)}$ aligns more and more with $u^{(1)}$ as $k \to \infty$, the method can also give us the eigenvector $u^{(1)}$. Many variations and refinements of the power method are found in the literature.

Algorithm

The algorithm for the power method as just described is as follows:

input n, A, x, M
output $0, x$
for $k = 1$ **to** M **do**
    $y \leftarrow Ax$
    $r \leftarrow \varphi(y)/\varphi(x)$
    $x \leftarrow y$
    output k, x, r
end do

Here φ is some linear functional.

In a practical realization of the power method, it is advisable to introduce a normalization of the vectors $x^{(k)}$, since otherwise they may converge to 0 or become unbounded. Thus, we can modify the iteration as follows:

input n, A, x, M
output $0, x$
for $k = 1$ **to** M **do**
    $y \leftarrow Ax$
    $r \leftarrow \varphi(y)/\varphi(x)$
    $x \leftarrow y/\|y\|$
    output k, x, r
end do

The norm used here can be any convenient one: the ℓ_∞-norm $\|x\|_\infty = \max_{1 \leqq j \leqq n} |x_j|$, for example. The ratios r are the same as in the unnormalized version of the algorithm. (Problem 2 asks for a proof of this assertion.)

Example 1 Use the power method on a matrix and an initial vector as follows:

$$A = \begin{bmatrix} 6 & 5 & -5 \\ 2 & 6 & -2 \\ 2 & 5 & -1 \end{bmatrix} \qquad x = (-1, 1, 1)^T$$

Solution The normalized version of the algorithm was programmed, and the linear functional φ was taken to be $\varphi(x) = x_2$. The normalized vectors $x^{(k)}$ and the ratios r_k are shown here for a few values of k:

$$
\begin{array}{llll}
k = 0 & x^{(0)} = (\,-1.00000,\ \ 1.00000,\ \ 1.00000\,) & & \\
k = 1 & x^{(1)} = (\,-1.00000,\ \ 0.33333,\ \ 0.33333\,) & r_0 = & 2.0 \\
k = 2 & x^{(2)} = (\,-1.00000, -0.11111, -0.11111\,) & r_1 = & -\ 2.0 \\
k = 3 & x^{(3)} = (\,-1.00000, -0.40741, -0.40741\,) & r_2 = & 22.0 \\
k = 4 & x^{(4)} = (\,-1.00000, -0.60494, -0.60494\,) & r_3 = & 8.9091 \\
\vdots & \vdots & \vdots & \\
k = 6 & x^{(6)} = (\,-1.00000, -0.82442, -0.82442\,) & r_5 = & 6.71508 \\
\vdots & \vdots & \vdots & \\
k = 28 & x^{(28)} = (\,-1.00000, -0.99998, -0.99998\,) & r_{27} = & 6.00007
\end{array}
$$

The leading eigenvalue of A is 6, and an eigenvector is $(1, 1, 1)^T$. ■

Our analysis of the power method is valid when some or all of the eigenvalues are complex, and under our assumptions the algorithm computes λ_1 and an associated eigenvector.

Aitken Acceleration

If the ratios r_k are regarded as approximations to λ_1, then there is some interest in estimating the errors $|r_k - \lambda_1|$. The pertinent results here are contained in Problems 3 and 4. It turns out that

$$r_{k+1} - \lambda_1 = (c + \delta_k)(r_k - \lambda_1)$$

for a constant c satisfying $|c| < 1$ and for numbers δ_k that converge to 0. In the terminology of Section 1.3, this implies that the sequence $[r_k]$ converges linearly to λ_1. Because of this additional knowledge of how the errors behave, a general procedure known as **Aitken acceleration** can be used. From the given sequence $[r_k]$, we construct another sequence $[s_k]$ by means of the formula

$$s_k = \frac{r_k r_{k+2} - r_{k+1}^2}{r_{k+2} - 2r_{k+1} + r_k}$$

This sequence converges faster than the original, in accordance with the following general result.

THEOREM 1 **Aitken Acceleration Theorem** *Let $[r_n]$ be a sequence of numbers that converges to a limit r. Then the new sequence*

$$s_n = \frac{r_n r_{n+2} - r_{n+1}^2}{r_{n+2} - 2r_{n+1} + r_n} \qquad (n \geqq 0)$$

converges to r faster if $r_{n+1} - r = (c + \delta_n)(r_n - r)$ with $|c| < 1$ and $\lim_{n\to\infty} \delta_n = 0$. Indeed, $(s_n - r)/(r_n - r) \to 0$ as $n \to \infty$.

Proof Define the error sequence $h_n \equiv r_n - r$. Then a short calculation reveals that

$$s_n = \frac{(r + h_n)(r + h_{n+2}) - (r + h_{n+1})^2}{(r + h_{n+2}) - 2(r + h_{n+1}) + (r + h_n)}$$

$$= r + \frac{h_n h_{n+2} - h_{n+1}^2}{h_{n+2} - 2h_{n+1} + h_n}$$

Use the hypothesis $h_{n+1} = (c + \delta_n)h_n$ to obtain $h_{n+2} = (c + \delta_{n+1})(c + \delta_n)h_n$ and

$$s_n - r = \frac{h_n(c + \delta_{n+1})(c + \delta_n)h_n - (c + \delta_n)^2 h_n^2}{(c + \delta_{n+1})(c + \delta_n)h_n - 2(c + \delta_n)h_n + h_n}$$

$$= h_n \frac{(c + \delta_{n+1})(c + \delta_n) - (c + \delta_n)^2}{(c + \delta_{n+1})(c + \delta_n) - 2(c + \delta_n) + 1}$$

It is now clear that $\lim_{n\to\infty}(s_n - r)/h_n = 0$, since in the previous equation the numerator converges to 0 and the denominator converges to $(c - 1)^2$, which is not 0. ■

It is important to stop the Aitken acceleration process soon after it produces apparently stationary values because subtractive cancellation in the formula will eventually spoil the results.

Before discussing other variations of the power method, we shall prove an elementary fact about eigenvalues.

THEOREM 2 *If λ is an eigenvalue of A and if A is nonsingular, then λ^{-1} is an eigenvalue of A^{-1}.*

Proof Let $Ax = \lambda x$, with $x \neq 0$. Then $x = A^{-1}(\lambda x) = \lambda A^{-1}x$. Hence, $A^{-1}x = \lambda^{-1}x$ and λ^{-1} is an eigenvalue of A^{-1}. ■

Inverse Power Method

Theorem 2 suggests a means of computing the *smallest* eigenvalue of A. Suppose that the eigenvalues of A can be arranged as follows:

$$|\lambda_1| \geqq |\lambda_2| \geqq \cdots \geqq |\lambda_{n-1}| > |\lambda_n| > 0$$

This hypothesis implies that A is nonsingular since 0 is not an eigenvalue. The eigenvalues of A^{-1} are the numbers λ_j^{-1}, and they are arranged like this:

$$|\lambda_n^{-1}| > |\lambda_{n-1}^{-1}| \geqq \cdots \geqq |\lambda_1^{-1}| > 0$$

Consequently, we can compute λ_n^{-1} by applying the power method to A^{-1}. It is *not* a good idea to compute the inverse, A^{-1}, first and then use $x^{(k+1)} = A^{-1}x^{(k)}$. Rather, we obtain $x^{(k+1)}$ by solving the equation

$$Ax^{(k+1)} = x^{(k)}$$

This can be done efficiently by using the Gaussian elimination method. It is only necessary to carry out the factorization phase of Gaussian elimination once and then repeat the solution phase by changing the right-hand sides from $x^{(0)}$ to $x^{(1)}$ to $x^{(2)}$, and so on. This is the **inverse power method**.

Example 2 Illustrate the inverse power method with the same matrix used in Example 1. Its LU-factorization is

$$\begin{bmatrix} 6 & 5 & -5 \\ 2 & 6 & -2 \\ 2 & 5 & -1 \end{bmatrix} = \begin{bmatrix} 1 & 0 & 0 \\ \frac{1}{3} & 1 & 0 \\ \frac{1}{3} & \frac{10}{13} & 1 \end{bmatrix} \begin{bmatrix} 6 & 5 & -5 \\ 0 & \frac{13}{3} & -\frac{1}{3} \\ 0 & 0 & \frac{12}{13} \end{bmatrix}$$

Solution We began with the vector $x = (3,\ 7,\ -13)^T$ and computed 25 steps in the process. In each step, we obtain $x^{(k+1)}$ by solving $Ux^{(k+1)} = L^{-1}x^{(k)}$. Then a ratio is computed and printed—namely, $r_k = x_1^{(k+1)}/x_1^{(k)}$. Before we proceed to the next step, $x^{(k+1)}$ is normalized (that is, divided by its ℓ_∞-norm). Here is some of the output:

$$\begin{array}{lll}
k = 0 & x^{(0)} = (\ \ 3.00000,\ \ 7.00000,\ -13.00000\ ) & \\
k = 1 & x^{(1)} = (\ -0.80165,\ -0.00826,\ -1.00000\ ) & r_0 = -5.8889 \\
k = 2 & x^{(2)} = (\ -0.95089,\ -0.01774,\ -1.00000\ ) & r_1 = 1.19759 \\
k = 3 & x^{(3)} = (\ -0.98759,\ -0.00712,\ -1.00000\ ) & r_2 = 1.02750 \\
k = 4 & x^{(4)} = (\ -0.99688,\ -0.00223,\ -1.00000\ ) & r_3 = 1.00446 \\
\vdots & \vdots & \vdots \\
k = 6 & x^{(6)} = (\ -0.99980,\ -0.00017,\ -1.00000\ ) & r_5 = 1.00012 \\
\vdots & \vdots & \vdots \\
k = 11 & x^{(11)} = (-1.00000,\ \ 0.00000,\ -1.00000\ ) & r_{10} = 1.00000
\end{array}$$

The remaining iterates showed no change. ■

At this stage, we have outlined the *power method* for computing the largest eigenvalue of A and the *inverse power method* for computing the smallest eigenvalue of A. By considering the **shifted matrix** $A - \mu I$, we can arrive at a procedure for computing the eigenvalue of A closest to a given value μ. Suppose that one eigenvalue of A, say λ_k, satisfies the inequality $0 < |\lambda_k - \mu| < \varepsilon$, where μ is a prescribed complex number. Suppose that all other eigenvalues of A satisfy the inequality $|\lambda_j - \mu| > \varepsilon$. Since the eigenvalues of $A - \mu I$ are the numbers $\lambda_j - \mu$, the inverse power method can be applied to $A - \mu I$, resulting in an approximate value for $(\lambda_k - \mu)^{-1}$. Here again the sequence of vectors required will be obtained from solving the equations $(A - \mu I)x^{(k+1)} = x^{(k)}$. The Gaussian decomposition of $A - \mu I$ would be computed once in this algorithm. Since this procedure computes

$z = (\lambda_k - \mu)^{-1}$, we can recover λ_k from the formula $\lambda_k = z^{-1} + \mu$. This algorithm is called the **shifted inverse power method**.

In a similar manner, we can compute the eigenvalue, say λ_k, farthest from a given value μ. Suppose there is a positive ε such that $|\lambda_k - \mu| > \varepsilon$ for some eigenvalue λ_k of A, and for all other eigenvalues λ_j of A, we have $0 < |\lambda_j - \mu| < \varepsilon$. The power method applied to $(A - \mu I)$ computes $z = \lambda_k - \mu$. Hence, we obtain $\lambda_k = z + \mu$.

Summary

We summarize these results in the following table:

Method	Equation	Computes
power	$x^{(k+1)} = Ax^{(k)}$	largest eigenvalue, λ_1
inverse power	$Ax^{(k+1)} = x^{(k)}$	smallest eigenvalue, λ_n
shifted power	$x^{(k+1)} = (A - \mu I)x^{(k)}$	eigenvalue farthest from μ
shifted inverse power	$(A - \mu I)x^{(k+1)} = x^{(k)}$	eigenvalue closest to μ

PROBLEMS 5.1

1. Let A be an $n \times n$ matrix that has a linearly independent set of n eigenvectors, $\{u^{(1)}, u^{(2)}, \dots, u^{(n)}\}$. Let $Au^{(i)} = \lambda_i u^{(i)}$, and let P be the matrix whose columns are the vectors $u^{(1)}, u^{(2)}, \dots, u^{(n)}$. What is $P^{-1}AP$?
2. Show that if the normalized and unnormalized versions of the power method are started at the same initial vector, then the values of r in the two algorithms will be the same.
3. In the power method, let $r_k = \varphi(x^{(k+1)})/\varphi(x^{(k)})$. We know that $\lim_{k\to\infty} r_k = \lambda_1$. Show that the relative errors obey

$$\frac{r_k - \lambda_1}{\lambda_1} = \left(\frac{\lambda_2}{\lambda_1}\right)^k c_k$$

where the numbers c_k form a bounded sequence.

4. (Continuation) Show that $r_{k+1} - \lambda_1 = (c + \delta_k)(r_k - \lambda_1)$, where $|c| < 1$ and $\lim_{n\to\infty} \delta_k = 0$, so that Aitken acceleration is applicable. Assume $|\lambda_2| > |\lambda_3|$.
5. Prove that in Aitken acceleration, $(s_n - r)/(r_{n+2} - r) \to 0$ as $n \to \infty$, provided that $c \neq 0$.
6. Count the number of multiplications and/or divisions involved in carrying out m steps of the basic (unnormalized) power method.
7. In the normalized power method, show that if $\lambda_1 > 0$, then the vectors $x^{(k)}$ converge to an eigenvector.
8. Devise a simple modification of the power method to handle the following case: $\lambda_1 = -\lambda_2 > |\lambda_3| \geqq |\lambda_4| \geqq \cdots \geqq |\lambda_n|$.
9. What can you prove about Aitken acceleration if the sequence $[r_n]$ satisfies only the hypothesis $|r_{n+1} - r| \leqq c|r_n - r|$ with $0 < c < 0.2$?

10. Let the eigenvalues of A satisfy $\lambda_1 > \lambda_2 > \cdots > \lambda_n$ (all real, but not necessarily positive). What value of the parameter β should be used in order for the power method to converge most rapidly to λ_1 when applied to $A + \beta I$?

11. Prove that $I - AB$ has the same eigenvalues as $I - BA$, if either A or B is nonsingular.

12. If the power method is applied to a real matrix with a real starting vector, what will happen if a dominant eigenvalue is complex? Does the theory outlined in the text apply?

13. What is the characteristic polynomial of this matrix?

$$\begin{bmatrix} a_1 & a_2 & a_3 & \cdots & a_{n-1} & a_n \\ 1 & 0 & 0 & \cdots & 0 & 0 \\ 0 & 1 & 0 & \cdots & 0 & 0 \\ \vdots & \vdots & \ddots & \ddots & \vdots & \vdots \\ 0 & 0 & 0 & \ddots & 0 & 0 \\ 0 & 0 & 0 & \cdots & 1 & 0 \end{bmatrix}$$

14. Prove that for any complex number λ and any $n \times n$ matrix A,

$$\dim\{x : Ax = \lambda x\} = n - \operatorname{rank}(A - \lambda I)$$

15. Let $A = LU$, where L is unit lower triangular and U is upper triangular. Put $B = UL$ and show that B and A have the same eigenvalues.

16. Suppose that A has a row, say row k, such that $\sum_{j=1,\, j\neq k}^{n} |a_{kj}| = 0$. Let B denote the matrix obtained by removing row k and column k from A. Show that a_{kk} is an eigenvalue of A and that the remaining eigenvalues of A are eigenvalues of B.

17. An $n \times n$ matrix is said to be **defective** if its eigenvectors do not span $\mathbb{R}^n$. Show that the matrix $A = \begin{bmatrix} 1 & 1 \\ 0 & 1 \end{bmatrix}$ is defective.

18. Prove that if an $n \times n$ matrix has n distinct eigenvalues, then it is not defective.

19. Prove that the converse of Problem 18 is not true.

20. In some experiments to compute eigenvalues of real matrices, it was observed that real eigenvalues occurred frequently. Prove that a real $n \times n$ matrix must have at least one real eigenvalue if n is odd.

21. (Continuation) A real polynomial can be factored into real quadratic and linear factors. The quadratic factors may have complex roots. Compute the probability that a real quadratic, $x^2 + ax + b$, will have imaginary roots, assuming that the coefficients a and b are uniformly distributed random variables chosen from the square $|a| \leqq \rho$, $|b| \leqq \rho$. Show that this probability converges to 0 as $\rho \to \infty$ and converges to $1/2$ as $\rho \to 0$. What does this suggest about the eigenvalues of real matrices?

22. To compute $x^{(k)} = A^k x$ for high values of k, one can perform repeated squarings of matrices:

$$A \to A^2 \to A^4 \to A^8 \to \cdots \to A^{2^m}$$

Then a single matrix-vector product produces $x^{(2^m)} = A^{2^m} x$. Compare this procedure to the ordinary power method. Count the number of multiplications required by each method in arriving at $x^{(2^m)}$. Prove that for large m, the squaring method is always more economical. Devise a means of avoiding overflow by scaling the successive powers of A.

23. Determine an approximate value for the spectral radius, $\rho(A)$, of the following matrix by taking two iterations of the power method using the infinity norm for φ. Use the initial vector $(1, 1, 1)^T$. (A norm is not a linear functional.)

$$A = \begin{bmatrix} 2 & 0 & -1 \\ -2 & -10 & 0 \\ -1 & -1 & 4 \end{bmatrix}$$

24. Prove that when φ is a norm, the ratios r_k in the power method converge to $|\lambda_1|$.

25. Show that a preferable form of the Aitken acceleration for computation is

$$s_k = r_k - \frac{(\Delta r_k)^2}{\Delta^2 r_k}$$

where

$$\Delta r_k = r_{k+1} - r_k \qquad \Delta^2 r_k = \Delta r_{k+1} - \Delta r_k$$

These are **forward differences**, which are discussed in Chapter 6.

26. For each j and k in the range $1 \leqq j, k \leqq n$, let f_{jk} be a linear function of λ. (A **linear function** has the form $\lambda \to \alpha\lambda + \beta$.) How would you determine the values of λ for which $\det\big(f_{jk}(\lambda)\big) = 0$? How many such values are there in general? Where are they located in the complex plane?

COMPUTER PROBLEMS 5.1

1. Employ the power method using the matrix in Example 1 with starting vector $(1, 2, 3)^T$. Take 100 steps and explain the apparent convergence at an early stage that is followed by later convergence to a different value.

2. Write a subroutine or procedure to apply M steps of the normalized power method on a given $n \times n$ matrix A, starting with a given vector x. Incorporate Aitken acceleration. In each step, print the current vector $x^{(k)}$, the current ratio

r_k, and the current *accelerated* value s_k as defined in the text. Test the procedure on $A - \mu I$, where

(i) $A = \begin{bmatrix} 5 & 4 & 1 & 1 \\ 4 & 5 & 1 & 1 \\ 1 & 1 & 4 & 2 \\ 1 & 1 & 2 & 4 \end{bmatrix}$ and $\mu = 0, 3, 6, 11$

(ii) $A = \begin{bmatrix} 2 & 3 & 4 \\ 7 & -1 & 3 \\ 1 & -1 & 5 \end{bmatrix}$ and $\mu = 0, 5, 10$

3. Write a computer program for the inverse power method and test it on some matrices of your own choosing.

4. **(a)** Write a computer program to reproduce the results given in Example 1 on your computer system. Modularize your program into a number of subroutines or procedures to compute each major segment of the algorithm. (For example, you may wish to construct routines for **(i)** a matrix times a vector, **(ii)** dot product, **(iii)** replace one vector by another, **(iv)** norm of a vector, **(v)** normalize a vector, and others.)

(b) Add Aitken acceleration to your code and rerun it. Write up your conclusions from this project.

5. **(a)** Repeat part (a) of Computer Problem 4 for Example 2.

(b) Repeat part (b) of Computer Problem 4 for Example 2.

6. Write and test a computer program to compute the eigenvalue farthest from a given complex number. Test the routine on the example matrix in this section.

7. Construct an example to show that Aitken acceleration will produce meaningless results if it is not stopped at an appropriate stage.

5.2 Schur's and Gershgorin's Theorems

We continue our review of the basic eigenvalue theory of matrices. Recall that two matrices A and B are **similar** to each other if there exists a nonsingular matrix P such that $B = PAP^{-1}$. The importance of this concept derives from the theorem that two matrices representing the same linear transformation with respect to different bases are similar to each other.

THEOREM 1 *Similar matrices have the same eigenvalues.*

Proof Let A and B be similar matrices, that is,

$$B = PAP^{-1}$$

We shall see that A and B have the same characteristic polynomial. Indeed,

$$\begin{aligned}\det(B - \lambda I) &= \det(PAP^{-1} - \lambda I)\\ &= \det[P(A - \lambda I)P^{-1}]\\ &= \det P \det(A - \lambda I) \det P^{-1}\\ &= \det(A - \lambda I)\end{aligned}$$

Here we invoked two basic facts: The determinant of a product of two matrices is the product of their determinants, and the determinant of the inverse of a matrix is the reciprocal of the determinant of the matrix. ■

Schur's Factorization

Theorem 1 suggests a strategy for finding the eigenvalues of A: Transform A to a matrix B by a similarity transformation, $B = PAP^{-1}$, and compute the eigenvalues of B. If B is simpler than A, the computation of its eigenvalues may be easier. Specifically, suppose that B is *triangular*. Then the eigenvalues of B (and the eigenvalues of A) are simply the diagonal elements of B. This brings us naturally to the important theorem of Schur, according to which the preceding strategy is always possible (theoretically). Recall that a matrix U is **unitary** if $UU^* = I$. Here U^* denotes the conjugate transpose of $U : (U^*)_{ij} = \overline{U}_{ji}$.

THEOREM 2 **Schur's Theorem** *Every square matrix is unitarily similar to a triangular matrix.*

Proof We proceed by induction on n, the order of the matrix A. The theorem is trivial for $n = 1$. Suppose now that the theorem has been proved for all matrices of order $n - 1$, and consider a matrix A of order n. Let λ be an eigenvalue of A, and let x be an accompanying eigenvector. There is no loss of generality in assuming that $\|x\|_2 = 1$. As usual, let $e^{(1)}$ denote the standard unit vector $e^{(1)} = (1, 0, \dots, 0)^T$. Let $\beta = x_1/|x_1|$ if $x_1 \neq 0$, and put $\beta = 1$ if $x_1 = 0$. By Lemma 2 that follows, there is a unitary matrix U such that $Ux = \beta e^{(1)}$. Since U is unitary, $U^{-1} = U^*$ and $\beta^{-1}x = U^*e^{(1)}$ so that

$$UAU^*e^{(1)} = UA\beta^{-1}x = \beta^{-1}\lambda Ux = \lambda e^{(1)}$$

This proves that the first column of UAU^* is $\lambda e^{(1)}$. Let $\widetilde{A}$ denote the matrix obtained from UAU^* by deleting the first row and column. By the induction hypothesis, there is a unitary matrix Q of order $n-1$ such that $Q\widetilde{A}Q^*$ is triangular. The unitary matrix that reduces A to triangular form is

$$V = \begin{bmatrix} 1 & 0 \\ 0 & Q \end{bmatrix} U$$

because

$$
\begin{aligned}
VAV^* &= \begin{bmatrix} 1 & 0 \\ 0 & Q \end{bmatrix} UAU^* \begin{bmatrix} 1 & 0 \\ 0 & Q^* \end{bmatrix} \\
&= \begin{bmatrix} 1 & 0 \\ 0 & Q \end{bmatrix} \begin{bmatrix} \lambda & w \\ 0 & \widetilde{A} \end{bmatrix} \begin{bmatrix} 1 & 0 \\ 0 & Q^* \end{bmatrix} \\
&= \begin{bmatrix} 1 & 0 \\ 0 & Q \end{bmatrix} \begin{bmatrix} \lambda & wQ^* \\ 0 & \widetilde{A}Q^* \end{bmatrix} = \begin{bmatrix} \lambda & wQ^* \\ 0 & Q\widetilde{A}Q^* \end{bmatrix}
\end{aligned}
$$

In this equation, w is a row vector of order $n-1$, and the 0's represent null vectors of order $n-1$. Since the matrix

$$
\begin{bmatrix} 1 & 0 \\ 0 & Q \end{bmatrix}
$$

is unitary, the proof is complete. ■

The reader will observe that the proof just given is not constructive because it appeals to the existence of eigenvalues (and uses them) but does not address the question of how to compute them.

COROLLARY 1 *Every square matrix is similar to a triangular matrix.*

COROLLARY 2 *Every Hermitian matrix is unitarily similar to a diagonal matrix.*

Proof If A is **Hermitian**, then $A = A^*$. Let U be a unitary matrix such that UAU^* is upper triangular. Then $(UAU^*)^*$ is lower triangular. But

$$
(UAU^*)^* = U^{**}A^*U^* = UAU^*
$$

Thus, the matrix UAU^* is both upper and lower triangular; hence, it must be a diagonal matrix. ■

LEMMA 1 *The matrix $I - vv^*$ is unitary if and only if $\|v\|_2^2 = 2$ or $v = 0$.*

Proof For the matrix $U = I - vv^*$ to be unitary, we require

$$
\begin{aligned}
I &= UU^* \\
&= (I - vv^*)(I - vv^*) \\
&= I - 2vv^* + vv^*vv^* \\
&= I - 2vv^* + (v^*v)(vv^*) \\
&= I - (2 - v^*v)vv^*
\end{aligned}
$$

(Here we moved the scalar quantity v^*v to the front.) The necessary and sufficient condition on v is that $v^*v = 2$ or $vv^* = 0$. ■

LEMMA 2 *Let x and y be two vectors such that $\|x\|_2 = \|y\|_2$ and $\langle x, y\rangle$ is real. Then there exists a unitary matrix U of the form $I - vv^*$ such that $Ux = y$.*

Proof If $x = y$, let $v = 0$. If $x \neq y$, an inspired guess leads us to try $v = \alpha(x - y)$, with $\alpha = \sqrt{2}/\|x - y\|_2$. This choice of v leads to

$$\begin{aligned} Ux - y &= (I - vv^*)x - y = x - vv^*x - y \\ &= x - y - \alpha^2(x - y)(x^* - y^*)x \\ &= (x - y)[1 - \alpha^2(x^*x - y^*x)] \end{aligned}$$

This last expression turns out to be 0 because the number inside the brackets vanishes. To see this, we use the hypotheses $x^*x = y^*y$ and $y^*x = x^*y$ to compute:

$$\begin{aligned} 1 - \alpha^2(x^*x - y^*x) &= 1 - \frac{1}{2}\alpha^2(x^*x + x^*x - y^*x - y^*x) \\ &= 1 - \frac{1}{2}\alpha^2(x^*x + y^*y - y^*x - x^*y) \\ &= 1 - \frac{1}{2}\alpha^2(x^* - y^*)(x - y) \\ &= 1 - \frac{1}{2}\alpha^2\|x - y\|_2^2 = 0 \end{aligned}$$

■

Example 1 Here is a concrete case of the Schur factorization, $UAU^* = T$, in which U is unitary and T is upper triangular.

Solution

$$\begin{bmatrix} 0.36 & 0.48 & 0.80 \\ 0.48 & 0.64 & -0.60 \\ 0.80 & -0.60 & 0.00 \end{bmatrix} \begin{bmatrix} 361 & 123 & -180 \\ 148 & 414 & -240 \\ -92 & 169 & 65 \end{bmatrix} \begin{bmatrix} 0.36 & 0.48 & 0.80 \\ 0.48 & 0.64 & -0.60 \\ 0.80 & -0.60 & 0.00 \end{bmatrix}$$
$$= \begin{bmatrix} 125 & 380 & -125 \\ 0 & 465 & 1250 \\ 0 & 0 & 250 \end{bmatrix}$$

■

If an eigenvalue λ of an $n \times n$ matrix A is known, then the proof of Schur's Theorem shows how to produce an $(n - 1) \times (n - 1)$ matrix $\widetilde{A}$ whose eigenvalues are the same as those of A, except for λ. This procedure is known as **deflation**.

The formalization of the **deflation process** for eigenvalues is as follows:

(i) Obtain an eigenvector x corresponding to a known eigenvalue λ.

(ii) Define $\beta = x_1/|x_1|$ if $x_1 \neq 0$, and let $\beta = 1$ otherwise.

(iii) Define $\alpha = \sqrt{2}/\|x - \beta e^{(1)}\|_2$, $v = \alpha\left(x - \beta e^{(1)}\right)$, and $U = I - vv^*$.

(iv) Let $\widetilde{A}$ be the matrix obtained from UAU^* by omitting the first row and column.

An analogous deflation process was discussed in Section 3.5 in finding the roots of a polynomial p. After one root, say ξ, has been found, we can divide $p(x)$ by $x - \xi$ to obtain a polynomial of lower degree with the same roots (except for ξ).

Most numerical methods for calculating eigenvalues compute one eigenvalue at a time. Any such method can be combined with the deflation procedure to compute as many eigenvalues of a matrix as we wish. In practice, this strategy must be used cautiously because the successive eigenvalues may be infected with more and more roundoff error.

Localizing Eigenvalues

Many theorems in the literature describe in a coarse manner where the eigenvalues of a matrix are situated in the complex plane. The most famous of these **localization** theorems is the following:

THEOREM 3 **Gershgorin's Theorem** *The spectrum of an $n \times n$ matrix A (that is, the set of its eigenvalues) is contained in the union of the following n disks, D_i, in the complex plane:*

$$D_i = \left\{ z \in \mathbb{C} : |z - a_{ii}| \leqq \sum_{\substack{j=1 \\ j \neq i}}^{n} |a_{ij}| \right\} \qquad (1 \leqq i \leqq n)$$

Proof Let λ be any element of the spectrum of A. Select a vector x such that $Ax = \lambda x$ and $\|x\|_\infty = 1$. Let i be an index for which $|x_i| = 1$. Since $(Ax)_i = \lambda x_i$, we have

$$\lambda x_i = \sum_{j=1}^{n} a_{ij} x_j$$

Therefore,

$$(\lambda - a_{ii}) x_i = \sum_{\substack{j=1 \\ j \neq i}}^{n} a_{ij} x_j$$

Taking absolute values and using the triangle inequality and $|x_j| \leqq 1 = |x_i|$, we have

$$|\lambda - a_{ii}| \leqq \sum_{\substack{j=1 \\ j \neq i}}^{n} |a_{ij}||x_j| \leqq \sum_{\substack{j=1 \\ j \neq i}}^{n} |a_{ij}|$$

Thus, $\lambda \in D_i$. ■

Example 2 The Gershgorin disks for the matrix

$$A = \begin{bmatrix} -1+i & 0 & \frac{1}{4} \\ \frac{1}{4} & 1 & \frac{1}{4} \\ 1 & 1 & 3 \end{bmatrix}$$

are shown in Figure 5.1. We can deduce from the figure that all eigenvalues of A satisfy the inequality $\frac{1}{2} \leqq |\lambda| \leqq 5$.

THEOREM 4 *If the matrix A is diagonalized by the similarity transformation $P^{-1}AP$, and if B is any matrix, then the eigenvalues of $A + B$ lie in the union of the disks*

$$\{\lambda \in \mathbb{C} : |\lambda - \lambda_i| \leqq \kappa_\infty(P)\|B\|_\infty\}$$

where $\lambda_1, \lambda_2, \ldots, \lambda_n$ are the eigenvalues of A, and $\kappa_\infty(P)$ is the condition number of P.

Proof We can establish a somewhat more precise result. If $P^{-1}AP = D$, the diagonal of the diagonal matrix D consists of $\lambda_1, \lambda_2, \ldots, \lambda_n$. Using "sp" for **spectrum**, we have

$$\text{sp}(A + B) = \text{sp}[P^{-1}(A + B)P] = \text{sp}(D + P^{-1}BP)$$

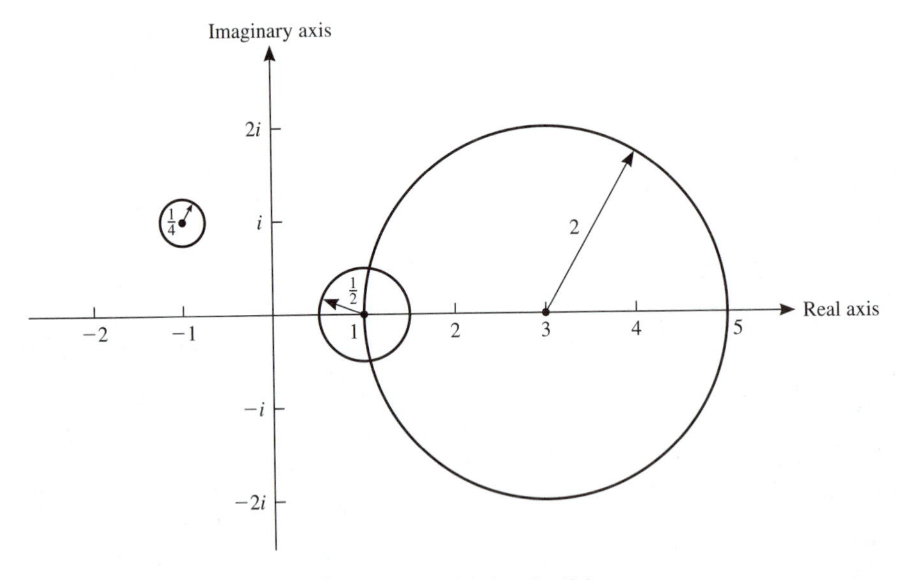

FIGURE 5.1 Gershgorin disks

By applying Gershgorin's Theorem to $D + C$, with $C = P^{-1}BP$, we conclude that the spectrum of $A + B$ is in the union of the Gershgorin disks of $D + C$, which are

$$\left\{\lambda \in \mathbb{C} : |\lambda - \lambda_i - c_{ii}| \leqq \sum_{\substack{j=1 \\ j \neq i}}^{n} |d_{ij} + c_{ij}| = \sum_{\substack{j=1 \\ j \neq i}}^{n} |c_{ij}|\right\}$$

By the triangle inequality and the definition of $\|C\|_\infty$, we have

$$|\lambda - \lambda_i| \leqq |\lambda - \lambda_i - c_{ii}| + |c_{ii}| \leqq |c_{ii}| + \sum_{\substack{j=1 \\ j \neq i}}^{n} |c_{ij}|$$

$$\leqq \|C\|_\infty \leqq \|P^{-1}\|_\infty \|B\|_\infty \|P\|_\infty = \kappa_\infty(P)\|B\|_\infty \qquad \blacksquare$$

Theorem 4 indicates that when a matrix A is perturbed, its eigenvalues will be perturbed by an amount not exceeding $\kappa_\infty(P)\|B\|_\infty$, where B is the perturbing matrix and $\kappa_\infty(P)$ is the condition number of P computed from the ℓ_∞-matrix norm. (See Section 4.4.)

For a Hermitian matrix (that is, $A^* = A$), the matrix P can be chosen to be *unitary*; this is Corollary 2 of Schur's Theorem. Hence, in this case the rows of P are vectors satisfying $\|x\|_2 = 1$. It follows that $\|P\|_\infty \leqq \sqrt{n}$. This is true for P^{-1} also, and thus $\kappa_\infty(P) \leqq n$. Hence, for any matrix B, the eigenvalues of $A + B$ lie in the union of disks

$$\{\lambda \in \mathbb{C} : |\lambda - \lambda_i| \leqq n\|B\|_\infty\}$$

where $\lambda_1, \lambda_2, \ldots, \lambda_n$ are the eigenvalues of the Hermitian matrix A.

PROBLEMS 5.2

1. Find the Schur factorizations of

$$\begin{bmatrix} 3 & 8 \\ -2 & 3 \end{bmatrix} \quad \text{and} \quad \begin{bmatrix} 4 & 7 \\ 1 & 12 \end{bmatrix}$$

2. Prove that the eigenvalues of A lie in the intersection of the two sets D and E defined by

$$D = \bigcup_{i=1}^{n} \left\{ z \in \mathbb{C} : |z - a_{ii}| \leqq \sum_{\substack{j=1 \\ j \neq i}}^{n} |a_{ij}| \right\}$$

$$E = \bigcup_{i=1}^{n} \left\{ z \in \mathbb{C} : |z - a_{ii}| \leqq \sum_{\substack{j=1 \\ j \neq i}}^{n} |a_{ji}| \right\}$$

3. Prove that if λ is an eigenvalue of A, then there is a nonzero vector x such that $x^T A = \lambda x^T$. (Here x^T denotes a **row vector**.)

4. Prove that if A is Hermitian, then the deflation technique in the text will produce a Hermitian matrix.

5. Prove that if one column in the matrix A, say the jth column, satisfies $a_{ij} = 0$ for $i \neq j$, then a_{jj} is an eigenvalue of A.

6. A **normal** matrix is one that commutes with its adjoint: $AA^* = A^*A$. Prove that if A is normal, then so is $A - \lambda I$ for any scalar λ.

7. Suppose that A is normal and that x and y are eigenvectors of A corresponding to different eigenvalues. Prove that $x^*y = 0$.

8. Prove that if A is normal, then A and A^* have the same eigenvectors.

9. Prove that if A is normal, then the condition $AB = BA$ implies that $A^*B = BA^*$.

10. Prove that if x and y are points in $\mathbb{C}^n$ having the same Euclidean norm, then there is a unitary matrix U such that $Ux = y$.

11. Prove or disprove: If $\{x_1, x_2, \dots, x_k\}$ and $\{y_1, y_2, \dots, y_k\}$ are orthonormal sets in $\mathbb{C}^n$, then there is a unitary matrix U such that $Ux_i = y_i$ for $1 \leqq i \leqq k$.

12. Prove that if $(I - vv^*)x = y$ for some triple of vectors v, x, y, then $\langle x, y\rangle$ is real.

13. Find the precise conditions on a pair of vectors u and v in order that $I - uv^*$ be unitary.

14. Prove that if Q is a unitary matrix, then

$$\|A\|_2 = \|QA\|_2 = \|AQ\|_2$$

Note: $\|A\|_2^2 = \rho(AA^T)$.

15. Prove that if A is a diagonal matrix, then

$$\|A\|_2 = \max_{1 \leqq i \leqq n} |a_{ii}|$$

16. Prove that for any square matrix A, $\|A\|_2^2 \leqq \|A^*A\|_2$.

17. Let A_j denote the jth column of A. Prove that $\|A_j\|_2 \leqq \|A\|_2$. Is this true for all subordinate matrix norms?

18. The **trace** of a matrix A is $\text{tr}(A) = \sum_{i=1}^{n} a_{ii}$. Prove that the trace of a matrix equals the sum of its eigenvalues. (Schur's Theorem may be useful here.)

19. Prove that if $\lambda_1, \lambda_2, \dots, \lambda_n$ are the eigenvalues of A, then the trace of A^m is

$$\text{tr}(A^m) = \lambda_1^m + \lambda_2^m + \cdots + \lambda_n^m$$

20. Prove that if the eigenvalues of A satisfy $|\lambda_1| > |\lambda_i|$ for $i = 2, 3, \dots, n$, then

$$\lambda_1 = \lim_{m \to \infty} \frac{\text{tr}(A^{m+1})}{\text{tr}(A^m)}$$

21. Prove that AB and BA have the same eigenvalues if A and B are square matrices.

22. Prove that if $a^2 + b^2 = 1$ and $c^2 + d^2 = 1$, then the following matrix is unitary:

$$\begin{bmatrix} ad & ac & b \\ bd & bc & -a \\ c & -d & 0 \end{bmatrix}$$

Notice that for arbitrary θ and φ we can let $a = \sin\theta$, $b = \cos\theta$, $c = \sin\varphi$, and $d = \cos\varphi$.

23. Prove that if x is a vector such that $\|x\|_2 = 1$, then there is a unitary matrix whose first column is x.

24. Let $\|x\|_2 = 1$ and put $U = I - 2xx^*$. Show that $U^2 = I$.

25. Let A be $n \times n$, let B be $m \times m$, and let C be $n \times m$. Prove that if C has rank m and if $AC = CB$, then

$$\text{sp}(B) \subseteqq \text{sp}(A)$$

26. If $x^*x = 2$, what is $(I - xx^*)^{-1}$?

27. Let $x^*x = 1$ and determine whether $I - xx^*$ is invertible.

28. Prove or disprove: If A is a square matrix, then there is a unitary Hermitian matrix U such that UAU is triangular.

29. Without computing them, prove that the eigenvalues of the matrix

$$A = \begin{bmatrix} 6 & 2 & 1 \\ 1 & -5 & 0 \\ 2 & 1 & 4 \end{bmatrix}$$

satisfy the inequality $1 \leqq |\lambda| \leqq 9$.

30. Show that the imaginary parts of the eigenvalues of

$$\begin{bmatrix} 3 & \frac{1}{3} & \frac{2}{3} \\ 1 & -4 & 0 \\ \frac{1}{2} & \frac{1}{2} & -1 \end{bmatrix}$$

all lie in the interval $[-1, 1]$.

31. Find the Schur factorization $UAU^* = T$ for the matrix

$$A = \begin{bmatrix} 2.888 & 0.984 & -1.440 \\ 1.184 & 3.312 & -1.920 \\ -0.160 & 2.120 & -0.200 \end{bmatrix}$$

Hint: The unitary matrix in Example 1 should serve.

32. (See Problems 24, 26, and 27.) Prove that $I - xx^*$ is singular if and only if $x^*x = 1$, and find the inverse in all nonsingular cases.

33. If $|a_{ii} - \lambda| > \sum_{j=1,\, j\neq i}^{n} |a_{ij}|$, then the matrix $A - \lambda I$ is diagonally dominant. Using this idea, prove Gershgorin's Theorem.

34. Let $U = I - \lambda uu^*$, where u is a given vector. Find all the complex values of λ for which U is unitary.

35. Suppose that the matrix A is similar to a *nearly diagonal* matrix B. Can we conclude that the eigenvalues of A are close to the diagonal elements of B? Prove a theorem about this, and investigate the question when B is *nearly triangular*.

36. Let A be an $n \times n$ matrix and let $D_1, D_2, \ldots, D_n$ be its Gershgorin disks. Suppose that an eigenvalue λ lies in D_k but not in any of the other disks. Let x be an eigenvector corresponding to λ. Show that $|x_k| > |x_i|$ for $i \neq k$.

37. **(a)** Sketch the Gershgorin disks for the matrix

$$\begin{bmatrix} 0 & 2 & -1 \\ -2 & -10 & 0 \\ -1 & -1 & 4 \end{bmatrix}$$

and give a bound for the spectral radius, $\rho(A)$.

(b) Determine an upper and lower bound for $\rho(A)$, using $\|A\|_1$, where

$$\begin{bmatrix} 1 & 0 & 1 \\ 1 & 4 & 0 \\ 0 & 0 & 3 \end{bmatrix}$$

and then repeat using Gershgorin's Theorem.

38. Use Gershgorin's Theorem to prove that a diagonally dominant matrix does not have 0 as an eigenvalue and is therefore nonsingular.

39. Prove that $\det(I + xx^*) = 1 + x^*x$. *Hint:* There is a unitary matrix that maps x into a multiple of $e_1 = (1, 0, 0, \ldots, 0)^T$.

---

*5.3 Orthogonal Factorizations and Least-Squares Problems

In this section, we continue to work with complex vectors and matrices. At the beginning of this chapter, we reviewed basic algebra in the complex space $\mathbb{C}^n$. The inner product $\langle \cdot, \cdot \rangle$ in this space allows us to define the concept of orthogonality. An indexed set of vectors $[v_1, v_2, \ldots, v_k]$ in $\mathbb{C}^n$ is said to be **orthogonal** if $\langle v_i, v_j \rangle = 0$ whenever $i \neq j$. If $\langle v_i, v_j \rangle = \delta_{ij}$, the set is said to be **orthonormal**. By taking $i = j$, we see that this implies $\|v_i\|_2 = 1$ for each i. If the v_i's are the columns of an $n \times k$ matrix A, then the orthonormality is expressed by the equation $A^*A = I$.

Basic Concepts

Suppose now that $[v_1, v_2, \ldots, v_n]$ is an orthonormal basis for $\mathbb{C}^n$. Each element $x \in \mathbb{C}^n$ has a unique representation in the form

$$x = \sum_{i=1}^{n} c_i v_i$$

for appropriate complex scalars c_i. On taking the inner product of both sides of this equation with some v_j, we obtain

$$\langle x, v_j \rangle = \left\langle \sum_{i=1}^{n} c_i v_i, v_j \right\rangle = \sum_{i=1}^{n} c_i \langle v_i, v_j \rangle = \sum_{i=1}^{n} c_i \delta_{ij} = c_j$$

This establishes that for all $x \in \mathbb{C}^n$,

$$x = \sum_{i=1}^{n} \langle x, v_i \rangle v_i \tag{1}$$

The term $\langle x, v_i \rangle v_i$ is referred to as the **component** of x in the direction v_i. A sketch of a typical situation in $\mathbb{R}^2$ is shown in Figure 5.2.

An **inner-product space** is an abstract algebraic system of which $\mathbb{C}^n$ is one embodiment. It is a linear space (over the complex field) in which an inner product has been defined subject to these axioms:

(i) $\langle x, x \rangle > 0 \qquad \text{if } x \neq 0$

(ii) $\langle \alpha x + \beta y, z \rangle = \alpha \langle x, z \rangle + \beta \langle y, z \rangle \qquad \alpha, \beta \in \mathbb{C}$

(iii) $\langle x, y \rangle = \overline{\langle y, x \rangle}$

The definitions of norm, orthogonality, and orthonormality are as given in $\mathbb{C}^n$. An inner-product space need not be finite-dimensional. It is easily proved from the preceding axioms that $\langle x, y + z \rangle = \langle x, y \rangle + \langle x, z \rangle$ and that $\langle x, \alpha y \rangle = \overline{\alpha} \langle x, y \rangle$.

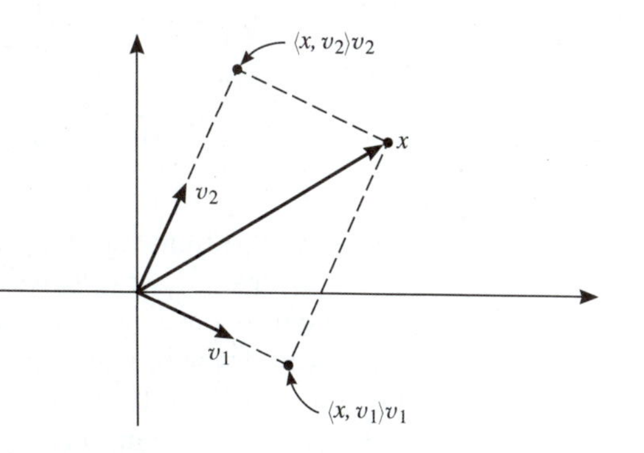

FIGURE 5.2 Orthogonal components of a vector

The **Pythagorean rule** is valid in any inner-product space: If $\langle x, y\rangle = 0$, then

$$\|x + y\|_2^2 = \|x\|_2^2 + \|y\|_2^2$$

This follows from writing

$$\|x + y\|_2^2 = \langle x + y, x + y\rangle = \langle x, x\rangle + \langle y, x\rangle + \langle x, y\rangle + \langle y, y\rangle = \|x\|_2^2 + \|y\|_2^2$$

Gram-Schmidt Process

The classic **Gram-Schmidt process** can be used to obtain orthonormal systems in any inner-product space. To describe it, we suppose that a linearly independent sequence of vectors is given in an inner-product space: $[x_1, x_2, \ldots]$. (This sequence can be finite or infinite.) We generate an orthonormal sequence $[u_1, u_2, \ldots]$ by the formula

$$u_k = \left(\left\|x_k - \sum_{i<k}\langle x_k, u_i\rangle u_i\right\|_2\right)^{-1}\left(x_k - \sum_{i<k}\langle x_k, u_i\rangle u_i\right) \qquad (k \geqq 1) \tag{2}$$

THEOREM 1 *The* **Gram-Schmidt sequence** $[u_1, u_2, \ldots]$ *has the property that* $\{u_1, u_2, \ldots, u_k\}$ *is an orthonormal base for the linear span of* $\{x_1, x_2, \ldots, x_k\}$ *for* $k \geqq 1$.

Proof We proceed by induction on k. For $k = 1$, we have from Equation (2) $u_1 = \left(\|x_1\|_2\right)^{-1}x_1$. Hence, the set $\{u_1\}$ is orthonormal, and its linear span is identical with the linear span of $\{x_1\}$. Notice that $\|x_1\|_2 \neq 0$ because the set $\{x_1, x_2, \ldots\}$ is linearly independent.

Suppose, as an induction hypothesis, that $\{u_1, u_2, \ldots, u_{k-1}\}$ is an orthonormal base for $\text{span}\{x_1, x_2, \ldots, x_{k-1}\}$. Let

$$v = x_k - \sum_{i<k}\langle x_k, u_i\rangle u_i \tag{3}$$

Then v is orthogonal to $u_1, u_2, \ldots, u_{k-1}$ because for $j < k$,

$$\begin{aligned}\langle v, u_j\rangle &= \langle x_k, u_j\rangle - \sum_{i<k}\langle x_k, u_i\rangle\langle u_i, u_j\rangle \\ &= \langle x_k, u_j\rangle - \sum_{i<k}\langle x_k, u_i\rangle\delta_{ij} = \langle x_k, u_j\rangle - \langle x_k, u_j\rangle = 0\end{aligned}$$

If $v = 0$, we see from Equation (3) that $x_k \in \text{span}\{u_1, u_2, \ldots, u_{k-1}\}$. Using the induction hypothesis, we conclude that $x_k \in \text{span}\{x_1, x_2, \ldots, x_{k-1}\}$, contradicting the assumed linear independence of $\{x_1, x_2, \ldots, x_k\}$. Consequently, $v \neq 0$, and u_k is well defined by the expression $(\|v\|_2)^{-1}v$. Since u_k is of norm one, the set $\{u_1, u_2, \ldots, u_k\}$ is orthonormal. The induction hypothesis together with Equation (3) shows that v (and u_k) is in $\text{span}\{x_1, x_2, \ldots, x_k\}$. Therefore, we see that $\text{span}\{u_1, u_2, \ldots, u_k\} \subseteqq \text{span}\{x_1, x_2, \ldots, x_k\}$. Since both $\{u_1, u_2, \ldots, u_k\}$ and $\{x_1, x_2,$

$\dots, x_k\}$ are linearly independent (both sets span a k-dimensional space), we have $\text{span}\{u_1, u_2, \dots, u_k\} = \text{span}\{x_1, x_2, \dots, x_k\}$. ■

If the Gram-Schmidt process is applied to the columns of a matrix, we can interpret the result as a matrix factorization. In this case, the inner products that arise in the computation can be saved in a matrix, which will be one of the factors. We apply the process to the columns $A_1, A_2, \dots, A_n$ of an $m \times n$ matrix A, arriving after n steps at an $m \times n$ matrix B whose columns form an orthonormal set. Here is the **Gram-Schmidt algorithm**:

for $j = 1$ **to** n **do**
  for $i = 1$ **to** $j - 1$ **do**
    $t_{ij} \leftarrow \langle A_j, B_i \rangle$
  end do
  $C_j \leftarrow A_j - \sum_{i<j} t_{ij} B_i$
  $t_{jj} \leftarrow \|C_j\|_2$
  $B_j \leftarrow t_{jj}^{-1} C_j$
end do

THEOREM 2 *The Gram-Schmidt process, when applied to the columns of an $m \times n$ matrix A of rank n, produces a factorization*

$$A = BT \tag{4}$$

in which B is an $m \times n$ matrix with orthonormal columns and T is an $n \times n$ upper triangular matrix with positive diagonal.

Proof First, observe that the preceding algorithm is actually carrying out the Gram-Schmidt process encapsulated in Equation (2). Next, we complete the definition of the T-matrix by setting $t_{ij} = 0$ when $i > j$. By Theorem 1, the columns of B form an orthonormal set of n vectors in $\mathbb{R}^m$, and each A_j is a linear combination of $B_1, B_2, \dots, B_j$. In fact, by Equation (1),

$$\begin{aligned} A_j &= \sum_{i=1}^{j} \langle A_j, B_i \rangle B_i = \sum_{i=1}^{j-1} \langle A_j, B_i \rangle B_i + \langle A_j, B_j \rangle B_j \\ &= \sum_{i=1}^{j-1} t_{ij} B_i + \langle A_j, B_j \rangle B_j \end{aligned} \tag{5}$$

Now, we stop to compute

$$\begin{aligned} \langle A_j, B_j \rangle &= \left\langle C_j + \sum_{i<j} t_{ij} B_i,\ B_j \right\rangle = \langle C_j, B_j \rangle \\ &= \langle C_j, C_j \rangle / t_{jj} = t_{jj} \end{aligned} \tag{6}$$

When the result from Equation (6) is used in Equation (5), we obtain

$$A_j = \sum_{i=1}^{j} t_{ij} B_i = \sum_{i=1}^{n} t_{ij} B_i \qquad (1 \leqq j \leqq n) \tag{7}$$

This is another way of expressing Equation (4). ■

Modified Gram-Schmidt Algorithm

Experience has shown (Rice [1966]) that a certain rearrangement of the Gram-Schmidt process generally has superior numerical properties. The **modified Gram-Schmidt algorithm** goes as follows:

for $k = 1$ **to** n **do**
  $A_k \leftarrow \left(\|A_k\|_2 \right)^{-1} A_k$
  for $j = k + 1$ **to** n **do**
    $A_j \leftarrow A_j - \langle A_j, A_k \rangle A_k$
  end do
end do

We can see that the components of A_j in the direction of the base vector A_k are being removed as soon as possible, and one at a time. In this algorithm, there is considerable overwriting, and in the end the original set $\{A_1, A_2, \ldots, A_n\}$ has been *replaced* by an orthonormal set.

To avoid the square roots involved in computing $\|x\|_2$, the modified Gram-Schmidt algorithm is often given in the following form, which yields a slightly different factorization:

for $k = 1$ **to** n **do**
  $d_k \leftarrow \|A_k\|_2^2$
  $t_{kk} \leftarrow 1$
  for $j = k + 1$ **to** n **do**
    $t_{kj} \leftarrow d_k^{-1} \langle A_j, A_k \rangle$
    $A_j \leftarrow A_j - t_{kj} A_k$
  end do
end do

THEOREM 3 *If the modified Gram-Schmidt process is applied to the columns of an $m \times n$ matrix A of rank n, the transformed $m \times n$ matrix B has an orthogonal set of columns and satisfies*

$$A = BT$$

where T is a unit $n \times n$ upper triangular matrix whose elements t_{kj} (for $j > k$) are generated in the algorithm.

Proof To prove this theorem, we write the algorithm in such a way that quantities are not overwritten but saved. If this were not done, then in a formal proof A_j would be ambiguous, since it would denote different vectors at different stages in the algorithm. We relabel the original set as $\{A_1^{(1)}, A_2^{(1)}, \dots, A_n^{(1)}\}$. The algorithm now reads:

for $k = 1$ **to** n **do**
    $d_k \leftarrow \|A_k^{(k)}\|_2^2$
    for $j = k + 1$ **to** n **do**
        $t_{kj} \leftarrow d_k^{-1}\langle A_j^{(k)}, A_k^{(k)}\rangle$
        $A_j^{(k+1)} \leftarrow A_j^{(k)} - t_{kj}A_k^{(k)}$
    end do
    for $j = 1$ **to** k **do**
        $A_j^{(k+1)} \leftarrow A_j^{(k)}$
    end do
end do

(In this algorithm there is no overwriting.)

The following assertion will now be proved by induction on k:

$$\mathcal{A}(k): \text{ If } \min(i, j) < k, \text{ then } \langle A_i^{(k)}, A_j^{(k)}\rangle = d_i\delta_{ij}.$$

Observe that $\mathcal{A}(1)$ is vacuously true because no pairs satisfy $\min(i, j) < 1$. Now adopt $\mathcal{A}(k)$ as an induction hypothesis. The proof of $\mathcal{A}(k + 1)$ then proceeds as follows: Consider any pair (i, j) for which $\min(i, j) < k + 1$. By symmetry, we can assume $i \leqq j$. There are four cases to analyze.

(a) If $i < k$ and $j \leqq k$, then by $\mathcal{A}(k)$,

$$\langle A_i^{(k+1)}, A_j^{(k+1)}\rangle = \langle A_i^{(k)}, A_j^{(k)}\rangle = d_i\delta_{ij}$$

(b) If $i < k$ and $j > k$, then by $\mathcal{A}(k)$,

$$\begin{aligned}\langle A_i^{(k+1)}, A_j^{(k+1)}\rangle &= \langle A_i^{(k)}, A_j^{(k)} - t_{kj}A_k^{(k)}\rangle \\ &= \langle A_i^{(k)}, A_j^{(k)}\rangle - \bar{t}_{kj}\langle A_i^{(k)}, A_k^{(k)}\rangle \\ &= d_i\delta_{ij} - \bar{t}_{kj}d_i\delta_{ik} = 0\end{aligned}$$

(c) Letting $i = j = k$, we have

$$\langle A_k^{(k+1)}, A_k^{(k+1)}\rangle = \langle A_k^{(k)}, A_k^{(k)}\rangle = d_k$$

(d) Letting $i = k < j$, we have

$$\begin{aligned}\langle A_k^{(k+1)}, A_j^{(k+1)}\rangle &= \langle A_k^{(k)}, A_j^{(k)} - t_{kj}A_k^{(k)}\rangle = \langle A_k^{(k)}, A_j^{(k)}\rangle - \bar{t}_{kj}\langle A_k^{(k)}, A_k^{(k)}\rangle \\ &= \bar{t}_{kj}d_k - \bar{t}_{kj}d_k = 0\end{aligned}$$

By the preceding inductive proof, $\mathcal{A}(n)$ is true, and this means $\langle A_i^{(n)}, A_j^{(n)}\rangle = d_i\delta_{ij}$ if $i < n$. Thus, the set $\{A_1^{(n)}, A_2^{(n)}, \ldots, A_n^{(n)}\}$ is orthogonal. If B is the matrix whose jth column is $A_j^{(n)}$, then $B^*B = \text{diag}(d_1, d_2, \ldots, d_n)$. Now we set $t_{kk} = 1$ and $t_{kj} = 0$ if $k > j$. To see that $A = BT$, notice that the kth column of A is given by

$$\begin{aligned}A_k = A_k^{(1)} &= (A_k^{(1)} - A_k^{(2)}) + (A_k^{(2)} - A_k^{(3)}) + \cdots + (A_k^{(k-1)} - A_k^{(k)}) + A_k^{(k)} \\ &= t_{1k}A_1^{(1)} + t_{2k}A_2^{(2)} + \cdots + t_{k-1,k}A_{k-1}^{(k-1)} + t_{kk}A_k^{(k)} \\ &= t_{1k}A_1^{(n)} + t_{2k}A_2^{(n)} + \cdots + t_{kk}A_k^{(n)}\end{aligned}$$

■

Least-Squares Problem

An important application of the orthogonal factorizations being discussed here is to the **least-squares** problem for a linear system of equations. Consider a system of m equations in n unknowns written as

$$Ax = b \tag{8}$$

Here A is $m \times n$, x is $n \times 1$, and b is $m \times 1$. We shall assume that the rank of A is n; consequently, $m \geqq n$. Usually System (8) will have no solution because b will not ordinarily be situated in the n-dimensional subspace of $\mathbb{C}^m$ generated by the columns of A. In such cases it is often required to find an x that minimizes the norm of the residual vector, $b - Ax$. The least-squares "solution" of (8) is the vector x that makes $\|b - Ax\|_2$ a minimum. (Under the hypothesis we have made concerning the rank of A, this x will be unique.)

LEMMA 1 *If x is a point such that $A^*(Ax - b) = 0$, then x solves the least-squares problem.*

Proof Let y be any other point. Since $A^*(Ax - b) = 0$, we conclude that $b - Ax$ is orthogonal to the column space of A. Moreover, since $A(x - y)$ is in the column space of A, we have $\langle b - Ax, A(x - y)\rangle = 0$, and the Pythagorean rule gives

$$\|b - Ay\|_2^2 = \|b - Ax + A(x - y)\|_2^2 = \|b - Ax\|_2^2 + \|A(x - y)\|_2^2 \geqq \|b - Ax\|_2^2$$ ■

If A has been factored in the form $A = BT$ as described in the preceding theorem, then the least-squares solution of the system $Ax = b$ will be the exact solution of the $n \times n$ system

$$Tx = (B^*B)^{-1}B^*b$$

This is verified by using Lemma 1:

$$A^*Ax = (BT)^*BTx = T^*B^*B(B^*B)^{-1}B^*b = T^*B^*b = A^*b$$

The matrix $(B^*B)^{-1}$ is $\text{diag}(d_1^{-1}, d_2^{-1}, \ldots, d_n^{-1})$, the numbers d_i being those computed in the modified Gram-Schmidt algorithm. As noted previously, this arrangement of the computations avoids the calculation of square roots.

Another approach to the least-squares problem associated with the system $Ax = b$ is to use Lemma 1 *directly*. Thus, the quantity $\|Ax - b\|_2$ will be a minimum if

$$A^*(Ax - b) = 0$$

If the $m \times n$ matrix A is of rank n, then A^*A is a nonsingular $n \times n$ matrix (Problem 2), and in this case there is exactly one least-squares solution; it is determined uniquely by solving the nonsingular $n \times n$ system of so-called **normal equations**:

$$A^*Ax = A^*b \tag{9}$$

The matrix A^*A is Hermitian and positive definite (Problem 3). The Cholesky factorization may therefore be used to solve (9). If the rank of A is less than n, Equation (9) will be consistent but it may have many solutions.

The direct use of the normal equations for solving a least-squares problem is very appealing because of its conceptual simplicity. However, it is regarded as one of the *least* satisfactory methods to use on this problem. One reason for this judgment is that the condition number of A^*A may be considerably worse than that of A. The phenomenon is illustrated by this example:

$$A = \begin{bmatrix} 1 & 1 & 1 \\ \varepsilon & 0 & 0 \\ 0 & \varepsilon & 0 \\ 0 & 0 & \varepsilon \end{bmatrix} \qquad A^*A = \begin{bmatrix} 1+\varepsilon^2 & 1 & 1 \\ 1 & 1+\varepsilon^2 & 1 \\ 1 & 1 & 1+\varepsilon^2 \end{bmatrix} \tag{10}$$

In A, the nonzero parameter ε is all that prevents the rank from being 1. In A^*A, it is only ε^2 that prevents the matrix from having rank 1. Thus, the nonsingularity of A^*A is a more delicate matter for small ε. Of course *in a computer*, we can have A of rank 3 and A^*A of rank 1.

Householder's *QR*-Factorization

We turn now to the most useful of the orthogonal factorizations—namely, one due to Alston Householder and known by his name. The objective is to factor an $m \times n$ matrix A into a product

$$A = QR$$

where Q is an $m \times m$ unitary matrix and R is an $m \times n$ upper triangular matrix. The factorization algorithm actually produces

$$Q^* A = R$$

and Q^* is built up step by step as a product of unitary matrices having the special form

$$\begin{bmatrix} I_k & 0 \\ 0 & I_{m-k} - vv^* \end{bmatrix}$$

These are called **reflections** or **Householder transformations**. We do not make any assumptions about the rank of A.

First, we determine $v \in \mathbb{C}^m$ so that $I - vv^*$ is unitary and so that $(I - vv^*)A$ will begin to look like R. Specifically, its first column should be of the correct form—namely, $(\beta, 0, 0, \dots, 0)^T$. Let the original first column of A be denoted by A_1. We want $(I - vv^*)A_1 = \beta e^{(1)}$, and, by the proof of Lemma 2 of Section 5.2, this is accomplished as follows: First, select a complex number β such that $|\beta| = \|A_1\|_2$ and such that $\langle A_1, \beta e^{(1)} \rangle$ is real. Then put $v = \alpha(A_1 - \beta e^{(1)})$ with $\alpha = \sqrt{2}/\|A_1 - \beta e^{(1)}\|_2$. This description admits two choices for β, and we select the one for which there is less cancellation in computing the first component of v. To understand how this is done, write the complex numbers β and a_{11} in polar form:

$$\beta = \|A_1\|_2 e^{i\varphi} \qquad a_{11} = |a_{11}| e^{i\theta}$$

Then we have

$$\langle A_1, \beta e^{(1)} \rangle = a_{11}\overline{\beta} = |a_{11}|\, \|A_1\|_2 e^{i(\theta - \varphi)}$$

This must be real, and so $\theta - \varphi$ should be either 0 or π. If we choose $\theta - \varphi = \pi$, the first component of v will have no subtractive cancellation since

$$v_1 = \alpha(a_{11} - \beta) = \alpha\left(|a_{11}|e^{i\theta} - |\beta| e^{i(\theta - \pi)}\right) = \alpha\left(|a_{11}| + |\beta|\right)e^{i\theta}$$

Thus, we define β by

$$\beta = -\|A_1\|_2\, e^{i\theta} = -\|A_1\|_2\, a_{11}/|a_{11}|$$

The algorithm for obtaining the matrix U in this first step is as follows:

$\beta \leftarrow -(a_{11}/|a_{11}|)\|A_1\|_2$
$y \leftarrow A_1 - \beta e^{(1)}$
$\alpha \leftarrow \sqrt{2}/\|y\|_2$
$v \leftarrow \alpha y$
$U \leftarrow I - vv^*$

The succeeding steps in the QR-factorization are similar to the first step. After k steps, we shall have multiplied A on the left by k unitary matrices and the result will be a matrix having its first k columns of the correct form; that is, they have 0's below the diagonal. The situation can be summarized by

$$U_k U_{k-1} \cdots U_1 A = \begin{bmatrix} J & H \\ 0 & W \end{bmatrix}$$

in which J is an upper triangular $k \times k$ matrix, 0 is an $(m-k) \times k$ null matrix, H is $k \times (n-k)$, and W is $(m-k) \times (n-k)$. By the previous analysis, there is a vector $v \in \mathbb{C}^{m-k}$ such that $I - vv^*$ is a unitary matrix of order $m-k$ and such that $(I - vv^*)W$ has 0's below the initial element in its first column. Now observe that

$$\begin{bmatrix} I & 0 \\ 0 & I - vv^* \end{bmatrix} \begin{bmatrix} J & H \\ 0 & W \end{bmatrix} = \begin{bmatrix} J & H \\ 0 & (I - vv^*)W \end{bmatrix}$$

The first factor on the left in this equation is unitary and is what we denote by U_{k+1}.

The process described terminates when the $(n-1)$th column of R has been put into proper form. At that stage, we have an equation $Q^*A = R$, where Q^* denotes the product of all the unitary matrices that have been used as factors. Since Q is unitary, $A = QR$, as we wished. This is **Householder's QR-factorization**. The equation $Q^* = U_{n-1}U_{n-2} \cdots U_1$ leads to $Q = U_1^* U_2^* \cdots U_{n-1}^*$. From the form of U_k,

$$U_k = \begin{bmatrix} I_{k-1} & 0 \\ 0 & I_{n-k+1} - vv^* \end{bmatrix}$$

we see that U_k is Hermitian ($U_k^* = U_k$), and hence,

$$Q = U_1 U_2 \cdots U_{n-1}$$

Example 1 We illustrate the QR-factorization with this matrix:

$$A = \begin{bmatrix} 63 & 41 & -88 \\ 42 & 60 & 51 \\ 0 & -28 & 56 \\ 126 & 82 & -71 \end{bmatrix}$$

Solution In the first step, we compute β, which can be taken to be $-\|A_1\|_2$ since A_1 is real:

$$\beta = -\|A_1\|_2 = -\|(63,\ 42,\ 0,\ 126)^T\|_2 = -147$$

Next, we compute α:

$$\alpha = \sqrt{2}/\|A_1 - \beta e^{(1)}\|_2 = \sqrt{2}/\|(210,\ 42,\ 0,\ 126)^T\|_2 = 1/21\sqrt{70}$$

The vector v is therefore given by

$$v = \alpha(A_1 - \beta e^{(1)}) = (10,\ 2,\ 0,\ 6)^T/\sqrt{70}$$

The first unitary factor is then

$$U_1 = I - vv^* = \frac{1}{35}\begin{bmatrix} -15 & -10 & 0 & -30 \\ -10 & 33 & 0 & -6 \\ 0 & 0 & 35 & 0 \\ -30 & -6 & 0 & 17 \end{bmatrix}$$

The product $U_1 A$ is now computed to be

$$U_1 A = \frac{1}{35}\begin{bmatrix} -5145 & -3675 & 2940 \\ 0 & 1078 & 2989 \\ 0 & -980 & 1960 \\ 0 & -196 & 1127 \end{bmatrix}$$

In the second step, the corresponding calculations are

$$\begin{aligned} \beta &= -\|(30.8,\ -28,\ -5.6)^T\|_2 = -42 \\ \alpha &= \sqrt{2}/\|(72.8,\ -28,\ -5.6)^T\|_2 = 0.018085 \\ v &= \alpha(1.3166,\ -0.50637,\ -0.10127)^T \\ I - vv^* &= \begin{bmatrix} -0.73333 & 0.66667 & 0.13333 \\ 0.66667 & 0.74359 & -0.05128 \\ 0.13333 & -0.05128 & 0.98974 \end{bmatrix} \end{aligned}$$

The second unitary factor is therefore

$$U_2 = \begin{bmatrix} 1. & 0. & 0. & 0. \\ 0. & -0.73333 & 0.66667 & 0.13333 \\ 0. & 0.66667 & 0.74359 & -0.05128 \\ 0. & 0.13333 & -0.05128 & 0.98974 \end{bmatrix}$$

and we have

$$U_2 U_1 A = \begin{bmatrix} -147. & -105. & 84. \\ 0. & -42. & -21. \\ 0. & 0. & 96.9231 \\ 0. & 0. & 40.3846 \end{bmatrix}$$

In the last step, the analogous computations are

$$\begin{aligned} \beta &= -\|(96.9231,\ 40.3846)^T\|_2 = -105 \\ \alpha &= \sqrt{2}/\|(201.9231,\ 40.3846)^T\|_2 = 0.0068677 \end{aligned}$$

$$v = (1.38675,\ 0.27735)^T$$

$$I - vv^* = \begin{bmatrix} -0.92308 & -0.38462 \\ -0.38462 & 0.92308 \end{bmatrix}$$

The third unitary factor is therefore

$$U_3 = \begin{bmatrix} 1. & 0. & 0. & 0. \\ 0. & 1. & 0. & 0. \\ 0. & 0. & -0.92308 & -0.38462 \\ 0. & 0. & -0.38462 & 0.92308 \end{bmatrix}$$

The upper triangular matrix R is then

$$R = \begin{bmatrix} -147. & -105. & 84. \\ 0. & -42. & -21. \\ 0. & 0. & -105. \\ 0. & 0. & 0. \end{bmatrix} = 21 \begin{bmatrix} -7 & -5 & 4 \\ 0 & -2 & -1 \\ 0 & 0 & -5 \\ 0 & 0 & 0 \end{bmatrix}$$

Finally,

$$Q^* = U_3 U_2 U_1 = \begin{bmatrix} -0.42857 & -0.28571 & 0. & -0.85714 \\ 0.09524 & -0.71429 & 0.66667 & 0.19048 \\ 0.47619 & -0.57143 & -0.66667 & -0.04762 \\ -0.76190 & -0.28571 & -0.33333 & 0.47619 \end{bmatrix}$$

$$= \frac{1}{21} \begin{bmatrix} -9 & -6 & 0 & -18 \\ 2 & -15 & 14 & 4 \\ 10 & -12 & -14 & -1 \\ -16 & -6 & -7 & 10 \end{bmatrix}$$

We can verify that $A = QR$; that is,

$$\begin{bmatrix} 63 & 41 & -88 \\ 42 & 60 & 51 \\ 0 & -28 & 56 \\ 126 & 82 & -71 \end{bmatrix} = \begin{bmatrix} -9 & 2 & 10 & -16 \\ -6 & -15 & -12 & -6 \\ 0 & 14 & -14 & -7 \\ -18 & 4 & -1 & 10 \end{bmatrix} \begin{bmatrix} -7 & -5 & 4 \\ 0 & -2 & -1 \\ 0 & 0 & -5 \\ 0 & 0 & 0 \end{bmatrix} \quad \blacksquare$$

PROBLEMS 5.3

1. Prove that if $x \neq y$ and $\langle x, y\rangle$ is real, then a unitary matrix U satisfying $Ux = y$ is given by $U = I - vu^*$, with $v = x - y$ and $u = 2v/\|v\|_2^2$. Explain why this is a better method for constructing the Householder transformations. Assume $\|x\|_2 = \|y\|_2$.
2. Prove that if A is an $m \times n$ matrix of rank n, then A^*A is nonsingular.
3. Prove that if A is an $m \times n$ matrix of rank n, then A^*A is Hermitian and positive definite.

4. Let $\{u_1, u_2, \ldots, u_n\}$ be an orthonormal base for an inner-product space X. Prove that for $x, y \in X$,

(a) $\|x\|_2^2 = \sum_{i=1}^{n} |\langle x, u_i\rangle|^2$

(b) $\langle x, y\rangle = \sum_{i=1}^{n} \langle x, u_i\rangle\overline{\langle y, u_i\rangle}$

5. Let $\{u_1, u_2, \ldots, u_n\}$ be an orthonormal base for a subspace U in an inner-product space X. Define $P : X \to U$ by the equation $Px = \sum_{i=1}^{n}\langle x, u_i\rangle u_i$. Prove:

(a) P is linear.

(b) P is idempotent ($P^2 = P$).

(c) $Px = x$ if $x \in U$.

(d) $\|Px\|_2 \leqq \|x\|_2$ for all $x \in X$.

The map P is called the **orthogonal projection** of X onto U.

6. What is the determinant of a unitary matrix? What is the determinant of an orthogonal matrix?

7. What is the necessary and sufficient condition for an orthogonal set to be linearly independent? Prove that any orthonormal set is linearly independent.

8. For fixed u and x, what value of t makes the expression $\|u - tx\|_2$ a minimum? (The answer is to be valid in the complex case.)

9. Prove that the matrix having elements $\langle x_i, y_j\rangle$ is unitary if $\{x_1, x_2, \ldots, x_n\}$ and $\{y_1, y_2, \ldots, y_n\}$ are orthonormal bases for $\mathbb{C}^n$.

10. A matrix A such that $A^2 = I$ is called an **involution** or is said to be **involutory**. Find the necessary and sufficient conditions on u and v in order that $I - uv^*$ be an involution.

11. Let $v^*v = 2$. Show that the partitioned matrix

$$\begin{bmatrix} I & 0 \\ 0 & I - vv^* \end{bmatrix}$$

is an involution. (See Problem 10.)

12. Is a product of involutory matrices involutory?

13. If U and V are unitary, does it follow that

$$\begin{bmatrix} U & 0 \\ 0 & V \end{bmatrix}$$

is unitary?

14. **(a)** Prove that the set of all unitary matrices (of fixed order) is a multiplicative group.

(b) Prove that the set is closed under the operations $A \to A^T$, A^*, and $\overline{A}$.

15. Use Householder's algorithm to find the QR-factorization of

$$\begin{bmatrix} 0 & -4 \\ 0 & 0 \\ -5 & -2 \end{bmatrix}$$

16. Prove that if Q is unitary, then for all x and y,

$$\|x\|_2 = \|Qx\|_2 \text{ and } \langle x, y\rangle = \langle Qx, Qy\rangle$$

Thus, Q (when considered as a transformation) preserves lengths, distances, and angles in Euclidean space. Compute $\|Q\|_2$ using the matrix norm subordinate to the Euclidean norm.

17. Prove that A is unitary if and only if its rows form an orthonormal system. (Here A is a square matrix.)

18. Let A be an $m \times n$ matrix, b an m-vector, and $\alpha > 0$. Using the Euclidean norm, define

$$F(x) = \|Ax - b\|_2^2 + \alpha\|x\|_2^2$$

Prove that $F(x)$ is a minimum when x is a solution of the equation

$$(A^T A + \alpha I)x = A^T b$$

Prove that when x is so defined,

$$F(x + h) = F(x) + (Ah)^T Ah + \alpha h^T h$$

19. Let D be a diagonal matrix and U a unitary matrix. Under what further hypotheses on D can we infer that DU is unitary?

20. If U and AU are unitary, what conclusion can be drawn about A?

21. Show that in solving the least-squares problem for the equation $Ax = b$, we can replace the normal equations by $CAx = Cb$, where C is any $n \times m$ matrix row-equivalent to A^T. *Hint:* Recall that two matrices G and H are row-equivalent if there is a nonsingular matrix F for which $G = FH$.

22. Let A be an $m \times n$ matrix of rank n. Let b be any point in $\mathbb{R}^m$. Show that the sets

$$K_\lambda = \left\{x \in \mathbb{R}^n : \|Ax - b\|_2 \leqq \lambda\right\}$$

are closed and bounded. (The norms can be arbitrary on $\mathbb{R}^m$ and $\mathbb{R}^n$.)

23. (Continuation) Assume the hypotheses in Problem 22. Prove that if $\lambda = 2\|b\|_2$, then

$$\inf_{x\in\mathbb{R}^n} \|Ax - b\|_2 = \inf_{x\in K_\lambda} \|Ax - b\|_2$$

24. (Continuation) If A is any $m \times n$ matrix of rank n, then the least-squares solution of $Ax = b$ satisfies the inequality

$$\|x\|_2 \leqq 2\|b\|_2 \, \|B\|_2$$

where B is any left inverse of A. Here the Euclidean vector norm is used in $\mathbb{R}^n$ and $\mathbb{R}^m$, and $\|B\|_2$ denotes the corresponding subordinate matrix norm.

25. Let A be an $m \times n$ matrix of unspecified rank. Let $b \in \mathbb{R}^m$, and let

$$\rho = \inf \left\{ \|Ax - b\| : x \in \mathbb{R}^n \right\}$$

Prove that this infimum is *attained*. In other words, prove the existence of an x for which $\|Ax - b\| = \rho$. In this problem, the norm is an arbitrary one defined on $\mathbb{R}^m$.

26. (Continuation) Adopt the assumptions in Problem 25 and prove that the equation $A^T Ax = A^T b$ has a solution. Do *not* make any assumption about the rank of A.

27. Prove that for any system of equations, $Ax = b$, the system $A^T Ax = A^T b$ has a solution. Make no assumption about the relative magnitudes of m and n or about the rank of A.

28. Give an example of vectors x and y for which $\|x + y\|_2^2 = \|x\|_2^2 + \|y\|_2^2$ and $\langle x, y \rangle \neq 0$.

29. Let $AB = Q$, where Q is $m \times n$, B is $n \times n$, and $Q^* Q = I$. Show that QQ^* is the orthogonal projection of $\mathbb{R}^m$ onto the range of A. (See Problem 5.)

30. Find the least-squares solution of the system

$$[x \; y] \begin{bmatrix} 3 & 2 & 1 \\ 2 & 3 & 2 \end{bmatrix} = \begin{bmatrix} 3 & 0 & 1 \end{bmatrix}$$

31. It is asserted that the solution of Problem 30 is $(29/21, -2/3)$. How can this be verified without solving for x and y?

32. Let A be an $(n+1) \times n$ matrix of rank n, and let z be a nonzero vector orthogonal to the columns of A. Show that the equation $Ax + \lambda z = b$ has a solution in x and λ. Show that the x-vector obtained in this way is the least-squares solution of the equation $Ax = b$.

33. The following example has been given by Noble and Daniel [1988] to indicate the effect of roundoff error in using the *unmodified Gram-Schmidt method*. Let ε be a small positive number such that $1 + \varepsilon$ and $3 + 2\varepsilon$ are machine numbers but $3 + 2\varepsilon + \varepsilon^2$ is computed to be $3 + 2\varepsilon$. Now let the Gram-Schmidt process be applied to the three vectors

$$v_1 = (1 + \varepsilon, \; 1, \; 1)^T \quad v_2 = (1, \; 1 + \varepsilon, \; 1)^T \quad v_3 = (1, \; 1, \; 1 + \varepsilon)^T$$

Verify that the supposedly orthonormal base produced by the computer will be

$$x_1 = [1/\sqrt{3+2\varepsilon}\,]v_1 \quad x_2 = [1/\sqrt{2}\,](-1, 1, 0)^T \quad x_3 = [1/\sqrt{2}\,](-1, 0, 1)^T$$

Notice that $\langle x_2, x_3\rangle = 1/2$.

34. Which elementary row operations on the system $Ax = b$ leave all least-squares solutions unchanged?

35. Let A be an $m \times n$ matrix in which $m > n$. Let A have the QR-factorization $A = QR$. Let Q' be the $m \times n$ matrix that results by deleting the last $m - n$ columns of Q. Let R' be the $n \times n$ matrix that results from deleting the last $m - n$ rows of R. Prove that $A = Q'R'$.

36. Provide details in the following method for obtaining the QR-factorization of an $m \times n$ matrix A. Denote the jth columns in A, Q, and R by A_j, Q_j, and R_j. Show that $A_j = \sum_{k=1}^{j} r_{kj} Q_k$ and that $r_{kj} = \langle A_j, Q_k\rangle$. Use this equation to determine Q_1 and r_{11}. Then show how to get Q_j and R_j once the following have been obtained: $Q_1, Q_2, \ldots, Q_{j-1}$ and $R_1, R_2, \ldots, R_{j-1}$.

37. Find the QR-factorization of the matrix

$$\begin{bmatrix} 3 & 2 & 3 \\ 4 & 5 & 6 \end{bmatrix}$$

38. Determine $\kappa_\infty(A^*A)$ using the matrices of Equation (10). What happens as $\varepsilon \to 0$?

39. Apply the Gram-Schmidt process to the three vectors $(3, 4, 0)$, $(1, 1, 1)$, and $(1, 2, 0)$, in that order.

COMPUTER PROBLEMS 5.3

1. Program both the Gram-Schmidt algorithm and the modified Gram-Schmidt algorithm. Then test the two to see which is better. The first test could involve a 20×10 matrix whose elements are random numbers uniformly distributed in the interval [0,1]. The second test could involve a 20×10 matrix whose elements are generated by some elementary function such as

$$a_{ij} = \left(\frac{2i - 21}{19}\right)^{j-1} \qquad (1 \leqq i \leqq 20,\ 1 \leqq j \leqq 10)$$

In each case, generate from A a matrix B whose columns should be orthonormal. Examine $B^T B$ to see how close it is to the identity matrix. For further information on such tests, see Rice [1966].

2. Write a subroutine or procedure to carry out the modified Gram-Schmidt algorithm. Input to the subroutine will be an $m \times n$ matrix A of rank n, and the output should be the matrices B and T.

3. (Continuation) Write a subroutine or procedure to solve a system of equations $Ax = b$ in the least-squares sense. This procedure should *call*—that is, refer to—the procedure in Problem 2.

4. Program and test the Householder QR-factorization.

*5.4 Singular-Value Decomposition and Pseudoinverses

The singular-value decomposition is another matrix factorization that has many applications. We begin our study of it with a theorem that describes its form and asserts its existence.

THEOREM 1 *An arbitrary complex $m \times n$ matrix A can be factored as*

$$A = PDQ$$

where P is an $m \times m$ unitary matrix, D is an $m \times n$ diagonal matrix, and Q is an $n \times n$ unitary matrix.

Proof The matrix A^*A is an $n \times n$ Hermitian matrix. It is also positive semidefinite since

$$x^*(A^*A)x = (Ax)^*(Ax) \geqq 0$$

It follows that the eigenvalues of A^*A are nonnegative. (See Problem 39.) Let them be denoted by $\sigma_1^2, \sigma_2^2, \dots, \sigma_n^2$. (In this list, each σ_i^2 is repeated according to its multiplicity as a root of the characteristic equation.) Furthermore, we order the σ_i's so that $\sigma_1^2, \sigma_2^2, \dots, \sigma_r^2$ are positive and $\sigma_{r+1}^2, \sigma_{r+2}^2, \dots, \sigma_n^2$ are 0. Let $\{u_1, u_2, \dots, u_n\}$ be an orthonormal set of eigenvectors for A^*A, arranged so that

$$A^*Au_i = \sigma_i^2 u_i$$

Then

$$\|Au_i\|_2^2 = u_i^*A^*Au_i = u_i^*\sigma_i^2 u_i = \sigma_i^2$$

This shows that $Au_i = 0$ when $i \geqq r + 1$. Observe that

$$r = \text{rank}(A^*A) \leqq \min\{\text{rank}(A^*), \text{rank}(A)\} \leqq \min\{m, n\}$$

We form an $n \times n$ matrix Q whose rows are $u_1^*, u_2^*, \dots, u_n^*$. Next, define

$$v_i = \sigma_i^{-1}Au_i \qquad (1 \leqq i \leqq r)$$

The v_i's form an orthonormal system because for $1 \leqq i, j \leqq r$, we have

$$v_i^*v_j = \sigma_i^{-1}(Au_i)^*\sigma_j^{-1}(Au_j) = (\sigma_i\sigma_j)^{-1}(u_i^*A^*Au_j) = (\sigma_i\sigma_j)^{-1}(u_i^*\sigma_j^2u_j) = \delta_{ij}$$

We select additional vectors v_i so that $\{v_1, v_2, \ldots, v_m\}$ is an orthonormal base for $\mathbb{C}^m$. Let P be the $m \times m$ matrix whose columns are $v_1, v_2, \ldots, v_m$. Let D be the $m \times n$ matrix having $\sigma_1, \sigma_2, \ldots, \sigma_r$ on its diagonal and 0's elsewhere. Then

$$A = PDQ$$

To prove this, we show that

$$P^*AQ^* = D$$

Now

$$(P^*AQ^*)_{ij} = v_i^* A u_j$$

This will be 0 when $j \geqq r + 1$. If $j \leqq r$, this term is $v_i^* \sigma_j v_j = \sigma_j \delta_{ij}$. ■

The numbers $\sigma_1, \sigma_2, \ldots, \sigma_n$ (taken to be nonnegative) are called the **singular values** of A. They are, then, the nonnegative square roots of the eigenvalues of A^*A. The factorization $A = PDQ$ in Theorem 1 is a **singular-value decomposition.**

Notice that in the proof, arbitrary choices are made in certain steps. For example, the ordering of $\sigma_1, \sigma_2, \ldots, \sigma_r$ is arbitrary, and the vectors $v_{r+1}, v_{r+2}, \ldots, v_m$ allow some choice. Hence, a matrix will always have many singular-value decompositions.

Example 1 Find a singular-value decomposition of the following matrix:

$$A = \begin{bmatrix} 7 & 0 & 0 & 0 \\ 0 & 3 & 0 & 0 \\ 0 & 0 & 0 & 0 \end{bmatrix}$$

Solution We have

$$A^*A = \begin{bmatrix} 49 & 0 & 0 & 0 \\ 0 & 9 & 0 & 0 \\ 0 & 0 & 0 & 0 \\ 0 & 0 & 0 & 0 \end{bmatrix}$$

Thus, $\sigma_1 = 7$, $\sigma_2 = 3$, $\sigma_3 = 0$, and $\sigma_4 = 0$. Following the proof of the theorem, we form these matrices:

$$P = \begin{bmatrix} 1 & 0 & 0 \\ 0 & 1 & 0 \\ 0 & 0 & 1 \end{bmatrix} \qquad D = \begin{bmatrix} 7 & 0 & 0 & 0 \\ 0 & 3 & 0 & 0 \\ 0 & 0 & 0 & 0 \end{bmatrix} \qquad Q = \begin{bmatrix} 1 & 0 & 0 & 0 \\ 0 & 1 & 0 & 0 \\ 0 & 0 & 1 & 0 \\ 0 & 0 & 0 & 1 \end{bmatrix}$$

The reader can verify that another singular-value decomposition of A is

$$\begin{bmatrix} 7 & 0 & 0 & 0 \\ 0 & 3 & 0 & 0 \\ 0 & 0 & 0 & 0 \end{bmatrix} = \begin{bmatrix} 0 & 1 & 0 \\ 1 & 0 & 0 \\ 0 & 0 & -1 \end{bmatrix} \begin{bmatrix} 3 & 0 & 0 & 0 \\ 0 & 7 & 0 & 0 \\ 0 & 0 & 0 & 0 \end{bmatrix} \begin{bmatrix} 0 & 1 & 0 & 0 \\ 1 & 0 & 0 & 0 \\ 0 & 0 & 0 & 1 \\ 0 & 0 & 1 & 0 \end{bmatrix}$$ ■

Example 2 Find a singular-value decomposition for the matrix

$$\begin{bmatrix} 0 & -1.6 & 0.6 \\ 0 & 1.2 & 0.8 \\ 0 & 0 & 0 \\ 0 & 0 & 0 \end{bmatrix}$$

Solution Again, we follow the proof of the theorem. We find that

$$A^*A = \begin{bmatrix} 0 & 0 & 0 \\ 0 & 4 & 0 \\ 0 & 0 & 1 \end{bmatrix}$$

and we put $\sigma_1 = 1, \sigma_2 = 2$, and $\sigma_3 = 0$. (One other ordering is possible.) Selecting eigenvectors and forming Q, we have (as one of several choices)

$$Q = \begin{bmatrix} 0 & 0 & 1 \\ 0 & -1 & 0 \\ 1 & 0 & 0 \end{bmatrix}$$

Then

$$v_1 = Au_1 = (0.6,\ 0.8,\ 0,\ 0)^*$$
$$v_2 = \frac{1}{2}Au_2 = (0.8,\ -0.6,\ 0,\ 0)^*$$

There is some freedom in the choice of v_3 and v_4. The simplest choice is

$$v_3 = (0,\ 0,\ 1,\ 0)^* \qquad v_4 = (0,\ 0,\ 0,\ 1)^*$$

Thus, one singular-value decomposition is

$$\begin{bmatrix} 0 & -1.6 & 0.6 \\ 0 & 1.2 & 0.8 \\ 0 & 0 & 0 \\ 0 & 0 & 0 \end{bmatrix} = \begin{bmatrix} 0.6 & 0.8 & 0 & 0 \\ 0.8 & -0.6 & 0 & 0 \\ 0 & 0 & 1 & 0 \\ 0 & 0 & 0 & 1 \end{bmatrix} \begin{bmatrix} 1 & 0 & 0 \\ 0 & 2 & 0 \\ 0 & 0 & 0 \\ 0 & 0 & 0 \end{bmatrix} \begin{bmatrix} 0 & 0 & 1 \\ 0 & -1 & 0 \\ 1 & 0 & 0 \end{bmatrix}$$ ■

Pseudoinverse

For an $m \times n$ matrix of the form

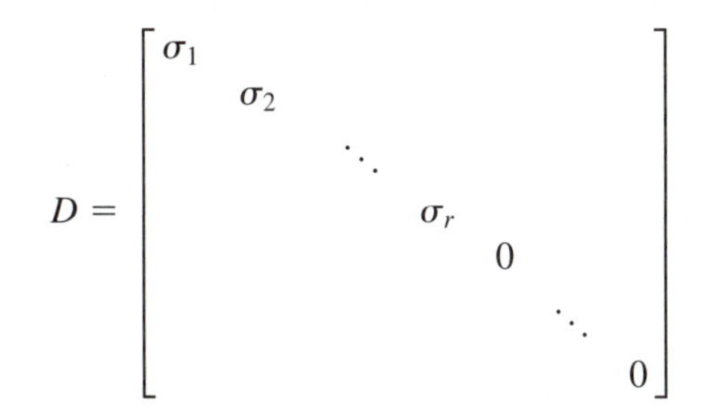

in which each σ_i is positive, we define its **pseudoinverse** to be the $n \times m$ matrix

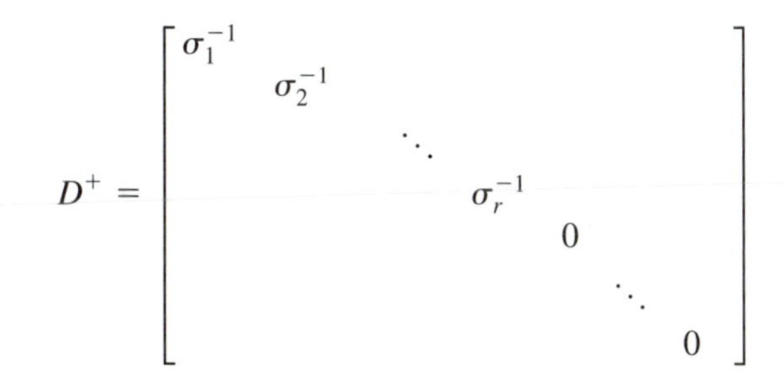

In all of these matrices, elements not displayed are 0's. The **pseudoinverse** of a general matrix A is defined by first taking a singular-value decomposition

$$A = PDQ$$

and then putting

$$A^+ = Q^* D^+ P^*$$

We shall see later that the pseudoinverse of a matrix is uniquely determined, although the singular-value decomposition is not unique.

Example 3 What is the pseudoinverse of the matrix A in Example 1?

Solution Since P and Q are identity matrices for this example, it follows immediately that

$$A^+ = \begin{bmatrix} 7^{-1} & 0 & 0 \\ 0 & 3^{-1} & 0 \\ 0 & 0 & 0 \\ 0 & 0 & 0 \end{bmatrix}$$

■

Example 4 Find the pseudoinverse of the matrix A in Example 2.

Solution The results of Example 2 give us

$$A^+ = \begin{bmatrix} 0 & 0 & 1 \\ 0 & -1 & 0 \\ 1 & 0 & 0 \end{bmatrix} \begin{bmatrix} 1 & 0 & 0 & 0 \\ 0 & 0.5 & 0 & 0 \\ 0 & 0 & 0 & 0 \end{bmatrix} \begin{bmatrix} 0.6 & 0.8 & 0 & 0 \\ 0.8 & -0.6 & 0 & 0 \\ 0 & 0 & 1 & 0 \\ 0 & 0 & 0 & 1 \end{bmatrix}$$

$$= \begin{bmatrix} 0 & 0 & 0 & 0 \\ -0.4 & 0.3 & 0 & 0 \\ 0.6 & 0.8 & 0 & 0 \end{bmatrix}$$

■

Inconsistent and Underdetermined Systems

The principal application of the pseudoinverse is to systems of equations that are inconsistent or have nonunique solutions. Consider then a system of equations written in matrix form as

$$Ax = b$$

where A is $m \times n$, x is $n \times 1$, and b is $m \times 1$. The **minimal solution** of this problem is defined as follows:

(a) If the system is consistent and has a unique solution, x, then the minimal solution is defined to be x.
(b) If the system is consistent and has a set of solutions, then the minimal solution is the element of this set having the least Euclidean norm.
(c) If the system is inconsistent and has a unique least-squares solution x, then the minimal solution is defined to be x.
(d) If the system is inconsistent and has a set of least-squares solutions, then the minimal solution is the element of this set having the least Euclidean norm.

An alternative definition proceeds as follows: Using the Euclidean norm, let

$$\rho = \inf \left\{ \|Ax - b\|_2 : x \in \mathbb{C}^n \right\}$$

Then the *minimal solution* of equation $Ax = b$ is the element of least norm in the set $K = \left\{ x : \|Ax - b\|_2 = \rho \right\}$. We see immediately that this encompasses all four cases described earlier. For example, if $\rho = 0$, we have cases **(i)** and **(ii)**, whereas cases **(iii)** and **(iv)** correspond to $\rho > 0$.

THEOREM 2 *The minimal solution of the equation $Ax = b$ is given by the pseudoinverse*

$$x = A^+ b$$

Proof Let a singular-value decomposition of A be $A = PDQ$. Let

$$c = P^* b \qquad \text{and} \qquad y = Qx$$

As x runs over $\mathbb{C}^n$, so does y because Q is *surjective*; that is, it maps $\mathbb{C}^n$ *onto* $\mathbb{C}^n$. Therefore,

$$\begin{aligned}\rho &= \inf_x \|Ax - b\|_2 = \inf_x \|PDQx - b\|_2 = \inf_x \|P^*(PDQx - b)\|_2 \\ &= \inf_x \|DQx - P^*b\|_2 = \inf_y \|Dy - c\|_2\end{aligned}$$

Now from the special nature of the matrix D, we have

$$\|Dy - c\|_2^2 = \sum_{i=1}^{r}(\sigma_i y_i - c_i)^2 + \sum_{i=r+1}^{m} c_i^2$$

This quantity is minimized by letting $y_i = c_i/\sigma_i$ for $1 \leqq i \leqq r$ and by permitting $y_{r+1}, y_{r+2}, \dots, y_n$ to be arbitrary. Thus,

$$\rho = \left(\sum_{i=r+1}^{m} c_i^2\right)^{1/2}$$

Among all the y-vectors that yield this minimum value ρ, the vector of least norm has $y_{r+1} = y_{r+2} = \cdots = y_n = 0$. This vector is given by

$$y = D^+c$$

The minimal solution of our problem is therefore

$$x = Q^*y = Q^*D^+c = Q^*D^+P^*b = A^+b \qquad \blacksquare$$

The pseudoinverse plays the same role for inconsistent or underdetermined systems as the inverse does for invertible systems. It should be noted that the minimal solution of any equation $Ax = b$ is unique because the set K is convex and has a unique element of least norm.

Example 5 Find the minimal solution of the system

$$\begin{cases} 0x - 1.6y + 0.6z = 5 \\ 0x + 1.2y + 0.8z = 7 \\ 0x + 0y + 0z = 3 \\ 0x + 0y + 0z = -2 \end{cases}$$

Solution The coefficient matrix is the A of Example 2. Its pseudoinverse was found in Example 4. Hence, the minimal solution is

$$A^+b = \begin{bmatrix} 0 & 0 & 0 & 0 \\ -0.4 & 0.3 & 0 & 0 \\ 0.6 & 0.8 & 0 & 0 \end{bmatrix} \begin{bmatrix} 5 \\ 7 \\ 3 \\ -2 \end{bmatrix} = \begin{bmatrix} 0.0 \\ 0.1 \\ 8.6 \end{bmatrix} \qquad \blacksquare$$

Penrose Properties

The pseudoinverse has some (but not all) of the properties of an inverse. For example, we cannot expect $A^+A = I$ to be true, if $n > m$, because the ranks of A^+, A, and A^+A are at most m, whereas I is $n \times n$. However, equations such as $AA^+A = A$ *are* true for arbitrary A. The next theorem concerns four such equations in a general setting.

THEOREM 3 *Corresponding to any matrix A there exists at most one matrix X having these four properties:*

(i) $AXA = A$
(ii) $XAX = X$
(iii) $(AX)^* = AX$
(iv) $(XA)^* = XA$

Proof Let X and Y be two matrices having properties **(i)–(iv)**. Then by systematic use of these properties as indicated, we have

$$
\begin{aligned}
X &= XAX && \textbf{(ii)} \\
&= XAYAX && \textbf{(i)} \\
&= XAYAYAYAX && \textbf{(i)} \\
&= (XA)^*(YA)^*Y(AY)^*(AX)^* && \textbf{(iv) and (iii)} \\
&= A^*X^*A^*Y^*YY^*A^*X^*A^* && \\
&= (AXA)^*Y^*YY^*(AXA)^* && \\
&= A^*Y^*YY^*A^* && \textbf{(i)} \\
&= (YA)^*Y(AY)^* && \\
&= YAYAY && \textbf{(iv) and (iii)} \\
&= YAY && \textbf{(ii)} \\
&= Y && \textbf{(ii)}
\end{aligned}
$$

■

This theorem was given by R. Penrose [1955], and the conditions **(i)–(iv)** are now known as the **Penrose properties**.

THEOREM 4 *The pseudoinverse of a matrix has the four Penrose properties. Hence, each matrix has a unique pseudoinverse.*

Proof Let A be any matrix and let its singular-value decomposition be

$$A = PDQ$$

Then

$$A^+ = Q^*D^+P^*$$

If A is $m \times n$, then so is D, and D has the form

$$D_{ij} = \begin{cases} \sigma_i & \text{if } i = j \leqq r \\ 0 & \text{otherwise} \end{cases}$$

From this we can prove that

$$DD^+D = D$$

To do so, we write

$$(DD^+D)_{ij} = \sum_{\nu=1}^{n} D_{i\nu} \sum_{\mu=1}^{m} D^+_{\nu\mu} D_{\mu j}$$

The right-hand side will be 0 unless $i \leqq r$ and $j \leqq r$ because of the presence of terms $D_{i\nu}$ and $D_{\mu j}$. Thus, we assume $i \leqq r$ and $j \leqq r$ and continue, simplifying the right-hand side to

$$\sum_{\nu=1}^{r} D_{i\nu} \sum_{\mu=1}^{r} D^+_{\nu\mu} D_{\mu j} = \sigma_i \sum_{\mu=1}^{r} D^+_{i\mu} D_{\mu j} = \sigma_i \sigma_i^{-1} D_{ij} = D_{ij}$$

By similar reasoning, we prove that D^+ has the remaining three Penrose properties relative to D. Then it is a simple matter to prove these four properties for A^+. For example, the first property is proved as follows:

$$\begin{aligned} AA^+A &= PDQQ^*D^+P^*PDQ \\ &= PDD^+DQ \\ &= PDQ = A \end{aligned}$$

Proofs of the remaining properties are left as Problems 28 and 29. ■

Singular-Value Decomposition Theorems

THEOREM 5 *Let A have the singular-value decomposition $A = PDQ$, as described in the proof of Theorem 1. Then we have:*

(i) *The* rank *of A is r.*
(ii) *$\{u_{r+1}, u_{r+2}, \ldots, u_n\}$ is an orthonormal base for the null space of A.*
(iii) *$\{v_1, v_2, \ldots, v_r\}$ is an orthonormal base for the range of A.*
(iv) $\|A\|_2 = \max_{1 \leqq i \leqq n} |\sigma_i|$

Proof Since P and Q are nonsingular, the rank of A is the same as the rank of D, which is obviously r. If $r < i \leqq n$, then $Au_i = 0$, as shown in the proof of Theorem 1. Since A has rank r, the null space of A has dimension $n - r$. Hence, $\{u_{r+1}, u_{r+2}, \ldots, u_n\}$ is a basis for the null space. Since the rank of A is r, the range of A has dimension

r. As shown in the proof of Theorem 1, $v_i = \sigma_i^{-1}Au_i$ and $\{v_1, v_2, \dots, v_r\}$ is an orthonormal base for the range of the matrix A. Since P and Q are unitary matrices, they act as isometries (norm-preserving maps) on $\mathbb{C}^m$ and $\mathbb{C}^n$, respectively. Hence,

$$\begin{aligned}
\|A\|_2 &= \sup\left\{\|Ax\|_2 : \|x\|_2 = 1\right\} \\
&= \sup\left\{\|PDQx\|_2 : \|x\|_2 = 1\right\} \\
&= \sup\left\{\|Dy\|_2 : \|y\|_2 = 1\right\} \\
&= \sup\left\{\sqrt{(\sigma_1 y_1)^2 + \cdots + (\sigma_n y_n)^2} : \|y\|_2 = 1\right\} \\
&= \left[\sup\left\{\sigma_1^2 y_1^2 + \cdots + \sigma_n^2 y_n^2 : \sum_{i=1}^n y_i^2 = 1\right\}\right]^{1/2} \\
&= \left[\max_{1 \leqq i \leqq n} \sigma_i^2\right]^{1/2} = \max_{1 \leqq i \leqq n} |\sigma_i| \qquad \blacksquare
\end{aligned}$$

A geometric interpretation of the singular-value decomposition is inherent in the proof of Theorem 1. A more easily visualized interpretation is given in the following theorem.

THEOREM 6 *Let L be a linear transformation from $\mathbb{C}^m$ to $\mathbb{C}^n$. Then there are orthonormal bases $\{u_1, u_2, \dots, u_m\}$ for $\mathbb{C}^m$ and $\{v_1, v_2, \dots, v_n\}$ for $\mathbb{C}^n$ such that*

$$Lu_i = \begin{cases} \sigma_i v_i & \text{if } 1 \leqq i \leqq \min(m, n) \\ 0 & \text{if } \min(m, n) < i \leqq m \end{cases}$$

Proof First, let $\{e_1, e_2, \dots, e_m\}$ be the familiar basis for $\mathbb{C}^m$, and $\{\bar{e}_1, \bar{e}_2, \dots, \bar{e}_n\}$ be the one for $\mathbb{C}^n$. Define the $m \times n$ matrix A by the equation $Le_i = \sum_{j=1}^n A_{ij}\bar{e}_j$, $1 \leqq i \leqq m$. Select a singular-value decomposition for A, say $A = PDQ$. Let the rows of P^* be denoted by $u_1, u_2, \dots, u_m$. Let the columns of Q^* be denoted by $v_1, v_2, \dots, v_m$. Then, as we shall see, $Lu_i = \sigma_i v_i$, where σ_i are the diagonal elements of the $m \times n$ matrix D. Here is the calculation that proves this. Fix k in the range $1, 2, \dots, m$. Then

$$\begin{aligned}
Lu_k &= L\left(\sum_{i=1}^m \langle u_k, e_i\rangle e_i\right) = \sum_{i=1}^m \langle u_k, e_i\rangle Le_i \\
&= \sum_{i=1}^m (P^*)_{ki} \sum_{j=1}^n A_{ij}\bar{e}_j = \sum_{j=1}^n \sum_{i=1}^m (P^*)_{ki} A_{ij}\bar{e}_j \\
&= \sum_{j=1}^n (P^*A)_{kj}\bar{e}_j = \sum_{j=1}^n (P^*A)_{kj} \sum_{s=1}^n \langle \bar{e}_j, v_s\rangle v_s \\
&= \sum_{s=1}^n \sum_{j=1}^n (P^*A)_{kj}(Q^*)_{js} v_s = \sum_{s=1}^n (P^*AQ^*)_{ks} v_s
\end{aligned}$$

Now the equation $A = PDQ$ leads to $P^*AQ^* = D$. The matrix D is a diagonal $m \times n$ matrix having the singular values of A, $\sigma_1, \sigma_2, \ldots, \sigma_\ell$, on its diagonal, with $\ell = \min(m, n)$. Hence,

$$Lu_k = \sum_{s=1}^{n} D_{ks} v_s = \begin{cases} \sigma_k v_k & \text{if } k \leqq \ell \\ 0 & \text{if } k > \ell \end{cases}$$

■

PROBLEMS 5.4

1. Find the singular-value decompositions of these matrices:

$$\begin{bmatrix} 4 & 0 & 0 \\ 0 & 0 & 0 \\ 0 & 0 & 7 \\ 0 & 0 & 0 \end{bmatrix} \qquad [2 \quad 1] \qquad \begin{bmatrix} 5 \\ -4 \end{bmatrix}$$

2. Find the minimal solutions of these systems of equations:

(a) $x_1 + x_2 = b_1$

(b) $\begin{cases} x_1 = b_1 \\ x_1 = b_2 \\ x_1 = b_3 \end{cases}$

(c) $\begin{cases} 4x_1 = b_1 \\ 0x_1 = b_2 \\ 7x_3 = b_3 \\ 0x_2 = b_4 \end{cases}$

3. Find A^+ in the case that AA^* is invertible.

4. Find A^+ in the case that $A^*A = I$.

5. Find A^+ in the case that A is Hermitian and **idempotent**; that is, $A^* = A$ and $A^2 = A$.

6. Prove that if A is Hermitian, then so is A^+.

7. If A is Hermitian and positive definite, what is its singular-value decomposition?

8. Prove that the pseudoinverse is a discontinuous function of a matrix. *Hint:* Compute the pseudoinverse of the matrix

$$\begin{bmatrix} 1 & 0 \\ 0 & \varepsilon \\ 0 & 0 \end{bmatrix}$$

9. If A is Hermitian, what is the relationship between its eigenvalues and its singular values?

10. Prove these properties of the pseudoinverse:

(a) $A^{++} = A$

(b) $A^{+*} = A^{*+}$

11. Prove these properties of the pseudoinverse:

(a) $(AA^*)^+ = A^{+*}A^+$

(b) $A^+ = A^*(AA^*)^+$

12. Show by a suitable example that in general $(AB)^+ \neq B^+A^+$.

13. Prove:

> **Theorem** *If A is an $m \times n$ matrix of rank r, with $m \geqq n \geqq r$, then A can be factored as $A = VSU$, in which V is an $m \times r$ matrix with orthonormal columns, S is a nonsingular $r \times r$ diagonal matrix, and U is an $r \times n$ matrix with orthonormal rows.*

This has been termed the **economical version** of the **singular-value decomposition**.

14. Refer to the proof of Theorem 1 and prove that

$$A = \sum_{j=1}^{r} \sigma_j v_j u_j^*$$

15. Let A be an $m \times n$ matrix of rank r, with $m \geqq n \geqq r$ and singular-value decomposition $A = PDQ$. Prove that the system of equations $Ax = b$ is consistent if and only if $(P^*b)_i = 0$ for $r < i \leqq m$.

16. Let A be an $m \times n$ matrix of rank r. Define u_i, v_i, and σ_i as in the proof of Theorem 1. Show that

$$A^+ = \sum_{j=1}^{r} \sigma_j^{-1} u_j v_j^*$$

(Compare with Problem 14.)

17. Use the result of Problem 14 to show that if the singular-value decomposition of A is available, then Ax can be computed at a cost of $(n + m + 1)r$ multiplications and $r(n + m - 1) - m$ additions. Compare this to the straightforward multiplication, which costs nm multiplications and $(n - 1)m$ additions.

18. Prove that if A is Hermitian and positive semidefinite, then its eigenvalues are identical with its singular values.

19. Prove that if two matrices are unitarily equivalent, then their singular values are the same. (Two matrices A and B are **unitarily equivalent** if $A = UBV$ for suitable unitary matrices U and V.)

20. Let A be an $n \times n$ matrix having singular values $\sigma_1, \sigma_2, \ldots, \sigma_n$. Prove that the determinant of A is $\det(A) = \pm\sigma_1\sigma_2 \cdots \sigma_n$.

21. Let $\|A\|_2$ denote the matrix norm subordinate to the Euclidean vector norm. Let the singular values of A be $\sigma_1 \geqq \sigma_2 \geqq \cdots \geqq \sigma_n$. Show that $\|A\|_2 = \sigma_1$.

22. Let A be a square matrix with singular-value decomposition $A = PDQ$. Prove that the characteristic polynomial of A is $\pm \det(D - \lambda P^*Q^*) = 0$.

23. Let A be an $m \times n$ matrix, and let X be an $n \times m$ matrix that has the four Penrose properties relative to A. Prove that the minimal solution to the system $Ax = b$ is Xb.

24. Prove that if A is real, then it has a real singular-value decomposition and a real pseudoinverse.

25. If the sum in Problem 14 is truncated with k summands, we obtain an approximation to A. Prove that the error in so doing satisfies

$$\left\| A - \sum_{i=1}^{k} \sigma_i v_i u_i^* \right\|_2 = \sigma_{k+1}$$

where the matrix norm subordinate to the Euclidean vector norm has been used.

26. Prove that the squares of the elements in $v_i u_i^*$ sum to 1. (Notation is as in the proof of Theorem 1.)

27. Suppose that $A = UDV$, where U is $m \times m$ unitary, V is $n \times n$ unitary, and D is $m \times n$ diagonal. Show that $|d_{ii}|^2$ $(1 \leqq i \leqq n)$ are the eigenvalues of A^*A.

28. Prove the three remaining Penrose properties for D^+.

29. (Continuation) Complete the proof of Theorem 4.

30. Prove that if the $m \times n$ matrix A has rank n, then $A^+ = (A^*A)^{-1}A^*$.

31. Prove that the pseudoinverse of a diagonal $m \times n$ matrix is a diagonal $n \times m$ matrix.

32. Find the pseudoinverse of an arbitrary $m \times 1$ matrix and of an arbitrary $1 \times n$ matrix.

33. Show that the orthogonal projection of $\mathbb{C}^m$ onto the column space of A is AA^+.

34. Prove that if A is symmetric, then so is A^+.

35. Prove that the equation $(AB)^+ = B^+A^+$ holds if A and B are of full rank. An $m \times n$ matrix A is said to have full rank if $\text{rank}(A) = \min(m, n)$.

36. Find the pseudoinverse of uv^*, where u and v are members of $\mathbb{C}^n$.

37. Find the pseudoinverse of an $m \times n$ matrix all of whose elements are 1's.

38. Use the Penrose Theorem to prove that if B is an $m \times r$ matrix, if C is an $r \times n$ matrix, and if B, C, and BC all have rank r, then $(BC)^+ = C^*(CC^*)^{-1}(B^*B)^{-1}B^*$.

39. Prove that the eigenvalues of a positive semidefinite matrix are nonnegative.

COMPUTER PROBLEM 5.4

1. Write a computer program to find the minimal solution of a system of equations, $Ax = b$, using the singular-value decomposition and pseudoinverse.

*5.5 The *QR*-Algorithm of Francis for the Eigenvalue Problem

In Section 5.2, we proved Schur's Theorem, according to which any square matrix is unitarily similar to a triangular matrix. Thus, a factorization of the type

$$UAU^* = T \tag{1}$$

is possible, where U is unitary and T is triangular. Since the eigenvalues of A and T are the same, and since the eigenvalues of a triangular matrix are simply its diagonal elements, we shall find the eigenvalues of A displayed on the diagonal of T. Although we know that the factorization in Equation (1) exists for any given matrix A, it is not a simple matter to compute it. Finding U must be as difficult as finding all the (complex) roots of a polynomial of degree n because in effect we are computing the roots of the characteristic polynomial of A. By Problem 3, every polynomial is (except for a scalar multiple) the characteristic polynomial of a matrix.

QR-Factorization

The *QR*-algorithm of Francis [1961] is an iterative procedure designed to reveal the eigenvalues of A by producing T in Equation (1). As may be inferred from its name, the algorithm uses the *QR*-factorization. Here, all matrices will be $n \times n$.

In Section 5.3, an algorithm was developed for producing a factorization

$$A = QR \tag{2}$$

where Q is unitary and R is upper triangular. Here, we need a slight refinement of this. Namely, we wish R to have a nonnegative diagonal. This is easily arranged. In fact, if Equation (2) is given and R does not have a nonnegative diagonal, we can define a certain diagonal unitary matrix D and replace Equation (2) by

$$A = (QD)(D^*R) = \widehat{Q}\widehat{R} \tag{3}$$

The definition of $D = \text{diag}(d_{ii})$ should be $d_{ii} = r_{ii}/|r_{ii}|$ if $r_{ii} \neq 0$ and $d_{ii} = 1$ if $r_{ii} = 0$. It is easily verified that D is unitary and that the matrix $\widehat{R} = D^*R$ has a nonnegative diagonal.

The *QR*-algorithm, in its basic form, is as follows:

$A_1 \leftarrow A$
for $k = 1$ **to** M **do**
  QR-factorization: $A_k = Q_k R_k$, where Q_k is unitary and
    R_k is upper triangular with nonnegative diagonal
  $A_{k+1} \leftarrow R_k Q_k$
end do

If the circumstances are favorable, the diagonal of A_k converges (as $k \to \infty$) to a vector whose components are the eigenvalues of A.

In practice, the basic QR-algorithm is combined with several additional procedures to economize the arithmetic and speed the convergence. These will be discussed presently. First, we observe that all the matrices A_k produced in the algorithm are unitarily similar to A. This follows from the equation

$$A_k = Q_k R_k = (Q_k R_k)(Q_k Q_k^*) = Q_k A_{k+1} Q_k^*$$

Second, we note that if the matrix A is *real*, then the subsequent matrices A_k will also be real. Thus, if A has some nonreal eigenvalues, we can only expect, under the best of circumstances, that A_k will converge to a "triangular" matrix with 2×2 submatrices on its diagonal.

Reduction to Upper Hessenberg Form

To economize on the amount of arithmetic involved in the QR-iteration, the matrix A is first reduced to **upper Hessenberg** form by means of unitary similarity transforms. An upper Hessenberg matrix is a matrix H in which $h_{ij} = 0$ when $i > j + 1$. Thus, H has the form

$$H = \begin{bmatrix} * & * & * & * & \cdots & * & * & * & * \\ * & * & * & * & \cdots & * & * & * & * \\ 0 & * & * & * & \cdots & * & * & * & * \\ 0 & 0 & * & * & \cdots & * & * & * & * \\ \vdots & \vdots & \vdots & \ddots & \ddots & \vdots & \vdots & \vdots & \vdots \\ 0 & 0 & 0 & 0 & \ddots & * & * & * & * \\ 0 & 0 & 0 & 0 & \cdots & * & * & * & * \\ 0 & 0 & 0 & 0 & \cdots & 0 & * & * & * \\ 0 & 0 & 0 & 0 & \cdots & 0 & 0 & * & * \end{bmatrix}$$

The reduction of A to H by unitary similarity transforms uses an algorithm of Householder. In the first step, the first column is brought into the proper form. In the second step, the second column is brought into proper form, and so on. Let us describe the kth step. At the beginning of step k, columns 1 to $k - 1$ will have the correct form for an upper Hessenberg matrix. Let us partition the partially reduced matrix in the following way, where the dimensions have been indicated:

$$\begin{array}{c} \\ k \\ n-k \end{array} \begin{array}{c} \begin{array}{cc} k & n-k \end{array} \\ \begin{bmatrix} B & C \\ D & E \end{bmatrix} \end{array}$$

Here B is a $k \times k$ upper Hessenberg matrix. The matrix D is $(n - k) \times k$ and has 0's everywhere but in its kth column. The matrix C is $k \times (n - k)$, and the matrix E is $(n - k) \times (n - k)$. The latter two matrices have no special structure. Let U be any unitary matrix of order $n - k$. Then

$$\begin{bmatrix} I & 0 \\ 0 & U \end{bmatrix} \begin{bmatrix} B & C \\ D & E \end{bmatrix} \begin{bmatrix} I & 0 \\ 0 & U^* \end{bmatrix} = \begin{bmatrix} B & CU^* \\ UD & UEU^* \end{bmatrix} \tag{4}$$

We shall select U so that UD will have in its kth column a vector $(\beta, 0, 0, \dots, 0)^T$. Notice that the form of the matrix D is

$$D = \begin{bmatrix} 0 & \cdots & 0 & d_1 \\ 0 & \cdots & 0 & d_2 \\ \vdots & \ddots & \vdots & \vdots \\ 0 & \cdots & 0 & d_{n-k} \end{bmatrix}$$

It therefore suffices to determine U so that

$$Ud = \beta e^{(1)} \quad \text{where} \quad d = \begin{bmatrix} d_1 \\ d_2 \\ \vdots \\ d_{n-k} \end{bmatrix}$$

and where $e^{(1)}$ denotes the first standard unit vector in $\mathbb{C}^{n-k}$. In the product UD, the first $k - 1$ columns will be 0's automatically, and the kth column will have 0's below the element β.

Using Lemma 2 in Section 5.2, we see that β should be chosen so that $\langle d, \beta e^{(1)} \rangle$ is real and $\|\beta e^{(1)}\|_2 = \|d\|_2$. We put

$$U = I - vv^* \quad \text{where} \quad v = \alpha\left(d - \beta e^{(1)}\right) \tag{5}$$

having chosen $\beta = -(d_1/|d_1|)\|d\|_2$ and $\alpha = \sqrt{2}/\|d - \beta e^{(1)}\|_2$. (These details come from the proof of Lemma 2 and from the explanation of the Householder factorization in Section 5.3.)

Example 1 Reduce the following matrix to upper Hessenberg form by means of unitary similarity transforms:

$$A = \begin{bmatrix} \begin{bmatrix} 1 \end{bmatrix} & \begin{bmatrix} 2 & 3 & 4 \end{bmatrix} \\ \begin{bmatrix} 4 \\ 2 \\ 4 \end{bmatrix} & \begin{bmatrix} 5 & 6 & 7 \\ 1 & 5 & 0 \\ 2 & 1 & 0 \end{bmatrix} \end{bmatrix}$$

Solution Notice that we have already partitioned A in the way required by the first step in the procedure outlined previously. Initially, we let $d = (4, 2, 4)^T$. Since the first component is real, we can simply let $\beta = -\|d\|_2 = -6$ and $\alpha = 1/\sqrt{60}$. Using

Formula (5), we obtain

$$v = \frac{1}{\sqrt{60}}(10,\ 2,\ 4)^T$$

Then the first U-matrix is

$$U = I - vv^* = \frac{1}{15}\begin{bmatrix} -10 & -5 & -10 \\ -5 & 14 & -2 \\ -10 & -2 & 11 \end{bmatrix}$$

The first step is completed by carrying out the multiplication indicated in Equation (4). The result is

$$UAU^* = \begin{bmatrix} \begin{bmatrix} 1 & -5 \\ -6 & \frac{385}{45} \end{bmatrix} & \begin{bmatrix} \frac{72}{45} & \frac{54}{45} \\ -\frac{163}{45} & \frac{34}{45} \end{bmatrix} \\ \begin{bmatrix} 0 & \frac{62}{45} \\ 0 & \frac{259}{45} \end{bmatrix} & \begin{bmatrix} \frac{677}{152} & -\frac{311}{225} \\ -\frac{536}{225} & -\frac{352}{225} \end{bmatrix} \end{bmatrix}$$

$$= \begin{bmatrix} \begin{bmatrix} 1. & -5. \\ -6. & 8.5556 \end{bmatrix} & \begin{bmatrix} 1.6 & 1.2 \\ -3.6222 & 0.75556 \end{bmatrix} \\ \begin{bmatrix} 0. & 1.3778 \\ 0. & 5.7556 \end{bmatrix} & \begin{bmatrix} 3.0089 & -1.3822 \\ -2.3822 & -1.5644 \end{bmatrix} \end{bmatrix}$$

In step 2, the current partially reduced matrix is partitioned in blocks as shown in the previous equation. The computations lead to

$$d = (1.3778,\quad 5.7555)^T$$

$$\beta = -5.9182$$

$$\alpha = 0.15218$$

$$v = (1.1103,\quad 0.87590)^T$$

$$U = \begin{bmatrix} -0.23280 & -0.97252 \\ -0.97252 & 0.23280 \end{bmatrix}$$

Carrying out Equation (4) again, we obtain the final upper Hessenberg matrix

$$H = \begin{bmatrix} 1. & -5. & -1.5395 & -1.2767 \\ -6. & 8.5556 & 0.10848 & 3.6986 \\ 0. & -5.9182 & -2.1689 & -1.1428 \\ 0. & 0. & -0.14276 & 3.6133 \end{bmatrix}$$

Although the calculations were carried out on a computer like the `Marc-32` with seven significant digits, the results have been rounded off for display here. ■

Shifted QR-Factorization

The next technique that is combined with the basic QR-algorithm is a repeated **origin shift**. Before we discuss that technique, we give an indication of why it is necessary.

Example 2 We show the result of applying the basic QR-algorithm to the 4×4 matrix H in Example 1. Thus, a sequence of matrices A_k is generated by the basic iteration $A_k = Q_k R_k$, $A_{k+1} = R_k Q_k$. Some of the results are shown, rounded to five significant figures.

Solution

$$A_1 \leftarrow \text{Hessenberg}\,(A)$$

$$A_2 = \begin{bmatrix} 10.135 & 1.9821 & -0.75082 & 5.5290 \\ 6.7949 & -2.8402 & 0.52664 & 1.2616 \\ 0. & 0.19692 & 1.5057 & 1.7031 \\ 0. & 0. & 1.7508 & 2.1994 \end{bmatrix}$$

$$\vdots$$

$$A_{10} = \begin{bmatrix} 11.105 & -4.7599 & 3.8826 & -4.0296 \\ -0.00045570 & -3.8487 & -0.72647 & 1.2553 \\ 0. & -0.068658 & 3.5669 & 0.16324 \\ 0. & 0. & 0. & 0.17645 \end{bmatrix}$$

$$\vdots$$

$$A_{20} = \begin{bmatrix} 11.106 & -4.7403 & 3.9060 & -4.0296 \\ 0. & -3.8526 & -0.68985 & 1.2559 \\ 0. & -0.032156 & 3.5706 & 0.15706 \\ 0. & 0. & 0. & 0.17645 \end{bmatrix}$$

We observe that the hoped-for upper triangular form is not being rapidly achieved since the element in the (3,2)-position is far from 0. The sluggish convergence to an upper triangular matrix evident in this example is rather troublesome, although one eigenvalue (namely, 0.17645) has been determined very accurately by the tenth step. Another eigenvalue, 11.106, is also determined quickly, but two others in the middle can only be estimated to two or three significant figures at the 20th step. ■

The slow convergence of the basic algorithm is alleviated by *shifts* performed on the successive matrices, with a shift defined as replacement of a matrix A by $A - zI$. The **shifted QR-algorithm** proceeds in the following manner:

$A_1 \leftarrow$ Hessenberg (A)
for $k = 1$ **to** M **do**
  given A_k, compute scalar z_k as explained below
  QR-factorization: $A_k - z_k I = Q_k R_k$

$$A_{k+1} \leftarrow R_k Q_k + z_k I$$
end do

If the scalar z_k in the algorithm is taken to be the lower right diagonal element of A_k, then the iteration should rapidly produce a vector of the form $(0, 0, \dots, 0, \alpha)^T$ in the last row. The number α is then an eigenvalue of A. The best way to proceed thereafter is to **deflate** the matrix by dropping the last row and column in a manner explained in Section 5.2, and repeat the entire process on smaller and smaller matrices. An initial reduction to Hessenberg form is advisable for large matrices to reduce the burden of computation.

LEMMA 1 *Let A be a matrix in partitioned form*

$$A = \begin{bmatrix} B & C \\ 0 & E \end{bmatrix}$$

in which B and E are square matrices. Then the spectrum of A is the union of the spectra of B and E.

Proof The equation $Ax = \lambda x$, in partitioned form, is

$$\begin{bmatrix} B & C \\ 0 & E \end{bmatrix} \begin{bmatrix} u \\ v \end{bmatrix} = \lambda \begin{bmatrix} u \\ v \end{bmatrix} \tag{6}$$

or, equivalently,

$$\begin{cases} Bu + Cv = \lambda u \\ \phantom{Bu + {}} Ev = \lambda v \end{cases}$$

If λ is an eigenvalue of A, then (6) has a nontrivial solution $(u, v)^T$. If $v \neq 0$, then λ is an eigenvalue of E. If $v = 0$, then $u \neq 0$, and λ is an eigenvalue of B. This proves that $\text{sp}\,(A) \subseteqq \text{sp}\,(B) \cup \text{sp}\,(E)$.

Conversely, if λ is an eigenvalue of B, and if u is a corresponding (nonzero) eigenvector, then $(u, 0)^T$ will solve (6). If λ is an eigenvalue of E but not an eigenvalue of B, then let v be a nonzero vector satisfying $Ev = \lambda v$. Next, solve the equation $(B - \lambda I)u = -Cv$. This can be done since λ is not an eigenvalue of B. Then the vector $(u, v)^T$ solves (6). This proves that $\text{sp}\,(B) \cup \text{sp}\,(E) \subseteqq \text{sp}\,(A)$. ■

Example 3 Apply the shifted QR-algorithm to the Hessenberg matrix H of Example 1.

Solution The results of five shifted QR-factorizations and then deflation of the matrices are shown here:

$$A \rightarrow H$$

$$H \rightarrow A_5 = \begin{bmatrix} 2.6141 & -10.087 & -2.4480 & -2.4727 \\ -5.5345 & 4.6668 & 3.5719 & 2.8753 \\ 0. & -0.28730 & 0.14546 & 0.10900 \\ 0. & 0. & 0. & 3.5736 \end{bmatrix}$$

$$\text{deflate}(A_5) \rightarrow \widetilde{A}_5 = \begin{bmatrix} 11.001 & -5.0329 & -4.1730 \\ -0.30955 & -3.7507 & 1.3719 \\ 0. & 0. & 0.17645 \end{bmatrix}$$

$$\text{deflate}(\widetilde{A}_5) \rightarrow \widehat{A}_5 = \begin{bmatrix} 11.106 & -4.7234 \\ 0. & -3.8556 \end{bmatrix}$$

Clearly, the computed eigenvalues are 3.5736, 0.17645, 11.106, and −3.8556. They are accurate to the number of decimal digits shown. ■

Elementary Row and Column Operations

An alternative (and somewhat simpler) method can be used to reduce a matrix to Hessenberg form. This method uses elementary row and column operations. Each step premultiplies the current matrix by a matrix E_i and postmultiplies by E_i^{-1}; thus, $A \leftarrow E_i A E_i^{-1}$. The inverses E_i^{-1} are easily determined. (See Problem 5 in Section 4.1.) The rows and columns are processed one at a time, proceeding from left to right. Example 4 indicates how this is done.

Example 4 We want to reduce the following matrix to upper Hessenberg form with similarity transforms:

$$A = \begin{bmatrix} -3 & 3 & 7 & 2 \\ 1 & 2 & 3 & -5 \\ 2 & -1 & 0 & 3 \\ 4 & 2 & -2 & 4 \end{bmatrix}$$

Solution First, if we do not *pivot*, we would subtract multiples of row 2 from rows 3 and 4 to create the desired 0's. However, as in Gaussian elimination, we want the multipliers to be small. Hence, we interchange rows 2 and 4 to bring the *stronger* pivot into the correct position. (Here we use only row pivoting rather than scaled row pivoting.) Then it is necessary to interchange columns 2 and 4 to have a similarity transformation. (To simplify the process, we actually do the interchanges here and do not use an index array as was done in the Gaussian elimination algorithm in Section 4.3.) The resulting matrices are

$$A \rightarrow \begin{bmatrix} -3 & 3 & 7 & 2 \\ 4 & 2 & -2 & 4 \\ 2 & -1 & 0 & 3 \\ 1 & 2 & 3 & -5 \end{bmatrix} \rightarrow \begin{bmatrix} -3 & 2 & 7 & 3 \\ 4 & 4 & -2 & 2 \\ 2 & 3 & 0 & -1 \\ 1 & -5 & 3 & 2 \end{bmatrix}$$

Next subtract $1/2$ of row 2 from row 3, and $1/4$ of row 2 from row 4. Follow this by the inverse column operations; that is, *add* $1/2$ of column 3 to column 2, and $1/4$ of column 4 to column 2. (See Problem 3 in Section 4.1.) The results are

$$\begin{bmatrix} -3 & 2 & 7 & 3 \\ 4 & 4 & -2 & 2 \\ 0 & 1 & 1 & -2 \\ 0 & -6 & \frac{7}{2} & \frac{3}{2} \end{bmatrix} \to \begin{bmatrix} -3 & \frac{11}{2} & 7 & 3 \\ 4 & 3 & -2 & 2 \\ 0 & \frac{3}{2} & 1 & -2 \\ 0 & -\frac{17}{4} & \frac{7}{2} & \frac{3}{2} \end{bmatrix}$$

$$\to \begin{bmatrix} -3 & \frac{25}{4} & 7 & 3 \\ 4 & \frac{7}{2} & -2 & 2 \\ 0 & 1 & 1 & -2 \\ 0 & -\frac{31}{8} & \frac{7}{2} & \frac{3}{2} \end{bmatrix}$$

In processing the second column, we find pivoting is necessary again. Carrying out the appropriate row operations and inverse column operations brings us to the following two matrices, the last of which is the one we want:

$$\begin{bmatrix} -3 & \frac{25}{4} & 7 & 3 \\ 4 & \frac{7}{2} & -2 & 2 \\ 0 & -\frac{31}{8} & \frac{7}{2} & \frac{3}{2} \\ 0 & 1 & 1 & -2 \end{bmatrix} \to \begin{bmatrix} -3 & \frac{25}{4} & 3 & 7 \\ 4 & \frac{7}{2} & 2 & -2 \\ 0 & -\frac{31}{8} & \frac{3}{2} & \frac{7}{2} \\ 0 & 1 & -2 & 1 \end{bmatrix}$$

$$\to \begin{bmatrix} -3 & \frac{25}{4} & -\frac{37}{8} & 3 \\ 4 & \frac{7}{2} & -\frac{39}{4} & 2 \\ 0 & -\frac{31}{8} & \frac{3}{2} & \frac{7}{2} \\ 0 & 0 & -\frac{50}{31} & \frac{59}{31} \end{bmatrix}$$ ■

PROBLEMS 5.5

1. Let A be an $n \times n$ upper Hessenberg matrix having a 0 in position $A_{k,k-1}$. Show that the spectrum of A is the union of the spectra of the two submatrices A_{ij} $(1 \leqq i, j < k)$ and A_{ij} $(k < i, j \leqq n)$.
2. Show that in the QR-algorithm we have $A_{k+1} = Q_k^* A_k Q_k$. From this, prove that the QR-factoring of A^k is $(Q_1 Q_2 \cdots Q_k)(R_k R_{k-1} \cdots R_1)$.
3. Let $a_0, a_1, \ldots, a_{n-1}$ be arbitrary complex numbers. Put

$$p(t) = a_0 + a_1 t + a_2 t^2 + \cdots + a_{n-1} t^{n-1} + t^n$$

The **companion matrix** of this polynomial is defined to be the $n \times n$ matrix

$$A = \begin{bmatrix} 0 & 1 & 0 & 0 & \cdots & 0 & 0 \\ 0 & 0 & 1 & 0 & \cdots & 0 & 0 \\ 0 & 0 & 0 & 1 & \cdots & 0 & 0 \\ \vdots & \vdots & \vdots & \ddots & \ddots & \vdots & \vdots \\ 0 & 0 & 0 & 0 & \ddots & 1 & 0 \\ 0 & 0 & 0 & 0 & \cdots & 0 & 1 \\ -a_0 & -a_1 & -a_2 & -a_3 & \cdots & -a_{n-2} & -a_{n-1} \end{bmatrix}$$

Prove that $(-1)^n p$ is the characteristic polynomial of A. *Hint:* Use induction to show that $\det(A - \lambda I) = (-1)^n p_n(\lambda)$. To reduce the determinant of $A - \lambda I$, expand it by the elements of its nth column.

4. For the matrix A in Problem 3, prove that $A^n = -\sum_{i=0}^{n-1} a_i A^i$. *Hint:* A direct proof is possible, or one can use Problem 3 and the Cayley-Hamilton Theorem.

5. Prove that for any set of n integers, there is an $n \times n$ integer matrix whose spectrum is the given set of integers.

6. Find the eigenvalues of the matrix $\begin{bmatrix} -1 & -4 & 1 \\ -1 & -2 & -5 \\ 5 & 4 & 3 \end{bmatrix}$.

7. Prove that in the shifted QR-algorithm A_{k+1} is unitarily similar to A_k.

8. The problem of finding values of λ for which the equation $Ax = \lambda Bx$ has nontrivial solutions is the **generalized eigenvalue problem**. Show that if B is nonsingular, the problem can be recast as a usual eigenvalue problem.

9. (Continuation) Show that if we find a generalized eigenvalue μ in the problem $Ax = \mu(B + tA)x$, then an eigenvalue for the problem $Ax = \lambda Bx$ can be easily computed provided $\mu t \neq 1$.

10. Select integers p and q such that $1 \leqq p < q \leqq n$. Select complex numbers α and β such that $|\alpha|^2 + |\beta|^2 = 1$. Let U be the matrix that is the $n \times n$ identity matrix except for these four elements: $U_{pp} = U_{qq} = \alpha$ and $U_{pq} = -U_{qp} = \beta$. Prove that U is unitary. Assume $\alpha\overline{\beta}$ is real.

11. Let A be a real matrix having upper triangular block structure

$$A = \begin{bmatrix} A_{11} & A_{12} & A_{13} & \cdots & A_{1n} \\ 0 & A_{22} & A_{23} & \cdots & A_{2n} \\ 0 & 0 & A_{33} & \cdots & A_{3n} \\ \vdots & \vdots & \vdots & \ddots & \vdots \\ 0 & 0 & 0 & \cdots & A_{nn} \end{bmatrix}$$

in which each A_{ii} is a 2×2 matrix. Give a simple procedure for computing the eigenvalues of A, including proofs.

12. Prove or disprove: If U is unitary, R is upper triangular, and UR is upper Hessenberg, then U is upper Hessenberg.

13. Refer to Problem 11 and prove that A is singular if and only if at least one of A_{ii} is singular.

14. Prove that if T is upper triangular and invertible, then the same is true of T^{-1}.

15. Prove that if A is upper Hessenberg and T is upper triangular, then AT and TA are upper Hessenberg.

16. Prove that if T is upper triangular and AT is upper Hessenberg, then TA is upper Hessenberg. (Assume that A and T are $n \times n$ matrices, not necessarily invertible.)

17. In the reduction of an $n \times n$ matrix to upper Hessenberg form as outlined in Problem 16 and in Example 4, approximately how many multiplications are required?

COMPUTER PROBLEMS 5.5

1. Write computer subroutines or procedures for reducing a matrix to Hessenberg form via Equation (4) and apply it to the matrix of Example 1.

2. (Continuation) Add computer subroutines or procedures for carrying out the basic *QR*-algorithm and reproduce the results of Example 2.

3. (Continuation) Modify the computer subroutines or procedures so that they carry out the shifted *QR*-algorithm. Then reproduce the results of Example 3. Also, use your routines to find the eigenvalues (343, 294, and $147 \pm 196i$) of the matrix

$$\begin{bmatrix} 190 & 66 & -84 & 30 \\ 66 & 303 & 42 & -36 \\ 336 & -168 & 147 & -112 \\ 30 & -36 & 28 & 291 \end{bmatrix}$$

4. Write a procedure or subroutine to reduce a matrix to upper Hessenberg form by using a sequence of similarity transformations as in Example 4. Each similarity transformation should follow the algorithm outlined in this example—that is, elementary row operations followed by the inverse elementary column operation. Test on the matrix of Example 4. Apply to the matrix of Example 1 and compare to the procedure given in the text.

Chapter Six
Approximating Functions

In this chapter, the problem of representing functions within a computer is discussed. Several different subproblems will be considered. They differ according to the type of function being represented, whether it is known at relatively few points or at many (or all) points. The representation chosen (a polynomial, a spline function, a continued fraction, and so on) also determines the nature of the theory. We begin with the case of polynomials, which is the oldest and simplest.

6.1 Polynomial Interpolation

In this section, we solve the following problem: We are given a table of $n + 1$ data points (x_i, y_i):

x	x_0	x_1	x_2	$\cdots$	x_n
y	y_0	y_1	y_2	$\cdots$	y_n

and we seek a polynomial p of lowest possible degree for which

$$p(x_i) = y_i \qquad (0 \leqq i \leqq n)$$

Such a polynomial is said to **interpolate** the data. Here is the theorem that governs this problem.

THEOREM 1 *If $x_0, x_1, \ldots, x_n$ are distinct real numbers, then for arbitrary values $y_0, y_1, \ldots, y_n$ there is a unique polynomial p_n of degree at most n such that*

$$p_n(x_i) = y_i \qquad (0 \leqq i \leqq n)$$

Proof Let us prove the uniqueness or *unicity* first. Suppose there were two such polynomials, p_n and q_n. Then the polynomial $p_n - q_n$ would have the property $(p_n - q_n)(x_i) = 0$ for $0 \leqq i \leqq n$. Since the degree of $p_n - q_n$ can be at most n, this polynomial can have at most n zeros if it is not the 0 polynomial. Since the x_i are distinct, $p_n - q_n$ has $n + 1$ zeros; it must therefore be 0. Hence, $p_n \equiv q_n$.

For the existence part of the theorem, we proceed inductively. For $n = 0$, the existence is obvious since a constant function p_0 (polynomial of degree $\leqq 0$) can be chosen so that $p_0(x_0) = y_0$. Now suppose that we have obtained a polynomial p_{k-1} of degree $\leqq k - 1$ with $p_{k-1}(x_i) = y_i$ for $0 \leqq i \leqq k - 1$. We try to construct p_k in the form

$$p_k(x) = p_{k-1}(x) + c(x - x_0)(x - x_1) \cdots (x - x_{k-1}) \tag{1}$$

Note that this is unquestionably a polynomial of degree at most k. Furthermore, p_k interpolates the data that p_{k-1} interpolates because

$$p_k(x_i) = p_{k-1}(x_i) = y_i \qquad (0 \leqq i \leqq k - 1)$$

Now we determine the unknown coefficient c from the condition $p_k(x_k) = y_k$. This leads to the equation

$$p_{k-1}(x_k) + c(x_k - x_0)(x_k - x_1) \cdots (x_k - x_{k-1}) = y_k \tag{2}$$

Equation (2) can certainly be solved for c because the factors multiplying c are not 0. (Why?) ■

Newton Form of the Interpolation Polynomial

Before we attempt to write an algorithm that carries out the recursive process in this proof, we need to make some observations. First, the polynomials $p_0, p_1, \ldots, p_n$ constructed in the proof have the property that each p_k is obtained simply by adding a single term to p_{k-1}. Thus, at the end of the process, p_n will be a sum of terms, and each $p_0, p_1, \ldots, p_{n-1}$ will be clearly visible in the expression for p_n. Each p_k has the form

$$p_k(x) = c_0 + c_1(x - x_0) + c_2(x - x_0)(x - x_1) + \cdots + c_k(x - x_0)\cdots(x - x_{k-1}) \quad \textbf{(3)}$$

The compact form of this is

$$p_k(x) = \sum_{i=0}^{k} c_i \prod_{j=0}^{i-1} (x - x_j) \quad \textbf{(4)}$$

Here the convention has been adopted that $\prod_{j=0}^{m}(x - x_j) = 1$ whenever $m < 0$. The first few cases of Equation (4) are

$$\begin{aligned} p_0(x) &= c_0 \\ p_1(x) &= c_0 + c_1(x - x_0) \\ p_2(x) &= c_0 + c_1(x - x_0) + c_2(x - x_0)(x - x_1) \end{aligned}$$

and so on. These polynomials are called the **interpolation polynomials in Newton's form**.

To evaluate $p_k(x)$, assuming that the coefficients $c_0, c_1, \ldots, c_k$ are known, an efficient method called **nested multiplication** or **Horner's algorithm** is used. This can be explained most easily for an arbitrary expression of the form

$$u = \sum_{i=0}^{k} c_i \prod_{j=0}^{i-1} d_j = c_0 + c_1 d_0 + c_2 d_0 d_1 + \cdots + c_k d_0 d_1 \cdots d_{k-1} \quad \textbf{(5)}$$

The idea is to write it in the form

$$u = (\cdots(((c_k)d_{k-1} + c_{k-1})d_{k-2} + c_{k-2})d_{k-3} + \cdots + c_1)d_0 + c_0 \quad \textbf{(6)}$$

The algorithm to compute u can be developed as follows: We denote the quantities in the parentheses by $u_k, u_{k-1}, \ldots, u_0$, starting with the innermost parentheses. Thus,

$$\begin{aligned} u_k &\leftarrow c_k \\ u_{k-1} &\leftarrow u_k d_{k-1} + c_{k-1} \\ u_{k-2} &\leftarrow u_{k-1} d_{k-2} + c_{k-2} \\ &\vdots \\ u_0 &\leftarrow u_1 d_0 + c_0 \end{aligned}$$

Since only u_0 is wanted, we can write algorithmically

$u \leftarrow c_k$
for $i = k - 1$ **to** 0 **step** -1 **do**
    $u \leftarrow ud_i + c_i$
end do

Going back to the polynomial given by Equation (3) or (4), we use the following algorithm to obtain $u = p_k(t)$ for a prescribed value of t:

$u \leftarrow c_k$
for $i = k - 1$ **to** 0 **step** -1 **do**
    $u \leftarrow (t - x_i)u + c_i$
end do

Now we can write an algorithm for computing the coefficients c_i in Equation (4). Notice that in Equation (2) the coefficient c is what we later have labeled c_k. The formula for c_k is then

$$c_k = \frac{y_k - p_{k-1}(x_k)}{(x_k - x_0)(x_k - x_1) \cdots (x_k - x_{k-1})}$$

An algorithm to compute $c_0, c_1, \ldots, c_n$ from the table of values for $x_0, x_1, \ldots, x_n$ and $y_0, y_1, \ldots, y_n$ follows:

$c_0 \leftarrow y_0$
for $k = 1$ **to** n **do**
    $d \leftarrow x_k - x_{k-1}$
    $u \leftarrow c_{k-1}$
    for $i = k - 2$ **to** 0 **step** -1 **do**
        $u \leftarrow u(x_k - x_i) + c_i$
        $d \leftarrow d(x_k - x_i)$
    end do
    $c_k \leftarrow (y_k - u)/d$
end do

The inner loop computes $p_{k-1}(x_k)$ and $(x_k - x_0)(x_k - x_1) \cdots (x_k - x_{k-1})$, the former by a straightforward modification of the algorithm discussed previously.

The algorithm just given serves a didactic purpose only. It illustrates the important process of translating a constructive existence proof into an algorithm suitable for a computer program. However, there is a more efficient procedure that achieves the same result. The alternative method uses **divided differences** to compute the coefficients $c_0, c_1, \ldots, c_k$ in Equation (4). This method will be presented in Section 6.2.

The preceding algorithm was programmed and given a simple test as follows: We define the polynomial

$$p_3(x) = 4x^3 + 35x^2 - 84x - 954 \tag{7}$$

Four values of this function are given here:

x	5	−7	−6	0
y	1	−23	−54	−954

(8)

This set of values was given as input to the computer routine, and it correctly computed the coefficients $c_0 = 1, c_1 = 2, c_2 = 3, c_3 = 4$. These are the coefficients in the Newton form of the polynomial in Equation (7); namely,

$$p_3(x) = 1 + 2(x-5) + 3(x-5)(x+7) + 4(x-5)(x+7)(x+6)$$

Lagrange Form of the Interpolation Polynomial

We now present an alternative form for the interpolating polynomial p associated with a table of data points (x_i, y_i) for $0 \leqq i \leqq n$. It is important to understand that there is one and only one interpolating polynomial of degree $\leqq n$ associated with the data (assuming, of course, that the $n+1$ abscissas x_i are *distinct*). However, the possibility certainly exists for expressing this polynomial in different forms and for arriving at it by different algorithms.

The alternative method will express p in the form

$$p(x) = y_0\ell_0(x) + y_1\ell_1(x) + \cdots + y_n\ell_n(x) = \sum_{k=0}^{n} y_k\ell_k(x) \tag{9}$$

Here $\ell_0, \ell_1, \ldots, \ell_n$ are polynomials that depend on the nodes $x_0, x_1, \ldots, x_n$ but *not* on the *ordinates* $y_0, y_1, \ldots, y_n$. Since the ordinates could be all 0 except for 1 occupying the ith position, we see that

$$\delta_{ij} = p_n(x_j) = \sum_{k=0}^{n} y_k\ell_k(x_j) = \sum_{k=0}^{n} \delta_{ki}\ell_k(x_j) = \ell_i(x_j)$$

(Recall that the **Kronecker delta** $\delta_{ki} = 1$ if $k = i$ and $\delta_{ki} = 0$ if $k \neq i$.) We can easily arrive at a set of polynomials having this property.

Let us consider ℓ_0. It is to be a polynomial of degree n that takes the value 0 at $x_1, x_2, \ldots, x_n$ and the value 1 at x_0. Clearly, ℓ_0 must be of the form

$$\ell_0(x) = c(x-x_1)(x-x_2)\cdots(x-x_n) = c\prod_{j=1}^{n}(x-x_j)$$

The value of c is obtained by putting $x = x_0$, so that

$$1 = c\prod_{j=1}^{n}(x_0 - x_j)$$

and

$$c = \prod_{j=1}^{n} (x_0 - x_j)^{-1}$$

Hence,

$$\ell_0(x) = \prod_{j=1}^{n} \frac{x - x_j}{x_0 - x_j}$$

Each ℓ_i is obtained by similar reasoning, and the general formula is then

$$\ell_i(x) = \prod_{\substack{j=0 \\ j \neq i}}^{n} \frac{x - x_j}{x_i - x_j} \qquad (0 \leqq i \leqq n) \tag{10}$$

For the set of nodes $x_0, x_1, \ldots, x_n$, these polynomials are known as the **cardinal functions**. With the cardinal polynomials in hand, Equation (9) gives the **Lagrange form** of the interpolating polynomials.

Example 1 What are the cardinal functions and Lagrange form of the interpolating polynomial for the data in Table (8)?

Solution The nodes are 5, −7, −6, 0. Hence, the cardinal functions are

$$\ell_0(x) = \frac{(x + 7)(x + 6)x}{(5 + 7)(5 + 6)5}$$

$$\ell_1(x) = \frac{(x - 5)(x + 6)x}{(-7 - 5)(-7 + 6)(-7)}$$

$$\ell_2(x) = \frac{(x - 5)(x + 7)x}{(-6 - 5)(-6 + 7)(-6)}$$

$$\ell_3(x) = \frac{(x - 5)(x + 7)(x + 6)}{(0 - 5)(0 + 7)(0 + 6)}$$

The interpolating polynomial is

$$p_3(x) = \ell_0(x) - 23\ell_1(x) - 54\ell_2(x) - 954\ell_3(x)$$

This may not look like the p_3 in Equation (7), but it must be identical to it *as a function* even though its form is quite different. ■

Example 2 Find an interpolation formula for the two-point table

x	x_0	x_1
y	y_0	y_1

Solution The Lagrange interpolation formula gives us

$$p(x) = y_0\left(\frac{x - x_1}{x_0 - x_1}\right) + y_1\left(\frac{x - x_0}{x_1 - x_0}\right) \qquad \blacksquare$$

Still other algorithms for polynomial interpolation have various advantages and disadvantages. Since there is one and only one polynomial of degree $\leqq n$ that takes prescribed values at $n+1$ given (distinct) points, these algorithms produce the same polynomial in different forms. For example, we can require our polynomial to be expressed in powers of x:

$$p(x) = a_0 + a_1x + a_2x^2 + \cdots + a_nx^n \tag{11}$$

The interpolation conditions, $p(x_i) = y_i$ for $0 \leqq i \leqq n$, lead to a system of $n+1$ linear equations for determining $a_0, a_1, \ldots, a_n$. This system has the form

$$\begin{bmatrix} 1 & x_0 & x_0^2 & \cdots & x_0^n \\ 1 & x_1 & x_1^2 & \cdots & x_1^n \\ 1 & x_2 & x_2^2 & \cdots & x_2^n \\ \vdots & \vdots & \vdots & \ddots & \vdots \\ 1 & x_n & x_n^2 & \cdots & x_n^n \end{bmatrix} \begin{bmatrix} a_0 \\ a_1 \\ a_2 \\ \vdots \\ a_n \end{bmatrix} = \begin{bmatrix} y_0 \\ y_1 \\ y_2 \\ \vdots \\ y_n \end{bmatrix} \tag{12}$$

The coefficient matrix in Equation (12) is called a **Vandermonde matrix**. It is nonsingular because the system has a unique solution for any choice of $y_0, y_1, \ldots, y_n$ (Theorem 1). The determinant of the Vandermonde matrix is therefore nonzero for distinct nodes $x_0, x_1, \ldots, x_n$. A formula for it is given in Problem 34. However, the Vandermonde matrix is often *ill conditioned*, and the coefficients a_i may therefore be inaccurately determined by solving System (12). (See Gautschi [1984].) Furthermore, the amount of work involved to obtain the polynomial in Equation (11) is excessive. Therefore, this approach is *not* recommended.

If we let $x = x_i$ for $0 \leqq i \leqq n$ in the Lagrange polynomial (9), we obtain the following $n+1$ linear equations:

$$\begin{bmatrix} \ell_0(x_0) & \ell_1(x_0) & \cdots & \ell_n(x_0) \\ \ell_0(x_1) & \ell_1(x_1) & \cdots & \ell_n(x_1) \\ \vdots & \vdots & \ddots & \vdots \\ \ell_0(x_n) & \ell_1(x_n) & \cdots & \ell_n(x_n) \end{bmatrix} \begin{bmatrix} y_0 \\ y_1 \\ \vdots \\ y_n \end{bmatrix} = \begin{bmatrix} p(x_0) \\ p(x_1) \\ \vdots \\ p(x_n) \end{bmatrix}$$

The coefficient matrix reduces to the identity matrix I, and the solution is $y_i = p(x_i)$ for $0 \leqq i \leqq n$, as it should be. Since there can be only one interpolating polynomial for the $n+1$ distinct points $x_0, x_1, \ldots, x_n$, we obtain an identity satisfied by any polynomial of degree at most n:

$$p(x) = \sum_{i=0}^{n} p(x_i)\ell_i(x)$$

For numerical work, it is probably best to use the Newton form of the interpolation polynomial. This can be combined with the divided difference algorithm (discussed in Section 6.2) for computing the required coefficients. On the other hand, the Lagrange form of the interpolating polynomial can be written down at once, since the coefficients in the Lagrange formula are the given y_i. This fact will be useful in Chapter 7 when quadrature formulas are constructed.

A situation where the Lagrange form would have the advantage over the Newton form would be if one had a single set of fixed x_i-nodes with many different y_i-values associated with them, such as from experimental data. Here the cardinal functions would remain the same for each case.

Another advantage of the Newton form is that if more data points are added to the interpolation problem, the coefficients already computed will not have to be changed. Thus in Equation (3), c_0 depends on the point (x_0, y_0). Then c_1 depends on the two points (x_0, y_0) and (x_1, y_1), and so on. The Newton form easily accommodates additional data to be interpolated.

The efficient algorithm of nested multiplication applies to the Newton form and to the form in Equation (11). This is a clear advantage over the Lagrange form. However, there are algorithms for evaluating the Lagrange form efficiently. See Werner [1984] and the references given there. See also Problems 31 and 32.

The Error in Polynomial Interpolation

We present now some theorems that pertain to the discrepancy between a function and a polynomial interpolant to it.

THEOREM 2 *Let f be a function in $C^{n+1}[a, b]$, and let p be the polynomial of degree $\leqq n$ that interpolates the function f at $n+1$ distinct points $x_0, x_1, \ldots, x_n$ in the interval $[a, b]$. To each x in $[a, b]$ there corresponds a point ξ_x in (a, b) such that*

$$f(x) - p(x) = \frac{1}{(n+1)!} f^{(n+1)}(\xi_x) \prod_{i=0}^{n} (x - x_i) \tag{13}$$

Proof If x is one of the nodes of interpolation x_i, the assertion is obviously true since both sides of Equation (13) reduce to 0. So, let x be any point other than a node. Put

$$w(t) \equiv \prod_{i=0}^{n} (t - x_i) \qquad \phi \equiv f - p - \lambda w$$

where λ is the real number that makes $\phi(x) = 0$. (Remember, x is fixed.) Thus,

$$\lambda = \frac{f(x) - p(x)}{w(x)}$$

Now $\phi \in C^{n+1}[a, b]$, and ϕ vanishes at the $n+2$ points $x, x_0, x_1, \ldots, x_n$. By Rolle's Theorem, ϕ' has at least $n+1$ distinct zeros in (a, b). Similarly, ϕ'' has at least n distinct zeros in (a, b). If this argument is repeated, we conclude eventually that

$\phi^{(n+1)}$ has at least one zero, say ξ_x, in (a, b). Now

$$\phi^{(n+1)} = f^{(n+1)} - p^{(n+1)} - \lambda w^{(n+1)} = f^{(n+1)} - (n+1)!\,\lambda$$

Hence,

$$0 = \phi^{(n+1)}(\xi_x) = f^{(n+1)}(\xi_x) - (n+1)!\,\lambda = f^{(n+1)}(\xi_x) - (n+1)!\frac{f(x) - p(x)}{w(x)}$$

This is Equation (13) in disguise. ■

Example 3 If the function $f(x) = \sin x$ is approximated by a polynomial of degree 9 that interpolates f at ten points in the interval $[0, 1]$, how large is the error on this interval?

Solution To answer this question, we use Equation (13) in Theorem 2. Obviously, $|f^{(10)}(\xi_x)| \leqq 1$ and $\prod_{i=0}^{9} |x - x_i| \leqq 1$. So, for all x in $[0, 1]$,

$$|\sin x - p(x)| \leqq \frac{1}{10!} < 2.8 \times 10^{-7}$$ ■

Chebyshev Polynomials

In Theorem 2, there is a term that can be *optimized* by choosing the nodes in a special way. An analysis of this problem was first given by the great mathematician Chebyshev (1821–1894). The optimization process leads naturally to a system of polynomials called **Chebyshev polynomials**, and we begin with their definition and basic properties.

The Chebyshev polynomials (of the first kind) are defined recursively as follows:

$$\begin{cases} T_0(x) = 1 \qquad T_1(x) = x \\ T_{n+1}(x) = 2xT_n(x) - T_{n-1}(x) \qquad (n \geqq 1) \end{cases}$$

The explicit forms of the next few T_n are readily calculated:

$$\begin{aligned} T_2(x) &= 2x^2 - 1 \\ T_3(x) &= 4x^3 - 3x \\ T_4(x) &= 8x^4 - 8x^2 + 1 \\ T_5(x) &= 16x^5 - 20x^3 + 5x \\ T_6(x) &= 32x^6 - 48x^4 + 18x^2 - 1 \end{aligned}$$

Figure 6.1 show the graphs of these first few Chebyshev polynomials.

These polynomials arose when Chebyshev (Чебышев) was studying the motion of linkages in a steam locomotive. Since then they have become a primary ingredient in applied mathematics. The Chebyshev polynomials have many marvelous properties. (See Rivlin [1990].)

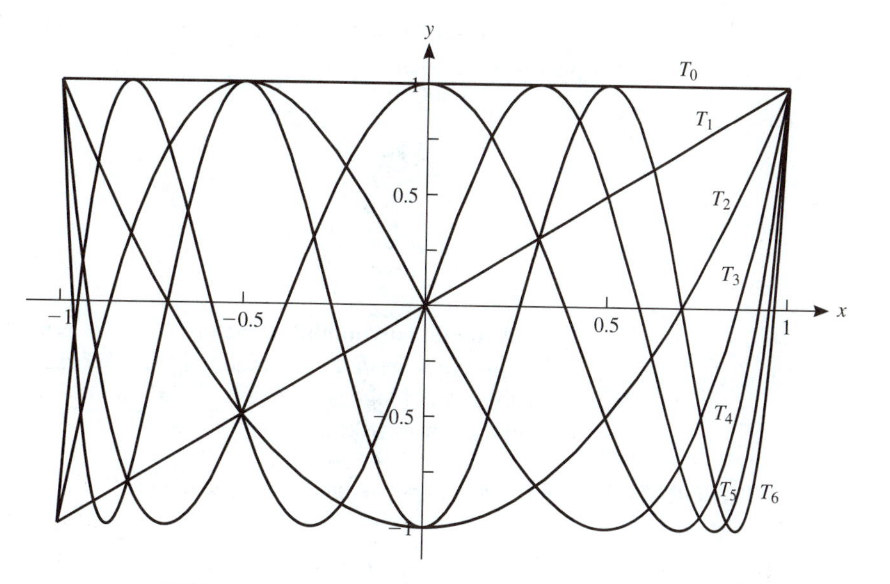

FIGURE 6.1 Chebyshev polynomials $T_n(x)$ for $0 \leqq n \leqq 6$

THEOREM 3 *For x in the interval $[-1, 1]$, the Chebyshev polynomials have this closed-form expression:*

$$T_n(x) = \cos(n \cos^{-1} x) \qquad (n \geqq 0)$$

Proof Recall the addition formula for the cosine:

$$\cos(A + B) = \cos A \cos B - \sin A \sin B$$

From this we obtain

$$\begin{aligned} \cos(n + 1)\theta &= \cos\theta \cos n\theta - \sin\theta \sin n\theta \\ \cos(n - 1)\theta &= \cos\theta \cos n\theta + \sin\theta \sin n\theta \end{aligned}$$

On adding these two equations and rearranging, we obtain

$$\cos(n + 1)\theta = 2\cos\theta \cos n\theta - \cos(n - 1)\theta \tag{14}$$

Now let $\theta = \cos^{-1} x$ and $x = \cos\theta$. Equation (14) shows that the functions f_n defined by

$$f_n(x) = \cos(n \cos^{-1} x)$$

obey this system of equations:

$$\begin{cases} f_0(x) = 1 \qquad f_1(x) = x \\ f_{n+1}(x) = 2x f_n(x) - f_{n-1}(x) \qquad (n \geqq 1) \end{cases}$$

It follows that $f_n = T_n$ for all n. ■

From the formula in Theorem 3, we obtain further properties of the Chebyshev polynomials, such as

$$|T_n(x)| \leqq 1 \qquad (-1 \leqq x \leqq 1)$$

$$T_n\left(\cos\frac{j\pi}{n}\right) = (-1)^j \qquad (0 \leqq j \leqq n)$$

$$T_n\left(\cos\frac{2j-1}{2n}\pi\right) = 0 \qquad (1 \leqq j \leqq n)$$

A **monic polynomial** is one in which the term of highest degree has a coefficient of unity. From the definition of the Chebyshev polynomials, we see that in $T_n(x)$ the term of highest degree is $2^{n-1}x^n$ for $n > 0$. Therefore, $2^{1-n}T_n$ is a monic polynomial for $n > 0$.

THEOREM 4 *If p is a monic polynomial of degree n, then*

$$\|p\|_\infty = \max_{-1 \leqq x \leqq 1} |p(x)| \geqq 2^{1-n}$$

Proof We proceed by contradiction. Suppose that

$$|p(x)| < 2^{1-n} \qquad (|x| \leqq 1)$$

Let $q = 2^{1-n}T_n$ and $x_i = \cos(i\pi/n)$. As noted above, q is a monic polynomial of degree n. Then

$$(-1)^i p(x_i) \leqq |p(x_i)| < 2^{1-n} = (-1)^i q(x_i)$$

Consequently,

$$(-1)^i \left[q(x_i) - p(x_i)\right] > 0 \qquad (0 \leqq i \leqq n)$$

This shows that the polynomial $q - p$ oscillates in sign $n + 1$ times on the interval $[-1, 1]$. It therefore must have at least n roots in $(-1, 1)$. But this is not possible because $q - p$ has degree at most $n - 1$. (Remember that q and p are monic, and the term x^n will not appear in $q - p$.) ■

Choosing the Nodes

In Theorem 2, assume that the interpolation nodes are in the interval $[-1, 1]$. If x is in this same interval, then ξ_x also will be in that interval. Hence, we can deduce that

$$\max_{|x| \leqq 1} |f(x) - p(x)| \leqq \frac{1}{(n+1)!} \max_{|x| \leqq 1} |f^{(n+1)}(x)| \max_{|x| \leqq 1} \left|\prod_{i=0}^{n}(x - x_i)\right|$$

By Theorem 4, we have (for any set of nodes)

$$\max_{|x| \leqq 1} \left|\prod_{i=0}^{n}(x - x_i)\right| \geqq 2^{-n}$$

This minimum value will be attained if $\prod_{i=0}^{n}(x - x_i)$ is the monic multiple of T_{n+1}—that is, $2^{-n}T_{n+1}$. The nodes then will be the roots of T_{n+1}. These are

$$x_i = \cos\left(\frac{2i+1}{2n+2}\pi\right) \qquad (0 \leqq i \leqq n)$$

These considerations establish the next result.

THEOREM 5 *If the nodes x_i are the roots of the Chebyshev polynomial T_{n+1}, then the error formula in Theorem 2 yields (for $|x| \leqq 1$)*

$$|f(x) - p(x)| \leqq \frac{1}{2^n(n+1)!} \max_{|t| \leqq 1} |f^{(n+1)}(t)|$$

The Convergence of Interpolating Polynomials

If a continuous function f is prescribed on an interval $[a, b]$, and if interpolating polynomials p_n of higher and higher degree (with evenly spaced nodes) are constructed for f, then the natural expectation is that these polynomials will converge to f uniformly on $[a, b]$. That is, we expect the quantity

$$\|f - p_n\|_\infty = \max_{a \leqq x \leqq b} |f(x) - p_n(x)|$$

to converge to 0 as $n \longrightarrow \infty$. We have seen examples, such as $f(x) = \sin x$, where this is true. But it must be recognized that $\sin x$ is far from being a typical continuous function. It is of class C^∞ on the real line and, as a function of a complex variable, it is an **entire function**, which means that it has no singularities at all in (the finite part of) the complex plane.

The surprising state of affairs is that for most continuous functions, the quantity $\|f - p_n\|_\infty$ will not converge to 0. The first example of this was exhibited by Meray in 1884. He took as nodes the n ***n*th roots of unity** in the complex plane. These are points evenly spaced on the unit circle $|z| = 1$, as shown in Figure 6.2 for $n = 6$. We denote these nodes by $\omega_1, \omega_2, \ldots, \omega_n$. Now consider the problem of

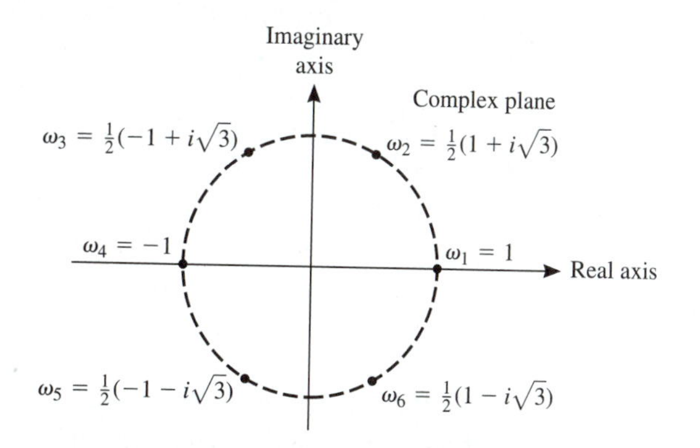

FIGURE 6.2 The 6 sixth roots of unity

interpolating the function $f(z) = 1/z$ at the nodes $\omega_1, \omega_2, \ldots, \omega_n$. A unique polynomial p_{n-1} of degree $n-1$ solves this problem. Not surprisingly, it is $p_{n-1}(z) = z^{n-1}$ because

$$p_{n-1}(\omega_j) = \omega_j^{n-1} = \frac{1}{\omega_j}\omega_j^n = \frac{1}{\omega_j} = f(\omega_j) \qquad (1 \leqq j \leqq n)$$

Let us measure the discrepancy between f and p_{n-1} on the unit circle. We have

$$\begin{aligned}\|f - p_{n-1}\|_\infty &= \max_{|z|=1} |f(z) - p_{n-1}(z)| = \max_{|z|=1} |z^{-1} - z^{n-1}| \\ &= \max_{|z|=1} \frac{1}{|z|} |1 - z^n| = 2\end{aligned}$$

(As z traverses the circle $|z| = 1$, the same is true of z^n, and when $z^n = -1$ we obtain 2 in the preceding calculation.) Thus, when $n \to \infty$, p_n and f in this example are always two units apart.

Another example, this time in the real domain, was given by Runge in 1901. This is the function $f(x) = (x^2 + 1)^{-1}$ on the interval $[-5, 5]$. If interpolating polynomials p_n are constructed for this function using equally spaced nodes in $[-5, 5]$, it is found that the sequence $\|f - p_n\|_\infty$ is not bounded. A computer experiment to see this behavior is outlined in the next section. Even for modest values of n (say $n = 15$), we see wild oscillations in the polynomial p_n. This phenomenon has become a standard example in textbooks and papers and is called the **Runge function** or **Runge example**. See, for example, Epperson [1987].

Subsequent work by other mathematicians such as Faber disclosed the following very general result in 1914:

THEOREM 6 *For any prescribed system of nodes*

$$a \leqq x_0^{(n)} < x_1^{(n)} < \cdots < x_n^{(n)} \leqq b \qquad (n \geqq 0) \tag{15}$$

there exists a continuous function f on $[a, b]$ such that the interpolating polynomials for f using these nodes fail to converge uniformly to f.

This matter is a little more subtle than it seems at first because we have the following positive result:

THEOREM 7 *If f is a continuous function on $[a, b]$, then there is a system of nodes as in Equation (15) such that the polynomials p_n of interpolation to f at these nodes satisfy* $\lim_{n\to\infty} \|f - p_n\|_\infty = 0$.

This theorem is obtained by joining two other powerful theorems: the **Weierstrass Approximation Theorem** and the **Chebyshev Alternation Theorem**. We turn now to the first of these two results. (The second appears in Section 6.9.)

THEOREM 8 **Weierstrass Approximation Theorem** *If f is continuous on $[a, b]$ and if $\varepsilon > 0$, then there is a polynomial p satisfying $|f(x) - p(x)| \leqq \varepsilon$ on the interval $[a, b]$.*

Proof It suffices to prove this theorem for the special interval [0, 1], since the change of variable $x = a + t(b - a)$ can be used to go back and forth between $[a, b]$ and [0, 1]. Of course, a polynomial of a linear function of t is a polynomial in t. If $f \in C[0, 1]$, the sequence of Bernstein polynomials $B_n f$, as explained later, converges uniformly to f. ■

Now, let us discuss the **Bernstein polynomials**, introduced by Serge Bernstein in 1912. They are given by

$$(B_n f)(x) = \sum_{k=0}^{n} f\left(\frac{k}{n}\right) g_{nk}(x) \quad \text{where} \quad g_{nk}(x) = \binom{n}{k} x^k (1 - x)^{n-k}$$

In the definition of g_{nk}, the term $\binom{n}{k}$ is a **binomial coefficient**, defined by

$$\binom{n}{k} = \begin{cases} \dfrac{n!}{k!\,(n-k)!} & 0 \leqq k \leqq n \\ 0 & \text{otherwise} \end{cases}$$

Why do we care about the Bernstein polynomials? We use them primarily to give an elementary proof of the Weierstrass Theorem. However, Bernstein polynomials are also used in computer-aided design. (Recently, there has been a tendency to replace them by B-splines in this application.)

It is essential to interpret B_n as a linear operator that maps elements of $C[0, 1]$ into other elements. The linearity of B_n is expressed by the equation

$$B_n(af + bg) = aB_n f + bB_n g \qquad \left(a, b \in \mathbb{R}; \quad f, g \in C[0, 1]\right)$$

which is easily verified. Another important property of these operators is that they are *positive*. This means that if $f \geqq 0$, then $B_n f \geqq 0$. The truth of this follows at once from the fact that $g_{nk} \geqq 0$ on the interval [0, 1]. (We regard all functions as being defined only on the interval [0, 1].)

The properties of linearity and positivity together are extremely powerful, as indicated in the following theorem.

THEOREM 9 **Bohman-Korovkin Theorem** *Let L_n $(n \geqq 1)$ be a sequence of positive linear operators defined on $C[a, b]$ and taking values in the same space. If $\|L_n f - f\|_\infty \to 0$ for the three functions $f(x) = 1, x, x^2$, then the same is true for all $f \in C[a, b]$.*

Proof In the space $C[a, b]$, we can take absolute values, denoting by $|f|$ that function whose value at x is $|f(x)|$. Notice also that if L is a positive linear operator, we have this implication:

$$f \geqq g \Rightarrow f - g \geqq 0 \Rightarrow L(f - g) \geqq 0 \Rightarrow Lf - Lg \geqq 0 \Rightarrow Lf \geqq Lg$$

A further property is that $L(|f|) \geqq |Lf|$, for $|f| \geqq f$ and $|f| \geqq -f$, from which $L(|f|) \geqq Lf$ and $L(|f|) \geqq -Lf$. Now set $h_k(x) = x^k$ for $k = 0, 1, 2$. Also define functions α_n, β_n, and γ_n by putting

$$\alpha_n = L_n h_0 - h_0 \qquad \beta_n = L_n h_1 - h_1 \qquad \gamma_n = L_n h_2 - h_2$$

The hypothesis of the theorem asserts that

$$\|\alpha_n\|_\infty \to 0 \qquad \|\beta_n\|_\infty \to 0 \qquad \|\gamma_n\|_\infty \to 0$$

(Here and for the remainder of the proof we are using the infinity norm.) For the main part of the proof, let f be any element of $C[a, b]$. Let $\varepsilon > 0$. We want to prove that there exists an integer m such that

$$n \geqq m \Rightarrow \|L_n f - f\|_\infty < 3\varepsilon$$

Since f is continuous on a compact interval, it is uniformly continuous. Consequently, there exists a positive δ such that for all x and y in $[a, b]$,

$$|x - y| < \delta \Rightarrow |f(x) - f(y)| < \varepsilon$$

Observe that with $c = 2\|f\|_\infty/\delta^2$, we have

$$|x - y| \geqq \delta \Rightarrow |f(x) - f(y)| \leqq 2\|f\|_\infty \leqq 2\|f\|_\infty \frac{(x-y)^2}{\delta^2} = c(x-y)^2$$

Thus, for all x and y in $[a, b]$, we have

$$|f(x) - f(y)| \leqq \varepsilon + c(x - y)^2$$

This inequality can be written in the form

$$|f - f(y)h_0| \leqq \varepsilon h_0 + c[h_2 - 2yh_1 + y^2 h_0]$$

To verify this, simply substitute x and use the definitions of h_k. From remarks made at the beginning of the proof, we infer that

$$|L_n f - f(y)L_n h_0| \leqq \varepsilon L_n h_0 + c[L_n h_2 - 2yL_n h_1 + y^2 L_n h_0]$$

This is an inequality among functions, and we are at liberty to substitute y into it:

$$\begin{aligned}
&|(L_n f)(y) - f(y)(L_n h_0)(y)| \\
&\quad \leqq \varepsilon(L_n h_0)(y) + c\big[(L_n h_2)(y) - 2y(L_n h_1)(y) + y^2(L_n h_0)(y)\big] \\
&\quad = \varepsilon\big[1 + \alpha_n(y)\big] + c\Big[y^2 + \gamma_n(y) - 2y\big(y + \beta_n(y)\big) + y^2\big(1 + \alpha_n(y)\big)\Big] \\
&\quad = \varepsilon + \varepsilon\alpha_n + c\gamma_n(y) - 2cy\beta_n(y) + cy^2\alpha_n(y) \\
&\quad \leqq \varepsilon + \varepsilon\|\alpha_n\|_\infty + c\|\gamma_n\|_\infty + 2c\|h_1\|_\infty\,\|\beta_n\|_\infty + c\|h_2\|_\infty\,\|\alpha_n\|_\infty
\end{aligned}$$

Select m so that the final right-hand side of this inequality is less than 2ε whenever $n \geqq m$. Then for $n \geqq m$, we have

$$\|L_n - f \cdot L_n h_0\|_\infty \leqq 2\varepsilon$$

Finally, we write

$$\begin{aligned}\|L_n f - f\|_\infty &\leqq \|L_n f - f \cdot L_n h_0\|_\infty + \|f \cdot L_n h_0 - f \cdot h_0\|_\infty \\ &\leqq 2\varepsilon + \|f\|_\infty \|\alpha_n\|_\infty\end{aligned}$$

If necessary, increase m so that for $n \geqq m$ we have $\|f\|_\infty \|\alpha_n\|_\infty < \varepsilon$. Then the last term above is no greater than 3ε. ■

In the proof of the Weierstrass Theorem, a remaining detail is to show that $B_n h_k \longrightarrow h_k$ for $k = 0, 1, 2$. Here $h_k(x) = x^k$. For h_0, we use the Binomial Theorem to write

$$(B_n h_0)(x) = \sum_{k=0}^{n} g_{nk}(x) = \sum_{k=0}^{n} \binom{n}{k} x^k (1-x)^{n-k} = \big(x + (1-x)\big)^n = 1$$

For h_1, we have, with the help of Problem 35,

$$\begin{aligned}(B_n h_1)(x) &= \sum_{k=0}^{n} \left(\frac{k}{n}\right) \binom{n}{k} x^k (1-x)^{n-k} = \sum_{k=1}^{n} \binom{n-1}{k-1} x^k (1-x)^{n-k} \\ &= x \sum_{k=0}^{n-1} \binom{n-1}{k} x^k x^{n-1-k} = x\end{aligned}$$

Lastly, for $k = 2$, we have, using Problem 35 twice,

$$\begin{aligned}(B_n h_2)(x) &= \sum_{k=0}^{n} \left(\frac{k}{n}\right)^2 \binom{n}{k} x^k (1-x)^{n-k} = \sum_{k=1}^{n} \left(\frac{k}{n}\right) \binom{n-1}{k-1} x^k (1-x)^{n-k} \\ &= \sum_{k=1}^{n} \left(\frac{n-1}{n}\,\frac{k-1}{n-1} + \frac{1}{n}\right) \binom{n-1}{k-1} x^k (1-x)^{n-k} \\ &= \frac{n-1}{n} x^2 \sum_{k=2}^{n} \binom{n-2}{k-2} x^{k-2} (1-x)^{n-k} + \frac{x}{n} \\ &= \frac{n-1}{n} x^2 + \frac{x}{n} \longrightarrow x^2\end{aligned}$$

PROBLEMS 6.1

1. Find the polynomials of least degree that interpolate these sets of data:

(a)

x	3	7
y	5	-1

(b)

x	7	1	2
y	146	2	1

(c)

x	3	7	1	2
y	10	146	2	1

(d)

x	3	7	1	2
y	12	146	2	1

(e)

x	1.5	2.7	3.1	-2.1	-6.6	11.0
y	0.0	0.0	0.0	1.0	0.0	0.0

2. The polynomial p of degree $\leqq n$ that interpolates a given function f at $n+1$ prescribed nodes is uniquely defined. Hence, there is a mapping $f \longmapsto p$. Denote this mapping by L and show that

$$Lf = \sum_{i=0}^{n} f(x_i)\ell_i$$

Show that L is **linear**; that is, $L(af + bg) = aLf + bLg$.

3. Prove that the algorithm for computing the coefficients c_i in the Newton form of the interpolating polynomial involves n^2 *long* operations (multiplications and divisions).

4. Refer to Problem 2 and define another mapping G by the formula

$$Gf = \sum_{i=0}^{n} f(x_i)\ell_i^2$$

Prove that Gf is a polynomial of degree at most $2n$, that Gf interpolates f at the nodes, and that Gf is nonnegative whenever f is nonnegative.

5. Discuss the problem of determining a polynomial of degree at most 2 for which $p(0)$, $p(1)$, and $p'(\xi)$ are prescribed, ξ being any preassigned point.

6. Prove that the mapping L in Problem 2 has the property that $Lq = q$ for every polynomial q of degree at most n.

7. (Continuation) Prove that $\sum_{i=0}^{n} \ell_i(x) = 1$ for all x.

8. (Continuation) If p is a polynomial of degree $\leqq n$ that interpolates the function f at $x_0, x_1, \dots, x_n$ (distinct points), then

$$f(x) - p(x) = \sum_{i=0}^{n} [f(x) - f(x_i)]\ell_i(x)$$

9. Prove that if g interpolates the function f at $x_0, x_1, \ldots, x_{n-1}$ and if h interpolates f at $x_1, x_2, \ldots, x_n$, then the function

$$g(x) + \frac{x_0 - x}{x_n - x_0}[g(x) - h(x)]$$

interpolates f at $x_0, x_1, \ldots, x_n$. Notice that h and g need not be polynomials.

10. Prove that the coefficient of x^n in the polynomial p of Equation (9) is

$$\sum_{i=0}^{n} y_i \prod_{\substack{j=0 \\ j \neq i}}^{n} (x_i - x_j)^{-1}$$

11. Prove that for any polynomial q of degree $\leqq n - 1$,

$$\sum_{i=0}^{n} q(x_i) \prod_{\substack{j=0 \\ j \neq i}}^{n} (x_i - x_j)^{-1} = 0$$

12. Determine whether the algorithm

$x \leftarrow a_n b_n$
for $i = 1$ **to** n **do**
  $x \leftarrow (x + a_{n-i}) b_i$
end do

computes

$$x = \sum_{i=0}^{n} a_i \prod_{j=0}^{i} b_j$$

13. Prove that if we take *any* set of 23 nodes in the interval $[-1, 1]$ and interpolate the function $f(x) = \cosh x$ with a polynomial p of degree 22, then the relative error $|p(x) - f(x)|/|f(x)|$ is no greater than 5×10^{-16} on $[-1, 1]$.

14. Let p be a polynomial of degree $\leqq n - 1$ that interpolates the function $f(x) = \sinh x$ at any set of n nodes in the interval $[-1, 1]$, subject only to the condition that one of the nodes is 0. Prove that the relative error obeys this inequality on $[-1, 1]$:

$$|p(x) - f(x)| \leqq \frac{2^n}{n!}|f(x)|$$

15. What is the final value of v in the algorithm shown?

$v \leftarrow c_{i-1}$
for $j = i$ **to** n **do**

$v \leftarrow vx + c_j$
end do

What is the number of additions and subtractions involved in this algorithm?

16. Write an efficient algorithm for evaluating

$$u = \sum_{i=1}^{n} \prod_{j=1}^{i} d_j$$

17. Suppose that p is a polynomial of degree greater than n that interpolates f at $n + 1$ nodes. What can you discover about $f(x) - p(x)$?

18. Prove or disprove: If n is a divisor of m, then each zero of T_n is a zero of T_m.

19. Find a polynomial that assumes these values:

x	1	2	0	3
y	3	2	-4	5

20. Prove that for $x \geqq 1$, $T_n(x) = \cosh(n \cosh^{-1} x)$. *Hint:* Imitate the proof of Theorem 3.

21. Write the Lagrange and Newton interpolating polynomials for these data:

x	2	0	3
$f(x)$	11	7	28

22. Find the Lagrange and Newton forms of the interpolating polynomial for these data:

x	-2	0	1
$f(x)$	0	1	-1

Write both polynomials in the form $a + bx + cx^2$ in order to verify that they are identical as functions.

23. Consider the data

x	$-\sqrt{\frac{3}{5}}$	0	$\sqrt{\frac{3}{5}}$
$f(x)$	$f\left(-\sqrt{\frac{3}{5}}\right)$	$f(0)$	$f\left(\sqrt{\frac{3}{5}}\right)$

What are the Newton interpolation polynomial and the Lagrange interpolation polynomial for these data?

24. The formula for the leading coefficient in T_n is 2^{n-1}. What is the formula for the coefficient of x^{n-2}? What about x^{n-1}?

25. Find the interpolating polynomial for the table

x	1	3	2	6
y	-2	-22	-1	-37

26. The equation $x - 9^{-x} = 0$ has a solution in $[0, 1]$. Find the interpolation polynomial on $x_0 = 0, x_1 = 0.5, x_2 = 1$ for the function on the left side of the equation. By setting the interpolation polynomial equal to 0 and solving the equation, find an approximate solution to the equation.

27. If we interpolate the function $f(x) = e^{x-1}$ with a polynomial p of degree 12 using 13 nodes in $[-1, 1]$, what is a good upper bound for $|f(x) - p(x)|$ on $[-1, 1]$?

28. Let p_k be the polynomial of degree $\leqq k$ such that $p_k(x_i) = y_i$ for $0 \leqq i \leqq k$. Prove that $p_k = p_{k-1}$ if and only if $p_{k-1}(x_k) = y_k$.

29. Devise a method for solving an equation $f(x) = 0$ that gives the correct root in $n + 1$ steps if f^{-1} is a polynomial of degree n in a neighborhood of the root sought. Here f^{-1} is an inverse function: $f^{-1}(f(x)) = x$.

30. Prove the following: If g is a function (not necessarily a polynomial) that interpolates a function f at nodes $x_0, x_1, \ldots, x_{n-1}$, and if h is a function such that $h(x_i) = \delta_{in}$ $(0 \leqq i \leqq n)$, then for some c the function $g + ch$ interpolates f at $x_0, x_1 \ldots, x_n$.

31. Refer to the Lagrange interpolation process and define

$$w_i = \prod_{\substack{j=0 \\ j \neq i}}^{n} (x_i - x_j)^{-1}$$

Show that if x is not a node, then the interpolating polynomial can be evaluated by the formula

$$p(x) = \left[\sum_{i=0}^{n} y_i w_i (x - x_i)^{-1} \right] \Bigg/ \left[\sum_{i=0}^{n} w_i (x - x_i)^{-1} \right]$$

This is called the **barycentric form** of the Lagrange interpolation process.

32. Show that the evaluation of p in Problem 31 is stable in the sense that if the w_i are incorrectly computed, we still have the interpolation property: $\lim_{x \to x_k} p(x) = y_k$ $(0 \leqq k \leqq n)$.

33. Let E be an $(n + 1)$–dimensional vector space of functions defined on a domain D. Let $x_0, x_1, \ldots, x_n$ be distinct points in D. Show that the interpolation problem

$$f(x_i) = y_i \qquad (0 \leqq i \leqq n) \qquad f \in E$$

has a unique solution for any choice of ordinates y_i if and only if no element of E other than 0 vanishes at all the points $x_0, x_1, \ldots, x_n$.

34. Prove that

$$\det \begin{vmatrix} 1 & x_0 & x_0^2 & \cdots & x_0^n \\ 1 & x_1 & x_1^2 & \cdots & x_1^n \\ 1 & x_2 & x_2^2 & \cdots & x_2^n \\ \vdots & \vdots & \vdots & \ddots & \vdots \\ 1 & x_n & x_n^2 & \cdots & x_n^n \end{vmatrix} = \prod_{0 \leqq j < k \leqq n} (x_k - x_j)$$

$$= (x_n - x_0)(x_n - x_1)(x_n - x_2) \cdots (x_n - x_{n-1}) \cdots$$
$$(x_3 - x_0)(x_3 - x_1)(x_3 - x_2)(x_2 - x_0) \cdots$$
$$(x_2 - x_1)(x_1 - x_0)$$

35. Prove that

$$\left(\frac{k}{n}\right)\binom{n}{k} = \binom{n-1}{k-1}$$

36. The functions $1/(1 + x^2)$ and e^{-x^2} have a similar appearance. Do they behave similarly in the interpolation process for equally spaced nodes?

37. The first U.S. postage stamp was issued in 1885 with a value of 2 cents. In 1917, it was raised to 3 cents but then returned to 2 cents in 1919. In 1932, it was upped to 3 cents again where it remained for 26 years. Then there was a series of increases every few years as follows: 1958 = 4 cents, 1963 = 5 cents, 1968 = 6 cents, 1971 = 8 cents, 1974 = 10 cents, 1978 = 15 cents, 1981 = 18 cents in March and 20 cents in October, 1985 = 22 cents, 1988 = 25 cents, 1991 = 29 cents, and 1995 = 32 cents. Determine the Newton interpolation polynomial for these data. Based on this, when will it cost \$1 to mail a letter? When will it cost \$10?

6.2 Divided Differences

In the preceding section, we discussed the problem of *interpolating* a function by a polynomial. We return to that problem now. Let f be a function whose values are known or computable at a set of points (*nodes*) $x_0, x_1, \ldots, x_n$. We assume in this section that these points are *distinct*, but they need not be *ordered* on the real line. We know that there exists a unique polynomial p of degree at most n that interpolates f at the $n + 1$ nodes:

$$p(x_i) = f(x_i) \qquad (0 \leqq i \leqq n) \tag{1}$$

Of course, the polynomial p can be constructed as a linear combination of the basic polynomials $1, x, x^2, \ldots, x^n$. As discussed in the previous section, this basis is not recommended, and we prefer to use a basis appropriate to the Newton form of the interpolating polynomial:

$$
\begin{aligned}
q_0(x) &= 1 \\
q_1(x) &= (x - x_0) \\
q_2(x) &= (x - x_0)(x - x_1) \\
q_3(x) &= (x - x_0)(x - x_1)(x - x_2) \\
&\ \vdots \\
q_n(x) &= (x - x_0)(x - x_1)(x - x_2)\cdots(x - x_{n-1})
\end{aligned}
$$

These lead to the Newton form

$$p(x) = \sum_{j=0}^{n} c_j q_j(x)$$

The interpolation conditions give rise to a linear system of equations for the determination of the unknown coefficients c_j:

$$\sum_{j=0}^{n} c_j q_j(x_i) = f(x_i) \qquad (0 \leqq i \leqq n) \tag{2}$$

In this system of equations, the coefficient matrix is an $(n+1) \times (n+1)$ matrix A whose elements are

$$a_{ij} = q_j(x_i) \qquad (0 \leqq i, j \leqq n) \tag{3}$$

The matrix A is lower triangular because

$$
\begin{aligned}
q_j(x) &= \prod_{k=0}^{j-1}(x - x_k) \\
q_j(x_i) &= \prod_{k=0}^{j-1}(x_i - x_k) = 0 \qquad \text{if } i \leqq j-1
\end{aligned}
\tag{4}
$$

For example, consider the case of three nodes with

$$
\begin{aligned}
p_2(x) &= c_0 q_0(x) + c_1 q_1(x) + c_2 q_2(x) \\
&= c_0 + c_1(x - x_0) + c_2(x - x_0)(x - x_1)
\end{aligned}
$$

Setting $x = x_0$, $x = x_1$, and $x = x_2$, we have a lower triangular system

$$\begin{bmatrix} 1 & 0 & 0 \\ 1 & (x_1 - x_0) & 0 \\ 1 & (x_2 - x_0) & (x_2 - x_0)(x_2 - x_1) \end{bmatrix} \begin{bmatrix} c_0 \\ c_1 \\ c_2 \end{bmatrix} = \begin{bmatrix} f(x_0) \\ f(x_1) \\ f(x_2) \end{bmatrix} \tag{5}$$

In solving Equation (2) for $c_0, c_1, \ldots, c_n$, we can start at the top and work down, computing the coefficients c_j in the order given by their subscripts. In this process, we see that c_0 depends on only $f(x_0)$, that c_1 depends on $f(x_0)$ and $f(x_1)$, and so on. Thus, c_n depends on f at $x_0, x_1, \ldots, x_n$. The notation

$$c_n = f[x_0, x_1, \ldots, x_n] \tag{6}$$

was adopted many years ago to signify this dependency. We are therefore *defining* the symbol $f[x_0, x_1, \ldots, x_n]$ to be the coefficient of q_n when $\sum_{k=0}^{n} c_k q_k$ interpolates f at $x_0, x_1, \ldots, x_n$. Since

$$q_n(x) = (x - x_0)(x - x_1) \cdots (x - x_{n-1}) = x^n + \text{lower-order terms}$$

we can also say that $f[x_0, x_1, \ldots, x_n]$ is the coefficient of x^n in the polynomial of degree at most n that interpolates f at $x_0, x_1, \ldots, x_n$. In all of the preceding description, n can take on arbitrary values. The expressions $f[x_0, x_1, \ldots, x_n]$ are called **divided differences** of f.

Explicit formulas for the first few divided differences will now be given. First, $f[x_0]$ is the coefficient of x^0 in the polynomial of degree 0 interpolating f at x_0. Thus, we must have

$$f[x_0] = f(x_0) \tag{7}$$

The quantity $f[x_0, x_1]$ is the coefficient of x in the polynomial of degree at most 1 interpolating f at x_0 and x_1. Since that polynomial is

$$p(x) = f(x_0) + \frac{f(x_1) - f(x_0)}{x_1 - x_0}(x - x_0) \tag{8}$$

we see that the coefficient of $q_1(x)$ is

$$f[x_0, x_1] = \frac{f(x_1) - f(x_0)}{x_1 - x_0} \tag{9}$$

This gives a hint as to why the term **divided difference** was adopted. A divided difference table with the following form can be displayed:

$$\begin{array}{lll} x_0 & f(x_0) & f[x_0, x_1] \\ x_1 & f(x_1) & \end{array}$$

and the interpolation polynomial is easily formed from it:

$$p(x) = f(x_0) + f[x_0, x_1](x - x_0)$$

Formulas (7) and (9) can also be obtained by solving for c_0 and c_1 in System (5) because $c_0 = f[x_0]$ and $c_1 = f[x_0, x_1]$, in accordance with Equation (6). Equation (1) allows us to write the Newton interpolating polynomial in the form

$$p(x) = \sum_{k=0}^{n} c_k q_k(x) = \sum_{k=0}^{n} f[x_0, x_1, \ldots, x_k] \prod_{j=0}^{k-1} (x - x_j) \tag{10}$$

The correctness of Equation (10) is proved by the following reasoning: If we *truncate* the sum in Equation (10) at the value $k = m$, we obtain $\sum_{k=0}^{m} c_k q_k(x)$. We know that this is the polynomial of degree at most m that interpolates f at $x_0, x_1, \ldots, x_m$. Hence, $c_m = f[x_0, x_1, \ldots, x_m]$. This is true for all m in the range $0 \leqq m \leqq n$.

Higher-Order Divided Differences

For computing the higher-order divided differences, the following theorem can be used.

THEOREM 1 *Divided differences satisfy the equation*

$$f[x_0, x_1, \ldots, x_n] = \frac{f[x_1, x_2, \ldots, x_n] - f[x_0, x_1, \ldots, x_{n-1}]}{x_n - x_0} \tag{11}$$

Proof First, let p_k denote the polynomial of degree at most k that interpolates f at the nodes $x_0, x_1, \ldots, x_k$. We shall be needing p_n and p_{n-1}. Let q denote the polynomial of degree at most $n - 1$ that interpolates f at $x_1, x_2, \ldots, x_n$. Then we have

$$p_n(x) = q(x) + \frac{x - x_n}{x_n - x_0}\left[q(x) - p_{n-1}(x)\right] \tag{12}$$

This equation is proved by noting first that on both sides of the equation there stands a polynomial of degree at most n. Then we verify that the values of these polynomials on the right and the left are the same at the points $x_0, x_1, \ldots, x_n$. Hence, the polynomials must be identical. Now let us examine the coefficient of x^n on the left and right sides of Equation (12). These coefficients must be equal, and so we arrive at Equation (11). ■

Theorem 1 gives us these particular formulas:

$$f[x_0, x_1] = \frac{f[x_1] - f[x_0]}{x_1 - x_0}$$

$$f[x_0, x_1, x_2] = \frac{f[x_1, x_2] - f[x_0, x_1]}{x_2 - x_0}$$

$$\vdots$$

In these formulas, $x_0, x_1, x_2, \ldots$ can be interpreted as independent variables. Because of that, we also have equations such as

$$f[x_i, x_{i+1}, \ldots, x_{i+j}] = \frac{f[x_{i+1}, x_{i+2}, \ldots, x_{i+j}] - f[x_i, x_{i+1}, \ldots, x_{i+j-1}]}{x_{i+j} - x_i} \tag{13}$$

Here $f[x_i]$, $f[x_i, x_{i+1}]$, $f[x_i, x_{i+1}, x_{i+2}]$, $f[x_i, x_{i+1}, x_{i+2}, x_{i+3}]$, and so on are differences of order 0, 1, 2, 3, and so forth, respectively.

If a table of function values $(x_i, f(x_i))$ is given, we can construct from it a table of divided differences. This is customarily laid out in the following form, where differences of orders 0, 1, 2, and 3 are shown in each successive column:

x_0	$f[x_0]$	$f[x_0, x_1]$	$f[x_0, x_1, x_2]$	$f[x_0, x_1, x_2, x_3]$
x_1	$f[x_1]$	$f[x_1, x_2]$	$f[x_1, x_2, x_3]$	
x_2	$f[x_2]$	$f[x_2, x_3]$		
x_3	$f[x_3]$			

The information to the left of the vertical line is given, and the quantities on the right are to be computed. Formula (11) is used to do this. The recursive nature of Formula (11) dictates the triangular form of the divided difference table. For example, the data given do not allow us to compute $f[x_3, x_4]$, $f[x_2, x_3, x_4]$, and so on.

By comparing Equations (10) and (11), we see that the coefficients required in the Newton interpolating polynomial occupy the top row in the divided difference table.

Example 1 Compute a divided difference table for these function values:

x	3	1	5	6
$f(x)$	1	-3	2	4

(14)

Solution We arrange the given table vertically and compute divided differences by the use of Formula (11), arriving at

3	1	2	$-3/8$	$7/40$
1	-3	$5/4$	$3/20$	
5		2	2	
6	4			

■

Example 2 Find the Newton interpolating polynomial for the function values in Table (14).

Solution $p(x) = 1 + 2(x - 3) - \frac{3}{8}(x - 3)(x - 1) + \frac{7}{40}(x - 3)(x - 1)(x - 5)$ ■

Algorithm for Divided Differences

An algorithm for computing a divided difference table can be very efficient and is recommended as the best means for producing an interpolating polynomial. Let us change the notation so that our divided difference table has the entries shown here:

$$
\begin{array}{ll|llllll}
x_0 & c_{00} & c_{01} & c_{02} & c_{03} & \cdots & c_{0,n-1} & c_{0,n} \\
x_1 & c_{10} & c_{11} & c_{12} & c_{13} & \cdots & c_{1,n-1} & \\
x_2 & c_{20} & c_{21} & c_{22} & c_{23} & \iddots & & \\
\vdots & \vdots & \vdots & \vdots & \iddots & & & \\
\vdots & \vdots & \vdots & \iddots & & & & \\
x_{n-1} & c_{n-1,0} & c_{n-1,1} & & & & & \\
x_n & c_{n0} & & & & & &
\end{array}
$$

Again, the vertical line separates the data (on the left) from the entries to be computed. It is clear that we have set

$$
c_{ij} = f[x_i, x_{i+1}, \dots, x_{i+j}]
$$

An algorithm is obtained from a direct translation of Equation (13):

```
for j = 1 to n do
    for i = 0 to n − j do
        c_ij ← (c_i+1,j−1 − c_i,j−1)/(x_i+j − x_i)
    end do
end do
```

In this algorithm, the numbers c_{i0} (which are input) are the values of the function f at the points x_i. They are also the values that the interpolating polynomial will have at those points. The interpolating polynomial, of course, is

$$
\begin{aligned}
p(x) &= c_{00} + c_{01}(x - x_0) + c_{02}(x - x_0)(x - x_1) + \cdots \\
&\quad + c_{0n}(x - x_0)(x - x_1) \cdots (x - x_{n-1}) \\
&= \sum_{i=0}^{n} c_{0i} \prod_{j=0}^{i-1} (x - x_j)
\end{aligned}
$$

If the divided difference algorithm is being used only to compute the coefficients in the Newton interpolating polynomial, then another algorithm can be designed that uses less storage space in the computer. A singly subscripted variable can be used. Let it be denoted by $d = [d_0, d_1, \dots, d_n]$. At the beginning, we put the given function values $f(x_0), f(x_1), \dots, f(x_n)$ into d. Observe that d_0 is already the first of the desired coefficients for the Newton interpolating polynomial. Next, we compute the first column of divided differences, putting them in positions $d_1, d_2, \dots, d_n$. After this has been done, d_0 will still have its original value, but d_1

will have become the next of the required coefficients in the polynomial. We continue this pattern, being careful to store the new divided differences in the *bottom part* of the d-vector so as not to disturb the *top part*, where the final values are gradually being created. Here is the algorithm:

for $i = 0$ **to** n **do**
  $d_i \leftarrow f(x_i)$
end do
for $j = 1$ **to** n **do**
  for $i = n$ **to** j **step** -1 **do**
    $d_i \leftarrow (d_i - d_{i-1})/(x_i - x_{i-j})$
  end do
end do

At the conclusion of this algorithm the vector d contains the coefficients of the polynomial:

$$p(x) = \sum_{i=0}^{n} d_i \prod_{j=0}^{i-1} (x - x_j)$$

In the problems, the efficiency of this procedure is compared to that of the procedure discussed in Section 6.1.

Divided Difference Properties

We conclude this section with some of the nice properties of divided differences.

THEOREM 2 *The divided difference is a symmetric function of its arguments. Thus, if $(z_0, z_1, \ldots, z_n)$ is a permutation of $(x_0, x_1, \ldots, x_n)$, then*

$$f[z_0, z_1, \ldots, z_n] = f[x_0, x_1, \ldots, x_n] \tag{15}$$

Proof The divided difference on the left side of Equation (15) is the coefficient of x^n in the polynomial of degree at most n that interpolates f at the points $z_0, z_1, \ldots, z_n$. The divided difference on the right is the coefficient of x^n in the polynomial of degree at most n that interpolates f at the points $x_0, x_1, \ldots, x_n$. These two polynomials are, of course, the same. ■

THEOREM 3 *Let p be the polynomial of degree at most n that interpolates a function f at a set of $n + 1$ distinct nodes, $x_0, x_1, \ldots, x_n$. If t is a point different from the nodes, then*

$$f(t) - p(t) = f[x_0, x_1, \ldots, x_n, t] \prod_{j=0}^{n} (t - x_j) \tag{16}$$

Proof First, let q be the polynomial of degree at most $n+1$ that interpolates f at the nodes $x_0, x_1, \ldots, x_n, t$. We know that q is obtained from p by adding one term. In fact,

$$q(x) = p(x) + f[x_0, x_1, \ldots, x_n, t] \prod_{j=0}^{n} (x - x_j)$$

Since $q(t) = f(t)$, we obtain at once (by letting $x = t$)

$$f(t) = p(t) + f[x_0, x_1, \ldots, x_n, t] \prod_{j=0}^{n} (t - x_j)$$ ■

THEOREM 4 *If f is n times continuously differentiable on $[a, b]$ and if $x_0, x_1, \ldots, x_n$ are distinct points in $[a, b]$, then there exists a point ξ in (a, b) such that*

$$f[x_0, x_1, \ldots, x_n] = \frac{1}{n!} f^{(n)}(\xi) \qquad \textbf{(17)}$$

Proof First, let p be the polynomial of degree at most $n-1$ that interpolates f at the nodes $x_0, x_1, \ldots, x_{n-1}$. By Theorem 2 in Section 6.1, there exists a point ξ in (a, b) such that

$$f(x_n) - p(x_n) = \frac{1}{n!} f^{(n)}(\xi) \prod_{j=0}^{n-1} (x_n - x_j) \qquad \textbf{(18)}$$

By Theorem 3 in this section,

$$f(x_n) - p(x_n) = f[x_0, x_1, \ldots, x_n] \prod_{j=0}^{n-1} (x_n - x_j) \qquad \textbf{(19)}$$

By comparing Equations (18) and (19), we deduce Equation (17). ■

Hermite-Gennochi Formula

A formula for divided differences, called the **Hermite-Gennochi** formula, is needed in many situations. It asserts that the divided difference $f[x_0, x_1, \ldots, x_n]$ is the integral of $f^{(n)}(u_0 x_0 + \cdots + u_n x_n)$ over the n-dimensional simplex. This **simplex** is the set S_n in $\mathbb{R}^{n+1}$ described as follows:

$$S_n = \left\{ (u_0, u_1, \ldots, u_n) \in \mathbb{R}^{n+1} : u_i \geqq 0, \quad \sum_{i=0}^{n} u_i = 1 \right\}$$

We shall prove the Hermite-Gennochi formula by induction on n, and therefore begin with the case $n = 1$. We have

$$\begin{aligned} S_1 &= \{(u_0, u_1) : u_0 \geqq 0,\ u_1 \geqq 0,\ u_0 + u_1 = 1\} \\ &= \{(u_0, 1 - u_0) : 0 \leqq u_0 \leqq 1\} \end{aligned}$$

Hence, the integral in question can be written as

$$\begin{aligned} \int_{S_1} f'(u_0 x_0 + u_1 x_1)\, du &= \int_0^1 f'\big(u_0 x_0 + (1 - u_0) x_1\big)\, du_0 \\ &= \int_0^1 f'\big(x_1 + u_0(x_0 - x_1)\big)\, du_0 \\ &= \int_0^1 \frac{d}{du_0}\big[f\big(x_1 + u_0(x_0 - x_1)\big)\big] \frac{du_0}{x_0 - x_1} \\ &= \frac{1}{x_0 - x_1} f\big(x_1 + u_0(x_0 - x_1)\big)\bigg|_{u_0=0}^{u_0=1} \\ &= \frac{1}{x_0 - x_1}\big\{f(x_0) - f(x_1)\big\} = f[x_0, x_1] \end{aligned}$$

For the inductive step, let us assume the correctness of the formula for $n - 1$ and prove it for n. Define

$$I(x_0, x_1, \ldots, x_n) = \int_{S_n} f^{(n)}(u_0 x_0 + \cdots + u_n x_n)\, du$$

Since $\sum_{i=0}^n u_i = 1$, we have $u_0 = 1 - \sum_{i=1}^n u_i$, and consequently,

$$\begin{aligned} I(x_0, x_1, \ldots, x_n) &= \int_{S_n} f^{(n)}\big(x_0 + u_1(x_1 - x_0) + \cdots + u_n(x_n - x_0)\big)\, du \\ &= \int_0^1 \int_0^{1-u_1} \cdots \int_0^{1-u_1-\cdots-u_{n-1}} f^{(n)}\big(x_0 + u_1(x_1 - x_0) \\ &\qquad + \cdots + u_n(x_n - x_0)\big)\, du_n \cdots du_1 \end{aligned}$$

The innermost integration is carried out as in the case $n = 1$; that is, we write

$$\begin{aligned} &\int_0^{1-u_1-\cdots-u_{n-1}} \frac{d}{du_n}\Big[f^{(n-1)}\big(x_0 + u_1(x_1 - x_0) + \cdots + u_n(x_n - x_0)\big)\Big] \frac{du_n}{x_n - x_0} \\ &= \frac{1}{x_n - x_0}\Big[f^{(n-1)}\big(x_0 + u_1(x_1 - x_0) + \cdots + u_n(x_n - x_0)\big)\Big]\bigg|_{u_n=0}^{u_n=1-u_1-\cdots-u_{n-1}} \end{aligned}$$

$$= \frac{1}{x_n - x_0}\left[f^{(n-1)}\left(x_0 + \sum_{i=1}^{n-1} u_i(x_i - x_0) + \left(1 - \sum_{i=1}^{n-1} u_i\right)(x_n - x_0)\right)\right.$$
$$\left. - f^{(n-1)}\left(x_0 + \sum_{i=1}^{n-1} u_i(x_i - x_0)\right)\right]$$
$$= \frac{1}{x_n - x_0}\left[f^{(n-1)}\left(x_n + \sum_{i=1}^{n-1} u_i(x_i - x_n)\right) - f^{(n-1)}\left(x_0 + \sum_{i=1}^{n-1} u_i(x_i - x_0)\right)\right]$$

Thus our integral I becomes

$$I(x_0, x_1, \ldots, x_n)$$
$$= \frac{1}{x_n - x_0}\int_0^1 \int_0^{1-u_1} \cdots \int_0^{1-u_1-\cdots-u_{n-2}} \left[f^{(n-1)}\left(x_n + \sum_{i=1}^{n-1} u_i(x_i - x_n)\right)\right.$$
$$\left. - f^{(n-1)}\left(x_0 + \sum_{i=1}^{n-1} u_i(x_i - x_0)\right)\right] du_{n-1} \cdots du_1$$
$$= \frac{1}{x_n - x_0}\left[I(x_n, x_1, \ldots, x_{n-1}) - I(x_0, x_1, \ldots, x_n)\right]$$

By the induction hypothesis, this last expression is

$$\frac{1}{x_n - x_0}\{f[x_n, x_1, \ldots, x_{n-1}] - f[x_0, x_1, \ldots, x_n]\}$$

By the recursive formula for divided differences and the invariance of divided differences under permutation of the arguments, the preceding expression is none other than $f[x_0, x_1, \ldots, x_n]$.

PROBLEMS 6.2

1. Prove Equation (12) by the process outlined in the text.
2. Prove that if f is continuous, then $f[x_0, x_1, \ldots, x_n]$ is continuous on the open set in $\mathbb{R}^{n+1}$ where the components of the vector $(x_0, x_1, \ldots, x_n)$ are distinct.
3. Let $f \in C^n[a, b]$. Prove that if $x_0 \in (a, b)$ and if $x_1, x_2, \ldots, x_n$ all converge to x_0, then $f[x_0, x_1, \ldots, x_n]$ will converge to $f^{(n)}(x_0)/n!$.
4. Prove that if f is a polynomial of degree k, then for $n > k$,
$$f[x_0, x_1, \ldots, x_n] = 0$$
5. Prove that if p is a polynomial of degree at most n, then
$$p(x) = \sum_{i=0}^{n} p[x_0, x_1, \ldots, x_i] \prod_{j=0}^{i-1} (x - x_j)$$

6. Show that the divided differences are linear maps on functions. That is, prove the equation

$$(\alpha f + \beta g)[x_0, x_1, \ldots, x_n] = \alpha f[x_0, x_1, \ldots, x_n] + \beta g[x_0, x_1, \ldots, x_n]$$

7. The divided difference $f[x_0, x_1]$ is analogous to a first derivative, as indicated in Theorem 4. Does it have a property analogous to $(fg)' = f'g + fg'$?

8. Using the functions ℓ_i defined in Section 6.1 and based on nodes $x_0, x_1, \ldots, x_n$, show that for any f,

$$\sum_{i=0}^{n} f(x_i)\ell_i(x) = \sum_{i=0}^{n} f[x_0, x_1, \ldots, x_i] \prod_{j=0}^{i-1}(x - x_j)$$

9. (Continuation) Prove this formula:

$$f[x_0, x_1, \ldots, x_n] = \sum_{i=0}^{n} f(x_i) \prod_{\substack{j=0 \\ j \neq i}}^{n} (x_i - x_j)^{-1}$$

10. Compare the efficiency of the divided difference algorithm to the procedure described in Section 6.1 for computing the coefficients in a Newton interpolating polynomial.

11. Use Cramer's rule in matrix theory to prove that

$$f[x_0, x_1, \ldots, x_n] = \begin{vmatrix} 1 & x_0 & x_0^2 & \cdots & x_0^{n-1} & f(x_0) \\ 1 & x_1 & x_1^2 & \cdots & x_1^{n-1} & f(x_1) \\ \vdots & \vdots & \vdots & & \vdots & \vdots \\ 1 & x_n & x_n^2 & \cdots & x_n^{n-1} & f(x_n) \end{vmatrix} \div \begin{vmatrix} 1 & x_0 & x_0^2 & \cdots & x_0^n \\ 1 & x_1 & x_1^2 & \cdots & x_1^n \\ \vdots & \vdots & \vdots & & \vdots \\ 1 & x_n & x_n^2 & \cdots & x_n^n \end{vmatrix}$$

12. For the particular function $f(x) = x^m$, $m \in \mathbb{N}$, show that

$$f[x_0, x_1 \ldots, x_n] = \begin{cases} 1 & \text{if } n = m \\ 0 & \text{if } n > m \end{cases}$$

13. (See Problem 7.) Prove the **Leibniz formula**

$$(fg)[x_0, x_1, \ldots, x_n] = \sum_{k=0}^{n} f[x_0, x_1, \ldots, x_k] g[x_k, x_{k+1}, \ldots, x_n]$$

14. Write the equation in Problem 9 in this form:

$$f[x_0, x_1 \ldots, x_n] = \sum_{i=0}^{n} \alpha_i f(x_i) \quad \text{where} \quad \alpha_i = \prod_{\substack{j=0 \\ j \neq i}}^{n} (x_i - x_j)^{-1}$$

Prove that if the x_i's are ordered thus:

$$x_0 < x_1 < x_2 < \cdots < x_n$$

then the α_i's alternate in sign.

15. (Continuation) Prove that

$$\sum_{i=0}^{n} \alpha_i x_i^n = 1 \quad \text{and} \quad \sum_{i=0}^{n} \alpha_i = \begin{cases} 1 & \text{if } n = 0 \\ 0 & \text{if } n > 0 \end{cases}$$

16. Let $f(x) = 1/x$ and prove that

$$f[x_0, x_1, \dots, x_n] = (-1)^n \prod_{i=0}^{n} x_i^{-1}$$

17. Find the Newton interpolating polynomial for these data:

x	1	3/2	0	2
$f(x)$	3	13/4	3	5/3

18. Prove that if f is a polynomial, then the divided difference $f[x_0, x_1, \dots, x_n]$ is a polynomial in the variables $x_0, x_1, \dots, x_n$.

19. Show that if u is any function that interpolates f at $x_0, x_1, \dots, x_{n-1}$, and if v is a function that interpolates f at $x_1, x_2, \dots, x_n$, then the function

$$[(x_n - x)u(x) + (x - x_0)v(x)] / (x_n - x_0)$$

interpolates f at $x_0, x_1, \dots, x_n$.

20. (Continuation) Consider the array

$$\begin{array}{cc|ccc} x_0 & y_0 & a_0 & b_0 & c_0 \\ x_1 & y_1 & a_1 & b_1 & \\ x_2 & y_2 & a_2 & & \\ x_3 & y_3 & & & \end{array}$$

in which, for some fixed x, the a_i, b_i, and c_i are computed by the formulas

$$a_i = \left[(x_{i+1} - x)y_i + (x - x_i)y_{i+1}\right] / (x_{i+1} - x_i)$$
$$b_i = [(x_{i+2} - x)a_i + (x - x_i)a_{i+1}] / (x_{i+2} - x_i)$$
$$c_i = [(x_{i+3} - x)b_i + (x - x_i)b_{i+1}] / (x_{i+3} - x_i)$$

Show that c_0 is the value of the cubic interpolating polynomial at x.

21. (Continuation) Generalize the algorithm suggested in Problem 20 to compute $p_n(x)$ for any n. This is known as **Neville's algorithm**.

22. Determine the Newton interpolating polynomial for this table:

x	0	1	2	7
y	51	3	1	201

23. The polynomial $p(x) = 2 - (x + 1) + x(x + 1) - 2x(x + 1)(x - 1)$ interpolates the first four points in the table:

x	-1	0	1	2	3
y	2	1	2	-7	10

By adding one additional term to p, find a polynomial that interpolates the whole table.

24. Write the Newton interpolating polynomial for these data:

x	4	2	0	3
$f(x)$	63	11	7	28

25. Establish an iteration method for solving $f(x) = 0$ as follows: Let q_n be the interpolating quadratic polynomial for the following data and let x_{n+1} be the zero of q_n closest to x_n:

$f(x_n)$	$f(x_{n-1})$	$f(x_{n-2})$
x_n	x_{n-1}	x_{n-2}

26. Prove that for $h > 0$,

$$f(x + 2h) - 2f(x + h) + f(x) = h^2 f''(\xi)$$

for some ξ in the interval $(x, x + 2h)$.

27. Prove

$$m!\, f[0, 1, \ldots, m] = \sum_{j=0}^{m} (-1)^{m-j} \binom{m}{j} f(j)$$

Hint: Use induction. In the middle of your proof, you will have to stop and prove the identity that is the basis for Pascal's triangle:

$$\binom{m}{j-1} + \binom{m}{j} = \binom{m+1}{j}$$

28. Suppose that the Hermite-Gennochi formula were adopted as the *definition* of divided differences. On that basis, derive the recursive formula, Equation (11).

29. Extend the result in Problem 2 by proving that if $f^{(n)}$ is continuous, then $f[x_0, x_1, \dots, x_n]$ is continuous everywhere in $\mathbb{R}^{n+1}$.

30. In Theorem 4, does ξ depend continuously on $x_0, x_1, \dots, x_n$? Answer the same question for $f^{(n)}(\xi)$.

COMPUTER PROBLEMS 6.2

1. For $n = 5$, 10, and 15, find the Newton interpolating polynomial p_n for the function $f(x) = 1/(1 + x^2)$ on the interval $[-5, 5]$. Use equally spaced nodes. In each case, compute $f(x) - p_n(x)$ for 30 equally spaced points in $[-5, 5]$ in order to see the divergence of p_n from f.

2. Program the algorithm described in Problem 25 and test it.

6.3 Hermite Interpolation

The term **Hermite interpolation** refers to the interpolation of a function and some of its derivatives at a set of nodes. (A more precise definition will be given later.) When a distinction is being made between this type of interpolation and the simpler type (in which no derivatives are interpolated), the latter is often called **Lagrange interpolation**.

Basic Concepts

An instructive and useful example of Hermite interpolation is the following: We require a polynomial of least degree that interpolates a function f and its *derivative* f' at two distinct points, say x_0 and x_1. The polynomial sought will satisfy these four conditions:

$$p(x_i) = f(x_i) \qquad p'(x_i) = f'(x_i) \qquad (i = 0, 1)$$

Since there are four conditions, it seems reasonable to look for a solution in Π_3, the linear space of all polynomials of degree at most 3. An element of Π_3 has four coefficients at our disposal. Rather than writing $p(x)$ in terms of $1, x, x^2, x^3$, however, let us write

$$p(x) = a + b(x - x_0) + c(x - x_0)^2 + d(x - x_0)^2(x - x_1)$$

since this will simplify the work. This leads to

$$p'(x) = b + 2c(x - x_0) + 2d(x - x_0)(x - x_1) + d(x - x_0)^2$$

The four conditions on p can now be written in the form

$$
\begin{aligned}
f(x_0) &= a \\
f'(x_0) &= b \\
f(x_1) &= a + bh + ch^2 \qquad (h = x_1 - x_0) \\
f'(x_1) &= b + 2ch + dh^2
\end{aligned}
$$

Obviously, a and b are obtained at once. Then c can be determined from the third equation, in which the terms involving a and b should be transferred to the left side. Finally, d can be determined from the fourth equation. Hence, the problem is solvable—no matter what the values of $f(x_i)$ and $f'(x_i)$ may be.

In general, if values of a function f and some of its derivatives are to be interpolated by a polynomial, we shall encounter some difficulties because the linear systems of equations (from which we expect to compute the coefficients in the polynomial) may be *singular*. A simple example will illustrate this.

Example 1 Find a polynomial p that assumes these values: $p(0) = 0$, $p(1) = 1$, $p'(\frac{1}{2}) = 2$.

Solution Since there are three conditions, we try a quadratic:

$$p(x) = a + bx + cx^2$$

The condition $p(0) = 0$ leads to $a = 0$. The other two conditions lead to

$$
\begin{aligned}
1 &= p(1) &= b + c \\
2 &= p'(\tfrac{1}{2}) &= b + c
\end{aligned}
$$

Thus, no quadratic solves our problem. Notice that the coefficient matrix is singular. If we now try a cubic polynomial for the same problem:

$$p(x) = a + bx + cx^2 + dx^3$$

we discover that there exists a solution but it is not unique. We notice that $a = 0$ as before. The remaining conditions are

$$
\begin{aligned}
1 &= b + c + d \\
2 &= b + c + \frac{3}{4}d
\end{aligned}
$$

The solution of this system is $d = -4$ and $b + c = 5$. ■

The *general* problem of this type obviously has some intriguing difficulties associated with it. The topic is known as **Birkhoff interpolation**, and a large corpus of recent research is devoted to it. The reader wishing to explore further is referred to the monograph of Lorentz, Jetter, and Riemenschneider [1983].

A large class of interpolation problems having unique solutions will now be discussed. The problems in this restricted class are the ones generally known as **Hermite interpolation**. In a Hermite problem, it is assumed that whenever a derivative $p^{(j)}(x_i)$ is to be prescribed (at a node x_i), then $p^{(j-1)}(x_i)$, $p^{(j-2)}(x_i)$, $\dots$, $p'(x_i)$, and $p(x_i)$ will also be prescribed. We choose our notation so that at node x_i, k_i interpolatory conditions are prescribed. Notice that k_i may vary with i. Let the nodes be $x_0, x_1, \dots, x_n$, and suppose that at node x_i these interpolation conditions are given:

$$p^{(j)}(x_i) = c_{ij} \qquad (0 \leqq j \leqq k_i - 1,\ 0 \leqq i \leqq n) \tag{1}$$

The total number of conditions on p is denoted by $m + 1$, and therefore

$$m + 1 = k_0 + k_1 + \cdots + k_n \tag{2}$$

THEOREM 1 *There exists a unique polynomial p in Π_m fulfilling the Hermite interpolation conditions in Equation* (1).

Proof The polynomial p is sought in the space Π_m, and it therefore has $m + 1$ coefficients. The number of interpolatory conditions that are imposed on p by Equation (1) is also $m + 1$. Thus, we have a *square* system of $m + 1$ equations in $m + 1$ unknowns to solve, and we wish to be assured that the coefficient matrix is nonsingular. To prove that a square matrix A is nonsingular, it suffices to prove that the homogeneous equation $Au = 0$ has *only* the 0 solution ($u = 0$). In the interpolation problem under discussion, the *homogeneous* problem is to find $p \in \Pi_m$ such that

$$p^{(j)}(x_i) = 0 \qquad (0 \leqq j \leqq k_i - 1,\ 0 \leqq i \leqq n)$$

Such a polynomial has a zero of multiplicity k_i at x_i ($0 \leqq i \leqq n$) and must therefore be a multiple of the polynomial q given by

$$q(x) = \prod_{i=0}^{n} (x - x_i)^{k_i}$$

Observe, however, that q is of degree

$$m + 1 = \sum_{i=0}^{n} k_i$$

whereas p is to be of degree at most m. We therefore conclude that $p = q = 0$. ■

Example 2 What happens in Hermite interpolation when there is only one node?

Solution In this case, we require a polynomial p of degree k, say, for which

$$p^{(j)}(x_0) = c_{0j} \qquad (0 \leqq j \leqq k)$$

The solution is the Taylor polynomial

$$p(x) = c_{00} + c_{01}(x - x_0) + \frac{c_{02}}{2!}(x - x_0)^2 + \cdots + \frac{c_{0k}}{k!}(x - x_0)^k$$ ■

Newton Divided Difference Method

Now let us explain how the Newton divided difference method can be extended to solve Hermite interpolation problems. We begin with a simple case in which a quadratic polynomial p is sought taking prescribed values:

$$p(x_0) = c_{00} \qquad p'(x_0) = c_{01} \qquad p(x_1) = c_{10} \tag{3}$$

We write the divided difference table in this form:

$$\begin{array}{cc|cc} x_0 & c_{00} & c_{01} & ? \\ x_0 & c_{00} & ? & \\ x_1 & c_{10} & & \end{array}$$

The question marks stand for entries that are not yet computed. Observe that x_0 appears twice in the argument column since two conditions are being imposed on p at x_0. Note further that the prescribed value of $p'(x_0)$ has been placed in the column of first-order divided differences. This is in accordance with the equation

$$\lim_{x \to x_0} f[x_0, x] = \lim_{x \to x_0} \frac{f(x) - f(x_0)}{x - x_0} = f'(x_0)$$

This equation justifies our defining

$$f[x_0, x_0] \equiv f'(x_0)$$

The remaining entries in the divided difference table can be computed in the usual way. The difficulty to be expected when the nodes are repeated occurs only at the entry c_{01}, and the value of c_{01} has been supplied by the *data* rather than being computed. The entries denoted by question marks are then computed in the usual way:

$$p[x_0, x_1] = \frac{p(x_1) - p(x_0)}{x_1 - x_0} = \frac{c_{10} - c_{00}}{x_1 - x_0} \tag{4}$$

and

$$p[x_0, x_0, x_1] = \frac{p[x_0, x_1] - p[x_0, x_0]}{x_1 - x_0} = \frac{c_{10} - c_{00}}{(x_1 - x_0)^2} - \frac{c_{01}}{x_1 - x_0} \tag{5}$$

The interpolating polynomial is written in the usual way:

$$p(x) = p(x_0) + p[x_0, x_0](x - x_0) + p[x_0, x_0, x_1](x - x_0)^2 \tag{6}$$

A direct verification that this polynomial solves the problem in Equation (3) is suggested in Problem 5.

Returning to the example used at the beginning of this section, we can obtain the interpolation polynomial

$$\begin{aligned} p(x) = f(x_0) &+ f'(x_0)(x - x_0) + f[x_0, x_0, x_1](x - x_0)^2 \\ &+ f[x_0, x_0, x_1, x_1](x - x_0)^2(x - x_1) \end{aligned}$$

directly from the following divided difference table:

$$\begin{array}{ll|lll} x_0 & f(x_0) & f'(x_0) & f[x_0, x_0, x_1] & f[x_0, x_0, x_1, x_1] \\ x_0 & f(x_0) & f[x_0, x_1] & f[x_0, x_1, x_1] & \\ x_1 & f(x_1) & f'(x_1) & & \\ x_1 & f(x_1) & & & \end{array}$$

Divided differences in which all arguments are identical are defined in accordance with Theorem 4 in Section 6.2. That theorem asserts the existence of a point ξ such that

$$f[x_0, x_1, \dots, x_k] = \frac{1}{k!} f^{(k)}(\xi)$$

Here it must be assumed that $f^{(k)}$ exists and is continuous in the smallest interval containing the nodes $x_0, x_1, \dots, x_k$. The point ξ will lie in the same interval. If the length of that interval shrinks to 0, we obtain in the limit

$$f[x_0, x_0, \dots, x_0] = \frac{1}{k!} f^{(k)}(x_0) \tag{7}$$

Notice that when $k \geqq 2$, we must be careful to include the factor $1/k!$.

Example 3 Use the extended Newton divided difference algorithm to determine a polynomial that takes these values:

$$p(1) = 2 \qquad p'(1) = 3 \qquad p(2) = 6 \qquad p'(2) = 7 \qquad p''(2) = 8$$

Solution We put the data in the divided difference array as follows, using question marks to signify that quantities are to be computed. The final result is on the right.

$$\begin{array}{cc|cccc} 1 & 2 & 3 & ? & ? & ? \\ 1 & 2 & ? & ? & ? & \\ 2 & 6 & 7 & 4 & & \\ 2 & 6 & & & & \\ 2 & 6 & & & & \end{array} \qquad \begin{array}{cc|cccc} 1 & 2 & 3 & 1 & 2 & -1 \\ 1 & 2 & 4 & 3 & 1 & \\ 2 & 6 & 7 & 4 & & \\ 2 & 6 & & & & \\ 2 & 6 & & & & \end{array}$$

Notice that in the third row, a second difference of 4 is inserted in accordance with Formula (7) in the case $k = 2$. When the array is completed, the numbers in the top row (excluding the node) are the coefficients in the interpolating polynomial:

$$p(x) = 2 + 3(x - 1) + (x - 1)^2 + 2(x - 1)^2(x - 2) - (x - 1)^2(x - 2)^2 \quad \blacksquare$$

Lagrange Form

In principle, formulas akin to the Lagrange interpolation formula can be developed for Hermite interpolation. We shall present one such formula that pertains to an important special case. As before, let the nodes be $x_0, x_1, \dots, x_n$, and let us assume that at each node a function value and the first derivative have been prescribed. The polynomial p that we seek must satisfy these equations:

$$p(x_i) = c_{i0} \qquad p'(x_i) = c_{i1} \qquad (0 \leqq i \leqq n) \tag{8}$$

In analogy with the Lagrange formula, we write

$$p(x) = \sum_{i=0}^{n} c_{i0} A_i(x) + \sum_{i=0}^{n} c_{i1} B_i(x) \tag{9}$$

in which A_i and B_i are to be polynomials with certain properties. A moment's reflection reveals that the following properties would serve our purpose perfectly:

$$\begin{cases} A_i(x_j) = \delta_{ij} \\ A_i'(x_j) = 0 \end{cases} \qquad \begin{cases} B_i(x_j) = 0 \\ B_i'(x_j) = \delta_{ij} \end{cases}$$

Indeed, if A_i and B_i have these properties, then it is elementary to prove that p, as given in Equation (9), will have the interpolation properties in Equation (8). It can now be verified that with the aid of the functions

$$\ell_i(x) = \prod_{\substack{j=0 \\ j \neq i}}^{n} \frac{x - x_j}{x_i - x_j} \qquad (0 \leqq i \leqq n)$$

A_i and B_i can be defined as follows:

$$\begin{cases} A_i(x) = \left[1 - 2(x - x_i)\ell_i'(x_i)\right]\ell_i^2(x) & (0 \leqq i \leqq n) \\ B_i(x) = (x - x_i)\ell_i^2(x) & (0 \leqq i \leqq n) \end{cases}$$

Notice that each ℓ_i is of degree n, and therefore A_i and B_i are of degree $2n + 1$. Hence, p will be of degree at most $2n + 1$. This is as it should be because $2n + 2$ conditions have been placed on p in Equation (8).

The Lagrange form of the interpolation polynomial in Example 1 is given by

$$p(x) = f(x_0)A_0(x) + f(x_1)A_1(x) + f'(x_0)B_0(x) + f'(x_1)B_1(x)$$

where

$$\begin{aligned} A_0(x) &= [1 - 2(x - x_0)\ell_0'(x_0)]\ell_0^2(x) \\ A_1(x) &= [1 - 2(x - x_1)\ell_1'(x_1)]\ell_1^2(x) \\ B_0(x) &= (x - x_0)\ell_0^2(x) \\ B_1(x) &= (x - x_1)\ell_1^2(x) \end{aligned}$$

and

$$\ell_0(x) = \frac{x - x_1}{x_0 - x_1}$$
$$\ell_1(x) = \frac{x - x_0}{x_1 - x_0}$$
$$\ell_0'(x) = \frac{1}{x_0 - x_1}$$
$$\ell_1'(x) = \frac{1}{x_1 - x_0}$$

An error formula for this type of Hermite interpolation is given in the next theorem.

THEOREM 2 *Let $x_0, x_1, \ldots, x_n$ be distinct nodes in $[a, b]$ and let $f \in C^{2n+2}[a, b]$. If p is the polynomial of degree at most $2n + 1$ such that*

$$p(x_i) = f(x_i) \qquad p'(x_i) = f'(x_i) \qquad (0 \leqq i \leqq n)$$

then to each x in $[a, b]$ there corresponds a point ξ in (a, b) such that

$$f(x) - p(x) = \frac{f^{(2n+2)}(\xi)}{(2n+2)!} \prod_{i=0}^{n} (x - x_i)^2$$

Proof If x is a node, the formula is obviously correct. Now fix x, not a node. Define w and ϕ by

$$w(t) = \prod_{i=0}^{n} (t - x_i)^2 \qquad \phi = f - p - \lambda w$$

where λ is chosen so that $\phi(x) = 0$. Notice that ϕ has at least $n + 2$ zeros in $[a, b]$—namely, $x, x_0, x_1, \ldots, x_n$. By Rolle's Theorem, ϕ' has at least $n + 1$ zeros different from the points just enumerated. In addition, ϕ' vanishes at each node. Thus ϕ' has at least $2n + 2$ zeros in $[a, b]$. By Rolle's Theorem, ϕ'' has at least $2n + 1$ zeros in (a, b), and by repeating the argument, we conclude that $\phi^{(2n+2)}$ has a zero, say ξ, in (a, b). Thus,

$$0 = \phi^{(2n+2)}(\xi) = f^{(2n+2)}(\xi) - p^{(2n+2)}(\xi) - \lambda w^{(2n+2)}(\xi)$$

Since p is of degree at most $2n + 1$, we have $p^{(2n+2)} = 0$. Since $w(t)$ has the leading term t^{2n+2}, it follows that $w^{(2n+2)} = (2n + 2)!$. Using this information together with the value $\lambda = \left[f(x) - p(x)\right]/w(x)$ in the equation for $\phi^{(2n+2)}$ gives us

$$0 = f^{(2n+2)}(\xi) - \left[f(x) - p(x)\right](2n + 2)!/w(x)$$

which is another form of the equation to be proved. ■

Divided Differences with Repetitions

The remainder of this section is devoted to a discussion of divided differences when repetitions are permitted in the arguments. There are many ways to proceed, such as with a recursive definition (Braess [1984]), or using a definition with determinants (Schumaker [1981]), or by simply extending the definition previously given for $f[x_0, x_1, \ldots, x_n]$ in the case of distinct arguments. We follow the latter course, as given by Conte and de Boor [1980].

Recall that if the points $x_0, x_1, \ldots, x_n$ are distinct, then $f[x_0, x_1, \ldots, x_n]$ has been defined in Section 6.2 as the coefficient of x^n in the polynomial from Π_n that interpolates f at $x_0, x_1, \ldots, x_n$. Let us now define the meaning of interpolation when the list of points contains repetitions.

DEFINITION 1 *We say that f interpolates 0 at $x_0, x_1, \ldots, x_n$ if $f^{(k-1)}(\xi) = 0$ for each point ξ that occurs k or more times in the list $x_0, x_1, \ldots, x_n$.*

Thus, for example, we shall say that f interpolates 0 at 1, 3, 8, 1, 13, 1, 8 if

$$f(1) = f(3) = f(8) = f'(1) = f(13) = f''(1) = f'(8) = 0$$

It is easy to prove that a polynomial p interpolates 0 at $x_0, x_1, \ldots, x_n$ if and only if $p(x)$ contains the factor

$$q(x) = \prod_{j=0}^{n} (x - x_j)$$

If f and g are two functions such that $f - g$ interpolates 0 at $x_0, x_1, \ldots, x_n$, then we say that f interpolates g (or g interpolates f) at $x_0, x_1, \ldots, x_n$.

Using the terminology just explained, we can restate the theorem proved earlier in this section in the following form.

THEOREM 3 *Let $x_0, x_1, \ldots, x_m$ be a list of points in which no element is repeated more than k times. Let f belong to class C^{k-1} on an interval containing these points. Then there exists a unique polynomial in Π_m that interpolates f at the given points.*

The definition of $f[x_0, x_1, \ldots, x_n]$ in the general case is that it is the coefficient of x^n in the polynomial from Π_n that interpolates f at $x_0, x_1, \ldots, x_n$. If a number ξ occurs k times in the list, then this definition requires the existence of the derivative $f^{(k-1)}(\xi)$. In any contrary case, the divided difference does not exist (that is, is not defined). Having made this definition, we can now prove the validity of the *general* Newton interpolation formula

$$p(x) = \sum_{j=0}^{n} f[x_0, x_1, \ldots, x_j] \prod_{i=0}^{j-1} (x - x_i) \tag{10}$$

Recall that $\prod_{i=0}^{-1}(x - x_i) = 1$ by definition.

THEOREM 4 *If f is sufficiently differentiable so that the divided differences occurring in Equation (10) exist, then that equation gives the polynomial p in Π_n that interpolates f at $x_0, x_1, \dots, x_n$.*

Proof We proceed by induction on n. For $n = 0$, it is asserted that $f[x_0]$ is the polynomial in Π_0 that interpolates f at x_0. This is obviously true.

Suppose now that the polynomial q defined by

$$q(x) = \sum_{j=0}^{n-1} f[x_0, x_1, \dots, x_j] \prod_{i=0}^{j-1} (x - x_i)$$

interpolates f at $x_0, x_1, \dots, x_{n-1}$. Let p be the polynomial in Π_n that interpolates f at $x_0, x_1, \dots, x_n$. The previous theorems guarantee the existence and unicity of p. By the definition of the divided difference, the coefficient of x^n in p is $f[x_0, x_1, \dots, x_n]$. Hence, the polynomial

$$p(x) - f[x_0, x_1, \dots, x_n] \prod_{i=0}^{n-1} (x - x_i)$$

is of degree at most $n - 1$. It interpolates f at $x_0, x_1, \dots, x_{n-1}$, by Problems 8, 9, and 11. By the unicity of q, we must have therefore

$$p(x) - f[x_0, x_1, \dots, x_n] \prod_{i=0}^{n-1} (x - x_i) = q(x)$$

Hence,

$$\begin{aligned} p(x) &= q(x) + f[x_0, x_1, \dots, x_n] \prod_{i=0}^{n-1} (x - x_i) \\ &= \sum_{j=0}^{n} f[x_0, x_1, \dots, x_j] \prod_{i=0}^{j-1} (x - x_i) \end{aligned}$$

■

The final issue to be settled is whether the divided differences in their general form obey the recurrence relation

$$f[x_0, x_1, \dots, x_n] = \left\{ f[x_1, x_2, \dots, x_n] - f[x_0, x_1, \dots, x_{n-1}] \right\} / (x_n - x_0)$$

Let us assume that $x_0 \leqq x_1 \leqq \cdots \leqq x_n$. This involves no loss of generality because the generalized divided differences are symmetric functions of their arguments. (The proof of this fact given in Section 6.2 is valid in the general case.) The recurrence relation will obviously fail in the case $x_n = x_0$. But this equality implies that $x_0 = x_1 = \cdots = x_n$, and so this is precisely the case in which we shall want to use the formula

$$f[x_0, x_0, \dots, x_0] = \frac{1}{n!} f^{(n)}(x_0)$$

In all other cases the recursive formula is valid. We now prove this formally.

THEOREM 5 *Let $x_0 \leqq x_1 \leqq \cdots \leqq x_n$. Then the divided differences obey this recursive formula:*

$$f[x_0, x_1, \dots, x_n] = \begin{cases} \dfrac{f[x_1, x_2, \dots, x_n] - f[x_0, x_1, \dots, x_{n-1}]}{x_n - x_0} & \text{if } x_n \neq x_0 \\ f^{(n)}(x_0)/n! & \text{if } x_n = x_0 \end{cases} \tag{11}$$

Proof We proceed by induction on n. The polynomial in Π_1 that interpolates f at x_0 and x_1 is

$$p(x) = \begin{cases} (x - x_0)[f(x_1) - f(x_0)]/(x - x_0) + f(x_0) & (x_0 \neq x_1) \\ f'(x_0)(x_1 - x_0) + f(x_0) & (x_0 = x_1) \end{cases}$$

In this polynomial, the coefficient of x is

$$f[x_0, x_1] = \begin{cases} \{f[x_1] - f[x_0]\}/(x_1 - x_0) & (x_0 \neq x_1) \\ f'(x_0) & (x_0 = x_1) \end{cases}$$

This is in accordance with Equation (11).

Now assume that Equation (11) is valid for all n from 1 to $m - 1$, inclusive. Let $x_0 \leqq x_1 \leqq \cdots \leqq x_m$ be a list of nodes, and consider the polynomial p in Π_m that interpolates f at these points. If $x_m = x_0$, then all the points $x_0, x_1, \dots, x_m$ are the same. In this case, by Problem 7, p is the Taylor polynomial of f at x_0 given by

$$p(x) = \sum_{k=0}^{m} \frac{1}{k!} f^{(k)}(x_0)(x - x_0)^k$$

The coefficient of x^m in $p(x)$ is $f^{(m)}(x_0)/m!$, and this establishes Equation (11) when $n = m$ and $x_m = x_0$. In the other case, the argument used to prove Theorem 1 in Section 6.2 is valid. ■

PROBLEMS 6.3

1. Use the extended Newton divided difference method to obtain a quartic polynomial that takes these values:

x	0	1	2
$p(x)$	2	−4	44
$p'(x)$	−9	4	

2. (Continuation) Find a quintic polynomial that takes the values given in Problem 1 and, in addition, satisfies $p(3) = 2$. *Hint:* Add a suitable term to the polynomial found in Problem 1.

3. Obtain a formula for the polynomial p of least degree that takes these values:

$$p(x_i) = y_i \qquad p'(x_i) = 0 \qquad (0 \leqq i \leqq n)$$

4. What condition will have to be placed on the nodes x_0 and x_1 if the interpolation problem

$$p(x_i) = c_{i0} \qquad p''(x_i) = c_{i2} \qquad (i = 0, 1)$$

is to be solvable by a cubic polynomial (for arbitrary c_{ij})?

5. Show that the polynomial given in Equation (6), with the divided differences in Equations (4) and (5), fulfills the conditions in Equation (3).

6. Verify the properties claimed for the functions A_i and B_i.

7. Prove that the Taylor polynomial

$$p(x) = \sum_{j=0}^{k-1} \frac{1}{j!} f^{(j)}(x_0)(x - x_0)^j$$

interpolates f at $x_0, x_0, \ldots, x_0$ (k repetitions).

8. Prove that a polynomial interpolates 0 at $x_0, x_1, \ldots, x_n$ (repetitions permitted) if and only if it contains the factor $\prod_{j=0}^{n}(x - x_j)$.

9. Prove that if f interpolates g at $x_0, x_1, \ldots, x_n$ and if h interpolates 0 at these points, then $f \pm ch$ interpolates g at these points.

10. Fix $x_0, x_1, \ldots, x_n$ and show that the set of functions that interpolate 0 at these points is an *algebra*; that is, it is closed under addition, multiplication, and multiplication by scalars.

11. Prove that if f interpolates 0 at the nodes $x_0, x_1, \ldots, x_n$, then it interpolates 0 at the nodes $x_0, x_1, \ldots, x_{n-1}$.

12. Let $x_0 < x_1 < \cdots < x_n$, and let f be continuously differentiable. Show that

$$\frac{\partial}{\partial x_i} f[x_0, x_1, \ldots, x_n] = f[x_0, x_1, \ldots, x_i, x_i, x_{i+1}, \ldots, x_n]$$

13. Write out in detail and then simplify the functions A_i and B_i occurring in Equation (9), taking $n = 2$.

14. Derive a formula for $\ell_i'(x)$.

15. (Use Rolle's Theorem.) Let $f \in C^n[\alpha, \beta]$. Suppose that f has a zero of multiplicity m at α and a root of multiplicity k at β, where $m \geqq 1$, $k \geqq 1$, and $m + k - 1 = n$. Prove that $f^{(n)}$ has at least one zero in (α, β).

16. Consider the polynomial

$$p(t) = b - (b-a)\left[3\left(\frac{b-t}{b-a}\right)^2 - 2\left(\frac{b-t}{b-a}\right)^3\right]$$

Show that $|p'(t)| \leqq p'((a+b)/2) = 3/2$, $p(a) = a$, $p(b) = b$, $p'(a) = 0$, and $p'(b) = 0$.

6.4 Spline Interpolation

A **spline function** consists of polynomial pieces on subintervals joined together with certain continuity conditions. Formally, suppose that $n+1$ points $t_0, t_1, \ldots, t_n$ have been specified and satisfy $t_0 < t_1 < \cdots < t_n$. These points are called **knots.** Suppose also that an integer $k \geqq 0$ has been prescribed. A **spline function of degree** k having knots $t_0, t_1, \ldots, t_n$ is a function S such that:

(i) On each interval $[t_{i-1}, t_i)$, S is a polynomial of degree $\leqq k$.
(ii) S has a continuous $(k-1)$st derivative on $[t_0, t_n]$.

Hence, S is a continuous piecewise polynomial of degree at most k having continuous derivatives of all orders up to $k-1$.

Splines of degree 0 are piecewise constants. A spline of degree 0 can be given explicitly in the form

$$S(x) = \begin{cases} S_0(x) = c_0 & x \in [t_0, t_1) \\ S_1(x) = c_1 & x \in [t_1, t_2) \\ \quad\vdots & \quad\vdots \\ S_{n-1}(x) = c_{n-1} & x \in [t_{n-1}, t_n] \end{cases}$$

The intervals $[t_{i-1}, t_i)$ do not intersect each other, and so no ambiguity arises in defining such a function at the knots. A typical spline of degree 0 with six knots is shown in Figure 6.3.

Figure 6.4 shows the graph of a typical spline function of degree 1 with nine knots. A function such as this can be defined explicitly by

$$S(x) = \begin{cases} S_0(x) = a_0x + b_0 & x \in [t_0, t_1) \\ S_1(x) = a_1x + b_1 & x \in [t_1, t_2) \\ \quad\vdots & \quad\vdots \\ S_{n-1}(x) = a_{n-1}x + b_{n-1} & x \in [t_{n-1}, t_n] \end{cases}$$

If the knots t_i and the coefficients a_i, b_i are all prescribed, then the value of S at x is obtained by first identifying the subinterval $[t_i, t_{i+1})$ that contains x. The spline

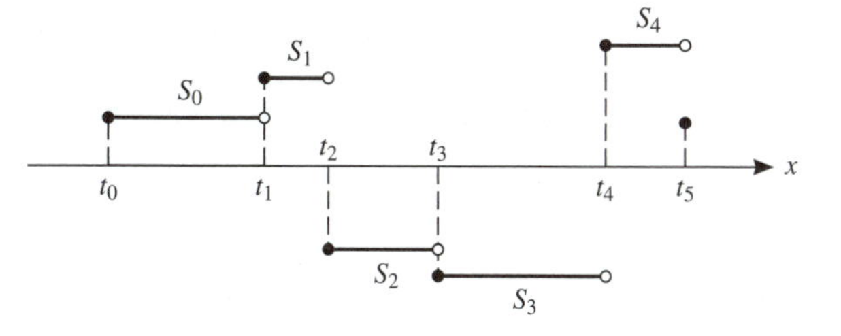

FIGURE 6.3 A spline of degree 0

function can be defined on the entire real line. For convenience, we can use the expression $a_0x + b_0$ on the interval $(-\infty, t_1)$ and the expression $a_{n-1}x + b_{n-1}$ on the interval $[t_{n-1}, \infty)$. The function S is continuous, and so the piecewise polynomials match up at the nodes; that is, $S_i(t_{i+1}) = S_{i+1}(t_{i+1})$. A pseudocode to evaluate the first-degree spline $S(x)$ follows:

input $(t_i), (a_i), (b_i), x, n$
for $i = 1$ **to** $n - 1$ **do**
    if $x < t_i$ **then**
        $S(x) = a_{i-1}x + b_{i-1}$
        output $S(x)$
        exit loop
    end if
end do
$S(x) = a_{n-1}x + b_{n-1}$
output $S(x)$

Cubic Splines

We shall develop more thoroughly the theory and construction of the **cubic splines** ($k = 3$) since these are often used in practice. We assume that a table of values has been given:

x	t_0	t_1	$\cdots$	t_n
y	y_0	y_1	$\cdots$	y_n

(1)

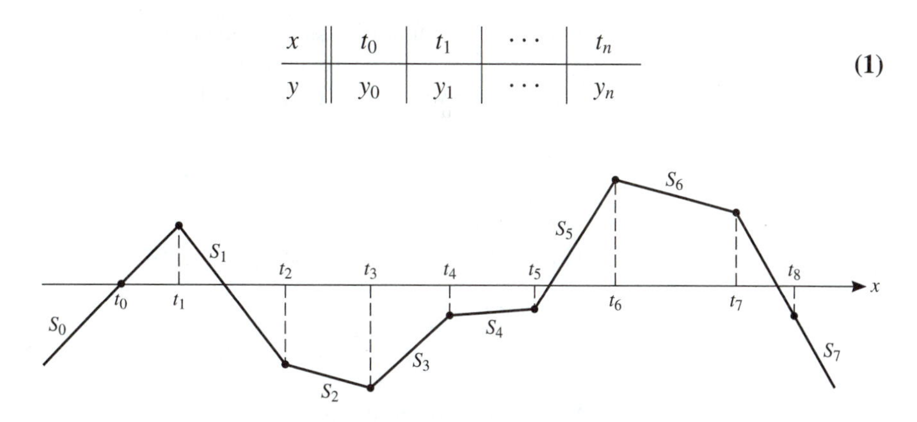

FIGURE 6.4 A spline of degree 1

and that a cubic spline S is to be constructed to interpolate the table. On each interval $[t_0, t_1]$, $[t_1, t_2], \ldots, [t_{n-1}, t_n]$, S is given by a different cubic polynomial. Let S_i be the cubic polynomial that represents S on $[t_i, t_{i+1}]$. Thus,

$$S(x) = \begin{cases} S_0(x) & x \in [t_0, t_1] \\ S_1(x) & x \in [t_1, t_2] \\ \vdots & \vdots \\ S_{n-1}(x) & x \in [t_{n-1}, t_n] \end{cases} \tag{2}$$

The polynomials S_{i-1} and S_i interpolate the same value at the point t_i and therefore

$$S_{i-1}(t_i) = y_i = S_i(t_i) \qquad (1 \leqq i \leqq n-1) \tag{3}$$

Hence, S is automatically continuous. Moreover, S' and S'' are assumed to be continuous, and these conditions will be used in the derivation of the cubic spline function.

Does the continuity of S, S', and S'' provide enough conditions to define a cubic spline? There are $4n$ coefficients in the piecewise cubic polynomial since there are four coefficients in each of the n cubic polynomials. On each subinterval $[t_i, t_{i+1}]$, there are two interpolation conditions, $S(t_i) = y_i$ and $S(t_{i+1}) = y_{i+1}$, giving $2n$ conditions. The continuity of S gives no additional conditions since it has already been counted in the interpolation conditions. The continuity of S' gives one condition at each interior knot, $S'_{i-1}(t_i) = S'_i(t_i)$, accounting for $n-1$ additional conditions. Similarly, the continuity of S'' gives another $n-1$ conditions. Thus, there are altogether $4n-2$ conditions for determining $4n$ coefficients. Two *degrees of freedom* remain, and there are various ways of using them to advantage.

Now we derive the equation for $S_i(x)$ on the interval $[t_i, t_{i+1}]$. First we define the numbers $z_i = S''(t_i)$. Clearly, z_i exist for $0 \leqq i \leqq n$ and satisfy

$$\lim_{x \downarrow t_i} S''(x) = z_i = \lim_{x \uparrow t_i} S''(x) \qquad (1 \leqq i \leqq n-1) \tag{4}$$

because S'' is continuous at each interior knot. Since S_i is a cubic polynomial on $[t_i, t_{i+1}]$, S''_i is a linear function satisfying $S''_i(t_i) = z_i$ and $S''_i(t_{i+1}) = z_{i+1}$ and therefore is given by the straight line between z_i and z_{i+1}:

$$S''_i(x) = \frac{z_i}{h_i}(t_{i+1} - x) + \frac{z_{i+1}}{h_i}(x - t_i) \tag{5}$$

where $h_i \equiv t_{i+1} - t_i$. If this is integrated twice, the result is S_i itself:

$$S_i(x) = \frac{z_i}{6h_i}(t_{i+1} - x)^3 + \frac{z_{i+1}}{6h_i}(x - t_i)^3 + C(x - t_i) + D(t_{i+1} - x) \tag{6}$$

where C and D are constants of integration. (To verify, differentiate (6) twice to obtain (5).) The interpolation conditions $S_i(t_i) = y_i$ and $S_i(t_{i+1}) = y_{i+1}$ can now be imposed on S_i to determine C and D. The result is

$$S_i(x) = \frac{z_i}{6h_i}(t_{i+1} - x)^3 + \frac{z_{i+1}}{6h_i}(x - t_i)^3 + \left(\frac{y_{i+1}}{h_i} - \frac{z_{i+1}h_i}{6}\right)(x - t_i) + \left(\frac{y_i}{h_i} - \frac{z_i h_i}{6}\right)(t_{i+1} - x) \tag{7}$$

Equation (7) is also verified easily. Simply let $x = t_i$ and $x = t_{i+1}$ to see that the interpolation conditions are fulfilled. Once the values of $z_0, z_1, \ldots, z_n$ have been determined, Equations (2) and (7) can be used to evaluate $S(x)$ for any x in the interval $[t_0, t_n]$.

To determine $z_1, z_2, \ldots, z_{n-1}$, we use the continuity conditions for S'. At the interior knots t_i, we must have $S'_{i-1}(t_i) = S'_i(t_i)$. Equation (7) gives us $S'_i(x)$ by differentiating. Then substitution of $x = t_i$ and simplification lead to

$$S'_i(t_i) = -\frac{h_i}{3}z_i - \frac{h_i}{6}z_{i+1} - \frac{y_i}{h_i} + \frac{y_{i+1}}{h_i} \tag{8}$$

Analogously, using Equation (7) to obtain S_{i-1}, we have

$$S'_{i-1}(t_i) = \frac{h_{i-1}}{6}z_{i-1} + \frac{h_{i-1}}{3}z_i - \frac{y_{i-1}}{h_{i-1}} + \frac{y_i}{h_{i-1}} \tag{9}$$

When the right-hand sides of Equations (8) and (9) are set equal to each other, the result can be written as

$$h_{i-1}z_{i-1} + 2(h_i + h_{i-1})z_i + h_i z_{i+1} = \frac{6}{h_i}(y_{i+1} - y_i) - \frac{6}{h_{i-1}}(y_i - y_{i-1}) \tag{10}$$

This equation is used only for $i = 1, 2, \ldots, n-1$. (Why?) It then gives a system of $n-1$ linear equations for the $n+1$ unknowns $z_0, z_1, \ldots, z_n$. We can select z_0 and z_n arbitrarily and solve the resulting system of equations to obtain $z_1, z_2, \ldots, z_{n-1}$. One excellent choice is $z_0 = z_n = 0$. The resulting spline function is called a **natural cubic spline**. (This choice of spline is supported by a theorem to be proved later.)

The linear system of equations (10) for $1 \leqq i \leqq n-1$ with $z_0 = 0$ and $z_n = 0$ is symmetric, tridiagonal, diagonally dominant, and of the form

$$\begin{bmatrix} u_1 & h_1 & & & & \\ h_1 & u_2 & h_2 & & & \\ & h_2 & u_3 & h_3 & & \\ & & \ddots & \ddots & \ddots & \\ & & & h_{n-3} & u_{n-2} & h_{n-2} \\ & & & & h_{n-2} & u_{n-1} \end{bmatrix} \begin{bmatrix} z_1 \\ z_2 \\ z_3 \\ \vdots \\ z_{n-2} \\ z_{n-1} \end{bmatrix} = \begin{bmatrix} v_1 \\ v_2 \\ v_3 \\ \vdots \\ v_{n-2} \\ v_{n-1} \end{bmatrix}$$

where

$$
\begin{aligned}
h_i &= t_{i+1} - t_i \\
u_i &= 2(h_i + h_{i-1}) \\
b_i &= \frac{6}{h_i}(y_{i+1} - y_i) \\
v_i &= b_i - b_{i-1}
\end{aligned}
$$

It can be solved by a special algorithm (Gaussian elimination without scaled row pivoting) as follows:

input $n, (t_i), (y_i)$
for $i = 0$ **to** $n - 1$ **do**
  $h_i \leftarrow t_{i+1} - t_i$
  $b_i \leftarrow 6(y_{i+1} - y_i)/h_i$
end do
$u_1 \leftarrow 2(h_0 + h_1)$
$v_1 \leftarrow b_1 - b_0$
for $i = 2$ **to** $n - 1$ **do**
  $u_i \leftarrow 2(h_i + h_{i-1}) - h_{i-1}^2/u_{i-1}$
  $v_i \leftarrow b_i - b_{i-1} - h_{i-1}v_{i-1}/u_{i-1}$
end do
$z_n \leftarrow 0$
for $i = n - 1$ **to** 1 **step** -1 **do**
  $z_i \leftarrow (v_i - h_i z_{i+1})/u_i$
end do
$z_0 \leftarrow 0$
output (z_i)

A subprogram or procedure can be easily written based on the preceding algorithm. It would accept as input an array of nodes (t_i) and an array of corresponding function values (y_i). The routine would compute and return an array of the (z_i) values.

Since there are divisions by u_i in this algorithm, we need to prove that $u_i \neq 0$. We show by induction that $u_i > h_i > 0$. For $i = 1$, this is clear since $u_1 = 2(h_0 + h_1)$. If $u_{i-1} > h_{i-1}$, then $u_i > h_i$ since

$$u_i = 2(h_i + h_{i-1}) - \frac{h_{i-1}^2}{u_{i-1}} > 2(h_i + h_{i-1}) - h_{i-1} > h_i = t_{i+1} - t_i > 0$$

After the coefficients $z_0, z_1, \ldots, z_n$ have been determined, any value of the cubic spline function (2) can be computed from Equation (7). Given any x, it is necessary first to determine which of the intervals

$$(-\infty, t_1), \quad [t_1, t_2), \quad \ldots, \quad [t_{n-2}, t_{n-1}), \quad [t_{n-1}, \infty)$$

contains x. By convention, we use S_0 not only on $[t_0, t_1]$ but also on $(-\infty, t_0)$. Likewise, we use S_{n-1} on $[t_{n-1}, t_n]$ and on (t_n, ∞). The interval containing x is determined by testing, in order,

$$x - t_{n-1}, \quad x - t_{n-2}, \quad \ldots, \quad x - t_1$$

to see whether any are nonnegative. If one of these is nonnegative, we take the first such occurrence, say $x - t_i$. Then $x - t_i \geqq 0$ but $x - t_{i+1} < 0$, so $t_i \leqq x < t_{i+1}$. If all the tested terms are negative, then we have $x \in (-\infty, t_1]$. From the index i so determined, the desired polynomial S_i could be evaluated at the given x value using Equation (7). However, we can rewrite (7) in a more efficient *nested* form by using Problem 4:

$$S_i(x) = y_i + (x - t_i)\Big[C_i + (x - t_i)\big[B_i + (x - t_i)A_i\big]\Big] \tag{11}$$

where

$$\begin{aligned} A_i &= \frac{1}{6h_i}(z_{i+1} - z_i) \\ B_i &= \frac{z_i}{2} \\ C_i &= -\frac{h_i}{6}z_{i+1} - \frac{h_i}{3}z_i + \frac{1}{h_i}(y_{i+1} - y_i) \end{aligned}$$

With these explanations in mind, the reader should have no trouble writing a subprogram or procedure that carries out the calculation in Equation (11). It would take as input the number n, the array of nodes (t_i), the array of function values (y_i), and the (z_i) that came from the previous routine. It also would take a single real number x as input and return the value of $S(x)$.

Here are some of the results from a simple computer test for two routines based on the algorithms just given. We sample the function $f(x) = \sqrt{x}$ at 11 knots equally spaced in the interval [0, 2.25], interpolate with a cubic spline function S, and print the differences $E(x) \equiv S(x) - f(x)$ at 37 points:

x	$\lvert E(x)\rvert$
0.0000	0.0
0.0625	$1.07321E - 01$
0.1250	$7.52666E - 02$
0.1875	$3.32617E - 02$
0.2500	0.0
⋮	⋮
1.7500	0.0
1.8125	$3.64780E - 05$
1.8750	$6.36578E - 05$
1.9375	$5.85318E - 08$

x	$\lvert E(x)\rvert$
2.0000	0.0
2.0625	$1.14083E-04$
2.1250	$2.12312E-04$
2.1875	$2.04682E-04$
2.2500	0.0

The output shows that the error, E, is 0 at each knot. For points other than knots, $|E|$ is no greater than 0.11. The greatest values of $|E|$ occur in $[t_0, t_1]$. (Why?)

We now present a theorem to the effect that the natural cubic spline produces the *smoothest possible* interpolating function. The word **smooth** is given a technical meaning in the theorem.

THEOREM 1 *Let f'' be continuous in $[a, b]$ and let $a = t_0 < t_1 < \cdots < t_n = b$. If S is the natural cubic spline interpolating f at the knots t_i for $0 \leqq i \leqq n$, then*

$$\int_a^b [S''(x)]^2\,dx \leqq \int_a^b [f''(x)]^2\,dx$$

Proof Let $g \equiv f - S$. Then $g(t_i) = 0$ for $0 \leqq i \leqq n$ and

$$\int_a^b (f'')^2\,dx = \int_a^b (S'')^2\,dx + \int_a^b (g'')^2\,dx + 2\int_a^b S''g''\,dx$$

The proof will be complete if we can show that

$$\int_a^b S''g''\,dx \geqq 0$$

We establish this in the following analysis, using integration by parts, the condition $S''(t_0) = S''(t_n) = 0$, and the fact that S''' is constant (say c_i) on $[t_{i-1}, t_i]$.

$$\begin{aligned}
\int_a^b S''g''\,dx &= \sum_{i=1}^n \int_{t_{i-1}}^{t_i} S''g''\,dx \\
&= \sum_{i=1}^n \left\{ (S''g')(t_i) - (S''g')(t_{i-1}) - \int_{t_{i-1}}^{t_i} S'''g'\,dx \right\} \\
&= -\sum_{i=1}^n \int_{t_{i-1}}^{t_i} S'''g'\,dx = -\sum_{i=1}^n c_i \int_{t_{i-1}}^{t_i} g'\,dx \\
&= -\sum_{i=1}^n c_i[g(t_i) - g(t_{i-1})] = 0
\end{aligned}$$

■

Recall that the **curvature** of a curve described by the equation $y = f(x)$ is the quantity

$$|f''(x)|\left[1 + \left\{f'(x)\right\}^2\right]^{-3/2}$$

If the nonlinear bracketed term is dropped, $|f''(x)|$ is left as an approximation to the curvature. In natural cubic spline interpolation, we are finding a curve with minimal (approximate) curvature over an interval because the quantity $\int_a^b [f''(x)]^2\,dx$ is being minimized.

At one step in the above proof, we note that there is a "collapsing" sum:

$$\sum_{i=1}^{n} \left[(S''g')(t_i) - (S''g')(t_{i-1})\right] = (S''g')(b) - (S''g')(a)$$

Our proof will still be valid if this last expression is nonnegative. Putting in $g = f - S$, we arrive at the condition

$$S''(b)[f'(b) - S'(b)] \geqq S''(a)[f'(a) - S'(a)]$$

For example, if instead of assuming that S is the *natural* cubic spline interpolant, we assume the end conditions $S'(a) = f'(a)$ and $S'(b) = f'(b)$, then the conclusion is true.

Tension Splines

In some data-fitting problems, it is useful to have a parameter τ available called the **tension**. When τ is given a large value, the curve passing through the data points will have high tension. This can be interpreted as a force that stretches the curve tightly among the data points, as shown in Figures 6.5(a) and (b). When τ is given

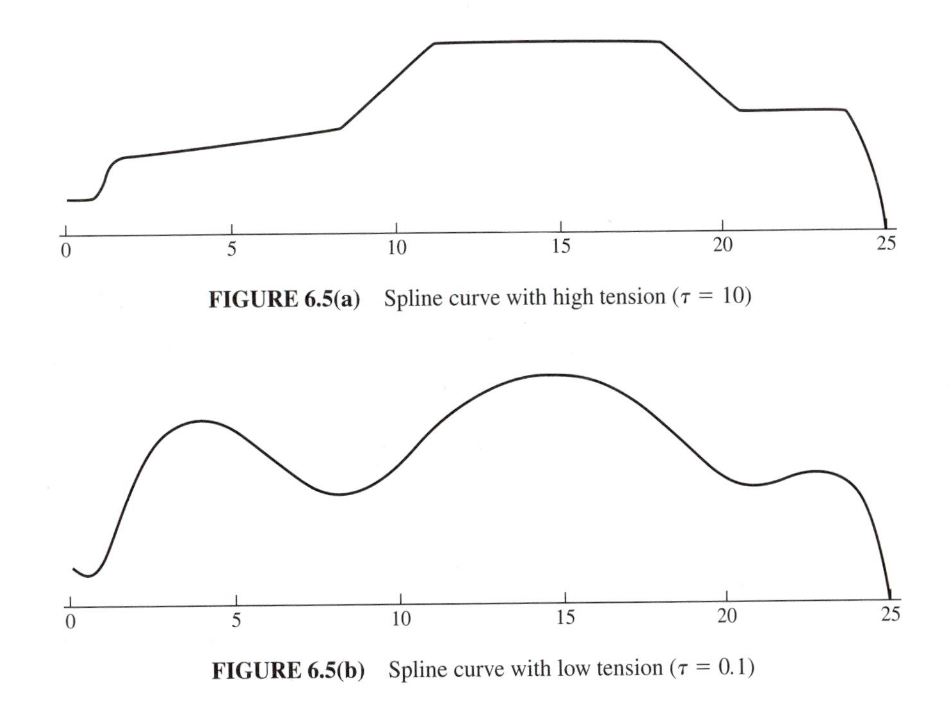

FIGURE 6.5(a) Spline curve with high tension ($\tau = 10$)

FIGURE 6.5(b) Spline curve with low tension ($\tau = 0.1$)

a low value, the curve will assume more nearly the shape of the interpolating cubic spline. When $\tau \to +\infty$, the curve approaches the piecewise linear function—that is, a spline of degree 1.

One mathematical model of a curve such as that just described is as follows. As before, we have knots

$$t_0 < t_1 < \cdots < t_n$$

and data y_i given at each t_i. The **tension spline** that we seek is a function f having these properties:

(i) $f \in C^2[t_0, t_n]$
(ii) $f(t_i) = y_i \quad (0 \leqq i \leqq n)$
(iii) On each open interval (t_{i-1}, t_i), f satisfies $f^{(4)} - \tau^2 f'' = 0$.

Thus, f has two continuous derivatives globally, interpolates the given data, and satisfies a certain differential equation in each subinterval. It is clear that this prescription yields a cubic spline when $\tau = 0$ because the solutions of the equation $f^{(4)} = 0$ are cubic polynomials.

To determine f, we proceed as in the case of the cubic splines. Hence, we set $z_i \equiv f''(t_i)$ and write down the conditions f must satisfy on the interval $[t_i, t_{i+1}]$:

$$\begin{array}{cc} & f^{(4)} - \tau^2 f'' = 0 \\ f(t_i) = y_i & f(t_{i+1}) = y_{i+1} \\ f''(t_i) = z_i & f''(t_{i+1}) = z_{i+1} \end{array}$$

One can verify that the solution of this **two-point boundary-value problem** is

$$\begin{aligned} f(x) = {} & \left\{z_i \sinh[\tau(t_{i+1} - x)] + z_{i+1} \sinh[\tau(x - t_i)]\right\} / [\tau^2 \sinh(\tau h_i)] \\ & + (y_i - z_i/\tau^2)(t_{i+1} - x)/h_i + (y_{i+1} - z_{i+1}/\tau^2)(x - t_i)/h_i \end{aligned} \qquad \textbf{(12)}$$

After the coefficients z_i have been determined, this equation will be used to compute values of f on the interval $[t_i, t_{i+1}]$. All of this is analogous to the case of cubic splines.

In order that f have C^2 global smoothness, the conditions

$$\lim_{x \downarrow x_i} f'(x) = \lim_{x \uparrow x_i} f'(x) \qquad (1 \leqq i \leqq n - 1)$$

must be imposed at the interior knots. The tedious calculations involved in this are not given here. They proceed as in the case of the cubic splines. The result is a tridiagonal system of equations for the unknowns $z_0, z_1, \ldots, z_n$ that can be written in the form

$$\alpha_{i-1} z_{i-1} + (\beta_{i-1} + \beta_i) z_i + \alpha_i z_{i+1} = \gamma_i - \gamma_{i-1} \qquad (1 \leqq i \leqq n - 1) \qquad \textbf{(13)}$$

with these abbreviations:

$$\begin{aligned}
\alpha_i &= 1/h_i - \tau/\sinh(\tau h_i)\\
\beta_i &= \tau\cosh(\tau h_i)/\sinh(\tau h_i) - 1/h_i\\
\gamma_i &= \tau^2(y_{i+1} - y_i)/h_i
\end{aligned}$$

It will be observed that two additional conditions are needed to determine the z-vector. As in the cubic spline case, one possibility is to specify $z_0 = z_n = 0$.

The steps in obtaining a tension spline f to fit data (t_i, y_i) are as follows:

(i) Test to be sure that $t_0 < t_1 < \cdots < t_n$.
(ii) Compute $h_i, \alpha_i, \beta_i, \gamma_i$ for $0 \leqq i \leqq n - 1$.
(iii) Set $z_1 = z_n = 0$.
(iv) Solve the tridiagonal system (13) for z_i for $2 \leqq i \leqq n - 1$.
(v) Use Formula (12) to compute values of f on the interval $[t_i, t_{i+1}]$.

The unsophisticated approach we have outlined here can be used for interesting experimentation. For example, it led to the two curves shown in Figures 6.5(a) and (b). In the first we let $\tau = 10$, and in the second $\tau = 0.1$. The data were the same and consisted of prescribed function values at the integer points $t_i = i$ for $0 \leqq i \leqq 25$.

Tension splines were introduced by Schweikert [1966]. Related papers are by Cline [1974a, b] and Pruess [1976, 1978]. Software for computing curves and surfaces with the aid of tension splines has been developed by Cline. An alternative to the *spline in tension* is the **taut spline** of de Boor [1984]. These functions are ordinary cubic splines with additional knots (and data) placed in regions where the curve is desired to change direction abruptly. The advantage of the taut splines is that no new computer programs are needed, and the computational burden involved in using hyperbolic functions is avoided.

Theory of Higher-Degree Natural Splines

In the final part of this section, we present some of the theory of higher-degree natural splines. Natural splines exist for odd degrees only, and the degree will be denoted here by $2m + 1$. When $m = 1$, one has the natural cubic splines, as developed previously. For the general theory, it is convenient to proceed in a slightly different manner. As before, a set of knots is given:

$$t_0 < t_1 < \cdots < t_n$$

A natural spline of degree $2m + 1$ is a function $S \in C^{2m}(\mathbb{R})$ that reduces to a polynomial of degree $\leqq 2m + 1$ in each interval $[t_0, t_1], [t_1, t_2], \ldots, [t_{n-1}, t_n]$ and to a polynomial of degree at most m in $(-\infty, t_0)$ and (t_n, ∞). A natural cubic spline, under this definition, must reduce to linear polynomials in $(-\infty, t_0)$ and (t_n, ∞).

The linear space of all natural splines of degree $2m + 1$ with $n + 1$ knots $t_0, t_1, \ldots, t_n$ is denoted by $\mathcal{N}^{2m+1}(t_0, t_1, \ldots, t_n)$ or, for brevity, $\mathcal{N}_n^{2m+1}$.

It is convenient to use the so-called **truncated power function**, denoted by x_+^n. This is defined to be x^n if $x \geqq 0$, and defined to be 0 if $x < 0$. This function belongs to the continuity class C^{n-1}.

THEOREM 2 *Every member of $\mathcal{N}_n^{2m+1}$ has a representation*

$$S(x) = \sum_{i=0}^{m} a_i x^i + \sum_{i=0}^{n} b_i (x - t_i)_+^{2m+1} \tag{14}$$

with $\sum_{i=0}^{n} b_i t_i^j = 0$ for $0 \leqq j \leqq m$.

Proof In the interval $(-\infty, t_0)$, S is a polynomial p_0 of degree at most m. This polynomial determines the coefficients a_i. In the interval (t_0, t_1), S reduces to a polynomial p_1 of degree $2m + 1$. The continuity conditions at t_0 are

$$p_0^{(j)}(t_0) = p_1^{(j)}(t_0) \qquad (0 \leqq j \leqq 2m)$$

By Taylor's Theorem, p_1 can be written in the form

$$\begin{aligned} p_1(x) &= \sum_{j=0}^{2m+1} \frac{1}{j!} p_1^{(j)}(t_0)(x - t_0)^j \\ &= \sum_{j=0}^{2m} \frac{1}{j!} p_0^{(j)}(t_0)(x - t_0)^j + b_0(x - t_0)^{2m+1} \\ &= p_0(x) + b_0(x - t_0)^{2m+1} \end{aligned}$$

This equation shows that on $(-\infty, t_1)$, we have

$$S(x) = p_0(x) + b_0(x - t_0)_+^{2m+1}$$

(We note in passing that $S^{(j)}(t_0) = 0$ for $m < j \leqq 2m$.) The argument used at t_0 is now repeated at each node to obtain the remaining terms $b_i(x - t_i)_+^{2m+1}$. On the interval (t_n, ∞), S must reduce to a polynomial of degree $\leqq m$. Hence, in this interval,

$$0 = S^{(m+1)}(x) = \sum_{i=0}^{n} b_i (2m + 1)(2m) \cdots (m + 1)(x - t_i)^m$$

From this equation, with the help of the Binomial Theorem, we obtain

$$0 = \sum_{i=0}^{n} b_i (x - t_i)^m = \sum_{i=0}^{n} b_i \sum_{j=0}^{m} \binom{m}{j} x^{m-j}(-t_i)^j$$

Here we have a polynomial of degree at most m in x. Its coefficients must vanish, and so $\sum_{i=0}^{n} b_i t_i^j = 0$ for $j = 0, 1, \ldots, m$. ■

THEOREM 3 *Let the knots $t_0 < t_1 < \cdots < t_n$ be given, and let $0 \leqq m \leqq n$. There is a unique natural spline of degree $2m+1$ taking prescribed values at the knots.*

Proof By Theorem 2, a natural spline has the form

$$S(x) = \sum_{j=0}^{m} a_j x^j + \sum_{j=0}^{n} b_j (x - t_j)_+^{2m+1}$$

If the values prescribed for S are λ_i, then (in accordance with Theorem 2) the interpolation problem requires us to solve the system

$$\begin{cases} S(t_i) \equiv \displaystyle\sum_{j=0}^{m} a_j t_i^j + \sum_{j=0}^{n} b_j (t_i - t_j)_+^{2m+1} = \lambda_i & (0 \leqq i \leqq n) \\ \displaystyle\sum_{j=0}^{m} b_j t_j^i = 0 & (0 \leqq i \leqq m) \end{cases}$$

This is a square system of $m+n+2$ equations in $m+n+2$ unknowns. To show that it is nonsingular, it suffices to prove that the corresponding homogeneous problem has only the 0-solution. Suppose therefore that $S(t_i) = 0$ for $0 \leqq i \leqq n$. We shall prove that

$$I \equiv \int_a^b \left[S^{(m+1)}(x)\right]^2 dx = 0 \tag{15}$$

where $a = t_0$ and $b = t_n$. Integration by parts yields

$$\begin{aligned} I &= S^{(m+1)}(x) S^{(m)}(x) \Big|_a^b - \int_a^b S^{(m)}(x) S^{(m+2)}(x)\,dx \\ &= -\int_a^b S^{(m)}(x) S^{(m+2)}(x)\,dx \end{aligned}$$

Here we exploited the fact that S is a polynomial of degree at most m in $(-\infty, a)$, and hence $S^{(m+1)}(a) = 0$. Likewise, $S^{(m+1)}(b) = 0$. This argument is repeated until we arrive at

$$I = (-1)^m \int_a^b S^{(1)}(x) S^{(2m+1)}(x)\,dx$$

Now $S^{(2m+1)}$ is piecewise constant, and so

$$I = (-1)^m \sum_{i=1}^{n} \int_{t_{i-1}}^{t_i} c_i S'(x)\,dx = (-1)^m \sum_{i=1}^{n} c_i \left[S(t_i) - S(t_{i-1})\right] = 0$$

This proves Equation (15). From it we infer that $S^{(m+1)} \equiv 0$. Hence, S is a polynomial of degree at most m. Since S has zeros at $t_0, t_1, \ldots, t_n$ and $n + 1 > m$, we see that $S = 0$. ■

The analogue of Theorem 1 follows.

THEOREM 4 *Let $m \leqq n$ and $f \in C^{m+1}[a, b]$. Let S be the natural spline of degree $2m + 1$ that interpolates f at knots $a = t_0 < t_1 < \cdots < t_n = b$. Then*

$$\int_a^b \left[S^{(m+1)}(x)\right]^2 dx \leqq \int_a^b \left[f^{(m+1)}(x)\right]^2 dx$$

Proof Proceed as in Theorem 1. Let $g = f - S$. Then $g(t_i) = 0$ for $0 \leqq i \leqq n$. By repeated integration by parts, we can prove that

$$\int_a^b g^{(m+1)}(x) S^{(m+1)}(x)\, dx = 0$$

Then the proof is completed by writing

$$\begin{aligned}
\int_a^b \left[f^{(m+1)}(x)\right]^2 dx &= \int_a^b \left[S^{(m+1)}(x) + g^{(m+1)}(x)\right]^2 dx \\
&= \int_a^b \left[S^{(m+1)}(x)\right]^2 dx + 2\int_a^b S^{(m+1)}(x) g^{(m+1)}(x)\, dx + \int_a^b \left[g^{(m+1)}(x)\right]^2 dx \\
&= \int_a^b \left[S^{(m+1)}(x)\right]^2 dx + \int_a^b \left[g^{(m+1)}(x)\right]^2 dx \\
&\geqq \int_a^b \left[S^{(m+1)}(x)\right]^2 dx
\end{aligned}$$

■

PROBLEMS 6.4

1. Refer to the algorithm for determining the values of z_i, and prove that for all $i = 1, 2, \ldots, n - 1$, we have $u_i z_i + h_i z_{i+1} - v_i = 0$.
2. (Continuation) Denote the left side of the equation in Problem 1 by E_i. Then $(h_i/u_i)E_i + E_{i+1} = 0$. Show that the latter equation can be reduced to Equation (10) by using the formulas in the algorithms. This will establish that the algorithm produces a solution to Equation (10).
3. Show that if we replace t_{i+1} by $t_i + h_i$ in Equation (6), the result is

$$S_i(x) = \frac{z_{i+1}}{6h_i}(x - t_i)^3 - \frac{z_i}{6h_i}(x - t_i - h_i)^3 + C(x - t_i) - D(x - t_i - h_i)$$

4. (Continuation) By expanding the term $(x - t_i - h_i)^3$ in the preceding equation and by using the correct values of C and D, show that Equation (11) in the text is correct.

5. Determine whether this is a quadratic spline function:

$$f(x) = \begin{cases} x & x \in (-\infty, 1] \\ -\frac{1}{2}(2-x)^2 + \frac{3}{2} & x \in [1, 2] \\ \frac{3}{2} & x \in [2, \infty) \end{cases}$$

6. Is the function in Problem 5 a cubic spline function?

7. Determine all the values of a, b, c, d, e for which the following function is a cubic spline:

$$f(x) = \begin{cases} a(x-2)^2 + b(x-1)^3 & x \in (-\infty, 1] \\ c(x-2)^2 & x \in [1, 3] \\ d(x-2)^2 + e(x-3)^3 & x \in [3, \infty) \end{cases}$$

Next, determine the values of the parameters so that the cubic spline interpolates this table:

x	0	1	4
y	26	7	25

8. Verify that Equations (7), (9), and (10) in the text are correct.

9. Using the development of the cubic splines as a model, derive the appropriate equations and algorithm to provide a quadratic spline interpolant to data (t_i, y_i) for $0 \leqq i \leqq n$, where $t_0 < t_1 < \cdots < t_n$. If Q is the spline interpolant, then the numbers $z_i = Q'(t_i)$ are well defined. Find the equations governing $z_0, z_1, \ldots, z_n$. You should discover that one of the z points can be arbitrary, say $z_0 = 0$.

10. Show that in the algorithm for solving the system (10), $u_i > h_i + h_{i-1}$ for all $i = 1, 2, \ldots, n-1$.

11. Determine the values of a, b, and c so that this is a cubic spline having knots 0, 1, and 2:

$$f(x) = \begin{cases} 3 + x - 9x^2 & x \in [0, 1] \\ a + b(x-1) + c(x-1)^2 + d(x-1)^3 & x \in [1, 2] \end{cases}$$

Next, determine d so that $\int_0^2 [f''(x)]^2\,dx$ is a minimum. Finally, find the value of d that makes $f''(2) = 0$ and explain why this value is different from the one previously determined.

12. Determine whether this is a cubic spline:

$$f(x) = \begin{cases} x^3 + x & x \leqq 0 \\ x^3 - x & x \geqq 0 \end{cases}$$

Show that

$$\lim_{x \uparrow 0} f''(x) = \lim_{x \downarrow 0} f''(x)$$

13. Determine whether the natural cubic spline that interpolates the table

x	0	1	2	3
y	1	1	0	10

is or is not the function

$$f(x) = \begin{cases} 1 + x - x^3 & x \in [0, 1] \\ 1 - 2(x-1) - 3(x-1)^2 + 4(x-1)^3 & x \in [1, 2] \\ 4(x-2) + 9(x-2)^2 - 3(x-2)^3 & x \in [2, 3] \end{cases}$$

14. Determine whether this function is a natural cubic spline:

$$f(x) = \begin{cases} 2(x+1) + (x+1)^3 & x \in [-1, 0] \\ 3 + 5x + 3x^2 & x \in [0, 1] \\ 11 + 11(x-1) + 3(x-1)^2 - (x-1)^3 & x \in [1, 2] \end{cases}$$

15. A theorem in calculus asserts that if a function is differentiable at a point, then it is necessarily continuous at that point. The reason for this is elementary: If the limit defining $f'(x)$ exists,

$$f'(x) = \lim_{h \to 0} \frac{f(x+h) - f(x)}{h}$$

then the limit of the numerator alone must be 0, and this implies the continuity of f at x. Suppose now that for a certain function f and a certain point x_0, we have

$$\lim_{x \downarrow x_0} f'(x) = \lim_{x \uparrow x_0} f'(x)$$

Can we conclude that f' is continuous at x_0?

16. (Continuation) In checking to determine whether a piecewise cubic function f is a cubic spline, does it suffice to verify the equation

$$\lim_{x \downarrow t} f''(x) = \lim_{x \uparrow t} f''(t)$$

for each knot t?

17. Find the natural cubic spline function whose knots are -1, 0, and 1 and that takes the values $S(-1) = 13$, $S(0) = 7$, and $S(1) = 9$.

18. Which properties of a natural cubic spline does the following function possess, and which properties does it not possess?

$$f(x) = \begin{cases} (x+1) + (x+1)^3 & x \in [-1, 0] \\ 4 + (x-1) + (x-1)^3 & x \in (0, 1] \end{cases}$$

19. Find a natural cubic spline function whose knots are -1, 0, and 1 and that takes these values:

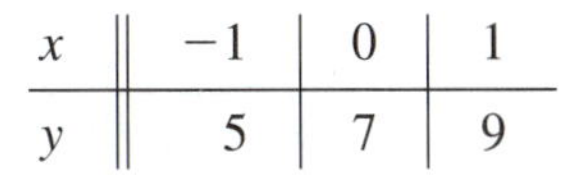

x	-1	0	1
y	5	7	9

20. Determine whether the coefficients a, b, c, and d exist so that the function

$$S(x) = \begin{cases} 1 - 2x & x \in (-\infty, -3] \\ a + bx + cx^2 + dx^3 & x \in [-3, 4] \\ 157 - 32x & x \in [4, +\infty) \end{cases}$$

is a natural cubic spline for the interval $[-3, 4]$.

21. Is the following function a natural cubic spline?

$$S(x) = \begin{cases} x^3 - 1 & x \in [-1, \frac{1}{2}] \\ 3x^3 - 1 & x \in [\frac{1}{2}, 1] \end{cases}$$

22. Repeat Problem 21 for the function

$$S(x) = \begin{cases} x^3 - 1 & x \in [-1, 0] \\ 3x^3 - 1 & x \in [0, 1] \end{cases}$$

23. Can a and b be defined so that the function

$$S(x) = \begin{cases} (x-2)^3 + a(x-1)^2 & x \in (-\infty, 2] \\ (x-2)^3 - (x-3)^2 & x \in [2, 3] \\ (x-3)^3 + b(x-2)^2 & x \in [3, +\infty) \end{cases}$$

is a natural cubic spline? Why or why not?

24. If S is a first-degree spline function that interpolates f at a sequence of knots $0 = t_0 < t_1 < \cdots < t_n = 1$, what is $\int_0^1 S(x)\,dx$?

25. What value of (a, b, c, d) makes this a cubic spline?

$$f(x) = \begin{cases} x^3 & x \in [-1, 0] \\ a + bx + cx^2 + dx^3 & x \in [0, 1] \end{cases}$$

26. Determine the value of (a, b, c) that makes the function

$$f(x) = \begin{cases} x^3 & x \in [0, 1] \\ \frac{1}{2}(x-1)^3 + a(x-1)^2 + b(x-1) + c & x \in [1, 3] \end{cases}$$

a cubic spline. Is it a natural cubic spline?

27. Let $t_0 < t_1 < \cdots < t_n$ and $-\infty < x < \infty$. What is the output value of k from the following algorithm?

```
for i = 1 to n do
    if x < t_i then
        k ← i
        exit loop
    end if
end do
```

28. How many conditions are needed to define a quadratic spline interpolation function $S(x)$ over $a = t_0 < t_1 < \cdots < t_{20} = b$? Does the continuity of $S'(x)$ provide all the needed conditions?

29. Prove Theorem 1 when the end conditions on S are changed to $S'(a) = f'(a)$ and $S'(b) = f'(b)$.

30. Develop a suitable procedure for finding the cubic spline interpolant when the end conditions specify values of $S'(t_0)$ and $S'(t_n)$.

31. Give the simplified version of natural cubic spline interpolation that pertains to the case of equally spaced knots.

32. Show that a spline function of degree 1 having knots $t_0 < t_1 < \cdots < t_n$ can be expressed in the form

$$S(x) = ax + b + \sum_{i=1}^{n-1} c_i|x - t_i|$$

33. Show that the function in Equation (12) solves the two-point boundary-value problem preceding Equation (12).

34. Show that Equation (13) follows from the global smoothness conditions, as asserted in the text.

35. Prove that the coefficient matrix in Equation (13) is diagonally dominant.

36. Using the notation and hypotheses of Theorem 4, prove that the inequality $\|f^{(m+1)}\| \geqq \|f^{(m+1)} - S^{(m+1)}\|$ is valid.

COMPUTER PROBLEMS 6.4

1. Draw a curve such as an oval or spiral on a sheet of graph paper. Select points in a more or less regular distribution along the curve and label them $t_0 = 1.0$, $t_1 = 2.0$, and so on. Read the x- and y-coordinates at each selected point to obtain a table of $x(t)$ and $y(t)$. Fit these functions by spline functions S and S^*. Then the formulas $x = S(t)$ and $y = S^*(t)$ give an approximate parametric representation of the curve. Plot the resulting curves for several test cases using an automatic plotter.

2. In the development of Problem 9, find a way to solve for $(z_0, z_1, \ldots, z_n)$ so that $\sum_{i=0}^{n} z_i^2$ is a minimum. That is, use this condition to remove the arbitrariness referred to in the problem. Incorporate this feature in your algorithm and test it on a computer.

3. Prove this formula:

$$\int_{t_i}^{t_{i+1}} S_i(x)\,dx = \frac{h_i}{2}(y_i + y_{i+1}) - \frac{h_i^3}{24}(z_i + z_{i+1})$$

Then write and test a program to compute

$$\int_{t_0}^{t_n} S(x)\,dx$$

4. Write and test a computer program to compute the cubic spline function S having prescribed knots $t_0 < t_1 < \cdots < t_n$ and satisfying these conditions:

$$\begin{cases} S(t_i) = y_i & (0 \leqq i \leqq n) \\ S''(t_0) = \alpha \\ S''(t_n) = \beta \end{cases}$$

5. Program the algorithm for tension splines and test it with a variety of values of the tension parameter τ.

6. Start with the silhouette of a car body as shown in Figure 6.6. Prepare a table of abscissas and ordinates for 10 to 20 points. Using values of the tension $\tau = 0.25$, 4, and 10, generate and plot values of the interpolating tension spline to see which one produces the most pleasing appearance.

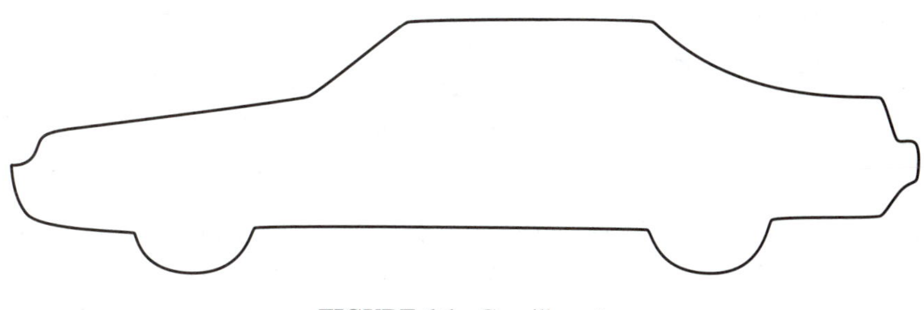

FIGURE 6.6 Car silhouette

7. Draw a script letter such as the one shown in Figure 6.7. Then reproduce it with the aid of cubic splines and a plotter. Proceed as follows: Select a modest number of points on the curve, say $n = 11$. Label these $t = 1, 2, \ldots, n$. For each point, obtain the corresponding x- and y-coordinates. Then fit $x = S_x(t)$

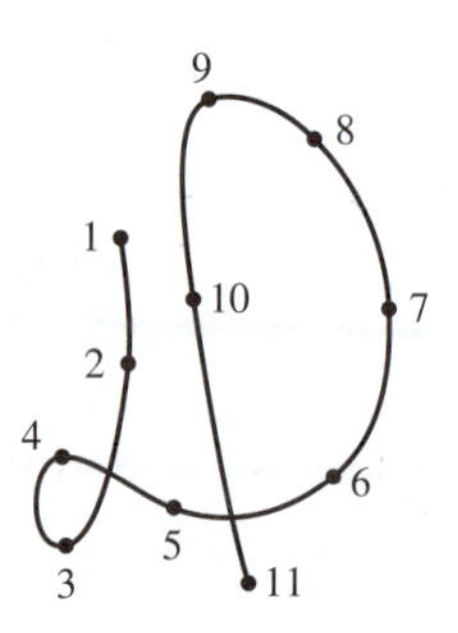

FIGURE 6.7 Script letter from 11 knots

and $y = S_y(t)$, using cubic spline interpolating functions S_x and S_y. This will produce a parametric representation of the original curve. Compute a large number of values of $S_x(t)$ and $S_y(t)$ to give to the plotter. To learn more about how spline curves are used in designing typefaces, the reader should consult Knuth [1979].

8. Interpret the results of the following numerical experiment and draw some conclusions.

(a) Define p to be the polynomial of degree 20 that interpolates the function $f(x) = (1 + 6x^2)^{-1}$ at 21 equally spaced nodes in the interval $[-1, 1]$. Include the endpoints as nodes. Print a table of $f(x)$, $p(x)$, and $f(x) - p(x)$ at 41 equally spaced points on the interval.

(b) Repeat the experiment using the **Chebyshev nodes** given by

$$x_i = \cos[(i-1)\pi/20] \qquad (1 \leqq i \leqq 21)$$

(c) With 21 equally spaced knots, repeat the experiment using a cubic interpolating spline.

*6.5 The *B*-Splines: Basic Theory

This section is devoted to a system of spline functions from which all other spline functions can be obtained by forming linear combinations. These splines provide bases for certain spline spaces and are therefore called ***B*-splines**. Once the knots are known, the *B*-splines are easily generated by recurrence relations and the algorithm is relatively simple. The *B*-splines are distinguished by their elegant theory and their model behavior in numerical calculations. Moreover, *B*-splines can be generalized.

We begin with a system of knots, t_i, on the real line. For practical purposes, only a finite set of knots is ever needed, but for the theoretical development it is much easier to suppose that the knots form an infinite set extending to $+\infty$ on the right and to $-\infty$ on the left:

$$\cdots < t_{-2} < t_{-1} < t_0 < t_1 < t_2 < \cdots$$

This knot sequence is assumed to be fixed throughout this section, and all of our splines will be based on it.

B-Splines of Degree 0

The B-splines of degree 0 are denoted by B_i^0 and have the appearance shown in Figure 6.8. The index i ranges over all the integers. The heavy dots on the graph indicate that we define $B_i^0(t_i) = 1$ and $B_i^0(t_{i+1}) = 0$. The formal definition is

$$B_i^0(x) = \begin{cases} 1 & \text{if } t_i \leqq x < t_{i+1} \\ 0 & \text{otherwise} \end{cases}$$

These B-splines form an infinite sequence, $\{B_i^0 : i \in \mathbb{Z}\}$. (Here $\mathbb{Z}$ denotes the set of all integers, positive, negative, or 0.) We observe some of their salient properties:

(i) The **support** of B_i^0, defined as the set of x where $B_i^0(x) \neq 0$, is the interval $[t_i, t_{i+1})$.
(ii) $B_i^0(x) \geqq 0$ for all i and all x.
(iii) B_i^0 is continuous from the right on the entire real line.
(iv) $\sum_{i=-\infty}^{\infty} B_i^0(x) = 1$ for all x.

This last equation is verified by selecting any $x \in \mathbb{R}$ and then determining the knot interval in which x lies, say $t_j \leqq x < t_{j+1}$; then $\sum_{i=-\infty}^{\infty} B_i^0(x) = B_j^0(x) = 1$.

A final remark about the splines B_i^0 is that they do form a basis for all splines of degree 0 based on the given knot sequence, provided that we standardize such splines to be continuous from the right. To verify this assertion, suppose that S is such a spline function. Then it is *piecewise constant* and is defined by a set of rules of the form

$$S(x) = c_i \quad \text{if} \quad t_i \leqq x < t_{i+1} \qquad (i \in \mathbb{Z})$$

It is apparent that $S(x) = \sum_{i=-\infty}^{\infty} c_i B_i^0(x)$. (Thus, we have a basis in the sense of Schauder: Each vector in the space has a *unique* representation as an *infinite* series $\sum_{i=-\infty}^{\infty} c_i B_i^0$.)

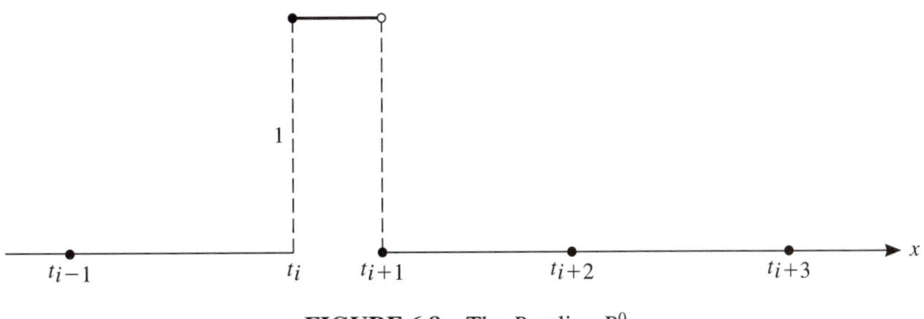

FIGURE 6.8 The B-spline B_i^0

The functions B_i^0 are the starting point for a recursive definition of all the higher-degree B-splines. The basic recurrence relation is

$$B_i^k(x) = \left(\frac{x - t_i}{t_{i+k} - t_i}\right) B_i^{k-1}(x) + \left(\frac{t_{i+k+1} - x}{t_{i+k+1} - t_{i+1}}\right) B_{i+1}^{k-1}(x) \qquad (k \geqq 1) \quad \textbf{(1)}$$

All the properties of the higher-order B-splines will follow from this recursive definition. By introducing some special linear functions

$$V_i^k(x) = \frac{x - t_i}{t_{i+k} - t_i} \quad \textbf{(2)}$$

we can write the recurrence relation in the following more elegant form:

$$B_i^k = V_i^k B_i^{k-1} + (1 - V_{i+1}^k) B_{i+1}^{k-1} \quad \textbf{(3)}$$

Since B_i^0 is a piecewise polynomial of degree 0, and since V_i^k is linear, B_i^1 is a piecewise polynomial of degree $\leqq 1$. The same reasoning shows that, in general, B_i^k will be a piecewise polynomial of degree $\leqq k$.

B-Splines of Degree 1

With the aid of Equation (1), we can give an explicit formula for $B_i^1(x)$ as follows:

$$\begin{aligned} B_i^1(x) &= \left(\frac{x - t_i}{t_{i+1} - t_i}\right) B_i^0(x) + \left(\frac{t_{i+2} - x}{t_{i+2} - t_{i+1}}\right) B_{i+1}^0(x) \\ &= \begin{cases} 0 & \text{if } x < t_i \text{ or } x \geqq t_{i+2} \\ \dfrac{x - t_i}{t_{i+1} - t_i} & \text{if } t_i \leqq x < t_{i+1} \\ \dfrac{t_{i+2} - x}{t_{i+2} - t_{i+1}} & \text{if } t_{i+1} \leqq x < t_{i+2} \end{cases} \end{aligned}$$

The graph of B_i^1 is shown in Figure 6.9.

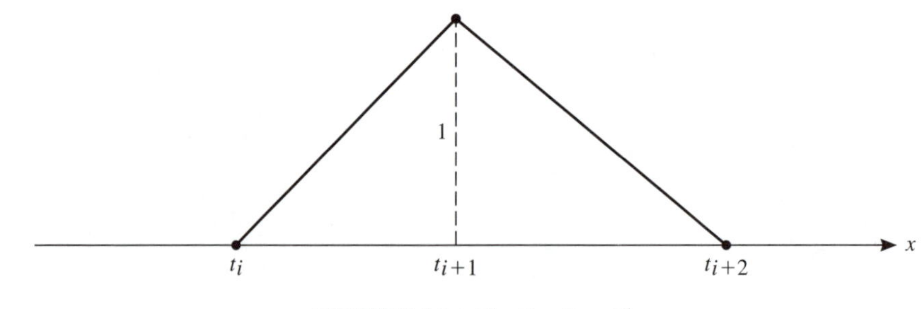

FIGURE 6.9 The B-spline B_i^1

Again, some observations can be made about the functions B_i^1:

(i) The support of B_i^1 is (t_i, t_{i+2}).
(ii) $B_i^1(x) \geqq 0$ for all i and all x.
(iii) B_i^1 is continuous and is differentiable at every point except t_i, t_{i+1}, and t_{i+2}.
(iv) $\sum_{i=-\infty}^{\infty} B_i^1(x) = 1$ for all x.

To verify this last equation, take any $x \in \mathbb{R}$. Since t_j converges to $+\infty$ when i increases, and it converges to $-\infty$ when i decreases, we can find an index j such that $t_j \leqq x < t_{j+1}$. Then $B_i^1(x) = 0$ for all i with the possible exception of $i = j$ or $i = j - 1$. Hence for this x,

$$\sum_{i=-\infty}^{\infty} B_i^1(x) = B_{j-1}^1(x) + B_j^1(x) = \frac{t_{j+1} - x}{t_{j+1} - t_j} + \frac{x - t_j}{t_{j+1} - t_j} = 1$$

Properties of *B*-Splines

Now in a sequence of lemmas, we shall develop the important properties of the family B_i^k $(i \in \mathbb{Z}, k \in \mathbb{N})$.

LEMMA 1 *If $k \geqq 1$ and $x \notin (t_i, t_{i+k+1})$, then $B_i^k(x) = 0$.*

Proof We have already observed that this is true for $k = 1$ but not for $k = 0$. If it is true for a certain index $k-1$, then it is true for k by the following reasoning: If $x \notin (t_i, t_{i+k+1})$, then $x \notin (t_i, t_{i+k})$ and $x \notin (t_{i+1}, t_{i+k+1})$. By the induction hypothesis, $B_i^{k-1}(x) = 0$ and $B_{i+1}^{k-1}(x) = 0$. From Equation (12), it follows that $B_i^k(x) = 0$. ■

LEMMA 2 *Let $k \geqq 0$. If $x \in (t_i, t_{i+k+1})$, then $B_i^k(x) > 0$.*

Proof We observed earlier that the assertion of Lemma 2 is true when $k = 0$ or $k = 1$. (It is true for $k = 1$ by the explicit formulas given previously for B_i^1.) Now assume that the assertion is true for an index $k - 1$, with $k \geqq 2$. This assumption and Lemma 1 imply that $B_i^{k-1}(x) \geqq 0$ for all x and for all i. Let $t_i < x < t_{i+k+1}$. Then the linear factors on the right side of Equation (1) are positive. By the induction hypothesis, $B_i^{k-1}(x) > 0$ in (t_i, t_{i+k}), and $B_{i+1}^{k-1}(x) > 0$ in (t_{i+1}, t_{i+k+1}). Since $k \geqq 2$, these two intervals overlap, and by Equation (1) we see that $B_i^k(x) > 0$. ■

Since we expect to use the B-splines B_i^k as a basis for all splines of degree k, we shall be interested in linear combinations of the form $\sum_{i=-\infty}^{\infty} c_i B_i^k(x)$.

LEMMA 3

$$\sum_{i=-\infty}^{\infty} c_i B_i^k = \sum_{i=-\infty}^{\infty} \left[c_i V_i^k + c_{i-1}(1 - V_i^k)\right] B_i^{k-1}$$

Proof We use Equation (3) and elementary series manipulations as follows:

$$\sum_{i=-\infty}^{\infty} c_i B_i^k = \sum_{i=-\infty}^{\infty} c_i \left[V_i^k B_i^{k-1} + (1 - V_{i+1}^k) B_{i+1}^{k-1} \right]$$

$$= \sum_{i=-\infty}^{\infty} c_i V_i^k B_i^{k-1} + \sum_{i=-\infty}^{\infty} c_{i-1}(1 - V_i^k) B_i^{k-1} \qquad \blacksquare$$

Numerical Procedure

In Lemma 3, the coefficients c_i can be constants or functions. The lemma therefore provides a means for evaluating a function given in the form

$$f(x) = \sum_{i=-\infty}^{\infty} C_i^k(x) B_i^k(x)$$

We suppose that the functions C_i^k are given; they of course may be constants. We define now

$$C_i^{k-1}(x) = C_i^k(x) V_i^k(x) + C_{i-1}^k(x)[1 - V_i^k(x)] \tag{4}$$

Then by Lemma 3 and Formula (4), we have

$$\sum_{i=-\infty}^{\infty} C_i^k(x) B_i^k(x) = \sum_{i=-\infty}^{\infty} C_i^{k-1}(x) B_i^{k-1}(x)$$

By repeating the argument for $k-1, k-2, \dots, 0$, we eventually obtain

$$\sum_{i=-\infty}^{\infty} C_i^k(x) B_i^k(x) = \sum_{i=-\infty}^{\infty} C_i^0(x) B_i^0(x)$$

As we know, the expression on the right is easily evaluated: For $t_j \leqq x < t_{j+1}$, its value is $C_j^0(x)$. We write Equation (4) in detail, using Equation (2), as follows:

$$C_i^{j-1}(x) = \left[(x - t_i) C_i^j(x) + (t_{i+j} - x) C_{i-1}^j(x)\right] / (t_{i+j} - t_i) \tag{5}$$

The preceding remarks lead to the following numerical procedure.

ALGORITHM 1 *If coefficients C_i^k are given, the spline function $S(x) = \sum_{i=-\infty}^{\infty} C_i^k B_i^k(x)$ for a given x can be computed as follows: Determine the index m such that $t_m \leqq x < t_{m+1}$.*

Compute the triangular array

$$\begin{array}{llll} C_m^k & C_m^{k-1} & \cdots\ C_m^1 & C_m^0 \\ C_{m-1}^k & C_{m-1}^{k-1} & \cdots\ C_{m-1}^1 & \\ \vdots & \vdots & \ddots & \\ C_{m-k+1}^k & C_{m-k+1}^{k-1} & & \\ C_{m-k}^k & & & \end{array}$$

using Equation (5). *Then* $S(x) = C_m^0$.

LEMMA 4 *For all k,*

$$\sum_{i=-\infty}^{\infty} B_i^k(x) = 1$$

Proof We can use the algorithm just described. We start with $\sum_{i=-\infty}^{\infty} C_i^k B_i^k(x)$, in which $C_i^k = 1$ for all i. We then fix x and apply Equation (4) to compute

$$\begin{aligned} C_i^{k-1} &= C_i^k V_i^k + C_{i-1}^k \left[1 - V_i^k\right] \\ &= V_i^k + 1 - V_i^k \\ &= 1 \end{aligned}$$

By repeating the argument, we discover that $C_i^j = 1$ for all i and for $j = k, k-1, \ldots, 0$. Hence,

$$\sum_{i=-\infty}^{\infty} B_i^k(x) = \sum_{i=-\infty}^{\infty} B_i^0(x) = 1 \qquad \blacksquare$$

Derivative and Integral of *B*-Splines

The next lemma gives an important formula for the derivative of B_i^k. It is convenient to use Formula (2) for V_i^k and to set

$$\alpha_i^k = \frac{1}{t_{i+k} - t_i} \tag{6}$$

We note that, with this notation,

$$\frac{d}{dx} V_i^k(x) = \alpha_i^k \tag{7}$$

Other useful formulas, which are trivial to verify, are these:

$$\alpha_i^k V_i^{k+1} = \alpha_i^{k+1} V_i^k \tag{8}$$

$$\alpha_{i+1}^k (1 - V_i^{k+1}) = \alpha_i^{k+1} (1 - V_{i+1}^k) \tag{9}$$

In both cases, the verification consists simply in using the definitions of α_i^k and V_i^k.

LEMMA 5 *For $k \geqq 2$,*

$$\frac{d}{dx} B_i^k(x) = \left(\frac{k}{t_{i+k} - t_i} \right) B_i^{k-1}(x) - \left(\frac{k}{t_{i+k+1} - t_{i+1}} \right) B_{i+1}^{k-1}(x) \tag{10}$$

When $k = 1$, the equation is true for all x except $x = t_i, t_{i+1}, t_{i+2}$.

Proof The proof is by induction, with the cases $k = 1$ and 2 left to the reader. We assume that the formula is true for a fixed k, and with that hypothesis, we establish the next case. The induction hypothesis is Equation (10), which is now rewritten in a compact form using notation previously explained:

$$\frac{d}{dx} B_i^k = k\alpha_i^k B_i^{k-1} - k\alpha_{i+1}^k B_{i+1}^{k-1} \tag{11}$$

From the basic recurrence relation, Equation (3), we have

$$\begin{aligned} \frac{d}{dx} B_i^{k+1} &= \frac{d}{dx} \left[V_i^{k+1} B_i^k + (1 - V_{i+1}^{k+1}) B_{i+1}^k \right] \\ &= V_i^{k+1} \frac{d}{dx} B_i^k + \alpha_i^{k+1} B_i^k + (1 - V_{i+1}^{k+1}) \frac{d}{dx} B_{i+1}^k - \alpha_{i+1}^{k+1} B_{i+1}^k \end{aligned} \tag{12}$$

Here we used Equation (7) twice. Next we replace the derivatives that appear on the right side of (12) by using the induction hypothesis, Equation (11). The result is

$$\begin{aligned} \frac{d}{dx} B_i^{k+1} &= V_i^{k+1} (k\alpha_i^k B_i^{k-1} - k\alpha_{i+1}^k B_{i+1}^{k-1}) + \alpha_i^{k+1} B_i^k \\ &\quad + (1 - V_{i+1}^{k+1})(k\alpha_{i+1}^k B_{i+1}^{k-1} - k\alpha_{i+2}^k B_{i+2}^{k-1}) - \alpha_{i+1}^{k+1} B_{i+1}^k \end{aligned} \tag{13}$$

By straightforward regrouping, we write this in the form

$$\begin{aligned} \frac{d}{dx} B_i^{k+1} &= \alpha_i^{k+1} B_i^k + k\alpha_i^k V_i^{k+1} B_i^{k-1} - \alpha_{i+1}^{k+1} B_{i+1}^k \\ &\quad - k\alpha_{i+2}^k (1 - V_{i+1}^{k+1}) B_{i+2}^{k-1} - k\alpha_{i+1}^k V_i^{k+1} B_{i+1}^{k-1} \\ &\quad + k\alpha_{i+1}^k (1 - V_{i+1}^{k+1}) B_{i+1}^{k-1} \end{aligned} \tag{14}$$

In this equation we shall make the following replacements—all justified by Equations (8) and (9):

$$k\alpha_i^k V_i^{k+1} B_i^{k-1} = k\alpha_i^{k+1} V_i^k B_i^{k-1} \tag{15}$$

$$-k\alpha_{i+2}^k (1 - V_{i+1}^{k+1}) B_{i+2}^{k-1} = -k\alpha_{i+1}^{k+1} (1 - V_{i+2}^k) B_{i+2}^{k-1} \tag{16}$$

$$\begin{aligned} &- k\alpha_{i+1}^k V_i^{k+1} B_{i+1}^{k-1} + k\alpha_{i+1}^k (1 - V_{i+1}^{k+1}) B_{i+1}^{k-1} \\ &= k\alpha_{i+1}^k (1 - V_i^{k+1}) B_{i+1}^{k-1} - k\alpha_{i+1}^k V_{i+1}^{k+1} B_{i+1}^{k-1} \\ &= k\alpha_i^{k+1} (1 - V_{i+1}^k) B_{i+1}^{k-1} - k\alpha_{i+1}^{k+1} V_{i+1}^k B_{i+1}^{k-1} \end{aligned} \tag{17}$$

The transformed version of Equation (14) is now

$$\frac{d}{dx}B_i^{k+1} = \alpha_i^{k+1}B_i^k + k\left[\alpha_i^{k+1}V_i^k B_i^{k-1} + \alpha_i^{k+1}(1 - V_{i+1}^k)B_{i+1}^{k-1}\right]$$
$$- \alpha_{i+1}^{k+1}B_{i+1}^k - k\left[\alpha_{i+1}^{k+1}V_{i+1}^k B_{i+1}^{k-1} + \alpha_{i+1}^{k+1}(1 - V_{i+2}^k)B_{i+2}^{k-1}\right] \quad \textbf{(18)}$$

By means of the basic recurrence relation (3), the bracketed terms in Equation (18) can be simplified, the result being

$$\frac{d}{dx}B_i^{k+1} = \alpha_i^{k+1}B_i^k + k\alpha_i^{k+1}B_i^k - \alpha_{i+1}^{k+1}B_{i+1}^k - k\alpha_{i+1}^{k+1}B_{i+1}^k$$
$$= (k+1)\alpha_i^{k+1}B_i^k - (k+1)\alpha_{i+1}^{k+1}B_{i+1}^k \quad \textbf{(19)}$$

Since this is Equation (11) with k replaced by $k + 1$, the induction step is complete. ■

LEMMA 6 *For $k \geqq 1$, the B-splines B_i^k belong to continuity class $C^{k-1}(\mathbb{R})$.*

Proof It is clear that B_i^1 is continuous: $B_i^1 \in C^0(\mathbb{R})$. Now let us assume that $B_i^k \in C^{k-1}(\mathbb{R})$. By Lemma 5, $(d/dx)B_i^{k+1} \in C^{k-1}(\mathbb{R})$ since this derivative is a linear combination of B_i^k and B_{i+1}^k. Hence, $B_i^{k+1} \in C^k(\mathbb{R})$. By the principle of induction, this completes the proof. ■

From the lemma concerning derivatives of B-splines, we obtain a useful formula:

$$\frac{d}{dx}\sum_{i=-\infty}^{\infty} c_i B_i^k(x) = k\sum_{i=-\infty}^{\infty}\left(\frac{c_i - c_{i-1}}{t_{i+k} - t_i}\right)B_i^{k-1}(x) \qquad (k \geqq 2) \quad \textbf{(20)}$$

This formula can be used for numerical differentiation, although numerical differentiation of noisy data is a very dubious undertaking. For $k = 1$, Equation (20) is true for all x except the knots.

LEMMA 7 $$\int_{-\infty}^{x} B_i^k(s)\,ds = \left(\frac{t_{i+k+1} - t_i}{k+1}\right)\sum_{j=i}^{\infty} B_j^{k+1}(x)$$

Proof We can verify that the derivatives of both sides of the equation are equal by using Equation (20) with

$$c_j = \begin{cases} 0 & \text{if } j < i \\ 1 & \text{if } j \geqq i \end{cases}$$

Then $c_j - c_{j-1}$ is 0 except when $j = i$, and we obtain

$$B_i^k(x) = \left(\frac{t_{i+k+1} - t_i}{k+1}\right)(k+1)\left(\frac{1}{t_{i+k+1} - t_i}\right)B_i^k(x)$$

In order to be certain that the functions on the two sides of the equation do not differ by a constant, note that they both reduce to 0 at $x = t_i$. ■

Additional Properties

If f is a function and K is a subset of its domain, then $f \mid K$ denotes the **restriction** of f to K. Thus,

$$(f \mid K)(x) = f(x) \qquad (x \in K)$$

This concept is useful in working with splines because each function $B_i^k \mid (t_j, t_{j+1})$ is a polynomial (more precisely, the restriction of a polynomial). When it is said that a set of functions f_i is **linearly independent on a set** K, this means that the set of restricted functions $f_i \mid K$ is linearly independent in the usual sense.

Now consider the B-splines $B_0^k, B_1^k, \ldots, B_k^k$. When these are restricted to any single interval between knots $(t_\nu, t_{\nu+1})$, the result is a set of polynomials of degree $\leqq k$. A surprising and useful fact is that these restrictions form a basis for the polynomial space Π_k if the interval is (t_k, t_{k+1}).

LEMMA 8 *The set of B-splines* $\{B_j^k, B_{j+1}^k, \ldots, B_{j+k}^k\}$ *is linearly independent on* (t_{k+j}, t_{k+j+1}).

Proof First consider the case $k = 0$. The lemma asserts that $\{B_j^0\}$ is linearly independent on the interval (t_j, t_{j+1}). This is obviously true. For the purposes of an inductive proof, let $k \geqq 1$, and assume that the lemma is correct for index $k - 1$. On the basis of this assumption, we shall prove the lemma for the index k. Let $S = \sum_{i=0}^{k} c_{j+i} B_{j+i}^k$, and suppose that $S \mid (t_{k+j}, t_{k+j+1}) = 0$. By Equation (20),

$$0 = S' \mid (t_{k+j}, t_{k+j+1}) = k \sum_{i=1}^{k} \frac{c_{j+i} - c_{j+i-1}}{t_{j+i+k} - t_{j+i}} B_{j+i}^{k-1} \mid (t_{k+j}, t_{k+j+1}) \tag{21}$$

In arriving at this equation, we used $B_{j+k+1}^{k-1} = 0$ and $B_j^{k-1} = 0$ on (t_{k+j}, t_{k+j+1}). By applying the induction hypothesis to $\{B_{j+1}^{k-1}, B_{j+2}^{k-1}, \ldots, B_{j+k}^{k-1}\}$, we conclude that this set is linearly independent on the interval (t_{k+j}, t_{k+j+1}). Therefore, in Equation (20) all the coefficients must be 0, and thus we have $c_0 = c_1 = \cdots = c_k$. If this common value is denoted by λ, we have $S(x) = \lambda$ on (t_{k+j}, t_{k+j+1}) by Lemma 4. (Observe that in Lemma 4, the only terms that are nonzero on the interval (t_{k+j}, t_{k+j+1}) are $B_j^k, B_{j+1}^k, \ldots, B_{j+k}^k$.) Since it has been assumed that S vanished on (t_{k+j}, t_{k+j+1}), we conclude that $\lambda = 0$. ■

LEMMA 9 *The set of B-splines* $\{B_{-k}^k, B_{-k+1}^k, \ldots, B_{n-1}^k\}$ *is linearly independent on* (t_0, t_n).

Proof Let $S = \sum_{i=-k}^{n-1} c_i B_i^k$, and suppose that $S \mid (t_0, t_n) = 0$. On the interval (t_0, t_1) only $B_{-k}^k, B_{-k+1}^k, \ldots, B_0^k$ are nonzero, and therefore

$$0 = S|(t_0, t_1) = \sum_{i=-k}^{0} c_i B_i^k \mid (t_0, t_1) \tag{22}$$

By Lemma 8, the set $\{B^k_{-k}, B^k_{-k+1}, \dots, B^k_0\}$ is linearly independent on (t_0, t_1). Hence, from Equation (22), we infer that $c_i = 0$ when $-k \leqq i \leqq 0$. If all the c_i's are 0, we have the desired conclusion. If not all the c_i's are 0, let j be the first index for which $c_j \neq 0$. By the prior work, $j \geqq 1$. Hence, $(t_j, t_{j+1}) \subseteqq (t_0, t_n)$. For any $x \in (t_j, t_{j+1})$, we obtain the contradiction

$$0 = S(x) = \sum_{i=j}^{n-1} c_i B^k_i(x) = c_j B^k_j(x) \neq 0$$

Hence, all the c_i's are 0. ■

PROBLEMS 6.5

1. Verify this formula for $B^2_j(t_i)$:

$$B^2_j(t_i) = \left(\frac{t_i - t_{i-1}}{t_{i+1} - t_{i-1}}\right) \delta_{i,j+1} + \left(\frac{t_{i+1} - t_i}{t_{i+1} - t_{i-1}}\right) \delta_{i,j+2}$$

2. Prove that if $t_m \leqq x < t_{m+1}$, then

$$\sum_{i=-\infty}^{\infty} c_i B^k_i(x) = \sum_{i=m-k}^{m} c_i B^k_i(x)$$

3. Let $h_i = t_{i+1} - t_i$. Show that if

$$c_{i-1} h_{i-1} + c_{i-2} h_i = y_i(h_i + h_{i-1}) \qquad (i \in \mathbb{Z})$$

then the spline function $S = \sum_{i=-\infty}^{\infty} c_i B^2_i$ will have the interpolation property $S(t_j) = y_j$ for all j.

4. Suppose that the knots are taken to be all the integers: $t_i = i\ (i \in \mathbb{Z})$. Show that $B^k_i(x) = B^k_0(x - t_i)$.

5. Count the number of multiplications, divisions, and additions/subtractions involved in the algorithm for computing $\sum_{i=-\infty}^{\infty} C^k_i B^k_i(x)$.

6. Verify Formula (20) for derivatives of splines.

7. Prove that

$$\int_{-\infty}^{\infty} B^k_i(x)\, dx = \frac{t_{i+k+1} - t_i}{k+1}$$

8. Prove that if $\sum_{i=-\infty}^{\infty} c_i B^k_i(x) = 0$ for all x, then $c_i = 0$ for all i.

9. Define functions of s by the formula

$$U^k_i(s) = (t_{i+1} - s)(t_{i+2} - s) \cdots (t_{i+k} - s)$$

For $k = 0$, let $U_i^0(s) = 1$. Prove that

$$U_i^k(s)V_i^k(x) + U_{i-1}^k(s)\left[1 - V_i^k(x)\right] = (x - s)U_i^{k-1}(s)$$

10. (Continuation) Prove that

$$\sum_{i=-\infty}^{\infty} U_i^k(s)B_i^k(x) = (x - s)\sum_{i=-\infty}^{\infty} U_i^{k-1}(s)B_i^{k-1}(x)$$

11. (Continuation) Prove **Marsden's Identity**:

$$\sum_{i=-\infty}^{\infty} U_i^k(s)B_i^k(x) = (x - s)^k$$

12. (Continuation) Prove that every polynomial of degree $\leqq k$ is expressible in the form $\sum_{i=-\infty}^{\infty} c_i B_i^k$.

13. Using the notation in the text, prove that B_i^2 is given by the formulas

$$B_i^2(x) = \begin{cases} V_i^2 V_i^1 & x \in [t_i, t_{i+1}) \\ V_i^k - V_{i+1}^1(V_i^2 - 1 + V_{i+1}^2) & x \in [t_{i+1}, t_{i+2}) \\ (1 - V_{i+1}^2)(1 - V_{i+2}^1) & \text{x} \in [t_{i+2}, t_{i+3}) \\ 0 & \text{elsewhere} \end{cases}$$

14. Verify these two equations:

$$\frac{d}{dx} B_i^1 = \frac{B_i^0}{t_{i+1} - t_i} - \frac{B_{i+1}^0}{t_{i+2} - t_{i+1}}$$

$$\frac{d}{dx} B_i^2 = \frac{2B_i^1}{t_{i+2} - t_i} - \frac{2B_{i+1}^1}{t_{i+3} - t_{i+1}}$$

15. Prove that

$$\sup_{-\infty<x<\infty} \left| \sum_{i=-\infty}^{\infty} c_i B_i^k(x) \right| \leqq \sup_{-\infty<i<\infty} |c_i|$$

16. Prove that if $\sup_i |t_{i+1} - t_i| \leqq m$, then

$$\int_{-\infty}^{\infty} \left| \sum_{i=-\infty}^{\infty} c_i B_i^k(x) \right| dx \leqq m \sum_{i=-\infty}^{\infty} |c_i|$$

17. Find an upper bound for

$$\int_{-\infty}^{\infty} \left[\sum_{i=-\infty}^{\infty} c_i B_i^k(x) \right]^2 dx$$

18. Prove that for $k \geqq 3$,

$$\frac{d^2}{dx^2} \sum_{i=-\infty}^{\infty} c_i B_i^k(x) = k(k-1) \sum_{i=-\infty}^{\infty} \left(\frac{c_i - c_{i-1}}{t_{i+k} - t_i} - \frac{c_{i-1} - c_{i-2}}{t_{i+k-1} - t_{i-1}} \right) \frac{B_i^{k-2}(x)}{t_{i+k-1} - t_i}$$

19. Prove that if $a_i = k^{-1}(t_{i+1} + t_{i+2} + \cdots + t_{i+k})$, then

$$\sum_{i=-\infty}^{\infty} a_i B_i^k(x) = x \qquad (k \geqq 1)$$

20. Give an example of a spline function $\sum_{i=-\infty}^{\infty} c_i B_i^k$ that is not the 0 function yet vanishes at all the knots.

21. Suppose that in the sequence of constants $C_{m-k}^k, C_{m-k+1}^k, \ldots, C_m^k$, two adjacent elements are the same. Show that the spline function

$$\sum_{i=-\infty}^{\infty} C_i^k B_i^k(x)$$

will be a polynomial of degree less than k on the interval (t_m, t_{m+1}).

22. Use the algorithm that follows Lemma 3 to prove that $\sum_{i=-\infty}^{\infty} c_i B_i^k$ is a polynomial of degree $\leqq k$ on (t_m, t_{m+1}).

23. Use the algorithm following Lemma 3 to give new proofs of Lemmas 1 and 2.

24. Prove formally that $B_i^k(x) > 0$ on (t_j, t_{j+1}) if and only if $j - k \leqq i \leqq j$.

25. Prove that Formula (10) in Lemma 5 is true for $k = 1$ except at the three knots t_i, t_{i+1}, t_{i+2}.

26. Find α and β in the formula

$$\text{support}\left(\prod_{i=0}^{r} B_i^k \right) = \bigcap_{i=0}^{r} \text{support}(B_i^k) = (\alpha, \beta)$$

27. What is the smallest interval on which $[B_j^k, B_{j+1}^k, \ldots, B_{n+k+j-1}^k]$ is linearly independent?

28. Prove that

$$\sum_{i=0}^{n} B_i^k(x) = 1 \qquad (t_k \leqq x \leqq t_{k+n})$$

29. Prove that

$$\sum_{i=0}^{n} B_i^k(x) > 0 \qquad (t_1 < x < t_{n+k+1})$$

COMPUTER PROBLEMS 6.5

1. Write and test a subroutine for computing approximate values of $g(x) = \int_a^x f(t)\,dt$ on an interval $[a, b]$. The function f is furnished by the user in a separate function subprogram. The method to be used is first to interpolate f on $[a, b]$ by a natural cubic spline, S, using n equally spaced knots, and then to use $g(x) \approx \int_a^x S(t)\,dt$. The user will specify a, b, and n.
2. Let knots $t_1, t_2, \ldots, t_{n+k+1}$ be specified, as well as n and k. Let coefficients $c_1, c_2, \ldots, c_n$ be given, and write $f(x) = \sum i = 1^n c_i B_i^k(x)$. Write a subroutine that gives the value of $f(x)$ for any real value of x.
3. (Challenging problem) Let the knots be the set of all integers. Write a program to compute the supremum norm of B_0^k for $1 \leqq k \leqq 100$. A short program will suffice. Test cases: for $k = 1, 5$, and 10, the values are 1, 0.55, and 0.410963.

*6.6 The *B*-Splines: Applications

In this section, we adopt the notation of the preceding section. An initial sequence of knots is prescribed

$$\cdots < t_{-2} < t_{-1} < t_0 < t_1 < t_2 < \cdots \tag{1}$$

and a system of *B*-splines, $B_i^k(x)$, is defined, based on the given knots. We want to relate these *B*-splines to the spline functions introduced at the beginning of Section 6.4. There, we considered functions that are globally of class C^{k-1} and are piecewise polynomials of degree $\leqq k$ on the n intervals $[t_0, t_1], [t_1, t_2], \ldots, [t_{n-1}, t_n]$. Let us denote the family of all such splines by S_n^k. The notation does not display the knots, which are fixed at the beginning. We regard the spline functions in S_n^k as having domain $[t_0, t_n]$. How are these functions related to the *B*-splines B_i^k? Throughout, we assume that $n \geqq 1$.

Basis for the Space S_n^k

In order that the domains of these functions be the same, we restrict the functions B_i^k to the interval $[t_0, t_n]$. The notation for the restricted functions is $B_i^k \mid [t_0, t_n]$.

THEOREM 1 *A basis for the space S_n^k is*

$$\left\{B_i^k \mid [t_0, t_n] : -k \leqq i \leqq n-1\right\} \tag{2}$$

Consequently, the dimension of S_n^k is $k + n$.

Proof First, it is obvious that the functions in Equation (2) belong to S_n^k because they are splines of degree k based on the given knot sequence in (1).

Next we prove that the dimension of S_n^k is at most $k + n$. This will be accomplished by exhibiting a set of $k + n$ functions that spans S_n^k. In fact, each element of S_n^k can be expressed in the form

$$S(x) = \sum_{i=0}^{k} a_i x^i + \sum_{i=1}^{n-1} b_i (x - t_i)_+^k \tag{3}$$

In this equation, the truncated powers are used:

$$(x - t_i)_+^k = \begin{cases} (x - t_i)^k & \text{if } x \geqq t_i \\ 0 & \text{if } x < t_i \end{cases}$$

In order to prove Equation (3), we start with the interval $[t_0, t_1]$ in which all the truncated powers are 0. On $[t_0, t_1]$, $S(x)$ is a polynomial of degree k, say p_0. Then we have $p_0(x) = \sum_{i=0}^{k} a_i x^i$. This determines all the coefficients a_i. In the interval $[t_1, t_2]$, $S(x)$ is another polynomial, say p_1. By the continuity conditions at t_1, we have

$$(p_1 - p_0)^{(r)}(t_1) = 0 \qquad (0 \leqq r \leqq k - 1)$$

Since $p_1 - p_0$ is of degree at most k, we conclude that $(p_1 - p_0)(x) = b_1(x - t_1)^k$ for some b_1. Therefore we can write

$$S(x) = \sum_{i=0}^{k} a_i x^i + b_1 (x - t_1)_+^k \qquad (t_0 \leqq x \leqq t_2)$$

This same argument is repeated at $t_2, t_3, \ldots, t_{n-1}$ to justify the remaining terms in Equation (3).

The third part of the proof is to apply Lemma 9 of Section 6.5 and conclude that the set in (2) is linearly independent. It is therefore a basis for S_n^k. ■

The proof of Theorem 1 shows that the functions

$$1, x, x^2, \ldots, x^k, (x - t_1)_+^k, (x - t_2)_+^k, \ldots, (x - t_{n-1})_+^k$$

form a basis for S_n^k. We should be cautious about using this basis for numerical work because it is very badly conditioned. Instead, one should use the basis of B-splines given in the statement of the theorem.

Interpolation Matrix

Spline functions can be used for interpolation at points other than the knots. Let a set of **nodes** be given: $x_1 < x_2 < \cdots < x_n$. We would like to interpolate arbitrary data given at the nodes by a spline function of the form $\sum_{j=1}^{n} c_j B_j^k$. For this to be possible, the **interpolation matrix** A given by

$$A_{ij} = B_j^k(x_i) \qquad (1 \leqq i, j \leqq n) \tag{4}$$

must be nonsingular. A beautiful theorem of Schoenberg and Whitney reveals the necessary and sufficient conditions for this nonsingularity—namely, that the diagonal of A should contain no 0 elements.

For the proof of this result, we follow de Boor [1976 and personal communications]. Several parts will be considered separately.

LEMMA 1 *If the matrix in Equation (4) is nonsingular, then $A_{ii} \neq 0$ for $1 \leqq i \leqq n$.*

Proof Suppose that $A_{rr} = 0$ for some r. Then $B_r^k(x_r) = 0$, and by Lemma 2 in Section 6.5, $x_r \notin (t_r, t_{r+k+1})$. Suppose first that $x_r \leqq t_r$. If $i \leqq r \leqq j$, then $x_i \leqq x_r \leqq t_r \leqq t_j$ and x_i is not in the support of B_j. Consequently, $A_{ij} = B_j^k(x_i) = 0$. The first r rows in A can be interpreted as vectors in $\mathbb{R}^{r-1}$ because $A_{ij} = 0$ for $j = r, r+1, \dots, n$. This set of rows is therefore linearly dependent, and A is singular.

In the other case, $x_r \geqq t_{r+k+1}$. If $i \geqq r \geqq j$, then $x_i \geqq x_r \geqq t_{r+k+1} \geqq t_{j+k+1}$, and $A_{ij} = B_j^k(x_i) = 0$. The columns numbered 1 to r form a linearly dependent set because their components numbered $r, r+1, \dots, n$ are all 0. Again we conclude that A is singular. ■

LEMMA 2 *If $k = 1$ and $t_i < x_i < t_{i+2}$ for $1 \leqq i \leqq n$, then A is nonsingular.*

Proof We use induction on n to establish this. If $n = 1$, the matrix A is a 1×1 matrix with sole element $A_{11} = B_1^1(x_1) \neq 0$.

Now suppose that Lemma 2 has been proved whenever the number of nodes is less than n. Consider then the case of n nodes, with $n \geqq 2$. If there is an index r for which $1 \leqq r \leqq n-1$ and $x_r \leqq t_{r+1}$, then for any pair (i, j) satisfying $i \leqq r < j$, we shall have $x_i \leqq x_r \leqq t_{r+1} \leqq t_j$ and $A_{ij} = B_j^1(x_i) = 0$. The matrix A will therefore have the form

$$A = \begin{bmatrix} C & 0 \\ E & D \end{bmatrix}$$

in which C is $r \times r$ and D is $(n-r) \times (n-r)$. The matrix A is invertible if and only if C and D are invertible. (See Problem 11.) The matrices C and D are just like A except for being smaller (because $r < n$ and $n - r < n$). Therefore they are invertible by the induction hypothesis.

A similar argument can be used if $x_r \geqq t_{r+1}$ for some index r in the range $2 \leqq r \leqq n$. In this case, if $j < r \leqq i$, then $A_{ij} = 0$. The structure of A is

$$A = \begin{bmatrix} C & E \\ 0 & D \end{bmatrix}$$

with C being $(r-1) \times (r-1)$ and D being $(n-r+1) \times (n-r+1)$.

The only case remaining is that in which $x_i > t_{i+1}$ for $1 \leqq i \leqq n-1$ and $x_i < t_{i+1}$ for $2 \leqq i \leqq n$. This can occur only if $n = 1$ (a case already considered) or if $n = 2$. In this special case, $n = 2$, $k = 1$, $x_1 > t_2$ and $x_2 < t_3$. Hence, $t_2 < x_1 < x_2 < t_3$. On the interval (t_2, t_3) we have $B_1^1(x) + B_2^1(x) = 1$. Thus the matrix A has the form

$$\begin{bmatrix} B_1^1(x_1) & B_2^1(x_1) \\ B_1^1(x_2) & B_2^1(x_2) \end{bmatrix} = \begin{bmatrix} \lambda & 1-\lambda \\ \mu & 1-\mu \end{bmatrix}$$

The determinant is $\lambda - \mu \equiv B_1^1(x_1) - B_1^1(x_2) > 0$. ■

The Schoenberg-Whitney Theorem follows.

THEOREM 2 **Schoenberg-Whitney Theorem** *Assume that $x_1 < x_2 < \cdots < x_n$. For the matrix A given by $A_{ij} = B_j^k(x_i)$ to be nonsingular, it is necessary and sufficient that there be no 0 elements on its diagonal.*

Proof The necessity has been proved in Lemma 1. The sufficiency, when $k = 0$, is trivial because the condition $B_i^0(x_i) \neq 0$ is equivalent to $t_i \leqq x_i < t_{i+1}$. Since the supports of the B-splines are disjoint, $B_j^0(x_i) = \delta_{ij}$.

For $k = 1$, the proof of sufficiency has been given in Lemma 2. Now we proceed by induction on k. Assume that the theorem has been proved for B-splines of degree $< k$. On that hypothesis, we prove it for splines of degree k. Here $k \geqq 2$ because of Lemma 2. Our proof will require another induction on n just as in the proof of Lemma 2, but we shall be less formal about it here. The point of the argument is that if $x_r \leqq t_{r+1}$ for some $r \in \{1, 2, \ldots, n-1\}$ or if $x_r \geqq t_{r+k}$ for some $r \in \{2, 3, \ldots, n\}$, then the matrix A will have a block structure such as in Lemma 2 and will be invertible by the induction hypothesis applied to two submatrices of A. We can therefore assume that

$$t_{i+1} < x_i < x_{i+1} < t_{i+k+1} \qquad (1 \leqq i \leqq n-1) \tag{5}$$

Suppose that A is singular. Then $Au = 0$ for some $u \neq 0$. Let $f = \sum_1^n u_j B_j^k$. We note that f vanishes at the $n+2$ points $t_1, x_1, x_2, \ldots, x_n, t_{n+k+1}$. Since $k \geqq 2$, f' exists and is continuous. Hence Rolle's Theorem is applicable, and we infer the existence of $n+1$ zeros, ξ_i, of f' arranged as follows:

$$t_1 < \xi_1 < x_1 < \xi_2 < x_2 < \cdots < x_{n-1} < \xi_n < x_n < \xi_{n+1} < t_{n+k+1}$$

Now f' is of the form $\sum_{j=1}^{n+1} v_j B_j^{k-1}$, as shown in Equation (20) of Section 6.5. We also have

$$\sum_{j=1}^{n+1} v_j B_j^{k-1}(\xi_i) = f'(\xi_i) = 0 \qquad (1 \leqq i \leqq n+1) \tag{6}$$

By using Equation (5) we see that ξ_i belongs to the support of B_i^{k-1} for $2 \leqq i \leqq n$.

Now several cases arise. In case 1, assume that $\xi_1 < t_{k+1}$ and that $\xi_{n+1} > t_{n+1}$. Then ξ_1 is in the support of B_1^{k-1} and ξ_{n+1} is in the support of B_{n+1}^{k-1}. Hence, by the induction hypothesis, the $(n+1) \times (n+1)$ matrix $(B_j^{k-1}(\xi_i))$ is nonsingular. The coefficients v_j in Equation (6) must therefore all be 0. The actual formula for v_j, as in Equation (20) of Section 6.5, is

$$v_j = k(u_j - u_{j-1})/(t_{j+k} - t_j)$$

We can use this formula for $1 \leqq j \leqq n+1$ if we agree to set $u_0 = u_{n+1} = 0$. The vanishing of the coefficients v_j implies that

$$u_1 - u_0 = u_2 - u_1 = \cdots = u_{n+1} - u_n = 0$$

Since $u_0 = u_{n+1} = 0$, we conclude that all $u_i = 0$.

In case 2, assume that $\xi_1 \geqq t_{k+1}$ and that $\xi_{n+1} > t_{n+1}$. Then $B_1^{k-1}(\xi_i) = 0$ for $1 \leqq i \leqq n+1$ because ξ_i is not in the support of B_1^{k-1}. In this case, we infer from Equation (6) that

$$\sum_{j=2}^{n+1} v_j B_j^{k-1}(\xi_i) = 0 \qquad (2 \leqq i \leqq n+1)$$

But ξ_i is in the support of B_i^{k-1} for $2 \leqq i \leqq n+1$, and so by the induction hypothesis we conclude that $v_j = 0$ for $2 \leqq j \leqq n+1$. As before, this leads to

$$u_2 - u_1 = u_3 - u_2 = \cdots = u_{n+1} - u_n = 0$$

and to $u_i = 0$ for $1 \leqq i \leqq n$ because $u_{n+1} = 0$.

In case 3, assume that $\xi_1 < t_{k+1}$ and that $\xi_n \leqq t_{n+1}$. This is similar to case 2. Now ξ_i is in the support of B_i^{k-1} for $1 \leqq i \leqq n$, and the same arguments as before lead to $u_i = 0$ for $1 \leqq i \leqq n$.

In case 4, assume that $\xi_1 \geqq t_{k+1}$ and $t_{n+1} \geqq \xi_{n+1}$. Now we have $B_1^{k-1}(\xi_i) = 0$ and $B_{n+1}^{k-1}(\xi_i) = 0$ for $1 \leqq i \leqq n+1$. Equation (6) gives us

$$\sum_{j=2}^{n} v_j B_j^{k-1}(\xi_i) = 0 \qquad (2 \leqq i \leqq n)$$

But ξ_i is in the support of B_i^{k-1} for $2 \leqq i \leqq n$, and by familiar reasoning, we conclude that

$$u_2 - u_1 = u_3 - u_2 = \cdots = u_n - u_{n-1} = 0$$

This shows that all the u_j are equal, say to λ. Hence, $f = \lambda \sum_1^n B_j^k$. Since $B_j^k \geqq 0$ and $B_1^k(x_1) > 0$, the equation $f(x_1) = 0$ shows that $\lambda = 0$. ■

Example 1 Let the knots be all the integers. If we wish to interpolate by linear combinations of $B_1^2, B_2^2, \ldots, B_5^2$, can we use the set of nodes $\{3.1, 3.5, 3.6, 6.1, 6.6\}$?

Solution In this case, the condition given by Theorem 2 can be written in the form

$$i < x_i < i + 3 \qquad (1 \leqq i \leqq 5)$$

One easily verifies that this condition *is* satisfied by the given nodes. ■

Existence

In Theorem 1, we saw that the spline space S_n^k has dimension $n + k$. The functions that make up S_n^k are regarded as having the domain $[t_0, t_n]$. If nodes $x_1, x_2, \ldots, x_{n+k}$ are chosen in $[t_0, t_n]$, what conditions must they satisfy so that interpolation by S_n^k will be possible?

THEOREM 3 *If nodes $x_1, x_2, \ldots, x_{n+k}$ are chosen in $[t_0, t_n]$ and satisfy*

$$t_{i-k-1} < x_i < t_i \qquad (1 \leqq i \leqq n + k)$$

then arbitrary data at these nodes can be interpolated by the spline space S_n^k.

Proof By Theorem 1, a basis for S_n^k is given by the set of functions $B_j^k \mid [t_0, t_n]$, with $-k \leqq j \leqq n - 1$. Relabel the nodes by putting $y_i = x_{i+1+k}$ for $-k \leqq i \leqq n - 1$. Then $y_i \in \text{supp}\,(B_i^k)$, and by Theorem 2, the matrix $B_j^k(y_i)$ is nonsingular. ■

If we assume that $B_i^k(x_i) \neq 0$ for $1 \leqq i \leqq n$ (which is the same as the assumption $t_i < x_i < t_{i+k+1}$), then our interpolating function can be taken to be

$$S(x) = \sum_{j=1}^{n} c_j B_j^k(x)$$

The coefficients in this equation are obtained by solving the following linear system of equations:

$$\sum_{j=1}^{n} B_j^k(x_i) c_j = f(x_i) \qquad (1 \leqq i \leqq n) \tag{7}$$

This solution to the interpolation problem is not necessarily the only one. We shall see this *nonuniqueness* now in considering the special case of interpolation at the knots. Suppose that $x_i = t_i$. The interpolation problem for these nodes requires the solution of the system

$$\sum_{j=-\infty}^{\infty} c_j B_j^k(t_i) = f(t_i) \qquad (1 \leqq i \leqq n) \tag{8}$$

The simplest case, $k = 0$, is solved immediately by using $B_j^0(t_i) = \delta_{ij}$. Hence, we have $c_j = f(t_j)$ for $1 \leqq j \leqq n$. The case $k = 1$ is solved in the same way because $B_j^1(t_i) = \delta_{i-1,j}$. Thus, $c_{i-1} = f(t_i)$ for $1 \leqq i \leqq n$.

To solve this interpolation problem when $k = 2$, we require the following fact (see Problem 1 in Section 6.5):

$$B_j^2(t_i) = \left(\frac{t_i - t_{i-1}}{t_{i+1} - t_{i-1}}\right)\delta_{i-1,j} + \left(\frac{t_{i+1} - t_i}{t_{i+1} - t_{i-1}}\right)\delta_{i-2,j} \tag{9}$$

Equation (9) implies immediately that each linear equation in System (8) will contain only two of the unknowns. Indeed, Equation (8) now reads as follows:

$$\left(\frac{t_{i+1} - t_i}{t_{i+1} - t_{i-1}}\right)c_{i-2} + \left(\frac{t_i - t_{i-1}}{t_{i+1} - t_{i-1}}\right)c_{i-1} = f(t_i) \qquad (1 \leqq i \leqq n) \tag{10}$$

This system contains n equations and $n + 1$ unknowns. We can solve it by assigning any value to c_{-1} and computing $c_0, c_1, \ldots, c_{n-1}$ recursively by means of Equation (10). Hence, there are *many* solutions to this interpolation problem.

If the case $k = 3$ is treated in the same manner, we obtain n equations with $n + 2$ unknowns. Here is system (8) for $k = 3$:

$$c_{i-3}B_{i-3}^3(t_i) + c_{i-2}B_{i-2}^3(t_i) + c_{i-1}B_{i-1}^3(t_i) = f(t_i) \qquad (1 \leqq i \leqq n) \tag{11}$$

The values $B_j^3(t_i)$ needed here can be computed by means of Theorem 4 in Section 6.5. The results are

$$\begin{aligned}
B_{i-3}^3(t_i) &= \frac{(t_{i+1} - t_i)^2}{(t_{i+1} - t_{i-2})(t_{i+1} - t_{i-1})} \\
B_{i-2}^3(t_i) &= \frac{(t_{i+2} - t_i)(t_i - t_{i-1})}{(t_{i+2} - t_{i-1})(t_{i+1} - t_{i-1})} + \frac{(t_i - t_{i-2})(t_{i+1} - t_i)}{(t_{i+1} - t_{i-2})(t_{i+1} - t_{i-1})} \\
B_{i-1}^3(t_i) &= \frac{(t_i - t_{i-1})^2}{(t_{i+2} - t_{i-1})(t_{i+1} - t_{i-1})}
\end{aligned}$$

As in the case $k = 2$, an easy solution to the system of equations is possible; namely, assign arbitrary values to c_{-2} and c_{-1}, then solve for $c_0, c_1, \ldots, c_{n-1}$ recursively, using Equation (11). This manner of solving system (11) is not actually recommended because the two extra parameters c_{-2} and c_{-1} are not being used to advantage. It is usual to impose one additional condition on the spline at the knots t_1 and t_n—that is, at *each end* of the interval of interest. For example, the *natural* spline conditions are defined to be $S''(t_1) = S''(t_n) = 0$.

The system of equations to be solved for the natural spline interpolant consists of Equation (11) together with a pair of additional equations that are given in Problems 8–10. For advice on spline interpolation and the arrangement of computations, the reader should consult de Boor [1984]. (In that book, B_j^k denotes a spline of degree $k - 1$.)

Noninterpolatory Approximation Methods

To illustrate *noninterpolatory* approximation methods, we turn to an elegant procedure introduced by Schoenberg [1967]. Given a function f, we define a spline function Sf by the equation

$$Sf = \sum_{i=-\infty}^{\infty} f(x_i)B_i^k \qquad x_i = \frac{1}{k}(t_{i+1} + \cdots + t_{i+k}) \tag{12}$$

When $k = 0$, we let $x_i = t_i$. In this case we have nothing new because Equation (12) is the interpolation scheme discussed previously. The same remark pertains to the case $k = 1$. But for higher values of k, the spline function Sf defined by Equation (12) does not interpolate f at any prescribed set of nodes. (Such operators have been termed **quasi-interpolation** operators.) The salient properties of this approximation scheme are these:

(i) If f is a linear function, then $Sf = f$.
(ii) For any linear function ℓ, $Sf - \ell$ has no more variations in sign than $f - \ell$.
(iii) If $f \geqq 0$, then $Sf \geqq 0$.
(iv) If $|f| \leqq M$, then $|Sf| \leqq M$.
(v) S is a linear operator: $S(\alpha f + \beta g) = \alpha Sf + \beta Sg$.

The interested reader should consult either Marsden [1970] or Schoenberg [1967].

Our next task is to investigate whether continuous functions can be approximated to arbitrary precision by splines. In situations like this, we expect to hold the degree k fixed while increasing the number of knots in order to obtain higher precision. The question addressed is whether any desired precision can be reached by increasing the density of knots; the theorem sought will be the spline analogue of the Weierstrass Approximation Theorem.

As usual, a set of knots is prescribed:

$$\cdots < t_{-2} < t_{-1} < t_0 < t_1 < t_2 < \cdots$$

The interval $[t_0, t_n]$ is denoted also by $[a, b]$, and we suppose that a function f is given on $[a, b]$. The function f is extended as shown in Figure 6.10. Thus, if f is continuous on $[a, b]$, its extension also will be continuous.

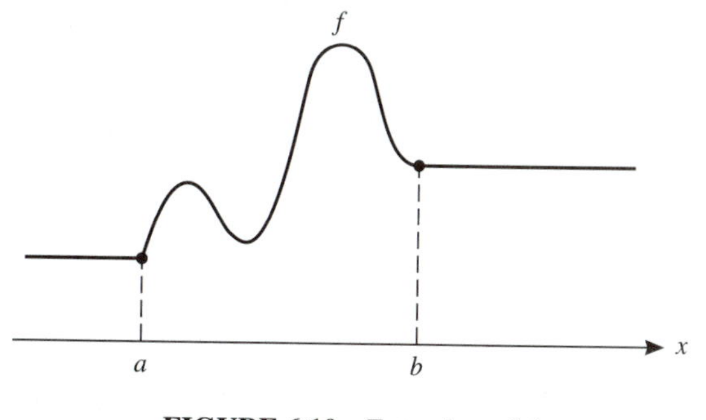

FIGURE 6.10 Extension of f

Whether f is continuous or not, we define its **modulus of continuity** by the equation

$$\omega(f;\delta) = \max_{|s-t| \leqq \delta} |f(s) - f(t)|$$

If f is a continuous function defined on an interval $[a, b]$, then it is uniformly continuous. This means that for any $\varepsilon > 0$, there is a $\delta > 0$ such that for all s and t in $[a, b]$,

$$|s - t| < \delta \Rightarrow |f(s) - f(t)| < \varepsilon$$

Hence, $\omega(f;\delta) \leqq \varepsilon$. In other words, for a continuous function f on a closed and bounded interval, the modulus of continuity $\omega(f;\delta)$ converges to 0 as δ converges to 0.

If f' exists, is continuous, and satisfies $|f'(x)| \leqq M$, then by the Mean-Value Theorem,

$$|f(s) - f(t)| = |f'(\xi)|\,|s - t| \leqq M|s - t|$$

Hence, $\omega(f, \delta) \leqq M\delta$.

The next step is to introduce a spline function that approximates f in a simple manner. For this purpose we choose

$$g = \sum_{i=-\infty}^{\infty} f(t_{i+2})B_i^k \tag{13}$$

With the aid of this function we can now prove the following result.

THEOREM 4 *If f is a function on $[t_0, t_n]$, then the spline function g in Equation (13) satisfies*

$$\max_{t_0 \leqq x \leqq t_n} |f(x) - g(x)| \leqq k\omega(f;\delta)$$

where $\delta = \max_{-k \leqq i \leqq n+1} |t_i - t_{i-1}|$ and $\omega(f;\cdot)$ is the modulus of continuity of f.

Proof Recall that $B_i^k \geqq 0$ and $\sum_{i=-\infty}^{\infty} B_i^k = 1$. Then with g as in Equation (13),

$$\begin{aligned}
|g(x) - f(x)| &= \left| \sum_{i=-\infty}^{\infty} f(t_{i+2})B_i^k(x) - f(x) \sum_{i=-\infty}^{\infty} B_i^k(x) \right| \\
&= \left| \sum_{i=-\infty}^{\infty} \left[f(t_{i+2}) - f(x) \right] B_i^k(x) \right| \\
&\leqq \sum_{i=-\infty}^{\infty} |f(t_{i+2}) - f(x)| B_i^k(x)
\end{aligned}$$

Let $x \in [t_j, t_{j+1}] \subseteqq [a, b]$. On the interval $[t_j, t_{j+1}]$, only $B^k_{j-k}, B^k_{j-k+1}, \ldots, B^k_j$ are *active*. Hence,

$$|g(x) - f(x)| \leqq \sum_{i=j-k}^{j} |f(t_{i+2}) - f(x)| B^k_i(x)$$
$$\leqq \max_{j-k \leqq i \leqq j} |f(t_{i+2}) - f(x)|$$

For i in the range $j - k \leqq i \leqq j$, we have

$$t_{i+2} - x \leqq t_{j+2} - t_j = (t_{j+2} - t_{j+1}) + (t_{j+1} - t_j) \leqq 2\delta$$
$$x - t_{i+2} \leqq t_{j+1} - t_{j-k+2} = (t_{j+1} - t_j) + \cdots + (t_{j-k+3} - t_{j-k+2}) \leqq k\delta$$

By the definition of the modulus of continuity and by Problem 20,

$$|f(t_{i+2}) - f(x)| \leqq \omega(f; k\delta) \leqq k\omega(f; \delta)$$ ■

Distance from a Function to a Spline Space

The preceding result can be stated in terms of the distance from f to the space S^k_n. The distance from a function f to a subspace G in a normed space is defined by

$$\operatorname{dist}(f, G) = \inf_{g \in G} \|f - g\|$$

Let us use the norm defined by

$$\|f\| = \max_{a \leqq x \leqq b} |f(x)|$$

Theorem 4 asserts that

$$\operatorname{dist}(f, S^k_n) \leqq k\omega(f; \delta) \tag{14}$$

If f is continuous, then

$$\lim_{\delta \downarrow 0} \omega(f; \delta) = 0$$

Hence, as the density of the knots is increased, the upper bound in Equation (14) will approach 0.

For functions possessing some derivatives, more can be said.

THEOREM 5 *Let $r < k < n$. If $f \in C^r[t_0, t_n]$, then (with δ as in Theorem 4)*

$$\operatorname{dist}(f, S^k_n) \leqq k^r \delta^r \|f^{(r)}\|$$

Proof Let g be any element of S_n^k. By Theorem 4,

$$\operatorname{dist}(f, S_n^k) = \operatorname{dist}(f - g, S_n^k) \leqq k\omega(f - g; \delta) \leqq k\delta \, \|f' - g'\|$$

As g runs over S_n^k, g' runs over all of S_n^{k-1}. (See Problem 13.) Thus, by taking an infimum on g in the previous inequality, we obtain

$$\operatorname{dist}(f, S_n^k) \leqq k\delta \operatorname{dist}(f', S_n^{k-1})$$

This argument can be repeated $r - 2$ times, yielding

$$\begin{aligned}\operatorname{dist}(f, S_n^k) &\leqq k^{r-1}\delta^{r-1} \operatorname{dist}(f^{(r-1)}, S_n^{k+1-r}) \\ &\leqq k^r \delta^{r-1} \omega(f^{(r-1)}; \delta) \\ &\leqq k^r \delta^r \, \|f^{(r)}\|\end{aligned}$$

■

PROBLEMS 6.6

1. Show that the matrix $B_j^k(x_i)$ that arises in B-spline interpolation is *banded*. Specifically, if $B_j^k(x_j) \neq 0$ and $x_j < x_{j+1}$ for each j, then each row and column of the matrix contains at most $2k + 1$ nonzero elements.
2. Let $k = 2$ and $f(x) = x$ in Equation (12). Show that $Sf = f$.
3. Prove that

$$x^2 = \sum_{i=-\infty}^{\infty} t_{i+1}t_{i+2}B_i^2(x)$$

4. Let $k = 2$ and $f(x) = 1$ in Equation (12). Show that $Sf = f$.
5. Prove property **(v)** of the Schoenberg approximation process.
6. Prove property **(i)** of the Schoenberg process in the case $k = 2$. *Hint:* Problems 2, 4, and 5 are helpful.
7. Prove properties **(iii)** and **(iv)** of the Schoenberg process.
8. Prove this theorem:

Theorem *If* $S = \sum_{j=-\infty}^{\infty} c_j B_j^3$, *then* $S'' = \sum_{j=-\infty}^{\infty} e_j B_j^1$ *with*

$$e_j = \frac{6}{t_{j+2} - t_j}\left(\frac{c_j - c_{j-1}}{t_{j+3} - t_j} - \frac{c_{j-1} - c_{j-2}}{t_{j+2} - t_{j-1}}\right)$$

9. (Continuation) Prove that if $S = \sum_{j=-\infty}^{\infty} c_j B_j^3$, then $S''(t_i) = e_{i-1}$, where e_j is defined in Problem 8.
10. (Continuation) Prove that the natural cubic spline interpolant for the function f is $\sum_{j=-2}^{n-1} c_j B_j^3$, where the coefficients satisfy Equation (11) in the text and

these *natural* end conditions corresponding to $i = 1$ and $i = n$:

$$(t_{i+2} - t_{i-1})c_{i-3} - (t_{i+2} + t_{i+1} - t_{i-1} - t_{i-2})c_{i-2} + (t_{i+1} - t_{i-2})c_{i-1} = 0$$

11. Let C and D be square matrices. Prove that

$$\begin{bmatrix} C & 0 \\ E & D \end{bmatrix}$$

is nonsingular if and only if C and D are nonsingular.

12. Let K be any subset of $\mathbb{R}$. Prove that in order for $\{B_1^k, \dots, B_n^k\}$ to be linearly independent on K, it is necessary and sufficient that $K \bigcap (t_i, t_{i+k+1})$ be nonempty for $1 \leqq i \leqq n$.

13. Use Theorem 1 to prove that the derivative operator is surjective from S_n^k to S_n^{k-1}.

14. Prove Lemma 8 of Section 6.5 as a corollary of the Schoenberg-Whitney Theorem.

15. Prove Lemma 9 of Section 6.5 as a corollary of the Schoenberg-Whitney Theorem.

16. Is this a valid version of the Schoenberg-Whitney Theorem?

> **Theorem** *If the nodes x_i are not necessarily ordered, then the nonsingularity of the matrix $(B_j^k(x_i))$ is equivalent to the assertion that each interval (t_i, t_{i+k+1}) for $1 \leqq i \leqq n$ contains at least one node.*

17. Use the method of proof in Theorem 4 to prove a similar result for the Schoenberg operator.

18. (Continuation) Refine the result obtained in Problem 17 in the case of equally spaced knots.

19. Prove Theorem 1 by showing that only $n + k$ elements of the sequence B_i^k have nonzero values on $[t_0, t_n]$.

20. Prove that $\omega(f; k\delta) \leqq k\omega(f; \delta)$.

21. Prove property **(i)** of the Schoenberg process for all values of k. It suffices, because of linearity, to prove that $Sf = f$ if $f(x) = 1$ or if $f(x) = x$. The first of these is very easy. To prove the second, show that $(Sf)' = 1$ by using Equation (20) in Section 6.5.

COMPUTER PROBLEM 6.6

1. Let $k = 3$ and take knots to be the integers. For $n = 5, 10, 15,$ and 20, estimate the conditioning of the matrix $(f_j(x_i))$ when f_j represents the basis for S_k^n consisting of powers and truncated powers (as in the proof of Theorem 1). Take x_i to be equally spaced points in $[t_0, t_n]$.

6.7 Taylor Series

In this short section, we want to illustrate (and emphasize) the utility of Taylor series as a mechanism for approximation. Taylor's Theorem is, of course, useful for functions that possess a certain number of continuous derivatives. However, it is not very useful for empirical data or for functions that have only a few derivatives. For functions to which Taylor's Theorem is applicable, one should not overlook the possibility of representing them efficiently by a *Taylor polynomial*.

Recall that if f is a function having a continuous $(n+1)$st derivative on an interval $(c-\delta, c+\delta)$, then

$$f(x) = p_n(x) + E_n(x)$$

where p_n is a polynomial of degree $\leqq n$ and E_n is a **remainder function**. These are given by

$$p_n(x) = \sum_{k=0}^{n} \frac{1}{k!} f^{(k)}(c)(x-c)^k$$

$$E_n(x) = \frac{1}{(n+1)!} f^{(n+1)}(\xi_x)(x-c)^{n+1} \qquad |\xi_x - c| < \delta$$

An important special case occurs when $c = 0$; this is the **Maclaurin series**.

By using Taylor's Theorem together with an analysis of $E_n(x)$ as $n \to \infty$, we obtain Taylor series for many important functions such as

$$\cos x = \sum_{k=0}^{\infty} (-1)^k \frac{x^{2k}}{(2k)!} \qquad (-\infty < x < \infty) \tag{1}$$

$$\frac{1}{x} = \sum_{k=0}^{\infty} (-1)^k (x-1)^k \qquad (0 < x < 2) \tag{2}$$

The series that occur in these illustrations are **power series**. Here is a theorem about the convergence of a power series.

THEOREM 1 *For every power series*

$$\sum_{k=0}^{\infty} a_k (x-c)^k$$

there is a number r in the range $[0, \infty]$ such that the series diverges for $|x-c| > r$ and converges for $|x-c| < r$.

The *number* r (it may be $+\infty$) is called the **radius of convergence** of the series. It can often be computed by the **ratio test**. (See Problems 2 and 3.) The radius of convergence of the cosine series, Equation (1), is $+\infty$; that of the series in Equation (2) is $+1$.

For applications the following theorem is important.

THEOREM 2 *Let r be the radius of convergence of $\sum_{k=0}^{\infty} a_k(x-c)^k$. Then the equation*

$$f(x) = \sum_{k=0}^{\infty} a_k(x-c)^k$$

defines a function that is continuously differentiable in the interval $|x-c| < r$. Furthermore,

$$f'(x) = \sum_{k=0}^{\infty} k a_k(x-c)^{k-1}$$

and this series has r for its radius of convergence. Finally, if $|b-c| < r$ and $|x-c| < r$, then

$$\int_b^x f(t)\,dt$$

can be obtained by integrating the f-series term by term. The resulting series has r for its radius of convergence.

In summary, this theorem says that we can integrate and differentiate a power series term by term within its interval of convergence.

As an illustration of how this theorem works, we consider one of the **higher transcendental functions**, the **sine integral**. This is defined by the formula

$$S(x) \equiv \int_0^x \frac{\sin t}{t}\,dt$$

This integration is not amenable to the usual techniques of elementary calculus. Instead, one can proceed as follows:

$$\sin t = \sum_{k=0}^{\infty} (-1)^k \frac{t^{2k+1}}{(2k+1)!}$$

$$\frac{\sin t}{t} = \sum_{k=0}^{\infty} (-1)^k \frac{t^{2k}}{(2k+1)!}$$

$$\int_0^x \frac{\sin t}{t}\,dt = \sum_{k=0}^{\infty} (-1)^k \frac{1}{(2k+1)!} \int_0^x t^{2k}\,dt$$

$$S(x) = \sum_{k=0}^{\infty} (-1)^k \frac{x^{2k+1}}{(2k+1)!(2k+1)}$$

The series we have obtained for $S(x)$ is rapidly convergent for small x. Thus, $S(1)$ can be computed to 20 decimal places of accuracy by taking only ten terms. (See Problem 36.)

Another similar example is

$$\int_0^x e^{t^2}\,dt = \sum_{k=0}^{\infty} \frac{x^{2k+1}}{(2k+1)k!} \tag{3}$$

PROBLEMS 6.7

1. Show that $\sum_{k=0}^{\infty} a_k x^k$ and $\sum_{k=0}^{\infty} a_k(x-c)^k$ have the same radius of convergence. For p a natural number, what about $\sum_{k=0}^{\infty} a_k x^{k+p}$?
2. (**Ratio test**) If $\lim_{n\to\infty} |A_{n+1}/A_n| < 1$, then $\sum_{k=0}^{\infty} A_k$ converges. Use the test to show that the cosine series in Equation (1) converges for all real x.
3. (**Ratio test**, continued) If $\lim_{n\to\infty} |A_{n+1}/A_n| > 1$, then $\sum_{k=0}^{\infty} A_k$ diverges. Use this fact together with Problem 2 to find the radius of convergence of

$$\sum_{k=0}^{\infty} (-1)^k (x-1)^k$$

4. Find the radius of convergence of

$$\sum_{k=0}^{\infty} k!\,x^k$$

5. If $f(x) = \sum_{k=0}^{\infty} a_k(x-c)^k$ and if the radius of convergence is r, then f possesses derivatives of all orders in the interval $|x-c| < r$. Furthermore,

$$f^{(n)}(x) = \sum_{k=n}^{\infty} \frac{a_k k!}{(k-n)!}(x-c)^{k-n} \qquad \left(|x-c| < r\right)$$

(Use Theorem 2 to prove this.)

6. Find a power series for

$$\frac{2}{\sqrt{\pi}} \int_0^x e^{-t^2}\,dt$$

Note: This function is known as the **error function** and is denoted by $\operatorname{erf}(x)$. It plays a role in statistics.

7. Find a power series for the function

$$f(x) = \int_0^x \frac{e^t - 1}{t}\,dt$$

Use the series you have obtained to compute $f(1)$ with three significant figures. (In this example, that means two decimal places.) Finally, prove that the sum of all the terms you did not include is so small that it will not affect the answer. *Cultural note:* The function f is an important one in applied mathematics. For further information, consult Abramowitz and Stegun [1964, chap. 5].

8. Show that the function $f(x) = \sum_{k=0}^{\infty} x^{2k}/(k!\,2^k)$ solves the differential equation $y' = xy$.

9. Let $a_0 = 1$ and $a_n = a_{n-1}/[2(n+1)]$ for $n \geqq 1$. What is the radius of convergence of $\sum_{k=0}^{\infty} a_k x^k$?

10. Let $f(x)$ be the function whose series is in Problem 9. What is the power series for $f'(x)$? A recurrence relation for the coefficients suffices.

11. Using the ratio test (see Problems 2 and 3), prove that if $\lim_{n\to\infty} |a_n/a_{n+1}|$ exists or is $+\infty$, then it is the radius of convergence of $\sum_{k=0}^{\infty} a_k(x-c)^k$.

12. Let $a_0 = 1$ and $a_{n+1} = [2 + (-1)^n]a_n$ for $n \geqq 1$. What is the radius of convergence of $\sum_{k=0}^{\infty} a_k(x-c)^k$?

13. If r and r' are the radii of convergence of $\sum_{k=0}^{\infty} a_k x^k$ and $\sum_{k=0}^{\infty} b_k x^k$, respectively, what is the radius of convergence of $\sum_{k=0}^{\infty}(a_k + b_k)x^k$?

14. A function related to the **dilogarithm** is defined by

$$f(x) = -\int_0^x \frac{\ln(1-t)}{t}\,dt \qquad (-\infty < x \leqq 1)$$

Find the Maclaurin series for f and determine its radius of convergence. How would you compute $f(-2)$? What about $f(0.001)$?

15. Obtain a series for $\ln x$ by integrating the series in Equation (2).

16. Show how the series for $\sin x$ can be obtained in two ways by differentiating or by integrating the series in Equation (1).

17. Obtain a power series for $\tan^{-1} x$ as follows: Start with $(1+x)^{-1} = \sum_{k=0}^{\infty}(-x)^k$, replace x by x^2, and then integrate the terms in the resulting equation. Compare this method to the alternative procedure of computing the successive derivatives of $\tan^{-1} x$ and obtaining the Taylor series.

18. Criticize the following analysis and correct it:

$$\int_0^x \frac{1-\cos t}{t}\,dt = \int_0^x \left[\frac{1}{t} - \frac{\cos t}{t}\right] dt = \int_0^x \left[\frac{1}{t} - \sum_{k=0}^{\infty} \frac{(-1)^k t^{2k-1}}{(2k)!}\right] dt$$

$$= \ln x - \sum_{k=0}^{\infty} \frac{(-1)^k x^{2k}}{2k(2k)!}$$

Hint: There are several errors.

19. Find the Maclaurin series for the function

$$f(x) = \int_0^x \frac{1-\cos t}{t}\,dt$$

Cultural note: This function is sometimes called the **cosine integral** and is denoted by $\text{Cin}(x)$. The nomenclature has not been standardized.

20. Find the Maclaurin series for the **Fresnel integral**:

$$\varphi(x) = \int_0^x \sin t^2 \, dt$$

21. (**Reciprocal function**) Suppose $f(x) = \sum_{k=0}^{\infty} a_k x^k$, with $a_0 = 1$. Then, in some neighborhood of 0, $1/f(x)$ is well defined. Assume a series $\sum_{k=0}^{\infty} b_k x^k$ for the reciprocal function, and determine the b_k's recursively from the fact that the product of the two series is 1.

22. (Continuation) Use the result of Problem 21 to find the Maclaurin series for $x/(e^x - 1)$.

23. (Continuation) Show that the coefficients b_k of Problem 21 have the property $b_3 = b_5 = b_7 = \cdots = 0$. *Cultural note:* The coefficients $k!b_k$ are called the **Bernoulli numbers** and are often denoted by $B_0, B_1, \ldots, B_k$.

24. (Continuation) Show that the Bernoulli numbers B_k in Problem 23 have the property

$$\sum_{k=0}^{n-1} \binom{n}{k} B_k = 0 \qquad (n > 1)$$

Show that with $B_0 = 1$, this equation can be used to compute $B_1, B_2, \ldots$ and that $B_0 = 1$, $B_1 = -\frac{1}{2}$, $B_2 = \frac{1}{6}$, $B_3 = 0$, $B_4 = -\frac{1}{30}$.

25. Prove that if r is the radius of convergence of $\sum_{k=0}^{\infty} a_k x^k$, then the radius of the derivative series, $\sum_{k=0}^{\infty} k a_k x^{k-1}$, is also r.

26. Show that

$$\frac{x+1}{x-1} = 1 - 2\sum_{k=0}^{\infty} x^k \qquad (|x| < 1)$$

27. Find a power series about 0 for the function $\csc x - 1/x$.

28. Find the Maclaurin series for the function f defined by the equation

$$-\ln(1 - x) f(x) = \frac{x}{1 - x}$$

29. Find a series in powers of $(x - 1)$ for the function

$$f(x) = \int_1^x \frac{e^t - e}{t - 1} \, dt$$

Use the resulting series to compute $f(0)$ correct to three significant figures.

30. Establish Equation (3).

31. What is the radius of convergence of the series $\sum_{k=0}^{\infty} k(2x)^k$?

32. What is the radius of convergence of $\sum_{k=0}^{\infty} c_k x^k$ if $c_0 = 1$ and $c_n = (n/3)c_{n-1}$ for $n > 1$?
33. What is the radius of convergence of the series $\sum_{k=0}^{\infty} c_k x^k$, where $c_0 = 1$ and $c_{n+1} = (3n+3)c_n/(n+5)$?
34. Find the first three terms in the power series in x for $e^{\cos x}$. *Hint:* Let $z = 1 - \cos x$.
35. Find a power series for

$$\int_0^x e^{-t}\, dt$$

36. Show that only ten terms in the series for the sine integral $S(1)$ are needed to obtain 20 decimal places of accuracy. How many are needed for 25 decimal places?

COMPUTER PROBLEM 6.7

1. Write a subprogram or procedure that computes the sine integral with ten decimal places of precision for any x in $[-1, 1]$. Let the number of terms in the series depend on x. Program an error return for x out of range.

*6.8 Best Approximation: Least-Squares Theory

One of the classic problems of best approximation can be stated as follows: A continuous function f is defined on an interval $[a, b]$. For a fixed n, we ask for a polynomial p of degree at most n that deviates as little as possible from f. The **deviation** can be measured by the expression

$$\max_{a \leqq x \leqq b} |f(x) - p(x)|$$

One recognizes immediately that this is quite different from simply interpolating f at some fixed set of nodes, and it is different from simply truncating the Taylor series of f, if it has one. In fact, we have here an **extremum problem** whose solution is not at all evident.

Existence

The general linear problem of best approximation encompasses the one just described. Here we consider any normed linear space E and a subspace G in E. For any $f \in E$, the distance from f to G is defined to be the quantity

$$\text{dist}(f, G) = \inf_{g \in G} \|f - g\|$$

This measures the absolute minimum deviation that we can hope to achieve in approximating the vector f by an element of G. If an element g of G has the property

$$\|f - g\| = \text{dist}(f, G)$$

then g *achieves* this minimum deviation and is termed a **best approximation** of f from G. The meaning of *best approximation* thus depends on the norm chosen for the problem.

In the classic problem mentioned above, the normed space E is the space $C[a, b]$ of all continuous functions defined on $[a, b]$, and the norm is

$$\|f\| = \max_{a \leqq x \leqq b} |f(x)| \qquad f \in C[a, b] \tag{1}$$

The subspace G would then be the space Π_n of all polynomials of degree $\leqq n$, interpreted as a subspace of $C[a, b]$. Section 6.9 is devoted to this particular type of approximation.

In the general problem of best approximation, one of the main issues is whether there exists a best approximation to f. Here is an important existence theorem.

THEOREM 1 *If G is a finite-dimensional subspace in a normed linear space E, then each point of E possesses at least one best approximation in G.*

Proof Let f be an element of E. Any candidate, g, for a best approximation to f must compete with the 0 element of G; hence,

$$\|f - g\| \leqq \|f - 0\| = \|f\|$$

We can therefore confine our search to the set

$$K = \{g \in G : \|g - f\| \leqq \|f\|\}$$

The set K is closed and bounded. Since G is finite dimensional, K is compact. Since the function $g \longmapsto \|f - g\|$ is continuous, we can invoke the theorem that a continuous real-valued function on a compact set attains its infimum. ■

In general, best approximations are not unique. An easy example occurs in approximating $f(x) = \cos x$ on the interval $[0, \pi/2]$ by a function $g(x) = \lambda x$, where λ is a constant at our disposal. Figure 6.11 shows several best approximations when the norm in Equation (1) is adopted.

Inner-Product Spaces

Actually obtaining best approximations for specific E, G, and f can be a difficult task. It usually entails the solution of a *nonlinear* system of equations. There is one important case, however, when we need only solve some *linear* equations. This is the case in which E is an inner-product space. Recall that a real inner-product space

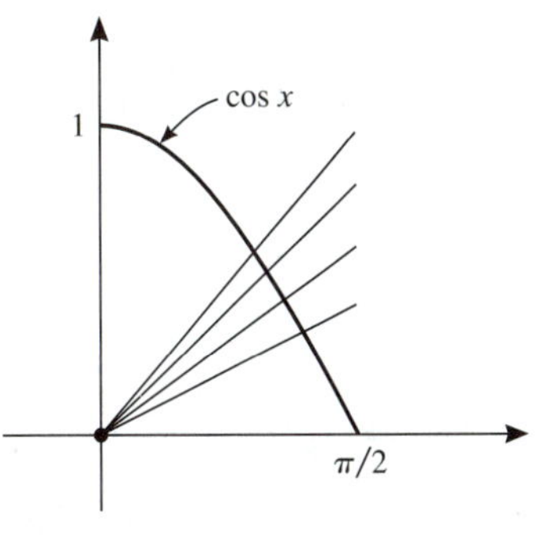

FIGURE 6.11 Approximating $\cos x$ by λx

is a linear space E in which an inner product and a norm have been introduced with these properties:

(i) $\langle f, h\rangle = \langle h, f\rangle$
(ii) $\langle f, \alpha h + \beta g\rangle = \alpha\langle f, h\rangle + \beta\langle f, g\rangle$
(iii) $\langle f, f\rangle > 0 \quad \text{if} \quad f \neq 0$
(iv) $\|f\| = \sqrt{\langle f, f\rangle}$

Two important inner-product spaces are $\mathbb{R}^n$, with

$$\langle x, y\rangle = \sum_{i=1}^{n} x_i y_i$$

and $C_w[a, b]$, the space of continuous functions on $[a, b]$, with

$$\langle f, g\rangle = \int_a^b f(x)\, g(x)\, w(x)\, dx$$

where w is a fixed continuous positive function. In any inner-product space, we write $f \perp g$ if $\langle f, g\rangle = 0$. We write $f \perp G$ if $f \perp g$ for all $g \in G$.

LEMMA 1 *In an inner-product space, we have*

(a) $\langle \sum_{i=1}^{n} a_i f_i, g\rangle = \sum_{i=1}^{n} a_i \langle f_i, g\rangle$
(b) $\|f + g\|^2 = \|f\|^2 + 2\langle f, g\rangle + \|g\|^2$
(c) *If* $f \perp g$, *then* $\|f + g\|^2 = \|f\|^2 + \|g\|^2$
(d) $|\langle f, g\rangle| \leqq \|f\|\, \|g\|$
(e) $\|f + g\|^2 + \|f - g\|^2 = 2\|f\|^2 + 2\|g\|^2$

Proof Part **(a)** is proved by induction, using axioms **(i)** and **(ii)**. Part **(b)** uses axioms **(i)** and **(ii)** and the definition in **(iv)**:

$$\begin{aligned}\|f + g\|^2 &= \langle f + g, f + g\rangle = \langle f, f\rangle + \langle f, g\rangle + \langle g, f\rangle + \langle g, g\rangle \\ &= \|f\|^2 + 2\langle f, g\rangle + \|g\|^2\end{aligned}$$

Part **(c)**, the **Pythagorean law**, follows at once from part **(b)**. Part **(d)** is the **Schwarz inequality**. To prove it, suppose that it fails for the pair (f, g):

$$|\langle f, g\rangle| > \|f\|\,\|g\|$$

Since $\langle f, 0\rangle = 0$, it is clear that $g \neq 0$. By homogeneity, we can assume $\|g\| = 1$, so that $|\langle f, g\rangle| > \|f\|$. By homogeneity, we can assume $\langle f, g\rangle = 1$, so that $\|f\| < 1$. Then a contradiction arises:

$$0 \leqq \|f - g\|^2 = \|f\|^2 - 2\langle f, g\rangle + \|g\|^2 = \|f\|^2 - 1$$

Part **(e)** is a straightforward computation based on part **(b)**. ■

Once the Schwarz inequality is available, the triangle inequality for the norm is proved by writing

$$\begin{aligned} \|f + g\|^2 &= \|f\|^2 + 2\langle f, g\rangle + \|g\|^2 \\ &\leqq \|f\|^2 + 2\|f\|\,\|g\| + \|g\|^2 \\ &= (\|f\| + \|g\|)^2 \end{aligned}$$

THEOREM 2 *Let G be a subspace in an inner-product space E. For $f \in E$ and $g \in G$, these properties are equivalent:*

(i) *g is a best approximation to f in G*
(ii) *$f - g \perp G$*

Proof If $f - g \perp G$, then for any $h \in G$ we have, by the Pythagorean law,

$$\|f - h\|^2 = \|(f - g) + (g - h)\|^2 = \|f - g\|^2 + \|g - h\|^2 \geqq \|f - g\|^2$$

For the converse, suppose that g is a best approximation to f. Let $h \in G$ and $\lambda > 0$. Then

$$\begin{aligned} 0 &\leqq \|f - g + \lambda h\|^2 - \|f - g\|^2 \\ &= \|f - g\|^2 + 2\lambda\langle f - g,\ h\rangle + \lambda^2\|h\|^2 - \|f - g\|^2 \\ &= \lambda\left\{2\langle f - g,\ h\rangle + \lambda\|h\|^2\right\} \end{aligned}$$

Letting $\lambda \downarrow 0$, we obtain $\langle f - g,\ h\rangle \geqq 0$. The same inequality must be satisfied by $-h$, and therefore $\langle f - g,\ h\rangle \leqq 0$. Hence, $\langle f - g,\ h\rangle = 0$. Since h was arbitrary in G, $f - g \perp G$. ■

Notice that a consequence of the proof is the *unicity* of a best approximation in the special setting.

Normal Equations

Example 1 Use Theorem 2 to determine the best approximation of the function $f(x) = \sin x$ by a polynomial $g(x) = c_1x + c_2x^3 + c_3x^5$ on the interval $[-1, 1]$, using the norm

$$\|f\| = \left\{ \int_{-1}^{1} [f(x)]^2\,dx \right\}^{1/2}$$

Solution The optimal function g has the property $f - g \perp G$, where G is the space generated by $g_1(x) = x$, $g_2(x) = x^3$, and $g_3(x) = x^5$. Thus, $\langle g - f, g_i\rangle = 0$ is required for $i = 1, 2, 3$. These equations can be written

$$c_1\langle g_1, g_i\rangle + c_2\langle g_2, g_i\rangle + c_3\langle g_3, g_i\rangle = \langle f, g_i\rangle \qquad (i = 1, 2, 3)$$

and are called the **normal equations** in this problem. Putting in the details, we have

$$\begin{cases} c_1 \displaystyle\int_{-1}^{1} x^2\,dx + c_2 \int_{-1}^{1} x^4\,dx + c_3 \int_{-1}^{1} x^6\,dx = \int_{-1}^{1} x \cdot \sin x\,dx \\ c_1 \displaystyle\int_{-1}^{1} x^4\,dx + c_2 \int_{-1}^{1} x^6\,dx + c_3 \int_{-1}^{1} x^8\,dx = \int_{-1}^{1} x^3 \sin x\,dx \\ c_1 \displaystyle\int_{-1}^{1} x^6\,dx + c_2 \int_{-1}^{1} x^8\,dx + c_3 \int_{-1}^{1} x^{10}\,dx = \int_{-1}^{1} x^5 \sin x\,dx \end{cases}$$

After all the integrals have been worked out numerically, we have this 3×3 system of equations to solve:

$$\begin{bmatrix} \frac{1}{3} & \frac{1}{5} & \frac{1}{7} \\ \frac{1}{5} & \frac{1}{7} & \frac{1}{9} \\ \frac{1}{7} & \frac{1}{9} & \frac{1}{11} \end{bmatrix} \begin{bmatrix} c_1 \\ c_2 \\ c_3 \end{bmatrix} = \begin{bmatrix} \alpha - \beta \\ -3\alpha + 5\beta \\ 65\alpha - 101\beta \end{bmatrix}$$

where $\alpha = \sin 1$ and $\beta = \cos 1$. Using numerical values for α and β, we solve, obtaining $c_1 \approx -0.99998$, $c_2 \approx -0.16652$, and $c_3 \approx 0.00802$. Recall that we discussed the normal equations and numerical difficulties associated with them in Section 5.3. Also, this coefficient matrix is an example of the ill-conditioned Hilbert matrix that was encountered in Section 2.3. Thus, the basis chosen for G is a poor one for numerical work. ■

Orthonormal Systems

There is another way of approaching approximation problems in an inner-product space, and this is by means of orthonormal systems. We say that a finite or infinite sequence of vectors $f_1, f_2, \ldots$ in an inner-product space is **orthogonal** if

$$\langle f_i, f_j\rangle = 0 \qquad (i \neq j)$$

The set is said to be **orthonormal** if for all i and j,

$$\langle f_i, f_j \rangle = \delta_{ij}$$

Such systems are very well adapted to approximation because of the following result.

THEOREM 3 *Let $\{g_1, g_2, \ldots, g_n\}$ be an orthonormal system in an inner-product space E. The best approximation of f by an element $\sum_{i=1}^n c_i g_i$ is obtained if and only if $c_i = \langle f, g_i \rangle$.*

Proof Let G be the subspace generated by $g_1, g_2, \ldots, g_n$. Notice that the best approximation, $\sum_{i=1}^n c_i g_i$, is characterized as in Theorem 2 by the condition

$$f - \sum_{i=1}^{n} c_i g_i \perp G$$

Thus, we need only verify that this condition is equivalent to the condition $c_i = \langle f, g_i \rangle$. Now, $f - \sum_{i=1}^n c_i g_i$ will be orthogonal to G if and only if it is orthogonal to each basis vector g_j. Computing the required inner products, we have (for $1 \leqq j \leqq n$):

$$\left\langle f - \sum_{i=1}^{n} c_i g_i,\ g_j \right\rangle = \langle f, g_j \rangle - \sum_{i=1}^{n} c_i \langle g_i, g_j \rangle$$
$$= \langle f, g_j \rangle - c_j = 0 \quad \blacksquare$$

This is the strategy suggested by these considerations: If it is desired to approximate elements of E by elements of a subspace G, first obtain an orthonormal base $\{g_1, g_2, \ldots, g_n\}$ for G. Then the best approximation of f is $\sum_{i=1}^n \langle f, g_i \rangle g_i$.

As an illustration of this strategy, we reconsider the previous example

$$\sin x \approx c_1 x + c_2 x^3 + c_3 x^5$$

It is known that an orthonormal base for our three-dimensional subspace is provided by three Legendre polynomials as follows:

$$g_1(x) = x\big/\sqrt{2/3}$$
$$g_2(x) = (5x^3 - 3x)\big/\sqrt{2/7}$$
$$g_3(x) = (63x^5 - 70x^3 + 15x)\big/\sqrt{2/11}$$

The solution to our problem is then the polynomial $\sum_{i=1}^3 c_i g_i$, where $c_i = \langle f, g_i \rangle$. Thus,

$$c_1 = \sqrt{3/2} \int_{-1}^{1} x \cdot \sin x \, dx = 2\sqrt{3/2}\,(\alpha - \beta)$$

$$c_2 = \sqrt{7/2}\int_{-1}^{1} \sin x(5x^3 - 3x)\,dx = 2\sqrt{7/2}\,(-18\alpha + 28\beta)$$

$$c_3 = \sqrt{11/2}\int_{-1}^{1} \sin x(63x^5 - 70x^3 + 15x)\,dx = 2\sqrt{11/2}\,(4320\alpha - 6728\beta)$$

where $\alpha = \sin 1$ and $\beta = \cos 1$. Using numerical values for α and β, we have the approximate solution $c_1 \approx 0.73771$, $c_2 \approx -6.7391 \times 10^{-2}$, and $c_3 \approx 2.2904 \times 10^{-3}$. Since the orthonormal base is already available, these results can be obtained with less effort using this second method. Notice that the coefficients are of modest size; our basis is well conditioned.

Generalized Pythagorean Law and Bessel's Inequality

The next result is a generalized Pythagorean law.

LEMMA 2 *If $[g_1, g_2, \ldots, g_n]$ is orthogonal, then*

$$\left\|\sum_{i=1}^{n} a_i g_i\right\|^2 = \sum_{i=1}^{n} a_i^2\,\|g_i\|^2$$

Proof If $n = 1$, it is obvious. If the equation is correct for n, then it is correct for $n+1$ because $a_{n+1}g_{n+1}$ is orthogonal to $\sum_{i=1}^{n} a_i g_i$, and the elementary case of the Pythagorean law applies:

$$\begin{aligned}\left\|\sum_{i=1}^{n+1} a_i g_i\right\|^2 &= \left\|\sum_{i=1}^{n} a_i g_i\right\|^2 + \|a_{n+1}g_{n+1}\|^2 \\ &= \sum_{i=1}^{n} a_i^2\|g_i\|^2 + a_{n+1}^2\|g_{n+1}\|^2\end{aligned}$$ ■

The following result is **Bessel's inequality**.

LEMMA 3 *If $[g_1, g_2, \ldots, g_n]$ is orthonormal, then*

$$\sum_{i=1}^{n} |\langle f, g_i\rangle|^2 \leqq \|f\|^2$$

Proof Let $g^* = \sum_{i=1}^{n}\langle f, g_i\rangle g_i$. By Theorem 3, g^* is the best approximation to f in the space G generated by the g_i. By Theorem 2, $f - g^* \perp G$. By the Pythagorean law (applied twice),

$$\|f\|^2 = \|f - g^*\|^2 + \|g^*\|^2 \geqq \|g^*\|^2 = \sum_{i=1}^{n} |\langle f, g_i\rangle|^2$$ ■

The Gram-Schmidt Process

We turn our attention now to the question of how to obtain orthonormal bases. The **Gram-Schmidt process** is described in the following theorem. (Recall that the Gram-Schmidt process was discussed in Chapter 5 for the matrix case.)

THEOREM 4 *Let $\{v_1, v_2, \dots, v_n\}$ be a basis for a subspace U in an inner-product space. Define recursively*

$$u_i = \left(\left\| v_i - \sum_{j=1}^{i-1} \langle v_i, u_j \rangle u_j \right\|\right)^{-1} \left(v_i - \sum_{j=1}^{i-1} \langle v_i, u_j \rangle u_j \right) \qquad (i = 1, 2, \dots, n)$$

Then $\{u_1, u_2, \dots, u_n\}$ is an orthonormal base for U.

Proof Denote by U_i the space spanned by $\{v_1, v_2, \dots, v_i\}$. We shall prove by induction on i that:

(i) $\{u_1, u_2, \dots, u_i\} \subseteqq U_i$
(ii) $\{u_1, u_2, \dots, u_i\}$ is orthonormal

For $i = 1$, we see at once that $u_1 \in U_1$ and that $\|u_1\| = 1$. Now suppose that **(i)** and **(ii)** are true for the index $i - 1$. Then $\{u_1, u_2, \dots, u_{i-1}\} \subseteqq U_{i-1}$ and $v_i \notin U_{i-1}$. Thus v_i is not a linear combination of $u_1, u_2, \dots, u_{i-1}$, and the denominator in the definition of u_i is not 0. It follows that u_i is well defined and belongs to U_i. Hence, $\{u_1, u_2, \dots, u_i\} \subseteqq U_i$. Since $\sum_{j=1}^{i-1} \langle v_i, u_j \rangle u_j$ is the best approximation of v_i in U_{i-1} (by Theorem 3), we have $u_i \perp U_{i-1}$ by Theorem 2. Hence, $u_i \perp \{u_1, u_2, \dots, u_{i-1}\}$. Obviously, $\|u_i\| = 1$. ■

A striking simplification occurs when the Gram-Schmidt process is applied to the monomial functions $1, x, x^2, \dots$ (in their natural order). In the following theorem, any inner product can be used as long as it has the property that for any three functions,

$$\langle fg, h \rangle = \langle f, gh \rangle$$

This is obviously valid for the commonly used inner product:

$$\langle f, g \rangle = \int_a^b f(x)\, g(x)\, w(x)\, dx$$

THEOREM 5 *The sequence of polynomials defined inductively as follows is orthogonal:*

$$p_n(x) = (x - a_n) p_{n-1}(x) - b_n p_{n-2}(x) \qquad (n \geqq 2)$$

with $p_0(x) = 1$, $p_1(x) = x - a_1$, and

$$a_n = \langle x p_{n-1}, p_{n-1} \rangle / \langle p_{n-1}, p_{n-1} \rangle$$
$$b_n = \langle x p_{n-1}, p_{n-2} \rangle / \langle p_{n-2}, p_{n-2} \rangle$$

Proof It is clear from the formulas that each p_n is a monic polynomial of degree n and is therefore not 0. Hence, the denominators in the formulas are not 0. Now we show by induction on n that $\langle p_n, p_i \rangle = 0$ for $i = 0, 1, \ldots, n-1$. For $n = 0$, there is nothing to prove. For $n = 1$, we note that by the definition of a_1,

$$\langle p_1, p_0 \rangle = \big\langle (x - a_1)p_0, p_0 \big\rangle = \langle xp_0, p_0 \rangle - a_1 \langle p_0, p_0 \rangle = 0$$

Now assume the validity of our assertion for an index $n - 1$, where $n \geqq 2$. Then

$$\langle p_n, p_{n-1} \rangle = \langle xp_{n-1}, p_{n-1} \rangle - a_n \langle p_{n-1}, p_{n-1} \rangle - b_n \langle p_{n-2}, p_{n-1} \rangle = 0$$
$$\langle p_n, p_{n-2} \rangle = \langle xp_{n-1}, p_{n-2} \rangle - a_n \langle p_{n-1}, p_{n-2} \rangle - b_n \langle p_{n-2}, p_{n-2} \rangle = 0$$

For any $i = 0, 1, \ldots, n-3$, we have

$$\begin{aligned} \langle p_n, p_i \rangle &= \langle xp_{n-1}, p_i \rangle - a_n \langle p_{n-1}, p_i \rangle - b_n \langle p_{n-2}, p_{n-i} \rangle \\ &= \langle p_{n-1}, xp_i \rangle \\ &= \langle p_{n-1}, p_{i+1} + a_{i+1} p_i + b_{i+1} p_{i-1} \rangle = 0 \end{aligned}$$

In the last step, we used the recurrence formula to represent xp_i. If $i = 0$, we should write instead $xp_0 = p_1 + a_1 p_0$. ■

Example 2 Use Theorem 5, with the inner product $\int_{-1}^{1} f(x)\,g(x)\,dx$, to obtain the **Legendre polynomials**.

Solution The first few calculations proceed as follows:

$$\begin{aligned} p_0(x) &= 1 \\ a_1 &= \langle xp_0, p_0 \rangle / \langle p_0, p_0 \rangle = 0 \\ p_1(x) &= x \\ a_2 &= \langle xp_1, p_1 \rangle / \langle p_1, p_1 \rangle = 0 \\ b_2 &= \langle xp_1, p_0 \rangle / \langle p_0, p_0 \rangle = \frac{1}{3} \\ p_2(x) &= x^2 - \frac{1}{3} \end{aligned}$$

The next three Legendre polynomials are

$$p_3(x) = x^3 - \frac{3}{5}x$$

$$p_4(x) = x^4 - \frac{6}{7}x^2 + \frac{3}{35}$$

$$p_5(x) = x^5 - \frac{10}{9}x^3 + \frac{5}{21}x$$

■

Example 3 Show that the **Chebyshev polynomials** form an orthogonal system on $[-1, 1]$ when the following inner product is used:

$$\langle f, g \rangle = \int_{-1}^{1} f(x)\, g(x) \frac{dx}{\sqrt{1 - x^2}}$$

Solution By making the change of variable $x = \cos\theta$, we put the inner product into the form

$$\langle f, g \rangle = \int_{0}^{\pi} f(\cos\theta)\, g(\cos\theta)\, d\theta$$

Since $T_n(x) = \cos(n \cos^{-1} x)$, we have (if $n \neq m$)

$$\begin{aligned} \langle T_n, T_m \rangle &= \int_0^{\pi} \cos n\theta \cos m\theta \, d\theta \\ &= \frac{1}{2} \int_0^{\pi} \left[\cos(n+m)\theta + \cos(n-m)\theta \right] d\theta \\ &= \frac{1}{2} \left[\frac{\sin(n+m)\theta}{n+m} + \frac{\sin(n-m)\theta}{n-m} \right]_0^{\pi} = 0 \end{aligned}$$

■

Algorithm

If a polynomial is given in the form $u = \sum_{i=0}^{n} c_i p_i$, where the polynomials p_i are as described in Theorem 5, then the evaluation of $u(x)$ can be accomplished efficiently as follows:

$d_{n+2} \leftarrow 0;\ d_{n+1} \leftarrow 0$
for $k = n$ **to** 0 **step** -1 **do**
  $d_k \leftarrow c_k + (x - a_{k+1}) d_{k+1} - b_{k+2} d_{k+2}$
end do

Then $u(x) = d_0$.

Here is the proof that the algorithm is valid:

$$\begin{aligned} u(x) &= \sum_{k=0}^{n} c_k p_k(x) \\ &= \sum_{k=0}^{n} \left[d_k - (x - a_{k+1}) d_{k+1} + b_{k+2} d_{k+2} \right] p_k(x) \\ &= d_0 p_0(x) + d_1 \left[p_1(x) - (x - a_1) p_0(x) \right] \\ &\quad + \sum_{k=2}^{n} d_k \left[p_k(x) - (x - a_k) p_{k-1}(x) + b_k p_{k-2}(x) \right] \\ &= d_0 \end{aligned}$$

THEOREM 6 *The polynomials p_n described in Theorem 5 have this property: p_n is the monic polynomial of degree n for which $\|p_n\|$ is a minimum.*

Proof An arbitrary monic polynomial of degree n can be represented as $p_n - \sum_{i=0}^{n-1} c_i p_i$. The norm of this function will be a minimum if

$$p_n - \sum_{i=0}^{n-1} c_i p_i \perp \Pi_{n-1}$$

This orthogonality relation is achieved by taking all $c_i = 0$ because $p_n \perp \Pi_{n-1}$. ■

If $[u_1, u_2, \ldots]$ is an orthonormal system in an inner-product space E, then a family of projection operators, P_n, is defined by the equation

$$P_n f = \sum_{i=1}^{n} \langle f, u_i \rangle u_i$$

THEOREM 7 *The operators P_n have these properties:*

(i) *P_n maps E linearly into the subspace U_n generated by $u_1, u_2, \ldots, u_n$.*

(ii) *Each P_n is a projection; that is, $P_n^2 = P_n$.*

(iii) *Each P_n is an orthogonal map in the sense that $f - P_n f \perp U_n$.*

(iv) *$P_n f$ is the best approximation of f in U_n.*

(v) *Each P_n is self-adjoint: $\langle P_n f, g \rangle = \langle f, P_n g \rangle$.*

Proof The proof is left to the reader as Problem 18. ■

The Gram Matrix

Approximation problems in inner-product spaces can be solved in an elementary fashion without using orthonormal bases. If $\{u_1, u_2, \ldots, u_n\}$ is any basis for a subspace U, and if we wish to compute a best approximation in U for an element f, we can proceed as follows: In order that an element $u \in U$ be the best approximation to f, it is necessary and sufficient that $u - f \perp U$ by Theorem 2. An equivalent condition is that $\langle u - f, u_i \rangle = 0$ for $1 \leqq i \leqq n$. On setting $u = \sum_{j=1}^{n} c_j u_j$, we find the condition

$$\sum_{j=1}^{n} c_j \langle u_j, u_i \rangle = \langle f, u_i \rangle \qquad (1 \leqq i \leqq n)$$

These are the **normal equations** for the problem. They constitute a system of n linear equations in the n unknowns $c_1, c_2, \ldots, c_n$. The coefficient matrix is called a **Gram** matrix; its elements are $G_{ij} = \langle u_i, u_j \rangle$.

LEMMA 4 *If $[u_1, u_2, \ldots, u_n]$ is linearly independent, then its Gram matrix is nonsingular.*

Proof By the Gram-Schmidt process, we can find an $n \times n$ matrix B such that the vectors

$$v_i = \sum_{s=1}^{n} B_{is} u_s$$

form an orthonormal set. Then

$$\delta_{ij} = \langle v_i, v_j \rangle = \left\langle \sum_s B_{is} u_s, \sum_r B_{jr} u_r \right\rangle = \sum_s \sum_r B_{is} \langle u_s, u_r \rangle B_{rj}^T$$

This equation is of the form $I = BGB^T$. It follows that G is nonsingular. ∎

It is impossible to resist giving another proof of this result. The approximation problem described above has a unique solution, u, by the remark following Theorem 2. The expression of u in terms of the basis, $u = \sum c_i u_i$, is unique also because this is a property of any basis. It follows that our problem has a unique solution, $(c_1, c_2, \ldots, c_n)$. Since this is true for any f, the coefficient matrix G must be nonsingular.

Lemma 4 assures us that the normal equations have a unique solution. Thus, in principle, any basis for the subspace U can be used to solve the approximation problem. In practice, the condition number of the Gram matrix must be considered. Recall the remarks made in Section 2.3 about the Hilbert matrix. This is the Gram matrix for the functions $u_j(x) = x^{j-1}$ when the inner product $\langle f, g \rangle = \int_0^1 f(x) g(x)\, dx$ is used. From the standpoint of conditioning, the most satisfactory bases are the orthonormal ones because their Gram matrices are identity matrices.

PROBLEMS 6.8

1. Find the best approximation to $\sin x$ by a function $u(x) = \lambda x$ on the interval $[0, \pi/2]$ using the supremum norm. *Hint:* Draw a sketch. The function $\sin t - \lambda t$ should have a maximum at a point ξ in $(0, \pi/2)$, and it should have a minimum of the same magnitude at $\pi/2$. These conditions should determine ξ and λ.

2. Solve Problem 1 using the usual quadratic norm.

3. Suppose that we wish to approximate an even function by a polynomial of degree $\leqq n$ using the norm $\|f\| = \left\{\int_{-1}^{1} |f(x)|^2\, dx\right\}^{1/2}$. Prove that the best approximation is also even. Generalize.

4. Let $p_0, p_1, p_2, \ldots$ be a sequence of polynomials such that (for each n) p_n has exact degree n. Show that the sequence is linearly independent.

5. Prove the **Parseval identity**:

$$\langle f, g \rangle = \sum_{i=1}^{n} \langle f, u_i \rangle \langle g, u_i \rangle$$

which is valid if f and g are in the span of the orthonormal set $[u_1, u_2, \ldots, u_n]$.

6. Show that the notorious **Hilbert matrix**, with elements

$$a_{ij} = (1 + i + j)^{-1} \qquad (0 \leqq i, j \leqq n)$$

is a Gram matrix for the functions $1, x, x^2, \ldots, x^{n-1}$.

7. In a vector space with basis $\{v_1, v_2, \ldots, v_n\}$, any other basis is obtained by a linear transformation

$$u_j = \sum_{i=1}^{n} a_{ij} v_i \qquad (1 \leqq j \leqq n)$$

in which the coefficient matrix is nonsingular. Show that the matrix that arises in this way from the Gram-Schmidt process is upper triangular.

8. In the three-term recurrence relation for the orthogonal polynomials, assume that the inner product is $\langle f, g\rangle = \int_{-a}^{a} f(x)\, g(x)\, w(x)\, dx$, where w is an even function. Prove that $a_n = 0$ for all n. Prove that p_n is even if n is even and that p_n is odd if n is odd.

9. Let $\{v_1, v_2, \ldots, v_n\}$ be an orthogonal set of vectors in an inner-product space. What choice of coefficients produces a minimum value in $\|f - \sum_{i=1}^{n} c_i v_i\|$? Don't overlook the possibility that some of the v's may be 0.

10. In the algorithm for computing a linear combination of orthogonal polynomials, show that at most $2n - 1$ multiplications are required. In the case of the Chebyshev polynomials, n multiplications suffice.

11. In the three-term recurrence formula for orthogonal polynomials, prove that b_n is positive by establishing that $b_n = \|p_{n-1}\|^2/\|p_{n-2}\|^2$.

12. Prove that for the Legendre polynomials, the coefficients in the three-term recurrence are $a_n = 0$ and $b_n = (n-1)^2/[(2n-1)(2n-3)]$.

13. How would the three-term recurrence formula for orthogonal polynomials have to be changed if it were desired to produce an *orthonormal* system?

14. Using the inner product $\langle u, v\rangle = \int_{-1}^{1} u(x)\, v(x)\, dx$, let the Gram-Schmidt process be applied to the sequence of functions $x \longmapsto (x^2 - 1)x^k$, $k = 0, 1, 2, \ldots$. Prove that if the resulting orthonormal sequence is renormalized to form a sequence of *monic* polynomials, then the latter satisfy a three-term recurrence relation of the form

$$q_{n+1}(x) = x q_n(x) - b_n q_{n-1}(x)$$

Give a formula for b_n. Determine the first three q-polynomials.

15. Prove that an orthogonal set of nonzero elements is necessarily linearly independent.

16. Let A be a linear transformation on an inner-product space. Assume that A is **self-adjoint**, which means that $\langle Af, g\rangle = \langle f, Ag\rangle$ for all f and g. Prove that the solutions of the equation $Af = \lambda f$ corresponding to different values of λ are mutually orthogonal.

17. Let $[u_1, u_2, \ldots]$ be an orthonormal sequence in an inner-product space. Prove that for any f in the space, the Fourier coefficients $\langle f, u_n \rangle$ are **square summable**: $\sum_{n=1}^{\infty} \langle f, u_n \rangle^2 < \infty$.

18. Prove Theorem 7. *Suggestion:* The order (a), (d), (c), (b), (e) is quite efficient.

19. Let $\widetilde{T}_n$ be the monic multiple of T_n. Find the three-term recurrence relation satisfied by $\widetilde{T}_0, \widetilde{T}_1, \ldots$.

20. Find a formula for $\text{dist}(f, G)$, where G is the subspace spanned by an orthonormal set $[g_1, g_2, \ldots, g_n]$.

21. Derive these Legendre polynomials:

$$p_3(x) = x^3 - \frac{3}{5}x$$

$$p_4(x) = x^4 - \frac{6}{7}x^2 + \frac{3}{35}$$

$$p_5(x) = x^5 - \frac{10}{9}x^3 + \frac{5}{21}x$$

22. Using Theorem 5 directly, find p_0, p_1, p_2, p_3 for $[a, b] = [0, 1]$ and $w(x) = 1$.

23. (Continuation) Determine p_3 in the form $p_3 = x^3 + Bx^2 + Cx + D$ by making p_3 orthogonal to $\prod_2$. Verify your results by the use of Problem 22.

24. Devise an algorithm for computing $\sum_{i=0}^{n} c_i p_i$ that computes, in order, each partial sum, $\sum_{i=0}^{k} c_i p_i$, for $k = 0, 1, 2, \ldots, n$. Assume that the coefficients a_k and b_k in Theorem 5 are known.

*6.9 Best Approximation: Chebyshev Theory

In this section, we work with the space $C(X)$ of all continuous real-valued functions defined on a given topological space X. We assume that X is a compact Hausdorff space. The reader who wishes to avoid considerations of general topology may take X to be a closed and bounded set in the real space $\mathbb{R}^k$—for example, an interval $[a, b]$ in $\mathbb{R}$.

The space $C(X)$ becomes a normed space (indeed a *Banach* space) if we define the norm to be

$$\|f\| = \max_{x \in X} |f(x)|$$

This norm is used throughout this section.

An important problem of best approximation in the space $C(X)$ is as follows: An element f is given in $C(X)$, and a finite-dimensional subspace G is given in $C(X)$. We want to approximate f as well as possible by an element of G. Hence (as in the preceding section) we define

$$\text{dist}(f, G) = \inf_{g \in G} \|f - g\|$$

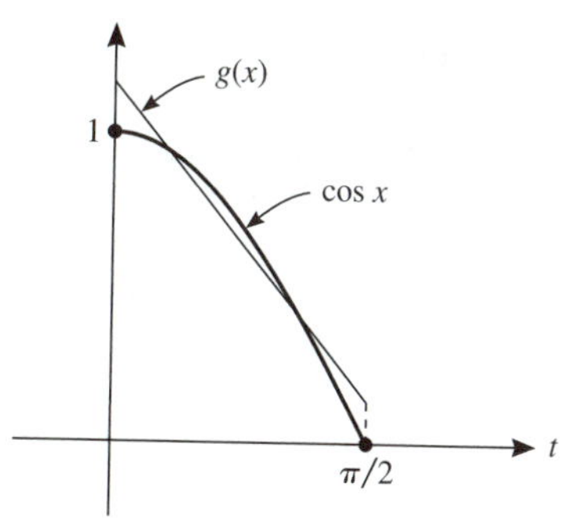

FIGURE 6.12 Approximating $\cos x$ on the interval $[0, \pi/2]$

and ask whether there exists a *best* approximation—that is, an element $g \in G$ such that

$$\|f - g\| = \text{dist}(f, G)$$

This question was answered affirmatively in Section 6.8 (Theorem 1). We can therefore address the problem of determining best approximations in this setting. To give some idea of what to expect, let us consider an example.

Example 1 Describe the best approximation of the function $f(x) = \cos x$ on the interval $\left[0, \frac{\pi}{2}\right]$ by an element of Π_1.

Solution Elements of Π_1 are linear functions, whose graphs are straight lines. A graph of $\cos x$ and a typical linear function not too far from $\cos x$ is shown in Figure 6.12.

Is the linear function shown a *best* approximation? No, it could be lowered, thus reducing the maximum deviation. Also its slope could be adjusted. For the best approximation, there will be *three* points where the maximum deviation occurs. These will apparently be the points $0, \frac{\pi}{2}$, and a point ξ interior to the interval. Denoting the maximum deviation by δ and the linear function by $g(x)$, we have

$$\begin{cases} g(0) - f(0) = \delta \\ g(\xi) - f(\xi) = -\delta \\ g(\pi/2) - f(\pi/2) = \delta \\ g'(\xi) - f'(\xi) = 0 \end{cases}$$

These four equations serve to determine δ, ξ, and the two coefficients in the equation $g(x) = c_1 + c_2 x$. ■

Characterizing Best Approximations

Before going on to the *characterization* of best approximations, we make an observation.

LEMMA 1 *For an element $g \in G$, these properties are equivalent:*

(i) *g is a best approximation to f.*
(ii) *0 is a best approximation to $f - g$.*

Proof If **(i)** is true, then for all $h \in G$,

$$\|f - g\| \leqq \|f - g - h\|$$

This implies that 0 is a best approximation of $f - g$. Conversely, if **(ii)** is true, then the preceding inequality is valid for all $h \in G$, and (since $g + h$ can be any element of G) **(i)** is true. ∎

Thus, we need to understand exactly which elements f of $C(X)$ have 0 for a best approximation in G—that is, have the property $\|f\| = \text{dist}(f, G)$. For $f \in C(X)$, we define its **critical set** to be

$$\text{crit}(f) = \{x \in X : |f(x)| = \|f\|\}$$

THEOREM 1 **Kolmogorov's Characterization Theorem** *For an element f and a subspace G in $C(X)$, these properties are equivalent:*

(i) $\|f\| = \text{dist}(f, G)$
(ii) *No element of G has the same signs as f on* $\text{crit}(f)$.

Proof Suppose that **(i)** is false. Then for some $g \in G$, $\|f - g\| < \|f\|$. For $x \in \text{crit}(f)$ we put $\sigma(x) = \text{sgn}\, f(x)$ and write

$$\sigma(x)[f(x) - g(x)] \leqq |f(x) - g(x)| \leqq \|f - g\| < \|f\| = |f(x)| = \sigma(x)f(x)$$

Thus, $\sigma(x)g(x) > 0$, and $g(x)$ has the same sign as $f(x)$ on $\text{crit}(f)$.

Now assume that **(ii)** is false. Let $g(x)f(x) > 0$ on $\text{crit}(f)$. There is no loss of generality in supposing that $\|g\| = 1$. Since $\text{crit}(f)$ is a compact set and gf is a continuous function, there is a positive ε such that $g(x)f(x) > \varepsilon$ on $\text{crit}(f)$. Put

$$\mathcal{O} = \{x \in X \;:\; g(x)f(x) > \varepsilon\}$$

Then $\mathcal{O}$ is an open set containing $\text{crit}(f)$. Its complement is a compact set *disjoint* from $\text{crit}(f)$. Hence,

$$\rho \equiv \max\{|f(x)| : x \in X \setminus \mathcal{O}\} < \|f\|$$

Now we shall try to approximate f by λg, where the coefficient λ is to be judiciously chosen. Let us see what is required. First, for the points in $\mathcal{O}$, we shall want this inequality to be valid pointwise:

$$(f - \lambda g)^2 = f^2 - 2\lambda g f + \lambda^2 g^2 \leqq \|f\|^2 - 2\lambda\varepsilon + \lambda^2 = \|f\|^2 - \lambda(2\varepsilon - \lambda) < \|f\|^2$$

This will be true if $0 < \lambda < 2\varepsilon$, as is easily verified. For the remaining points—that is, those in $X \setminus \mathcal{O}$—we shall want

$$|f - \lambda g| \leqq |f| + \lambda|g| \leqq \rho + \lambda < \|f\|$$

This is true if $0 < \lambda < \|f\| - \rho$. Thus, if λ is correctly chosen, $\|f - \lambda g\| < \|f\|$, showing that **(i)** is false. ■

By using Lemma 1 and Theorem 1, we obtain the next corollary.

COROLLARY 1 *Let f be an element of $C(X)$, G a subspace of $C(X)$, and g^* an element of G. In order that g^* be a best approximation of f in G, it is necessary and sufficient that no element g in G satisfy $g(x)[f(x) - g^*(x)] > 0$ on the set $\{x : |f(x) - g^*(x)| = \|f - g^*\|\}$.*

COROLLARY 2 *In order that an element $g \in \Pi_1$ be a best approximation to an $f \in C[a, b]$, it is necessary and sufficient that the function $f - g$ assume the value $\pm\|f - g\|$ with alternating sign in at least three points of $[a, b]$.*

Proof By the Characterization Theorem applied to $f - g$, the characteristic property is that no element of Π_1 can have the same signs as $f - g$ on crit$(f - g)$. There must be at least three points in crit$(f - g)$ at which the values of $f - g$ alternate in sign because otherwise there would be a point ξ in $[a, b]$ such that the points satisfying $f(x) - g(x) = \|f - x\|$ were on one side of ξ and the points satisfying $f(x) - g(x) = -\|f - g\|$ were on the other. Then a linear function vanishing at ξ would exist having the same signs as $f - g$ on crit$(f - g)$. ■

COROLLARY 3 *Let X be a closed and bounded set in $\mathbb{R}^2$. Let G be the subspace of $C(X)$ consisting of linear functions*

$$g(x, y) = a + bx + cy$$

In order that g be a best approximation to an element $f \in C(X)$, it is necessary and sufficient that the critical set of $f - g$ contain one of the three patterns in Figure 6.13.

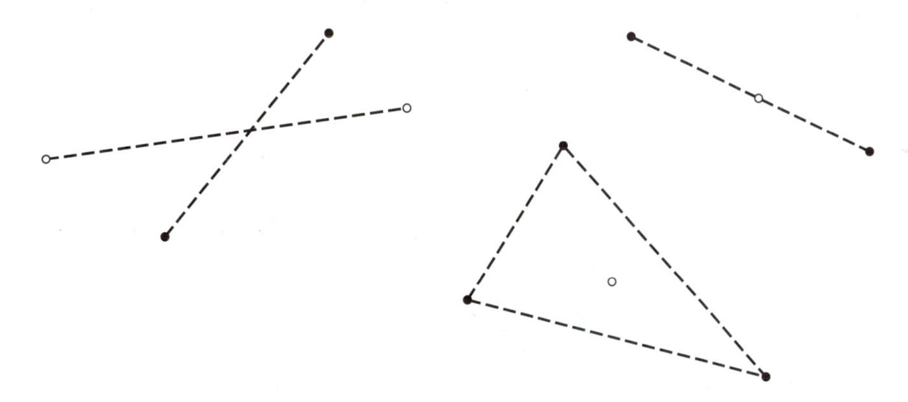

FIGURE 6.13 Critical points in a best approximation

Convexity

A set K in a linear space is said to be **convex** if it contains every line segment connecting two points of K. Formally, this is expressed by the implication:

$$\left.\begin{array}{l} u, v \in K \\ 0 \leqq \theta \leqq 1 \end{array}\right\} \Rightarrow \theta u + (1-\theta)v \in K$$

A convex set and a nonconvex set in $\mathbb{R}^2$ are shown in Figure 6.14.

Linear combinations of vectors of the form $\sum_{i=1}^{k} \theta_i u_i$ are called **convex combinations** when $\sum_{i=1}^{k} \theta_i = 1$ and $\theta_i \geqq 0$. The set of all convex combinations of points selected from a given set S is called the **convex hull** of S. Thus,

$$\mathrm{co}(S) = \left\{ \sum_{i=1}^{k} \theta_i u_i : k \in \mathbb{N},\ u_i \in S,\ \theta_i \geqq 0,\ \sum_{i=1}^{k} \theta_i = 1 \right\}$$

LEMMA 2 *Let K be a closed convex set in $\mathbb{R}^n$ (or any Hilbert space). Then K contains a unique point of minimum norm. Furthermore, these properties of K are equivalent:*

(i) $0 \notin K$
(ii) *There is a vector v such that $\langle v, u \rangle > 0$ for all $u \in K$.*

Proof Let $\rho = \inf\{\|u\| : u \in K\}$, and select $u_j \in K$ so that $\|u_j\| \to \rho$. The sequence $[u_j]$ is a Cauchy sequence because the parallelogram law yields

$$\begin{aligned} \|u_i - u_j\|^2 &= 2\|u_i\|^2 + 2\|u_j\|^2 - \|u_i + u_j\|^2 \\ &= 2\|u_i\|^2 + 2\|u_j\|^2 - 4\|(u_i + u_j)/2\|^2 \\ &\leqq 2\|u_i\|^2 + 2\|u_j\|^2 - 4\rho^2 \to 2\rho^2 + 2\rho^2 - 4\rho^2 = 0 \end{aligned}$$

By the completeness of the space, the Cauchy sequence converges, say $u_j \to u$. Then $u \in K$ because K is closed. Also $\|u\| = \rho$ by continuity. The uniqueness is proved by an argument like the one above: If u and u' are two elements of norm ρ

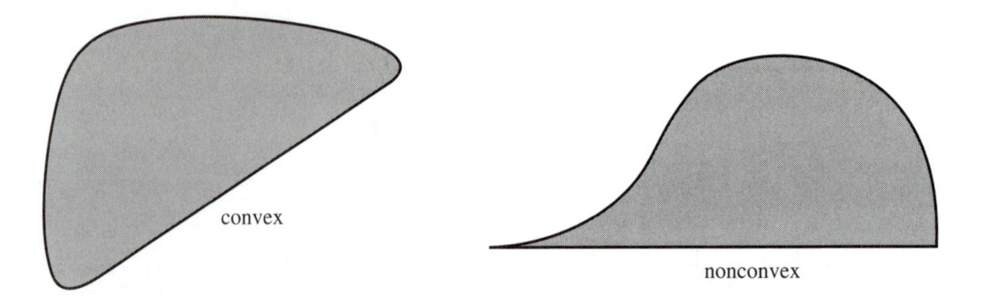

FIGURE 6.14 A convex set and a nonconvex set

in K, then

$$\|u - u'\|^2 \leqq 2\|u\|^2 + 2\|u'\|^2 - 4\rho^2 = 0$$

If **(ii)** is true, then **(i)** follows at once. For the converse, suppose that **(i)** is true. Let v be the point of least norm in K. We have, for arbitrary $u \in K$ and $\theta \in (0, 1)$,

$$0 \leqq \|\theta u + (1-\theta)v\|^2 - \|v\|^2 = \|\theta(u-v) + v\|^2 - \|v\|^2 = \theta^2\|u-v\|^2 + 2\theta\langle u-v, v\rangle$$

This shows that

$$0 \leqq \theta\|u - v\|^2 + 2\langle u - v, v\rangle$$

Letting $\theta \downarrow 0$, we conclude that $\langle u - v, v\rangle \geqq 0$, whence $\langle v, u\rangle \geqq \langle v, v\rangle > 0$. ■

THEOREM 2 **Carathéodory's Theorem** *Let S be a subset of an n-dimensional linear space. Each point of the convex hull of S is a convex combination of at most $n + 1$ points of S.*

Proof Let p be a point in the convex hull of S. There is no loss of generality in taking p to be 0. Then we have $0 = \sum_{i=1}^k \theta_i u_i$ for appropriate $\theta_i \in [0, 1]$, $u_i \in S$, $\sum_{i=1}^k \theta_i = 1$. We can assume that a minimal such representation of 0 has been chosen (that is, k is as small as possible). It follows that $\theta_i > 0$ for $i = 1, \dots, k$. If $k \leqq n + 1$, the proof is finished. If $k > n + 1$, then find a nontrivial linear dependence $\sum_{i=2}^k \lambda_i u_i = 0$. Put $\lambda_1 = 0$. Observe that for all real t, $\sum_{i=1}^k (\theta_i + t\lambda_1)u_i = 0$. Let

$$\phi(t) = \min_{1 \leqq i \leqq k} (\theta_i + t\lambda_i)$$

The function ϕ is continuous and $\phi(0) > 0$. For some t, $\phi(t) < 0$. Hence there is a particular t_0 for which $\phi(t_0) = 0$. Putting $\theta_i' = \theta_i + t_0\lambda_i$, we have $\theta_i' \geqq 0$, $\min \theta_i' = 0$, and $\theta_1' = \theta_1 > 0$. Hence in the expression $\sum_{i=1}^k \theta_i' u_i = 0$, one term has $\theta_i' = 0$. Dividing by $\Sigma\, \theta_i'$, we express 0 as a convex combination of $k - 1$ elements of S, thus contradicting the assumed minimality of k. ■

LEMMA 3 *The convex hull of a compact set in a finite-dimensional normed linear space is compact.*

Proof Let S be a compact set in an n-dimensional space X. Consider the set V of all $(2n + 2)$-tuples

$$v \equiv (\theta_0, \theta_1, \dots, \theta_n, u_0, u_1, \dots, u_n)$$

in which $\theta_i \in \mathbb{R}$, $u_i \in S$, $\theta_i \geqq 0$, and $\sum_{i=0}^n \theta_i = 1$. These $(2n+2)$-tuples are points in $\mathbb{R}^{n+1} \times X \times X \times \cdots \times X$. This space has dimension $n + 1 + (n+1)n = (n+1)^2$, and our set V is closed and bounded; hence, V is compact. Define $f(v) = \sum_{i=0}^n \theta_i u_i$. Then f maps V into co(S). By Carathéodory's Theorem, f is *surjective* (onto). It is also continuous. Since the continuous image of a compact set is compact, we conclude that co(S) is compact. ■

THEOREM 3 **Theorem on Linear Inequalities** *For a compact set S in $\mathbb{R}^n$, these properties are equivalent:*

(i) *There is a v such that $\langle v, u \rangle > 0$ for all $u \in S$.*
(ii) *0 is not in the convex hull of S.*

Proof If **(ii)** is false, then we can write $0 = \sum_{i=1}^{k} \theta_i u_i$ for appropriate $\theta_i > 0$ and $u_i \in S$. For any $v \in \mathbb{R}^n$, we have

$$0 = \langle v, 0 \rangle = \sum_{i=1}^{k} \theta_i \langle v, u_i \rangle$$

It is clear that not all the numbers $\langle v, u_i \rangle$ can be positive. Hence **(i)** is false.

For the converse, assume that **(ii)** is true. By Lemma 3, co(S) is compact. Lemma 2 can then be applied to conclude that **(i)** is true. ■

Chebyshev Solution of Linear Equations

COROLLARY 4 *Let A be an $m \times n$ matrix, $b \in \mathbb{R}^m$, $x \in \mathbb{R}^n$, $\sigma_i = \operatorname{sgn}(Ax - b)_i$, and $I = \{i : |(Ax - b)_i| = \|Ax - b\|_\infty\}$. In order that x minimize the norm $\|Ax - b\|_\infty$, it is necessary and sufficient that 0 lie in the convex hull of the set $\{\sigma_i A_i : i \in I\}$, where A_i denotes the ith row of A.*

Proof In seeking to minimize $\|Ax - b\|_\infty$, we are looking for a linear combination of the columns of A that is as close as possible to b. We interpret all column vectors as functions on the set $T = \{1, 2, \ldots, m\}$. The element being approximated is b, and the subspace G of approximants is the column space of A. By the Kolmogorov Theorem, the solution x is characterized by the fact that no element of G has the signs σ_i on I. Equivalently, the system $\sigma_i (Av)_i > 0$, $i \in I$, is inconsistent. Since this can be written $\langle \sigma_i A_i, v \rangle > 0$, $i \in I$, Theorem 3 gives an equivalent condition; namely, $0 \in \operatorname{co}\{\sigma_i A_i : i \in I\}$. ■

Example 2 Use Corollary 4 to determine whether $x = (2, 3)^T$ is the Chebyshev solution of the system

$$\begin{cases} 5x_1 - 7x_2 = 14 \\ 3x_1 + x_2 = 8 \\ x_1 - 9x_2 = -23 \\ 3x_1 - 1x_2 = 6 \\ 6x_1 - x_2 = 6 \end{cases}$$

Solution The residuals, $r_i = \langle A_i, x \rangle - b_i$, are 3, 1, -2, -3, 3. Hence the σ_i are $+1$, $+1$, -1, -1, $+1$. The critical set is $I = \{1, 4, 5\}$. The corollary asserts that x is a Chebyshev

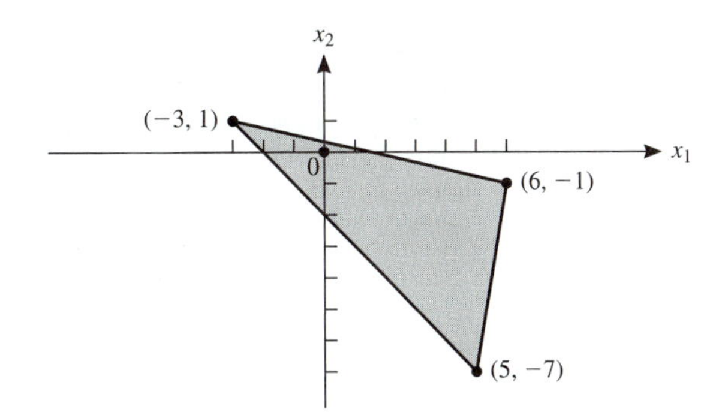

FIGURE 6.15 Zero in the hull of three points

solution if and only if

$$0 \in \text{co}\{\sigma_1 A_1,\ \sigma_4 A_4,\ \sigma_5 A_5\}$$

These three vectors can be quickly plotted to see that this circumstance is actually true. See Figure 6.15. ■

Further Characterization Theorems

In the next theorems, we use some special notation. If G is an n-dimensional subspace in $C(X)$, and if $\{g_1, g_2, \ldots, g_n\}$ is a basis for G, we put

$$\vec{g}(x) = \big(g_1(x), g_2(x), \ldots, g_n(x)\big) \qquad (x \in X)$$

This is interpreted as a point in $\mathbb{R}^n$.

For any $x \in X$ we define a linear functional $\widehat{x}$ on $C(X)$ by means of the equation

$$\widehat{x}(f) = f(x) \qquad \big(f \in C(X)\big)$$

The linearity of $\widehat{x}$ is a consequence of a simple calculation:

$$\widehat{x}(\alpha f + \beta g) = (\alpha f + \beta g)(x) = \alpha f(x) + \beta g(x) = \alpha\widehat{x}(f) + \beta\widehat{x}(g)$$

THEOREM 4 *Let G be an n-dimensional subspace in $C(X)$, and let f be an element of $C(X)$. These statements are equivalent:*

(i) $\|f\| = \text{dist}(f, G)$.
(ii) *No element of G has the same signs as $f(x)$ on* crit(f).
(iii) 0 *lies in the convex hull of the set* $\{f(x)\vec{g}(x) : x \in \text{crit}(f)\}$.
(iv) *There is a functional $\sum_{i=1}^{k} \lambda_i\, \widehat{x}_i$ that annihilates G and satisfies the conditions $x_i \in \text{crit}(f)$, $\lambda_i f(x_i) > 0$, and $k \leqq n + 1$.*

Proof The implication **(i)** ⇒ **(ii)** is part of the Kolmogorov Theorem proved earlier.

For the implication **(ii)** ⇒ **(iii)**, assume **(ii)**. Let $\{g_1, g_2, \ldots, g_n\}$ be any basis for G. Then we *cannot* find $c_1, c_2, \ldots, c_n$ such that $\sum_{i=1}^{n} c_i f(x) g_i(x) > 0$ on crit(f). Since this inequality can be written as $\langle c, f(x)\vec{g}(x)\rangle > 0$, the Theorem on Linear Inequalities implies that 0 lies in the convex hull of the point set

$$\{f(x)\vec{g}(x) : x \in \text{crit}(f)\}$$

For the implication **(iii)** ⇒ **(iv)**, assume **(iii)**. By Carathéodory's Theorem, we can write

$$0 = \sum_{i=1}^{k} \theta_i f(x_i)\vec{g}(x_i)$$

where $k \leqq n+1$, $\theta_i > 0$, and $x_i \in \text{crit}(f)$. Putting $\lambda_i = \theta_i f(x_i)$, we have $\lambda_i f(x_i) > 0$ and $0 = \sum_{i=1}^{k} \lambda_i \vec{g}(x_i)$. From the definition of $\vec{g}$, this last equation shows that for $j = 1, \ldots, n$, $0 = \sum_{i=1}^{k} \lambda_i g_j(x_i) = (\sum_{i=1}^{k} \lambda_j \widehat{x}_i)(g_j)$. Since the functions g_j span G, we see that the functional $\sum_{i=1}^{k} \lambda_i \widehat{x}_i$ annihilates G.

For the implication **(iv)** ⇒ **(i)**, assume **(iv)**. If $h \in G$, then

$$\|f\| \sum_{i=1}^{k} |\lambda_i| = \sum_{i=1}^{k} \lambda_i f(x_i) = \sum_{i=1}^{k} \lambda_i \big[f(x_i) - h(x_i)\big] \leqq \|f - h\| \sum_{i=1}^{k} |\lambda_i|$$

Hence, $\|f\| \leqq \|f - h\|$, and **(i)** is true. ■

Example 3 Let $X = [0, 1]$ and let G be generated by $g_1(x) = 1 - 2x^2$ and $g_2(x) = x - x^2$. Show that any $f \in C[0, 1]$ that satisfies $\|f\| = f(0) = f(1)$ must have the property $\|f\| = \text{dist}(f, G)$.

Solution Use part **(iv)** of Theorem 4. For $g \in G$, we notice that $g(0) + g(1) = 0$. (Simply test the two basis functions for this property.) Thus we can take $x_1 = 0$, $x_2 = 1$, $\lambda_1 = 1$, and $\lambda_2 = 1$ in the theorem. ■

Haar Subspaces

DEFINITION 1 *An n-dimensional subspace G in C(X) is called a* **Haar subspace** *if no element of G (except 0) has n or more zeros in X.*

Example 4 The subspace Π_{n-1} of polynomials having degree $< n$ is a Haar subspace in $C(X)$ whenever X is a subset of $\mathbb{R}$.

The next result shows that Haar subspaces are ideally suited for interpolation problems.

LEMMA 4 *An n-dimensional subspace G in C(X) is a Haar subspace if and only if for arbitrary real numbers $\lambda_1, \lambda_2, \ldots, \lambda_n$ and for arbitrary distinct points $x_1, x_2, \ldots, x_n$ in X, there is a unique element g in G such that $g(x_i) = \lambda_i$ $(1 \leqq i \leqq n)$.*

Proof Select a basis $\{g_1, g_2, \dots, g_n\}$ for G. The interpolation problem then is to determine $c_1, c_2, \dots, c_n$ so that

$$\sum_{j=1}^{n} c_j g_j(x_i) = \lambda_i \qquad (1 \leqq i \leqq n)$$

If this is solvable for all choices of λ_i, then the matrix having elements $g_j(x_i)$ is nonsingular, and the corresponding homogeneous problem (with all $\lambda_i = 0$) can have only the 0 solution ($c_i = 0$ for all i). Thus, no linear combination of $g_1, g_2, \dots, g_n$ can vanish at the points $x_1, x_2, \dots, x_n$. This is the Haar property of G because the points $x_1, x_2, \dots, x_n$ are arbitrary. Since the argument is reversible, the proof is complete. ■

For Haar subspaces, best approximations can be described in the following manner.

THEOREM 5 *Let G be an n-dimensional Haar subspace in $C(X)$, and let $f \in C(X)$. A necessary and sufficient condition for $\|f\| = \text{dist}(f, G)$ is that there exists a functional of the form $\sum_{i=1}^{n+1} \lambda_i \widehat{x}_i$ that annihilates G and satisfies $x_i \in \text{crit}(f)$ and $\lambda_i f(x_i) > 0$.*

Proof Theorem 4 gives this same necessary and sufficient condition with $n + 1$ replaced by k. The equation $\sum_{i=1}^{k} \lambda_i g(x_i) = 0$ cannot be valid nontrivially unless $k \geqq n + 1$. In order to verify this, let $k \leqq n$ and use the interpolating property of G (Lemma 4) to select $g \in G$ such that $g(x_i) = \lambda_i$. Consequently, we would have $0 = \sum_{i=1}^{k} \lambda_i g(x_i) = \sum_{i=1}^{k} \lambda_i^2$. ■

Unicity of Best Approximations

THEOREM 6 **Strong Unicity Theorem** *Let G be a finite-dimensional Haar subspace in $C(X)$, and let f be an element of $C(X)$ such that $\|f\| = \text{dist}(f, G)$. Then there is a positive constant γ (depending on f) such that for all $g \in G$, $\|f + g\| \geqq \|f\| + \gamma\|g\|$.*

Proof Let n be the dimension of G. By Theorem 5, there exist points $x_0, x_1, \dots, x_n$ in $\text{crit}(f)$ and positive coefficients θ_i such that $\sum_{i=0}^{n} \theta_i \sigma_i g(x_i) = 0$ for all $g \in G$, with $\sigma_i = \text{sgn}\, f(x_i)$. Let h be an element of norm 1 in G. Since $\sum_{i=0}^{n} \theta_i \sigma_i h(x_i) = 0$, at least one of the numbers $\sigma_i h(x_i)$ is positive. Thus, $\max_i \sigma_i h(x_i) > 0$. Since this expression is continuous and the surface of the unit cell of G is compact, we infer that

$$\gamma \equiv \inf_h \max_i \sigma_i h(x_i) > 0$$

If $g \in G$ and $g \neq 0$, then for some i, $\sigma_i g(x_i)/\|g\| \geqq \gamma$. Hence,

$$\|f + g\| \geqq \sigma_i f(x_i) + \sigma_i g(x_i) \geqq \|f\| + \gamma\|g\|$$ ■

COROLLARY 5 *If G is a finite-dimensional Haar subspace in $C(X)$, then each element of $C(X)$ has a unique best approximation in G.*

Proof Let g be a best approximation to an element f in $C(X)$. Then $f - g$ has 0 for a best approximation in G. If $h \in G$ and $h \neq 0$, then

$$\|f - g + h\| \geqq \|f - g\| + \gamma\|h\| > \|f - g\|$$

by Theorem 6. ■

THEOREM 7 **Continuity Theorem** *Let G be a finite-dimensional Haar subspace in $C(X)$. Let $A : C(X) \to G$ be the map such that $\|f - Af\| = \text{dist}(f, G)$. Then for each f there is a positive number $\lambda(f)$ such that*

$$\|Af - Ah\| \leqq \lambda(f)\|f - h\| \qquad h \in C(X)$$

Proof By the Strong Unicity Theorem, there exists a positive number $\gamma(f)$ such that for all $g \in G$,

$$\|f - g\| \geqq \|f - Af\| + \gamma(f)\|Af - g\|$$

Letting $g = Ah$, we then have

$$\begin{aligned}
\gamma(f)\|Af - Ah\| &\leqq \|f - Ah\| - \|f - Af\| \\
&\leqq \|f - h\| + \|h - Ah\| - \|f - Af\| \\
&\leqq \|f - h\| + \|h - Af\| - \|f - Af\| \\
&\leqq \|f - h\| + \|h - f\| + \|f - Af\| - \|f - Af\| \\
&= 2\|f - h\|
\end{aligned}$$

■

Chebyshev's Alternation Theorem

LEMMA 5 *Let G be an n-dimensional Haar subspace in $C[a, b]$. Suppose that we have $n + 1$ points such that $a \leqq x_0 < x_1 < \cdots < x_n \leqq b$ and that $\sum_{i=0}^{n} \lambda_i g(x_i) = 0$ for every $g \in G$. If $\sum_{i=0}^{n} |\lambda_i| \neq 0$, then the λ's alternate in sign: $\lambda_i \lambda_{i-1} < 0$ for $i = 1, 2, \ldots, n$.*

Proof For any $j \in 1, 2, \ldots, n$ we carry out the following argument: By Lemma 4, there exists a unique element g_j in G having this interpolation property:

$$\begin{cases} g_j(x_i) = 0 & \text{for } 0 \leqq i \leqq j - 2 \text{ and for } j + 1 \leqq i \leqq n \\ g_j(x_j) = 1 \end{cases}$$

Then we have

$$0 = \sum_{i=0}^{n} \lambda_i g_j(x_i) = \lambda_{j-1} g_j(x_{j-1}) + \lambda_j \tag{1}$$

Now $g_j(x_{j-1}) > 0$ because if $g_j(x_{j-1}) = 0$ or if $g_j(x_{j-1}) < 0$, we can conclude that g_j has n zeros, contrary to the Haar property. Equation (1) now shows that if any

λ_j is 0, then all are 0, contrary to hypothesis. Then Equation (1) shows that λ_j and λ_{j-1} must have opposite signs. ■

THEOREM 8 **Chebyshev's Alternation Theorem** *Let G be an n-dimensional Haar subspace in $C[a, b]$, and let X be a closed set in $[a, b]$. For an element $f \in C(X)$, these properties are equivalent:*

(i) $\|f\| = \text{dist}(f, G)$; *that is, f has 0 as a best approximation in G.*
(ii) *There exist points $x_0 < x_1 < \cdots < x_n$ in X such that $f(x_{i-1})f(x_i) = -\|f\|^2$ for $1 \leqq i \leqq n$.*

Proof By Theorem 7, **(i)** is equivalent to

(iii) there exist points $x_0, \ldots, x_n$ in crit(f) and coefficients $\lambda_0, \ldots, \lambda_n$ such that

$$x_0 < x_1 < \cdots < x_n, \quad \sum_{i=0}^{n} \lambda_i g(x_i) = 0 \text{ for all } g \in G, \text{ and } \lambda_i f(x_i) > 0$$

If **(iii)** is true, then by Lemma 5, the coefficients λ_i must alternate in sign. Since $\lambda_i f(x_i) > 0$, the numbers $f(x_i)$ must alternate in sign. Hence, **(ii)** is true.

Conversely, if **(ii)** is true, it is clear that each x_i is a critical point of f, and the signs of $f(x_i)$ alternate. Suppose for definiteness that sgn $f(x_i) = (-1)^i$. If **(i)** is false, then for some $g \in G$, $\|f - g\| < \|f\|$. Hence,

$$(-1)^i \left[f(x_i) - g(x_i) \right] \leqq \|f - g\| < \|f\| = (-1)^i f(x_i)$$

Thus $(-1)^i g(x_i) > 0$, and consequently g must have at least n zeros. This contradicts the Haar property of G. Notice that these zeros of g are certainly in $[a, b]$ but not necessarily in X. That is why the Haar condition is needed on $[a, b]$. ■

COROLLARY 6 *Let G be an n-dimensional Haar subspace in $C[a, b]$. Let $f \in C[a, b]$ and $g \in G$. In order that g be the best approximation of f in G, it is necessary and sufficient that there exist points $x_0 < x_1 < \cdots < x_n$ in $[a, b]$ such that*

$$f(x_i) - g(x_i) = (-1)^i c\|f - g\| \qquad (0 \leqq i \leqq n, \quad |c| = 1)$$

Algorithms

In the remainder of this section, we discuss the numerical problems of Chebyshev approximation and outline some of the algorithms available. We take as a basic problem the minimization of an expression

$$\Delta(c) = \left\| f - \sum_{i=1}^{n} c_i g_i \right\|_\infty = \sup_{x \in X} \left| f(x) - \sum_{i=1}^{n} c_i g_i(x) \right|$$

In this problem, the functions $f, g_1, g_2, \ldots, g_n$ have been prescribed in $C(X)$, and we wish to determine the coefficient vectors $c = (c_1, c_2, \ldots, c_n)$ that make $\Delta(c)$ as small as possible.

One of the basic algorithms for this problem is an iterative one that requires in each step the solution of a similar problem of more elementary nature. This algorithm is known as the **Remez first algorithm**; we describe it now.

Let us denote by G the subspace generated by $\{g_1, g_2, \dots, g_n\}$. At the kth step of the algorithm, a finite subset X_k is given in X. This finite set is used to define a seminorm in $C(X)$ via the equation

$$\|f\|_k = \max_{x \in X_k} |f(x)|$$

At this step it is necessary to determine (by methods to be discussed later) an element h_k in G to minimize the seminorm

$$\|f - h_k\|_k$$

This is another Chebyshev approximation problem, but it involves only the finite point set X_k. Next, a point x_k is selected in X so that

$$|f(x_k) - h_k(x_k)| = \|f - h_k\|$$

This point is now adjoined to the set X_k to form the next set X_{k+1}, and the process is repeated. The vector h_k obtained at the kth step is usually a good starting point in the search for h_{k+1} in the next step.

THEOREM 9 *If the initial seminorm $\|\ \|_1$ in the Remez first algorithm is a true norm on G, then $\lim_{k\to\infty} \|f - h_k\| = \operatorname{dist}(f, G)$. The sequence $[h_k]$ has cluster points, and each of these is a best approximation to f.*

Proof Directly from the definitions of the norms and the inclusions $X_1 \subseteqq X_2 \subseteqq \cdots$ we have, for $1 \leqq k \leqq i$ and $g \in G$,

$$\|f - g\|_1 \leqq \|f - g\|_k \leqq \|f - g\|_i \leqq \|f - g\|$$

From the preceding inequality we infer that

$$\|f - h_k\|_1 \leqq \|f - h_k\|_k \leqq \|f - h_i\|_i \leqq \operatorname{dist}(f, G)$$

This shows that the sequence $\|h_k\|_1$ is bounded. Since all norms are equivalent on the finite-dimensional space G, the sequence $\|h_k\|$ is bounded. Hence some subsequence of the h_k's converges to a point h^*. Given $\varepsilon > 0$, let k be chosen so that $\|h_k - h^*\| < \varepsilon$. Select $i > k$ so that $\|h_i - h^*\| < \varepsilon$. Then we have

$$\begin{aligned}
\operatorname{dist}(f, G) &\leqq \|f - h^*\| \leqq \|f - h_k\| + \|h_k - h^*\| \\
&\leqq |f(x_k) - h_k(x_k)| + \varepsilon \\
&\leqq \|f - h_k\|_i + \varepsilon \\
&\leqq \|f - h_i\|_i + \|h_i - h^*\|_i + \|h^* - h_k\|_i + \varepsilon \\
&\leqq \operatorname{dist}(f, G) + 3\varepsilon
\end{aligned}$$

Since ε was arbitrary, we conclude that $\|f - h^*\| = \text{dist}(f, G)$. Thus any cluster point of the sequence $[h_k]$ is a best approximation to f in G.

It remains to be proved that the sequence $d_k = \|f - h_k\|$ converges to $\text{dist}(f, G)$. This sequence is bounded and has convergent subsequences. Let $d_{k_i} \to d^*$. Let h' now denote a cluster point of the subsequence $[h_{k_i}]$. By the first part of our proof, $d^* = \|f - h'\| = \text{dist}(f, G)$. This shows that the sequence $[d_k]$ has only one cluster point—namely, $\text{dist}(f, G)$. Consequently, we have the convergence $d_k \to \text{dist}(f, G)$. ■

It is useful to note that at each step of the Remez first algorithm, an upper and a lower bound are easily computed for the unknown number $d^* = \text{dist}(f, G)$. In fact, we have

$$\|f - h_k\|_k \leqq d^* \leqq \min_{1 \leqq i \leqq k} \|f - h_i\|$$

The lower bound converges monotonically upward to d^*, whereas the upper bound converges monotonically downward to d^*. The simpler upper bound $\|f - h_k\|$ also converges to d^* but not always monotonically.

Next we discuss a variation of this algorithm, called the **exchange method**, in which the sets X_k are restrained from growing larger by deleting one *old* element whenever a *new* element is added. In practical problems, it is usually this method that is used, although its validity depends on further assumptions being made concerning the basic functions $g_1, g_2, \dots, g_n$. The process of deleting one element and adding another is called an **exchange**. If the initial subset X_1 contains $n + 1$ elements and if an exchange occurs in every step of the algorithm, then each X_k will contain exactly $n + 1$ elements.

From the characterization theorem for best approximations, we know that at the kth step, the origin of $\mathbb{R}^n$ will lie in the convex hull of the $n + 1$ vectors

$$e_k(x)\big(g_1(x), g_2(x), \dots, g_n(x)\big) \qquad x \in X_k \tag{2}$$

where $e_k(x) = f(x) - h_k(x)$. With h_k in hand, the point referred to previously as x_k is chosen. That is, x_k is a critical point of the function $f - h_k$. Now one of the points in X_k is *replaced* by x_k. This exchange is done in such a way that the origin of $\mathbb{R}^n$ remains in the convex hull of the vectors described in (2). Thus, the vector

$$e_k(x_k)\big(g_1(x_k), g_2(x_k), \dots, g_n(x_k)\big)$$

will replace one of the previous set of $n + 1$ vectors. The theorem that governs the exchange and reveals how to do it is presented next.

THEOREM 10 **Exchange Theorem** *Let $\{u_0, u_1, \dots, u_{n+1}\}$ be a set of $n + 2$ points in $\mathbb{R}^n$. If 0 lies in the convex hull of $\{u_0, u_1, \dots, u_n\}$, then for some $k \leqq n$ this assertion remains true when u_k is replaced by u_{n+1}.*

Proof If it is possible to write $0 = \sum_{i=0}^{n} \theta_i u_i$ with $\theta_i \geqq 0$, $\sum_{i=0}^{n} \theta_i = 1$, and $\min_i \theta_i = 0$, then do so and select k so that $\theta_k = 0$. Clearly we can replace u_k by u_{n+1} in the equation $0 = \sum_{i=0}^{n} \theta_i u_i$.

Suppose that it is *not* possible to write $0 = \sum_{i=0}^{n} \theta_i u_i$ with $\theta_i \geqq 0, \sum_{i=0}^{n} \theta_i = 1$, and $\min_i \theta_i = 0$. By Carathéodory's Theorem, the vectors $u_0, u_1, \ldots, u_n$ must span an n-dimensional space, which of course must be $\mathbb{R}^n$. Hence, we can find $\lambda_0, \ldots, \lambda_n$ such that $u_{n+1} = \sum_{i=0}^{n} \lambda_i u_i$. Also there exist $\theta_i > 0$ such that $0 = \sum_{i=0}^{n} \theta_i u_i$. Select k so that $\lambda_k/\theta_k \geqq \lambda_i/\theta_i$ for all i. For $0 \leqq i \leqq n$, put $\theta_i' = \lambda_k \theta_i - \lambda_i \theta_k$. Also let $\theta_{n+1}' = \theta_k$. Observe that $\theta_k' = 0$. Now we have

$$\begin{aligned}\sum_{i=0}^{n+1} \theta_i' u_i &= \sum_{i=0}^{n} \theta_i' u_i + \theta_{n+1}' u_{n+1} = \sum_{i=0}^{n} (\lambda_k \theta_i - \lambda_i \theta_k) u_i + \theta_k u_{n+1} \\ &= \lambda_k \sum_{i=0}^{n} \theta_i u_i - \theta_k \sum_{i=0}^{n} \lambda_i u_i + \theta_k u_{n+1} \\ &= \lambda_k(-\theta_k u_k) - \theta_k(u_{n+1} - \lambda_k u_k) + \theta_k u_{n+1} = 0\end{aligned}$$

Furthermore, $\theta_i' \geqq 0$ because $\theta_{n+1}' = \theta_k > 0$, and for $i \leqq n$,

$$\theta_i' = \theta_i \theta_k (\lambda_k/\theta_k - \lambda_i \theta_i) \geqq 0$$

If we divide the equation $\sum \theta_i' u_i = 0$ by $\sum \theta_i'$, we see that 0 is expressed as a convex combination of $u_0, \ldots, u_{n+1}$, with u_k omitted. ■

PROBLEMS 6.9

1. Solve for c_1, c_2, δ, and ξ in Example 1 of this section.
2. Find the best approximation of $\sqrt{x}$ by a first-degree polynomial on the interval $[0, 1]$.
3. Show that the subspaces in $C[0, 1]$ spanned by these sets are Haar subspaces:
   (i) $\{1, x^2, x^3\}$ **(ii)** $\{1, e^x, e^{2x}\}$ **(iii)** $\{(x+2)^{-1}, (x+3)^{-1}, (x+4)^{-1}\}$
4. Show that the subspaces in $C[-1, 1]$ spanned by these sets are not Haar subspaces:
   (i) $\{1, x^2, x^3\}$ **(ii)** $\{|x|, |x-1|\}$ **(iii)** $\{e^x, x+1\}$
5. In the space $C[0, 1]$, consider the subspace Π_0 of polynomials of degree 0 (that is, constant functions). Using the quantities

   $$M(f) = \max_{0 \leqq x \leqq 1} f(x) \quad \text{and} \quad m(f) = \min_{0 \leqq x \leqq 1} f(x)$$

   describe the best approximation of an element f in $C[0, 1]$ by an element of Π_0.
6. Let $u_0, u_1, \ldots, u_n$ be $n+1$ points in $\mathbb{R}^n$ such that it is not possible to write $0 = \sum_{i=0}^{n} \theta_i u_i$ with $\theta \geqq 0$, $\theta \neq 0$, and $\min_i \theta_i = 0$. Show that each set of n vectors chosen from $\{u_0, u_1, \ldots, u_n\}$ is a basis for $\mathbb{R}^n$ or give a counterexample.
7. Let A be an $n \times (n+2)$ matrix. Show that if the system

   $$Ax = 0 \quad x \geqq 0 \quad x \neq 0 \quad x_{n+2} = 0 \qquad x \in \mathbb{R}^{n+2}$$

is consistent, then for some $k < n + 2$ so is the system

$$Ax = 0 \quad x \geqq 0 \quad x \neq 0 \quad x_k = 0 \qquad x \in \mathbb{R}^{n+2}$$

8. Prove that the quadratic polynomial of best approximation to the function $\cosh x$ on the interval $[-1, 1]$ is $a + bx^2$, where $b = \cosh 1 - 1$ and a is obtained by solving this pair of equations simultaneously for a and t:

$$\begin{aligned} 2a &= 1 + \cosh t - t^2 b \\ \sinh t &= 2tb \end{aligned}$$

9. Prove that the convex hull of a set is convex and that it is the smallest convex set containing the original set.

10. Prove that in a normed linear space every closed ball $\{f : \|f - g\| \leqq r\}$ is convex.

11. Give an example of a convex set whose complement is bounded.

12. Given a line in the plane, $ax + by + c = 0$, with $a^2 + b^2 > 0$, provide a complete description of the set of points on the line whose distance from the origin is a minimum, using the ℓ_∞-norm to define distance.

13. Let f be a continuous function of (x, y) in the square $0 \leqq x \leqq 1, 0 \leqq y \leqq 1$. Describe the best approximation of f by a continuous function of x alone.

14. Do the three functions

$$g_0(x, y) = 1 \qquad g_1(x, y) = x \qquad g_2(x, y) = y$$

generate a Haar subspace in $C(\mathbb{R}^2)$?

15. Prove that the set of functions $\{1, x, x^2, \ldots, x^{n-1}, f\}$ generates a Haar subspace on $[a, b]$ if $f^{(n)}(x) > 0$ on $[a, b]$.

16. Let $f = a_0T_0 + a_1T_1 + \cdots + a_{n+1}T_{n+1}$, where the T_k are Chebyshev polynomials. Prove that the best approximation of f in the sup-norm on $[-1, 1]$ in the space Π_n is $a_0T_0 + a_1T_1 + \cdots + a_nT_n$.

*6.10 Interpolation in Higher Dimensions

The problem of finding smooth interpolants for functions of several variables is a difficult one that has attracted much attention, both in the past and currently. The multivariate case shows some unusual features that are not present in the univariate case, and these features are already apparent when the number of variables is only two. Therefore, very little is lost in restricting the discussion to the *bivariate* case (two independent variables), at least at the beginning.

Interpolation Problem

The central problem that we shall discuss is as follows: A set of interpolation points (or *nodes*) is given in the xy-plane. These can be denoted by

$$(x_1, y_1) \quad (x_2, y_2) \quad \dots \quad (x_n, y_n) \tag{1}$$

We assume that these n points are *distinct*. With each point (x_i, y_i) there is associated a real number, c_i, and our objective is to find a smooth and easily computed function F such that

$$F(x_i, y_i) = c_i \qquad (1 \leqq i \leqq n)$$

It is understood that F will be defined on all of $\mathbb{R}^2$, or at least on some large domain that includes the nodes. In the preceding description, the terms *smooth* and *easily computed* have only informal or intuitive meanings.

Cartesian Product and Grid

The interpolation problem just described can sometimes be solved by a *tensor product* of univariate interpolation methods. Let us dispose of this case first. The procedure is limited to the situation in which the set of nodes, denoted here by $\mathcal{N}$, is a **Cartesian product:**

$$\mathcal{N} = \{x_1, x_2, \dots, x_p\} \times \{y_1, y_2, \dots, y_q\}$$

Thus, $\mathcal{N}$ is the set of *all* pairs (x_i, y_j) where x_i is chosen from the first set and y_j from the second. In other words,

$$\mathcal{N} = \{(x_i, y_j) : 1 \leqq i \leqq p,\ 1 \leqq j \leqq q\} \tag{2}$$

Observe that the notation in Equation (2) is more convenient in this case than the notation in (1). An example of Equation (2), in which $p = 4$ and $q = 3$, is shown in Figure 6.16. Such an array of nodes is often called a **Cartesian grid**. For

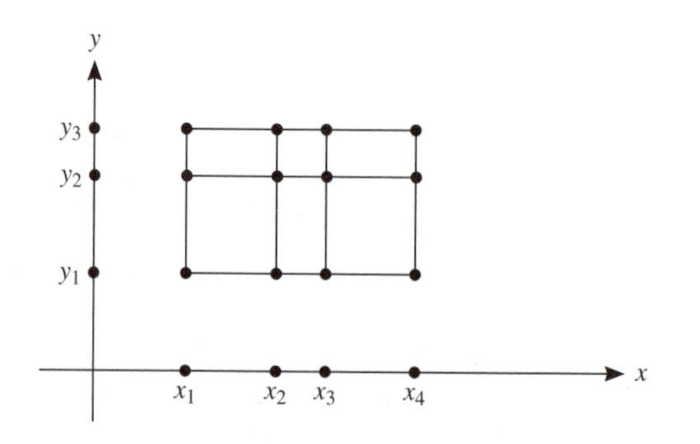

FIGURE 6.16 A Cartesian grid of nodes

convenience, we have numbered the x-points from left to right and the y-points from bottom to top, although this is not necessary.

Suppose that we have a linear interpolation scheme for nodes $x_1, x_2, \ldots, x_p$. This will be a **univariate process**. We want to think of this as a linear operator P of the form

$$(Pf)(x) = \sum_{i=1}^{p} f(x_i)\, u_i(x) \tag{3}$$

in which the functions u_i have the **cardinal property**

$$u_i(x_j) = \delta_{ij} \qquad (1 \leqq i, j \leqq p) \tag{4}$$

For example, in ordinary polynomial interpolation, these functions u_i are given by a familiar formula from Section 6.1:

$$u_i(x) = \prod_{\substack{j=1 \\ j \neq i}}^{p} \frac{x - x_j}{x_i - x_j} \qquad (1 \leqq i \leqq p) \tag{5}$$

Notice that the operator P can be extended in a trivial manner to operate on functions of *two* or more variables. Thus, if f is a function of (x, y), we can write

$$(\overline{P}f)(x, y) = \sum_{i=1}^{p} f(x_i, y)\, u_i(x) \tag{6}$$

One can see immediately that $\overline{P}f$ is a function of two variables that interpolates f on the *vertical* lines

$$L_i = \{(x_i, y) : -\infty < y < \infty\} \qquad (1 \leqq i \leqq p) \tag{7}$$

Suppose that another operator is available for interpolation at the nodes $y_1, y_2, \ldots, y_q$. We write

$$(Qf)(y) = \sum_{i=1}^{q} f(y_i)\, v_i(y) \tag{8}$$

where the functions v_i are any convenient ones having the cardinal property

$$v_i(y_j) = \delta_{ij} \qquad (1 \leqq i, j \leqq q) \tag{9}$$

Again, Q can be extended to operate on bivariate functions by use of the equation

$$(\overline{Q}f)(x, y) = \sum_{i=1}^{q} f(x, y_i)\, v_i(y) \tag{10}$$

The function $\overline{Q}f$ interpolates f on all the *horizontal* lines

$$L^i = \{(x, y_i) : -\infty < x < \infty\} \qquad (1 \leqq i \leqq q) \tag{11}$$

Boolean Sum

Two useful bivariate interpolation operators can now be constructed from $\overline{P}$ and $\overline{Q}$; they are the **product** $\overline{P}\,\overline{Q}$ and the **Boolean sum** $\overline{P} \oplus \overline{Q}$, defined by

$$\overline{P} \oplus \overline{Q} = \overline{P} + \overline{Q} - \overline{P}\,\overline{Q} \tag{12}$$

More detailed formulas for these operators are easily derived from the definitions of $\overline{P}$ and $\overline{Q}$. Thus,

$$\begin{aligned}(\overline{P}\,\overline{Q}f)(x, y) = \overline{P}(\overline{Q}f)(x, y) &= \sum_{i=1}^{p} (\overline{Q}f)(x_i, y)\, u_i(x) \\ &= \sum_{i=1}^{p} \sum_{j=1}^{q} f(x_i, y_j)\, v_j(y)\, u_i(x)\end{aligned} \tag{13}$$

Since $v_j(y_k)\, u_i(x_\ell) = \delta_{jk}\delta_{i\ell}$, we see without difficulty that $\overline{P}\,\overline{Q}f$ is a function that interpolates f at all the nodes (x_i, y_j). The **tensor product** notation $P \otimes Q$ is also used for the operator $\overline{P}\,\overline{Q}$.

In the same way, a formula for $\overline{P} \oplus \overline{Q}$ is

$$\begin{aligned}\left[(\overline{P} \oplus \overline{Q})f\right](x, y) &= (\overline{P}f)(x, y) + (\overline{Q}f)(x, y) - (\overline{P}\,\overline{Q}f)(x, y) \\ &= \sum_{i=1}^{p} f(x_i, y)\, u_i(x) + \sum_{j=1}^{q} f(x, y_j)\, v_j(y) \\ &\quad - \sum_{i=1}^{p} \sum_{j=1}^{q} f(x_i, y_j)\, u_i(x)\, v_j(y)\end{aligned} \tag{14}$$

It is left as Problem 6 to prove that the function $(\overline{P} \oplus \overline{Q})f$ interpolates f on all the horizontal and vertical lines L_i $(1 \leqq i \leqq p)$ and L^j $(1 \leqq j \leqq q)$.

Example 1 Give a formula for a polynomial in two variables that assumes the following values:

(x, y)	(1, 1)	(2, 1)	(4, 1)	(5, 1)	(1, 3)	(2, 3)	(4, 3)
$f(x, y)$	1.7	−4.1	−3.2	4.9	6.1	−4.2	2.3

(5, 3)	(1, 4)	(2, 4)	(4, 4)	(5, 4)
7.5	−5.9	3.8	−1.7	2.5

Solution Observe first that the nodes form a Cartesian grid, and the tensor product method is applicable. The functions u_i and v_i are given by Equation (5). In this example, they are as follows:

$$u_1(x) = \frac{x-2}{1-2} \cdot \frac{x-4}{1-4} \cdot \frac{x-5}{1-5} = -\frac{1}{12}(x-2)(x-4)(x-5)$$

$$u_2(x) = \frac{1}{6}(x-1)(x-4)(x-5)$$

$$u_3(x) = -\frac{1}{6}(x-1)(x-2)(x-5)$$

$$u_4(x) = \frac{1}{12}(x-1)(x-2)(x-4)$$

$$v_1(y) = \frac{y-3}{1-3} \cdot \frac{y-4}{1-4} = \frac{1}{6}(y-3)(y-4)$$

$$v_2(y) = -\frac{1}{2}(y-1)(y-4)$$

$$v_3(y) = \frac{1}{3}(y-1)(y-3)$$

The polynomial interpolant is then

$$\begin{aligned} F(x, y) = u_1(x)&\left[1.7\, v_1(y) + 6.1\, v_2(y) - 5.9\, v_3(y)\right] \\ &+ u_2(x)\left[-4.1\, v_1(y) - 4.2\, v_2(y) + 3.8\, v_3(y)\right] \\ &+ u_3(x)\left[-3.2\, v_1(y) + 2.3\, v_2(y) - 1.7\, v_3(y)\right] \\ &+ u_4(x)\left[4.9\, v_1(y) + 7.5\, v_2(y) + 2.5\, v_3(y)\right] \end{aligned} \tag{15}$$

■

Tensor Product

If the function F in Example 1 is written as a sum of terms $x^i y^j$, the following 12 terms appear

$$1,\ x,\ x^2,\ x^3,\ y,\ xy,\ x^2y,\ x^3y,\ y^2,\ xy^2,\ x^2y^2,\ x^3y^2 \tag{16}$$

Thus, we are interpolating by means of a 12-dimensional subspace of bivariate polynomials. The proper notation for this subspace is $\Pi_3 \otimes \Pi_2$. This is the **tensor product** of two linear spaces and consists of all functions of the form

$$(x, y) \longmapsto \sum_{i=1}^{m} a_i(x) b_i(y)$$

in which $a_i \in \Pi_3$ and $b_i \in \Pi_2$. (The sum can have any number of terms.) It is not difficult to prove that a basis for this space consists of the functions in (16).

It is to be emphasized that the theory just outlined applies to general functions u_i and v_i, not just to polynomials. All that is needed is the cardinality property. (In an abstract theory, one works directly with the operators P and Q; their detailed structure does not enter the analysis.)

In the tensor product method of polynomial interpolation, the general case will involve bivariate polynomials from the space $\Pi_{p-1} \otimes \Pi_{q-1}$, where p and q are the numbers of points that figure in Equation (2). A basis for this space is given by the functions

$$(x, y) \longmapsto x^i y^j \qquad (0 \leqq i \leqq p-1,\ 0 \leqq j \leqq q-1) \tag{17}$$

A generic element of the space is then of the form

$$(x, y) \longmapsto \sum_{i=0}^{p-1} \sum_{j=0}^{q-1} c_{ij} x^i y^j$$

The **degree** of a term $x^i y^j$ is defined to be $i + j$. Thus, the space $\Pi_{p-1} \otimes \Pi_{q-1}$ will contain one basis element of degree $p + q - 2$—namely, $x^{p-1} y^{q-1}$. But it will not contain all terms of degree $p+q-2$. For example, a term such as $x^p y^{q-2}$ will not be present. The degree of a polynomial in (x, y) is defined to be the largest degree of the terms present in the polynomial. The space of all bivariate polynomials of degree at most k will be denoted here by $\Pi_k(\mathbb{R}^2)$. A typical element of $\Pi_k(\mathbb{R}^2)$ is a function of the form

$$(x, y) \longmapsto \sum_{i=0}^{k} \sum_{j=0}^{k-i} c_{ij} x^i y^j = \sum_{0 \leqq i+j \leqq k} c_{ij} x^i y^j \tag{18}$$

THEOREM 1 *A basis for $\Pi_k(\mathbb{R}^2)$ is the set of functions*

$$(x, y) \longmapsto x^i y^j \qquad (0 \leqq i + j \leqq k)$$

Proof It is clear that this set spans $\Pi_k(\mathbb{R}^2)$, and it is only necessary to prove its linear independence. Suppose therefore that the function in Equation (18) is 0. If y is assigned a fixed value, say $y = y_0$, then the equation

$$\sum_{i=0}^{k} \left(\sum_{j=0}^{k-i} c_{ij} y_0^j \right) x^i = 0$$

exhibits an apparent linear dependence among the functions $x \longmapsto x^i$. Since this set of functions is linearly independent, we conclude that

$$\sum_{j=0}^{k-i} c_{ij} y_0^j = 0 \qquad (0 \leqq i \leqq k)$$

In this equation, y_0 can be *any* point. By the linear independence of the set of functions

$$y \longmapsto y^j \qquad (0 \leqq j \leqq k)$$

we conclude that $c_{ij} = 0$ for all i and j. ■

COROLLARY 1 *The dimension of* $\Pi_k(\mathbb{R}^2)$ *is* $\frac{1}{2}(k+1)(k+2)$.

Proof The basis elements of $\Pi_k(\mathbb{R}^2)$ given in Theorem 1 can be arrayed as follows:

$$\begin{array}{lllll} x^k & & & & \\ x^{k-1} & x^{k-1}y & & & \\ x^{k-2} & x^{k-2}y & x^{k-2}y^2 & & \\ \vdots & \vdots & \vdots & \ddots & \\ x^0 & x^0 y & x^0 y^2 & \cdots & x^0 y^k \end{array}$$

The number of basis elements is thus

$$1 + 2 + 3 + \cdots + (k+1) = \frac{1}{2}(k+1)(k+2)$$ ■

Recall that in the one-variable case, Π_k can be used for interpolation at *any* set of $k + 1$ nodes in $\mathbb{R}$. It is natural to expect that for two variables, $\Pi_k(\mathbb{R}^2)$ can be used to interpolate at any set of $n \equiv \frac{1}{2}(k+1)(k+2)$ nodes. This expectation is not fulfilled, however, and a simple example will show this. Suppose that $k = 1$, so that $n = 3$. A generic element of $\Pi_1(\mathbb{R}^2)$ has the form

$$c_0 + c_1 x + c_2 y$$

If we attempt to solve an interpolation problem with three nodes (x_i, y_i), we are led to a linear system whose coefficient determinant is

$$\begin{vmatrix} 1 & x_1 & y_1 \\ 1 & x_2 & y_2 \\ 1 & x_3 & y_3 \end{vmatrix}$$

Since this determinant represents twice the area of a triangle whose vertices are the given nodes, the determinant will be zero when the nodes are colinear. In that case the interpolation problem will be (in general) insoluble.

Geometry

The preceding considerations indicate that the *geometry* of the node set $\mathcal{N}$ will determine whether interpolation by $\Pi_k(\mathbb{R}^2)$ is possible on $\mathcal{N}$. Of course, the number of nodes should be $n = \frac{1}{2}(k+1)(k+2)$. Some theorems concerning this question will be given here to illustrate what is known. The following theorem is due to Gasca and Maeztu [1982].

THEOREM 2 *Interpolation of arbitrary data by the subspace $\Pi_k(\mathbb{R}^2)$ is possible on a set of $\frac{1}{2}(k+1)(k+2)$ nodes if the nodes lie on lines $L_0, L_1, \ldots, L_k$ in such a way that (for each i) L_i contains exactly $i+1$ nodes.*

Proof Let $\mathcal{N}$ denote the set of nodes. Assuming the hypothesis of the theorem, we have $\#(\mathcal{N} \bigcap L_i) = i+1$, in which the symbol # is used to signify the number of elements in a set. The sets $\mathcal{N} \bigcap L_i$ must be pairwise disjoint; if they were not, the following contradiction would arise:

$$\#\mathcal{N} < \sum_{i=0}^{k} \#\left(\mathcal{N} \bigcap L_i\right) = \sum_{i=0}^{k} (i+1) = \frac{1}{2}(k+1)(k+2)$$

Since the number of nodes is equal to the dimension of the space $\Pi_k(\mathbb{R}^2)$, it suffices to prove that the homogeneous interpolation problem has only the 0 solution. Accordingly, let $p \in \Pi_k(\mathbb{R}^2)$, and suppose that $p(x, y) = 0$ for each point (x, y) in $\mathcal{N}$. For each i, let ℓ_i be a linear function describing L_i:

$$L_i = \{(x, y) : \ell_i(x, y) = 0\} \qquad (0 \leqq i \leqq k)$$

Notice that $p^2 + \ell_k^2$ has at least $k+1$ zeros—namely, the points of $\mathcal{N} \bigcap L_k$. Bézout's Theorem (given below) allows us to conclude that ℓ_k is a divisor of p. This argument can be repeated because $(p/\ell_k)^2 + \ell_{k-1}^2$ has at least k zeros, and ℓ_{k-1} must be a divisor of p/ℓ_k. After k steps in this argument, the conclusion is drawn that p is divisible by $\ell_1\ell_2 \cdots \ell_k$. Thus p is a *scalar* multiple of $\ell_1\ell_2 \cdots \ell_k$ because p is of degree at most k. Since p vanishes on $\mathcal{N} \bigcap L_0$ while $\ell_1\ell_2 \cdots \ell_k$ does not, p must be 0. ■

The **Theorem of Bézout** states that if $p \in \Pi_k(\mathbb{R}^2)$, if $q \in \Pi_m(\mathbb{R}^2)$, and if $p^2 + q^2$ has more than km zeros, then p and q must have a common nonconstant factor. Because this theorem is limited to $\mathbb{R}^2$, the same is true of Theorem 2. An algorithmic proof of Theorem 2, not requiring Bézout's Theorem, is given later.

Example 2 The sets of nodes in Figure 6.17 are suitable for interpolation by the space $\Pi_2(\mathbb{R}^2)$.

A theorem closely related to Theorem 2 can be given in higher-dimensional spaces, $\mathbb{R}^d$. We appeal to Problem 5, according to which the dimension of $\Pi_k(\mathbb{R}^d)$ is $\binom{d+k}{k}$. The following result is from Chung and Yao [1977].

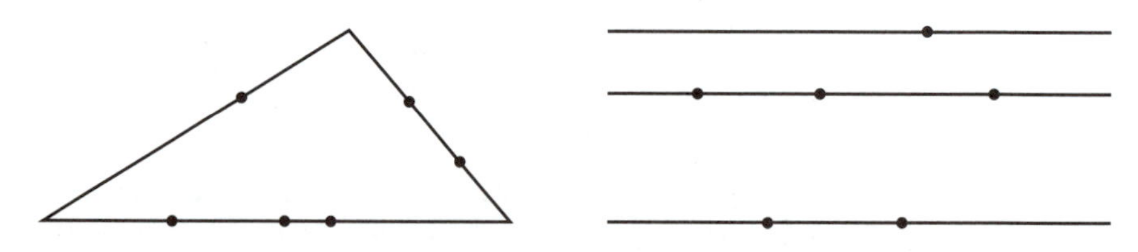

FIGURE 6.17 Node sets for interpolation by $\Pi_2(\mathbb{R}^2)$

THEOREM 3 *Let k and d be given, and set $n = \binom{d+k}{k}$. Let a set of n nodes $z_1, z_2, \dots, z_n$ be given in $\mathbb{R}^d$. If there exist hyperplanes H_{ij} in $\mathbb{R}^d$, with $1 \leqq i \leqq n$ and $1 \leqq j \leqq k$, such that*

$$z_j \in \bigcup_{\nu=1}^{k} H_{i\nu} \Longleftrightarrow j \neq i \qquad (1 \leqq i, j \leqq n) \tag{19}$$

then arbitrary data on the node set can be interpolated by polynomials in $\Pi_k(\mathbb{R}^d)$.

Proof Each hyperplane is the zero set of a nonzero linear function, and we write

$$H_{ij} = \left\{z \in \mathbb{R}^d : \ell_{ij}(z) = 0\right\}$$

where $\ell_{ij} \in \Pi_1(\mathbb{R}^d)$. Define the functions

$$q_i(z) = \prod_{j=1}^{k} \ell_{ij}(z) \qquad (1 \leqq i \leqq n)$$

Now z_i does not belong to any of the hyperplanes $H_{i1}, H_{i2}, \dots, H_{ik}$ by Condition (19), and therefore $\ell_{ij}(x_i) \neq 0$ for $1 \leqq j \leqq k$. This proves that $q_i(z_i) \neq 0$.

Again, by Condition (19), if $j \neq i$, then $z_j \in H_{i\nu}$ for some ν, and consequently $\ell_{i\nu}(z_j)$ and $q_i(z_j)$ are 0. Now if we set $p_i(z) = q_i(z)/q_i(z_i)$, we shall have the cardinality property, $p_i(z_j) = \delta_{ij}$. Since $p_i \in \Pi_k(\mathbb{R}^n)$, we have a Lagrangian formula for interpolating a function f at the nodes by a polynomial of degree k:

$$P(z) = \sum_{i=1}^{n} f(z_i)p_i(z) \qquad \blacksquare$$

A node configuration satisfying the hypotheses of Theorem 3 is shown in Figure 6.18. Here the dimension is $d = 2$, the degree is $k = 2$, and the number of nodes is $n = 6$.

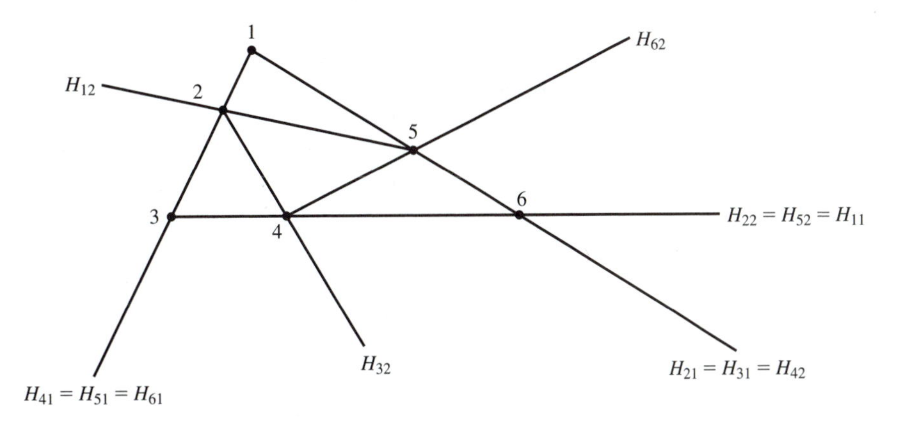

FIGURE 6.18 Illustration of Theorem 3

A very weak result concerning polynomial interpolation at arbitrary sets of nodes follows.

THEOREM 4 *The space $\Pi_k(\mathbb{R}^2)$ is capable of interpolating arbitrary data on any set of $k + 1$ distinct nodes in $\mathbb{R}^2$.*

Proof If the nodes are (x_i, y_i), with $0 \leqq i \leqq k$, we select a linear function

$$\ell(x, y) = ax + by + c$$

such that the $k + 1$ numbers $t_i = \ell(x_i, y_i)$ are all different. (See Problem 16.) If f is the function to be interpolated, we find $p \in \Pi_k(\mathbb{R})$ such that $p(t_i) = f(x_i, y_i)$. Then $p \circ \ell \in \Pi_k(\mathbb{R}^2)$ and

$$(p \circ \ell)(x_i, y_i) = p\big(\ell(x_i, y_i)\big) = p(t_i) = f(x_i, y_i) \qquad \blacksquare$$

A function of the form $f \circ \ell$, with $\ell \in \Pi_1(\mathbb{R}^2)$, has been termed a **ridge function**. Its graph is a ruled surface because $f \circ \ell$ remains constant on each line $\ell(x, y) = \lambda$. Theorem 4 has an obvious extension to $\mathbb{R}^d$.

A Newtonian Scheme

For the practical implementation of any interpolation method, it is advantageous to have an algorithm like Newton's procedure in univariate polynomial interpolation. Recall that one feature of the Newton scheme is that from a polynomial p interpolating f at nodes $x_1, x_2, \ldots, x_n$, we can easily obtain a polynomial p^* interpolating f at nodes $x_1, x_2, \ldots, x_{n+1}$ by adding one term to p. Indeed, we put

$$\begin{aligned} q(x) &= (x - x_1)(x - x_2) \cdots (x - x_n) \\ p^*(x) &= p(x) + cq(x) \\ c &= \big[f(x_{n+1}) - p(x_{n+1})\big]/q(x_{n+1}) \end{aligned}$$

The advantage of this algorithm is that an interpolating polynomial can be constructed step by step, adding one new interpolation node and one new term to p in each stage.

The abstract form of this procedure is as follows: Let X be a set and f a real-valued function defined on X. Let $\mathcal{N}$ be a set of nodes. If p is any function that interpolates f on $\mathcal{N}$, and if q is any function that vanishes on $\mathcal{N}$, then a function p^* interpolating f on $\mathcal{N} \bigcup \{\xi\}$ can be obtained in the form $p^* = p + cq$, provided that $q(\xi) \neq 0$.

A more general version of this strategy deals with sets of nodes. Let q be a function from X to $\mathbb{R}$ and let Z be its 0 set. If p interpolates f on $\mathcal{N} \bigcap Z$ and r interpolates $(f - p)/q$ on $\mathcal{N} \setminus Z$, then $p + qr$ interpolates f on $\mathcal{N}$.

The procedure just outlined can be used to give an algorithmic proof of Theorem 2. To begin, select $p_k \in \Pi_k(\mathbb{R}^2)$ that interpolates f on $\mathcal{N} \bigcap L_k$. (Use Theorem 4.) Proceeding inductively downward, suppose that p_i has been found in $\Pi_k(\mathbb{R}^2)$ and interpolates f on all the nodes in $L_k \bigcup L_{k-1} \bigcup \cdots \bigcup L_i$. We shall attempt to construct p_{i-1} in the **Newton form**

$$p_{i-1} = p_i + r\ell_k\ell_{k-1}\cdots\ell_i$$

It is clear that p_{i-1} will still interpolate f at the nodes in $L_k \bigcup L_{k-1} \bigcup \cdots \bigcup L_i$, since the term added to p_i vanishes on this set. In order to make p_{i-1} interpolate f on the nodes in L_{i-1}, we write

$$f(x) = p_i(x) + r(x)(\ell_k\ell_{k-1}\cdots\ell_i)(x) \qquad x \in \mathcal{N} \bigcap L_{i-1}$$

from which we infer that r should interpolate $(f - p_i)/(\ell_k\ell_{k-1}\cdots\ell_i)$ on $\mathcal{N} \bigcap L_{i-1}$. By Theorem 4, there is an $r \in \Pi_{i-1}(\mathbb{R}^2)$ that does so. Finally, observe that $p_{i-1} \in \Pi_k(\mathbb{R}^2)$ because r is of degree $i - 1$ and $(\ell_k\ell_{k-1}\cdots\ell_i)$ is of degree $k - i + 1$. This algorithm was given by Micchelli [1986a].

It is an interesting fact that *no* n-dimensional subspace in $C(\mathbb{R}^2)$ can serve for interpolation at arbitrary sets of n nodes (except in the trivial case of $n = 1$). This was probably first noticed by Haar in 1918, and his argument goes like this. Suppose that n functions $u_1, u_2, \ldots, u_n$ are given in $C(\mathbb{R}^2)$. Let n nodes in $\mathbb{R}^2$ be given, say $p_i = (x_i, y_i)$. If we wish to interpolate at these nodes using the base functions u_i, we shall have to solve a linear system whose determinant is

$$D = \begin{vmatrix} u_1(p_1) & u_2(p_1) & \cdots & u_n(p_1) \\ u_1(p_2) & u_2(p_2) & \cdots & u_n(p_2) \\ \vdots & \vdots & \ddots & \vdots \\ u_1(p_n) & u_2(p_n) & \cdots & u_n(p_n) \end{vmatrix}$$

This determinant may be nonzero for the given set of nodes. However, let the first two of the nodes undergo a continuous motion in $\mathbb{R}^2$ in such a way that during the motion these two points never coincide, nor do they coincide with any of the other nodes, yet at the end of the motion they have exchanged their original positions. By the rules of determinants, the determinant D will have changed sign (because rows 1 and 2 are interchanged). By continuity, D assumed the value 0 during the continuous motion described. Hence, D will *sometimes* be 0, even for distinct nodes. The fact that two nodes can move in $\mathbb{R}^2$ and exchange places without ever being coincident is characteristic of $\mathbb{R}^2, \mathbb{R}^3, \ldots$ but not of $\mathbb{R}^1$. This explains why interpolation in $\mathbb{R}^2, \mathbb{R}^3, \ldots$ must be approached somewhat differently from $\mathbb{R}^1$. What is usually done is to fix the nodes *first* and *then* to ask what subspaces of interpolating functions are suitable.

Shepard Interpolation

A very general method of this type (in which the subspace depends on the nodes) is known as **Shepard interpolation**, after its originator, Shepard [1968]. Let the (distinct) nodes be listed as

$$p_i = (x_i, y_i) \qquad (1 \leqq i \leqq n) \tag{20}$$

We shall use p and q to denote generic elements in $\mathbb{R}^2$ because this will make the extension to $\mathbb{R}^3, \mathbb{R}^4, \ldots$ conceptually transparent. Next, we select a real-valued function ϕ on $\mathbb{R}^2 \times \mathbb{R}^2$ subject to the sole condition that

$$\phi(p, q) = 0 \quad \text{if and only if} \quad p = q \tag{21}$$

Examples that come to mind are $\phi(p, q) = \|p - q\|$ and $\phi(p, q) = \|p - q\|^2$. Next, we set up some cardinal functions in exact analogy with the Lagrange formulas in univariate approximation. This is done as follows:

$$u_i(p) = \prod_{\substack{j=1 \\ j \neq i}}^{n} \frac{\phi(p, p_j)}{\phi(p_i, p_j)} \qquad (1 \leqq i \leqq n) \tag{22}$$

It is easy to see that these functions have the cardinality property

$$u_i(p_j) = \delta_{ij} \qquad (1 \leqq i, j \leqq n)$$

This is a consequence of the hypothesis in (21). It follows that an interpolant to f at the given nodes is provided by the function

$$F = \sum_{i=1}^{n} f(p_i) u_i \tag{23}$$

Example 3 What are the formulas for Shepard interpolation when $\|p - q\|^2$ is used for $\phi(p, q)$?

Solution Let $p_i = (x_i, y_i)$, $p = (x, y)$, and

$$\phi(p, p_j) = \|p - p_j\|^2 = (x - x_j)^2 + (y - y_j)^2$$

Then

$$F(x, y) = \sum_{i=1}^{n} f(x_i, y_i) \prod_{\substack{j=1 \\ j \neq i}}^{n} \frac{(x - x_j)^2 + (y - y_j)^2}{(x_i - x_j)^2 + (y_i - y_j)^2}$$ ■

An algorithm to compute $F(x, y)$ in Example 3 goes as follows:

input n, x, y
input $(x_i, y_i, c_i) \quad (1 \leqq i \leqq n)$
for $i = 1$ **to** n **do**
  $d_i = (x - x_i)^2 + (y - y_i)^2$

```
    for j = 1 to n do
        d_ij = (x_i − x_j)^2 + (y_i − y_j)^2
    end do
end do
```

Then

$$F(x, y) = \sum_{i=1}^{n} c_i \prod_{\substack{j=1 \\ j \neq i}}^{n} d_j / d_{ij}$$

(No attempt has been made to make the algorithm efficient. It contains much duplication.)

Another version of Shepard's method starts with the additional assumption on ϕ that it is a *nonnegative* function. Next, let

$$v_i(p) = \prod_{\substack{j=1 \\ j \neq i}}^{n} \phi(p, p_j) \qquad v(p) = \sum_{i=1}^{n} v_i(p) \qquad w_i(p) = v_i(p)/v(p) \tag{24}$$

By our assumptions on ϕ, we have $v_i(p_j) = 0$ if $i \neq j$ and $v_i(p) > 0$ for all points except $p_1, \ldots, p_{i-1}, p_{i+1}, \ldots, p_n$. It follows that $v(p) > 0$ and that w_i is well defined. By the construction, $w_i(p_j) = \delta_{ij}$ and $0 \leqq w_i(p) \leqq 1$. Furthermore, $\sum_{i=1}^{n} w_i(p) = 1$. The interpolation process is given by the equation

$$F = \sum_{i=1}^{n} f(p_i) w_i = \sum_{i=1}^{n} f(p_i) v_i / v \tag{25}$$

This process has two favorable properties not possessed by the previous version; namely, if the data are nonnegative, then the interpolant F will be a nonnegative function, and if f is a constant function, then $F = f$. These two properties give evidence that the interpolant F inherits certain characteristics of the function being interpolated. On the other hand, if ϕ is differentiable, then F will exhibit a flat spot at each node. This is because $0 \leqq w_i \leqq 1$ and $w_i(p_j) = \delta_{ij}$, so that the nodes are extrema (maximum points or minimum points) of each w_i. Thus the partial derivatives of w_i are 0 at each node, and consequently the same is true of F.

An important case of Shepard interpolation arises when the function ϕ is given by a power of the Euclidean distance:

$$\phi(x, y) = \|x - y\|^{\mu} \qquad (\mu > 0)$$

Here x and y can be points in $\mathbb{R}^s$. It will be seen that ϕ is differentiable if $\mu > 1$ but not if $0 < \mu \leqq 1$. It suffices to examine the simpler function $g(x) = \|x\|^{\mu}$ at the questionable point $x = 0$. The directional derivative of g at 0 is obtained by differentiating the function $G(t) = \|tu\|^{\mu}$, where u is a unit vector defining the direction. Since $G(t) = |t|^\mu$, the derivative at $t = 0$ does not exist when $0 < \mu \leqq 1$, but for $\mu > 1$, $G'(0) = 0$.

The formula for w_i can be given in two ways:

$$w_i(x) = \prod_{\substack{j=1 \\ j \neq i}}^{n} \|x - x_j\|^\mu \Big/ \sum_{k=1}^{n} \prod_{\substack{j=1 \\ j \neq k}}^{n} \|x - x_j\|$$

$$w_i(x) = \|x - x_i\|^{-\mu} \Big/ \sum_{j=1}^{n} \|x - x_j\|^{-\mu}$$

The second equation must be used with care since the right side assumes the indeterminate form ∞/∞ at x_i.

A *local* multivariate interpolation method of Franke and Little is designed so that the datum at one node will have a very small influence on the interpolating function at points far from that node. Given nodes (x_i, y_i), $1 \leqq i \leqq n$, we introduce functions

$$g_i(x, y) = \left(1 - r_i^{-1}\sqrt{(x - x_i)^2 + (y - y_i)^2}\right)_+^\mu$$

The subscript + indicates that when the quantity inside the parentheses is negative, it is replaced by 0. This will occur if (x, y) is far from the node (x_i, y_i). The parameter μ influences the smoothness of the function. The parameter r_i controls the support of g_i. Thus, $g_i(x, y) = 0$ if (x, y) is more than r units distant from (x_i, y_i).

If r_i is chosen to be the distance from (x_i, y_i) to the nearest neighboring node, then $g_i(x_j, y_j) = \delta_{ij}$. In this case, we interpolate an arbitrary function f by means of the function

$$\sum_{i=1}^{n} f(x_i, y_i) g_i(x, y)$$

Triangulation

Another general strategy for interpolating functions given on $\mathbb{R}^2$ begins by creating a **triangulation**. Informally, this means that triangles are drawn by joining nodes. In the end, we shall have a family of triangles, $T_1, T_2, \ldots, T_m$. We consider this collection of triangles to be the triangulation. These rules must be satisfied:

(i) Each interpolation node must be the vertex of some triangle T_i.
(ii) Each vertex of a triangle in the collection must be a node.
(iii) If a node belongs to a triangle, it must be a vertex of that triangle.

The effect of rule **(iii)** is to disallow the construction shown in Figure 6.19.

The simplest type of interpolation on a triangulation is the piecewise linear function that interpolates a function f at all the vertices of all triangles. In any

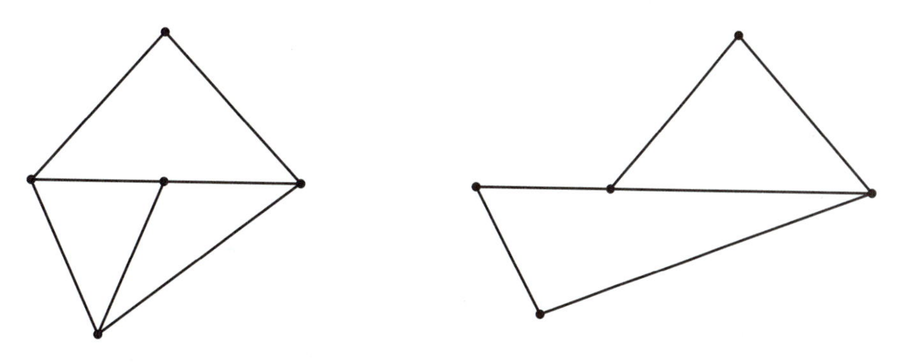

FIGURE 6.19 Illegal triangulations

triangle, T_i, a linear function will be prescribed:

$$\ell_i(x, y) = a_i x + b_i y + c_i \qquad (x, y) \in T_i$$

The coefficients in ℓ_i are uniquely determined by the prescribed function values at the vertices of T_i. This can be seen as an application of Theorem 2 because L_1 in that theorem can be taken to be one side of the triangle, and L_0 can be a line parallel to L_1 containing the vertex not on L_1.

Let us consider the situation shown in Figure 6.20. The line segment joining (x_2, y_2) to (x_3, y_3) is common to both triangles. This line segment can be represented as

$$\{t(x_2, y_2) + (1 - t)(x_3, y_3) : 0 \leqq t \leqq 1\}$$

The variable t can be considered to be the coordinate for the points on the line segment. The linear function ℓ_1, when restricted to this line segment, will be a

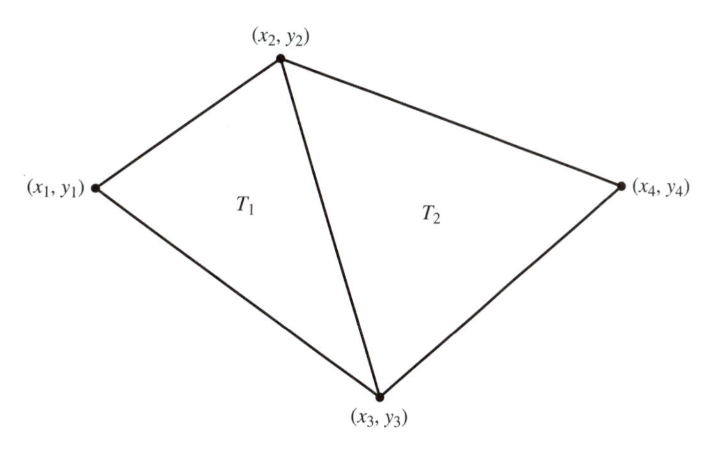

FIGURE 6.20 Triangulation

linear function of the single variable t—namely,

$$a_1\big(tx_2 + (1-t)x_3\big) + b_1\big(ty_2 + (1-t)y_3\big) + c_1$$

or

$$(a_1x_2 - a_1x_3 + b_1y_2 - b_1y_3)t + (a_1x_3 + b_1y_3 + c_1)$$

This linear function of t is completely determined by the interpolation conditions at (x_2, y_2) and (x_3, y_3). The same remarks pertain to the linear function ℓ_2. Thus ℓ_1 and ℓ_2 agree on this line segment, and the piecewise linear function defined on $T_1 \bigcup T_2$ is continuous. This proves the following result.

THEOREM 5 *Let $\{T_1, T_2, \dots, T_m\}$ be a triangulation in the plane. The piecewise linear function taking prescribed values at all the vertices of all the triangles T_i is continuous.*

Consider next the use of piecewise quadratic functions on a triangulation. In each triangle, T_i, a quadratic polynomial will be prescribed:

$$q_i(x, y) = a_1x^2 + a_2xy + a_3y^2 + a_4x + a_5y + a_6$$

Six conditions will be needed to determine the six coefficients. One such set of conditions consists of values at the vertices of the triangle and the midpoints of the sides. Again, an application of Theorem 2 shows that this interpolation is always uniquely possible. Indeed, in that theorem, L_2 can be one side of the triangle, L_1 can be the line passing through two midpoints not on L_2, and L_0 can be a line containing the remaining vertex but no other node. (See Figure 6.21.) Reasoning as before, we see that the global piecewise quadratic function will be continuous because the three prescribed function values on the side of a triangle determine the quadratic function of one variable on that side.

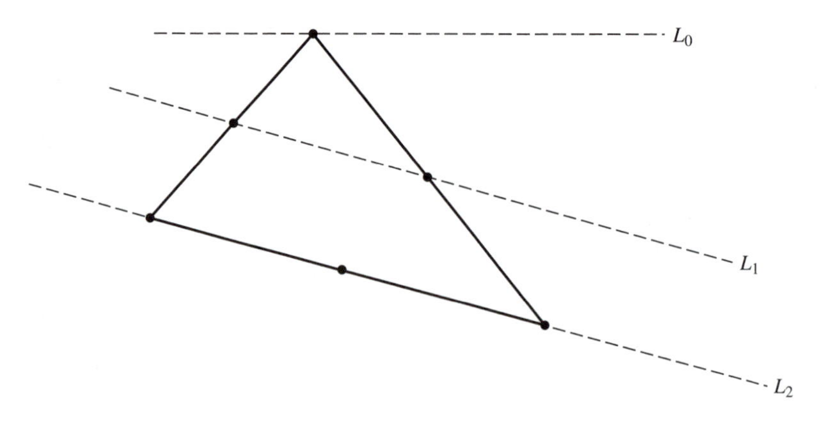

FIGURE 6.21 Applying Theorem 2

Moving Least Squares

Another versatile method of smoothing and interpolating multivariate functions is called **moving least squares**. First it is explained in a general setting, and then some specific examples will be given.

We start with a set X that is the domain of the functions involved. For example, X can be $\mathbb{R}$, or $\mathbb{R}^2$, or a subset of either. Next, a set of nodes $\{x_1, x_2, \ldots, x_n\}$ is given. These are the points at which a certain function f has been sampled. Thus, the values $f(x_i)$ are known for $1 \leqq i \leqq n$. For purposes of approximation, we select a set of functions $u_1, u_2, \ldots, u_m$. These are real-valued functions defined on X. The number m will usually be very small relative to n.

In the familiar least-squares method, a set of nonnegative weights $w_i \geqq 0$ is given. We try to find coefficients $c_1, c_2, \ldots, c_m$ to minimize the expression

$$\sum_{i=1}^{n} \left[f(x_i) - \sum_{j=1}^{m} c_j u_j(x_i) \right]^2 w_i$$

This is the sum of the squares of the residuals. If we write

$$\langle f, g \rangle = \sum_{i=1}^{n} f(x_i)\, g(x_i)\, w_i \qquad \|f\| = \sqrt{\langle f, f \rangle}$$

then the theory of approximation in an inner-product space is applicable, and the solution to the minimization problem is characterized by the orthogonality condition

$$f - \sum_{j=1}^{m} c_j u_j \perp u_i \qquad (1 \leqq i \leqq m)$$

This leads to the **normal equations**

$$\sum_{j=1}^{m} c_j \langle u_j, u_i \rangle = \langle f, u_i \rangle \qquad (1 \leqq i \leqq m)$$

How does the moving-least-squares method differ from the procedure just outlined? The weights w_i are now allowed to be functions of x. The formalism of the usual least-squares method can be retained, although the following notation may be better:

$$\langle f, g \rangle_x = \sum_{i=1}^{n} f(x_i)\, g(x_i)\, w_i(x)$$

The normal equations now should be written in the form

$$\sum_{j=1}^{m} c_j(x) \langle u_j, u_i \rangle_x = \langle f, u_i \rangle_x$$

and the final approximating function will be

$$g(x) = \sum_{j=1}^{m} c_j(x)\, u_j(x)$$

The computations necessary to produce this function will be quite formidable if m is large because the normal equations change with x. For this reason, m is usually no greater than 10.

The weight functions can be used to achieve several desirable effects. First, if $w_i(x)$ is "strong" at x_i, the function g will nearly interpolate f at x_i. In the limiting case, $w_i(x_i) = +\infty$ and $g(x_i) = f(x_i)$. If $w_i(x)$ decreases rapidly to 0 when x moves away from x_i, then the nodes far from x_i will have little effect on $g(x_i)$.

A choice for w_i that achieves these two objectives in a space $\mathbb{R}^d$ is

$$w_i(x) = \|x - x_i\|^{-2}$$

where any norm can be used, although the Euclidean norm is usual.

If the moving-least-squares procedure is used with a single function, $u_1(x) \equiv 1$, and with weight functions like the one just mentioned, then Shepard's method will result. To see that this is so, write the normal equation for this case, with $c_1(x) = c(x)$, $u_1(x) = u(x) = 1$:

$$c(x)\langle u, u\rangle_x = \langle f, u\rangle_x$$

The approximating function will be

$$\begin{aligned} g(x) &= c(x)u(x) = c(x) = \langle f, u\rangle_x / \langle u, u\rangle_x \\ &= \sum_{i=1}^{n} f(x_i) w_i(x) \Big/ \sum_{j=1}^{n} w_j(x) \end{aligned}$$

If $w_i(x) = \|x - x_i\|^{-2}$, then after the singularities are removed, $w_i / \sum_{j=1}^{n} w_j$ has the cardinal property: It takes the value 1 at x_i and the value 0 at all other nodes.

Interpolation by Multiquadrics

Another multivariate interpolation process is one proposed by R. L. Hardy [1971] that uses as its basic functions these so-called **multiquadrics:**

$$z_i(p) = \left\{\|p - p_i\|^2 + c^2\right\}^{1/2} \qquad (1 \leqq i \leqq n)$$

Here the norm is Euclidean, and c is a parameter that Hardy suggested to be set equal to 0.8 times the average distance between nodes. In interpolating with these functions, we need to know that the coefficient matrix $(z_i(p_j))$ is not singular. This has been proved by Micchelli [1986b].

Some further references on multivariate interpolation are Chui [1988], Hartley [1976], Micchelli [1986a], Franke [1982], and Lancaster and Salkauskas [1986].

References on Shepard's method of interpolation are Shepard [1968], Gordon and Wixom [1978], Newman and Rivlin [1983], Barnhill, Dube, and Little [1983], and Farwig [1986].

PROBLEMS 6.10

1. Give an algorithm for determining whether a set $\mathcal{N}$ of points in $\mathbb{R}^2$ can be expressed as a Cartesian product, as in Equation (2). If the factorization is possible, the algorithm should deliver the factors.

2. For each of these inclusions, either prove that it is true or prove that it is false:

   (a) $\Pi_n \otimes \Pi_m \subseteqq \Pi_{n+m}(\mathbb{R}^2)$

   (b) $\Pi_k(\mathbb{R}^2) \subseteqq \Pi_n \otimes \Pi_m$, where $k = \max(n, m)$

   (c) $\Pi_k(\mathbb{R}^2) \subseteqq \Pi_n \otimes \Pi_m$, where $k = \min(n, m)$

3. Prove that $\dim \Pi_k(\mathbb{R}^3) = \frac{1}{6}(k+1)(k+2)(k+3)$. Here the space consists of all polynomials of degree at most k in three variables.

4. This problem and the next deal with polynomials in d variables. Label the variables with subscripts and put them into a vector called x:

$$x = (x_1, x_2, \ldots, x_d) \in \mathbb{R}^d$$

   Now we require the concept of a **multi-index**. This is a d-tuple of nonnegative integers:

$$\alpha = (\alpha_1, \alpha_2, \ldots, \alpha_d) \in \mathbb{Z}_+^d$$

   We define

$$x^\alpha = x_1^{\alpha_1} x_2^{\alpha_2} x_3^{\alpha_3} \cdots x_d^{\alpha_d}$$

   This is called a **multinomial** and is the basic building block for polynomials in d variables. The **order** of the multi-index α is defined by the equation

$$|\alpha| = \alpha_1 + \alpha_2 + \cdots + \alpha_d$$

   Notice that what is called the **degree** of the multinomial x^α is nothing but $|\alpha|$. Prove that $x^\alpha x^\beta = x^{\alpha+\beta}$. Prove that any polynomial in d variables having degree $\leqq k$ can be written in the form

$$\sum_{|\alpha| \leqq k} c_\alpha x^\alpha$$

5. Let $\Pi_k(\mathbb{R}^d)$ be the set of all polynomials in d variables having degree at most k. Show that the dimension of this space is $\binom{d+k}{k}$.

6. Prove the assertion in the text (preceding Example 1) concerning the interpolation property of the operator $\overline{P} \oplus \overline{Q}$.

7. Show that interpolation by $\Pi_2(\mathbb{R}^2)$ is not generally possible at a set of six nodes if these nodes lie on an ellipse, a parabola, a hyperbola, or a pair of straight lines.

8. What type of bivariate polynomial would be suitable for interpolation on the sets of nodes in Figure 6.22?

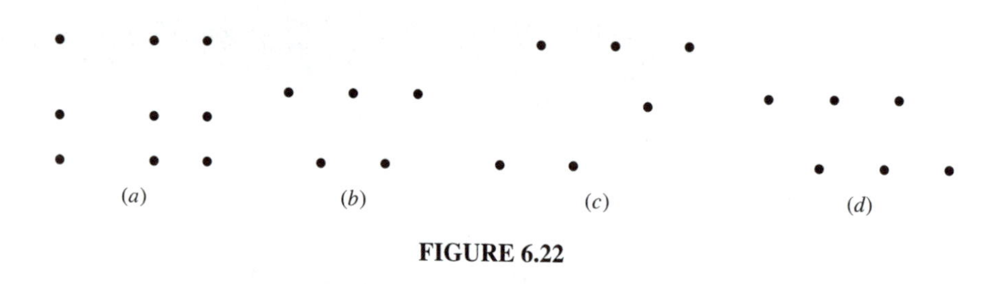

FIGURE 6.22

9. Prove that a basis for $\Pi_n \otimes \Pi_m$ is given by the functions

$$(x, y) \longmapsto x^i y^j \qquad (0 \leqq i \leqq n\,,\ 0 \leqq j \leqq m)$$

10. Give the formulas for the second version of Shepard interpolation using $\phi(p, q) = \|p - q\|_\infty^2$. Recall that this norm is $\|(x, y)\| = \max\{|x|, |y|\}$.

11. A polynomial in two variables having the form

$$F(x, y) = \sum_{0 \leqq i+j \leqq k} c_{ij} x^i y^j$$

is said to be of **degree at most** k. If there is a coefficient $c_{ij} \neq 0$ with $i + j = k$, we say that F is of **degree** k. Prove that the product of a polynomial in two variables of degree k with a polynomial in two variables of degree m is a polynomial of degree $m + k$.

12. Consider Shepard interpolation in two variables with n nodes and let the function ϕ be given by $\phi(p, q) = \|p - q\|^2$. What polynomial space $\Pi_k(\mathbb{R}^2)$ contains all the cardinal functions? (Give the least k.)

13. Let $f \in \Pi_k(\mathbb{R})$ and $\ell \in \Pi_1(\mathbb{R}^2)$. Prove that $f \circ \ell \in \Pi_k(\mathbb{R}^2)$. Is every element of $\Pi_k(\mathbb{R}^2)$ obtained in this way?

14. Prove that interpolation of arbitrary data at a set of four nodes is possible by a polynomial

$$p(x, y) = a + bx + cy + dxy$$

provided that the nodes are vertices of a rectangle whose sides are parallel to the coordinate axes.

15. Prove that interpolation by a linear polynomial

$$p(x, y) = ax + by + c$$

is always possible at the vertices of a triangle. Give two proofs, one of which is based on Theorem 3.

16. Prove that if n distinct points x_i are given in $\mathbb{R}^d$, then there is a vector b such that the n numbers $t_i = \langle x_i, b \rangle$ are all different. (This is needed for Theorem 4.)

17. If U and V are vector spaces of functions on X and Y, respectively, then $U \otimes V$ consists of all functions w that have a representation as a finite sum

$$w(x, y) = \sum_{i=1}^{n} u_i(x)v_i(y) \qquad u_i \in U, \ v_i \in V$$

Prove that if U and V are finite dimensional, then $\dim(U \otimes V) = \dim U \cdot \dim V$.

18. Prove that for an operator L of the form

$$Lf = \sum_{i=1}^{n} f(x_i)w_i$$

these properties are equivalent:

(i) For all f, $\min f(x_i) \leqq Lf \leqq \max f(x_i)$

(ii) $w_i \geqq 0$ and $\sum_{i=1}^{n} w_i = 1$

19. Is the following an accurate summary of the rules for a triangulation? *The set of nodes is exactly the same as the set of vertices of the triangles.*

COMPUTER PROBLEM 6.10

1. Write and test an efficient code to carry out Shepard's method. The user will specify nodes (x_i, y_i), ordinates c_i, and a list of points at which the Shepard interpolant is to be evaluated. If possible, obtain graphic output.

---

*6.11 Continued Fractions

Many of the special functions that occur in the applications of mathematics are defined by infinite processes, such as series, integrals, and iterations. The continued fraction is one of these infinite processes. An example of a continued fraction is this one, due to Lambert in 1770:

$$\tan^{-1} x = \cfrac{x}{1 + \cfrac{x^2}{3 + \cfrac{4x^2}{5 + \cfrac{9x^2}{7 + \cfrac{16x^2}{9 + \cdots}}}}} \qquad \bigl(|x| < 1\bigr) \tag{1}$$

This can also be written as

$$\tan^{-1} x = \frac{x}{1+}\,\frac{x^2}{3+}\,\frac{4x^2}{5+}\,\frac{9x^2}{7+}\,\frac{16x^2}{9+}\cdots \qquad \bigl(|x| < 1\bigr) \tag{2}$$

The right side of this equation represents a limit in the following way: The expression in Equation (2) is terminated after n terms to give a well-defined function f_n:

$$f_n(x) = \frac{x}{1+}\,\frac{x^2}{3+}\,\frac{4x^2}{5+}\cdots\frac{(n-1)^2x^2}{2n-1} \qquad (n \geqq 2) \tag{3}$$

This is called the nth *convergent* of the continued fraction. Equation (2) is defined to mean

$$\tan^{-1} x = \lim_{n\to\infty} f_n(x) \qquad \bigl(|x| < 1\bigr) \tag{4}$$

If we assume the validity of this equation, we have an alternative method of computing values of the arctangent function. Whether the method is practical depends on the *rapidity* of convergence in Equation (4). To judge this numerically, let us compute $\tan^{-1}(1/\sqrt{3}) = \pi/6 \approx 0.5235987756$ by means of the sequence $f_n(1/\sqrt{3})$ for $n \geqq 2$. Here are the results:

n	$f_n\bigl(1/\sqrt{3}\bigr)$
2	0.519615
3	0.523892
4	0.523577
5	0.523600
6	0.523599
7	0.523599

This list shows that six decimal places of precision have been obtained at the sixth convergent.

Recursive Formulas

The task of evaluating a continued fraction is not as easy as evaluating a series. In the case of an infinite series, say $\sum_{k=1}^{\infty} a_k$, we compute the partial sums $S_n = \sum_{k=1}^{n} a_k$

by means of the formula $S_{n+1} = S_n + a_{n+1}$. Let us consider the analogous problem for a continued fraction:

$$C = \frac{a_1}{b_1+}\ \frac{a_2}{b_2+}\ \frac{a_3}{b_3+}\ \cdots \tag{5}$$

We want to discover a recursive algorithm for computing the successive convergents:

$$C_n = \frac{a_1}{b_1+}\ \frac{a_2}{b_2+}\ \frac{a_3}{b_3+}\ \cdots\ \frac{a_{n-1}}{b_{n-1}+}\ \frac{a_n}{b_n} \tag{6}$$

We introduce for temporary use the functions

$$f_n(x) = \frac{a_1}{b_1+}\ \frac{a_2}{b_2+}\ \frac{a_3}{b_3+}\ \cdots\ \frac{a_{n-1}}{b_{n-1}+}\ \frac{a_n}{b_n + x} \tag{7}$$

Obviously then

$$C_n = f_n(0)$$

To obtain $f_n(x)$ from $f_{n-1}(x)$ we observe first

$$f_n(x) = f_{n-1}\left(\frac{a_n}{b_n + x}\right)$$

By direct calculation, we have

$$f_1(x) = \frac{a_1}{b_1 + x}$$

$$f_2(x) = \frac{a_1 b_2 + a_1 x}{b_1 b_2 + a_2 + b_1 x}$$

$$f_3(x) = \frac{a_1 b_2 b_3 + a_1 a_3 + (a_1 b_2)x}{b_1 b_2 b_3 + a_2 b_3 + b_1 a_3 + (b_1 b_2 + a_2)x}$$

The pattern that emerges here suggests that

$$f_n(x) = \frac{A_n + A_{n-1}x}{B_n + B_{n-1}x} \qquad (n \geqq 1) \tag{8}$$

where

$$\begin{cases} A_0 = 0, \ A_1 = a_1 \\ A_n = b_n A_{n-1} + a_n A_{n-2} \qquad (n \geqq 2) \end{cases} \tag{9}$$

and

$$\begin{cases} B_0 = 1, \ B_1 = b_1 \\ B_n = b_n B_{n-1} + a_n B_{n-2} \qquad (n \geqq 2) \end{cases} \tag{10}$$

THEOREM 1 *If the sequences $\{a_n\}_{n=1}^{\infty}$ and $\{b_n\}_{n=1}^{\infty}$ are given, and if the sequences $\{A_n\}_{n=0}^{\infty}$ and $\{B_n\}_{n=0}^{\infty}$ are defined by Formulas (9) and (10), then*

$$C_n = \frac{A_n}{B_n} \qquad (n \geqq 1) \tag{11}$$

Proof The validity of Equation (8) for indices 1, 2, and 3 has already been verified. Now suppose that Equation (8) is true for indices $1, 2, \ldots, n-1$. Then

$$\begin{aligned}
f_n(x) &= f_{n-1}\left(\frac{a_n}{b_n + x}\right) \\
&= \frac{A_{n-1} + A_{n-2}\, a_n/(b_n + x)}{B_{n-1} + B_{n-2}\, a_n/(b_n + x)} \\
&= \frac{A_{n-1}(b_n + x) + A_{n-2}\, a_n}{B_{n-1}(b_n + x) + B_{n-2}\, a_n} \\
&= \frac{A_{n-1}\, b_n + A_{n-2}\, a_n + A_{n-1}\, x}{B_{n-1}\, b_n + B_{n-2}\, a_n + B_{n-1}\, x} \\
&= \frac{A_n + A_{n-1}\, x}{B_n + B_{n-1}\, x}
\end{aligned}$$

This establishes Equation (8) by induction. Then, of course,

$$C_n = f_n(0) = \frac{A_n}{B_n}$$

■

The recursive formulas developed here form the basis for an efficient algorithm. For example, the numerical results for $\tan^{-1}(1/\sqrt{3})$, at the beginning of this section, were computed using this recursive relation with $a_n = (n-1)^2 x^2$ and $b_n = 2n - 1$, starting with $a_1 = x$ and $b_1 = 1$.

Conversion of Series to Continued Fractions

Many important special functions that arise in applied mathematics have expansions in continued fractions. Sources of information are Abramowitz and Stegun [1964] and the textbooks by Khovanskii [1963], Perron [1929], and Wall [1948]. We shall indicate here one procedure by which a continued fraction can be obtained from a series.

THEOREM 2

$$\sum_{k=1}^{\infty} \frac{1}{x_k} = \frac{1}{x_1 -}\ \frac{x_1^2}{x_1 + x_2 -}\ \frac{x_2^2}{x_2 + x_3 -} \cdots \frac{x_{n-1}^2}{x_{n-1} + x_n -} \cdots \tag{12}$$

Proof This can be proved by induction and is left as Problem 17 for the reader. ■

To illustrate how this theorem can be used, we construct a continued fraction from the Maclaurin series for arctan x:

$$\arctan x = x - \frac{x^3}{3} + \frac{x^5}{5} - \frac{x^7}{7} + \cdots$$

$$= \frac{1}{x^{-1}} + \frac{1}{-3x^{-3}} + \frac{1}{5x^{-5}} + \frac{1}{-7x^{-7}} + \cdots$$

$$= \frac{1}{x^{-1}-} \frac{(x^{-1})^2}{x^{-1}+(-3x^{-3})-} \frac{(-3x^{-3})^2}{(-3x^{-3})+(5x^{-5})-} \frac{(5x^{-5})^2}{(5x^{-5})+(-7x^{-7})-} \cdots$$

$$= \frac{x}{1+} \frac{-x^2}{x^2-3+} \frac{-9x^2}{-3x^2+5+} \frac{-25x^2}{5x^2-7-} \cdots$$

Notice that if both the numerator and denominator of one of the component fractions in a continued fraction are multiplied by the same quantity, then the numerator of the next component fraction is affected as well. (Why?)

Equations (9), (10), and (11) can be used to approximate arctan x for a given value of x when

$$a_n = -(2n-3)^2x^2 \quad \text{and} \quad b_n = (-1)^n[(2n-3)x^2 - (2n-1)]$$

for $n \geqq 2$. However, the algorithm for computing arctan x via this continued fraction is far more complicated than one based on computing the partial sums in the Maclaurin series, and it produces the same sequence of numbers. Observe that the Lambert continued fraction (1) for arctan x is not the same as the one derived from the Maclaurin series. To investigate this further, we compute $\arctan(1/\sqrt{3})$ using the partial sums in the series:

n	$f_n(1/\sqrt{3})$
1	0.577350
2	0.513200
3	0.526030
4	0.522976
5	0.523767
6	0.523551
7	0.523612
8	0.523595
9	0.523600
10	0.523598
11	0.523599
12	0.523599

A comparison of this table with the one produced from the continued fraction (1) shows that the continued fraction converges faster than the series for this example.

PROBLEMS 6.11

1. Show that

$$\frac{1}{1+}\,\frac{1}{1+}\,\frac{1}{1+}\,\cdots = \frac{1}{2}(\sqrt{5}-1)$$

Hint: Set x equal to the continued fraction, and look at $1/x$.

2. Show that

$$\sqrt{x} = 1 + 2\left(\frac{v}{1+}\,\frac{v}{1+}\,\frac{v}{1+}\,\cdots\right)$$

where $v = \frac{1}{4}(x-1)$.

3. Show that

$$\sqrt{b^2+a} = b + \frac{a}{2b+}\,\frac{a}{2b+}\,\frac{a}{2b+}\,\cdots$$

4. For the continued fraction in Problem 1, show that

$$C_n = \frac{A_n}{A_{n+1}}$$

5. For the continued fraction in Problem 2, show that

$$C_n = \frac{vA_n}{A_{n+1}}$$

6. Compare two methods for computing values of the function

$$f(x) = \int_0^x e^{-t^2}\,dt$$

namely, the Taylor series expansion given in Problem 6 of Section 6.7 and the continued fraction

$$f(x) = \frac{\sqrt{\pi}}{2} - \frac{1}{2}e^{-x^2}\left(\frac{1}{x+}\,\frac{1}{2x+}\,\frac{2}{x+}\,\frac{3}{2x+}\,\frac{4}{x+}\,\cdots\right)$$

7. Show that every real number in the interval $0 < x < 1$ can be expressed as a continued fraction (possibly terminating) in the form

$$x = \frac{1}{b_1+}\,\frac{1}{b_2+}\,\frac{1}{b_3+}\,\cdots$$

where each b_i is a positive integer.

8. Prove that if

$$x = \frac{a_1}{b_1+}\,\frac{a_2}{b_2+}\,\frac{a_3}{b_3+}\,\cdots$$

with the a_i and b_i all positive, then the convergents of the fraction are alternately greater than and less than x.

9. Show that f_n is increasing if and only if $A_{n-1}B_n > A_nB_{n-1}$, using the notation of Equation (8).

10. Count the number of long operations involved in computing C_n from A_n/B_n by the use of Formulas (9) and (10).

11. Show that any two successive convergents in the continued fraction of Equation (11) obey this equation:

$$\frac{A_n}{B_n} - \frac{A_{n-1}}{B_{n-1}} = (-1)^{n-1}\frac{a_1a_2\cdots a_n}{B_nB_{n-1}}$$

12. Use the result of Problem 11 to prove Problem 8.

13. Prove that if the numbers b_i are all positive and $a_i = 1$, then $B_n > \min(1, b_1)$ for all n.

14. Prove that if $b_i > 0$ and $a_i = 1$, then B_nB_{n-1} is increasing as a function of n.

15. Prove that if $b_i \geqq 1 = a_i$, then the continued fraction in Equation (11) converges.

16. Find the recursive formula for B_n if

$$\frac{a_1}{b_1+}\ \frac{a_2}{b_2+}\ \cdots = \frac{1}{B_1+}\ \frac{1}{B_2+}\ \cdots$$

17. Prove Theorem 2 in this section. *Hint:* Use induction.

18. Show that

$$e^x = \frac{1}{1-}\ \frac{x}{x+1-}\ \frac{x}{x+2-}\ \frac{2x}{x+3-}\ \frac{3x}{x+4-}\ \cdots$$

19. Show that

$$\frac{1}{1+}\ \frac{2}{2+}\ \frac{3}{3+}\ \cdots = \frac{1}{e-1}$$

Hint: Use Problem 18.

20. Show that the functions

$$g_n(x) = \frac{1}{x+}\ \frac{2}{x+}\ \frac{3}{x+}\ \cdots\ \frac{n}{x}$$

can be generated recursively by this algorithm:

$$g_n(x) = \frac{p_n(x)}{q_n(x)}$$

where

$$\begin{cases} p_0(x) = 0, \quad p_1(x) = 1 \\ p_{n+1}(x) = xp_n(x) + (n+1)p_{n-1}(x) \end{cases}$$

and

$$\begin{cases} q_0(x) = 1, \quad q_1(x) = x \\ q_{n+1}(x) = xq_n(x) + (n+1)q_{n-1}(x) \end{cases}$$

21. Assuming that the following continued fraction converges, find its value:

$$\frac{1}{6+}\,\frac{1}{6+}\,\frac{1}{6+}\,\frac{1}{6+}\cdots$$

22. Find the value of x:

$$x = 1 + \frac{1}{1+}\,\frac{1}{2+}\,\frac{1}{1+}\,\frac{1}{2+}\cdots$$

23. Find the value of x:

$$x = 2 + \frac{1}{4+}\,\frac{1}{4+}\,\frac{1}{4+}\cdots$$

24. If

$$f_n(x) = \frac{2}{1+}\,\frac{4}{2+}\,\frac{6}{3+}\cdots\frac{2n}{n+x}$$

how would you obtain f_{n+1} from f_n?

25. What is the value of $\sqrt{6+\sqrt{6+\sqrt{6+\cdots}}}$?

26. Assume that the continued fraction

$$\frac{1}{2x+}\,\frac{1}{2x+}\,\frac{1}{2x+}\cdots \qquad (x > 0)$$

converges. Determine a closed-form expression for it in terms of x.

COMPUTER PROBLEMS 6.11

1. Write a program for computing $\sqrt{x}$ by means of the equation in Problem 2. Compute $\sqrt{10}$, $\sqrt{100}$, $\sqrt{1000}$, and $\sqrt{10000}$ by printing a table of the first 50 convergent values for each.
2. Write a program for computing $\arctan(1/\sqrt{3})$ without using subscripted variables, and compare to the results given in the text.

3. Write a computer program to evaluate the continued fraction for $\arctan x$, as given in Equation (1). Use the following elementary approach: Given n, compute $f_n(x)$ in Equation (3) by starting at the right side and forming the appropriate fractions. Test your program by computing $\pi^{-1}\arctan(\sqrt{3})$ with $n = 5, 10, 15, 20$.

---

*6.12 Trigonometric Interpolation

First, we recall the salient facts about interpolation by ordinary algebraic polynomials. If $n+1$ function values are given in a table

x	x_0	x_1	x_2	$\cdots$	x_n
y	y_0	y_1	y_2	$\cdots$	y_n

(1)

then there exists a unique polynomial p of degree $\leqq n$ that *interpolates* the data. In other words,

$$p(x_j) = y_j \qquad (0 \leqq j \leqq n) \tag{2}$$

It is assumed that the points $x_0, x_1, \ldots, x_n$ are all different from each other, but no restriction is placed on the data y_j. The polynomial p is chosen from the linear space Π_n consisting of all polynomials of degree $\leqq n$. One basis for Π_n is given by the functions $b_k(x) = x^k$ for $0 \leqq k \leqq n$.

Fourier Series

For representing *periodic* phenomena, the algebraic polynomial space Π_n is, of course, not appropriate. Trigonometric functions will be much more suitable. One must know the period of the function in question before selecting the basic trigonometric functions. For convenience, let us assume that the function being interpolated is periodic with period 2π. Then the functions $1, \cos x, \cos 2x, \ldots$ and $\sin x, \sin 2x, \sin 3x, \ldots$ are appropriate. One of the basic theorems from Fourier analysis states that if f is 2π-periodic and has a continuous first derivative, then its Fourier series

$$\frac{a_0}{2} + \sum_{k=1}^{\infty}(a_k \cos kx + b_k \sin kx) \tag{3}$$

converges uniformly to f. The Fourier coefficients in the series are computed from the formulas

$$a_k = \frac{1}{\pi}\int_{-\pi}^{\pi} f(t)\cos kt\, dt \tag{4}$$

$$b_k = \frac{1}{\pi}\int_{-\pi}^{\pi} f(t)\sin kt\, dt \tag{5}$$

The cited theorem assures us that it is reasonable to approximate 2π-periodic functions with the sines and cosines listed above.

Complex Fourier Series

Much of this theory can be expressed in a more elegant form by using complex exponentials. We recall **Euler's formula**:

$$e^{i\theta} = \cos\theta + i\sin\theta \tag{6}$$

in which $i^2 = -1$. The Fourier series is given by

$$f(x) \sim \sum_{k=-\infty}^{\infty} \widehat{f}(k)e^{ikx} \tag{7}$$

where

$$\widehat{f}(k) = \frac{1}{2\pi}\int_{-\pi}^{\pi} f(t)e^{-ikt}\,dt \tag{8}$$

If f is a real function, then its Fourier series, as given in Equation (3), is the real part of the *complex* Fourier series given in Equation (7). Indeed, we have, from Equations (4), (5), (6), and (8),

$$\widehat{f}(k) = \frac{1}{2\pi}\int_{-\pi}^{\pi} f(t)[\cos kt - i\sin kt]\,dt = \frac{1}{2}(a_k - ib_k) \qquad (k \geqq 0) \tag{9}$$

Now we invoke the following theorem.

THEOREM 1 *Given the real sequences* $[a_k]_{k=0}^{\infty}$ *and* $[b_k]_{k=1}^{\infty}$, *define*

$$b_0 = 0 \qquad a_{-k} = a_k \qquad b_{-k} = -b_k \qquad c_k = \frac{1}{2}(a_k - ib_k)$$

Then

$$\frac{a_0}{2} + \sum_{k=1}^{n}(a_k\cos kx + b_k\sin kx) = \sum_{k=-n}^{n} c_k e^{ikx} \tag{10}$$

Proof The series on the right can be written as

$$\frac{1}{2}\sum_{k=-n}^{n}(a_k - ib_k)(\cos kx + i\sin kx) \tag{11}$$

The imaginary part of this series is 0, as shown by the following calculation:

$$\frac{1}{2}\sum_{k=-n}^{n}[a_k \sin kx - b_k \cos kx]$$

$$= \frac{1}{2}\sum_{k=1}^{n}[a_{-k}\sin(-kx) - b_{-k}\cos(-kx)] - \frac{b_0}{2} + \frac{1}{2}\sum_{k=1}^{n}[a_k \sin kx - b_k \cos kx]$$

$$= \frac{1}{2}\sum_{k=1}^{n}[-a_k \sin kx + b_k \cos kx] + \frac{1}{2}\sum_{k=1}^{n}[a_k \sin kx - b_k \cos kx] = 0$$

The real part of the series in (11) is

$$\frac{1}{2}\sum_{k=-n}^{n}[a_k \cos kx + b_k \sin kx]$$

$$= \frac{1}{2}\sum_{k=1}^{n}[a_{-k}\cos(-kx) + b_{-k}\sin(-kx)] + \frac{a_0}{2} + \frac{1}{2}\sum_{k=1}^{n}[a_k \cos kx + b_k \sin kx]$$

$$= \frac{a_0}{2} + \sum_{k=1}^{n}(a_k \cos kx + b_k \sin kx)$$

■

Inner Product, Pseudo-Inner Product, and Pseudonorm

The functions E_k defined by $E_k(x) = e^{ikx}$ (with $k = 0, \pm 1, \pm 2, \ldots$) form an orthonormal system of functions in the complex Hilbert space $L_2[-\pi, \pi]$ provided that we define the inner product to be

$$\langle f, g \rangle = \frac{1}{2\pi}\int_{-\pi}^{\pi} f(x)\overline{g(x)}\,dx$$

This means that $\langle E_k, E_n \rangle = 0$ when $n \neq k$ and that $\langle E_k, E_k \rangle = 1$. The calculation supporting this statement when $n \neq k$ is

$$\langle E_k, E_n \rangle = \frac{1}{2\pi}\int_{-\pi}^{\pi} E_k(x)\overline{E_n(x)}\,dx = \frac{1}{2\pi}\int_{-\pi}^{\pi} e^{ikx}e^{-inx}\,dx$$

$$= \frac{1}{2\pi}\int_{-\pi}^{\pi} e^{i(k-n)x}\,dx = \frac{1}{2\pi}\frac{e^{i(k-n)x}}{i(k-n)}\bigg|_{x=-\pi}^{x=\pi} = 0$$

Obviously, $\langle E_k, E_k \rangle = 1$ and the functions E_k form an orthonormal sequence.

It will be convenient to use the following inner-product notation:

$$\langle f, g \rangle_N = \frac{1}{N}\sum_{j=0}^{N-1} f(2\pi j/N)\overline{g(2\pi j/N)} \tag{12}$$

This function is not a true inner product because the condition $\langle f, f \rangle_N = 0$ does not imply that $f = 0$. It implies only that $f(x)$ takes the value 0 at each node $2\pi j/N$.

The other properties of a (complex) inner product are valid. They are:

(i) $\langle f, f\rangle_N \geqq 0$
(ii) $\langle f, g\rangle_N = \overline{\langle g, f\rangle_N}$
(iii) $\langle \alpha f + \beta g, h\rangle_N = \alpha\langle f, h\rangle_N + \beta\langle g, h\rangle_N$

Along with this **pseudo-inner product**, there is a **pseudonorm** defined by

$$\|f\|_N = \sqrt{\langle f, f\rangle_N}$$

We have $\|f\|_N = 0$ if and only if $f(2\pi j/N) = 0$ for $0 \leqq j \leqq N-1$.

For interpolation, the following theorem plays a crucial role.

THEOREM 2 *For any $N \geqq 1$, we have*

$$\langle E_k, E_m\rangle_N = \begin{cases} 1 & \text{if } k-m \text{ is divisible by } N \\ 0 & \text{otherwise} \end{cases} \tag{13}$$

Proof The expression in question can be written as

$$\frac{1}{N}\sum_{j=0}^{N-1} E_k\left(\frac{2\pi j}{N}\right)\overline{E_m\left(\frac{2\pi j}{N}\right)} = \frac{1}{N}\sum_{j=0}^{N-1}\left[e^{2\pi i(k-m)/N}\right]^j$$

If $k-m$ is divisible by N, then $(k-m)/N$ is an integer and $e^{2\pi i(k-m)/N} = 1$. Hence, each summand is 1 and their average is 1. On the other hand, if $k-m$ is not divisible by N, then $e^{2\pi i(k-m)/N} \neq 1$, and in this case we can apply the standard formula for the sum of a geometric series:

$$\sum_{j=0}^{N-1}\lambda^j = \frac{\lambda^N - 1}{\lambda - 1} \qquad (\lambda \neq 1)$$

The result is

$$\frac{e^{2\pi i(k-m)} - 1}{e^{2\pi i(k-m)/N} - 1} = 0$$

■

Exponential Polynomials

An **exponential polynomial** of degree at most n is any function of the form

$$P(x) = \sum_{k=0}^{n} c_k e^{ikx} = \sum_{k=0}^{n} c_k E_k(x) = \sum_{k=0}^{n} c_k (e^{ix})^k$$

The last expression in this equation explains the source of the terminology because it shows P to be a polynomial of degree $\leqq n$ in the variable e^{ix}. Interpolation by exponential polynomials is summarized in the next two results.

THEOREM 3 *Let E_k be the function $E_k(x) = e^{ikx}$. Then $\{E_0, E_1, \dots, E_{N-1}\}$ is orthonormal with respect to the inner product $\langle \cdot, \cdot \rangle_N$ defined in Equation (12).*

COROLLARY 1 *The exponential polynomial that interpolates a prescribed function f at equally spaced nodes $x_j = 2\pi j/N$ is given by the equations*

$$P = \sum_{k=0}^{N-1} c_k E_k \quad \text{with} \quad c_k = \langle f, E_k \rangle_N \tag{14}$$

Proof Using the given formula for c_k, we compute the value of the exponential polynomial at an arbitrary node, say x_ν. The result is

$$\begin{aligned}
\sum_{k=0}^{N-1} c_k E_k(x_\nu) &= \sum_{k=0}^{N-1} \langle f, E_k \rangle_N E_k(x_\nu) \\
&= \sum_{k=0}^{N-1} \frac{1}{N} \sum_{j=0}^{N-1} f(x_j)\, \overline{E_k(x_j)}\, E_k(x_\nu) \\
&= \sum_{j=0}^{N-1} f(x_j) \frac{1}{N} \sum_{k=0}^{N-1} \overline{E_j(x_k)}\, E_\nu(x_k) \\
&= \sum_{j=0}^{N-1} f(x_j) \langle E_\nu, E_j \rangle_N \\
&= f(x_\nu)
\end{aligned}$$

■

Example 1 Use Corollary 1 to obtain an explicit formula for the interpolant when $N = 2$.

Solution In this case, P is the exponential polynomial of degree 1 that interpolates f at the nodes 0 and π. Formula (14) gives us

$$\begin{aligned}
P(x) &= \frac{1}{2}\big[f(0) + f(\pi)\big] + \frac{1}{2}\big[f(0) + f(\pi)e^{-i\pi}\big]e^{ix} \\
&= \frac{1}{2}\big[f(0) + f(\pi)\big] + \frac{1}{2}\big[f(0) - f(\pi)\big]e^{ix}
\end{aligned}$$

■

COROLLARY 2 *If $n < N$, then the exponential polynomial $\sum_{k=0}^{n} c_k E_k$ that best approximates f in the least-squares sense on the finite set*

$$x_j = 2\pi j/N \qquad (0 \leqq j \leqq N-1)$$

is obtained with $c_k = \langle f, E_k \rangle_N$.

Proof By Theorem 3, the functions $x \longmapsto e^{ikx}$ form an orthonormal system with respect to the inner product defined in Equation (12). Now use Theorem 3 of Section 6.8 to complete the proof. ■

In Corollary 1, the choice of words suggests that the exponential polynomial described there is unique. To verify this, suppose that $\sum_{k=0}^{N-1} a_k E_k$ is an exponential polynomial that interpolates f at $x_0, x_1, \ldots, x_{N-1}$ (where $x_j = 2\pi j/N$). Then

$$\sum_{k=0}^{N-1} a_k E_k(x_j) = f(x_j) \qquad (0 \leqq j \leqq N-1)$$

If we multiply both sides of this equation by $E_n(-x_j)$ and sum with respect to j, the result is

$$\sum_{k=0}^{N-1} a_k \sum_{j=0}^{N-1} E_k(x_j)\, E_n(-x_j) = \sum_{j=0}^{N-1} f(x_j)\, E_n(-x_j)$$

By Equation (12), this implies that

$$\sum_{k=0}^{N-1} a_k \langle E_k, E_n \rangle_N = \langle f, E_n \rangle_N$$

Since $\langle E_k, E_n \rangle_N = \delta_{kn}$, we conclude that

$$a_n = \langle f, E_n \rangle_N = c_n$$

PROBLEMS 6.12

1. Using the notation of Theorem 3 and its corollary, show that if an exponential polynomial $g(x) = \sum_{k=0}^{N-1} a_k E_k(x)$ assumes the value 0 at each node x_j, then the coefficients a_k are all 0.
2. (Continuation) Use the result of Problem 1 to give another proof that the interpolating function in Corollary 1 is unique.
3. Prove that $E_k E_n = E_{k+n}$ and that $\overline{E}_k = E_{-k}$.
4. Prove that if f and g are functions such that

$$f(x_j) = \langle g, E_j \rangle_n \qquad (x_j = 2\pi j/n)$$

then $g(x_j) = n\langle f, E_j \rangle_n$.

5. By taking real and imaginary parts in a suitable exponential equation, prove that

$$\frac{1}{n}\sum_{j=0}^{n-1} \cos \frac{2\pi jk}{n} = \begin{cases} 1 & \text{if } k \text{ divides } n \\ 0 & \text{otherwise} \end{cases}$$

$$\frac{1}{n}\sum_{j=0}^{n-1} \sin \frac{2\pi jk}{n} = 0$$

6. Show that the inner product defined in Equation (12) satisfies the three properties **(i)**, **(ii)**, and **(iii)** following that equation. Why is $\|\cdot\|_N$ not a norm?

*6.13 Fast Fourier Transform

Fourier transforms are used to decompose a signal into its constituent frequencies. This is analogous to a prism separating white light into its component bands of colored light. In the same way that sunglasses reduce the glare of white light by permitting only the passage of the softer green light, the Fourier transform can be used to modify a signal to achieve a desired effect. Analyzing the component frequencies of a signal or a system, Fourier series and transforms find use in a wide variety of applications such as aircraft and spacecraft guidance, digital signal processing, medical imaging, oil and gas exploration, and the solution of differential equations. See, for example, Briggs and Henson [1995] or Walker [1992].

This section is devoted to the computational aspects of trigonometric interpolation. In particular, an algorithm called the **fast Fourier transform** will be developed for the efficient determination of the coefficients in Equation (14) of Section 6.12. We follow the exposition of Stoer and Bulirsch [1980].

Suppose that the coefficients $c_0, c_1, \ldots, c_{N-1}$ are defined as in Corollary 1 of Section 6.12. We write

$$c_k = \frac{1}{N}\sum_{j=0}^{N-1} f(x_j)(\lambda^k)^j \qquad (\lambda = e^{-2\pi i/N})$$

Thus, c_k is the result of evaluating a polynomial of degree $N-1$ at the point λ^k; it can be computed at a cost of approximately N multiplications and N additions. Since there are N coefficients c_k to be computed, the total cost of constructing the interpolating exponential polynomial is $\mathcal{O}(N^2)$ operations, in this straightforward approach.

The fast Fourier transform will bring this computing cost down to a more reasonable amount—namely, $N \log_2 N$. The table shows what this means for the large values of N that are commonly encountered in signal processing:

N	N^2	$N \log_2 N$
1024	1048576	10240
4096	16777216	49152
16384	268435456	229375

THEOREM 1 *Let p and q be exponential polynomials of degree $\leqq n-1$ such that, for the points $x_j = \pi j/n$, we have*

$$p(x_{2j}) = f(x_{2j}) \qquad q(x_{2j}) = f(x_{2j+1}) \qquad (0 \leqq j \leqq n-1) \tag{1}$$

Then the exponential polynomial of degree $\leqq 2n - 1$ that interpolates f at the points $x_0, x_1, \ldots, x_{2n-1}$ is given by

$$P(x) = \frac{1}{2}(1 + e^{inx})p(x) + \frac{1}{2}(1 - e^{inx})q(x - \pi/n) \tag{2}$$

Proof Since p and q have degrees $\leqq n - 1$, whereas e^{inx} is of degree n, it is clear that P has degree $\leqq 2n - 1$. It remains to be shown that P interpolates f at the nodes. We have, for $0 \leqq j \leqq 2n - 1$,

$$P(x_j) = \frac{1}{2}\left[1 + E_n(x_j)\right]p(x_j) + \frac{1}{2}\left[1 - E_n(x_j)\right]q(x_j - \pi/n)$$

Notice that

$$E_n(x_j) = e^{\pi i n j/n} = e^{\pi i j} = \begin{cases} +1 & j \text{ even} \\ -1 & j \text{ odd} \end{cases}$$

Thus for even j, we infer that $P(x_j) = p(x_j) = f(x_j)$, whereas for odd j, we have

$$P(x_j) = q(x_j - \pi/n) = q(x_{j-1}) = f(x_j) \qquad \blacksquare$$

THEOREM 2 *Let the coefficients of the polynomials described in Theorem 1 be as follows:*

$$p = \sum_{j=0}^{n-1} \alpha_j E_j$$

$$q = \sum_{j=0}^{n-1} \beta_j E_j$$

$$P = \sum_{j=0}^{2n-1} \gamma_j E_j$$

Then, for $0 \leqq j \leqq n - 1$,

$$\gamma_j = \frac{1}{2}\alpha_j + \frac{1}{2}e^{-ij\pi/n}\beta_j \tag{3}$$

$$\gamma_{j+n} = \frac{1}{2}\alpha_j - \frac{1}{2}e^{-ij\pi/n}\beta_j \tag{4}$$

Proof We shall be using Equation (2) and shall require a formula for $q(x - \pi/n)$:

$$q\left(x - \frac{\pi}{n}\right) = \sum_{j=0}^{n-1} \beta_j E_j\left(x - \frac{\pi}{n}\right)$$
$$= \sum_{j=0}^{n-1} \beta_j e^{ij(x-\pi/n)} = \sum_{j=0}^{n-1} \beta_j e^{-i\pi j/n} E_j(x)$$

Thus, from Equation (2),

$$P(x) = \frac{1}{2}\left[1 + E_n(x)\right]p(x) + \frac{1}{2}\left[1 - E_n(x)\right]q\left(x - \frac{\pi}{n}\right)$$
$$P = \frac{1}{2}\sum_{j=0}^{n-1}\left\{(1 + E_n)\alpha_j E_j + (1 - E_n)\beta_j e^{-i\pi j/n} E_j\right\}$$
$$= \frac{1}{2}\sum_{j=0}^{n-1}\left\{(\alpha_j + \beta_j e^{-i\pi j/n})E_j + (\alpha_j - \beta_j e^{-i\pi j/n})E_{n+j}\right\}$$

The formulas for the coefficients γ_j can now be read from this equation. ■

Example 1 Use Theorem 2 to find P when $n = 1$.

Solution Formulas (3) and (4) give us

$$\gamma_0 = \frac{1}{2}(\alpha_0 + \beta_0) \qquad \gamma_1 = \frac{1}{2}(\alpha_0 - \beta_0)$$

Hence, P is given by

$$P = \gamma_0 E_0 + \gamma_1 E_1 = \frac{1}{2}(\alpha_0 + \beta_0) + \frac{1}{2}(\alpha_0 - \beta_0)E_1$$

Putting in the values of α_0 and β_0, we obtain the result of Example 1 in Section 6.12:

$$P(x) = \frac{1}{2}\left[f(0) + f(\pi)\right] + \frac{1}{2}\left[f(0) - f(\pi)\right]e^{ix}$$ ■

Analysis

For the further analysis, let $R(n)$ denote the minimum number of multiplications necessary to compute the coefficients in an interpolating exponential polynomial for the set of points $\{2\pi j/n : 0 \leqq j \leqq n - 1\}$.

THEOREM 3 *The function R obeys the inequality*

$$R(2^m) \leqq m2^m$$

Proof We begin by establishing the inequality

$$R(2n) \leqq 2R(n) + 2n \tag{5}$$

By Theorem 2, the coefficients γ_j needed in the polynomial P can be obtained from the coefficients in p and q at the cost of $2n$ multiplications. Indeed, we require n multiplications to compute $\frac{1}{2}\alpha_j$ for $0 \leqq j \leqq n-1$, and another n multiplications to compute $(\frac{1}{2}e^{-ij\pi/n})\beta_j$ for $0 \leqq j \leqq n-1$. (In the latter, we assume that the factors $\frac{1}{2}e^{-ij\pi/n}$ have already been made available.) Since the coefficients $\alpha_0, \alpha_1, \ldots, \alpha_{n-1}$ cost $R(n)$ multiplications, and since the same is true of $\beta_0, \beta_1, \ldots, \beta_{n-1}$, we obtain a total cost for P of at most $2R(n) + 2n$ multiplications.

The inequality in the theorem is proved by induction. Consider the case $m = 0$. We wish to interpolate f at the point $x_0 = 0$ by an exponential polynomial of degree 0. The solution is the constant $f(0)$, and no multiplications are required. Thus the assertion of the theorem is true for $m = 0$. Now proceed by induction, using Equation (5). The calculation for the inductive step (m to $m+1$) is

$$\begin{aligned} R(2^{m+1}) &= R(2 \cdot 2^m) \leqq 2R(2^m) + 2 \cdot 2^m \\ &\leqq 2(m2^m) + 2^{m+1} = (m+1)2^{m+1} \end{aligned}$$

■

As a consequence of Theorem 3, we see that if N is a power of 2, say 2^m, then the cost of computing the interpolating exponential polynomial obeys the inequality $R(N) \leqq N \log_2 N$. The algorithm that carries out repeatedly the procedure in Theorem 1 is the fast Fourier transform.

The content of Theorem 1 can be interpreted in terms of two linear operators, L_n and T_h. For any f, let $L_n f$ denote the exponential polynomial of degree $n-1$ that interpolates f at the nodes $2\pi j/n$ for $0 \leqq j \leqq n-1$. Let T_h be a **translation operator** defined by

$$(T_h f)(x) = f(x+h)$$

We know from Corollary 1 in Section 6.12 that

$$L_n f = \sum_{k=0}^{n-1} \langle f, E_k \rangle_n E_k$$

Furthermore, in Theorem 1,

$$\begin{aligned} P &= L_{2n} f \\ p &= L_n f \\ q &= L_n T_{\pi/n} f \end{aligned}$$

The conclusion of Theorems 1 and 2 is that $L_{2n}f$ can be obtained efficiently from $L_n f$ and $L_n T_{\pi/n} f$. Of course, this is true for $n = 1, 2, \ldots$.

The notation just introduced allows Theorem 1 to be expressed in this nice form:

$$L_{2n}f = \frac{1}{2}(1 + E_n)L_n f + \frac{1}{2}(1 - E_n)T_{-\pi/n}L_n T_{\pi/n} f$$

Our goal now is to establish one version of the fast Fourier transform algorithm for computing $L_N f$, where $N = 2^m$.

DEFINITION 1 *Having fixed the function f and the index $N = 2^m$, we define*

$$P_k^{(n)} = L_{2^n} T_{2k\pi/N} f \qquad (0 \leqq n \leqq m,\ 0 \leqq k \leqq 2^{m-n} - 1) \tag{6}$$

An alternative description of $P_k^{(n)}$ is as the exponential polynomial of degree $2^n - 1$ that interpolates f in the following way:

$$P_k^{(n)}\left(\frac{2\pi j}{2^n}\right) = f\left(\frac{2\pi k}{N} + \frac{2\pi j}{2^n}\right) \qquad (0 \leqq j \leqq 2^n - 1) \tag{7}$$

By Problem 4, the set of nodes $(2\pi k/N) + (2\pi j/2^n)$ corresponding to one value of k is disjoint from the set corresponding to another value of k. A straightforward application of Theorem 1 shows that

$$P_k^{(n+1)}(x) = \frac{1}{2}(1 + e^{i2^n x})P_k^{(n)}(x) + \frac{1}{2}(1 - e^{i2^n x})P_{k+2^{m-n-1}}^{(n)}\left(x - \frac{\pi}{2^n}\right) \tag{8}$$

We can illustrate in a **tree diagram** how the exponential polynomials $P_k^{(n)}$ are related. Suppose that our objective is to compute $P_0^{(3)}$. In accordance with Equation (8), this function can be easily obtained from $P_0^{(2)}$ and $P_1^{(2)}$. Each of these, in turn, can be easily obtained from four polynomials of lower order, and so on. Figure 6.23 shows the connections.

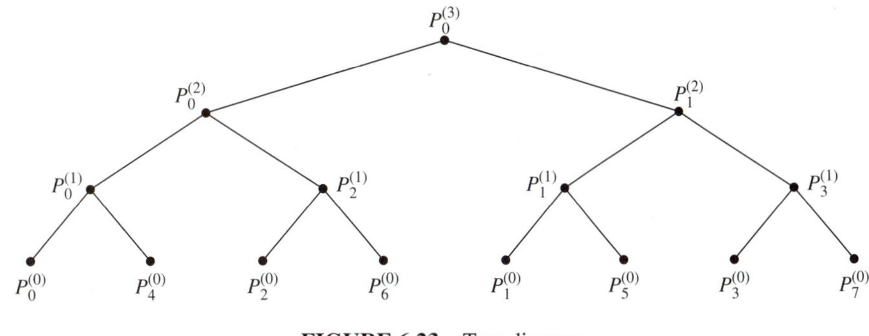

FIGURE 6.23 Tree diagram

Algorithm

Denote the coefficients of $P_k^{(n)}$ by $A_{kj}^{(n)}$. Here $0 \leqq n \leqq m$, $0 \leqq k \leqq 2^{m-n} - 1$, and $0 \leqq j \leqq 2^n - 1$. We have

$$P_k^{(n)}(x) = \sum_{j=0}^{2^n-1} A_{kj}^{(n)} E_j(x) = \sum_{j=0}^{2^n-1} A_{kj}^{(n)} e^{ijx}$$

By Theorem 2, the following equations are valid:

$$\begin{aligned} A_{kj}^{(n+1)} &= \frac{1}{2}\left[A_{kj}^{(n)} + e^{-ij\pi/2^n} A_{k+2^{m-n-1},j}^{(n)}\right] \\ A_{k,j+2^n}^{(n+1)} &= \frac{1}{2}\left[A_{kj}^{(n)} - e^{-ij\pi/2^n} A_{k+2^{m-n-1},j}^{(n)}\right] \end{aligned}$$

For a fixed n, the array $A^{(n)}$ requires N storage locations in memory because $0 \leqq k \leqq 2^{m-n} - 1$ and $0 \leqq j \leqq 2^n - 1$. One way to carry out the computations is to use two linear arrays of length N, one to hold $A^{(n)}$ and the other to hold $A^{(n+1)}$. At the next stage, one array will contain $A^{(n+1)}$ and the other $A^{(n+2)}$. Let us call these arrays C and D. The two-dimensional array $A^{(n)}$ is stored in C by the rule

$$C(2^n k + j) \leftarrow A_{kj}^{(n)} \qquad (0 \leqq k \leqq 2^{m-n} - 1,\ 0 \leqq j \leqq 2^n - 1)$$

Likewise, the array $A^{(n+1)}$ is stored in D by the rule

$$D(2^{n+1} k + j) \leftarrow A_{kj}^{(n+1)} \qquad (0 \leqq k \leqq 2^{m-n-1} - 1,\ 0 \leqq j \leqq 2^{n+1} - 1)$$

The factors $Z(j) = e^{-2\pi ij/N}$ are computed at the beginning and stored. Then we use the fact that $e^{-ij\pi/2^n} = Z(j2^{m-n-1})$. Here is the **fast Fourier transform algorithm**:

input m
$N \leftarrow 2^m$
$w \leftarrow e^{-2\pi i/N}$
for $k = 0$ **to** $N - 1$ **do**
    $Z(k) \leftarrow w^k$
    $C(k) \leftarrow f(2\pi k/N)$
end do
for $n = 0$ **to** $m - 1$ **do**
    for $k = 0$ **to** $2^{m-n-1} - 1$ **do**
        for $j = 0$ **to** 2^{n-1} **do**
            $u \leftarrow C(2^n k + j)$
            $v \leftarrow Z(j2^{m-n-1}) C(2^n k + 2^{m-1} + j)$
            $D(2^{n+1} k + j) \leftarrow (u + v)/2$
            $D(2^{n+1} k + j + 2^n) \leftarrow (u - v)/2$
        end do
    end do

```
        for j = 0 to N − 1 do
            C(j) ← D(j)
        end do
    end do
    output C(0), C(1), ..., C(N)
```

By scrutinizing the pseudocode, we can verify the bound $m2^m$ for the number of multiplications involved, as given by Theorem 3. Notice that in the nested loop of the code, n takes on m values; then k takes on 2^{m-n-1} values, and j takes on 2^n values. In this part of the code, there is really just one command involving a multiplication—namely, the one in which v is computed. This command will be encountered a number of times equal to the product $m \times 2^{m-n-1} \times 2^n = m2^{m-1}$. At an earlier point in the code, the computation of the Z-array involves $2^m - 1$ multiplications. On any binary computer, a multiplication by $1/2$ need not be counted because it is accomplished by subtracting 1 from the exponent of the floating-point number.

Example 2 Let $f = \sum_{k=0}^{7}(k+1)E_k$, and reconstruct f by means of the fast Fourier transform, using samples of f at eight equally spaced points.

Solution The algorithm given can be used with $m = 3$. A subroutine or procedure for computing values of f directly from the formula goes as follows:

```
input x
z ← cos x + i sin x
d ← 8
for k = 1 to 7 do
    d ← dz + 8 − k
end do
output d
```

When the programs were run on a 32-bit machine, the initial content of the C-array was as follows (rounded to five figures):

$$
\begin{aligned}
C_0 &= 36.000 & C_4 &= -4.0000 \\
C_1 &= -4.0000 - 9.6569i & C_5 &= -4.0000 + 1.6568i \\
C_2 &= -4.0000 - 4.0000i & C_6 &= -4.0000 + 4.0000i \\
C_3 &= -4.0000 - 1.6569i & C_7 &= -4.0000 + 9.6569i
\end{aligned}
$$

The final contents of the C-array was

$$(1.0000,\ 2.0000,\ 3.0000,\ 4.0000,\ 5.0000,\ 6.0000,\ 7.0000,\ 8.0000) \quad \blacksquare$$

The preceding algorithm serves a didactic purpose only. Codes to be used in *production* computation should have further refinements. For example, with additional programming, we can dispense with the D-array. Furthermore, whenever the factor $Z(j)$ is $+1$ or -1, the product $Z(j)C(k)$ should be programmed simply as $C(k)$ or $-C(k)$. Finally, a more versatile code can be written that can cope with values of N other than powers of 2.

Special codes can be written for functions f taking only real values at real arguments. Other codes can be written to produce cosine series or sine series alone. Consult the references cited at the end of this section for various ramifications of this topic.

Aliasing and Nyquist Frequency

As we attempt to reconstruct a function f from *samples* $f(x_j)$, the question arises of what limitations there may be in the process. Obviously, a finite number of samples cannot convey all the information contained in a function that can be an arbitrary element in an infinite-dimensional linear space. Thus, two different continuous functions may appear to be the same from the evidence obtained from a finite number of samples. This phenomenon is known as **aliasing**. To investigate this, let us suppose that f is represented by its Fourier series:

$$f = \sum_{k=-\infty}^{\infty} \langle f, E_k \rangle E_k \tag{9}$$

Here we have used the following notation:

$$E_k(x) = e^{ikx} \qquad \langle f, g \rangle = \frac{1}{2\pi} \int_{-\pi}^{\pi} f(x)\,\overline{g(x)}\,dx$$

By contrast, the interpolating exponential polynomial of degree $\leqq N-1$ is expressed as

$$P = \sum_{k=0}^{N-1} \langle f, E_k \rangle_N E_k \tag{10}$$

Here the notation is

$$\langle f, g \rangle_N = \frac{1}{N} \sum_{j=0}^{N-1} f(x_j)\,\overline{g(x_j)} \qquad x_j = 2\pi j/N$$

To compare the coefficients in the two equations (9) and (10), we write

$$\langle f, E_m \rangle_N = \left\langle \sum_{k=-\infty}^{\infty} \langle f, E_k \rangle E_k, E_m \right\rangle_N = \sum_{k=-\infty}^{\infty} \langle f, E_k \rangle \langle E_k, E_m \rangle_N$$

By Theorem 2 of Section 6.12, $\langle E_k, E_m \rangle_N$ is 0 except in those cases when $k - m$ is a multiple of N. Thus, if we let $k - m = \nu N$, with $\nu = 0, \pm 1, \pm 2, \ldots$, then we obtain

$$\langle f, E_m \rangle_N = \sum_{\nu=-\infty}^{\infty} \langle f, E_{m+\nu N} \rangle$$

Writing these terms in order of decreasing importance, we have

$$\langle f, E_m \rangle_N = \langle f, E_m \rangle + \langle f, E_{m+N} \rangle + \langle f, E_{m-N} \rangle + \cdots$$

This shows that the coefficient of E_m in the interpolating function (10) has not only the wanted term $\langle f, E_m \rangle$ from (9) but also terms that properly belong to higher frequencies, $\langle f, E_{m \pm \nu N} \rangle$ for $\nu = 1, 2, 3, \ldots$. These terms are unwanted and create distortions in the reconstructed signal. The first of these unwanted components is $\langle f, E_N \rangle$, which belongs to E_N in the Fourier series. But E_N is not present in the interpolating function of (10). The term E_N represents the lowest frequency component that is *present* in the Fourier series but *absent* in the interpolating function. This lowest frequency is known as the **Nyquist frequency**.

Computing Values of an Exponential Polynomial

The fast Fourier transform can also be used to evaluate an exponential polynomial at a set of equally spaced points. Suppose that such a polynomial is given in the form

$$p(x) = \sum_{j=0}^{n-1} a_j E_j(x)$$

To evaluate p at the points

$$t - \frac{2k\pi}{n} \qquad (0 \leqq k \leqq n-1)$$

we write $x_k = 2k\pi/n$ and

$$p(t - x_k) = \sum_{j=0}^{n-1} a_j E_j(t - x_k) = \sum_{j=0}^{n-1} a_j e^{ij(t-x_k)}$$

$$= \sum_{j=0}^{n-1} a_j e^{ijt}\, \overline{E_k(x_j)} = n\langle g, E_k \rangle_n$$

where g is a function such that

$$g(x_j) = a_j e^{ijt} \qquad (0 \leqq j \leqq n-1)$$

Thus, we apply the fast Fourier transform to g, obtaining the coefficients $\langle g, E_k \rangle_n$. When these are multiplied by n, we have the values $p(t - x_k)$.

Some sources of information on the fast Fourier transform are Davis and Rabinowitz [1984], Cooley, Lewis, and Welch [1967], Bloomfield [1976], Briggs and Henson [1995], Brigham [1974], Conte and de Boor [1980], Kahaner [1970, 1978], Lanczos [1966], Kahaner, Moler, and Nash [1989], Elliott and Rao [1982], Nussbaumer [1982], and Scheid [1988].

PROBLEMS 6.13

1. Prove that $T_h E_j$ is a scalar multiple of E_j.

2. Show that f and $f + \lambda E_k$ are indistinguishable if they are interpolated at points $x_j = 2\pi j/N$, with k a multiple of N. Thus, to detect all components of f having frequencies $\leqq \omega$, the sampling frequency should be greater than ω. This is another aspect of the aliasing effect.

3. **(Aliasing)** Two functions can appear to be the same if we know only samples of the functions taken at discrete points. Show that these two functions appear to be identical when sampled at the integer values of x:

$$f(x) = \cos\left[(n-\alpha)\pi x + \beta\right] \qquad g(x) = \cos\left[(n+\alpha)\pi x - \beta\right]$$

4. Prove that if $k \neq r$, then the node sets on the right side of Equation (7) are disjoint.

5. Establish the validity of Equation (8).

6. Prove this equivalent form of Theorem 2:

$$\begin{aligned}\langle f, E_j\rangle_{2n} &= (u_j + v_j)/2 \\ \langle f, E_{j+n}\rangle_{2n} &= (u_j - v_j)/2 \\ u_j = \langle f, E_j\rangle_n \qquad & v_j = e^{-ij\pi/n}\langle T_{\pi/n}E_j\rangle_n\end{aligned}$$

where $T_{\pi/n}$ is the translation operator defined by $(T_{\pi/n}f)(x) = f(x + \pi/n)$.

7. This problem and the next three give a matrix formulation of the fast Fourier transform. Fix an index n and a function f. Let u and v be column vectors having components

$$u_j = f\left(\frac{2\pi j}{n}\right) \qquad v_j = \langle f, E_j\rangle_n \qquad (0 \leqq j \leqq n-1)$$

Prove that $v = Au$, where A is the matrix with components $A_{jk} = n^{-1}e^{-2\pi i jk/n}$ $(0 \leqq j, k \leqq n-1)$.

8. (Continuation) Prove that the matrix A contains only n different elements. Prove that $n^{1/2}A$ is a unitary matrix. Prove that A is symmetric.

9. (Continuation) Using the Euclidean norm on $\mathbb{C}^n$, show that $\|v\| = n^{-1/2}\|u\|$.

10. (Continuation) Retain the notation of the preceding three problems, and write $A = A^{(n)}$ to show the dependence on n. If $u \in \mathbb{C}^{2n}$, put $u' = (u_0, u_2, \ldots, u_{2n-2})$ and $u'' = (u_1, u_3, \ldots, u_{2n-1})$. Prove that for $0 \leqq k \leqq n-1$,

$$(A^{(2n)}u)_k = \frac{1}{2}(A^{(n)}u')_k + \frac{1}{2}e^{-ik\pi/n}(A^{(n)}u'')_k$$

$$(A^{(2n)}u)_{n+k} = \frac{1}{2}(A^{(n)}u')_k - \frac{1}{2}e^{-ik\pi/n}(A^{(n)}u'')_k$$

COMPUTER PROBLEM 6.13

1. Write and test a suitable fast Fourier transform code to compute the values of $p(x) = \sum_{j=0}^{N-1} a_j E_j(x)$, assuming that $N = 2^m$ and that the coefficients a_j have been prescribed.

6.14 Adaptive Approximation

The distinguishing feature of *adaptive* approximation is that the domain of the function is repeatedly subdivided in order to obtain more accurate approximations on smaller subdomains. The resulting global approximation is then defined *piecewise* by patching together many local approximations. It is clear that spline functions with free knots are well suited for this type of approximation.

First-Degree Spline

We shall illustrate the principles with an algorithm that finds a first-degree spline function as an approximation to a given function f on a fixed interval $[a, b]$. It is assumed that an error tolerance, $\varepsilon > 0$, has been prescribed and that the result of our work is to be a first-degree spline function S that satisfies the inequality

$$|f(x) - S(x)| \leqq \varepsilon \quad \text{for} \quad a \leqq x \leqq b \tag{1}$$

The spline function S is piecewise linear and is chosen to interpolate f at the knots. The linear function that interpolates f at two points α and β is given by

$$\ell(f, \alpha, \beta; x) = \frac{f(\alpha)}{\beta - \alpha}(\beta - x) + \frac{f(\beta)}{\beta - \alpha}(x - \alpha)$$

If a set of knots is given, say

$$a = t_0 < t_1 < t_2 < \cdots < t_n = b$$

then the first-degree interpolating spline for f is defined by the formula

$$S(x) = \begin{cases} \ell(f, t_0, t_1; x) & t_0 \leqq x \leqq t_1 \\ \ell(f, t_1, t_2; x) & t_1 \leqq x \leqq t_2 \\ \vdots & \vdots \\ \ell(f, t_{n-1}, t_n; x) & t_{n-1} \leqq x \leqq t_n \end{cases}$$

If S meets the error criterion in Inequality (1), then our objective has been met. If, however,

$$\|f - S\|_\infty = \max_{a \leqq x \leqq b} |f(x) - S(x)| > \varepsilon$$

then the error tolerance ε has been exceeded at one or more points in the interval. Let ξ be a point where the error is worst; that is,

$$|f(\xi) - S(\xi)| = \|f - S\|_\infty$$

Let i be the index for which $t_{i-1} \leqq \xi \leqq t_i$. On this interval, the current spline function S is unsatisfactory. Thus, the difference $|f(x) - \ell(f, t_{i-1}, t_i; x)|$ exceeds ε at some points of the interval $[t_{i-1}, t_i]$. We therefore subdivide this interval by introducing ξ as a new knot. Instead of the single linear function $\ell(f, t_{i-1}, t_i; x)$ on $[t_{i-1}, t_i]$, we shall now have two linear functions—namely, $\ell(f, t_{i-1}, \xi; x)$ on $[t_{i-1}, \xi]$ and $\ell(f, \xi, t_i; x)$ on $[\xi, t_i]$. In this way, we pass from one set of knots to another set, and from one spline approximation to a better one. The process is repeated until the spline function meets the error criterion.

Algorithm

In writing an algorithm for this process, it is desirable to give high priority to efficiency. Thus at each stage of the process, we should store not only the current **knot array** $t = [t_0, t_1, \dots, t_n]$ but also the corresponding **ordinate array** $y = [y_0, y_1, \dots, y_n]$, where $y_i = f(t_i)$. Also, we should store the **deviation array** $d = [d_1, d_2, \dots, d_n]$ with $d_i = \max_{t_{i-1} \leqq x \leqq t_i} |f(x) - \ell(f, t_{i-1}, t_i; x)|$ and the array of **maximum deviation points** $c = [c_1, c_2, \dots, c_n]$ consisting of the points where these maximum deviations occur. Here d_i is the maximum deviation in the interval $[t_{i-1}, t_i]$, and c_i is a point in that interval where the maximum occurs. In terms of these stored quantities, we have

$$\begin{aligned} \ell(f, t_{i-1}, t_i; x) &= \left[y_{i-1}(t_i - x) + y_i(x - t_{i-1})\right] / (t_i - t_{i-1}) \\ \|f - S\|_\infty &= \max\{d_1, d_2, \dots, d_n\} \\ d_i &= |f(c_i) - \ell(f, t_{i-1}, t_i; c_i)| \qquad t_{i-1} \leqq c_i \leqq t_i \end{aligned}$$

At the nth step, the arrays t, y, d, and c will be available from prior work. Now a test is performed to see whether the present knot sequence is satisfactory. If it is not, an index i is selected for which $d_i = \max\{d_1, d_2, \dots, d_n\}$. Since c_i is a point in $[t_{i-1}, t_i]$ at which the deviation $|f(x) - S(x)|$ is a maximum, we can take c_i to be the new knot (denoted previously by ξ). In the next step, two new intervals will be present in our partition of $[a, b]$—namely, $[t_{i-1}, c_i]$ and $[c_i, t_i]$. All the stored data with indices $i, i+1, \dots, n$ are shifted *to the right* in memory, and t_i becomes c_i. In other words, the knots to the right of the new knot are relabeled, as well as the ordinates, deviations, and maximum deviation points, and the new knot is inserted into the array of knots. Then the interval containing the new knot is split at the new knot, and new points of maximum deviation over each of these subintervals are computed. In this way, a set of knots is generated with the maximum deviation over each subinterval within the specified tolerance, if possible. Here is an algorithm to do all this, starting with $t_0 = a$ and $t_1 = b$:

input a, b, ε, M
$t_0 \leftarrow a;\ t_1 \leftarrow b;\ y_0 \leftarrow f(t_0);\ y_1 \leftarrow f(t_1)$
call Max (f, t_0, t_1, c_1, d_1)
for $n = 1$ **to** $M - 1$ **do**
    select i so that $d_i = \max\{d_1, d_2, \dots, d_n\}$
    if $d_i \leqq \varepsilon$ **exit loop**
    for $j = n$ **to** $i + 1$ **step** -1 **do**
        $t_{j+1} \leftarrow t_j$
        $y_{j+1} \leftarrow y_j$
        $d_{j+1} \leftarrow d_j$
        $c_{j+1} \leftarrow c_j$
    end do
    $t_{i+1} \leftarrow t_i$
    $y_{i+1} \leftarrow y_i$
    $t_i \leftarrow c_i$
    $y_i \leftarrow f(c_i)$
    call Max$(f, t_{i-1}, t_i, c_i, d_i)$
    call Max$(f, t_i, t_{i+1}, c_{i+1}, d_{i+1})$
end do
output $n, (t_0, t_1, \dots, t_n), (y_0, y_1, \dots, y_n), (d_1, d_2, \dots, d_n)$

The algorithm starts with just two knots, $t_0 = a$ and $t_1 = b$. Next, there is a determination of the maximum deviation d_1 on the interval $[t_0, t_1]$. If d_1 is larger than the allowable tolerance ε, then the point c_1 where the maximum deviation occurred becomes a knot. The knot t_1 is relabeled t_2, and c_1 becomes t_1. The maximum deviations are then computed on $[t_0, t_1]$ and $[t_1, t_2]$. If either exceeds ε, the process is continued. Eventually either all deviations fall below ε (success) or the upper limit on the number of steps (M) is exceeded (failure). Enough storage must be provided to allow M components in the arrays t, y, d, and c.

In this algorithm, there are three references to a subroutine or procedure Max. The purpose of Max (f, α, β, c, d) is to compute the maximum value of $|f(x) - \ell(f, \alpha, \beta; x)|$ on the interval $[\alpha, \beta]$, storing the maximum value in d and storing the appropriate point at which this occurs in c. In practice, the values of c and d need not be computed to high precision. A rough way to do this is to sample the values of $|f(x) - \ell(f, \alpha, \beta; x)|$ at, say, 11 points across the interval, and to accept one of these as the desired point. Here is an algorithm to do just this:

procedure Max(f, α, β, c, d)
$k \leftarrow 10$
$h \leftarrow (\beta - \alpha)/k$
for $i = 0$ **to** k **do**
    $z_i \leftarrow f(\alpha + ih)$
end do
for $i = 1$ **to** $k - 1$ **do**
    $z_i \leftarrow |z_i - \big(iz_k + (k - i)z_0\big)/k|$
end do

```
d ← 0
for i = 1 to k − 1 do
    if z_i > d then
        d ← z_i
        c ← α + ih
    end if
end do
return
```

In the foregoing algorithm for `Max`, we have first stored $f(\alpha + ih)$ in z_i. After $z_0, z_1, \dots, z_k$ have been computed, we have this equation:

$$\ell(f, \alpha, \beta; x) = \frac{z_0}{kh}(\beta - x) + \frac{z_k}{kh}(x - \alpha)$$

where $h = (\beta - \alpha)/k$. For $x = \alpha + ih$, we have (because $\beta = \alpha + kh$)

$$\ell(f, \alpha, \beta; \alpha + ih) = \frac{z_0}{kh}(k - i)h + \frac{z_k}{kh}ih = \big(iz_k + (k - i)z_0\big)/k$$

Then the memory location z_i is used (line 8 in the algorithm) to store

$$\big|f(\alpha + ih) - \ell(f, \alpha, \beta; \alpha + ih)\big|$$

Example 1 A computer program incorporating the adaptive approximation method described above was tested using the function $f(x) = \sqrt{x}$ on the interval [0, 1] and with $\varepsilon = 10^{-2}$. As a result, ten computed knots were generated, clustered near the singularity at 0 (that is, a point with an infinite derivative). See Figure 6.24.

Solution The ten knots, the values of f, and the maximum deviations were:

knots	value of f	deviations
0.00	0.000	
0.000729	0.027	0.007
0.00243	0.049	0.002
0.0081	0.09	0.003
0.027	0.16	0.005
0.09	0.3	0.01
0.174	0.417	0.005
0.3	0.548	0.004
0.58	0.762	0.009
1.0	1.000	0.008

If ε is reduced to 10^{-3}, then 32 knots are generated by this adaptive procedure. ■

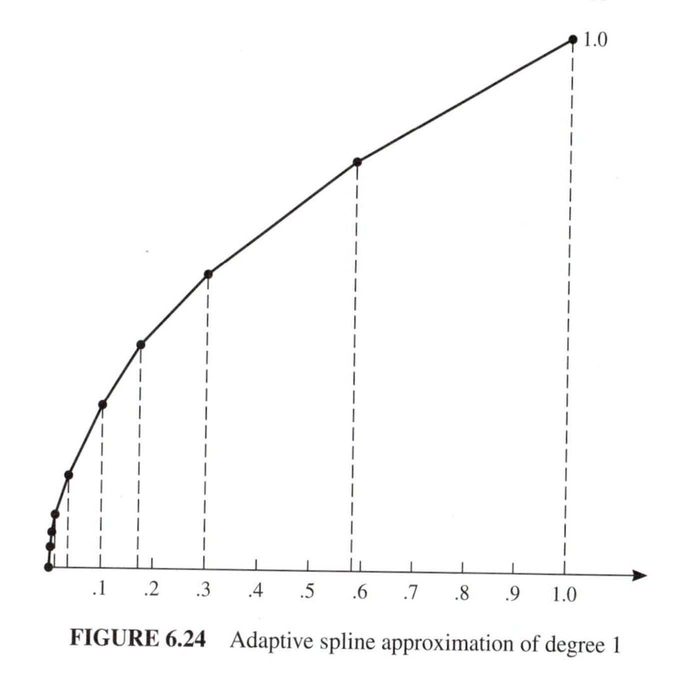

FIGURE 6.24 Adaptive spline approximation of degree 1

General Case

The principles illustrated by the first-degree adaptive spline algorithm can be abstracted so that they apply to more general cases. The essential ingredient in the process is a local approximation operator A that operates on a function f and an interval $[\alpha, \beta]$ and produces an approximation to f on the given interval. We write $A(f, \alpha, \beta; x)$ for the value at x of this approximating function. The operation $\ell(f, \alpha, \beta; x)$ used in the first-degree spline procedure is such an operator.

At each step in the adaptive process, a set of subintervals of $[a, b]$ will be given. These intervals are nonoverlapping and cover $[a, b]$. If $[\alpha, \beta]$ is one of these subintervals, we say that it is *satisfactory* if

$$|f(x) - A(f, \alpha, \beta; x)| \leqq \varepsilon \quad \text{on } [\alpha, \beta]$$

If all the subintervals are satisfactory, the objective is met by a global approximating function G that is made up from the functions $A(f, \alpha, \beta; x)$ on various subintervals. (Notice that there is no guarantee that G will be continuous.) If a subinterval $[\alpha, \beta]$ is not satisfactory, it is subdivided in some standard way. For example, the midpoint $\xi = \frac{1}{2}(\alpha + \beta)$ can be added as a knot. An alternative is to take as the new knot a point ξ where the error is worst on $[\alpha, \beta]$. In any case, the interval $[\alpha, \beta]$ is replaced by two new intervals $[\alpha, \xi]$ and $[\xi, \beta]$. The process is then repeated.

PROBLEMS 6.14

1. If the procedure of Computer Problem 1 is used on a monotone increasing function, what lower bound can be given on the number of subintervals required? The answer will be in terms of $f(b) - f(a)$ and ε.

2. For the function $f(x) = \sqrt{x}$, show that the maximum of

$$|f(x) - \ell(f, \alpha, \beta; x)|$$

occurs at $x = \frac{1}{4}(\sqrt{\beta} + \sqrt{\alpha})^2$.

3. What will happen if the adaptive algorithm is applied (exactly as described in the text) to the function $f(x) = \sin 10x$ on the interval $[0, \pi]$?

COMPUTER PROBLEMS 6.14

1. Write and test an adaptive routine that uses piecewise constant functions.
2. Write a computer routine for the adaptive approximation procedure and apply it to the example in this section. Produce a computer plot of the results.
3. Modify the adaptive algorithm for use with cubic splines.
4. Modify the procedure `Max` so that it uses the y-array. Experiment with some other procedure for `Max`.

Chapter Seven
Numerical Differentiation and Integration

7.1 Numerical Differentiation and Richardson Extrapolation

If the values of a function f are given at a few points, say $x_0, x_1, \ldots, x_n$, can that information be used to estimate a derivative $f'(c)$ or an integral $\int_a^b f(x)\,dx$? The answer is a qualified *yes*.

Let us begin by observing that from the values $f(x_0), f(x_1), \ldots, f(x_n)$ alone it is impossible to infer very much about f unless we are informed also that f belongs to some relatively small family of functions. Thus, if f is allowed to range over the family of all continuous real-valued functions, the values $f(x_i)$ are almost useless. Figure 7.1 illustrates three continuous functions that take the same values at six points.

On the other hand, if we know that f is a polynomial of degree at most n, then the values at $n + 1$ points completely determine f, by the theory of interpolation in Section 6.1. In this case, we recover f precisely and can then compute $f'(c)$ or $\int_a^b f(x)\,dx$ with complete confidence. In most realistic situations, however, the information at hand will not fully determine f, and any numerical estimate of its

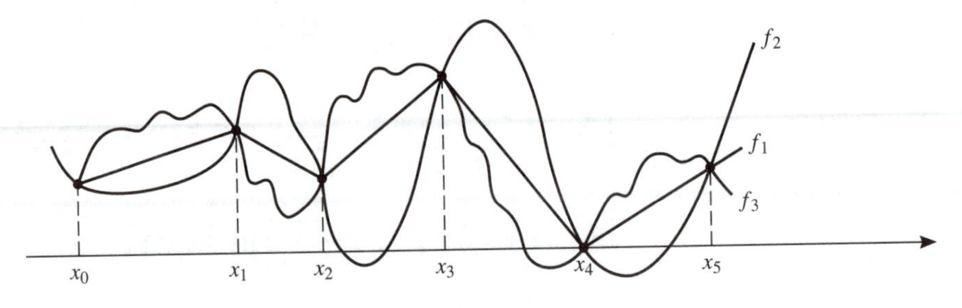

FIGURE 7.1 Three continuous functions through six points

derivative or integral should be viewed skeptically unless it is accompanied by some bound on the errors involved.

Numerical Differentiation

We illustrate these matters by examining a formula for numerical differentiation that emerges directly from the limit definition of $f'(x)$:

$$f'(x) \approx \frac{1}{h}\left[f(x+h) - f(x)\right] \tag{1}$$

For a linear function, $f(x) = ax + b$, the approximate Formula (1) is *exact*; that is, it yields the correct value of $f'(x)$ for any nonzero value of h. The formula may be exact in other cases too, but then only fortuitously. Let us therefore attempt to assess the error involved in this formula for numerical differentiation. The starting point is Taylor's Theorem in this form:

$$f(x+h) = f(x) + hf'(x) + \frac{h^2}{2}f''(\xi) \tag{2}$$

Here ξ is a point in the open interval between x and $x + h$. For the validity of Equation (2), f and f' should be continuous on the closed interval between x and $x + h$, and f'' should exist on the corresponding open interval. A rearrangement of Equation (2) yields

$$f'(x) = \frac{1}{h}\left[f(x+h) - f(x)\right] - \frac{h}{2}f''(\xi) \tag{3}$$

Equation (3) is more useful than Equation (1) because now on a large class of functions as described above, an error term is available along with the basic numerical formula. Notice that the error term in Equation (3) has two parts: a power of h and a factor involving some higher-order derivative of f. The latter gives us an indication of the class of functions to which the error estimate is applicable. The h-term in the error makes the entire expression converge to 0 as h approaches 0. The rapidity of this convergence will depend on the power of h. These remarks apply to many error estimates in numerical analysis: There will usually be a power of h and a factor telling us to what smoothness class the function must belong so that the estimate is valid.

Example 1 If Formula (1) is used to compute the derivative of $f(x) = \cos x$ at $x = \pi/4$ with $h = 0.01$, what is the answer and how accurate is it?

Solution Using a calculator, we find

$$f'(x) \approx \frac{1}{h}[f(x+h) - f(x)] = \frac{1}{0.01}[0.70000\,0476 - 0.70710\,6781]$$
$$= -0.71063\,051$$

The error term in Equation (3) can be estimated like this:

$$\left|\frac{h}{2}f''(\xi)\right| = 0.005|\cos\xi| \leqq 0.005$$

We can obtain a sharper bound by using the fact that $\pi/4 < \xi < \pi/4 + h$, so that $|\cos\xi| < 0.707107$. This gives the bound $0.00353\,55$. The actual error is

$$-\sin\frac{\pi}{4} + 0.71063\,051 = 0.00352\,3729$$ ■

The term $-(h/2)f''(\xi)$ in Equation (3) is called the **truncation error**. It is the error that arises because, at some stage in the derivation, a Taylor series has been truncated. In this case, the approximate Formula (1) was obtained by truncating the series

$$f(x+h) = f(x) + hf'(x) + \frac{h^2}{2}f''(x) + \frac{h^3}{3!}f'''(x) + \cdots$$

as follows:

$$f(x+h) \approx f(x) + hf'(x)$$

As we shall see presently, truncation error and roundoff error play equally important roles in the use of Formula (1) and others like it.

A glance at Equation (3) shows that in order to compute $f'(x)$ accurately, the step size h must be small. Therefore, let us perform an experiment in which h converges to 0 through a sequence of values, and the corresponding approximations to $f'(x)$ are computed. We select $f(x) = \tan^{-1} x$ and use the point $x = \sqrt{2}$. The result should be $f'(x) = (x^2+1)^{-1}$ at $\sqrt{2}$, which is $1/3$. Here is an algorithm for this task:

$f(x) := \tan^{-1} x$
input $s \leftarrow \sqrt{2};\ h \leftarrow 1;\ M \leftarrow 26$
$F_1 \leftarrow f(a)$
for $k = 0$ **to** M **do**
    $F_2 \leftarrow f(s+h)$
    $d \leftarrow F_2 - F_1$
    $r \leftarrow d/h$

```
    output k, h, F2, F1, d, r
    h ← h/2
end do
```

Some output from a 32-bit computer is shown here:

k	h	F_2	F_1	d	r
4	0.62×10^{-1}	0.97555095	0.95531660	0.02023435	0.32374954
12	0.24×10^{-3}	0.95539796	0.95531660	0.00008136	0.33325195
20	0.95×10^{-6}	0.95531690	0.95531660	0.00000030	0.31250000
24	0.60×10^{-7}	0.95531666	0.95531660	0.00000006	1.00000000
26	0.15×10^{-7}	0.95531660	0.95531660	0.00000000	0.00000000

In each line, d is computed as the difference $F_2 - F_1$, and r is computed as the ratio d/h. Because of **subtractive cancellation**, d has progressively fewer significant digits, until finally $d = 0$ and $r = 0$. The best value of r is obtained when $k = 12$, and it has four correct digits, if rounded to four decimal places. At this value of k, we notice that d has four significant digits. As k increases, the number of significant digits in d decreases, and of course r cannot have any more significant digits than d. Thus, roundoff error prevents us from getting accurate values when h is small. To have more precision in d, we would require more precision in F_1 and F_2, and that would require higher precision in the basic calculations. We can use multiple precision or carry out the calculations on a machine with a longer word length. The step at which the answers become worse will depend on the word length (or more exactly, the unit roundoff error) in the particular machine being used.

Numerical differentiation formulas find their most important application in the numerical solution of differential equations. A common stratagem there is to replace derivatives by approximations such as the one given in Equation (1). The precision of such numerical differentiation formulas is often judged simply by the power of h present in the error term. The higher the power of h the better because h is always a small number. In this assessment, Formula (2) fares poorly, as the error is $\mathcal{O}(h)$. A superior formula is

$$f'(x) \approx \frac{1}{2h}\left[f(x+h) - f(x-h)\right] \tag{4}$$

This is derived from two cases of Taylor's Theorem—namely,

$$f(x+h) = f(x) + hf'(x) + \frac{h^2}{2}f''(x) + \frac{h^3}{3!}f'''(\xi_1) \tag{5}$$

$$f(x-h) = f(x) - hf'(x) + \frac{h^2}{2}f''(x) - \frac{h^3}{3!}f'''(\xi_2) \tag{6}$$

Subtracting one of these equations from the other and rearranging, we obtain

$$f'(x) = \frac{1}{2h}\left[f(x+h) - f(x-h)\right] - \frac{h^2}{12}\left[f'''(\xi_1) + f'''(\xi_2)\right] \tag{7}$$

This is a more favorable result because of the h^2 term in the error. Notice, however, the presence of f''' in the error. This error term is applicable if f''' exists.

The error term in Equation (7) can be simplified if we make the slight additional assumption that the function f''' is continuous on $[x-h, x+h]$. Let M and m denote the greatest and least values of f''' on the interval $[x-h, x+h]$. Then $f'''(\xi_1)$, $f'''(\xi_2)$, and $c \equiv [f'''(\xi_1) + f'''(\xi_2)]/2$ all lie in the interval $[m, M]$. Since f''' is continuous, it assumes the value c at some point ξ in $[x-h, x+h]$. Hence,

$$f'''(\xi) = \frac{1}{2}\left[f'''(\xi_1) + f'''(\xi_2)\right]$$

When this expression is substituted in Equation (7), the result is

$$f'(x) = \frac{1}{2h}\left[f(x+h) - f(x-h)\right] - \frac{h^2}{6}f'''(\xi) \tag{8}$$

An important formula for second derivatives is obtained by extending Equations (5) and (6) by one more term and then adding the equations. After rearrangement and applying the method used previously, we have

$$f''(x) = \frac{1}{h^2}\left[f(x+h) - 2f(x) + f(x-h)\right] - \frac{h^2}{12}f^{(4)}(\xi) \tag{9}$$

for some $\xi \in (x-h, x+h)$. This formula is often used in the numerical solution of second-order differential equations.

Example 2 Use a computer to approximate $f'(x)$, where $f(x) = \tan^{-1} x$ and $x = \sqrt{2}$. Use Equation (8) with step size h approaching 0. Recall that the correct value is $1/3$.

Solution A suitable algorithm follows:

$f(x) := \tan^{-1} x$
input $s \leftarrow \sqrt{2};\ h \leftarrow 1;\ M \leftarrow 26$
for $k = 0$ **to** M **do**
  $F_2 \leftarrow f(s+h)$
  $F_1 \leftarrow f(s-h)$
  $d \leftarrow F_2 - F_1$
  $r \leftarrow d/(2h)$
  output k, h, F_2, F_1, d, r
  $h \leftarrow h/2$
end do

Some of the output from a 32-bit machine is shown here:

k	h	F_2	F_1	d	r
2	0.25	1.02972 674	0.86112 982	0.16859 692	0.33719 385
10	0.9765×10^{-3}	0.95564 199	0.95499 092	0.00065 106	0.33334 351
18	0.3815×10^{-5}	0.95531 786	0.95531 535	0.00000 250	0.32812 500
26	0.1490×10^{-7}	0.95531 660	0.95531 660	0.00000 000	0.00000 000

Again, we see that because of **subtractive cancellation**, there is a pronounced *deterioration* in accuracy as h approaches 0. The output begins to exhibit this phenomenon at $k = 9$. The values of F_1 and F_2 are very close to each other, and their difference, d, suffers a severe loss of significant digits. Eventually, as $h \to 0^+$, the values of F_1 and F_2 will be identical in the machine, and a value 0 will be computed as the derivative. This occurs at $k = 26$. This behavior will appear at different k values on computers with different word lengths. ■

The numerical differentiation of functions known only empirically is a risky procedure because errors present in the data will be magnified by the process. This is quite easily seen from Equation (8), for example. If the *sampling points* $x \pm h$ are accurately determined while the ordinates $f(x \pm h)$ are inaccurate, the errors in the ordinates will be multiplied by $1/(2h)$. Since h will be small, the influence of the errors will be large. (This phenomenon is *not* present in numerical integration.) Thus, numerical differentiation of *empirical* data should be avoided or undertaken with great caution.

Differentiation Via Polynomial Interpolation

A general approach to numerical differentiation and integration can be based on polynomial interpolation. Suppose that we have $n + 1$ values of a function f at points $x_0, x_1, \dots, x_n$. A polynomial that interpolates f at the nodes x_i can be written in the Lagrange form of Equation (9) in Section 6.1. Let us include the error term from Theorem 2 in that section, obtaining

$$f(x) = \sum_{i=0}^{n} f(x_i)\ell_i(x) + \frac{1}{(n+1)!} f^{(n+1)}(\xi_x) w(x) \tag{10}$$

Here we have written $w(x) = \prod_{i=0}^{n}(x - x_i)$. Taking derivatives in Equation (10), we have

$$f'(x) = \sum_{i=0}^{n} f(x_i)\ell_i'(x) + \frac{1}{(n+1)!} f^{(n+1)}(\xi_x) w'(x) + \frac{1}{(n+1)!} w(x) \frac{d}{dx} f^{(n+1)}(\xi_x)$$

If x is one of the nodes, say $x = x_\alpha$, then the preceding equation simplifies, since $w(x_\alpha) = 0$, and the result is

$$f'(x_\alpha) = \sum_{i=0}^{n} f(x_i)\ell_i'(x_\alpha) + \frac{1}{(n+1)!} f^{(n+1)}(\xi_{x_\alpha}) w'(x_\alpha)$$

This, in turn, can be simplified by computing $w'(x_\alpha)$. To do this, we note that

$$w'(x) = \sum_{i=0}^{n} \prod_{\substack{j=0 \\ j \neq i}}^{n} (x - x_j) \qquad \text{so} \qquad w'(x_\alpha) = \prod_{\substack{j=0 \\ j \neq \alpha}}^{n} (x_\alpha - x_j)$$

The final differentiation formula with error term is

$$f'(x_\alpha) = \sum_{i=0}^{n} f(x_i)\ell_i'(x_\alpha) + \frac{1}{(n+1)!} f^{(n+1)}(\xi_{x_\alpha}) \prod_{\substack{j=0\\ j\neq\alpha}}^{n} (x_\alpha - x_j) \qquad \textbf{(11)}$$

This formula is particularly well suited for nonequally spaced nodes.

Example 3 Give the explicit form of Equation (11) when $n = 2$ and $\alpha = 1$.

Solution The three cardinal functions for Lagrange interpolation in this case are

$$\begin{aligned}
\ell_0(x) &= \frac{(x - x_1)(x - x_2)}{(x_0 - x_1)(x_0 - x_2)} \\
\ell_1(x) &= \frac{(x - x_0)(x - x_2)}{(x_1 - x_0)(x_1 - x_2)} \\
\ell_2(x) &= \frac{(x - x_0)(x - x_1)}{(x_2 - x_0)(x_2 - x_1)}
\end{aligned}$$

Their derivatives are

$$\begin{aligned}
\ell_0'(x) &= \frac{2x - x_1 - x_2}{(x_0 - x_1)(x_0 - x_2)} \\
\ell_1'(x) &= \frac{2x - x_0 - x_2}{(x_1 - x_0)(x_1 - x_2)} \\
\ell_2'(x) &= \frac{2x - x_0 - x_1}{(x_2 - x_0)(x_2 - x_1)}
\end{aligned}$$

Evaluating at x_1, we obtain

$$\begin{aligned}
\ell_0'(x_1) &= \frac{x_1 - x_2}{(x_0 - x_1)(x_0 - x_2)} \\
\ell_1'(x_1) &= \frac{2x_1 - x_0 - x_3}{(x_1 - x_0)(x_1 - x_2)} \\
\ell_2'(x_1) &= \frac{x_1 - x_0}{(x_2 - x_0)(x_2 - x_1)}
\end{aligned}$$

The numerical differentiation formula with its error term is thus

$$\begin{aligned}
f'(x_1) = f(x_0) &\frac{x_1 - x_2}{(x_0 - x_1)(x_0 - x_2)} \\
&+ f(x_1)\frac{2x_1 - x_0 - x_2}{(x_1 - x_0)(x_1 - x_2)} \\
&+ f(x_2)\frac{x_1 - x_0}{(x_2 - x_0)(x_2 - x_1)} \qquad \textbf{(12)} \\
&+ \frac{1}{6} f'''(\xi_{x_1})(x_1 - x_0)(x_1 - x_2)
\end{aligned}$$

■

Example 4 In Example 3, what formula results if the nodes are equally spaced?

Solution Let $x_0 = x_1 - h$ and $x_2 = x_1 + h$. Then from Equation (12) we obtain (with $x = x_1$)

$$f'(x) = f(x-h)\left(\frac{-1}{2h}\right) + f(x+h)\left(\frac{1}{2h}\right) - \frac{1}{6}f'''(\xi_x)h^2$$

This is identical with Equation (8), which was previously derived. ■

Richardson Extrapolation

We shall now indicate how a procedure known as **Richardson extrapolation** can be used to coax more accuracy out of some numerical formulas. Consider Equations (5) and (6), extended with higher-order terms. We assume that $f(x)$ is represented by its Taylor series

$$f(x+h) = \sum_{k=0}^{\infty} \frac{1}{k!} h^k f^{(k)}(x) \tag{13}$$

$$f(x-h) = \sum_{k=0}^{\infty} \frac{1}{k!} (-1)^k h^k f^{(k)}(x) \tag{14}$$

If the second equation is subtracted from the first, all terms with an even value of k will cancel, leaving

$$f(x+h) - f(x-h) = 2hf'(x) + \frac{2}{3!}h^3 f'''(x) + \frac{2}{5!}h^5 f^{(5)}(x) + \cdots$$

A rearrangement yields

$$f'(x) = \frac{1}{2h}\left[f(x+h) - f(x-h)\right] - \left[\frac{1}{3!}h^2 f^{(3)}(x) + \frac{1}{5!}h^4 f^{(5)}(x) + \frac{1}{7!}h^6 f^{(7)}(x) + \cdots\right]$$

This equation has the form

$$L = \varphi(h) + a_2 h^2 + a_4 h^4 + a_6 h^6 + \cdots \tag{15}$$

where L stands for $f'(x)$ and $\varphi(h)$ stands for the numerical differentiation formula (4); that is,

$$\varphi(h) = \frac{1}{2h}\left[f(s+h) - f(s-h)\right]$$

where x is assigned a numerical value s at which the derivative is sought. The numerical procedure is designed to estimate L. The function φ is such that $\varphi(h)$ can be evaluated for $h > 0$, but it cannot be evaluated for $h = 0$. Thus, we can only let h *approach* 0 in our attempt to determine L. For each $h > 0$, the error is given by the series of terms $a_2 h^2 + a_4 h^4 + \cdots$. Assuming $a_2 \neq 0$, we see that the first term,

a_2h^2, is greater than the others when h is sufficiently small. We shall therefore look for a way of eliminating this dominant term, a_2h^2. Our analysis depends only on Equation (15) and will be applicable to other numerical processes; in particular, it is used in Section 7.4 with the trapezoid rule.

Write out Equation (15) with h replaced by $h/2$:

$$L = \varphi(h/2) + a_2h^2/4 + a_4h^4/16 + a_6h^6/64 + \cdots \tag{16}$$

The leading term a_2h^2 in the error series can be eliminated between Equations (15) and (16) by multiplying the latter by 4 and subtracting the former from it. Here is the work:

$$\begin{aligned} L &= \varphi(h) + a_2h^2 + a_4h^4 + a_6h^6 + \cdots \\ 4L &= 4\varphi(h/2) + a_2h^2 + a_4h^4/4 + a_6h^6/16 + \cdots \\ \hline 3L &= 4\varphi(h/2) - \varphi(h) - 3a_4h^4/4 - 15a_6h^6/16 - \cdots \end{aligned}$$

Thus, we have

$$L = \frac{4}{3}\varphi(h/2) - \frac{1}{3}\varphi(h) - a_4h^4/4 - 5a_6h^6/16 - \cdots \tag{17}$$

Equation (17) embodies the first step in Richardson extrapolation. It shows that a simple combination of $\varphi(h)$ and $\varphi(h/2)$ furnishes an estimate of L with accuracy $\mathcal{O}(h^4)$.

Example 5 Recompute the derivative in Example 2, incorporating Richardson extrapolation.

Solution A suitable algorithm is as follows:

$f(x) := \tan^{-1}(x)$
input $s \leftarrow \sqrt{2};\ h \leftarrow 1;\ M \leftarrow 30$
for $k = 0$ **to** M **do**
  $d_k \leftarrow \left[f(s+h) - f(s-h)\right]/(2h)$
  $h \leftarrow h/2$
end do
for $k = 1$ **to** M **do**
  $r_k \leftarrow d_k + (d_k - d_{k-1})/3$
end do
output $[k, d_k, r_k : 0 \leqq k \leqq M]$

A few lines of output are shown here.

k	d_k	r_k
2	0.33719 385	0.33333 480
4	0.33357 477	0.33333 364
8	0.33332 825	0.33332 571
16	0.33203 125	0.33138 022
26	0.00000 000	0.00000 000

A comparison of this output with that of Example 2 shows that two more digits of accuracy have been obtained.

The example just given is artificial because we already know the correct result, and we are using this knowledge to determine which value in the third column is most accurate. In a realistic situation, we would count the number of significant figures lost in computing d_k. Obviously, r_k cannot have more significant figures than d_k. The moral here is *not* to assume that r_k becomes more accurate as k increases indefinitely. ■

It has probably occurred to the reader that what was done to Equation (15) can now be done (with suitable modifications) to Equation (17). The work involved is as follows: Put

$$\psi(h) = \frac{4}{3}\varphi(h/2) - \frac{1}{3}\varphi(h)$$

in Equation (17). Then

$$\begin{aligned} L &= \psi(h) + b_4h^4 + b_6h^6 + \cdots \\ L &= \psi(h/2) + b_4h^4/16 + b_6h^6/64 + \cdots \end{aligned}$$

Now

$$\begin{aligned} L &= \psi(h) + b_4h^4 + b_6h^6 + \cdots \\ 16L &= 16\psi(h/2) + b_4h^4 + b_6h^6/4 + \cdots \\ \hline 15L &= 16\psi(h/2) - \psi(h) - 3b_6h^6/4 - \cdots \end{aligned}$$

Thus, we have

$$L = \frac{16}{15}\psi(h/2) - \frac{1}{15}\psi(h) - b_6h^6/20 - \cdots \tag{18}$$

Again, we can repeat this process by putting

$$\theta(h) = \frac{16}{15}\psi(h/2) - \frac{1}{15}\psi(h)$$

in Equation (18) so that

$$L = \theta(h) + c_6h^6 + c_8h^8 + \cdots$$

In a manner similar to that used above, it follows that

$$L = \frac{64}{63}\theta(h/2) - \frac{1}{63}\theta(h) - 3c_8h^8/252 - \cdots$$

As a matter of fact, any number of steps can be carried out to obtain formulas of increasing precision. The complete algorithm, allowing for M steps of Richardson extrapolation, is given next:

1. Select a convenient h (say $h = 1$) and compute the $M + 1$ numbers

$$D(n, 0) = \varphi(h/2^n) \qquad (0 \leqq n \leqq M)$$

2. Compute additional quantities by the formula

$$D(n, k) = \frac{4^k}{4^k - 1} D(n, k-1) - \frac{1}{4^k - 1} D(n-1, k-1) \tag{19}$$

Here $k = 1, 2, \ldots, M$ and $n = k, k+1, \ldots, M$.

Observe that $D(0, 0) = \varphi(h)$, $D(1, 0) = \varphi(h/2)$, and $D(1, 1) = \psi(h)$. The quantities $D(n, 1)$ conform to Equation (17) with h replaced by $h/2$ repeatedly. Similarly, $D(n, 2)$ corresponds to Equation (18), and so on. It is clear from our hypotheses and calculations that

$$D(n, 0) = L + \mathcal{O}(h^2)$$
$$D(n, 1) = L + \mathcal{O}(h^4)$$
$$D(n, 2) = L + \mathcal{O}(h^6)$$
$$D(n, 3) = L + \mathcal{O}(h^8)$$

The general result, proved in the next theorem, is that

$$D(n, k-1) = L + \mathcal{O}(h^{2k}) \qquad \text{as } h \to 0$$

At this point, let us reemphasize that the analysis can be applied to *any* numerical process that obeys Equation (15). The equation just preceding Equation (15) gives only one example of how it may arise in a specific numerical process.

If Equation (15) is valid, then the Richardson algorithm is effective, in the sense that successive columns in the $D(n, k)$-array will exhibit higher-order convergence. This is the content of the next theorem.

THEOREM 1 *The quantities $D(n, k)$ defined in the Richardson extrapolation algorithm obey an equation of the form*

$$D(n, k-1) = L + \sum_{j=k}^{\infty} A_{jk}(h/2^n)^{2j} \tag{20}$$

Proof When $k = 1$, we verify Equation (20) by using the definition of $D(n, 0)$ and Equation (15):

$$D(n, 0) = \varphi(h/2^n) = L - \sum_{j=1}^{\infty} a_{2j}(h/2^n)^{2j}$$

Thus, we can let $A_{j1} = -a_{2j}$. Now proceed by induction on k. We assume that Equation (20) is valid for some $k - 1$, and on that basis we prove it for k. From

Equations (19) and (20), we have

$$D(n,k) = \frac{4^k}{4^k - 1}\left[L + \sum_{j=k}^{\infty} A_{j,k}\left(\frac{h}{2^n}\right)^{2j}\right] - \frac{1}{4^k - 1}\left[L + \sum_{j=k}^{\infty} A_{j,k}\left(\frac{h}{2^{n-1}}\right)^{2j}\right]$$

This can be simplified to the following:

$$D(n,k) = L + \sum_{j=k}^{\infty} A_{j,k}\left[\frac{4^k - 4^j}{4^k - 1}\right]\left(\frac{h}{2^n}\right)^{2j} \tag{21}$$

Thus, $A_{j,k+1}$ should be defined by

$$A_{j,k+1} = A_{j,k}\left[\frac{4^k - 4^j}{4^k - 1}\right]$$

Notice that $A_{k,k+1} = 0$, and thus Equation (21) can be written as

$$D(n,k) = L + \sum_{j=k+1}^{\infty} A_{j,k+1}(h/2^n)^{2j}$$ ■

The formulas given for $D(n, 0)$ and $D(n, k)$ allow us to construct a triangular array:

$$\begin{array}{lllll}
D(0,0) & & & & \\
D(1,0) & D(1,1) & & & \\
D(2,0) & D(2,1) & D(2,2) & & \\
\vdots & \vdots & \vdots & \ddots & \\
D(M,0) & D(M,1) & D(M,2) & \cdots & D(M,M)
\end{array}$$

An algorithm to generate this triangular array is given next. In it, we refer to a function φ that must be made available to the code separately (that is, by a subprogram or procedure).

input h, M
for $n = 0$ **to** M **do**
  $D(n,0) \leftarrow \varphi(h/2^n)$
end do
for $k = 1$ **to** M **do**
  for $n = k$ **to** M **do**
    $D(n,k) \leftarrow D(n,k-1) + \big[D(n,k-1) - D(n-1,k-1)\big]/(4^k - 1)$
  end do
end do
output $D(n,k) \quad (0 \leqq n \leqq M, 0 \leqq k \leqq n)$

Example 6 Recompute the derivative in Example 5 using the complete Richardson extrapolation algorithm just described.

Solution Some output from this algorithm using Example 2 is shown here:

n	$D(n,0)$	$D(n,1)$	$D(n,2)$	$D(n,3)$	$D(n,4)$
0	0.39269 91				
1	0.34877 10	0.33412 83			
2	0.33719 38	0.33333 48	0.33328 19		
3	0.33429 81	0.33333 29	0.33333 28	0.33333 36	
4	0.33357 48	0.33333 36	0.33333 37	0.33333 37	0.33333 37

This algorithm obtains approximately the same precision as in Example 5. Eventually it too will produce meaningless results because of subtractive cancellation. ■

In recent years, software tools have been developed for automatic differentiation. Bischof, Carle, Khademi, and Mauer [1994] have written a software system called ADIFOR that evaluates derivatives by applying the chain rule to elementary instructions. One of their goals is to generate efficient derivative codes that require little human effort consistent with a *black-box* piece of software. For an introduction to techniques for evaluating derivatives of functions via computer programs, see Griewank and Corlis [1991] for example.

PROBLEMS 7.1

1. Carry out in all detail the derivation of Equation (11).

2. Let f be a polynomial of degree n. If the values of f at n points are known, can $f'(c)$ or $\int_a^b f(x)\,dx$ be estimated with confidence? Your answers will depend on c, a, and b.

3. Let a numerical process be described by

$$L = \varphi(h) + \sum_{j=1}^{\infty} a_j h^j$$

Explain how Richardson extrapolation will work in this case. Prove an analogue of Theorem 1 in this section for this case.

4. Give the explicit form of Equation (11) in these cases:

   (a) $n = 0$: $\alpha = 0$

   (b) $n = 1$: $\alpha = 0$ and $\alpha = 1$

   (c) $n = 2$: $\alpha = 0$ and $\alpha = 2$

5. Formula (9) for $f''(x)$ is often used in the numerical solution of differential equations. By adding the Taylor series for $f(x+h)$ and for $f(x-h)$, show that

the error in this formula has the form $\sum_{n=1}^{\infty} a_{2n}h^{2n}$. Determine the coefficients a_{2n} explicitly. Also derive the error term given in Equation (9).

6. Derive the following two formulas for approximating derivatives and show that they are both $\mathcal{O}(h^4)$ by establishing their error terms:

$$f'(x) \approx \frac{1}{12h}\left[-f(x+2h)+8f(x+h)-8f(x-h)+f(x-2h)\right]$$

$$f''(x) \approx \frac{1}{12h^2}\left[-f(x+2h)+16f(x+h)-30f(x)+16f(x-h)-f(x-2h)\right]$$

7. Derive the following two formulas for approximating the third derivative. Find their error terms. Which formula is more accurate?

$$f'''(x) \approx \frac{1}{h^3}\left[f(x+3h) - 3f(x+2h) + 3f(x+h) - f(x)\right]$$

$$f'''(x) \approx \frac{1}{2h^3}\left[f(x+2h) - 2f(x+h) + 2f(x-h) - f(x-2h)\right]$$

8. Carry out the instructions in Problem 7 for the following fourth-derivative formulas:

$$f^{(4)} \approx \frac{1}{h^4}\left[f(x+4h) - 4f(x+3h) + 6f(x+2h) - 4f(x+h) + f(x)\right]$$

$$f^{(4)} \approx \frac{1}{h^4}\left[f(x+2h) - 4f(x+h) + 6f(x) - 4f(x-h) + f(x-2h)\right]$$

9. Show that in Richardson extrapolation,

$$D(2, 2) = \frac{16}{15}\psi(h/2) - \frac{1}{15}\psi(h)$$

10. Show how to use Richardson extrapolation employing x_n and x_{n^2} if

$$L = x_n + a_1 n^{-1} + a_2 n^{-2} + a_3 n^{-3} + \cdots$$

11. Prove or disprove:

(a) If $L - x_n = \mathcal{O}(n^{-1})$, then $L - (2x_{2n} - x_n) = \mathcal{O}(n^{-2})$.

(b) If $L - x_n = \mathcal{O}(n^{-1})$, then $L - x_{n^2} = \mathcal{O}(n^{-2})$.

Discuss the numerical consequences of this problem.

12. Show how to use Richardson extrapolation if

$$L = \varphi(h) + a_1 h + a_3 h^3 + a_5 h^5 + \cdots$$

13. Suppose that $L = \lim_{h\to 0} f(h)$ and that $L - f(h) = c_6 h^6 + c_9 h^9 + \cdots$. What combination of $f(h)$ and $f(h/2)$ should be the best estimate of L?

14. Using Taylor series, derive the error term for the approximation

$$f'(x) \approx \frac{1}{2h}\left[-3f(x) + 4f(x+h) - f(x+2h)\right]$$

15. Derive a numerical differentiation formula of order $\mathcal{O}(h^4)$ by applying Richardson extrapolation to

$$f'(x) = \frac{1}{2h}\left[f(x+h) - f(x-h)\right] - \frac{h^2}{6}f'''(x) - \frac{h^4}{120}f^{(5)}(x) - \cdots$$

Give the error term of order $\mathcal{O}(h^4)$.

16. Using Taylor series expansions, derive the error term for the formula

$$f''(x) \approx \frac{1}{h^2}\left[f(x) - 2f(x+h) + f(x+2h)\right]$$

17. Establish a formula of the form

$$f''_n \approx \frac{1}{h^2}\left[Af_{n+3} + Bf_{n+2} + Cf_{n+1} + Df_n\right]$$

Here $f_{n+i} = f(x_n + ih)$.

18. Derive the approximation

$$f'(x_n) \approx \frac{3f(x_n) - 4f(x_{n-1}) + f(x_{n-2})}{3x_n - 4x_{n-1} + x_{n-2}}$$

and show that the error term is $\mathcal{O}(h^2)$ as $h \to 0$. Here $f_{n+i} = f(x_n + ih)$.

COMPUTER PROBLEMS 7.1

1. Program the algorithm in the text in which Richardson extrapolation is used repeatedly to estimate $f'(x)$. Test your program on the following:

(a) $\ln x$ at $x = 3$

(b) $\tan x$ at $x = \sin^{-1}(0.8)$

(c) $\sin(x^2 + \frac{1}{3}x)$ at $x = 0$

2. Write and test an algorithm for computing $f''(x)$ using Formula (9) together with repeated Richardson extrapolation.

3. In a typical use of Formula (1), the roundoff error (due mainly to subtractive cancellation) will behave like αh^{-1}. Test this assertion (and estimate α) on some examples. A computer program using multiple precision may be needed to simulate the computations.

4. Devise and test an algorithm for approximating f' and f'' using cubic splines.

5. Using the approximation in Problem 18, develop an iterative formula analogous to the secant method. Test the numerical behavior of this method on several functions, and compare with the secant method.

7.2 Numerical Integration Based on Interpolation

Numerical integration is the process of producing a numerical value for the integration of a function over a set. Here are some examples of integrals that might be calculated by suitable computer routines:

$$\int_0^2 e^{-x^2}\,dx$$

$$\int_0^1 \int_0^1 \sin(xye^x)\,dx\,dy$$

$$\int_0^1 \int_{x^2}^x \tan(xy^2)\,dy\,dx$$

$$\int_0^\pi \cos(3\cos\theta)\,d\theta$$

These integration problems are not amenable to the techniques learned in elementary calculus. Those techniques depends on **antidifferentiation**. Thus, to find the value of

$$\int_a^b f(x)\,dx$$

by calculus, we first must produce a function F with the property that $F' = f$. Then we have

$$\int_a^b f(x)\,dx = F(b) - F(a)$$

For example,

$$\int_1^4 x^2\,dx = \frac{1}{3}x^3\Big|_1^4 = \frac{1}{3}4^3 - \frac{1}{3} = 21$$

The function F given by $F(x) = (1/3)x^3$ is an antiderivative of the function f given by $f(x) = x^2$. Many elementary functions do not have simple antiderivatives. A good example is $f(x) = e^{x^2}$. An antiderivative of this function is

$$F(x) = \sum_{k=0}^{\infty} \frac{x^{2k+1}}{(2k+1)k!}$$

(See Equation (3) in Section 6.7 for the source of this equation.)

One powerful stratagem for computing the integral

$$\int_a^b f(x)\,dx \tag{1}$$

numerically is to replace f by another function g that approximates f well and is easily integrated. Then we simply say to ourselves that from $f \approx g$, it follows that

$$\int_a^b f(x)\,dx \approx \int_a^b g(x)\,dx$$

It should now occur to the reader that polynomials are good candidates for the function g, and indeed g can be a polynomial that *interpolates* to f at a certain set of nodes. Of course, a polynomial approximation to f can be obtained in other ways—for example, by truncating a Taylor series. Again, the reader should recall what was done along these lines in Section 6.7. Thus, for example,

$$\int_0^1 e^{x^2}\,dx \approx \int_0^1 \sum_{k=0}^{n} \frac{x^{2k}}{k!}\,dx = \sum_{k=0}^{n} \frac{1}{(2k+1)k!}$$

However, it is desirable to have general procedures that require only *evaluations* of the integrand. Methods based on interpolation fulfill this desideratum. Spline functions can also be used to interpolate f, and integrals of spline functions are easily computed.

Integration Via Polynomial Interpolation

We take up polynomial interpolation now. Suppose, then, that we want to evaluate the integral in (1). We can select nodes $x_0, x_1, \ldots, x_n$ in $[a, b]$ and set up a Lagrange interpolation process as in Section 6.1. Define

$$\ell_i(x) = \prod_{\substack{j=0\\ j\neq i}}^{n} \frac{x - x_j}{x_i - x_j} \qquad (0 \leqq i \leqq n)$$

These are the *fundamental polynomials* for interpolation. The polynomial of degree at most n that interpolates f at the nodes is

$$p(x) = \sum_{i=0}^{n} f(x_i)\ell_i(x) \tag{2}$$

Then, as mentioned previously, we simply write

$$\int_a^b f(x)\,dx \approx \int_a^b p(x)\,dx = \sum_{i=0}^{n} f(x_i) \int_a^b \ell_i(x)\,dx$$

In this way, we obtain a formula that can be used on *any* f. It reads as follows:

$$\int_a^b f(x)\,dx \approx \sum_{i=0}^{n} A_i f(x_i) \tag{3}$$

where

$$A_i = \int_a^b \ell_i(x)\,dx$$

A formula of the form (3) is called a **Newton-Cotes formula** if the nodes are equally spaced.

Trapezoid Rule

The simplest case results if $n = 1$ and the nodes are $x_0 = a$ and $x_1 = b$. The fundamental polynomials for interpolation are then

$$\ell_0(x) = \frac{b-x}{b-a} \qquad \ell_1(x) = \frac{x-a}{b-a}$$

Consequently,

$$A_0 = \int_a^b \ell_0(x)\,dx = \frac{1}{2}(b-a) = \int_a^b \ell_1(x)\,dx = A_1$$

The corresponding quadrature formula is

$$\int_a^b f(x)\,dx \approx \frac{b-a}{2}\big[f(a) + f(b)\big]$$

This is known as the **trapezoid rule**. It is exact for all $f \in \Pi_1$ (that is, polynomials of degree at most 1). Moreover, its error term is

$$-\frac{1}{12}(b-a)^3 f''(\xi)$$

where $\xi \in (a, b)$. This can be determined by integrating the error term in the polynomial approximation $f(x) - p_1(x) = f''(\xi_x)(x-a)(x-b)/2$, and using the Mean-Value Theorem for Integrals. We shall see the trapezoid rule again in Section 7.4 as a component of the Romberg integration method.

If the interval $[a, b]$ is partitioned like this:

$$a = x_0 < x_1 < \cdots < x_n = b$$

then the trapezoid rule can be applied to each subinterval. Here the nodes are not necessarily uniformly spaced. Thus, we obtain the **composite trapezoid rule**:

$$\int_a^b f(x)\,dx = \sum_{i=1}^{n} \int_{x_{i-1}}^{x_i} f(x)\,dx$$

$$\approx \frac{1}{2}\sum_{i=1}^{n}(x_i - x_{i-1})\big[f(x_{i-1}) + f(x_i)\big]$$

(A **composite rule** is one obtained by applying an integration formula for a single interval to each subinterval of a partitioned interval.) The composite trapezoid rule also arises if the integrand f is replaced by an interpolating spline function of degree 1—that is, a *broken line*.

With uniform spacing $h = (b-a)/n$ and $x_i = a+ih$, the **composite trapezoid rule** takes the form

$$\int_a^b f(x)\,dx \approx \frac{h}{2}\left[f(a) + 2\sum_{i=1}^{n-1} f(a+ih) + f(b)\right]$$

or

$$\int_a^b f(x)\,dx \approx h\sum_{i=0}^{n}{}'' f(a+ih) \tag{4}$$

where the double prime on the summation indicates that the first and last terms in the sum are to be halved. The error term for the composite trapezoid rule is

$$-\frac{1}{12}(b-a)h^2 f''(\xi)$$

where $\xi \in (a, b)$. It is obtained from summing the error term on each subinterval and using the fact that there is a point ξ in $[a, b]$ for which $f''(\xi) = (1/n)\sum_{i=1}^{n} f''(\xi_i)$, where $\xi_i \in (x_{i-1}, x_i)$ and $(1/n) = (b-a)/h$—that is, the average value.

Example 1 If we take $n = 2$ and $[a, b] = [0, 1]$ in the Newton-Cotes procedure, we obtain another formula:

$$\int_0^1 f(x)\,dx \approx \frac{1}{6}f(0) + \frac{2}{3}f\left(\frac{1}{2}\right) + \frac{1}{6}f(1) \tag{5}$$

Derive this rule by using Formula (3).

Solution The three fundamental polynomials for nodes 0, $\frac{1}{2}$, 1 are

$$\ell_0(x) = 2\left(x - \frac{1}{2}\right)(x-1) \qquad \ell_1(x) = -4x(x-1) \qquad \ell_2(x) = 2x\left(x - \frac{1}{2}\right)$$

Then

$$A_0 = \int_0^1 \ell_0(x)\,dx = \frac{1}{6}$$

and so forth. ■

Method of Undetermined Coefficients

From the manner in which Formula (3) was derived, we see at once that it is *exact* for all polynomials of degree $\leqq n$. On the other hand, suppose that Formula (3) is given to us, and we are told only that it is exact for all polynomials of degree $\leqq n$. Is the following equation true?

$$A_i = \int_a^b \ell_i(x)\,dx$$

The answer is *yes*, and we see this from the fact that Formula (3) should integrate any ℓ_j correctly. Hence,

$$\int_a^b \ell_j(x)\,dx = \sum_{i=0}^{n} A_i \ell_j(x_i) = A_j$$

Here, of course, we used two essential properties of the fundamental polynomials: ℓ_j is a polynomial of degree at most n and $\ell_i(x_j) = \delta_{ij}$.

The remarks just made enable us to arrive at formulas like (3) by the **method of undetermined coefficients**. To illustrate, let us derive Equation (5) in this way. We seek a formula

$$\int_0^1 f(x)\,dx \approx A_0 f(0) + A_1 f\left(\frac{1}{2}\right) + A_2 f(1)$$

that will be exact for all polynomials of degree $\leqq 2$. By using as trial functions the polynomials $f(x) = 1$, x, and x^2, in order, we get

$$\begin{aligned}
1 &= \int_0^1 dx &&= A_0 + A_1 + A_2 \\
\frac{1}{2} &= \int_0^1 x\,dx &&= \frac{1}{2}A_1 + A_2 \\
\frac{1}{3} &= \int_0^1 x^2\,dx &&= \frac{1}{4}A_1 + A_2
\end{aligned}$$

The solution of this system of three simultaneous equations is $A_0 = 1/6$, $A_1 = 2/3$, and $A_2 = 1/6$. Since the formula is linear, it will produce exact values of integrals for any quadratic polynomial, $f(x) = c_0 + c_1 x + c_2 x^2$.

Simpson's Rule

Similar calculations for an arbitrary interval $[a, b]$ lead to the familiar **Simpson's rule**:

$$\int_a^b f(x)\,dx \approx \frac{b-a}{6}\left[f(a) + 4f\left(\frac{a+b}{2}\right) + f(b)\right] \tag{6}$$

From the manner in which it was derived, we know that Simpson's rule is exact for all polynomials of degree $\leqq 2$. It is, unexpectedly, exact for all polynomials of degree $\leqq 3$. (See Problem 2.) The error term associated with Simpson's rule is

$$-\frac{1}{90}\left[(b-a)/2\right]^5 f^{(4)}(\xi)$$

for some $\xi \in (a, b)$. The derivation of this error term does *not* come from a straightforward integration of the error term in Theorem 2 of Section 6.1. Rather, we proceed as follows: If $h = (b-a)/2$, this numerical integration formula takes the form

$$\int_a^{a+2h} f(x)\,dx \approx \frac{h}{3}\left[f(a) + 4f(a+h) + f(a+2h)\right]$$

Using Taylor's Theorem from Section 1.1, we can write the right-hand side as

$$2hf(a) + 2h^2 f'(a) + \frac{4}{3}h^3 f''(a) + \frac{2}{3}h^4 f'''(a) + \frac{100}{3\cdot 5!}h^5 f^{(4)}(a) + \cdots$$

Now let

$$F(x) = \int_a^x f(t)\,dt$$

By the Fundamental Theorem of Calculus, $F' = f$, and with Taylor's Theorem we can write the left-hand side in Simpson's rule as $F(a+2h)$ or

$$2hf(a) + 2h^2 f'(a) + \frac{4}{3}h^3 f''(a) + \frac{2}{3}h^4 f'''(a) + \frac{32}{5!}h^5 f^{(4)}(a) + \cdots$$

Combining these two expansions, we have

$$\int_a^{a+2h} f(x)\,dx = \frac{h}{3}\left[f(a) + 4f(a+h) + f(a+2h)\right] - \frac{1}{90}h^5 f^{(4)}(a) - \cdots$$

and the error term follows.

A **composite Simpson's rule** using an even number of subintervals is often used. Let n be even, and set

$$x_i = a + ih \qquad h = (b-a)/n \qquad (0 \leqq i \leqq n)$$

Then

$$\int_a^b f(x)\,dx = \int_{x_0}^{x_2} f(x)\,dx + \int_{x_2}^{x_4} f(x)\,dx + \cdots + \int_{x_{n-2}}^{x_n} f(x)\,dx = \sum_{i=1}^{n/2} \int_{x_{2i-2}}^{x_{2i}} f(x)\,dx$$

Now let Simpson's rule (6) be applied to each subinterval, resulting in the formula

$$\int_a^b f(x)\,dx \approx \frac{h}{3} \sum_{i=1}^{n/2} \left[f(x_{2i-2}) + 4f(x_{2i-1}) + f(x_{2i}) \right]$$

The right-hand side of this formula should be computed as follows, to avoid repetition of terms:

$$\frac{h}{3} \left[f(x_0) + 2\sum_{i=2}^{n/2} f(x_{2i-2}) + 4\sum_{i=1}^{n/2} f(x_{2i-1}) + f(x_n) \right]$$

The error term for this formula is

$$-\frac{1}{180}(b-a)h^4 f^{(4)}(\xi)$$

for some $\xi \in (a, b)$.

General Integration Formulas

The procedure that led to the Newton-Cotes formulas can be used to produce more general integration formulas of the type

$$\int_a^b f(x)w(x)\,dx \approx \sum_{i=0}^{n} A_i f(x_i)$$

where w can be any fixed **weight function**. The only modification necessary is to replace Equation (3) by

$$A_i = \int_a^b \ell_i(x)w(x)\,dx$$

This is illustrated in the next example.

Example 2 Find a formula

$$\int_{-\pi}^{\pi} f(x)\cos x\,dx \approx A_0 f\left(-\frac{3}{4}\pi\right) + A_1 f\left(-\frac{1}{4}\pi\right) + A_2 f\left(\frac{1}{4}\pi\right) + A_3 f\left(\frac{3}{4}\pi\right)$$

that is exact when f is a polynomial of degree 3.

Solution Clearly, a polynomial of degree 3 is a linear combination of the four monomials 1, x, x^2, and x^3. Thus, we can determine the coefficients by substituting $f(x) = x^j$ $(0 \leqq j \leqq 3)$ and by solving the resulting four linear equations. By symmetry, $A_0 = A_3$ and $A_1 = A_2$; thus, the work can be simplified:

$$0 = \int_{-\pi}^{\pi} 1 \cos x \, dx = 2A_0 + 2A_1$$

$$-4\pi = \int_{-\pi}^{\pi} x^2 \cos x \, dx = 2A_0 \left(\frac{3}{4}\pi\right)^2 + 2A_1 \left(\frac{1}{4}\pi\right)^2$$

The solution is $A_1 = A_2 = -A_0 = -A_3 = 4/\pi$, and the formula is

$$\int_{-\pi}^{\pi} f(x) \cos x \, dx \approx \frac{4}{\pi}\left[-f\left(-\frac{3}{4}\pi\right) + f\left(-\frac{1}{4}\pi\right) + f\left(\frac{1}{4}\pi\right) - f\left(\frac{3}{4}\pi\right)\right] \quad \blacksquare$$

Change of Intervals

From a formula for numerical integration on one interval, we can derive a formula for any other interval by making a linear change of variable. If the first formula is exact for polynomials of a certain degree, the same will be true of the second. Let us see how this is accomplished.

Suppose that a numerical integration formula is given:

$$\int_c^d f(t) \, dt \approx \sum_{i=0}^{n} A_i f(t_i) \tag{7}$$

We do not care where this formula comes from; however, let us assume that it is exact for all polynomials of degree $\leqq m$. If a formula is needed for some other interval, say $[a, b]$, we first define a linear function λ of t such that if t traverses $[c, d]$, then $\lambda(t)$ will traverse $[a, b]$. The function λ is given explicitly by

$$\lambda(t) = \frac{b - a}{d - c} t + \frac{ad - bc}{d - c} \tag{8}$$

Now in the integral

$$\int_a^b f(x) \, dx$$

we make the change of variable $x = \lambda(t)$. Then $dx = \lambda'(t) \, dt = (b-a)(d-c)^{-1} \, dt$, and so we have

$$\begin{aligned} \int_a^b f(x) \, dx &= \frac{b - a}{d - c} \int_c^d f\big(\lambda(t)\big) \, dt \\ &\approx \frac{b - a}{d - c} \sum_{i=0}^{n} A_i f\big(\lambda(t_i)\big) \end{aligned}$$

Hence, we have

$$\int_a^b f(x)\,dx \approx \frac{b-a}{d-c}\sum_{i=0}^{n} A_i f\left(\frac{b-a}{d-c}t_i + \frac{ad-bc}{d-c}\right) \tag{9}$$

Observe that, because λ is linear, $f(\lambda(t))$ is a polynomial in t if f is a polynomial, and the degrees are the same. Hence, the new formula is exact for polynomials of degree m. For Simpson's rule, this procedure yields Equation (6) as a consequence of Equation (5) using $\lambda(t) = (b-a)t + a$.

Error Analysis

To assess the error involved in the numerical integration Formula (3), we use the error term in polynomial interpolation. Recall, from Equation (13) in Section 6.1, that if p is the polynomial of degree $\leqq n$ that interpolates f at $x_0, x_1, \ldots, x_n$, and if $f^{(n+1)}$ is continuous, then

$$f(x) - p(x) = \frac{1}{(n+1)!} f^{(n+1)}(\xi_x) \prod_{i=0}^{n} (x - x_i) \tag{10}$$

Consequently,

$$\int_a^b f(x)\,dx - \sum_{i=0}^{n} A_i f(x_i) = \frac{1}{(n+1)!}\int_a^b f^{(n+1)}(\xi_x)\prod_{i=0}^{n}(x - x_i)\,dx \tag{11}$$

If $|f^{(n+1)}(x)| \leqq M$ on $[a, b]$, then

$$\left|\int_a^b f(x)\,dx - \sum_{i=0}^{n} A_i f(x_i)\right| \leqq \frac{M}{(n+1)!}\int_a^b \prod_{i=0}^{n} |x - x_i|\,dx \tag{12}$$

The choice of nodes that makes the right-hand side of this inequality as small as possible is known to be

$$x_i = \frac{a+b}{2} + \frac{b-a}{2}\cos\left[\frac{(i+1)\pi}{n+2}\right] \qquad (0 \leqq i \leqq n)$$

If the interval is $[-1, 1]$, these nodes have the simpler form

$$x_i = \cos\left[\frac{(i+1)\pi}{n+2}\right] \qquad (0 \leqq i \leqq n)$$

These are the zeros of the function

$$U_{n+1}(x) = \frac{\sin[(n+2)\theta]}{\sin\theta} \qquad (x = \cos\theta) \tag{13}$$

The function U_{n+1} is known as a **Chebyshev polynomial of the second kind**. As defined in Equation (13), $U_{n+1}(x)$ is not a monic polynomial; that is, its leading coefficient is not 1. In fact, the coefficient of x^{n+1} in U_{n+1} is 2^{n+1}. Thus,

$$2^{-(n+1)}U_{n+1} = (x - x_0)(x - x_1)\cdots(x - x_n) \tag{14}$$

with $x_i = \cos[(i+1)\pi/(n+2)]$. A computation then shows that

$$\int_{-1}^{1} |(x - x_0)(x - x_1)\cdots(x - x_n)|\,dx = 2^{-n} \tag{15}$$

Thus, with these nodes and the proper coefficients A_i we have

$$\left| \int_{-1}^{1} f(x)\,dx - \sum_{i=0}^{n} A_i f(x_i) \right| \leqq \frac{M}{(n+1)!\,2^n} \tag{16}$$

THEOREM 1 *Among the monic polynomials p of degree n, the one for which $\int_{-1}^{1} |p(x)|\,dx$ is least is $2^{-n}U_n$.*

Proof First, we prove the orthogonality relation:

$$\int_{-1}^{1} U_m(x)\,\mathrm{sgn}\big[U_n(x)\big]\,dx = 0 \qquad (0 \leqq m < n)$$

Let I denote this integral. Make the change of variable $\cos\theta = x$ to get

$$\begin{aligned}
I &= \int_{0}^{\pi} \frac{\sin(m+1)\theta}{\sin\theta}\,\mathrm{sgn}\left[\frac{\sin(n+1)\theta}{\sin\theta}\right] \sin\theta\,d\theta \\
&= \sum_{k=0}^{n} (-1)^k \int_{k\varphi}^{(k+1)\varphi} \sin(m+1)\theta\,d\theta \qquad \varphi = \pi/(n+1) \\
&= (m+1)^{-1} \sum_{k=0}^{n} (-1)^{k+1}\big[\cos(m+1)(k+1)\varphi - \cos(m+1)k\varphi\big]
\end{aligned}$$

Let $\alpha = (m+1)\varphi + \pi$. Then (using $\mathrm{Re}(x)$ to signify real part of x) we have

$$\begin{aligned}
(m+1)I &= \sum_{k=0}^{n} \big[\cos(k+1)\alpha + \cos k\alpha\big] \\
&= \mathrm{Re}\left\{ \sum_{k=0}^{n} [e^{i\alpha(k+1)} + e^{i\alpha k}] \right\}
\end{aligned}$$

$$= \operatorname{Re}\left\{\frac{e^{i\alpha(n+2)} - e^{i\alpha} + e^{i\alpha n} - 1}{e^{i\alpha} - 1}\right\}$$

$$= \frac{\operatorname{Re}\left[(e^{-i\alpha} - 1)(e^{i\alpha(n+2)} - e^{i\alpha} + e^{i\alpha n} - 1)\right]}{|e^{i\alpha} - 1|^2}$$

Here the numerator is

$$\begin{aligned}\operatorname{Re}\left(e^{i\alpha n} - e^{i\alpha(n+2)}\right) &= \cos n\alpha - \cos(n+2)\alpha \\ &= \cos\left[(n+1)\alpha - \alpha\right] - \cos\left[(n+1)\alpha + \alpha\right] \\ &= \cos(k\pi - \alpha) - \cos(k\pi + \alpha) = 0\end{aligned}$$

where $k = m + n + 2$. To complete the proof, let p be a monic polynomial of degree n. It can be expressed as

$$p = 2^{-n}U_n + a_{n-1}U_{n-1} + \cdots + a_0U_0$$

Hence by the orthogonality relation,

$$\begin{aligned}\int_{-1}^{1} |p|\,dx &\geqq \int_{-1}^{1} p \operatorname{sgn}[U_n]\,dx = 2^{-n}\int_{-1}^{1} U_n \operatorname{sgn}[U_n]\,dx \\ &= 2^{-n}\int_{-1}^{1} |U_n|\,dx\end{aligned}$$

■

The subject of numerical integration can be studied further in Davis and Rabinowitz [1984], Krylov [1962], and Ghizetti and Ossiccini [1970].

PROBLEMS 7.2

1. Derive the Newton-Cotes formula for $\int_0^1 f(x)\,dx$ based on the nodes $0, \frac{1}{3}, \frac{2}{3}, 1$.
2. Prove (without using its error term) that Simpson's rule, Equation (6), correctly integrates all cubic polynomials.
3. Obtain Formula (6) from (5) by a suitable change of variable.
4. Verify that the following formula is exact for polynomials of degree $\leqq 4$:

$$\int_0^1 f(x)\,dx \approx \frac{1}{90}\left[7f(0) + 32f\left(\frac{1}{4}\right) + 12f\left(\frac{1}{2}\right) + 32f\left(\frac{3}{4}\right) + 7f(1)\right]$$

5. From the formula in Problem 4 obtain a formula for $\int_a^b f(x)\,dx$ that is exact for all polynomials of degree 4.
6. Calculate $\ln 2$ approximately by applying the formula in Problem 4 to

$$\int_0^1 \frac{dt}{t+1}$$

Compare your answer to the correct value and compute the error.

7. Calculate $\int_0^1 e^{x^2}\,dx$ to eight-decimal-place accuracy by use of the series in the text.

8. Find the formula

$$\int_0^1 f(x)\,dx \approx A_0 f(0) + A_1 f(1)$$

that is exact for all functions of the form $f(x) = ae^x + b\cos(\pi x/2)$.

9. Find a formula of the form

$$\int_0^{2\pi} f(x)\,dx = A_1 f(0) + A_2 f(\pi)$$

that is exact for any function having the form

$$f(x) = a + b\cos x$$

Prove that the resulting formula is exact for any function of the form

$$f(x) = \sum_{k=0}^{n} \left[a_k \cos(2k+1)x + b_k \sin kx\right]$$

10. Use the Lagrange interpolation polynomial to derive the formula of the form

$$\int_0^1 f(x)\,dx \approx Af\left(\frac{1}{3}\right) + Bf\left(\frac{2}{3}\right)$$

Transform the preceding formula to one for integration over $[a, b]$.

11. Using the polynomial of lowest order that interpolates $f(x)$ at x_1 and x_2, derive a numerical integration formula for

$$\int_{x_0}^{x_3} f(x)\,dx$$

Do not assume uniform spacing. Here $x_0 < x_1 < x_2 < x_3$.

12. Derive a formula for approximating

$$\int_1^3 f(x)\,dx$$

in terms of $f(0)$, $f(2)$, and $f(4)$. It should be exact for all f in Π_2.

13. Determine values for A, B, and C that make the formula

$$\int_0^2 x f(x)\,dx \approx Af(0) + Bf(1) + Cf(2)$$

exact for all polynomials of degree as high as possible. What is the maximum degree?

14. Derive the Newton-Cotes formula for

$$\int_0^1 f(x)\,dx$$

based on the Lagrange interpolation polynomial at the nodes -2, -1, and 0. Apply this result to evaluate the integral when $f(x) = \sin \pi x$.

15. Calculate

$$\int_0^{10^{-2}} \left(\frac{\sin x}{x}\right) dx$$

to seven-decimal-place accuracy using a series.

16. We intend to use $\int_0^1 p(x)\,dx$ as an estimate of $\int_0^1 f(x)\,dx$, where p is a polynomial of degree n that interpolates f at nodes $x_0, x_1, \dots, x_n$ in $[0, 1]$. Assume that $|f^{(n+1)}(x)| < M$ on $[0, 1]$. What upper bound can be given for the error if nothing is known about the location of the nodes? Can you find the best upper bound?

17. Determine the composite numerical integration rule based on the simple **right-side rectangle rule**:

$$\int_0^1 f(x)\,dx \approx f(1)$$

Assume unequal spacing $a = x_0 < x_1 < \cdots < x_n = b$.

18. Derive the composite rule for $\int_a^b f(x)\,dx$ based on the midpoint rule

$$\int_{-1}^1 f(x)\,dx \approx 2f(0)$$

Give formulas for unequal spacing and equal spacing of nodes.

19. (Continuation) The **midpoint rule** over the interval $[x_{i+1}, x_{i-1}]$ is given by

$$\int_{x_{i-1}}^{x_{i+1}} f(x)\,dx = (x_{i+1} - x_{i-1}) f(x_i)$$

Determine the composite midpoint rule over the interval $[a, b]$ with uniform spacing of $h = (b - a)/n$ such that $x_i = a + ih$ for $i = 0, 1, 2, \dots, n$ (n even).

20. Determine the integration rule for $\int_a^b f(x)\,dx$ based on the Gaussian quadrature rule

$$\int_{-1}^1 f(x)\,dx \approx f\left(-\frac{1}{\sqrt{3}}\right) + f\left(\frac{1}{\sqrt{3}}\right)$$

21. There are two Newton-Cotes formulas for $n = 2$ and $[a, b] = [0, 1]$; namely,

$$\int_0^1 f(x)\,dx \approx af(0) + bf\left(\frac{1}{2}\right) + cf(1)$$

$$\int_0^1 f(x)\,dx \approx \alpha f\left(\frac{1}{4}\right) + \beta f\left(\frac{1}{2}\right) + \gamma f\left(\frac{3}{4}\right)$$

Which is better?

22. Is there a formula of the form

$$\int_0^1 f(x)\,dx \approx \alpha\left[f(x_0) + f(x_1)\right]$$

that correctly integrates all quadratic polynomials?

23. Prove that if the formula

$$\int_{-1}^1 f(x)\,dx \approx \sum_{i=0}^n A_i f(x_i) \qquad (n \text{ even})$$

is exact for all polynomials of degree n and if the nodes are symmetrically placed about the origin, then the formula is exact for all polynomials of degree $n + 1$.

24. Let n be even. Show how the composite Simpson rule with $2n$ equally spaced nodes can be computed from the case of n equally spaced nodes with a minimum amount of additional work.

25. In Example 2, symmetry was used to simplify the calculations. Give a proof that $A_0 = A_3$ and $A_1 = A_2$.

26. Prove that the **Chebyshev polynomials** of the **second kind** are generated recursively by

$$\begin{cases} U_0(x) = 1 \qquad U_1(x) = 2x \\ U_{n+1} = 2xU_n - U_{n-1} \qquad (n \geqq 1) \end{cases}$$

27. (Continuation) Prove the orthogonality relation

$$\int_{-1}^1 U_n(x)U_m(x)\sqrt{1 - x^2}\,dx = \delta_{nm}\frac{\pi}{2}$$

28. (Continuation) Prove this relation between Chebyshev polynomials of the first and second kinds: $T_n' = nU_{n-1}$.

29. Establish error terms for the trapezoid rule and the composite trapezoid rule.

30. Fill in the details in establishing the error term for Simpson's rule and the composite Simpson's rule.

31. Determine the minimum number of subintervals needed to approximate

$$\int_1^2 (x + e^{-x^2})\,dx$$

to an accuracy of at least $\frac{1}{2} \times 10^{-7}$ using the trapezoid rule.

COMPUTER PROBLEMS 7.2

1. Write a computer program that computes $\int_0^x e^{-t^2}\,dt$ by summing an appropriate Taylor series until the individual terms fall below 10^{-8} in magnitude. Test your program by calculating the values of this integral for $x = 0.0, 0.1, 0.2, \ldots, 1.0$.

2. Write a computer program that estimates $\int_a^b f(x)\,dx$ by $\int_a^b S(x)\,dx$, where S is the natural cubic spline having knots $a + ih$ and interpolating f at these knots. Here $0 \leqq i \leqq n$ and $h = (b - a)/n$. First obtain a formula for

$$\int_{t_0}^{t_n} S(x)\,dx$$

starting with Equation (7) in Section 6.4. Then write the subprogram to compute this. Test your code on these well-known integrals:

(a) $\dfrac{4}{\pi}\displaystyle\int_0^1 (1 + x^2)^{-1}\,dx$

(b) $\dfrac{1}{\ln 3}\displaystyle\int_1^3 x^{-1}\,dx$

Use $n = 4, 8, 16$ in both examples. (See Computer Problem 3 in Section 6.4.)

3. Using a symbolic manipulation program, carry out the following exercises:

(a) Find the indefinite integral $\int (\cos x)^{-14}\,dx$.

(b) Find the definite integral $\int_0^1 \log(\log x)\,dx$.

(c) Obtain the numerical value of $\int_0^1 \sqrt{1 + \sin^3 x}\,dx$.

7.3 Gaussian Quadrature

In the preceding section, we saw how to create quadrature formulas of the type

$$\int_a^b f(x)\,dx \approx \sum_{i=0}^{n} A_i f(x_i) \tag{1}$$

that are *exact* for polynomials of degree $\leqq n$. In these formulas, the choice of nodes $x_0, x_1, \ldots, x_n$ was made *a priori*. Once the nodes were fixed, the coefficients were determined uniquely from the requirement that Formula (1) must be an equality for all $f \in \Pi_n$.

It is natural to ask whether some choices of nodes are better than others in Formula (1). For example, there may be a particular set of nodes for which the resulting coefficients A_i are all equal to each other. If $A_i = c$ for $0 \leqq i \leqq n$, then less arithmetic is involved in using Equation (1), since it now has this simpler form:

$$\int_a^b f(x)\,dx \approx c \sum_{i=0}^{n} f(x_i) \tag{2}$$

(This effects a reduction from $n+1$ to 1 in the number of multiplications.) Formulas of type (2) exist for only $n = 0, 1, 2, 3, 4, 5, 6$, and 8. They are known as **Chebyshev's quadrature formulas**. Here is the one corresponding to $n = 4$:

$$\int_{-1}^{1} f(x)\,dx \approx \frac{2}{5}\big[f(-\alpha) + f(-\beta) + f(0) + f(\beta) + f(\alpha)\big] \tag{3}$$

where

$$\alpha = \sqrt{\left(5 + \sqrt{11}\,\right)/12} \approx 0.83249\,74870\,00982$$

$$\beta = \sqrt{\left(5 - \sqrt{11}\,\right)/12} \approx 0.37454\,14095\,53581$$

The method of undetermined coefficients can be used to obtain the nodes α and β. Also, it can be shown that the formula is exact for polynomials of degree $\leqq 5$. Somewhat more complicated formulas with equal coefficients can also be given. As an example, we cite this one:

$$\int_{-1}^{1} f(x)\,dx \approx \frac{\pi}{n} \sum_{i=1}^{n} F\left(\cos \frac{2i-1}{2n}\pi\right) \tag{4}$$

in which $F(x) = f(x)\sqrt{1-x^2}$. This is known as **Hermite's quadrature formula**. It is exact for a certain linear space of dimension $2n$; namely,

$$G = \{pw \,:\, p \in \Pi_{2n-1}\} \qquad w(x) = \left(1 - x^2\right)^{-1/2}$$

It is not likely that any saving in computing effort will result when this formula is used. (Why?) On the other hand, with only n evaluations of f, we have a formula that is exact on a linear space of dimension $2n$. Pursuing this goal systematically leads us to the *Gaussian* quadrature formulas.

Gaussian Quadrature

The theory can be formulated for quadrature rules of a slightly more general form; namely,

$$\int_a^b f(x)w(x)\,dx \approx \sum_{i=0}^{n} A_i f(x_i) \tag{5}$$

where w is a fixed positive **weight function**. The case when $w(x) \equiv 1$ is, naturally, of special importance. Let us assume that Formula (5) is exact for $f \in \Pi_n$. As we learned in the preceding section, this is the case if and only if

$$A_i = \int_a^b w(x) \prod_{\substack{j=0 \\ j \neq i}}^{n} \frac{x - x_j}{x_i - x_j}\, dx \tag{6}$$

Since there are $n+1$ coefficients A_i and $n+1$ nodes x_i at our disposal with no *a priori* restrictions on them, we suspect that quadrature formulas of the form in Equation (5) can be discovered that will be exact for polynomials of degree $\leqq 2n + 1$. We now show that this is indeed true.

The idea, which originated with Carl Friedrich Gauss (1777–1855), is to use the variability of the nodes to force the quadrature Formulas (5) and (6) to be exact for all polynomials of degree $2n + 1$. Here is the theorem that discloses where the nodes should be placed.

THEOREM 1 *Let w be a positive weight function and let q be a nonzero polynomial of degree $n + 1$ that is w-**orthogonal** to Π_n in the sense that for any $p \in \Pi_n$, we have*

$$\int_a^b q(x)p(x)w(x)\, dx = 0 \tag{7}$$

If $x_0, x_1, \ldots, x_n$ are the zeros of q, then the quadrature Formula (5) with coefficients given by Equation (6) will be exact for all $f \in \Pi_{2n+1}$.

Proof Let $f \in \Pi_{2n+1}$. Divide f by q, obtaining a quotient p and remainder r. Then

$$f = qp + r \qquad (p, r \in \Pi_n)$$

Consequently, $f(x_i) = r(x_i)$. Using Equation (7) and the fact that Equation (5) is exact for elements of Π_n, we have

$$\int_a^b fw\, dx = \int_a^b rw\, dx = \sum_{i=0}^{n} A_i r(x_i) = \sum_{i=0}^{n} A_i f(x_i) \qquad \blacksquare$$

It turns out that the roots of q are *simple* roots and lie in the interior of the interval $[a, b]$. (In particular, they are real and not imaginary.) This follows immediately from the following theorem.

THEOREM 2 *Let w be a positive weight function in $C[a, b]$. Let f be a nonzero element of $C[a, b]$ that is w-orthogonal to Π_n. Then f changes sign at least $n + 1$ times on (a, b).*

Proof Since $1 \in \Pi_n$, we have $\int_a^b f(x)w(x)\, dx = 0$, and this shows that f changes sign at least once. Suppose that f changes sign only r times, with $r \leqq n$. Choose points t_i so that

$$a = t_0 < t_1 < t_2 < \cdots < t_r < t_{r+1} = b$$

and so that f is of one sign on each interval

$$(t_0, t_1), (t_1, t_2), \ldots, (t_r, t_{r+1})$$

The polynomial

$$p(x) = \prod_{i=1}^{r} (x - t_i)$$

has the same sign property, and therefore $\int_a^b f(x)p(x)w(x)\,dx \neq 0$. Since $p \in \Pi_n$, this is a contradiction. ■

If the weight function is $w(x) = 1$ and if the interval is $[-1, 1]$, we have the situation originally investigated by Gauss. Here are two of the formulas for this case, corresponding to $n = 1$ and $n = 4$:

$$\int_{-1}^{1} f(x)\,dx \approx f(-\alpha) + f(\alpha) \tag{8}$$

where $\alpha = 1/\sqrt{3}$ and

$$\int_{-1}^{1} f(x)\,dx \approx A_0 f(x_0) + A_1 f(x_1) + \cdots + A_4 f(x_4) \tag{9}$$

where

$$-x_0 = x_4 = \frac{1}{3}\sqrt{5 + 2\sqrt{10/7}} \approx 0.90617\,98459\,38664$$

$$-x_1 = x_3 = \frac{1}{3}\sqrt{5 - 2\sqrt{10/7}} \approx 0.53846\,93101\,05683$$

$$x_2 = 0.0$$

$$A_0 = A_4 = 0.3\big(0.7 + 5\sqrt{0.7}\big)/\big(2 + 5\sqrt{0.7}\big) \approx 0.23692\,68850\,56189$$

$$A_1 = A_3 = 0.3\big(-0.7 + 5\sqrt{0.7}\big)/\big(-2 + 5\sqrt{0.7}\big) \approx 0.47862\,86704\,99366$$

$$A_2 = 128/225 \approx 0.56888\,88888\,88889$$

By using the exact nodes and coefficients, we can compute these numbers to any desired precision. In fact, the nodes x_i are roots of the Legendre polynomials $p_2(x) = x^3 - \frac{1}{3}x$ and $p_5(x) = x^5 - \frac{10}{9}x^3 + \frac{5}{21}x$, respectively, which were determined in Example 2 of Section 6.8.

The nodes x_i and coefficients A_i for many integration formulas of various types are found in handbooks such as Abramowitz and Stegun [1964]. The tables found there cover values of n from 1 to 9 as well as $n = 11, 15, 19, 23, 31, 39, 47, 63, 79, 95$. For well-behaved integrands, modest accuracy is obtained with only a few function evaluations using Gaussian formulas, and even more accuracy can be obtained by using some of the high-order ones.

A short pseudocode is presented for approximating the integral

$$\int_a^b f(x)\,dx$$

by incorporating the preceding five-point Gaussian integration formula. Here we change intervals as described in Equation (9) of Section 7.2 and exploit the symmetry in Equation (8):

input
  $x_0 \leftarrow 0$
  $x_1 \leftarrow 0.53846\,93101\,05683$
  $x_2 \leftarrow 0.90617\,98459\,38664$
  $A_0 \leftarrow 0.56888\,88888\,88889$
  $A_1 \leftarrow 0.47862\,86704\,99366$
  $A_2 \leftarrow 0.23692\,68850\,56189$
$u \leftarrow (a + b)/2$
$S \leftarrow A_0 f(u)$
for $i = 1$ **to** 2 **do**
  $u \leftarrow ((b - a)x_i + a + b)/2$
  $v \leftarrow ((a - b)x_i + a + b)/2$
  $S \leftarrow S + A_i\left[f(u) + f(v)\right]$
end do
$S \leftarrow (b - a)S/2$
output S

The computation of the coefficients A_i in a Gaussian formula proceeds in the manner already indicated for non-Gaussian formulas once the nodes x_i have been determined. The nodes are, in turn, the roots of a certain polynomial. This polynomial depends on $n + 1$, and we now denote it by q_{n+1}. The polynomial q_{n+1} is uniquely defined by two conditions:

(i) q_{n+1} is a monic polynomial of degree $n + 1$.
(ii) q_{n+1} is w-orthogonal to Π_n.

The word **monic** means that in q_{n+1}, the coefficient of x^{n+1} is unity. The condition of w-**orthogonality** is that

$$\int_a^b q_{n+1}(x)p(x)w(x)\,dx = 0 \quad \text{for all} \quad p \in \Pi_n$$

These so-called **orthogonal polynomials** are useful in various branches of mathematics, and much is known about them. They can be generated by a recursive algorithm, as discussed in Section 6.8.

Example 1 Find the Gaussian quadrature rule when $[a, b] = [-1, 1]$, $w(x) = 1$, and $n = 2$.

Solution From Theorem 5 in Section 6.8, the orthogonal polynomials can be determined recursively as was done in Example 2 in that section. Hence,

$$q_0(x) = 1$$
$$q_1(x) = x$$
$$q_2(x) = x^2 - \frac{1}{3}$$
$$q_3(x) = x^3 - \frac{3}{5}x$$

The roots 0 and $\pm\sqrt{3/5}$ of q_3 are the nodes in the quadrature formula

$$\int_{-1}^{1} f(x)\,dx \approx \frac{5}{9}f\left(-\sqrt{\frac{3}{5}}\right) + \frac{8}{9}f(0) + \frac{5}{9}f\left(\sqrt{\frac{3}{5}}\right)$$

The constants $5/9$ and $8/9$ can be found by the method of undetermined coefficients. (See Problem 21.) ■

Convergence and Error Analysis

Some of the favorable characteristics of Gaussian quadrature formulas will now be discussed.

LEMMA 1 *In a Gaussian quadrature formula, the coefficients are positive, and moreover their sum is* $\int_a^b w(x)\,dx$.

Proof Fix n, and let q be a polynomial of degree $n+1$ that is w-orthogonal to Π_n, as in Theorem 1. The zeros of q are denoted by $x_0, x_1, \ldots, x_n$, and these are the nodes in the Gaussian formula (5). Let p be the polynomial $q(x)/(x - x_j)$ for some fixed j. Since p^2 is of degree at most $2n$, the Gaussian formula will be exact for it. Consequently,

$$0 < \int_a^b p^2(x)w(x)\,dx = \sum_{i=0}^{n} A_i p^2(x_i) = A_j p^2(x_j)$$

From this we conclude that $A_j > 0$. Since the Gaussian formula is exact for $f(x) \equiv 1$,

$$\int_a^b w(x)\,dx = \sum_{i=0}^{n} A_i$$ ■

A remarkable theorem of Stieltjes establishes the convergence of the Gaussian formulas when $n \longrightarrow \infty$. Fixing w and $[a, b]$, we have a Gaussian quadrature formula for each $n \in \{0, 1, 2, \ldots\}$. These we write now as

$$\int_a^b f(x)w(x)\,dx \approx \sum_{i=0}^{n} A_{ni} f(x_{ni}) \qquad (n \geqq 0) \tag{10}$$

THEOREM 3 *If f is continuous on $[a, b]$, then the approximations in Equation (10) converge to the integral as $n \to \infty$.*

Proof Let $\varepsilon > 0$. By the Weierstrass Approximation Theorem (Theorem 8 of Section 6.1), there is a polynomial p for which $|f(x) - p(x)| < \varepsilon$ on $[a, b]$. For any n such that $2n$ exceeds the degree of p, the nth Gaussian formula will integrate p correctly. Using elementary inequalities and Lemma 1, we have

$$\begin{aligned}
&\left| \int_a^b f(x)w(x)\,dx - \sum_{i=0}^{n} A_{ni} f(x_{ni}) \right| \\
&\quad \leqq \left| \int_a^b f(x)w(x)\,dx - \int_a^b p(x)w(x)\,dx \right| + \left| \sum_{i=0}^{n} A_{ni} p(x_{ni}) - \sum_{i=0}^{n} A_{ni} f(x_{ni}) \right| \\
&\quad \leqq \int_a^b |f(x) - p(x)| w(x)\,dx + \sum_{i=0}^{n} A_{ni} |p(x_{ni}) - f(x_{ni})| \\
&\quad \leqq \varepsilon \int_a^b w(x)\,dx + \varepsilon \sum_{i=0}^{n} A_{ni} = 2\varepsilon \int_a^b w(x)\,dx
\end{aligned}$$

■

THEOREM 4 *Consider a Gaussian formula with error term:*

$$\int_a^b f(x)w(x)\,dx = \sum_{i=0}^{n-1} A_i f(x_i) + E$$

For an $f \in C^{2n}[a, b]$, we have

$$E = \frac{f^{(2n)}(\xi)}{(2n)!} \int_a^b q^2(x)w(x)\,dx$$

where $a < \xi < b$ and $q(x) = \prod_{i=0}^{n-1}(x - x_i)$.

Proof By Hermite interpolation (Section 6.3), there is a polynomial p of degree at most $2n - 1$ such that

$$p(x_i) = f(x_i) \qquad p'(x_i) = f'(x_i) \qquad (0 \leqq i \leqq n - 1)$$

The error formula for this interpolation is

$$f(x) - p(x) = f^{(2n)}\big(\zeta(x)\big) q^2(x) / (2n)! \tag{11}$$

as given in Theorem 2 of Section 6.3. It follows that

$$\int_a^b f(x)w(x)\,dx - \int_a^b p(x)w(x)\,dx = \frac{1}{(2n)!}\int_a^b f^{(2n)}\big(\zeta(x)\big)q^2(x)w(x)\,dx$$

Now use the fact that p is of degree at most $2n-1$ to see that

$$\int_a^b p(x)w(x)\,dx = \sum_{i=0}^{n-1} A_i p(x_i) = \sum_{i=0}^{n-1} A_i f(x_i)$$

Furthermore, the Mean-Value Theorem for Integrals can be used to write

$$\int_a^b f^{(2n)}\big(\zeta(x)\big)q^2(x)w(x)\,dx = f^{(2n)}(\xi)\int_a^b q^2(x)w(x)\,dx$$

This requires continuity of $f^{(2n)}(\zeta(x))$, which can be inferred from Equation (11). The desired error formula results from making the substitution described. ■

Further references for Gaussian quadrature are Krylov [1962], Davis and Rabinowitz [1956, 1984], Stroud and Secrest [1966], and Abramowitz and Stegun [1956, 1964].

PROBLEMS 7.3

1. In Equation (2) let $n = 1$, $a = 0$, $b = 1$. Find all cases of the formula that are exact for $f \in \Pi_3$.
2. Solve Problem 1 with $n = 2$.
3. Derive the following Gaussian quadrature rule:

$$\int_{-1}^{1} f(x)\,dx \approx A_0 f(x_0) + A_1 f(x_1) + A_2 f(x_2) + A_3 f(x_3)$$

where

$$A_0 = A_2 = \frac{1}{2}\left(1 + \frac{1}{6}\sqrt{10/3}\right) \approx 0.34785\,48451\,37454$$

$$A_1 = A_3 = \frac{1}{2}\left(1 - \frac{1}{6}\sqrt{10/3}\right) \approx 0.65214\,51548\,62546$$

$$-x_0 = x_2 = \sqrt{\frac{1}{7}(3 - 4\sqrt{0.3})} \approx 0.86113\,63115\,94052$$

$$-x_1 = x_3 = \sqrt{\frac{1}{7}(3 + 4\sqrt{0.3})} \approx 0.33998\,10435\,848456$$

4. Verify the correctness of Equation (8) as a Gaussian formula.

5. Verify that the nodes in Equation (9) are correct.

6. We can determine a polynomial q of degree $n+1$ that is orthogonal to Π_n by writing

$$q(x) = x^{n+1} + c_1 x^n + \cdots + c_{n+1}$$

and imposing the conditions $\int_a^b q(x)x^k w(x)\,dx = 0$ for $0 \leqq k \leqq n$. The resulting system of $n+1$ equations in the $n+1$ unknowns $c_1, c_2, \ldots, c_{n+1}$ can then be solved. Carry out this process to obtain q_5 as needed in Problem 3. Do you think that this is a good way to obtain q?

7. **(a)** Find a formula of the form

$$\int_0^1 x f(x)\,dx \approx \sum_{i=0}^{n} A_i f(x_i)$$

with $n = 1$ that is exact for all polynomials of degree 3.

(b) Repeat with $n = 2$, making the formula exact on Π_5.

8. **(a)** Determine appropriate values of A_i and x_i so that the quadrature formula

$$\int_{-1}^1 x^2 f(x)\,dx \approx \sum_{i=0}^{n} A_i f(x_i)$$

will be correct when f is any polynomial of degree 3. Use $n = 1$.

(b) Repeat for f any polynomial of degree 5, using $n = 2$.

9. Find a quadrature formula

$$\int_{-1}^1 f(x)\,dx \approx c \sum_{i=0}^{2} f(x_i)$$

that is exact for all quadratic polynomials.

10. **(a)** If the integration formula

$$\int_{-1}^1 f(x)\,dx \approx f(\alpha) + f(-\alpha)$$

is to be exact for all quadratic polynomials, what value of α should be used? Answer the same question for all cubic polynomials.

Repeat part **(a)** for polynomials of the forms:

(b) $f(x) = a + bx + cx^3 + dx^4$

(c) $f(x) = a + \sum_{i=1}^{n} b_i x^{2i-1} + cx^{2n}$

11. For what value of α is this formula exact on Π_3?

$$\int_0^2 f(x)\,dx \approx f(\alpha) + f(2-\alpha)$$

12. Prove that if the interval is symmetric with respect to the origin and if w is an even function, then the Gaussian nodes will be symmetric and $A_i = A_{n-i}$ for $0 \leqq i \leqq n$.

13. Prove that every quadrature formula of the type

$$\int_a^b f(x)\,dx \approx \sum_{i=0}^{n} A_i f(x_i)$$

is exact on some infinite-dimensional subspace of $C[a, b]$.

14. Prove the following converse of Theorem 1:

Theorem *If*

$$\int_a^b f(x)w(x)\,dx = \sum_{i=0}^{n} A_i f(x_i)$$

for all $f \in \Pi_{2n+1}$, *then the polynomial* $\prod_{i=0}^{n}(x - x_i)$ *is orthogonal to* Π_n *on* $[a, b]$ *with respect to* w.

15. Determine the coefficients A_0, A_1, and A_2 that make the formula

$$\int_0^2 f(x)\,dx \approx A_0 f(0) + A_1 f(1) + A_2 f(2)$$

exact for all polynomials of degree 3.

16. Using only $f(0)$, $f'(-1)$, and $f''(1)$, compute an approximation to $\int_{-1}^{1} f(x)\,dx$ that is exact for all quadratic polynomials. Is the approximation exact for polynomials of degree 3? Why or why not?

17. If the formula

$$\int_0^1 x f(x)\,dx = \sum_{i=0}^{4} A_i f(x_i)$$

is correct for all $f \in \Pi_9$ (ninth-degree polynomials), then $x_0, x_1, \ldots, x_4$ must be the zeros of a fifth-degree polynomial q that has what properties?

18. If the formula

$$\int_1^2 (x^4 - 1) f(x)\,dx = A f(x_0) + B f(x_1) + C f(x_2)$$

is correct for all f that are polynomials of degree $\leqq 5$, then x_0, x_1, and x_2 must be roots of a polynomial q having what properties?

19. Which of these polynomials is orthogonal to Π_2 on the interval $[0, 1]$ with weight function $w(x) = 1$?

(i) $1 + x$

(ii) $x - \frac{1}{2}$

(iii) $x^2 - 3x + 1$

(iv) $35x^4 - 60x^2 + 32x - 3$

(v) $x^3 - 3x^2 + x - 1$

20. Find a nonzero polynomial that is orthogonal to Π_2 on the interval $[-1, 1]$ with respect to the weight function $1 + x^2$.

21. Consider a numerical integration rule of the form

$$\int_{-1}^{1} f(x)\,dx \approx Af\left(-\sqrt{\frac{3}{5}}\right) + Bf(0) + Cf\left(\sqrt{\frac{3}{5}}\right)$$

(a) What is the linear system that must be solved in the method of undetermined coefficients for finding A, B, and C? Solve for A, B, C.

(b) What three integrals must be evaluated in order to determine A, B, and C in a Newton-Cotes formula? Solve for A, B, C.

22. Show how the Gaussian quadrature rule

$$\int_{-1}^{1} f(x)\,dx \approx \frac{5}{9}f\left(-\sqrt{\frac{3}{5}}\right) + \frac{8}{9}f(0) + \frac{5}{9}f\left(\sqrt{\frac{3}{5}}\right)$$

can be used for $\int_a^b f(x)\,dx$. Apply this result to evaluate:

(a) $\displaystyle\int_0^{\pi/2} x\,dx$

(b) $\displaystyle\int_0^4 \frac{\sin t}{t}\,dt$

23. Given $f(0)$, $f'(-1)$, and $f''(1)$, compute an approximation to $\int_{-1}^{1} x^2 f(x)\,dx$ by the method of undetermined coefficients. Your formula should give exact results for all f in Π_2.

24. Find A, B, and C so that the numerical integration rule of the form

$$\int_{-1}^{1} xf(x)\,dx \approx Af(-1) + Bf(0) + Cf(+1)$$

is exact for all polynomials of maximum degree m. What is m?

25. Using the method of undetermined coefficients, find A, B, and C in the following rule, which should give exact results for polynomials of degree 2:

$$\int_{-3h}^{h} f(x)\,dx \approx h\big[Af(0) + Bf(-h) + Cf(-2h)\big]$$

26. (Continuation) How can a numerical integration rule of the form given in Problem 25 be used to obtain a rule for solving an ordinary differential equation of the following form?

$$\begin{cases} x' = f(t, x) \\ x(t_0) = x_0 \end{cases}$$

27. Find the coefficients and nodes of a Gaussian quadrature formula of the form

$$\int_0^1 x^4 f(x)\,dx \approx A_0 f(x_0) + A_1 f(x_1)$$

28. Prove that no Gaussian quadrature formula with n nodes can be exact on Π_{2n}.

29. What happens in the theory of Gaussian quadrature if the function w is not positive?

30. Use information from Section 7.2 to find the Gaussian formulas pertaining to the interval $[-1, 1]$ and the weight function $w(x) = \sqrt{1 - x^2}$. Problems 26 and 27 in Section 7.2 may be helpful.

31. Determine the nodes and weights for the Gaussian formula of the form

$$\int_{-1}^1 x^4 f(x)\,dx \approx A_0 f(x_0) + A_1 f(x_1)$$

32. Derive the exact values for the nodes and weights for each of the following formulas. Also determine the highest degree polynomial for which the formula is exact.

(a) Equation (3) **(b)** Equation (8) **(c)** Equation (9)

COMPUTER PROBLEM 7.3

1. Write a computer program to test whether Hermite's quadrature formula, Equation (4), with $n = 4$ is exact for all functions $p(x)(1 - x^2)^{-1/2}$, where $p \in \Pi_9$. It suffices to test the formula for the ten test functions $T_k(x)(1 - x^2)^{-1/2}$, where T_k is the kth Chebyshev polynomial defined by $T_k(x) = \cos(k \cos^{-1} x)$.

7.4 Romberg Integration

The algorithm to be described now goes by the name of Romberg, who first gave a recursive form for the method. We suppose that an approximate value is required for the integral

$$I = \int_a^b f(x)\,dx \tag{1}$$

The function f and the interval $[a, b]$ will remain fixed in the discussion.

Recursive Trapezoid Rule

Let the trapezoid rule for I using n subintervals of width $h = (b - a)/n$ be denoted by $T(n)$. From Equation (4) in Section 7.2, we have

$$T(n) = h \sum_{i=0}^{n}{}'' f(a + ih) = \frac{(b-a)}{n} \sum_{i=0}^{n}{}'' f\left(a + i\frac{(b-a)}{n}\right) \tag{2}$$

Here, the double prime on the summation sign signifies that the first and last terms are to be *halved*.

Example 1 What are the explicit formulas for $T(1)$, $T(2)$, $T(4)$, and $T(8)$ in the case when the interval is $[0, 1]$?

Solution Using Equation (2), we have

$$T(1) = \frac{1}{2}f(0) + \frac{1}{2}f(1)$$

$$T(2) = \frac{1}{4}f(0) + \frac{1}{2}\left[f\left(\frac{1}{2}\right)\right] + \frac{1}{4}f(1)$$

$$T(4) = \frac{1}{8}f(0) + \frac{1}{4}\left[f\left(\frac{1}{4}\right) + f\left(\frac{1}{2}\right) + f\left(\frac{3}{4}\right)\right] + \frac{1}{8}f(1)$$

$$T(8) = \frac{1}{16}f(0) + \frac{1}{8}\left[f\left(\frac{1}{8}\right) + f\left(\frac{1}{4}\right) + f\left(\frac{3}{8}\right) + f\left(\frac{1}{2}\right) + f\left(\frac{5}{8}\right) + f\left(\frac{3}{4}\right) + f\left(\frac{7}{8}\right)\right] + \frac{1}{16}f(1)$$

■

It is clear that if $T(2n)$ is to be computed, then we can take advantage of the work already done in the computation of $T(n)$. Only the terms present in $T(2n)$ but *not already present* in $T(n)$ must be computed. For example, from the preceding equations, we see that

$$T(2) = \frac{1}{2}T(1) + \frac{1}{2}\left[f\left(\frac{1}{2}\right)\right]$$

$$T(4) = \frac{1}{2}T(2) + \frac{1}{4}\left[f\left(\frac{1}{4}\right) + f\left(\frac{3}{4}\right)\right]$$

$$T(8) = \frac{1}{2}T(4) + \frac{1}{8}\left[f\left(\frac{1}{8}\right) + f\left(\frac{3}{8}\right) + f\left(\frac{5}{8}\right) + f\left(\frac{7}{8}\right)\right]$$

With $h = (b - a)/(2n)$, the general formula pertaining to any interval $[a, b]$ is as follows:

$$T(2n) = \frac{1}{2}T(n) + h\big[f(a + h) + f(a + 3h) + f(a + 5h) + \cdots + f(a + (2n - 1)h)\big]$$

or

$$T(2n) = \frac{1}{2}T(n) + h\sum_{i=1}^{n} f\big(a + (2i - 1)h\big) \tag{3}$$

We prove Equation (3) by first writing $T(2n)$ in terms of $T(n)$:

$$T(2n) = \frac{1}{2}T(n) + \left[T(2n) - \frac{1}{2}T(n)\right]$$

By using Equation (2), we can rewrite the bracketed expression in the form

$$T(2n) - \frac{1}{2}T(n) = h\sum_{i=0}^{2n}{}'' f(a + ih) - h\sum_{i=0}^{n}{}'' f(a + 2ih) = h\sum_{i=1}^{n} f\big(a + (2i - 1)h\big)$$

Since each term in the second summation is of the form $f(a + 2ih)$, the second summation cancels all even-numbered terms in the first summation. In the resulting sum, the range of i is obtained by noting that in the first summation the first odd term is $f(a + ih)$ so we want $2i - 1 = 1$, and the last odd term is $f(a + (2n - 1)h)$ so we want $2i - 1 = 2n - 1$. Hence, Equation (3) is established.

The integral I, Equation (1), now can be approximated by a recursive trapezoid rule. If there are 2^n uniform subintervals, then Equation (3) provides a recursive trapezoid rule:

$$T(2^n) = \frac{1}{2}T(2^{n-1}) + h_n\sum_{i=1}^{2^{n-1}} f\big(a + (2i - 1)h_n\big)$$

where

$$h_0 = b - a \qquad h_n = h_{n-1}/2 \qquad (n \geqq 1)$$

Romberg Algorithm

This formula is used in the Romberg algorithm. Letting $R(n, 0)$ denote the trapezoid estimate with 2^n subintervals, we have

$$\begin{cases} R(0,0) = \dfrac{1}{2}(b-a)[f(a)+f(b)] \\ R(n,0) = \dfrac{1}{2}R(n-1,0) + h_n \displaystyle\sum_{i=1}^{2^{n-1}} f\big(a+(2i-1)h_n\big) \end{cases} \tag{4}$$

The estimates $R(0, 0), R(1, 0), R(2, 0), \dots, R(M, 0)$ are computed for a modest value of M, and there are no duplicate function evaluations. In the remainder of the Romberg algorithm, additional quantities $R(n, m)$ are to be computed. All of these can be interpreted as estimates of the integral I. Further evaluations of the integrand f are not necessary after the element $R(M, 0)$ has been computed. The subsequent columns of the R-array for $n \geqq 1$ and $m \geqq 1$ are constructed from the formula

$$R(n,m) = R(n,m-1) + \frac{1}{4^m-1}\big[R(n,m-1) - R(n-1,m-1)\big] \tag{5}$$

This calculation is very simple. It is used to provide a final array of the form

$$\begin{array}{lllllll}
R(0,0) & & & & & & \\
R(1,0) & R(1,1) & & & & & \\
R(2,0) & R(2,1) & R(2,2) & & & & \\
R(3,0) & R(3,1) & R(3,2) & R(3,3) & & & \\
R(4,0) & R(4,1) & R(4,2) & R(4,3) & R(4,4) & & \\
\vdots & \vdots & \vdots & \vdots & \vdots & \ddots & \\
R(M,0) & R(M,1) & R(M,2) & R(M,3) & R(M,4) & \cdots & R(M,M)
\end{array}$$

Here is the pseudocode for the Romberg algorithm computed row-wise:

```
input a, b, M
h ← b − a
R(0, 0) ← ½(b − a)[f(a) + f(b)]
for n = 1 to M do
    h ← h/2
    R(n, 0) ← ½R(n − 1, 0) + h Σ_{i=1}^{2^{n−1}} f(a + (2i − 1)h)
    for m = 1 to n do
        R(n, m) ← R(n, m−1) + [R(n, m−1) − R(n−1, m−1)]/(4^m − 1)
    end do
end do
output R(n, m)   (0 ≦ n ≦ M, 0 ≦ m ≦ n)
```

Usually, only a moderate value of M is needed because the number of function values required is $2^M + 1$. A more sophisticated algorithm might include a procedure for stopping the calculations automatically when a certain prescribed error criterion is met.

Analysis

To explain the origin of Equation (5), we start with the **Euler-Maclaurin formula**:

$$\int_0^1 f(t)\,dt = \frac{1}{2}\big[f(0) + f(1)\big] + \sum_{k=1}^{m-1} A_{2k}\big[f^{(2k-1)}(0) - f^{(2k-1)}(1)\big] - A_{2m} f^{(2m)}(\xi_0) \tag{6}$$

where ξ_0 is some point between 0 and 1. It can be shown that Equation (6) is valid for any function $f \in C^{2m}[0, 1]$. A proof of this important formula is given in Section 7.7. The constants $k!\,A_k$ are known as **Bernoulli numbers**. They can be defined by the equation

$$\frac{x}{e^x - 1} = \sum_{k=0}^{\infty} A_k x^k$$

From the basic formula (6), we can obtain the following one in two steps: Apply Equation (6) to the function g defined by $g(t) = f(x_i + ht)$ with $h = x_{i+1} - x_i$ and, in the resulting integral, make the change of variable $t = (x - x_i)/h$:

$$\int_{x_i}^{x_{i+1}} f(x)\,dx = \frac{h}{2}\big[f(x_i) + f(x_{i+1})\big] + \sum_{k=1}^{m-1} A_{2k} h^{2k}\big[f^{(2k-1)}(x_i) - f^{(2k-1)}(x_{i+1})\big] - A_{2m} h^{2m+1} f^{(2m)}(\xi_i) \tag{7}$$

In Equation (7), we apply the operation $\sum_{i=0}^{2^n-1}$ to both sides. If $x_i = a + ih$ for $0 \leqq i \leqq 2^n$ and $h = (b - a)/2^n$, then

$$\int_a^b f(x)\,dx = \frac{h}{2}\sum_{i=0}^{2^n-1}\big[f(x_i) + f(x_{i+1})\big] + \sum_{k=1}^{m-1} A_{2k} h^{2k}\big[f^{(2k-1)}(a) - f^{(2k-1)}(b)\big] - A_{2m}(b - a) h^{2m} f^{(2m)}(\xi) \tag{8}$$

for some $\xi \in (a, b)$. In obtaining this equation, the error term has been handled as described in Problem 2. Equation (8) is interpreted as follows: The first term on the right is the trapezoidal estimate of I, Equation (1), using subintervals of size $h = (b - a)/2^n$. Thus, Equation (8) tells us that

$$I = R(n, 0) + c_2 h^2 + c_4 h^4 + c_6 h^6 + \cdots + c_{2m-2} h^{2m-2} + c_{2m} h^{2m} f^{(2m)}(\xi) \tag{9}$$

where $f \in C^{2m}[a, b]$ and $\xi \in (a, b)$. This equation is valid for arbitrary h, and the coefficients $c_2, c_4, \ldots, c_{2m}$ are independent of h. Since Equation (9) is a special case of Equation (15) of Section 7.1, we immediately have Equation (4) by the Richardson extrapolation analysis in Section 7.1.

To take full advantage of the Romberg algorithm, we shall want the function f to belong to class $C^{2m}[a, b]$, with as high a value of m as possible. This hypothesis justifies the use of the Euler-Maclaurin formula and implies that the quantities $R(n, m)$ converge to the integral of f with an error that is $\mathcal{O}(h^{2m})$, where $h = (b - a)2^{-n}$. What happens if f is assumed to be only continuous?

THEOREM 1 *If $f \in C[a, b]$, then each column in the Romberg array converges to the integral of f. Thus for each m,*

$$\lim_{n \to \infty} R(n, m) = \int_a^b f(x)\, dx$$

Proof We begin with the first column, which contains trapezoidal estimates of the integral, I. The trapezoid rule with k subintervals can be written in the form

$$h \sum_{i=0}^{k}{}'' f(a + ih) = \frac{1}{2} h \sum_{i=0}^{k-1} f(a + ih) + \frac{1}{2} h \sum_{i=1}^{k} f(a + ih)$$

The right side of this equation represents the average of two Riemann sums for I. Since $h = (b - a)/k$, the maximum width of subintervals converges to 0 as $k \to \infty$. Hence, by the theory of the Riemann integral, both Riemann sums converge to I. Of course, their average also converges to I. This proves that $\lim_{n \to \infty} R(n, 0) = I$. As for the second column, we note that

$$R(n, 1) = \frac{4}{3} R(n, 0) - \frac{1}{3} R(n - 1, 0)$$

from which we get

$$\lim_{n \to \infty} R(n, 1) = \frac{4}{3} I - \frac{1}{3} I = I$$

All subsequent columns can be analyzed in the same way. ∎

PROBLEMS 7.4

1. Derive Equation (7) starting with Equation (6).
2. Derive Equation (8) from Equation (7) and, in particular, justify the conversion from

$$h^{2m+1} \sum_{i=0}^{2^n - 1} f^{(2m)}(\xi_i) \qquad \text{to} \qquad (b - a) h^{2m} f^{(2m)}(\xi)$$

3. Establish the following equation in which $h = 1/2^n$:

$$I = \frac{4}{3}T\left(f, \frac{h}{2}\right) - \frac{1}{3}T(f, h) - \sum_{n=1}^{\infty} \frac{4^n - 1}{3(4^n)} c_{2n+2} h^{2n+2}$$

where

$$I = \int_0^1 f(x)\,dx \qquad \text{and} \qquad T(f, h) = h \sum_{i=0}^{2^n}{}'' f(ih)$$

4. Show that the second column in the Romberg array is the result of using Simpson's rule on f. (See Equation (6) in Section 7.2.)

5. Prove by induction that

$$I - R(n, m-1) = ah^{2m} + bh^{2m+2} + ch^{2m+4} + \cdots$$

6. Apply the Romberg algorithm to find $R(2, 2)$ for these integrals:

(a) $\displaystyle\int_1^3 \frac{dx}{x}$

(b) $\displaystyle\int_0^{\pi/2} \left(\frac{x}{\pi}\right)^2 dx$ (in terms of π)

7. Suppose that $S(f, h)$ is a quadrature rule for the integral I in Equation (1) and that the error series is $c_4h^4 + c_6h^6 + \cdots$. Combine $S(f, h)$ with $S(f, h/3)$ to find a more accurate approximation to I.

8. In the Romberg algorithm, $R(n, 0)$ is an estimate of $\int_a^b f(x)\,dx$ using the trapezoid rule with 2^n subintervals. How many evaluations of $f(x)$ are needed to compute $R(i, j)$ for $0 \leqq i \leqq N$ and $0 \leqq j \leqq N$?

9. If the trapezoid rule satisfied the equation

$$\int_a^b f(x)\,dx = T(f, h) + c_1h + c_2h^2 + c_3h^3 + \cdots$$

instead of Equation (9), then how would we have to modify Formula (5)?

10. In the Romberg algorithm, the elements in the second column satisfy

$$R(i, 1) = I + C_4h_i^4 + C_6h_i^6 + \cdots$$

where $I = \int_a^b f(x)\,dx$ and $h_i = (b-a)/2^i$. Derive the formula for computing elements in the third column and the first term in its error series.

11. Express $R(2, 2)$ (**Milne's rule**) in terms of elements in the first column of the Romberg array. Show that $R(3, 3)$ is *not* a Newton-Cotes formula, but $R(2, 2)$ is.

12. Show that Equation (3) follows immediately from the fact that

$$\sum_{0 \leqq i \leqq 2n} f(a+ih) - \sum_{\substack{0 \leqq i \leqq 2n \\ i \text{ even}}} f(a+ih) = \sum_{\substack{0 \leqq i \leqq 2n \\ i \text{ odd}}} f(a+ih)$$

COMPUTER PROBLEM 7.4

1. Write a subprogram to carry out the Romberg algorithm for a function f defined on an arbitrary interval $[a, b]$. The user will specify the number of rows to be computed in the array and will want to see the entire array when it has been computed. Write a main program and test your Romberg subprogram on these three examples:

(a) $\displaystyle\int_0^1 \frac{\sin x}{x}\,dx$

(b) $\displaystyle\int_{-1}^1 \frac{\cos x - e^x}{\sin x}\,dx$

(c) $\displaystyle\int_1^\infty (xe^x)^{-1}\,dx$

The routines for these integrals should be written to avoid serious loss of significance due to subtraction. Also, it is customary to define a function f at any questionable point x_0 by the equation $f(x_0) = \lim_{x \to x_0} f(x)$. If the limit exists, this method guarantees continuity of f at x_0. For the third example, make a suitable change of variable, such as $x = 1/t$. Compute seven rows in the Romberg array. Print the array in each case with a format that enables the convergence to be observed.

7.5 Adaptive Quadrature

Adaptive quadrature methods are intended to compute definite integrals by automatically taking into account the behavior of the integrand. Ideally, the user supplies only the integrand f, the interval $[a, b]$, and the accuracy ε desired for computing the integral

$$\int_a^b f(x)\,dx \tag{1}$$

The program then divides the interval into pieces of varying length so that numerical integration on the subintervals will produce results of acceptable precision.

Here a typical adaptive quadrature method will be developed. It will use Simpson's rule on subintervals together with an estimate of the errors involved.

Simpson's rule is given by Equation (6) in Section 7.2:

$$\int_a^b f(x)\,dx = S(a,b) - \frac{1}{90}\left[(b-a)/2\right]^5 f^{(4)}(\xi)$$

$$S(a,b) = \frac{b-a}{6}\left[f(a) + 4f\left(\frac{a+b}{2}\right) + f(b)\right] \tag{2}$$

for some $\xi \in (a,b)$. The main idea is that if Simpson's rule on a given subinterval is not sufficiently accurate, that interval will be divided into two equal parts, and Simpson's rule will be used on each half. This procedure will be repeated in an effort to obtain an approximation to the integral with the same accuracy over all subintervals involved. At the end, we shall have computed the integral with n applications of Simpson's rule,

$$\int_a^b f(x)\,dx = \sum_{i=1}^{n} \int_{x_{i-1}}^{x_i} f(x)\,dx = \sum_{i=1}^{n} (S_i + e_i) = \sum_{i=1}^{n} S_i + \sum_{i=1}^{n} e_i$$

where S_i is an approximation to the integral on the interval $[x_{i-1}, x_i]$ and e_i is the associated *local* error. If

$$|e_i| \leqq \varepsilon(x_i - x_{i-1})/(b-a) \tag{3}$$

then the *total* error will be bounded by

$$\left|\sum_{i=1}^{n} e_i\right| \leqq \sum_{i=1}^{n} |e_i| \leqq \frac{\varepsilon}{b-a}\sum_{i=1}^{n}(x_i - x_{i-1}) = \varepsilon$$

Thus, the local error criterion (3) leads to an absolute error bound

$$\left|\int_a^b f(x)\,dx - \sum_{i=1}^{n} S_i\right| \leqq \varepsilon$$

From Equation (2), the *basic* Simpson's rule on the interval $[u, v]$ is described by

$$\int_u^v f(x)\,dx = S(u,v) - \frac{1}{90}[(v-u)/2]^5 f^{(4)}(\xi_1) \tag{4}$$

for some $\xi_1 \in (u, v)$. If the interval of integration is divided into two equal subintervals at the midpoint $w = (u+v)/2$, then a more accurate value of the integral can be computed by using Simpson's rule on each subinterval. Here is the result of doing so:

$$\begin{aligned}\int_u^v f(x)\,dx &= \int_u^w f(x)\,dx + \int_w^v f(x)\,dx \\ &= S(u, w) - \frac{1}{90}[(w-u)/2]^5 f^{(4)}(\xi_2) + S(w, v) \\ &\quad - \frac{1}{90}[(v-w)/2]^5 f^{(4)}(\xi_3) \\ &= S^* + S^{**} - \frac{1}{90}\left(\frac{v-u}{2^2}\right)^5 \left[f^{(4)}(\xi_2) + f^{(4)}(\xi_3)\right] \\ &= S^* + S^{**} - \frac{1}{2^9}\cdot\frac{1}{90}(v-u)^5 f^{(4)}(\xi) \end{aligned} \tag{5}$$

In this calculation, we have set

$$S^* \equiv S(u, w) \qquad S^{**} \equiv S(w, v)$$

$$f^{(4)}(\xi) \equiv \frac{1}{2}\left[f^{(4)}(\xi_2) + f^{(4)}(\xi_3)\right] \tag{6}$$

where $\xi_2 \in (u, w)$, $\xi_3 \in (w, v)$, and $\xi \in (u, v)$. Equation (6) is justified if $f^{(4)}$ is assumed to be continuous. As is usual in such formulas, the error term cannot be estimated or bounded unless some knowledge of $f^{(4)}$ is available. For automatic computation, however, it is imperative to find a way of estimating the magnitude of $f^{(4)}(\xi)$. This will be accomplished by assuming that over small intervals, $f^{(4)}$ is *constant*. In particular, let us assume that $f^{(4)}(\xi_1) = f^{(4)}(\xi)$ in Equations (4) and (5). Then the terms involving $f^{(4)}$ can be eliminated by multiplying Equation (5) by $16/15$ and subtracting $1/15$ times Equation (4). The result is

$$\int_u^v f(x)\,dx \approx S^* + S^{**} + \frac{1}{15}\left[S^* + S^{**} - S(u, v)\right] \tag{7}$$

The groundwork has now been laid for an adaptive quadrature method. Suppose that the integral (1) is to be computed numerically with a prescribed tolerance ε. If $f^{(4)}$ exists and is slowly varying, then the approximation in Equation (7) will be satisfactory on small intervals. The adaptive algorithm starts by considering the interval $[a, b]$. For this interval (as well as for others that are considered subsequently), a vector having six components is constructed:

$$\left[a, h, f(a), f(a+h), f(a+2h), S\right] \qquad h = \frac{1}{2}(b-a) \tag{8}$$

This vector contains the Simpson estimate $S = S(a, b)$ as well as the data needed to compute it by the equation

$$S = \frac{h}{3}\left[f(a) + 4f(a+h) + f(a+2h)\right]$$

Now we compute $c = a + h$, $S^* = S(a, c)$, and $S^{**} = S(c, b)$. As explained previously, $S^* + S^{**}$ is a more refined estimate of the integral. To see whether it is good enough (that is, meets the tolerance ε), we shall use Equations (7) and (3). Hence, we test the inequality

$$\frac{1}{15}|S^* + S^{**} - S| < \varepsilon(2h)/(b - a) \tag{9}$$

If this is true, $S^* + S^{**} + [S^* + S^{**} - S]/15$ is accepted as the value of the integral (1), in accordance with Equation (7). If Inequality (9) fails, the interval is subdivided into two subintervals of equal length—namely, $[a, c]$ and $[c, b]$, where $c = (a + b)/2$. For each of these intervals, we construct vectors as described previously. The vector in (8) is discarded, and in its place we have two new vectors:

$$\begin{array}{ll} \left[a, h/2, f(a), f(y), f(c), S^*\right] & y = a + h/2 \\ \left[c, h/2, f(c), f(z), f(b), S^{**}\right] & z = c + h/2 \end{array}$$

Notice that only two new function evaluations, $f(y)$ and $f(z)$, are needed to compute $S^* = S(a, c)$ and $S^{**} = S(c, b)$ since all other quantities have already been computed, and they are now being stored in a particular format. The process that was applied to the vector in (8) will eventually be applied to every vector generated by the algorithm.

A description of the general step in the algorithm goes as follows: At any stage, there will be accumulated in a variable Σ the sum of integrals of f over certain subintervals. There is, at the same time, a *workstack* of vectors described earlier, each of which corresponds to an interval on which the integral of f has yet to be computed satisfactorily. One of these vectors, say

$$[u, h, \overline{u}, \overline{w}, \overline{v}, S]$$

is taken off the stack. It will have these properties:

$$\begin{array}{lll} w = u + h & v = u + 2h & \\ \overline{u} = f(u) & \overline{w} = f(w) & \overline{v} = f(v) \\ S = (\overline{u} + 4\overline{w} + \overline{v})h/3 & & \end{array}$$

Next, we replace h by $h/2$ and compute

$$\begin{array}{ll} y = u + h & z = u + 3h \\ \overline{y} = f(y) & \overline{z} = f(z) \\ S^* = (\overline{u} + 4\overline{y} + \overline{w})h/3 & \\ S^{**} = (\overline{w} + 4\overline{z} + \overline{v})h/3 & \end{array}$$

We wish to know whether $S^* + S^{**}$, as an estimate of $\int_u^v f(x)\,dx$, passes the error test, which is

$$|S^* + S^{**} - S| \leqq 30\varepsilon h/(b-a) \tag{10}$$

If the test (10) is passed, then $S^* + S^{**} + (S^* + S^{**} - S)/15$ is added to Σ, and a new vector is taken from the stack for processing. If the test fails, then two new vectors are added to the stack:

$$[u, h, \overline{u}, \overline{y}, \overline{w}, S^*]$$
$$[u + 2h, h, \overline{w}, \overline{z}, \overline{v}, S^{**}]$$

At this stage, another vector is removed from the stack and processed as before. The size of the stack is not allowed to exceed n vectors, where n is a parameter set by the user. This helps to prevent the algorithm from becoming infinite.

A pseudocode for this algorithm is given next. It follows the description just given. Some care is taken to save quantities that may be needed later. The vectors in the workstack are denoted by $v^{(1)}$, $v^{(2)}$, and so on. Each vector has six components:

$$v^{(k)} = [v_1^{(k)}, v_2^{(k)}, \ldots, v_6^{(k)}]$$

The first component, $v_1^{(k)}$, always represents the left-hand endpoint of an interval; $v_2^{(k)}$ is half the length of this interval; $v_3^{(k)}$, $v_4^{(k)}$, and $v_5^{(k)}$ are values of f at the left end, the middle point, and the right end of this interval, respectively; and $v_6^{(k)}$ is the value given by Simpson's rule for this interval.

The complete algorithm follows:

input a, b, ε, n
$\Delta \leftarrow b - a;\ \ \Sigma \leftarrow 0;\ \ h \leftarrow \Delta/2;\ \ c \leftarrow (a+b)/2;\ \ k \leftarrow 1$
$\overline{a} \leftarrow f(a)\ \ \overline{b} \leftarrow f(b)\ \ \overline{c} \leftarrow f(c)$
$S \leftarrow (\overline{a} + 4\overline{c} + \overline{b})h/3$
$v^{(1)} \leftarrow [a, h, \overline{a}, \overline{c}, \overline{b}, S]$
while $1 \leqq k \leqq n$
  $h \leftarrow v_2^{(k)}/2$
  $\overline{y} \leftarrow f(v_1^{(k)} + h)$
  $S^* \leftarrow (v_3^{(k)} + 4\overline{y} + v_4^{(k)})h/3$
  $\overline{z} \leftarrow f(v_1^{(k)} + 3h)$
  $S^{**} \leftarrow (v_4^{(k)} + 4\overline{z} + v_5^{(k)})h/3$
  if $|S^* + S^{**} - v_6^{(k)}| < 30\varepsilon h/\Delta$ **then**
    $\Sigma \leftarrow \Sigma + S^* + S^{**} + [S^* + S^{**} - v_6^{(k)}]/15$
    $k \leftarrow k - 1$
    if $k \leqq 0$ **then** output Σ; **exit**
  else
    if $k \geqq n$ **then output** failure; **exit**
      $\overline{v} \leftarrow v_5^{(k)}$

$v^{(k)} \leftarrow [v_1^{(k)}, h, v_3^{(k)}, \overline{y}, v_4^{(k)}, S^*]$
$k \leftarrow k + 1$
$v^{(k)} \leftarrow [v_1^{(k-1)} + 2h, h, v_5^{(k-1)}, \overline{z}, \overline{v}, S^{**}]$
end if
end do

In some programming languages and computers it is important to store the most frequently accessed quantities in the same column. If this is the case then the workstack vector could be stored in a two-dimensional array `V(I, K)` $\leftarrow v_i^{(k)}$.

Here we present the adaptive quadrature algorithm with an explicit stack, but it is essentially a recursive procedure. First, try a simple Simpson's rule—if the error is satisfactory, accept the answer; otherwise, split the interval in half and call the procedure recursively on each half. Using recursion, we see the concept of the algorithm more easily. Moreover, the recursive algorithm is simple to code since most programming languages support recursion. We leave this as Computer Problem 2.

PROBLEMS 7.5

1. The local error criterion (3) was used to obtain an *absolute* error bound. Try to establish a local error criterion that leads to a *relative* error bound

$$\left| \int_a^b f(x)\,dx - \sum_{i=1}^{n} S_i \right| \leqq \left| \int_a^b f(x)\,dx \right| \varepsilon$$

2. Establish an approximation similar to (7) using the trapezoid rule in the form

$$\int_u^v f(x)\,dv = T(u, v) - \frac{1}{12}(v - u)^3 f''(\xi)$$

where

$$T(u, v) = \frac{1}{2}(v - u)\big[f(u) + f(v)\big]$$

3. The composite trapezoid rule can be written as

$$I = T_m - \frac{1}{12}(v - u)h^2 f''(\xi)$$

where

$$I = \int_u^v f(x)\,dx$$

$$T_m = \frac{h}{2}\left[f(u) + 2\sum_{i=1}^{m-1} f(u + ih) + f(v)\right]$$

Show how T_m and T_{2m} can be combined to obtain a better approximation to I. Here m is the number of subintervals in the interval $[u, v]$.

4. The composite Simpson's rule can be written as

$$I = S_{2m} - \frac{1}{180}(v-u)h^4 f^{(4)}(\xi)$$

where

$$I = \int_u^v f(x)\,dx$$

$$S_{2m} = \frac{h}{3}\left[f(u) + 2\sum_{i=2}^{m} f(x_{2i-2}) + 4\sum_{i=1}^{m} f(x_{2i-1}) + f(v)\right]$$

and $h = (v-u)/(2m)$, $x_i = u + ih$ $(0 \leqq i \leqq 2m)$. Here there are $2m$ subintervals in the interval $[u, v]$. Show how S_{2m} and S_{4m} can be combined to obtain a better approximation to I.

5. In Section 7.4, the following relationship between the trapezoid rule over $[u, v]$ with m subintervals and that over $[u, v]$ with $2m$ subintervals was derived:

$$T_{2m} = \frac{1}{2}T_m + h\sum_{i=1}^{m} f(x_{2i-1})$$

where $h = (v-u)/(2m)$ and $x_i = u + ih$ $(0 \leqq i \leqq 2m)$. Can this be used to establish an adaptive scheme?

COMPUTER PROBLEMS 7.5

1. Program the adaptive algorithm and test it on these integrals:

(a) $\displaystyle\int_0^1 x^{1/2}\,dx$

(b) $\displaystyle\int_0^1 (1-x)^{1/2}\,dx$

(c) $\displaystyle\int_0^1 (1-x)^{1/4}\,dx$

2. Write and test the adaptive algorithm of this section as a recursive procedure.

*7.6 Sard's Theory of Approximating Functionals

A **linear functional** on a linear space is a linear map of that space into the scalar field (in this book, usually $\mathbb{R}$). For example, if the linear space is $C[a, b]$, an important linear functional φ is defined by

$$\varphi(f) = \int_a^b f(x)\,dx \qquad f \in C[a, b]$$

In numerical practice, the most basic functionals are **point evaluations**. Fixing x in $[a, b]$, we define a linear functional $\widehat{x}$ by the equation

$$\widehat{x}(f) = f(x) \qquad f \in C[a, b]$$

By taking linear combinations of point evaluations, we obtain more versatile functionals of the form ψ, defined by

$$\psi = \sum_{i=0}^{n} c_i \widehat{x}_i \qquad \text{that is} \qquad \psi(f) = \sum_{i=0}^{n} c_i f(x_i) \tag{1}$$

One can argue persuasively that in numerical work this is the most general type of functional that can be computed directly; other functionals (such as integration) must be replaced by approximating functionals like ψ. This is precisely what is done in the subject of numerical integration and numerical differentiation.

A coherent theory of approximating functionals was developed by Arthur Sard in the years 1940–1970. This theory involves natural splines in an interesting manner. The type of functional that we shall approximate is given by the equation

$$\varphi(f) = \sum_{i=0}^{N} \left\{ \int_a^b \alpha_i(x) f^{(i)}(x)\,dx + \sum_{j=1}^{n} \beta_{ij} f^{(i)}(z_{ij}) \right\} \tag{2}$$

The points z_{ij} lie in $[a, b]$, and the functions α_i are assumed to be piecewise continuous on $[a, b]$. The functional φ can be applied to any $f \in C^N[a, b]$.

We shall say that a functional φ **annihilates** a space W if $\varphi(f) = 0$ for each $f \in W$. The **Peano kernels** of a functional φ such as in Equation (2) are the functions

$$K_m(t) = \frac{1}{m!}\,\varphi_x\left[(x - t)_+^m\right]$$

where $m \geqq N$ and

$$(x - t)_+^m = \begin{cases} (x - t)^m & x \geqq t \\ 0 & x < t \end{cases}$$

and the notation φ_x indicates that the functional applies to $(x - t)_+^m$ as a function of x. The function x_+^m (obtained when $t = 0$) is called a **truncated power function**.

Example 1 Consider the functional φ defined by the equation

$$\varphi(f) = \int_0^{\pi} (\cos x) f'(x)\, dx$$

This illustrates Equation (2), with $N = 1$ and $a_1(x) = \cos x$. What is the Peano kernel, K_1, for φ?

Solution The definition of the Peano kernels can be used to compute them explicitly. Notice that

$$\frac{d}{dx}(x-t)_+^m = m(x-t)_+^{m-1} \qquad (m \geqq 1)$$

Consequently, in this example,

$$\begin{aligned} K_1(t) = \varphi_x\left[(x-t)_+^1\right] &= \int_0^{\pi} (\cos x) \frac{d}{dx}(x-t)_+\, dx \\ &= \int_0^{\pi} (\cos x)(x-t)_+^0\, dx \\ &= \int_t^{\pi} (\cos x)\, dx = -\sin t \end{aligned}$$

■

The following result was proved in 1905. (In it we use Π_m as the space of all polynomials of degree at most m.)

THEOREM 1 **Peano Kernel Theorem** *If the functional in Equation (2) annihilates Π_m, then for all $f \in C^{m+1}[a, b]$,*

$$\varphi(f) = \int_a^b K_m(t) f^{(m+1)}(t)\, dt \tag{3}$$

where $m \geqq N$ and $K_m(t)$ is the Peano kernel of φ.

Proof By Taylor's Theorem with integral remainder (Theorem 2 in Section 1.1),

$$f(x) = \sum_{k=0}^{m} \frac{1}{k!} f^{(k)}(a)(x-a)^k + r(x)$$

$$r(x) = \frac{1}{m!} \int_a^x f^{(m+1)}(t)\,(x-t)^m\, dt$$

Since φ annihilates Π_m, $\varphi(f) = \varphi(r)$. Now express r in the form

$$r(x) = \frac{1}{m!} \int_a^b f^{(m+1)}(t)\,(x-t)_+^m\, dt$$

From this, it follows that

$$\varphi(r) = \frac{1}{m!}\int_a^b f^{(m+1)}(t)\varphi_x\left[(x-t)_+^m\right] dt$$

This step involves passing the functional φ under the integral sign, which is justified by certain theorems in calculus that require the hypotheses we adopted for φ. ■

Example 2 An illustration given by Sard [1963, p. 31] is the functional

$$\varphi(f) = \int_0^1 f(x)x^{-1/2}\, dx$$

Find an approximate formula

$$\psi(f) = c_1 f(0) + c_2 f(1)$$

that is an exact substitute for φ on the polynomial space Π_1 and find its error.

Solution We require that $\varphi - \psi$ annihilate Π_1. Using the method of undetermined coefficients (as in Section 7.2), we first take $f(x) = 1$, getting

$$0 = \varphi(f) - \psi(f) = \int_0^1 x^{-1/2}\, dx - (c_1 + c_2) = 2 - c_1 - c_2$$

Taking $f(x) = x$, we obtain

$$0 = \varphi(f) - \psi(f) = \int_0^1 x^{1/2}\, dx - c_2 = \frac{2}{3} - c_2$$

Hence, $c_2 = 2/3$ and $c_1 = 4/3$. The Peano kernel K_1 for the functional $\varphi - \psi$ is

$$\begin{aligned}
(\varphi_x - \psi_x)(x-t)_+^1 &= \int_0^1 (x-t)_+ x^{-1/2}\, dx - \frac{4}{3}(0-t)_+ - \frac{2}{3}(1-t)_+ \\
&= \int_t^1 (x-t)x^{-1/2}\, dx - \frac{2}{3}(1-t) \\
&= \frac{4}{3}t(t^{1/2} - 1)
\end{aligned}$$

The Peano Kernel Theorem, applied to $\varphi - \psi$, now tells us that

$$\int_0^1 f(x)x^{-1/2}\, dx - \left[\frac{4}{3}f(0) + \frac{2}{3}f(1)\right] = \int_0^1 \frac{4}{3}t(t^{1/2} - 1)f''(t)\, dt$$

The right-hand side of this equation is an exact representation for the error when $f \in C^2[0, 1]$. Since the kernel is nonpositive on the interval $[0, 1]$, the Mean-Value

Theorem for Integrals (Section 1.1) gives

$$\int_0^1 \frac{4}{3}t(t^{1/2} - 1)f''(t)\,dt = f''(\xi)\int_0^1 \frac{4}{3}t(t^{1/2} - 1)\,dt = -\frac{2}{15}f''(\xi) \qquad \blacksquare$$

If φ and ψ are two functionals that agree on Π_m, then their difference will annihilate Π_m. The difference will have a Peano kernel, K_m, and by the **Cauchy-Schwarz inequality**,

$$|\varphi(f) - \psi(f)| \leqq \|K_m\|_2\,\|f^{(m+1)}\|_2 \tag{4}$$

The norm with subscript 2 is the usual L_2-norm on $[a, b]$. If there are parameters in ψ that are not fully determined by the requirement that $\varphi - \psi$ annihilate Π_m, then these parameters can be chosen to make the factor $\int_a^b K_m(t)^2\,dt$ a minimum. In this way, we obtain a functional ψ that is a best approximation to φ *in the sense of Sard*. Schoenberg discovered that such *best* formulas can be obtained in a rather simple way, and this important theorem is presented next. In the theorem, $\varphi \circ L$ is the composition of φ and L, so that $(\varphi \circ L)(f) = \varphi(Lf)$.

THEOREM 2 **Schoenberg Theorem** *Let φ be the linear functional given in Equation (2). Let knots be prescribed thus: $a = t_0 < t_1 < \cdots < t_n = b$, with $n > N$. Among all the functionals of the form $\sum_{i=0}^n c_i\widehat{t_i}$ that agree with φ on Π_m, the best approximation to φ in the sense of Sard is $\varphi \circ L$, where L is the linear operator producing natural spline interpolants of degree $2m + 1$ at the given knots.*

Proof Let ψ be of the form $\psi = \sum_{i=0}^n c_i\widehat{t_i}$, and assume that for any $p \in \Pi_m$, $\psi(p) = \varphi(p)$. Then $\varphi - \psi$ annihilates Π_m. Let K_m be its Peano kernel. Now $Lf = f$ if f is a natural spline of degree $2m + 1$ on the given knots. Since polynomials of degree $\leqq m$ are also such natural splines, $Lp = p$ for $p \in \Pi_m$. Thus, $\varphi - \varphi \circ L$ also annihilates Π_m. Let $\overline{K}_m$ be its Peano kernel. It is to be shown that

$$\int_a^b \left[\overline{K}_m(t)\right]^2 dt \leqq \int_a^b \left[K_m(t)\right]^2 dt \tag{5}$$

The Peano kernel of the functional

$$\theta = \varphi \circ L - \psi = (\varphi - \psi) - (\varphi - \varphi \circ L)$$

is $\overline{\overline{K}}_m = K_m - \overline{K}_m$ and is given by

$$\overline{\overline{K}}_m(t) = \frac{1}{m!}\,\theta_x\left[(x - t)_+^m\right] \tag{6}$$

If $\{s_0, s_1, \ldots, s_n\}$ is a cardinal basis for our natural spline space, then $s_i(t_j) = \delta_{ij}$ and L will have the form

$$Lf = \sum_{i=0}^{n} f(t_i)s_i$$

It follows that θ has the form

$$\theta(f) = \varphi(Lf) - \psi(f) = \sum_{i=0}^{n} f(t_i)\varphi(s_i) - \sum_{i=0}^{n} c_i f(t_i) = \sum_{i=0}^{n} \gamma_i f(t_i)$$

Thus, for $\overline{\overline{K}}_m$ we have the equation

$$\overline{\overline{K}}_m(t) = \frac{1}{m!}\sum_{i=0}^{n} \gamma_i (t_i - t)_+^m \tag{7}$$

Select a function g such that $g^{(m+1)} = \overline{\overline{K}}_m$. Then $g^{(2m+1)} = \overline{\overline{K}}_m^{(m)}$, and this is a step function or spline of degree 0. Consequently, g itself is a spline function of degree $2m+1$, with knots $t_0, t_1, \ldots, t_n$. In fact, g is a natural spline. To verify this, notice that by Equation (7), $\overline{\overline{K}}_m(t) = 0$ when $t \geqq b$. Hence, $g^{(m+1)}(t) = 0$ for $t \geqq b$. For $t \leqq a \leqq x$, we have from Equation (6),

$$\overline{\overline{K}}_m(t) = \frac{1}{m!}\,\theta_x\left[(x-t)^m\right]$$

Since θ annihilates Π_m, $\overline{\overline{K}}_m(t) = g^{(m+1)}(t) = 0$ when $-\infty < t \leqq a$. Since g is a natural spline, $Lg = g$. Hence,

$$\int_a^b \overline{K}_m \overline{\overline{K}}_m \, dt = \int_a^b \overline{K}_m g^{(m+1)} \, dt = (\varphi - \varphi \circ L)(g) = 0$$

Equation (5) can be established now by writing

$$\int_a^b K_m^2 \, dt = \int_a^b (\overline{K}_m + \overline{\overline{K}}_m)^2 \, dt = \int_a^b (\overline{K}_m^2 + 2\overline{K}_m\overline{\overline{K}}_m + \overline{\overline{K}}_m^2)\, dt$$
$$= \int_a^b \overline{K}_m^2 \, dt + \int_a^b \overline{\overline{K}}_m^2 \, dt \geqq \int_a^b \overline{K}^2 \, dt$$

In terms of the inner product for functions, these calculations show that we have $\langle \overline{K}_m, \overline{\overline{K}}_m \rangle = 0$, whence by the Pythagorean Law,

$$\|K_m\|_2^2 = \|\overline{K}_m + \overline{\overline{K}}_m\|_2^2 = \|\overline{K}_m\|_2^2 + \|\overline{\overline{K}}_m\|_2^2 \geqq \|\overline{K}_m\|_2^2 \qquad \blacksquare$$

Example 3 Let

$$\varphi(f) = \int_{-1}^{1} f(x)\, dx$$
$$\psi(f) = c_1 f(-1) + c_2 f(0) + c_3 f(1)$$

Considering only functionals ψ that agree with φ on Π_1, find the best approximation to φ in the sense of Sard.

Solution We require the natural spline interpolation operator L for this situation. One can verify that

$$(Lf)(x) = a_0 + a_1 x + b_0(x+1)_+^3 + b_1(x)_+^3 + b_2(x-1)_+^3$$

where

$$\begin{aligned} a_0 &= \left[-f(-1) + 6f(0) - f(1)\right]/4 \\ a_1 &= \left[-5f(-1) + 6f(0) - f(1)\right]/4 \\ b_0 &= b_2 = -b_1/2 = \left[f(-1) - 2f(0) + f(1)\right]/4 \end{aligned}$$

Then the best formula is given by

$$\begin{aligned} \psi(f) = \varphi(Lf) &= \int_{-1}^{1} (Lf)(x)\,dx \\ &= 2a_0 + 4b_0 + \frac{1}{4}b_1 \\ &= \frac{3}{8}f(-1) + \frac{5}{4}f(0) + \frac{3}{8}f(1) \end{aligned}$$

■

PROBLEMS 7.6

1. We seek an approximate formula

$$f'(x) \approx c_0 f(-1) + c_1 f(0) + c_2 f(1)$$

where x is a fixed point in $(-1, 1)$. Find the conditions on the coefficients that are necessary and sufficient for the formula to be correct on Π_1. Find the best approximation in the sense of Sard. Compute $\int_{-1}^{1} K_1(t)^2\,dt$ when the minimum is attained.

2. For Example 3, find the best constant in the error formula

$$|\varphi(f) - \psi(f)| \leqq c\,\|f''\|_2$$

3. Find the interpolating function f $(f(t_i) = \lambda_i,\ \ 0 \leqq i \leqq n)$ that minimizes the integral $\int_{t_0}^{t_n} (f'')^2\,dt$ assuming f'' is *piecewise* continuous.

4. Use the Peano Kernel Theorem on the functional

$$\varphi(f) = f(x+h) - f(x) - (h/2)\left[3f'(x) - f'(x-h)\right]$$

(x, h fixed) and prove that $\varphi(f) = (5/12)h^3 f'''(\xi)$ with $x - h < \xi < x + h$.

5. Use the Peano Kernel Theorem to obtain this result for Simpson's rule:

$$\int_0^2 f(x)\,dx = \frac{1}{3}\big[f(0) + 4f(1) + f(2)\big] - \frac{1}{90}f^{(4)}(\xi)$$

6. Use the Schoenberg Theorem of this section to prove that

$$\left|f(2) + \frac{1}{4}f(0) - \frac{7}{8}f(1) - \frac{3}{8}f(3)\right| \leqq \sqrt{\frac{5}{48}}\,\|f''\|_2$$

*7.7 Bernoulli Polynomials and the Euler-Maclaurin Formula

The Romberg integration algorithm depends for its validity on the Euler-Maclaurin formula, Equation (6) in Section 7.4. We derive this formula here, after establishing some properties of the Bernoulli polynomials.

Bernoulli Polynomials

The Bernoulli polynomials form an infinite sequence with many useful properties. The first few of these polynomials are

$$B_0(t) = 1$$

$$B_1(t) = t - \frac{1}{2}$$

$$B_2(t) = t^2 - t + \frac{1}{6}$$

$$B_3(t) = t^3 - \frac{3}{2}t^2 + \frac{1}{2}t$$

They are defined by the equation

$$\sum_{k=0}^{n} \binom{n+1}{k} B_k(t) = (n+1)t^n \tag{1}$$

Recall that

$$\binom{n}{m} = \frac{n!}{m!\,(n-m)!}$$

For example, if $n = 0$, then Equation (1) states that $B_0(t) = 1$. If $n = 1$, the equation states that $B_0(t) + 2B_1(t) = 2t$, from which we obtain $B_1(t) = t - \frac{1}{2}$. Since the coefficient of B_n in Equation (1) is $n + 1$, we can always solve for B_n in terms of

$B_0, B_1, \ldots, B_{n-1}$. The following theorem enumerates the properties of the Bernoulli polynomials that we shall require.

THEOREM 1 *The Bernoulli polynomials have these properties:*

(i) $B_n' = nB_{n-1} \qquad (n \geqq 1)$

(ii) $B_n(t+1) - B_n(t) = nt^{n-1} \qquad (n \geqq 2)$

(iii) $B_n(t) = \displaystyle\sum_{k=0}^{n} \binom{n}{k} B_k(0) t^{n-k}$

(iv) $B_n(1-t) = (-1)^n B_n(t)$

Proof Equation **(i)** will be proved by induction. The case $n = 1$ is verified from the fact that $B_1(t) = t - \frac{1}{2}$ and $B_0(t) = 1$. Now suppose that $B_k' = kB_{k-1}$ for $k = 1, 2, \ldots, n-1$. Then by differentiating in Equation (1) and using Equation (1) again, we obtain

$$\sum_{k=1}^{n} \binom{n+1}{k} B_k'(t) = n(n+1)t^{n-1} = (n+1)\sum_{k=0}^{n-1} \binom{n}{k} B_k(t) \tag{2}$$

By means of the induction hypothesis, Equation (2) can be put into the form

$$(n+1)B_n' + \sum_{k=1}^{n-1} \binom{n+1}{k} kB_{k-1} = (n+1)\sum_{k=0}^{n-1} \binom{n}{k} B_k$$

Now divide by $n+1$ and use the identity

$$\frac{k}{n+1}\binom{n+1}{k} = \binom{n}{k-1}$$

to obtain

$$B_n' + \sum_{k=1}^{n-1} \binom{n}{k-1} B_{k-1} = \sum_{k=0}^{n-1} \binom{n}{k} B_k$$

The cancellation of similar terms here leaves Equation **(i)**, which was to be proved.

To prove Equation **(ii)**, use Equation **(i)** repeatedly to obtain

$$\begin{aligned}
B_n''(t) &= n(n-1)B_{n-2}(t) \\
B_n'''(t) &= n(n-1)(n-2)B_{n-3}(t) \\
&\ \ \vdots \\
B_n^{(k)}(t) &= n(n-1)\cdots(n-k+1)B_{n-k}(t)
\end{aligned}$$

Thus, the Taylor series for B_n is

$$B_n(t+h) = \sum_{k=0}^{n} \frac{1}{k!} B_n^{(k)}(t) h^k = \sum_{k=0}^{n} \binom{n}{k} B_{n-k}(t) h^k \tag{3}$$

Now set $h = 1$. Use the identity

$$\binom{n}{k} = \binom{n}{n-k}$$

and Equation (1) to obtain Equation **(ii)**:

$$\begin{aligned} B_n(t+1) &= \sum_{k=0}^{n} \binom{n}{k} B_k(t) = B_n(t) + \sum_{k=0}^{n-1} \binom{n}{k} B_k(t) \\ &= B_n(t) + nt^{n-1} \end{aligned}$$

We obtain Equation **(iii)** by putting $t = 0$ and $h = t$ in Equation (3).

To derive Equation **(iv)**, use Equation **(ii)** with t replaced by $-t$. The result is

$$\begin{aligned} B_n(1-t) - B_n(-t) &= n(-1)^{n-1} t^{n-1} \\ &= (-1)^{n-1} \left[B_n(t+1) - B_n(t) \right] \end{aligned}$$

When this is written in the form

$$(-1)^n B_n(t+1) - B_n(-t) = (-1)^n B_n(t) - B_n(1-t) \equiv F(t)$$

we recognize it as an equation of the type $F(t+1) = F(t)$, which implies that F is periodic with period 1. Since F is a polynomial, it must be constant. Hence,

$$(-1)^n B_n(t) - B_n(1-t) = c_n \tag{4}$$

Now differentiate in Equation (4) and use Equation **(i)**:

$$(-1)^n B_n'(t) + B_n'(1-t) = (-1)^n n B_{n-1}(t) + n B_{n-1}(1-t) = 0$$

This is equivalent to Equation **(iv)**. ■

LEMMA 1 *The function $G(t) = B_{2n}(t) - B_{2n}(0)$ has no zeros in the open interval* $(0, 1)$.

Proof We begin by setting $t = 0$ in Equations **(ii)** and **(iv)** of Theorem 1. The result is

$$B_n(0) = B_n(1) = (-1)^n B_n(0)$$

Consequently, $B_3(0) = B_5(0) = B_7(0) = \cdots = 0$. Now suppose that G has a zero in (0,1). Since $G(0) = G(1) = 0$, we see by Rolle's Theorem that G' has two zeros in (0,1). Since B_{2n-1} is a multiple of G', B_{2n-1} has two zeros in (0,1). But B_{2n-1} vanishes also at 0 and 1, so B'_{2n-1} has three zeros in (0,1). Consequently B_{2n-2} has three zeros in (0,1). Continuing in this way, we conclude that B_k, for odd indices $k < 2n$, must have at least two zeros in (0,1). Hence, B_3 must have two zeros in (0,1) as well as the two zeros at 0 and 1. This is manifestly impossible because B_3 is a cubic polynomial. ■

THEOREM 2 **Euler-Maclaurin Formula** *If the function f possesses $2n$ continuous derivatives in $[0, 1]$, then*

$$\int_0^1 f(t)\,dt = \frac{1}{2}\big[f(0) + f(1)\big] - \sum_{k=0}^{n-1} \frac{b_{2k}}{(2k)!}\big[f^{(2k-1)}(1) - f^{(2k-1)}(0)\big] + R$$

where

$$\begin{cases} b_k = B_k(0) \\ R = -\dfrac{b_{2n}}{(2n)!} f^{(2n)}(\xi) \qquad (0 < \xi < 1) \end{cases}$$

Proof From Equation **(i)** of Theorem 1, we have

$$B_n(t) = \frac{1}{n+1} B'_{n+1}(t)$$

Using this formula and integration by parts, we write

$$\int_0^1 f(t)\,dt = \int_0^1 f(t)B_0(t)\,dt = B_1(t)f(t)\Big|_0^1 - \int_0^1 B_1(t)f'(t)\,dt \tag{5}$$

Since $B_1(1) = \frac{1}{2}$ and $B_1(0) = -\frac{1}{2}$, Equation (5) can be written in the form

$$\int_0^1 f(t)\,dt = \frac{1}{2}\big[f(1) + f(0)\big] - \int_0^1 B_1(t)f'(t)\,dt$$

The integration in this equation can also be carried out with an integration by parts, and the result of doing so is

$$\int_0^1 f(t)\,dt = \frac{1}{2}\big[f(1) + f(0)\big] - \frac{b_2}{2}\big[f'(1) - f'(0)\big] + \frac{1}{2}\int_0^1 B_2(t)f''(t)\,dt$$

This process can be continued. The formulas

$$\begin{aligned} B_n(0) = B_n(1) &= b_n \\ b_3 = b_5 = b_7 = \cdots &= 0 \end{aligned}$$

previously mentioned are helpful. After $2n$ steps, we shall have

$$\int_0^1 f(t)\,dt = \frac{1}{2}\big[f(1) + f(0)\big] - \sum_{k=1}^{n} \frac{b_{2k}}{(2k)!}\big[f^{(2k-1)}(1) - f^{(2k-1)}(0)\big] + \frac{1}{(2n)!}\int_0^1 B_{2n}(t) f^{(2n)}(t)\,dt$$

The last term in the sum can be expressed in the form

$$\frac{b_{2n}}{(2n)!}\big[f^{(2n-1)}(1) - f^{(2n-1)}(0)\big] = \frac{b_{2n}}{(2n)!}\int_0^1 f^{(2n)}(t)\,dt$$

This permits the formula to be written as in the statement of the theorem, except that the remainder now is

$$R = \frac{1}{(2n)!}\int_0^1 \big[B_{2n}(t) - b_{2n}\big] f^{(2n)}(t)\,dt \tag{6}$$

By Lemma 1, the function $B_{2n}(t) - b_{2n}$ does not change sign in $[0, 1]$. Hence, the Mean-Value Theorem for Integrals can be applied in Equation (6) to write R in the form

$$R = \frac{1}{(2n)!} f^{(2n)}(\xi) \int_0^1 \big[B_{2n}(t) - b_{2n}\big]\,dt$$

Since $B_{2n} = B'_{2n+1}/(2n + 1)$ and since $B_{2n+1}(t)$ is 0 at $t = 0$ and $t = 1$, we obtain finally

$$R = -\frac{b_{2n}}{(2n)!} f^{(2n)}(\xi)$$

■

PROBLEMS 7.7

1. Prove that the leading term in $B_n(t)$ is t^n and that the term t^{n-1} has coefficient $-n/2$.
2. Use Equation **(ii)** in Theorem 1 to prove that
$$1^p + 2^p + \cdots + n^p = \big[B_{p+1}(n + 1) - B_{p+1}(0)\big]/(p + 1)$$
3. Show that the functions $B_n(x) - B_n(0)$ have a simple zero at $\frac{1}{2}$ when n is odd.
4. Show that the numbers $B_0(0), B_2(0), B_4(0), \ldots$ alternate in sign.
5. Prove that for even n, B_n has at least two roots in $(0, 1)$, and for odd n, there is at least one root.
6. Prove that the binomial coefficients $\binom{n}{m}$ are integers. *Hint:* You could prove first Pascal's Law:
$$\binom{n}{m-1} + \binom{n}{m} = \binom{n+1}{m}$$

Chapter Eight

Numerical Solution of Ordinary Differential Equations

Introduction

This chapter concerns numerical problems involving ordinary differential equations. The central problem is to solve a single first-order equation when one point on the solution curve is given. Later sections are devoted to systems of equations, higher-order equations, and two-point boundary-value problems.

8.1 The Existence and Uniqueness of Solutions

Our model is an initial-value problem written in the form

$$\begin{cases} x' = f(t, x) \\ x(t_0) = x_0 \end{cases} \tag{1}$$

Here x is an unknown function of t that we hope to construct from the information given in Equations (1), where $x' = dx(t)/dt$. The second of the two equations in (1) specifies one particular value of the function $x(t)$. The first equation gives the slope of the curve x at any point t. Of course, the function f must be specified. For a concrete example, we can take

$$\begin{cases} x' = x \tan(t + 3) \\ x(-3) = 1 \end{cases} \tag{2}$$

We would like to determine $x(t)$ on an interval containing the initial point t_0. The *analytic* solution of this initial-value problem is $x(t) = \sec(t + 3)$, as we can easily verify. Since $\sec t$ becomes infinite at $t = \pm\pi/2$, our solution is valid only for $-\pi/2 < t + 3 < \pi/2$. The example in (2) is exceptional because it has a simple analytic solution from which numerical values are readily calculated. Typically, for problems of the type in (1), analytic solutions are *not* available and numerical methods must be used.

Existence

Will every initial-value problem of the form in (1) have a solution? No; some assumptions must be made about f, and even then we can only expect the solution to exist in a neighborhood of $t = t_0$. As an example of what could happen, consider

$$\begin{cases} x' = 1 + x^2 \\ x(0) = 0 \end{cases} \tag{3}$$

The solution curve starts at $t = 0$ with slope 1; that is, $x'(0) = 1$. Since the slope is positive, $x(t)$ is *increasing* near $t = 0$. Therefore, the expression $1 + x^2$ is also increasing. Hence, x' is increasing. Since x and x' are both increasing and are related by the equation $x' = 1 + x^2$, we can expect that at some finite value of t there will be no solution; that is, $x(t) = +\infty$. As a matter of fact, this occurs at $t = \pi/2$ because the analytic solution of (3) is $x(t) = \tan t$.

THEOREM 1 *If f is continuous in a rectangle R centered at (t_0, x_0), say*

$$R = \left\{(t, x) : |t - t_0| \leqq \alpha, \quad |x - x_0| \leqq \beta\right\} \tag{4}$$

then the initial-value problem (1) has a solution $x(t)$ for $|t - t_0| \leqq \min(\alpha, \beta/M)$, where M is the maximum of $|f(t, x)|$ in the rectangle R.

Example 1 Prove that the initial-value problem

$$\begin{cases} x' = (t + \sin x)^2 \\ x(0) = 3 \end{cases}$$

has a solution on the interval $-1 \leqq t \leqq 1$.

Solution In this example, $f(t, x) = (t + \sin x)^2$ and $(t_0, x_0) = (0, 3)$. In the rectangle

$$R = \{(t, x) : |t| \leqq \alpha,\ |x - 3| \leqq \beta\}$$

the magnitude of f is bounded by $|f(t, x)| \leqq (\alpha + 1)^2 \equiv M$. We want $\min(\alpha, \beta/M) \geqq 1$, and so we can let $\alpha = 1$. Then $M = 4$, and our objective is met by letting $\beta \geqq 4$. The existence theorem asserts that a solution of the initial-value problem exists on the interval $|t| \leqq \min(\alpha, \beta/M) = 1$. ■

Uniqueness

It can happen, even if f is continuous, that the initial-value problem (1) does not have a unique solution. A simple example of this phenomenon is given by the problem

$$\begin{cases} x' = x^{2/3} \\ x(0) = 0 \end{cases}$$

It is obvious that the 0 function, $x(t) \equiv 0$, is a solution of this problem. Another solution is the function $x(t) = \frac{1}{27}t^3$.

To prove that the initial-value problem (1) has a *unique* solution in a neighborhood of $t = t_0$, it is necessary to assume somewhat more about f. Here is the usual theorem on this.

THEOREM 2 *If f and $\partial f/\partial x$ are continuous in the rectangle R defined by (4), then the initial-value problem (1) has a unique solution in the interval $|t - t_0| < \min(\alpha, \beta/M)$.*

In both Theorems 1 and 2, the interval on the t-axis in which the solution is asserted to exist may be smaller than the base of the rectangle in which we have defined $f(t, x)$. Theorem 3 is of a different type that allows us to infer the existence and unicity of a solution on a prescribed interval $[a, b]$. (See Henrici [1962, p. 15].)

THEOREM 3 *If f is continuous in the strip $a \leqq t \leqq b$, $-\infty < x < \infty$ and satisfies there an inequality*

$$|f(t, x_1) - f(t, x_2)| \leqq L|x_1 - x_2| \tag{5}$$

then the initial-value problem (1) has a unique solution in the interval $[a, b]$.

Inequality (5) is called a **Lipschitz condition** in the second variable. For a function of one variable, such a condition would assert simply

$$|g(x_1) - g(x_2)| \leqq L|x_1 - x_2| \tag{6}$$

We see immediately that this condition is *stronger* than continuity because if x_2 approaches x_1, then the right-hand side in (6) approaches 0, and this forces $g(x_2)$ to approach $g(x_1)$. The condition (6) is weaker than having a bounded derivative. Indeed, if $g'(x)$ exists everywhere and does not exceed L in modulus, then by the Mean-Value Theorem,

$$|g(x_1) - g(x_2)| = |g'(\xi)|\,|x_1 - x_2| \leqq L|x_1 - x_2|$$

Example 2 Show that the function $g(x) = \sum_{i=1}^{n} a_i|x - w_i|$ satisfies a Lipschitz condition with the constant $L = \sum_{i=1}^{n} |a_i|$.

Solution

$$\begin{aligned}
|g(x_1) - g(x_2)| &= \left|\sum_{i=1}^{n} a_i|x_1 - w_i| - \sum_{i=1}^{n} a_i|x_2 - w_i|\right| \\
&= \left|\sum_{i=1}^{n} a_i\{|x_1 - w_i| - |x_2 - w_i|\}\right| \\
&\leqq \sum_{i=1}^{n} |a_i|\,\big|\,|x_1 - w_i| - |x_2 - w_i|\,\big| \\
&\leqq \sum_{i=1}^{n} |a_i||x_1 - x_2| = L|x_1 - x_2|
\end{aligned}$$

■

PROBLEMS 8.1

1. Find two solutions of the initial-value problem

$$\begin{cases} x' = x^{1/3} \\ x(0) = 0 \end{cases}$$

Hint: Try $x = ct^{\lambda}$, or observe that the equation is separable.

2. **(a)** Use Theorem 1 to predict in what interval a solution of the initial-value problem (3) exists. Find the largest interval.

   (b) Repeat part **(a)** for the initial-value problem (2).

3. Show that $x = -t^2/4$ and $x = 1 - t$ are solutions of the initial-value problem

$$\begin{cases} 2x' = \sqrt{t^2 + 4x} - t \\ x(2) = -1 \end{cases}$$

Why does this not contradict Theorem 2?

4. Solve the initial-value problem $x' = f(t, x)$, $x(0) = 0$ in these special cases:

(a) $f(t, x) = t^3$

(b) $f(t, x) = (1 - t^2)^{-1/2}$

(c) $f(t, x) = (1 + t^2)^{-1}$

(d) $f(t, x) = (t + 1)^{-1}$

5. Solve the initial-value problem $x' = f(t, x)$, $x(0) = 0$ in the following cases. Use the fact that $dt/dx = (dx/dt)^{-1}$ when $dx/dt \neq 0$.

(a) $f(t, x) = x^{-2}$

(b) $f(t, x) = 1 + x^2$

(c) $f(t, x) = (\sin x + \cos x)^{-1}$

6. Use Theorem 1 to show that the initial-value problem

$$\begin{cases} x' = \sqrt{|x|} \\ x(0) = 0 \end{cases}$$

has a solution on the entire real line.

7. Show by using Theorem 1 that the initial-value problem

$$\begin{cases} x' = \tan x \\ x(0) = 0 \end{cases}$$

has a solution in the interval $|t| < \pi/4$.

8. Let f be a continuous function of one variable, defined on all of $\mathbb{R}$. Let $M(r)$ denote the maximum of $|f(x)|$ for $|x| \leqq r$. If $M(r) = o(r)$ as $r \to \infty$, then the initial-value problem

$$\begin{cases} x' = f(x) \\ x(0) = 0 \end{cases}$$

has a solution on all of $\mathbb{R}$. Prove this assertion.

9. Prove that the initial-value problem

$$\begin{cases} x' = t^2 + e^x \\ x(0) = 0 \end{cases}$$

has a unique solution in the interval $|t| \leqq 0.351$.

10. Prove that if $f(t, x)$ is continuous and bounded in the domain $a \leqq t \leqq b$, $-\infty < x < \infty$, then the initial-value problem

$$\begin{cases} x' = f(t, x) \\ x(a) = \alpha \end{cases}$$

has a solution in the interval $a \leqq t \leqq b$.

11. Let R denote the rectangle in the tx-plane defined by $|t - t_0| \leqq \alpha$, $|x - x_0| \leqq \beta$. Let f be a continuous function defined on this rectangle and satisfying $\beta \geqq \alpha|f(t, x)|$. Prove that the initial-value problem $x' = f(t, x)$, $x(t_0) = x_0$ has a solution on the interval $|t - t_0| \leqq \alpha$.

12. Establish that the initial-value problem

$$\begin{cases} x' = 1 + x + x^2 \cos t \\ x(0) = 0 \end{cases}$$

has a solution in the interval $-1/3 \leqq t \leqq 1/3$.

13. Show that the initial-value problem

$$\begin{cases} x' = \sqrt{|x|} \\ x(0) = 0 \end{cases}$$

has two solutions, and indicate why Theorem 2 does not apply.

14. Prove that the problem

$$\begin{cases} x' = tx^{2/3} \\ x(0) = 1 \end{cases}$$

has a solution in the interval $-2 \leqq t \leqq 2$. Is there more than one solution?

15. Show that the initial-value problem

$$\begin{cases} x' = 2te^{-x} \\ x(0) = 0 \end{cases}$$

has a unique solution in the interval $-\infty < t < \infty$. Is there a choice of α and β in the existence theorem that allows this conclusion to be drawn?

16. Let $f(t, x)$ be continuous on the rectangle defined by $|t| \leqq 3$, $|x| \leqq 4$, and assume that $|f(t, x)| \leqq 7$ for all points in this rectangle. What is the largest interval in which the initial-value problem

$$\begin{cases} x' = f(t, x) \\ x(0) = 0 \end{cases}$$

must have a solution?

17. Find an interval in which we can be sure that the initial-value problem

$$\begin{cases} x' = \sec x \\ x(0) = 0 \end{cases}$$

has a unique solution.

18. Use each of Theorems 1, 2, and 3 to predict where this problem has a solution, and then solve the problem explicitly, comparing the theory to the fact:

$$\begin{cases} x' = x^2 \\ x(0) = 1 \end{cases}$$

19. Prove that the initial-value problem

$$\begin{cases} x' = 1 + x^2 \\ x(0) = 0 \end{cases}$$

has a solution in the interval $[-1, 1]$. Prove that this example does not satisfy the hypotheses of Theorem 3. Explain why this example does not contradict Theorem 3.

COMPUTER PROBLEM 8.1

1. Using a symbolic manipulation program, find the solution of the differential equation $x' + x = (1 + e^t)^{-1}$.

8.2 Taylor-Series Method

In the numerical solution of differential equations, we rarely expect to obtain the solution directly as a *formula* giving $x(t)$ as a function of t. Instead, we usually construct a table of function values of the form

t_0	t_1	t_2	t_3	$\cdots$	t_m
x_0	x_1	x_2	x_3	$\cdots$	x_m

(1)

Here, x_i is the computed approximate value of $x(t_i)$, our notation for the *exact* solution at t_i. From a table such as (1), a spline function or other approximating function can be constructed. However, most numerical methods for solving ordinary differential equations produce such a table first.

We consider again the initial-value problem

$$\begin{cases} x' = f(t, x) \\ x(t_0) = x_0 \end{cases} \tag{2}$$

where f is a prescribed function of two variables, and (t_0, x_0) is a single given point through which the solution curve should pass. A *solution* of (2) is a function $x(t)$ such that $dx(t)/dt = f(t, x(t))$ for all t in some neighborhood of t_0, and $x(t_0) = x_0$.

Illustration

For the Taylor-series method, it is necessary to assume that various partial derivatives of f exist. To illustrate the method, we take a concrete example:

$$\begin{cases} x' = \cos t - \sin x + t^2 \\ x(-1) = 3 \end{cases} \tag{3}$$

At the heart of the procedure is the Taylor series for x, which we write as

$$x(t + h) = x(t) + hx'(t) + \frac{h^2}{2!}x''(t) + \frac{h^3}{3!}x'''(t) + \frac{h^4}{4!}x^{(4)}(t) + \cdots \tag{4}$$

The derivatives appearing here can be obtained from the differential equation in (3). They are

$$\begin{aligned} x'' &= -\sin t - x'\cos x + 2t \\ x''' &= -\cos t - x''\cos x + (x')^2\sin x + 2 \\ x^{(4)} &= \sin t - x'''\cos x + 3x'x''\sin x + (x')^3\cos x \end{aligned}$$

At this point, our patience wears thin and we decide to use only terms up to and including h^4 in Formula (4). The terms that we have *not* included start with a term in h^5, and they constitute collectively the **truncation error** inherent in our procedure. The resulting numerical method is said to be of order 4. (The *order* of the Taylor-series method is n if terms up to and including $h^n x^{(n)}(t)/n!$ are used.) Notice that in differentiating terms such as $\sin x$ with respect to t, we must think of it as $d\{\sin[x(t)]\}/dt$ and employ the *chain rule* of differentiation. This, of course, accounts for the complexity of the formulas for $x'', x''', \ldots$. We could perform various substitutions to obtain formulas for $x'', x''', \ldots$ containing no derivatives of x on the right-hand side. It is not necessary to do this if the formulas are used in the order listed. They are recursive in nature.

Here is an algorithm to solve the initial-value problem (3), starting at $t = -1$ and progressing in steps of $h = 0.01$. We desire a solution in the t-interval $[-1, 1]$, and thus we must take 200 steps.

input $M \leftarrow 200;\ h \leftarrow 0.01;\ t \leftarrow -1.0;\ x \leftarrow 3.0$
output $0, t, x$
for $k = 1$ **to** M **do**
$\quad x' \leftarrow \cos t - \sin x + t^2$
$\quad x'' \leftarrow -\sin t - x'\cos x + 2t$
$\quad x''' \leftarrow -\cos t - x''\cos x + (x')^2\sin x + 2$
$\quad x^{(4)} \leftarrow \sin t + ((x')^3 - x''')\cos x + 3x'x''\sin x$
$\quad x \leftarrow x + h(x' + \frac{h}{2}(x'' + \frac{h}{3}(x''' + \frac{h}{4}(x^{(4)}))))$
$\quad t \leftarrow t + h$
$\quad$ **output** k, t, x
end do

What can be said of the errors in our computed solution? At each step, the **local truncation error** is $\mathcal{O}(h^5)$ since we have *not* included terms involving $h^5, h^6, \ldots$ from the Taylor series. Thus, as $h \to 0$, the behavior of the local errors should be similar to Ch^5. Unfortunately, we do not know C. But h^5 is 10^{-10} since $h = 10^{-2}$. So with good luck, the error in each step should be roughly of the magnitude 10^{-10}. After several hundred steps, these small errors could accumulate and spoil the numerical solution. At each step (except the first), the estimate x_k of $x(t_k)$ already contains errors, and further computations continue to add to these errors. These remarks should make us quite cautious about accepting blindly all the decimal digits in a numerical solution of a differential equation!

When the preceding algorithm was programmed and run, the solution at $t = 1$ was $x_{200} = 6.42194$. Here is a sample of the output from that computer program:

k	t	x
0	−1.00000	3.00000
1	−0.99000	3.01400
2	−0.98000	3.02803
3	−0.97000	3.04209
4	−0.96000	3.05617
⋮	⋮	⋮
196	0.96000	6.36566
197	0.97000	6.37977
198	0.98000	6.39386
199	0.99000	6.40791
200	1.00000	6.42194

In a subsequent computer run, the differential equation was integrated with this value of x_{200} as the *initial* condition and with $h = -0.01$. The result of this second computer run was $x_{200} \approx 3.00000$ at $t = -0.99999$. The close agreement with the original initial value leads us to think that the numerical solution is accurate to about six significant figures—the number of digits shown.

In the example just discussed, it is not difficult to estimate the local truncation error in each step of the numerical solution. To do this, we recall that the error term in the Taylor series (4) is of the form

$$E_n = \frac{1}{(n+1)!} h^{n+1} x^{(n+1)}(t + \theta h) \qquad (0 < \theta < 1)$$

This is the error that is present when the last power of h included in the sum is h^n. In the example, we have taken $n = 4$ and $h = 0.01$. One can estimate $x^{(5)}(t + \theta h)$ by a simple finite-difference approximation and arrive at

$$E_4 \approx \frac{1}{5!} h^5 \left[\frac{x^{(4)}(t+h) - x^{(4)}(t)}{h} \right] = \frac{h^4}{120} \left[x^{(4)}(t+h) - x^{(4)}(t) \right]$$

With a little extra programming, this estimate of E_4 can be incorporated into the algorithm and computed. We do not give the details here because it is Computer Problem 19. When this feature is added to the program, the computed output shows that our estimate of E_4 never exceeds 3.42×10^{-11} in absolute value. The program was run in double precision on a computer similar to the `Marc-32`.

Pros and Cons

What are the advantages and disadvantages of the Taylor-series method? For disadvantages, the method depends on repeated differentiation of the given differential equation (unless we intend to use only the method of order 1). Hence, the function $f(t, x)$ must possess partial derivatives in the region where the solution curve passes in the tx-plane. Such an assumption is, of course, not necessary for the existence of a solution. Also, the necessity of carrying out the preliminary analytic work is a decided drawback. For example, an error made at this stage might be overlooked and remain undetected. Finally, the various derivatives must be separately programmed.

Advantages are the conceptual simplicity of the procedure and its potential for very high precision. Thus, if we could easily obtain 20 derivatives of $x(t)$, then there is nothing to prevent us from using the method with order 20 (that is, terms up to and including the one involving h^{20}). With such a high order, the same accuracy can be obtained with a larger step size, say $h = 0.2$. Fewer steps are required to traverse the given interval, and this reduces the amount of computation. On the other hand, the calculations in each step are more burdensome. Some examples in which a high-order Taylor-series method can be used are among the problems.

In recent years, symbol-manipulating programs have become available for carrying out various routine mathematical calculations of a nonnumerical type. Differentiation and integration of rather complicated expressions can thus be turned over to the computer! Of course, these programs apply only to a restricted class of functions, but this class is broad enough to include all the functions that one encounters in the typical calculus textbook. With the use of such programs, the Taylor-series method of order 20, say, could be used without difficulty.

Errors

In solving a differential equation numerically, several types of errors arise. These are conveniently classified as follows:

(i) Local truncation error
(ii) Local roundoff error
(iii) Global truncation error
(iv) Global roundoff error
(v) Total error

The **local truncation error** is the error made in one step when we replace an infinite process by a finite one. In the Taylor-series method, we replace the infinite Taylor series for $x(t + h)$ by a partial sum. The local truncation error is inherent in any algorithm that we might choose and is quite independent of the roundoff error. **Roundoff error** is, of course, caused by the limited precision of our computing machines, and its magnitude depends on the word length of the computer (or on the number of bits in the mantissa of the floating-point machine numbers).

In the Taylor-series method, if we retain terms up to and including h^n in the series, then the local truncation error is the sum of all the remaining terms that we do *not* include. By Taylor's Theorem, these can be compressed into a single term of the form

$$\frac{h^{n+1}}{(n+1)!}x^{(n+1)}(\xi)$$

for some nearby point ξ. We say that the local truncation error is $\mathcal{O}(h^{n+1})$. An error of this sort is present in each step of the numerical solution. The accumulation of all these many local truncation errors gives rise to the **global truncation** error. Again, this error will be present even if all calculations are performed using exact arithmetic. It is an error that is associated with the method and is independent of the computer on which the calculations are performed. If the local truncation errors are $\mathcal{O}(h^{n+1})$, then the global truncation error must be $\mathcal{O}(h^n)$ because the number of steps necessary to reach an arbitrary point T, having started at t_0, is $(T - t_0)/h$.

The **global roundoff error** is the accumulation of local roundoff errors in prior steps. The **total error** is the sum of the global truncation error and the global roundoff error. If the global truncation error is $\mathcal{O}(h^n)$, we say that the numerical procedure is of **order** n.

Euler's Method

The Taylor-series method with $n = 1$ is called **Euler's method**. It looks like this:

$$x(t+h) = x(t) + hf(t, x)$$

This formula has the obvious advantage of *not* requiring any differentiation of f. This advantage is offset by the necessity of taking small values for h to gain acceptable precision. Still, the method serves as a useful example and is of great importance theoretically since existence theorems can be based on it. See Henrici [1962, pp. 15–25] for such an existence theorem.

Delay Differential Equations

A special type of differential equation, called a **delay differential equation** or a **differential equation with retarded argument**, arises in some practical problems. Population models and mixing problems often have this special feature, which is that the value of $x'(t)$ is related to values of the function x at previous values of t. For example, we might have

$$x'(t) = f\big(x(t-1)\big)$$

If we know the value of x at $t - 1$, the differential equation enables us to compute the value of $x'(t)$. To integrate the differential equation starting at $t = 0$, we shall need the history of $x(t)$ starting at $t = -1$. Thus, the values of $x(t)$ on the interval $[-1, 0]$ will have to be supplied to us as *initial values*.

An example of a specific and well-defined problem of this type might be

$$\begin{cases} x'(t) = x(t-1) & (t \geqq 0) \\ x(t) = t^2 & (-1 \leqq t \leqq 0) \end{cases} \tag{5}$$

The second of these equations gives the needed initial values for $x(t)$. If t is restricted to the interval $[0, 1]$, then $t - 1$ is in $[-1, 0]$, and hence

$$\begin{cases} x'(t) = x(t-1) = (t-1)^2 & (0 \leqq t \leqq 1) \\ x(0) = 0 \end{cases}$$

The solution of this is easily obtained by integrating and supplying an appropriate constant of integration:

$$x(t) = \frac{1}{3}(t-1)^3 + \frac{1}{3} \qquad (0 \leqq t \leqq 1)$$

If the solution is to be continued into the next interval $[1, 2]$, another step like the first can be taken. Thus, for t in $[1, 2]$, we have

$$\begin{cases} x'(t) = x(t-1) = \dfrac{1}{3}(t-2)^3 + \dfrac{1}{3} & (1 \leqq t \leqq 2) \\ x(1) = \dfrac{1}{3} \end{cases}$$

The solution of this is

$$x(t) = \frac{1}{12}(t-2)^4 + \frac{1}{3}t - \frac{1}{12} \qquad (1 \leqq t \leqq 2)$$

The solution can be continued indefinitely to the right by similar calculations.

For more complicated equations, such as

$$x'(t) = \sin\left[x(t-1)^3\right] + \log\left[x(t) + t^5\right]$$

one must resort to numerical procedures. The Taylor series method is available but is not without some drawbacks. Consider, for example,

$$\begin{cases} x'(t) = 2x(t-1) + x(t) & (t > 0) \\ x(t) = t^2 & (-1 \leqq t \leqq 0) \end{cases} \tag{6}$$

To compute the solution on the interval $[0, 1]$, one can proceed in steps of length h by using a short Taylor expansion:

$$x(t+h) = x(t) + hx'(t) + \frac{h^2}{2}x''(t) + \frac{h^3}{6}x'''(t)$$

As usual, in the Taylor-series method, we must provide formulas for x', x'', and x''' by using the differential equation. In this example, if we consider only the interval

[0, 1], we can use

$$x'(t) = 2x(t-1) + x(t) = 2(t-1)^2 + x(t)$$
$$x''(t) = 2x'(t-1) + x'(t) = 4(t-2)^2 + 2(t-1)^2 + x'(t)$$
$$x'''(t) = 2x''(t-1) + x''(t) = 8(t-3)^2 + 8(t-2)^2 + 2(t-1)^2 + x''(t)$$

The numerical solution produces values of $x(t)$ at discrete points in the interval [0, 1]. At the same time, we should store the values of $x'(t)$, $x''(t)$, and $x'''(t)$ at these same discrete points for use in the next interval. If we do not change the value of h, we can proceed in each interval in the same way using the appropriate saved values.

For the theory of delay differential equations, see the books by Driver [1977] and Diekmann, Van Gils, Verduyn Lunel, and Walther [1995]. Also, a code called DELSOL is described in Willé and Baker [1992].

PROBLEMS 8.2

1. In Computer Problem 3, it is necessary to compute derivatives of the function e^{-t^2}. Show that the nth derivative of this function is of the form $e^{-t^2}P_n(t)$, where the polynomials P_n are determined recursively from the formulas

$$\begin{cases} P_0 = 1 \\ P_{n+1}(t) = P_n'(t) - 2tP_n(t) \end{cases}$$

Show that, for example,

$$P_4(t) = 12 - 48t^2 + 16t^4$$
$$P_5(t) = -120t + 160t^3 - 32t^5$$

2. Verify that the function $x(t) = t^2/4$ solves the initial-value problem

$$\begin{cases} x' = \sqrt{x} \\ x(0) = 0 \end{cases}$$

Apply the Taylor-series method of order 1 and explain why the numerical solution differs from the solution $t^2/4$.

3. Compute $x(0.1)$ by solving the differential equation

$$\begin{cases} x' = -tx^2 \\ x(0) = 2 \end{cases}$$

with one step of the Taylor-series method of order 2. (Use a calculator.)

4. Using the ordinary differential equation

$$\begin{cases} x' = x^2 + xe^t \\ x(0) = 1 \end{cases}$$

and one step of the Taylor-series method of order 3, calculate $x(0.01)$.

5. Consider the ordinary differential equation

$$\begin{cases} 5tx' + x^2 = 2 \\ x(4) = 1 \end{cases}$$

Calculate $x(4.1)$ using one step of the Taylor-series method of order 2.

6. An **integral equation** is an equation involving an unknown function within an integration. For example, here is a typical integral equation (of a type known by the name **Volterra**):

$$x(t) = \int_0^t \cos\big(s + x(s)\big)\, ds + e^t$$

By differentiating this integral equation, obtain an equivalent initial-value problem for the unknown function.

7. If the Taylor-series method is used to solve an initial-value problem involving the differential equation

$$x' = \cos(tx)$$

what are the formulas for x'', x''', and $x^{(4)}$?

8. Let $x' = f(t, x)$. Determine x'', x''', and $x^{(4)}$ from this equation.

COMPUTER PROBLEMS 8.2

1. Write and test a computer program to solve the following differential equation with initial condition:

$$\begin{cases} x' = x + e^t + tx \\ x(1) = 2 \end{cases}$$

on the interval $[1, 3]$. Use the Taylor series of order 5 and $h = 0.01$.

2. Write and test a computer program for solving the following initial-value problem:

$$\begin{cases} x' = 1 + x^2 - t^3 \\ x(0) = -1 \end{cases}$$

Use the Taylor-series method of order 4, with h a binary machine number near 0.01. Find the solution in the interval $[0, 2]$. Account for any peculiar phenomenon in the solution.

3. Methods for solving initial-value problems can also be used to compute definite or indefinite integrals. For example, we can compute

$$\int_0^2 e^{-s^2}\, ds$$

by solving the initial-value problem

$$\begin{cases} x' = e^{-t^2} \\ x(0) = 0 \end{cases}$$

on the t-interval [0, 2]. Do this, using the Taylor-series method of order 4. From a table of the **error function**

$$\text{erf}(t) = \frac{2}{\sqrt{\pi}} \int_0^t e^{-s^2}\, ds$$

we obtain $x(2) \approx 0.88208\,13907$. (See Abramowitz and Stegun [1964, p. 311].)

4. (Continuation) Use Problem 1 to write a computer program for the sixth-order Taylor-series method applicable to the integral in Computer Problem 3. Test the code with $h = 0.01$ to see whether the value of $x(2)$ given in Computer Problem 3 is obtained. Print a table of the function $\frac{1}{2}\sqrt{\pi}\,\text{erf}(t)$ in steps of 0.01 from $t = 0$ to $t = 2$.

5. Solve the initial-value problem

$$\begin{cases} x' = 1 + x^2 \\ x(0) = 0 \end{cases}$$

on the interval [0, 1.56] using the Taylor-series method of order 4 with $h = 0.01$. Then use the computed value of $x(1.56)$ as the initial value to integrate back to $t = 0$. Compare the results and explain what happened.

6. The equation $\arctan(x/t) = \ln\sqrt{x^2 + t^2}$ defines x implicitly as a function of t. Verify that this implicit function is a solution of the initial-value problem

$$\begin{cases} x' = (t + x)/(t - x) \\ x(1) = 0 \end{cases}$$

Prepare a table of the function $x(t)$ on the interval [0, 2] with steps of ± 0.01. Use the Taylor-series method of order 4.

7. The function

$$\varphi(t) = \int_0^t \sin s^2\, ds$$

is known as a **Fresnel integral**. Prepare a table of this function on the interval [0, 10]. Use the Taylor-series method of order 5 with $h = 0.1$. If possible, obtain a computer plot of the function.

8. (Continuation) Show that the derivatives of the function $f(t) = \sin t^2$ are

$$f^{(n)}(t) = P_n(t)\sin t^2 + Q_n(t)\cos t^2$$

where P_n and Q_n are computed recursively as follows:

$$\begin{cases} P_0 = 1 \\ P_{n+1} = P_n' - 2tQ_n \end{cases} \qquad \begin{cases} Q_0 = 0 \\ Q_{n+1} = Q_n' + 2tP_n \end{cases}$$

Use this to generate a table of P_n and Q_n values for $n = 0$ to 6. Modify the code in Computer Problem 7 to become seventh order. Test as in Computer Problem 7, first with $h = 0.1$ and then with $h = 0.2$. Compare the value of $\varphi(10)$ obtained with $h = 0.2$ to the value obtained in Computer Problem 7.

9. The function

$$x(t) = \int_0^t (1 - k\sin^2\theta)^{1/2}\,d\theta$$

where k is a parameter in [0, 1], is an **elliptic integral of the second kind.** Make a table of this function when $k = 1/2$ for the interval $0 \leqq t \leqq \pi/2$ using the Taylor-series method of order 3 with $h = 0.01$.

10. Solve the initial-value problem

$$\begin{cases} x' = -\dfrac{3t^2x + x^2}{2t^3 + 3tx} \\ x(1) = -2 \end{cases}$$

by the Taylor-series method of order 2 on the interval [0, 1] with steps $h = -0.01$. Use the implicit solution $t^3x^2 + tx^3 + 4 = 0$ as a check on the computed solution. Also verify the correctness of the implicit solution.

11. Prepare a table of a **dilogarithm** function

$$f(x) = -\int_0^x \frac{\ln(1-t)}{t}\,dt$$

on the interval $-2 \leqq x \leqq 0$ by solving an appropriate initial-value problem. (Compare with Problem 14 of Section 6.7.)

12. The integral $\int \sqrt{1 + x^3}\,dx$ is one that cannot be obtained by the methods of elementary calculus. (It is an **elliptic integral.**) Prepare a table of the function

$$f(x) = \int_0^x \sqrt{1 + t^3}\,dt$$

on the interval $0 \leqq x \leqq 5$ by solving a suitable initial-value problem. Use the Taylor-series method of order 3 with $h = 1/64$.

13. Consider the initial-value problem $x' = 1 - xt^{-1}$, $x(2) = 2$. Prove that

$$\begin{cases} x'' = (1 - 2x')t^{-1} \\ x^{(n)} = -nx^{(n-1)}t^{-1} \qquad (n \geqq 3) \end{cases}$$

Write a computer program to solve this initial-value problem by the Taylor-series method of order 10. Test your program using $h = 1$ and the interval $[2, 20]$. What is the analytic solution of the problem? Compare it to your numerical solution. (See Conte and de Boor [1980].)

14. Write a computer program to solve the initial-value problem

$$\begin{cases} x' = t^2 + x^2 + 2tx \\ x(0) = 7 \end{cases}$$

using the Taylor-series method. A solution is wanted in the interval $-2 \leqq t \leqq 2$. Include terms up to and including h^3. Use step size 0.01.

15. Write a program to compute a table of values for the function

$$x(t) = \int_2^t \sin(u^3)\, du$$

We want to cover the interval $2 \leqq t \leqq 5$ in steps of 0.05. Do this by setting up an appropriate initial-value problem. Then carry out the numerical solution using the Taylor-series method of order 3. Estimate the accuracy of the resulting table.

16. Solve the integral equation in Problem 6 by using the Taylor-series method of order 3 on the equivalent initial-value problem. Obtain the solution on $[0, 3]$ using steps of 0.01.

17. Using a symbol-manipulating program, solve the initial-value problem given by Equation (3) in the text using the Taylor-series method of orders 4 and 20. Compare to the numerical results given in the text. Compare these two Taylor-series methods using various step sizes h. Determine how large h can be for the method of order 20 to obtain the same accuracy as the method of order 4 with $h = 0.01$.

18. (Continuation) Repeat for the initial-value problem in Computer Problem 6.

19. Reprogram the example in the text, including the estimate given for E_4.

20. Solve numerically the initial-value problem

$$\begin{cases} (x')^2 - 2tx' - x\cos t = 0 \\ x(0) = 0 \end{cases}$$

on the interval $0 \leqq t \leqq 1$. (The trivial solution is not the one wanted here!) What would you have done if $(x')^2$ had been $(x')^3$?

21. Solve Equation (5) numerically and compare to the true solution.

22. Carry out the numerical solution of Equation (6) using the Taylor-series method of order 3 over the interval $[0, 1]$ with step size $h = 0.01$.

8.3 Runge-Kutta Methods

The Taylor-series method of the preceding section has the drawback of requiring some analysis prior to programming it. Thus, if we wish to use the fourth-order Taylor-series method on the general problem

$$\begin{cases} x' = f(t, x) \\ x(t_0) = x_0 \end{cases} \tag{1}$$

we have to determine formulas for x'', x''', and $x^{(4)}$ by successive differentiation in (1). Then these functions will have to be programmed.

The Runge-Kutta methods avoid this difficulty, although they do imitate the Taylor-series method by means of clever combinations of values of $f(t, x)$. We illustrate by deriving a second-order Runge-Kutta procedure.

Second-Order Runge-Kutta Method

Let us begin with the Taylor series for $x(t + h)$:

$$x(t + h) = x(t) + hx'(t) + \frac{h^2}{2!}x''(t) + \frac{h^3}{3!}x'''(t) + \cdots \tag{2}$$

From the differential equation, we have

$$\begin{aligned} x'(t) &= f \\ x''(t) &= f_t + f_x x' = f_t + f_x f \\ x'''(t) &= f_{tt} + f_{tx} f + (f_t + f_x f) f_x + f(f_{xt} + f_{xx} f) \\ &\vdots \end{aligned}$$

Here subscripts denote partial derivatives, and the chain rule of differentiation is used repeatedly. The first three terms in Equation (2) can be written now in the form

$$\begin{aligned} x(t + h) &= x + hf + \frac{1}{2}h^2(f_t + ff_x) + \mathcal{O}(h^3) \\ &= x + \frac{1}{2}hf + \frac{1}{2}h[f + hf_t + hff_x] + \mathcal{O}(h^3) \end{aligned} \tag{3}$$

where x means $x(t)$, f means $f(t, x)$, and so on. We are able to eliminate the partial derivatives with the aid of the first few terms in the Taylor series in two variables (as given in Section 1.1):

$$f(t + h, x + hf) = f + hf_t + hff_x + \mathcal{O}(h^2)$$

Equation (3) can be rewritten as

$$x(t+h) = x + \frac{1}{2}hf + \frac{1}{2}hf(t+h, x+hf) + \mathcal{O}(h^3)$$

Hence, the formula for advancing the solution is

$$x(t+h) = x(t) + \frac{h}{2}f(t,x) + \frac{h}{2}f\big(t+h, x+hf(t,x)\big)$$

or equivalently,

$$x(t+h) = x(t) + \frac{1}{2}(F_1 + F_2) \tag{4}$$

where

$$\begin{cases} F_1 = hf(t,x) \\ F_2 = hf(t+h, x+F_1) \end{cases}$$

This formula can be used repeatedly to advance the solution one step at a time. It is called a **second-order Runge-Kutta method**. It is also known as **Heun's method**.

In general, second-order Runge-Kutta formulas are of the form

$$x(t+h) = x + w_1hf + w_2hf(t+\alpha h, x+\beta hf) + \mathcal{O}(h^3) \tag{5}$$

where w_1, w_2, α, and β are parameters at our disposal. Equation (5) can be rewritten with the aid of the Taylor series in two variables as

$$x(t+h) = x + w_1hf + w_2h[f + \alpha hf_t + \beta hff_x] + \mathcal{O}(h^3) \tag{6}$$

Comparing Equations (3) and (6), we see that we should impose these conditions:

$$\begin{cases} w_1 + w_2 = 1 \\ w_2\alpha = \frac{1}{2} \\ w_2\beta = \frac{1}{2} \end{cases} \tag{7}$$

One solution is $w_1 = w_2 = \frac{1}{2}$, $\alpha = \beta = 1$, which is the one corresponding to Heun's method in Equation (4). The system of equations (7) has solutions other than this one, such as the one obtained by letting $w_1 = 0$, $w_2 = 1$, $\alpha = \beta = \frac{1}{2}$. The resulting formula from (5) is called the **modified Euler method**:

$$x(t+h) = x(t) + F_2$$

where

$$\begin{cases} F_1 = hf(t, x) \\ F_2 = hf(t + \frac{1}{2}h,\ x + \frac{1}{2}F_1) \end{cases}$$

Compare this to the standard Euler method described in Section 8.2.

Fourth-Order Runge-Kutta Method

The higher-order Runge-Kutta formulas are very tedious to derive, and we shall not do so. The formulas are rather elegant, however, and are easily programmed once they have been derived. Here are the formulas for the *classical* **fourth-order Runge-Kutta method**:

$$x(t + h) = x(t) + \frac{1}{6}(F_1 + 2F_2 + 2F_3 + F_4) \tag{8}$$

where

$$\begin{cases} F_1 = hf(t, x) \\ F_2 = hf(t + \frac{1}{2}h,\ x + \frac{1}{2}F_1) \\ F_3 = hf(t + \frac{1}{2}h,\ x + \frac{1}{2}F_2) \\ F_4 = hf(t + h,\ x + F_3) \end{cases}$$

This is called a fourth-order method because it reproduces the terms in the Taylor series up to and including the one involving h^4. The error is therefore $\mathcal{O}(h^5)$. Exact expressions for the h^5 error term are available.

Example 1 Give an algorithm incorporating the Runge-Kutta method of order 4 for solving the following initial-value problem:

$$\begin{cases} x' = t^{-2}(tx - x^2) \\ x(1) = 2 \end{cases} \tag{9}$$

on the interval $[1, 3]$, using steps of $h = 1/128$.

Solution Since this is a numerical experiment to gauge the effectiveness of the procedure, a problem has been chosen with a known analytic solution. The solution of (9) is given by $x(t) = (\frac{1}{2} + \ln t)^{-1}t$. The values of the error are printed by the computer program.

input $M \leftarrow 256;\ t \leftarrow 1.0;\ x \leftarrow 2.0;\ h \leftarrow 0.0078125$
define $f(t, x) = (tx - x^2)/t^2$
define $u(t) = t/(\frac{1}{2} + \ln t)$
$e \leftarrow |u(t) - x|$

output $0, t, x, e$
for $k = 1$ **to** M **do**
  $F_1 \leftarrow hf(t, x)$
  $F_2 \leftarrow hf(t + \frac{1}{2}h, x + \frac{1}{2}F_1)$
  $F_3 \leftarrow hf(t + \frac{1}{2}h, x + \frac{1}{2}F_2)$
  $F_4 \leftarrow hf(t + h, x + F_3)$
  $x \leftarrow x + (F_1 + 2F_2 + 2F_3 + F_4)/6$
  $t \leftarrow t + h$
  $e \leftarrow |u(t) - x|$
  output k, t, x, e
end do

Some of the output from a computer program based on this algorithm follows:

k	t	x	e
0	1.00000	2.00000	
1	1.00781	1.98473	1.19×10^{-07}
2	1.01563	1.97016	0.
3	1.02344	1.95623	0.
4	1.03125	1.94293	1.19×10^{-07}
5	1.03906	1.93020	1.19×10^{-07}
6	1.04688	1.91802	0.
7	1.05469	1.90637	1.19×10^{-07}
8	1.06250	1.89521	1.19×10^{-07}
9	1.07031	1.88452	0.
⋮	⋮	⋮	⋮
249	2.94531	1.86387	7.15×10^{-07}
250	2.95313	1.86569	5.96×10^{-07}
251	2.96094	1.86750	7.15×10^{-07}
252	2.96875	1.86932	5.96×10^{-07}
253	2.97656	1.87115	4.77×10^{-07}
254	2.98438	1.87297	5.96×10^{-07}
255	2.99219	1.87480	5.96×10^{-07}
256	3.00000	1.87663	5.96×10^{-07}

■

Errors

We turn now to a discussion of **truncation error** in the Runge-Kutta procedure. Roughly described, the local truncation error is the error that arises in each step simply because our methods cannot take into account all the terms of a Taylor series. This error is inevitable and will be present even if the calculations are executed with *infinite* precision. In the case of the fourth-order Runge-Kutta method (discussed above), the formulas involve a local truncation error of $\mathcal{O}(h^5)$. This order of error corresponds exactly to the error in the Taylor-series method when terms up to h^4 are included in Taylor's formula.

At the very first step in the fourth-order Runge-Kutta procedure, a value of $x(t_0 + h)$ is computed by the algorithm. On the other hand, there is a correct solution, $x^*(t_0 + h)$, which we shall not know. The local truncation error in this step is, by definition,

$$x^*(t_0 + h) - x(t_0 + h)$$

The theory of the Runge-Kutta algorithm indicates that this truncation error behaves like Ch^5 for small values of h. Here C is a number independent of h but dependent on t_0 and on the function x^*. To estimate Ch^5, we assume that C does *not* change as t changes from t_0 to $t_0 + h$. Let v be the value of the approximate solution at $t_0 + h$ obtained by taking one step of length h from t_0. Let u be the approximate solution at $t_0 + h$ obtained by taking *two* steps of size $h/2$ from t_0. These are both *computable*. By the assumption made, we have

$$\begin{aligned} x^*(t_0 + h) &= v + Ch^5 \\ x^*(t_0 + h) &= u + 2C(h/2)^5 \end{aligned}$$

By subtraction, we obtain from these two equations:

$$\text{Local truncation error } = Ch^5 = (u - v)/(1 - 2^{-4})$$

Thus, the local truncation error is approximated by $u - v$.

In a computer realization of the Runge-Kutta method, the approximate truncation error can be occasionally monitored, by computing $|u - v|$, to ensure that it remains below a specified tolerance. If it does not, the step size can be decreased (usually halved) to improve the local truncation error. On the other hand, if the local truncation error is far below a permitted threshold, then the step size can be doubled.

As we saw in the derivation of the Runge-Kutta method of order 2, a number of parameters must be selected. A similar selection process occurs in establishing higher-order Runge-Kutta methods. Consequently, there is not just *one* Runge-Kutta method of each order, but a family of methods. As shown in the following table, the number of required function evaluations increases more rapidly than the order of the Runge-Kutta methods:

Number of function evaluations	1	2	3	4	5	6	7	8
Maximum order of Runge-Kutta method	1	2	3	4	4	5	6	6

Unfortunately, this makes the higher-order Runge-Kutta methods less attractive than the classical fourth-order method, since they are more expensive to use.

Adaptive Runge-Kutta-Fehlberg Method

In an effort to devise a procedure for automatically adjusting the step size in the Runge-Kutta method, Fehlberg [1969] turned to a fourth-order method with *five* function evaluations and a fifth-order method with *six* function evaluations. At first glance, this does not seem to be promising since his fourth-order method required more function evaluations than the classical method. Moreover, these two methods together would require a total of 11 function evaluations. However, Fehlberg was able to select the parameters in these methods to obtain two formulas of different orders with the same points for the function evaluations. Hence, only *six* function evaluations are required, and as a bonus we obtain an estimate for the local truncation error on which to base the control of the step size. The resulting **Runge-Kutta-Fehlberg method** is of order 5 and utilizes two formulas, having orders 4 and 5. These formulas give different approximate values of the solution, and we denote them by $x(t+h)$ and $\overline{x}(t+h)$:

$$x(t+h) = x(t) + \sum_{i=1}^{6} a_i F_i \tag{10}$$

$$\overline{x}(t+h) = x(t) + \sum_{i=1}^{6} b_i F_i \tag{11}$$

The quantities F_i are computed from formulas of the type

$$F_i = hf\left(t + c_i h,\ x + \sum_{j=1}^{i-1} d_{ij} F_j\right) \qquad (1 \leqq i \leqq 6) \tag{12}$$

Formula (10) is of the fifth order, and Formula (11) is of the fourth order. The difference, $e = x(t+h) - \overline{x}(t+h)$, can be interpreted as an estimate of the local truncation error associated with the less accurate Formula (11). Thus, e can be used to monitor the step size. Since Formula (10) is the more accurate, it is used to provide the output in the algorithm. Note that

$$e = x(t+h) - \overline{x}(t+h) = \sum_{i=1}^{6} (a_i - b_i) F_i \tag{13}$$

The values of the coefficients are given in the following table:

i	a_i	$a_i - b_i$	c_i	d_{ij}
1	$\frac{16}{135}$	$\frac{1}{360}$	0	0
2	0	0	$\frac{1}{4}$	$\frac{1}{4}$
3	$\frac{6656}{12825}$	$-\frac{128}{4275}$	$\frac{3}{8}$	$\frac{3}{32}, \frac{9}{32}$
4	$\frac{28561}{56430}$	$-\frac{2197}{75240}$	$\frac{12}{13}$	$\frac{1932}{2197}, -\frac{7200}{2197}, \frac{7296}{2197}$
5	$-\frac{9}{50}$	$\frac{1}{50}$	1	$\frac{439}{216}, -8, \frac{3680}{513}, -\frac{845}{4104}$
6	$\frac{2}{55}$	$\frac{2}{55}$	$\frac{1}{2}$	$-\frac{8}{27}, 2, -\frac{3544}{2565}, \frac{1859}{4104}, -\frac{11}{40}$

An adaptive routine using the preceding formulas attempts to keep the magnitude of the local truncation error, e, below a certain prescribed tolerance, δ. The solution is sought in a certain interval $[a, b]$, and an initial value $x(a) = \alpha$ is also given. An upper bound, M, on the number of steps would also be prescribed. Notice that the procedure is *conservative* because the local truncation error in a fourth-order method is being used to control the step size while the numerical solution is actually being computed with a fifth-order formula. An algorithm to carry out the procedure is given here. Since the local truncation error, as estimated by $x(t) - \overline{x}(t)$, is Ch^5, it seems sensible to double the step size when the doubling would lead to $C(2h)^5 < \delta/4$. Thus, we double h when the apparent local truncation error is less than $\delta/128$.

Another formula in common use to control the error per step is:

$$h \leftarrow 0.9h\left[\delta/|e|\right]^{1/(1+p)}$$

where p is the order of the first formula in the pair of Runge-Kutta formulas. (See, for example, Hull, Enright, Fellen, and Sedgwick [1972] and Shampine, Watts, and Davenport [1976].) The same formula is used for both increasing and decreasing the step size h. In one case, a rejected step size is recomputed, whereas in the other a new step size is established.

Here is an **adaptive Runge-Kutta-Fehlberg algorithm**:

```
input a, α, b, h, δ, M
t ← a;  x ← α
iflag ← 1
for k = 1 to M do
    d ← b − t
    if |d| ≦ |h| then
        iflag ← 0
        h ← d
    end if
    s ← t
    y ← x
    compute F1, F2, ..., F6      [Equation (12)]
    compute x      [Equation (10)]
    compute e      [Equation (13)]
        t ← t + h
        output k, t, x, e
    if iflag = 0 then stop
    if |e| ≧ δ then
            t ← s
            h ← h/2
            x ← y
            k ← k − 1
    else
        if |e| < δ/128 then h ← 2h
    end if
end do
```

Embedded Runge-Kutta procedures are composed of a pair of formulas of orders p and q (usually $q = p + 1$) that share the same function evaluation points. Generally they provide an efficient technique for solving nonstiff initial-value problems. Many higher-order embedded Runge-Kutta formulas have been derived in recent years. Examples of some of these are given in the problems. Although higher-order Runge-Kutta methods with complicated coefficients have been derived, the classic fourth-order formula remains the most popular except for adaptive schemes. Additional information on Runge-Kutta methods can be found in numerous sources—for example, Butcher [1987], Fehlberg [1969], Gear [1971], Jackson, Enright, and Hull [1978], Prince and Dormand [1981], Shampine and Gordon [1975], Thomas [1986], and Verner [1978].

PROBLEMS 8.3

1. Write out the second-order Runge-Kutta formulas when $\alpha = 2/3$.
2. If an initial-value problem involving the differential equation $x + 2tx' + xx' = 3$ is to be solved using a Runge-Kutta method, what function must be programmed?
3. Prove that the Runge-Kutta formula (8) is of order 4 in the special case that $f(t, x)$ is independent of x. Show that in this case the Runge-Kutta formula is equivalent to Simpson's rule. (See Equation (6) in Section 7.2.)
4. Derive the **modified Euler's method**:

$$x(t + h) = x(t) + hf\left(t + \frac{1}{2}h, x(t) + \frac{1}{2}hf\big(t, x(t)\big)\right)$$

   by performing Richardson's extrapolation on Euler's method using step sizes h and $h/2$. *Hint:* Assume the error term is Ch^2.
5. Derive the third-order Runge-Kutta formulas

$$x(t + h) = x(t) + \frac{1}{9}(2F_1 + 3F_2 + 4F_3)$$

   where

$$\begin{cases} F_1 = hf(t, x) \\ F_2 = hf(t + \frac{1}{2}h,\ x + \frac{1}{2}F_1) \\ F_3 = hf(t + \frac{3}{4}h,\ x + \frac{3}{4}F_2) \end{cases}$$

   Show that it agrees with the Taylor-series method of order 3 for the differential equation $x' = x + t$.
6. Prove that when the fourth-order Runge-Kutta method is applied to the problem $x' = \lambda x$, the formula for advancing this solution will be

$$x(t + h) = \left[1 + h\lambda + \frac{1}{2}h^2\lambda^2 + \frac{1}{6}h^3\lambda^3 + \frac{1}{24}h^4\lambda^4\right] x(t)$$

7. (Continuation) Prove that the local truncation error in Problem 6 is $\mathcal{O}(h^5)$.

COMPUTER PROBLEMS 8.3

1. Write a computer program to solve an initial-value problem $x' = f(t, x)$ with $x(t_0) = x_0$ on an interval $t_0 \leqq t \leqq t_m$ or $t_m \leqq t \leqq t_0$. Use the fourth-order Runge-Kutta method. Test it on this example:

$$\begin{cases} (e^t + 1)x' + xe^t - x = 0 \\ x(0) = 3 \end{cases}$$

Determine the *analytic* solution and compare it to the computed solution on the interval $-2 \leqq t \leqq 0$. Use $h = -0.01$.

2. Using various values of λ such as 5, -5, or -10, solve the following initial-value problem numerically using the fourth-order Runge-Kutta method:

$$\begin{cases} x' = \lambda x + \cos t - \lambda \sin t \\ x(0) = 0 \end{cases}$$

Compare the numerical solution to the analytic solution on the interval [0, 5]. Use step size $h = 0.01$. What effect does λ have on the numerical accuracy?

3. Program and test the Runge-Kutta-Fehlberg algorithm described in the text. For test cases, use the examples in Problems 8, 11, and 15 of Section 8.2.

4. Write and execute a program to solve this initial-value problem:

$$\begin{cases} x' = e^{xt} + \cos(x - t) \\ x(1) = 3 \end{cases}$$

Use the fourth-order Runge-Kutta formulas with $h = 0.01$. Stop the computation just *before* the solution overflows.

5. Solve $x' = x^2$, $x(0) = 1$ on the interval [0, 2] using the adaptive Runge-Kutta-Fehlberg procedure. Show that the true solution is $x(t) = 1/(1 - t)$. What happens in the algorithm near the discontinuity at $t = 1$?

6. Numerically compare the following fourth-order **Runge-Kutta-Gill method** with the classical Runge-Kutta method:

$$x(t + h) = x(t) + \frac{1}{6}\left[F_1 - \left(2 + \sqrt{2}\right)F_2 + \left(2 + \sqrt{2}\right)F_3 + F_4\right]$$

where

$$\begin{cases} F_1 = hf(t, x) \\ F_2 = hf(t + \frac{1}{2}h, x + \frac{1}{2}F_1) \\ F_3 = hf(t + \frac{1}{2}h, x + \frac{1}{2}\left(\sqrt{2} - 1\right)F_1 + \frac{1}{2}\left(2 - \sqrt{2}\right)F_2) \\ F_4 = hf(t + h, x - \frac{1}{2}\sqrt{2}F_2 + \frac{1}{2}\left(2 + \sqrt{2}\right)F_3) \end{cases}$$

7. Numerically compare the following **fifth-order Runge-Kutta method** with the classical Runge-Kutta method on a problem with a known solution:

$$x(t+h) = 7x(t) + \frac{1}{24}F_1 + \frac{5}{48}F_4 + \frac{27}{56}F_5 + \frac{125}{336}F_6$$

where

$$\begin{cases} F_1 = hf(t, x) \\ F_2 = hf\left(t + \frac{1}{2}h, x + \frac{1}{2}F_1\right) \\ F_3 = hf\left(t + \frac{1}{2}h, x + \frac{1}{4}F_1 + \frac{1}{4}F_2\right) \\ F_4 = hf\left(t + h, x - F_2 + 2F_3\right) \\ F_5 = hf\left(t + \frac{2}{3}h, x + \frac{7}{27}F_1 + \frac{10}{27}F_2 + \frac{1}{27}F_4\right) \\ F_6 = hf\left(t + \frac{1}{5}h, x + \frac{28}{625}F_1 - \frac{1}{5}F_2 + \frac{546}{625}F_3 + \frac{54}{625}F_4 - \frac{378}{625}F_5\right) \end{cases}$$

8. The **Runge-Kutta-Merson method** of order 5 is given by the coefficients in the following table. Write and test an adaptive routine based on this method.

i	a_i	$a_i - b_i$	c_i	d_{ij}
1	$\frac{1}{6}$	$\frac{1}{15}$	0	0
2	0	0	$\frac{1}{3}$	$\frac{1}{3}$
3	0	$-\frac{3}{10}$	$\frac{1}{3}$	$\frac{1}{6}, \frac{1}{6}$
4	$\frac{2}{3}$	$\frac{4}{15}$	$\frac{1}{2}$	$\frac{1}{8}, 0, \frac{3}{8}$
5	$\frac{1}{6}$	$-\frac{1}{30}$	1	$\frac{1}{2}, 0, -\frac{3}{2}, 2$

9. Write and test an adaptive routine based on the **fifth-order Runge-Kutta-Verner method** given by the coefficients in the following table (see Verner [1978]):

i	a_i	$a_i - b_i$	c_i	d_{ij}
1	$\frac{3}{80}$	$\frac{33}{640}$	0	0
2	0	0	$\frac{1}{18}$	$\frac{1}{18}$
3	$\frac{4}{25}$	$-\frac{132}{325}$	$\frac{1}{6}$	$-\frac{1}{12}, \frac{1}{4}$
4	$\frac{243}{1120}$	$\frac{891}{2240}$	$\frac{2}{9}$	$-\frac{2}{81}, \frac{4}{27}, \frac{8}{81}$
5	$\frac{77}{160}$	$-\frac{33}{320}$	$\frac{2}{3}$	$\frac{40}{33}, -\frac{4}{11}, -\frac{56}{11}, \frac{54}{11}$
6	$\frac{73}{700}$	$-\frac{73}{700}$	1	$-\frac{369}{73}, \frac{72}{73}, \frac{5380}{219}, -\frac{12285}{584}, \frac{2695}{1752}$
7	0	$\frac{891}{8329}$	$\frac{8}{9}$	$-\frac{8716}{891}, \frac{656}{297}, \frac{39520}{891}, -\frac{416}{11}, \frac{52}{27}$
8	0	$\frac{2}{35}$	1	$\frac{3015}{256}, -\frac{9}{4}, -\frac{4219}{78}, \frac{5985}{128}, -\frac{539}{394}, \frac{693}{3328}$

10. A pair of **embedded Runge-Kutta formulas** of orders 6 and 5 are given by the coefficients in the following table (see Prince and Dormand [1981]). Write and test an adaptive routine using them.

i	a_i	b_i	c_i	d_{ij}
1	$\frac{821}{10800}$	$\frac{61}{864}$	0	0
2	0	0	$\frac{1}{10}$	$\frac{1}{10}$
3	$\frac{19683}{71825}$	$\frac{98415}{321776}$	$\frac{2}{9}$	$-\frac{2}{81}, \frac{20}{81}$
4	$\frac{175273}{912600}$	$\frac{16807}{146016}$	$\frac{3}{7}$	$\frac{615}{1372}, -\frac{270}{343}, \frac{1053}{1372}$
5	$\frac{395}{3672}$	$\frac{1375}{7344}$	$\frac{3}{5}$	$\frac{3243}{5500}, -\frac{54}{55}, \frac{50949}{71500}, \frac{4998}{17875}$
6	$\frac{785}{2704}$	$\frac{1375}{5408}$	$\frac{4}{5}$	$-\frac{26492}{37125}, \frac{72}{55}, \frac{2808}{23375}, -\frac{24206}{37125}, \frac{338}{459}$
7	$\frac{3}{50}$	$-\frac{37}{1120}$	1	$\frac{5561}{2376}, -\frac{35}{11}, -\frac{24117}{31603}, \frac{899983}{200772}, -\frac{5225}{1836}, \frac{3925}{4056}$
8	0	$\frac{1}{10}$	1	$\frac{465467}{266112}, -\frac{2945}{1232}, -\frac{5610201}{14158144}, \frac{10513573}{3212352}, -\frac{424325}{205632}, \frac{376225}{454272}, 0$

11. Consider the initial-value problem $x' = -kx$, $x(0) = 1$ on the interval $[0, 1]$ with various values for the **decay constant** k. Show that the analytical solution is $x(t) = e^{-kt}$. For $k = 5$, compare the behavior of Euler's method and the Runge-Kutta-Gill method (given in Problem 10) using various values for the step size h such as 0.2, 0.4, 0.6, 0.8, and 1.0. Repeat using $k = 25$. For Euler's method the condition $0 \leqq hk \leqq 2$ must be satisfied in order to solve this problem correctly, whereas $0 \leqq hk < 2.8$ is needed for the Runge-Kutta-Gill method. These are the **regions of stability** for this problem using these methods. For a particular problem and method, we can either adjust the step size h to try to stay within the region of stability or alternatively use a higher-order method with a larger region of stability. As discussed by Thomas [1986], both of these methods become unworkable for large values of k.

8.4 Multistep Methods

The Taylor-series method and the Runge-Kutta method for solving the initial-value problem are **single-step methods** since they do not use any knowledge of prior values of $x(t)$ when the solution is being advanced from t to $t + h$. If $t_0, t_1, t_2 \dots, t_i$ are steps along the t-axis, then x_{i+1} (the approximate value of $x(t_{i+1})$) depends only on x_i, and knowledge of the approximate values $x_{i-1}, x_{i-2}, \dots, x_0$ is *not* used.

More efficient procedures can be devised if some prior values of the solution are utilized at each step. The principle involved here is as follows: We are solving the initial-value problem

$$\begin{cases} x' = f(t, x) \\ x(t_0) = x_0 \end{cases} \quad \textbf{(1)}$$

numerically. We have prescribed steps $t_0, t_1, t_2, \dots, t_n$ on the t-axis. (They will not always be equally spaced.) If the true solution is denoted by $x(t)$, then integrating

Equation (1) gives us

$$\int_{t_n}^{t_{n+1}} x'(t)\,dt = x(t_{n+1}) - x(t_n)$$

and then

$$x(t_{n+1}) = x(t_n) + \int_{t_n}^{t_{n+1}} f\big(t, x(t)\big)\,dt \tag{2}$$

The integral on the right can be approximated by a numerical quadrature scheme, and the result will be a formula for generating the approximate solution step by step.

Adams-Bashforth Formula

Suppose the resulting formula is of the following type:

$$x_{n+1} = x_n + af_n + bf_{n-1} + cf_{n-2} + \cdots \tag{3}$$

where f_i denotes $f(t_i, x_i)$. An equation of this type is called an **Adams-Bashforth formula**. Here is the **Adams-Bashforth formula of order 5**, based on equally spaced points $t_i = t_0 + ih$ for $0 \leqq i \leqq n$:

$$x_{n+1} = x_n + \frac{h}{720}\big[1901 f_n - 2774 f_{n-1} + 2616 f_{n-2} - 1274 f_{n-3} + 251 f_{n-4}\big] \tag{4}$$

How were these coefficients determined? We start with the intention of approximating the integral in Equation (2) as

$$\int_{t_n}^{t_{n+1}} f\big(t, x(t)\big)\,dt \approx h\big[Af_n + Bf_{n-1} + Cf_{n-2} + Df_{n-3} + Ef_{n-4}\big] \tag{5}$$

The coefficients A, B, C, D, E are determined by requiring that Equation (5) be exact whenever the integrand is a polynomial of degree $\leqq 4$. By Problem 6, there is no loss of generality in assuming $t_n = 0$ and $h = 1$ as sketched in Figure 8.1.

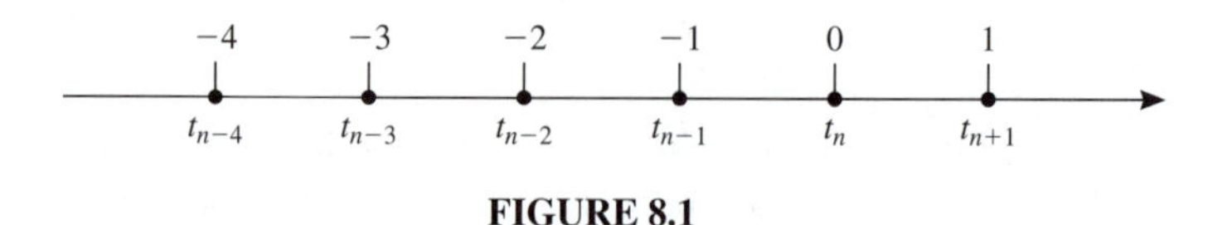

FIGURE 8.1

We take as a basis for Π_4 the following five polynomials:

$$
\begin{aligned}
p_0(t) &= 1 \\
p_1(t) &= t \\
p_2(t) &= t(t+1) \\
p_3(t) &= t(t+1)(t+2) \\
p_4(t) &= t(t+1)(t+2)(t+3)
\end{aligned}
$$

When these are substituted in the equation

$$\int_0^1 p_n(t)\,dt = Ap_n(0) + Bp_n(-1) + Cp_n(-2) + Dp_n(-3) + Ep_n(-4)$$

we obtain five equations for the determination of the coefficients A, B, C, D, E. This system of equations is

$$
\begin{cases}
A + B + C + D + E = 1 \\
-B - 2C - 3D - 4E = 1/2 \\
2C + 6D + 12E = 5/6 \\
-6D - 24E = 9/4 \\
24E = 251/30
\end{cases}
\tag{6}
$$

The coefficients of Formula (4) are obtained by back substitution.

The procedure just illustrated is called the **method of undetermined coefficients**. In principle, it can be used to obtain similar formulas of higher order and in a variety of other situations. (See Section 7.2.)

Alternatively, we could obtain the values of these unknown coefficients directly from Equation (4) by letting $x = p_n$ and using $f = p_n'$ since $x' = f$. A system of equations similar to (6) is obtained.

Adams-Moulton Formula

In numerical practice, the Adams-Bashforth formulas are rarely used by themselves. They are used in conjunction with other formulas to enhance the precision. To see how this might be possible, we return to Equation (2) and suppose that we use a numerical quadrature formula that involves f_{n+1}. Then Equation (3) will have the form

$$x_{n+1} = x_n + af_{n+1} + bf_n + cf_{n-1} + \cdots \tag{7}$$

Here is a formula of this type, known as the **Adams-Moulton formula of order 5**:

$$x_{n+1} = x_n + \frac{h}{720}\left[251f_{n+1} + 646f_n - 264f_{n-1} + 106f_{n-2} - 19f_{n-3}\right] \tag{8}$$

This formula also can be derived using the method of undetermined coefficients. Notice that it cannot be used directly to advance the solution because x_{n+1} occurs on both sides of the equation! Remember that f_i stands for $f(t_i, x_i)$, and so the term involving f_{n+1} can be computed only after x_{n+1} is known. However, a very satisfactory algorithm, called a **predictor-corrector method**, uses the Adams-Bashforth formula (4) to **predict** a tentative value for x_{n+1}, say x^*_{n+1}, and then the Adams-Moulton formula (8) to compute a *corrected* value of x_{n+1}. So in Formula (8), we evaluate f_{n+1} as $f(t_{n+1}, x^*_{n+1})$ using the predicted value x^*_{n+1} obtained from Formula (4).

In using this predictor-corrector method, a special procedure must be employed to start the method since initially only x_0 is known. Of course, a Runge-Kutta method is ideal for obtaining x_1, x_2, x_3, x_4. Usually formulas of the same order are used together. Therefore, fifth-order Runge-Kutta methods, such as those in Computer Problems 8–10 of Section 8.3, could be used in combination with the Adams-Bashforth formula (4) and the Adams-Moulton formula (8).

Still another method can be used to obtain the value of x_{n+1} in Equation (8). After all, that equation states that x_{n+1} is a fixed point of a certain mapping—namely, the one defined by

$$\phi(z) = \frac{251}{720} h\, f(t_{n+1}, z) + C$$

where C is composed of all the other terms in Equation (8). Therefore, the method of functional iteration suggests itself as a means to compute x_{n+1}. The theory of functional iteration, as developed in Section 3.4, tells us that the sequence determined by the equation

$$z_{k+1} = \phi(z_k) \qquad (k \geqq 0)$$

will converge to a fixed point of ϕ under appropriate hypotheses. In particular, Problem 31 of Section 3.4 can be used here. If ξ is the fixed point of ϕ (and thus the value of x_{n+1} that we seek), then we should start our iteration with a point z_0 in an interval centered at ξ in which $|\phi'(z)| < 1$. It is necessary to assume that ϕ' is continuous. In the case at hand,

$$\phi'(z) = \frac{251}{720} h \frac{\partial f(t_{n+1}, z)}{\partial z}$$

This can be made as small as we please by decreasing the step size, h. In practice, only one or two steps are required in this iteration to find the value of x_{n+1}.

Analysis of Linear Multistep Methods

In the remainder of this section, we take up the theory of linear multistep methods in general. The format of any such method is

$$a_k x_n + a_{k-1} x_{n-1} + \cdots + a_0 x_{n-k} = h\big[b_k f_n + b_{k-1} f_{n-1} + \cdots + b_0 f_{n-k}\big] \tag{9}$$

This is called a **k-step method**. The coefficients a_i and b_i are given. As before, x_i denotes an approximation to the solution at t_i, with $t_i = t_0 + ih$. The letter f_i denotes $f(t_i, x_i)$. Formula (9) is used to compute x_n, assuming that $x_0, x_1, \ldots, x_{n-1}$ are already known. Thus, we may assume that $a_k \neq 0$. The coefficient b_k can be 0. If $b_k = 0$, the method is said to be **explicit**. In this case, x_n can be computed directly from Equation (9) in an elementary fashion. If, however, $b_k \neq 0$, then on the right the term f_n involves the unknown x_n, and Equation (9) determines x_n implicitly. The method is then said to be **implicit**.

The accuracy of a numerical solution of a differential equation is largely determined by the **order** of the algorithm used. The order indicates how many terms in a Taylor-series solution are being simulated by the method. For example, the Adams-Bashforth method in Equation (4) is said to be of order 5 because it produces approximately the same accuracy as the Taylor-series method with terms h, h^2, h^3, h^4, h^5. The error can then be expected to be $\mathcal{O}(h^6)$ in each step of a solution using Equation (4). The intuitive idea of order will be made more precise in the following paragraphs.

In correspondence with the multistep method in Equation (9), we define a linear functional L by writing

$$Lx = \sum_{i=0}^{k} \left[a_i x(ih) - hb_i x'(ih)\right] \tag{10}$$

Here we let $k = n$ to simplify the notation and assume that the first value in Equation (9) begins at $t = 0$ rather than at $t = n - k$. The operation Lx can be applied to any function x that is differentiable. However, in the following analysis, we assume that x is represented by its Taylor series at $t = 0$. By using the Taylor series for x, one can express L in this form:

$$Lx = d_0 x(0) + d_1 hx'(0) + d_2 h^2 x''(0) + \cdots \tag{11}$$

To compute the coefficients d_i in Equation (11), we write the Taylor series for x and x':

$$x(ih) = \sum_{j=0}^{\infty} \frac{(ih)^j}{j!} x^{(j)}(0)$$

$$x'(ih) = \sum_{j=0}^{\infty} \frac{(ih)^j}{j!} x^{(j+1)}(0)$$

These series are then substituted in Equation (10). On rearranging the result in powers of h, we obtain an equation having the form of (11), with these values of d_i:

$$
\begin{cases}
d_0 = \sum_{i=0}^{k} a_i \\
d_1 = \sum_{i=0}^{k} (i a_i - b_i) \\
d_2 = \sum_{i=0}^{k} \left(\frac{1}{2} i^2 a_i - i b_i \right) \\
\quad \vdots \\
d_j = \sum_{i=0}^{k} \left(\frac{i^j}{j!} a_i - \frac{i^{j-1}}{(j-1)!} b_i \right) \qquad (j \geqq 1)
\end{cases}
\tag{12}
$$

Here we use $0! = 1$ and $i^0 = 1$, of course.

THEOREM 1 *These three properties of the multistep method* (9) *are equivalent:*

(i) $d_0 = d_1 = \cdots = d_m = 0$
(ii) $Lp = 0$ *for each polynomial p of degree* $\leqq m$.
(iii) Lx *is* $\mathcal{O}(h^{m+1})$ *for all* $x \in C^{m+1}$.

Proof If **(i)** is true, then Equation (11) has the form

$$
Lx = d_{m+1} h^{m+1} x^{(m+1)}(0) + \cdots \tag{13}
$$

If x is a polynomial of degree $\leqq m$, then $x^{(j)}(t) = 0$ for all $j > m$, and thus $Lx = 0$ from Equation (13). So **(i)** implies **(ii)**.

Assume that **(ii)** is true. If $x \in C^{m+1}$, then by Taylor's Theorem we can write $x = p + r$, where p is a polynomial of degree $\leqq m$ and r is a function whose first m derivatives vanish at 0. Since $Lp = 0$, Equation (11) gives us

$$
Lx = Lr = d_{m+1} h^{m+1} r^{(m+1)}(0) = \mathcal{O}(h^{m+1})
$$

and **(ii)** implies **(iii)**.

Finally, assume that **(iii)** is true. Then in Equation (11), we must have the condition $d_0 = d_1 = \cdots = d_m = 0$. Hence, **(iii)** implies **(i)**. ■

The **order** of the multistep method in Equation (9) is the unique natural number m such that

$$
d_0 = d_1 = \cdots = d_m = 0 \neq d_{m+1}
$$

Example 1 What is the order of the method described by this equation?

$$
x_n - x_{n-2} = \frac{1}{3} h (f_n + 4 f_{n-1} + f_{n-2})
$$

Solution The vector (a_0, a_1, a_2) is $(-1, 0, 1)$, and the vector (b_0, b_1, b_2) is $(\frac{1}{3}, \frac{4}{3}, \frac{1}{3})$. Thus, the d_i are

$$
d_0 = a_0 + a_1 + a_2 = 0
$$

$$
d_1 = -b_0 + (a_1 - b_1) + (2a_2 - b_2) = 0
$$

$$d_2 = \left(\frac{1}{2}a_1 - b_1\right) + (2a_2 - 2b_2) = 0$$

$$d_3 = \left(\frac{1}{6}a_1 - \frac{1}{2}b_1\right) + \left(\frac{4}{3}a_2 - 2b_2\right) = 0$$

$$d_4 = \left(\frac{1}{24}a_1 - \frac{1}{6}b_1\right) + \left(\frac{2}{3}a_2 - \frac{4}{3}b_2\right) = 0$$

$$d_5 = \left(\frac{1}{120}a_1 - \frac{1}{24}b_1\right) + \left(\frac{4}{15}a_2 - \frac{2}{3}b_2\right) = -\frac{1}{90}$$

The order of the method is 4. ■

Other characteristics being equal, we might prefer a method of high order to one of low order. If we desire to produce a k-step method of the form (9) of order $2k$, we simply write down the $2k + 1$ equations

$$d_0 = d_1 = \cdots = d_{2k} = 0$$

By Equation (12), this is a system of $2k + 1$ linear homogeneous equations in the $2k + 2$ unknowns a_i and b_i $(0 \leqq i \leqq k)$. By elementary linear algebra, this system has a nontrivial solution; Dahlquist [1956] proved that there is a solution with $a_k \neq 0$ (which is necessary in the method). We shall see in Section 8.5, however, that some characteristics of a multistep method other than its order *must* be taken into account. Chief among these is *stability*, defined in that section. Dahlquist also proved that a stable k-step method, as in Equation (9), cannot have an order greater than $k + 2$.

PROBLEMS 8.4

1. Derive the first-order (one-step) Adams-Moulton formula and verify that it is equivalent to the trapezoid rule. (See Section 7.2.)
2. Verify the correctness of the system of equations in (6).
3. Use the method of undetermined coefficients to derive Equation (8).
4. Use the method of undetermined coefficients to derive the fourth-order Adams-Bashforth formula

$$x_{n+1} = x_n + \frac{h}{24}\left[55f_n - 59f_{n-1} + 37f_{n-2} - 9f_{n-3}\right]$$

5. Derive the fourth-order Adams-Moulton formula

$$x_{n+1} = x_n + \frac{h}{24}\left[9f_{n+1} + 19f_n - 5f_{n-1} + f_{n-2}\right]$$

6. Prove that if every element of Π_m is correctly integrated by the formula

$$\int_0^1 f(x)\,dx \approx \sum_{i=-n}^{n} A_i f(i)$$

then the same is true of the formula

$$\int_{t_0}^{t_0+h} f(x)\,dx \approx h \sum_{i=-n}^{n} A_i f(t_0 + ih)$$

7. Derive the second-order Adams-Bashforth formula

$$x_{n+1} = x_n + h\left[\frac{3}{2}f_n - \frac{1}{2}f_{n-1}\right]$$

8. **(a)** Using the method of undetermined coefficients, determine A and B in the following Adams-Bashforth formula. (Do not change to a corresponding numerical integration formula.)

$$x_{n+1} = x_n + h[Af_n + Bf_{n-1}]$$

(b) Repeat part **(a)** by changing it to a corresponding numerical integration formula.

(c) Repeat the problem by integrating the Newton interpolation polynomial based on the nodes x_n and x_{n-1}.

9. **(a)** Use the method of undetermined coefficients to find A and B in the implicit Adams-Moulton formula of second order

$$x_{n+1} = x_n + h[Af_{n+1} + Bf_n]$$

(b) Use a numerical integration formula to derive the formula.

(c) Use an interpolation formula to derive the formula.

10. **(a)** Use the method of undetermined coefficients to derive a multistep formula of the form

$$x_{n+1} = x_n + h[Af_{n+1} + Bf_n + Cf_{n-1}]$$

(b) Repeat part **(a)** for

$$x_{n+1} = x_n + h[Af_n + Bf_{n-1} + Cf_{n-2}]$$

11. Derive an implicit multistep formula based on Simpson's rule (involving uniformly spaced points x_{n-1}, x_n, x_{n+1}) for numerically solving the ordinary differential equation $x' = f$.

12. The formula

$$x_{n+1} = (1-A)x_n + Ax_{n-1} + \frac{h}{12}\left[(5-A)x'_{n+1} + 8(1+A)x'_n + (5A-1)x'_{n-1}\right]$$

is known to be exact for all polynomials of degree m or less for all A. Determine A so that it will be exact for all polynomials of degree $m + 1$. Find A and m.

13. Compute the coefficients in a multistep method of the form

$$x_{n+1} = x_n + h[Af_n + Bf_{n-2} + Cf_{n-4}]$$

The formula should correctly integrate an equation $x' = f(t, x)$ when the right-hand side is of the form $f(t, x) = a + bt + ct^2$.

14. Find a method of order 4 in the form of Equation (9), taking $k = 2$.

15. Prove that the multistep method of Equation (9) is of order m if and only if $Lp_i = 0$ for $0 \leqq i \leqq m$ and $Lp_{m+1} \neq 0$, where $p_i(t) = t^i$ and L is as given in Equation (10).

16. Determine the order of this method by computing $d_0, d_1, \ldots$:

$$x_n = x_{n-3} + \frac{3}{8}h[f_n + 3f_{n-1} + 3f_{n-2} + f_{n-3}]$$

17. Determine the order of this method by the procedure in Problem 15:

$$x_n = x_{n-2} + 2hf_{n-1}$$

18. Let $a_0, a_1, \ldots, a_m$ be given, and assume that $\sum_{i=0}^{k} a_i = 0$. Do there necessarily exist $b_0, b_1, \ldots, b_m$ so that the multistep method having these coefficients will be of order at least m? (A theorem or an example is required.)

19. Prove that for Euler's method $d_0 = d_1 = 0$ and $d_j = 1/j!$ for $j \geqq 2$.

COMPUTER PROBLEMS 8.4

1. Write and test a subprogram or procedure for the fifth-order Adams-Bashforth-Moulton method coupled with a fifth-order Runge-Kutta method (Problems 15–17 in Section 8.3). Keep only the five most recent values of (t_i, x_i). Print statements should be included in the routine.

2. Write a computer program that calls the routine in Computer Problem 1 and solves the initial-value problem

$$\begin{cases} x' = (t - e^{-t})/(x + e^x) \\ x(0) = 0 \end{cases}$$

on the interval $[-1, 1]$. Use $h = 1/238$. Verify (analytically) that the true solution is given implicitly by the equation $x^2 - t^2 + 2e^x - 2e^{-t} = 0$. Use this equation in the program to provide a check on the computed solution. Use the Runge-Kutta method to get started.

3. Compute the solution of

$$\begin{cases} y' = -2xy^2 \\ y(0) = 1 \end{cases}$$

at $x = 1.0$ using $h = 0.25$ and the fourth-order Adams-Bashforth-Moulton method (Problems 4 and 5) together with the fourth-order Runge-Kutta method. Give the computed solution to five significant digits at 0.25, 0.5, 0.75, and 1.0. Compare your results to the exact solution $y = 1/(1 + x^2)$.

4. Find a program in your computer center program library that is suitable for solving the initial-value problem for a system of first-order ordinary differential equations. Use the program to solve this problem:

$$\begin{cases} dx/dt = \sin x + \cos(yt) & x(-1) = 2.37 \\ dy/dt = e^{-xt} + [\sin(yt)]/t & y(-1) = -3.48 \end{cases}$$

The solution is desired with eight-decimal-place accuracy in the interval $[-1, 4]$. The solution is to be printed at intervals of 0.1 on the t-axis.

(a) Obtain plots of the functions $x(t)$ and $y(t)$ for $-1 \leqq t \leqq 4$. Both curves should appear on the same plot. The maximum and minimum ordinates should be set as -4 and $+4$.

(b) Obtain a plot of the curve defined parametrically as

$$\{(x(t), y(t)) : -1 \leqq t \leqq 4\}$$

(Values of t will not appear on this plot.) The curve obtained here is called an **orbit** of the original system of differential equations. (The word **trajectory** is also used.)

*8.5 Local and Global Errors: Stability

The Adams-Bashforth and Adams-Moulton methods of Section 8.4 are only two examples of multistep methods for solving an initial-value problem

$$\begin{cases} x' = f(t, x) \\ x(t_0) = x_0 \end{cases} \tag{1}$$

The term **multistep** refers to the fact that in the computation of x_n, some previously computed terms, $x_{n-1}, x_{n-2}, \ldots$, are required. Let us review a few matters discussed in the preceding section.

Implicit/Explicit and Convergent Methods

Any multistep method can be described by an equation of the form

$$a_k x_n + a_{k-1} x_{n-1} + \cdots + a_0 x_{n-k} = h[b_k f_n + b_{k-1} f_{n-1} + \cdots + b_0 f_{n-k}] \tag{2}$$

For example, the fifth-order Adams-Moulton formula would read as follows:

$$x_n - x_{n-1} = h\left[\frac{251}{720}f_n + \frac{646}{720}f_{n-1} - \frac{264}{720}f_{n-2} + \frac{106}{720}f_{n-3} - \frac{19}{720}f_{n-4}\right] \tag{3}$$

In these equations, f_i stands for $f(t_i, x_i)$. In solving problem (1) using a formula such as (2), we assume that starting values $x_0, x_1, \ldots, x_{k-1}$ have been obtained by some other method. Then Equation (2) is used with $n = k, k+1, \ldots$ in succession. Observe that if $b_k \neq 0$, then Equation (2) contains the unknown ordinate x_n on *both* sides of the equation. In this case, the method is said to be **implicit**. If $b_k = 0$, the method is an **explicit** one. In our analysis, it is assumed that x_n satisfies Equation (2). In practice, x_n can be obtained by iteration, starting with a tentative value given by a *predictor* formula.

Associated with Equation (2) are two polynomials:

$$\begin{cases} p(z) = a_k z^k + a_{k-1}z^{k-1} + \cdots + a_0 \\ q(z) = b_k z^k + b_{k-1}z^{k-1} + \cdots + b_0 \end{cases} \tag{4}$$

The analysis will show that certain desirable properties of the multistep method depend on the location of the roots of the polynomials p and q.

The multistep method defined by Equation (2) is said to be **convergent** under the following circumstances: Imagine that numerical solutions of problem (1) are computed by using a variety of step sizes. We denote by $x(h, t)$ the approximate solution obtained by using step size h. As usual, the exact solution is written $x(t)$. Convergence means that

$$\lim_{h\to 0} x(h, t) = x(t) \qquad (t \text{ fixed}) \tag{5}$$

for all t in some interval $[t_0, t_m]$, provided only that the starting values obey the same equation; that is,

$$\lim_{h\to 0} x(h, t_0 + nh) = x_0 \qquad (0 \leqq n < k) \tag{6}$$

and that the function f satisfies the hypotheses of the basic existence theorem (Theorem 3 in Section 8.1).

Stability and Consistency

Two other terms that are used are *stable* and *consistent*. The method is **stable** if all roots of p lie in the disk $|z| \leqq 1$ and if each root of modulus 1 is simple. The method is **consistent** if $p(1) = 0$ and $p'(1) = q(1)$. The main theorem in this subject can now be stated.

THEOREM 1 *For the multistep method of Equation* (2) *to be convergent, it is necessary and sufficient that it be stable and consistent.*

Proof (*Stability is necessary.*) Suppose that the method is not stable. Then either p has a root λ satisfying $|\lambda| > 1$ or p has a root λ satisfying $|\lambda| = 1$ and $p'(\lambda) = 0$. In either case we consider a simple initial-value problem whose solution is $x(t) = 0$:

$$\begin{cases} x' = 0 \\ x(0) = 0 \end{cases} \tag{7}$$

The multistep method is governed by the equation

$$a_k x_n + a_{k-1} x_{n-1} + \cdots + a_0 x_{n-k} = 0 \tag{8}$$

This is a linear difference equation, one of whose solutions is $x_n = h\lambda^n$, where λ is one of the roots of p. If $|\lambda| > 1$, then for $0 \leqq n < k$ we have

$$|x(h, nh)| = h|\lambda^n| < h|\lambda|^k \to 0 \qquad (\text{as } h \to 0)$$

This establishes condition (6). But Equation (5) is violated because if $t = nh$, then $h = tn^{-1}$ and

$$|x(h, t)| = |x(h, nh)| = tn^{-1}|\lambda|^n \to \infty$$

On the other hand, if $|\lambda| = 1$ and $p'(\lambda) = 0$, then a solution of Equation (8) is $x_n = hn\lambda^n$. Condition (6) is fulfilled because if $0 \leqq n < k$, then

$$|x(h, nh)| = hn|\lambda|^n = hn < hk \to 0 \qquad (\text{as } h \to 0)$$

Condition (5) is violated because

$$|x(h, t)| = (tn^{-1})n|\lambda|^n = t \neq 0$$

■

Proof (*Consistency is necessary.*) Suppose that the method defined by Equation (2) is convergent. Consider the problem

$$\begin{cases} x' = 0 \\ x(0) = 1 \end{cases} \tag{9}$$

The exact solution is $x = 1$. Equation (2) again has the form (8). One solution is obtained by setting $x_0 = x_1 = \cdots = x_{k-1} = 1$ and then using (8) to generate the remaining values, $x_k, x_{k+1}, \ldots$. Since the method is convergent, $\lim_{n\to\infty} x_n = 1$. Putting this into Equation (8) yields the result $a_k + a_{k-1} + \cdots + a_0 = 0$ or, in other words, $p(1) = 0$.

Now consider the initial-value problem

$$\begin{cases} x' = 1 \\ x(0) = 0 \end{cases} \tag{10}$$

whose exact solution is $x = t$. Equation (8) becomes

$$a_k x_n + a_{k-1} x_{n-1} + \cdots + a_0 x_{n-k} = h\left[b_k + b_{k-1} + \cdots + b_0\right] \tag{11}$$

Since the method is convergent, it is stable by the preceding proof. Hence $p(1) = 0$ and $p'(1) \neq 0$. One solution of Equation (11) is given by $x_n = (n + k)h\gamma$, with $\gamma = q(1)/p'(1)$. Indeed, substitution of this into the left side of Equation (11) yields

$$
\begin{aligned}
h\gamma\big[a_k(n+k) + a_{k-1}(n+k-1) + \cdots + a_0 n\big]&\\
&= nh\gamma(a_k + a_{k-1} + \cdots + a_0) + h\gamma\big[ka_k + (k-1)a_{k-1} + \cdots + a_1\big]\\
&= nh\gamma p(1) + h\gamma p'(1)\\
&= h\gamma p'(1) = hq(1) = h\big[b_k + b_{k-1} + \cdots + b_0\big]
\end{aligned}
$$

Notice that the starting values in this numerical solution are consistent with the initial value $x(0) = 0$ because $\lim_{h\to 0}(n+k)h\gamma = 0$ for $n = 0, 1, \ldots, k-1$. Now the convergence condition demands that $\lim_{n\to\infty} x_n = t$ if $nh = t$. Hence, we have $\lim_{n\to\infty}(n+k)h\gamma = t$. Therefore, we conclude that $\gamma = 1$, or $p'(1) = q(1)$ since $\lim_{n\to\infty} kh = 0$. ■

The proof that stability and consistency together imply convergence is much more involved. The interested reader should consult the text by Henrici [1962, Section 5.3].

Milne Method

To illustrate Theorem 1, we analyze the **Milne method** defined by

$$
x_n - x_{n-2} = h\left[\frac{1}{3}f_n + \frac{4}{3}f_{n-1} + \frac{1}{3}f_{n-2}\right] \tag{12}
$$

This is an implicit method and is characterized by the two polynomials

$$
\begin{aligned}
p(z) &= z^2 - 1\\
q(z) &= \frac{1}{3}z^2 + \frac{4}{3}z + \frac{1}{3}
\end{aligned}
$$

Note that the roots of p are $+1$ and -1. They are simple roots. Furthermore, $p'(z) = 2z$, $p'(1) = 2$, and $q(1) = 2$. Thus, the conditions of consistency and stability are fulfilled. By the theorem, the Milne method is convergent.

Local Truncation Error

Our next task is to define and analyze the local truncation error that arises in using the multistep method (2). Suppose that Equation (2) is used to compute x_n, under the supposition that all previous values $x_{n-1}, x_{n-2}, \ldots$ are *correct*—that is, that $x_i = x(t_i)$ for $i < n$. Here $x(t)$ denotes the solution of the differential equation. The **local truncation error** is defined then to be $x(t_n) - x_n$. This error is due solely to our modeling of the differential equation by a difference equation. In this definition, roundoff error is not included; we assume that x_n has been computed with complete precision from the difference Equation (2). We want to prove that if the method has order m (as defined in Section 8.4), then the local truncation error will be $\mathcal{O}(h^{m+1})$. This is not true without qualification because our analysis presupposes certain smoothness in f and in the exact solution function $x(t)$.

THEOREM 2 *If the multistep method (2) is of order m, if $x \in C^{m+2}$, and if $\partial f/\partial x$ is continuous, then under the hypotheses of the preceding paragraph,*

$$x(t_n) - x_n = \left(\frac{d_{m+1}}{a_k}\right) h^{m+1} x^{(m+1)}(t_{n-k}) + \mathcal{O}(h^{m+2}) \tag{13}$$

(The coefficients d_k are defined in Section 8.4.)

Proof It suffices to prove the equation when $n = k$, since x_n can be interpreted as the value of a numerical solution that began at the point t_{n-k}. Using the linear functional L, Equation (10) in Section 8.4, we have

$$Lx = \sum_{i=0}^{k} \left[a_i x(t_i) - hb_i x'(t_i)\right] = \sum_{i=0}^{k} \left[a_i x(t_i) - hb_i f\left(t_i, x(t_i)\right)\right] \tag{14}$$

On the other hand, the numerical solution satisfies the equation

$$0 = \sum_{i=0}^{k} \left[a_i x_i - hb_i f(t_i, x_i)\right] \tag{15}$$

Since we have assumed that $x_i = x(t_i)$ for $i < k$, the result of subtracting Equation (15) from Equation (14) will be

$$Lx = a_k\left[x(t_k) - x_k\right] - hb_k\left[f\left(t_k, x(t_k)\right) - f(t_k, x_k)\right] \tag{16}$$

To the last term of Equation (16) we apply the Mean-Value Theorem, obtaining

$$\begin{aligned} Lx &= a_k\left[x(t_k) - x_k\right] - hb_k \frac{\partial f}{\partial x}(t_k, \xi)\left[x(t_k) - x_k\right] \\ &= \left[a_k - hb_k F\right]\left[x(t_k) - x_k\right] \end{aligned} \tag{17}$$

where $F = \partial f(t_k, \xi)/\partial x$ for some ξ between $x(t_k)$ and x_k. Now recall from Equation (13) in Section 8.4 that if the method being used is of order m, then Lx will have the form

$$Lx = d_{m+1} h^{m+1} x^{(m+1)}(t_0) + \mathcal{O}(h^{m+2}) \tag{18}$$

Combining Equations (17) and (18), we obtain (13). Here we can ignore $hb_k F$ in the denominator (see Problem 8). ■

Global Truncation Error

We now wish to establish a bound on the global truncation error in solving a differential equation numerically. At any given step in the solution process, say at t_n, the solution computed is denoted by x_n. Let us assume that all computations have been carried out with complete precision; that is, no roundoff error is involved. Of course, the true solution at t_n, $x(t_n)$, differs from the computed solution x_n since the latter has been obtained by formulas that approximate a Taylor series. The difference $x(t_n) - x_n$ is the **global truncation error**. It is *not* simply the sum of all local truncation errors that entered at previous points. To see that this is so, we must realize that each step in the numerical solution must use as its initial value the approximate ordinate computed at the preceding step. Since that ordinate is in error, the numerical process is in effect attempting to track the wrong solution curve. We therefore must begin the analysis by seeing how two solution curves differ if they are started with different initial conditions. In other words, we need to understand the effect of changing an initial value on the later ordinates of a solution curve. Figure 8.2 shows what we are trying to measure.

Consider the initial-value problem

$$\begin{cases} x' = f(t, x) \\ x(0) = s \end{cases} \tag{19}$$

We assume that $\partial f/\partial x$, denoted here by f_x, is continuous and satisfies the condition $f_x(t, x) \leqq \lambda$ in the region defined by $0 \leqq t \leqq T$ and $-\infty < x < \infty$. The solution of (19) is a function of t, but in order to show its dependence on the initial value s, we write it as $x(t; s)$. Define $u(t) = \partial x(t; s)/\partial s$. We can obtain a differential equation—the **variational equation**—for u by differentiating with respect to s in the initial-value problem (19). The result is

$$\begin{cases} u'(t) = f_x(t, x)u \\ u(0) = 1 \end{cases} \tag{20}$$

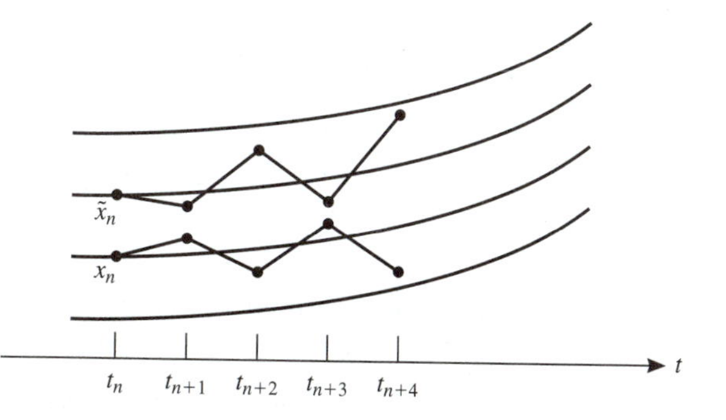

FIGURE 8.2 Effect of changing initial conditions

Example 1 Determine u explicitly in this initial-value problem:

$$\begin{cases} x' = x^2 \\ x(0) = s \end{cases} \tag{21}$$

Solution Here $f(t, x) = x^2$ and $f_x = 2x$. Hence the variational equation is

$$\begin{cases} u' = 2xu \\ u(0) = 1 \end{cases} \tag{22}$$

The solution of the initial-value problem (21) is $x(t) = s(1-st)^{-1}$. Hence, Equations (22) become

$$\begin{cases} u'(t) = 2s(1 - st)^{-1}u(t) \\ u(0) = 1 \end{cases}$$

and the solution of this is

$$u(t) = (1 - st)^{-2}$$

■

THEOREM 3 *If $f_x \leqq \lambda$, then the solution of the variational equation (20) satisfies the inequality*

$$|u(t)| \leqq e^{\lambda t} \qquad (t \geqq 0)$$

Proof From Equation (20), we have

$$u'/u = f_x = \lambda - \alpha(t) \tag{23}$$

in which $\alpha(t) \geqq 0$. On integrating Equation (23), we get

$$\log|u| = \lambda t - \int_0^t \alpha(\tau)\,d\tau = \lambda t - A(t)$$

in which $A(t)$ denotes the indicated integral. Since $t \geqq 0$, $A(t) \geqq 0$ and consequently $\log|u| \leqq \lambda t$. Thus, $|u| \leqq e^{\lambda t}$ since the exponential function is increasing. ■

THEOREM 4 *If the initial-value problem (19) is solved with initial values s and $s + \delta$, the solution curves at t differ by at most $|\delta|e^{\lambda t}$.*

Proof By the Mean-Value Theorem, by the definition of u, and by Theorem 3, we have

$$\begin{aligned} |x(t; s) - x(t; s + \delta)| &= \left| \frac{\partial}{\partial s} x(t, s + \theta\delta) \right| |\delta| \\ &= |u(t)|\,|\delta| \leqq |\delta| e^{\lambda t} \end{aligned}$$

■

THEOREM 5 *If the local truncation errors at $t_1, t_2, \ldots, t_n$ do not exceed δ in magnitude, then the global truncation error at t_n does not exceed $\delta(e^{n\lambda h} - 1)(e^{\lambda h} - 1)^{-1}$.*

Proof Let truncation errors of $\delta_1, \delta_2, \ldots$ be associated with the numerical solution at points $t_1, t_2, \ldots$. In computing x_2 there was an error of δ_1 in the initial condition, and by

Theorem 4 the effect of this at t_2 is at most $|\delta_1|e^{\lambda h}$. To this is added the truncation error at t_2. Thus the global truncation error at t_2 is at most

$$|\delta_1|e^{\lambda h} + |\delta_2|$$

The effect of this error at t_3 is (by Theorem 4) no greater than

$$\left(|\delta_1|e^{\lambda h} + |\delta_2|\right)e^{\lambda h}$$

To this is added the truncation error at t_3. Continuing in this way, we find that the global truncation error at t_n is no greater than

$$\sum_{k=1}^{n} |\delta_k|e^{(n-k)\lambda h} \leqq \delta \sum_{k=0}^{n-1} e^{k\lambda h} = \delta(e^{n\lambda h} - 1)(e^{\lambda h} - 1)^{-1} \qquad \blacksquare$$

THEOREM 6 *If the local truncation errors in the numerical solution are $\mathcal{O}(h^{m+1})$, then the global truncation error is $\mathcal{O}(h^m)$.*

Proof In Theorem 5, let δ be $\mathcal{O}(h^{m+1})$. Since $e^z - 1$ is $\mathcal{O}(z)$ and $nh = t$, we find a decrease of 1 in the order from using the formula in Theorem 5. $\blacksquare$

PROBLEMS 8.5

1. Discuss these multistep methods in the light of Theorem 1 of this section:
   (a) $x_n - x_{n-2} = 2hf_{n-1}$
   (b) $x_n - x_{n-2} = h\left[\frac{7}{3}f_{n-1} - \frac{2}{3}f_{n-2} + \frac{1}{3}f_{n-3}\right]$
   (c) $x_n - x_{n-1} = h\left[\frac{3}{8}f_n + \frac{19}{24}f_{n-1} - \frac{5}{24}f_{n-2} + \frac{1}{24}f_{n-3}\right]$
2. A method is said to be **weakly unstable** if p has a zero λ such that $\lambda \neq 1$, $|\lambda| = 1$, and $q(\lambda) < \lambda p'(\lambda)$. Show that the Milne method given by Equation (12) is weakly unstable.
3. Show that every multistep method in which $p(z) = z^k - z^{k-1}$ and $\sum_{i=0}^{k} b_i = 1$ is stable, consistent, convergent, and weakly stable.
4. Determine the numerical characteristics of the multistep method whose equation is

$$x_n + 4x_{n-1} - 5x_{n-2} = h\left[4f_{n-1} + 2f_{n-2}\right]$$

5. Is there any reason for distrusting this numerical scheme for solving $x' = f(t, x)$?

$$x_n - 3x_{n-1} + 2x_{n-2} = h\left[f_n + 2f_{n-1} + f_{n-2} - 2f_{n-3}\right]$$

   Explain.

6. Which of these multistep methods is *convergent*?

(a) $x_n - x_{n-2} = h(f_n - 3f_{n-1} + 4f_{n-2})$

(b) $x_n - 2x_{n-1} + x_{n-2} = h(f_n - f_{n-1})$

(c) $x_n - x_{n-1} - x_{n-2} = h(f_n - f_{n-1})$

(d) $x_n - 3x_{n-1} + 2x_{n-2} = h(f_n + f_{n-1})$

(e) $x_n - x_{n-2} = h(f_n - 3f_{n-1} + 2f_{n-2})$

7. A multistep method is said to be **strongly stable** if $p(1) = 0$, $p'(1) \neq 0$, and all other roots of p satisfy the inequality $|z| < 1$. Prove that a strongly stable method is convergent, using Theorem 1. Prove also that for any value of λ, a strongly stable method will solve the problem $x' = \lambda x$, $x(0) = 1$ without introducing any extraneous errors of exponential growth.

8. Prove that

$$\frac{Ah^{m+1} + \mathcal{O}(h^{m+2})}{B - Ch} = \frac{A}{B}h^{m+1} + \mathcal{O}(h^{m+2})$$

8.6 Systems and Higher-Order Ordinary Differential Equations

The standard form for a *system* of first-order differential equations is:

$$\begin{cases} x_1' = f_1(t, x_1, x_2, \dots, x_n) \\ x_2' = f_2(t, x_1, x_2, \dots, x_n) \\ \vdots \\ x_n' = f_n(t, x_1, x_2, \dots, x_n) \end{cases} \tag{1}$$

In this system, there are n unknown functions, $x_1, x_2, \dots, x_n$, to be determined. They are functions of the single independent variable t, and the notation x_i' denotes the derivative dx_i/dt. For a concrete example, consider this system, in which x and y have been written in place of x_1 and x_2:

$$\begin{cases} x' = x + 4y - e^t \\ y' = x + y + 2e^t \end{cases} \tag{2}$$

The *general* solution of System (2) is

$$\begin{cases} x = 2ae^{3t} - 2be^{-t} - 2e^t \\ y = ae^{3t} + be^{-t} + \frac{1}{4}e^t \end{cases} \tag{3}$$

where a and b are arbitrary constants. The purported solution can be verified, of course, by differentiation and substitution in System (2). Notice that the example is a *linear* system in the unknown functions x and y.

In a well-defined physical problem having presumably a *unique* solution, the system of differential equations would be accompanied by auxiliary conditions that serve to determine the arbitrary constants in the *general* solution. Thus, if System (2) were accompanied by initial conditions

$$x(0) = 4 \qquad y(0) = \tfrac{5}{4}$$

then the solution would be

$$\begin{cases} x = 4e^{3t} + 2e^{-t} - 2e^{t} \\ y = 2e^{3t} - e^{-t} + \frac{1}{4}e^{t} \end{cases} \tag{4}$$

The initial-value problem for the general system in (1) consists of the n differential equations together with a prescribed *initial value* for t, say $t = t_0$, and a specification of the value of each function x_i at t_0.

Vector Notation

A convenient vector notation can be used to rewrite System (1). We let X denote the column vector whose components are $x_1, x_2, \dots, x_n$. These components are functions of t. Hence, X is a mapping of $\mathbb{R}$ (or an interval in $\mathbb{R}$) to $\mathbb{R}^n$. Similarly, let F denote the column vector with components $f_1, f_2, \dots, f_n$. Each of these is a function on $\mathbb{R}^{n+1}$ (or a subset thereof), and so F is a mapping of $\mathbb{R}^{n+1}$ to $\mathbb{R}^n$. System (1) can be written now as

$$X' = F(t, X) \tag{5}$$

An initial-value problem for System (5) would also include numerical values for the vector $X(t_0)$, where t_0 is the initial value of t.

A differential equation of high order can be converted to a system of first-order equations. Suppose that a single differential equation is given in the form

$$y^{(n)} = f(t, y, y', y'', \dots, y^{(n-1)})$$

Here, of course, all derivatives are with respect to t: $y^{(i)} = d^i y/dt^i$. Next, we introduce new variables $x_1, x_2, \dots, x_n$ according to these definitions:

$$x_1 = y \qquad x_2 = y' \qquad x_3 = y'' \qquad \cdots \qquad x_n = y^{(n-1)}$$

The new variables satisfy the following system of first-order differential equations:

$$\begin{cases} x_1' = x_2 \\ x_2' = x_3 \\ x_3' = x_4 \\ \quad\vdots \\ x_n' = f(t, x_1, x_2, x_3, \ldots, x_n) \end{cases}$$

This is a system of the form presented in Equation (5).

To solve differential equations using widely available software, it is almost always necessary to convert the problem into a system such as (5). We illustrate this process with two examples.

Example 1 Convert the initial-value problem

$$\begin{cases} (\sin t)y''' + \cos(ty) + \sin(t^2 + y'') + (y')^3 = \log t \\ y(2) = 7 \\ y'(2) = 3 \\ y''(2) = -4 \end{cases} \tag{6}$$

into a system of first-order differential equations with initial values.

Solution We introduce new variables x_1, x_2, x_3 as follows: $x_1 = y$, $x_2 = y'$, $x_3 = y''$. The system of equations governing $X = (x_1, x_2, x_3)^T$ is

$$\begin{cases} x_1' = x_2 \\ x_2' = x_3 \\ x_3' = \left[\log t - x_2^3 - \sin(t^2 + x_3) - \cos(tx_1)\right] / \sin t \end{cases} \tag{7}$$

The initial conditions at $t = 2$ are $X = (7, 3, -4)^T$. ■

Systems of higher-order equations can be treated in a similar manner, as illustrated in the next example.

Example 2 Convert the system shown into a system of first-order equations:

$$\begin{cases} (x'')^2 + te^y + y' = x' - x \\ y'y'' - \cos(xy) + \sin(tx'y) = x \end{cases} \tag{8}$$

The initial conditions are omitted in this example.

Solution After introducing the new variables $x_1 = x$, $x_2 = x'$, $x_3 = y$, $x_4 = y'$, the problem can be written as

$$\begin{cases} x_1' = x_2 \\ x_2' = \left(x_2 - x_1 - x_4 - te^{x_3}\right)^{1/2} \\ x_3' = x_4 \\ x_4' = \left[x_1 - \sin(tx_2x_3) + \cos(x_1x_3)\right]/x_4 \end{cases} \tag{9}$$

■

Taylor-Series Method for Systems

The Taylor-series method, which was discussed in Section 8.2, can be applied to systems of first-order equations. We write a truncated Taylor series for each variable like this:

$$x_i(t+h) = x_i(t) + hx_i'(t) + \frac{h^2}{2!}x_i''(t) + \frac{h^3}{3!}x_i'''(t) + \cdots + \frac{h^n}{n!}x_i^{(n)}(t)$$

or in vector notation:

$$X(t+h) = X(t) + hX'(t) + \frac{h^2}{2!}X''(t) + \frac{h^3}{3!}X'''(t) + \cdots + \frac{h^n}{n!}X^{(n)}(t) \tag{10}$$

The derivatives appearing here can be obtained from the differential equation. Usually these derivatives must be computed in a particular order when used in a computer program. We must be sure that quantities needed at one step are available as the results from prior steps.

Example 3 Write the Taylor-series algorithm of order 3 for the following initial-value problem. Use $|h| = 0.1$ and compute the solution on the interval $-2 \leqq t \leqq 1$.

$$\begin{cases} x' = x + y^2 - t^3 & x(1) = 3 \\ y' = y + x^3 + \cos t & y(1) = 1 \end{cases} \tag{11}$$

Solution The higher derivatives required are

$$\begin{aligned} x'' &= x' + 2yy' - 3t^2 \\ y'' &= y' + 3x^2x' - \sin t \\ x''' &= x'' + 2yy'' + 2(y')^2 - 6t \\ y''' &= y'' + 6x(x')^2 + 3x^2x'' - \cos t \end{aligned}$$

A suitable algorithm to carry out the computation follows:

input $t \leftarrow 1;\ x \leftarrow 3;\ y \leftarrow 1;\ h \leftarrow -0.1;\ M \leftarrow 30$
output $0, t, x, y$
for $k = 1$ **to** M **do**
  $x' \leftarrow x + y^2 - t^3$
  $y' \leftarrow y + x^3 + \cos t$

$x'' \leftarrow x' + 2yy' - 3t^2$
$y'' \leftarrow y' + 3x^2x' - \sin t$
$x''' \leftarrow x'' + 2yy'' + 2(y')^2 - 6t$
$y''' \leftarrow y'' + 6x(x')^2 + 3x^2x'' - \cos t$
$x \leftarrow x + h(x' + \frac{1}{2}h(x'' + \frac{1}{3}h(x''')))$
$y \leftarrow y + h(y' + \frac{1}{2}h(y'' + \frac{1}{3}h(y''')))$
$t \leftarrow t + h$
output k, t, x, y
end do

The numerical output from the computer implementation of this algorithm is not given here, but a computer plot of the two functions $x(t)$ and $y(t)$ is shown in Figure 8.3. Since plotting routines are not at all standardized, no purpose would be served in showing the computer program by which the graphs were obtained. It suffices to remark that usually we must give the automatic plotter two arrays, say $\{t_i : 0 \leqq i \leqq m\}$ and $\{x_i : 0 \leqq i \leqq m\}$, from which it would then plot the points (t_i, x_i) on a Cartesian grid. Usually the plotter can be instructed to connect these points with straight lines. Hence, it is imperative that the points be ordered with this in mind. Usually, of course, $t_0 < t_1 < \cdots < t_m$. If the plotted points are close to each other, the resulting curve will appear to be a smooth curve and not a sequence of line segments. ■

From the theoretical standpoint, there is no loss of generality in assuming that the equations in System (1) do not contain t explicitly. We can write the system in the form

$$x_i' = f_i(x_0, x_1, \ldots, x_n)$$

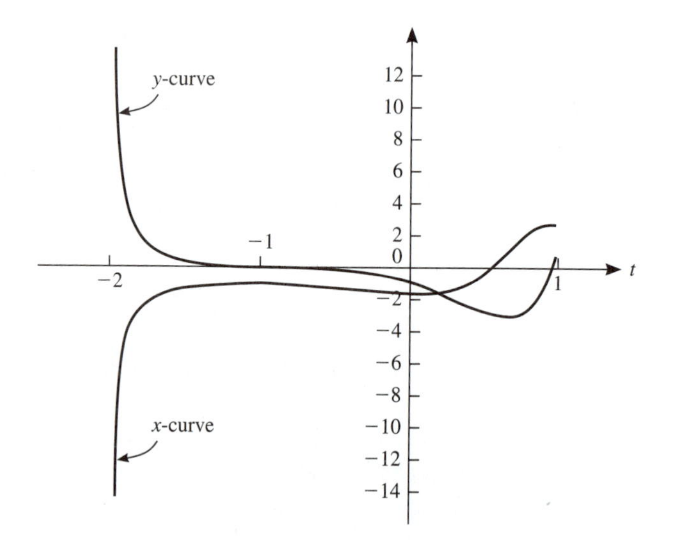

FIGURE 8.3 Solution curves for Example 3

by introducing a variable $x_0 = t$. The differential equation for the new variable is simply $x_0' = 1$. By use of this device, we can write the system of equations (5) as

$$X' = F(X) \tag{12}$$

without t appearing explicitly. Here $X = (x_0, x_1, \dots, x_n)^T$. A system in the form (12) is said to be **autonomous**.

Example 4 Convert the initial-value problem of Example 1 into a system in which t does not appear explicitly.

Solution Let x_1, x_2, x_3 be as in Example 1. Put $x_0 = t$. The new system is

$$\begin{cases} x_0' = 1 \\ x_1' = x_2 \\ x_2' = x_3 \\ x_3' = \left[\log x_0 - x_2^3 - \sin(x_0^2 + x_3) - \cos(x_0 x_1)\right] / \sin x_0 \end{cases} \tag{13}$$

and the initial condition is $X = (2, 7, 3, -4)^T$. ■

Other Methods for Systems

The Runge-Kutta procedure for systems of first-order equations is most easily written down in the case when our system is autonomous; that is, it has the form of Equation (12). The classical **fourth-order Runge-Kutta formulas**, in vector form, are

$$X(t + h) = X(t) + \frac{1}{6}(F_1 + 2F_2 + 2F_3 + F_4) \tag{14}$$

where

$$\begin{cases} F_1 = hF(X) \\ F_2 = hF(X + \frac{1}{2}F_1) \\ F_3 = hF(X + \frac{1}{2}F_2) \\ F_4 = hF(X + F_3) \end{cases}$$

The programming of a routine to carry out this procedure is left as Computer Problem 4. Similar formulas can be given for the Runge-Kutta-Fehlberg method discussed in Section 8.3.

Multistep methods can also be extended to apply to systems of equations. As an example, we give the vector form of the **Adams-Bashforth-Moulton predictor-corrector method** (4) and (8) from Section 8.4:

$$X^*(t + h) = X(t) + \frac{h}{720}\left[1901F\big(X(t)\big) - 2774F\big(X(t - h)\big) + 2616F\big(X(t - 2h)\big) - 1274F\big(X(t - 3h)\big) + 251F\big(X(t - 4h)\big)\right]$$

$$X(t+h) = X(t) + \frac{h}{720}\big[251F\big(X^*(t+h)\big) + 646F\big(X(t)\big)$$
$$- 264F\big(X(t-h)\big) + 106F\big(X(t-2h)\big) - 19F\big(X(t-3h)\big)\big]$$

As in the case of a single equation, a single-step procedure such as a fifth-order Runge-Kutta method (Computer Problems 7–9 in Section 8.3) could be used to provide starting values:

$$X(t_0+h) \qquad X(t_0+2h) \qquad X(t_0+3h) \qquad X(t_0+4h)$$

PROBLEMS 8.6

1. Find the general solution of the system

$$\begin{cases} x' = 3x - 4y + e^t \\ y' = x - y - e^t \end{cases}$$

   Hint: Try functions of the form e^t, te^t, and t^2e^t.

2. Write an autonomous system of first-order equations equivalent to

$$\begin{cases} x''' - \sin(x'') + e^t x' + 2t\cos x = 25 \\ x(0) = 5 \qquad x'(0) = 3 \qquad x''(0) = 7 \end{cases}$$

3. Write the third-order ordinary differential equation

$$\begin{cases} x''' + 2x'' - x' - 2x = e^t \\ x(8) = 3 \qquad x'(8) = 2 \qquad x''(8) = 1 \end{cases}$$

   as an autonomous system of first-order equations.

4. Convert the system of second-order ordinary differential equations

$$\begin{cases} x'' - x'y = 3y'x\log t \\ y'' - 2xy' = 5x'y\sin t \end{cases}$$

   into a system of first-order equations in which t does not appear explicitly.

5. Write

$$\begin{cases} y'' + yz = 0 \qquad y(0) = 1 \qquad y'(0) = 0 \\ z' + 2yz = 4 \qquad z(0) = 3 \end{cases}$$

   as a system of first-order equations with initial conditions. Use vector notation.

6. Write an autonomous system of first-order equations equivalent to

$$\begin{cases} x''' - \left[\sin x'' + e^t x'\right]^2 + \cos x = 0 \\ x(0) = 3 \qquad x'(0) = 4 \qquad x''(0) = 5 \end{cases}$$

7. Explain how to solve this problem numerically, assuming the availability of a computer program for solving the initial-value problem for a system of first-order equations:

$$\begin{cases} x'' = x\cos t + e^t x' + 3t^2 + 7 \\ x(1) = 5 \qquad x'(1) = 9 \end{cases}$$

8. Write this system of higher-order ordinary differential equations as a system of first-order equations:

$$\begin{cases} y_1^{(n)} = f_1(t, y_1, y_1', \dots, y_1^{(n-1)}) \\ y_2^{(n)} = f_2(t, y_2, y_2', \dots, y_2^{(n-1)}) \\ \quad\vdots \\ y_m^{(n)} = f_m(t, y_m, y_m', \dots, y_m^{(n-1)}) \end{cases}$$

9. Convert the following system of higher-order differential equations into a system of first-order equations in which t does not appear explicitly:

$$\begin{cases} x''' - 5tx''y'' + \ln(x')z = 0 \\ y'' - \sin(ty) + 7tx'' = 0 \\ z' + 16ty' - e^t zx' = 0 \end{cases}$$

10. Write Euler's method, Heun's method, and the modified Euler's method for a system of equations in the form of Equation (12).

COMPUTER PROBLEMS 8.6

1. Write a computer program to solve this initial-value problem using the Taylor-series method. Include terms in h, h^2, and h^3 and continue the solution to $t = 1$. Let $h = 0.01$.

$$\begin{cases} x_1' = t + x_1^2 + x_2 & x_1(-1) = 0.43 \\ x_2' = t^2 - x_1 + x_2^2 & x_2(-1) = -0.69 \end{cases}$$

2. (A challenging problem) Follow the instructions in Computer Problem 1 for this initial-value problem:

$$\begin{cases} x_1' = \sin x_1 + \cos(tx_2) & x_1(-1) = 2.37 \\ x_2' = t^{-1}\sin(tx_1) & x_2(-1) = -3.48 \end{cases}$$

The programming of x_2', x_2'', and x_2''' must be very carefully carried out because of the singularity at $t = 0$.

3. Solve numerically the system of differential equations

$$\begin{cases} x_1' = (-1 - 9c^2 + 12sc)x_1 + (12c^2 + 9sc)x_2 & x_1(0) = -12 \\ x_2' = (-12s^2 + 9s)x_1 + (-1 - 9s^2 - 12sc)x_2 & x_2(0) = 6 \end{cases}$$

in which $c = \cos 6t$ and $s = \sin 6t$. Integrate over the interval [0, 10] with step size $h = 0.01$. Verify that the correct solution is

$$\begin{cases} x_1 = e^{-13t}(s - 2c) \\ x_2 = e^{-13t}(2s + c) \end{cases}$$

Does your computed solution compare favorably with the true solution? This example has been given by Lambert [1973].

4. Write a subprogram or procedure that takes one step of length h in the fourth-order Runge-Kutta procedure. Make it capable of handling a system of n differential equations with $n \leqq 20$. Test your program by solving the following system on the interval $1 \leqq t \leqq 2$. Use $h = -0.01$.

$$\begin{cases} x' = x^{-2} + \log y + t^2 \\ y' = e^y - \cos x + (\sin t)x - (xy)^{-3} \\ x(2) = -2 \qquad y(2) = 1 \end{cases}$$

5. Solve and plot the resulting curve over the interval [0, 5] for the ordinary differential equation $x'' + 192x = 0$ with $x(0) = \frac{1}{6}$, $x'(0) = 0$. This corresponds to an **undamped spring-mass system**.

6. Write and test a subroutine or procedure for the adaptive Runge-Kutta-Fehlberg algorithm that can handle systems of first-order differential equations.

7. Write and test a subroutine or procedure that is capable of handing systems of first-order differential equations for the fifth-order Adams-Bashforth-Moulton method coupled with a fifth-order Runge-Kutta method (see Computer Problems 7–9 in Section 8.3).

*8.7 Boundary-Value Problems

In the preceding sections of this chapter, various methods have been considered for solving the initial-value problem. Thus, for example, the initial-value problem

$$\begin{cases} x'' = f(t, x, x') \\ x(a) = \alpha \qquad x'(a) = \beta \end{cases} \tag{1}$$

can be put into the form of a system of first-order equations by letting $x_1 = x$ and $x_2 = x'$:

$$\begin{cases} x_1' = x_2 & x_1(a) = \alpha \\ x_2' = f(t, x_1, x_2) & x_2(a) = \beta \end{cases} \tag{2}$$

System (2) can then be solved by one of the step-by-step methods described earlier.

The situation would be quite different, however, if the problem in (1) were changed to read

$$\begin{cases} x'' = f(t, x, x') \\ x(a) = \alpha \qquad x(b) = \beta \end{cases} \tag{3}$$

The difference lies in the specification of conditions at two different points, $t = a$ and $t = b$. The step-by-step procedures for the initial-value problem are not adapted to the solution of (3) since the numerical solution cannot begin without a *full complement* of initial values. In (3), we have a typical example of a **two-point boundary-value problem**. Such problems usually present difficulties of a greater magnitude than those of the initial-value problem.

Here is an example of a two-point boundary-value problem that can be solved without numerical work:

$$\begin{cases} x'' = -x \\ x(0) = 3 \qquad x(\frac{\pi}{2}) = 7 \end{cases} \tag{4}$$

We can find first the general solution of the differential equation, which is

$$x(t) = A \sin t + B \cos t \tag{5}$$

Then the constants A and B can be determined so that the boundary conditions are satisfied. Thus,

$$\begin{cases} 3 = x(0) = A \sin 0 + B \cos 0 = B \\ 7 = x(\frac{\pi}{2}) = A \sin \frac{\pi}{2} + B \cos \frac{\pi}{2} = A \end{cases}$$

The solution of (4) is then

$$x(t) = 7 \sin t + 3 \cos t$$

Further examples of this type are given in the problems.

The technique just illustrated is not effective if the general solution of the differential equation in (3) is unknown. Our interest here is in numerical methods by which any two-point boundary-value problem can be attacked.

Existence

Before considering numerical methods, a few comments about the existence question are appropriate. In general, we cannot assure the existence of a solution to (3) simply by assuming that f is a *nice* function. A simple example in support of this assertion is

$$\begin{cases} x'' = -x \\ x(0) = 3 \qquad x(\pi) = 7 \end{cases} \tag{6}$$

This is only slightly different from the problem in (4), but when we try to impose the boundary values on the general solution (5), we get the contradictory equations $3 = B$ and $7 = -B$. Thus, the problem in (6) has no solution.

Existence theorems for solutions of the two-point boundary-value problem (3) tend to be rather complicated, and we refer the reader to the books by Stoer and Bulirsch [1980] and Keller [1968]. One elegant result in Keller [1968, p. 108] is the following.

THEOREM 1 *The boundary-value problem*

$$\begin{cases} x'' = f(t, x) \\ x(0) = 0 \qquad x(1) = 0 \end{cases}$$

has a unique solution if $\partial f/\partial x$ is continuous, nonnegative, and bounded in the infinite strip defined by the inequalities $0 \leqq t \leqq 1$, $-\infty < x < \infty$.

Example 1 Show that this two-point boundary-value problem has a unique solution:

$$\begin{cases} x'' = (5x + \sin 3x)e^t \\ x(0) = x(1) = 0 \end{cases}$$

Solution Let us try to use Theorem 1. Then

$$\frac{\partial f}{\partial x} = (5 + 3\cos 3x)e^t$$

This is continuous in the strip $0 \leqq t \leqq 1$, $-\infty < x < \infty$. Furthermore, it is bounded above by $8e$ and is nonnegative, since $3\cos 3x \geqq -3$. The hypotheses of Theorem 1 are therefore fulfilled. ■

Changes of Variables

Although Theorem 1 applies to a special case, simple changes of variables can reduce more general problems to the special case. To do this, we begin by changing the t-interval. Suppose that the original problem is

$$\begin{cases} x'' = f(t, x) \\ x(a) = \alpha \qquad x(b) = \beta \end{cases} \tag{7}$$

where $x = x(t)$. A change of variable $t = a + (b - a)s$ is called for here. Notice that $s = 0$ corresponds to $t = a$, and $s = 1$ corresponds to $t = b$. Therefore, we define $y(s) = x(a + \lambda s)$ with $\lambda = b - a$. Then $y'(s) = \lambda x'(a + \lambda s)$ and $y''(s) = \lambda^2 x''(a + \lambda s)$. Likewise $y(0) = x(a) = \alpha$ and $y(1) = x(b) = \beta$. Thus, if x is a solution of (7), then y is a solution of the boundary-value problem

$$\begin{cases} y''(s) = \lambda^2 f\big(a + \lambda s,\ y(s)\big) \\ y(0) = \alpha \qquad y(1) = \beta \end{cases} \tag{8}$$

Conversely, if y is a solution of (8), then the function $x(t) = y((t-a)/(b-a))$ is a solution of (10).

THEOREM 2 *Consider these two-point boundary-value problems*

$$\text{(i)} \begin{cases} x'' = f(t, x) \\ x(a) = \alpha \qquad x(b) = \beta \end{cases} \qquad \text{(ii)} \begin{cases} y'' = g(t, y) \\ y(0) = \alpha \qquad y(1) = \beta \end{cases}$$

in which

$$g(p, q) = (b-a)^2 f\big(a + (b-a)p, q\big)$$

If y is a solution of **(ii)**, *then the function x defined by $x(t) = y((t-a)/(b-a))$ is a solution of* **(i)**. *Moreover, if x is a solution of* **(i)**, *then $y(t) = x(a + (b-a)t)$ is a solution of* **(ii)**.

Proof It is a simple verification as follows:

$$\begin{aligned}
x(a) &= y\left(\frac{a-a}{b-a}\right) = y(0) = \alpha \\
x(b) &= y\left(\frac{b-a}{b-a}\right) = y(1) = \beta \\
x'(t) &= y'\left(\frac{t-a}{b-a}\right)\frac{1}{b-a} \\
x''(t) &= y''\left(\frac{t-a}{b-a}\right)\frac{1}{(b-a)^2} \\
&= g\left(\frac{t-a}{b-a}, y\left(\frac{t-a}{b-a}\right)\right)\frac{1}{(b-a)^2} \\
&= (b-a)^2 f\left(a + (b-a)\frac{t-a}{b-a}, y\left(\frac{t-a}{b-a}\right)\right)\frac{1}{(b-a)^2} = f\big(t, x(t)\big) \quad \blacksquare
\end{aligned}$$

Example 2 By use of Theorem 2, we see that these two-point boundary-value problems are equivalent:

$$\begin{cases} x'' = \sin(tx) + x^2 \\ x(1) = 3 \qquad x(4) = 7 \end{cases}$$

$$\begin{cases} y'' = 16\{\sin[(4s+1)y] + y^2\} \\ y(0) = 3 \qquad y(1) = 7 \end{cases}$$

To reduce a two-point boundary-value problem

$$\begin{cases} y'' = g(t, y) \\ y(0) = \alpha \qquad y(1) = \beta \end{cases}$$

to one having homogeneous boundary values, we simply subtract from y a linear function that takes the values α and β at 0 and 1. Here is the theorem.

THEOREM 3 *Consider these two-point boundary-value problems:*

$$\textbf{(ii)} \begin{cases} y'' = g(t, y) \\ y(0) = \alpha \qquad y(1) = \beta \end{cases} \qquad \textbf{(iii)} \begin{cases} z'' = h(t, z) \\ z(0) = 0 \qquad z(1) = 0 \end{cases}$$

where

$$h(p, q) = g\big(p, q + \alpha + (\beta - \alpha)p\big)$$

If z solves **(iii)**, *then the function $y(t) = z(t) + \alpha + (\beta - \alpha)t$ solves* **(ii)**. *Moreover, if y solves* **(ii)**, *then $z(t) = y(t) - [\alpha + (\beta - \alpha)t]$ solves* **(iii)**.

Proof Again there is a straightforward verification:

$$\begin{aligned} y(0) &= z(0) + \alpha + (\beta - \alpha)0 = \alpha \\ y(1) &= z(1) + \alpha + (\beta - \alpha)1 = \beta \\ y''(t) &= z''(t) = h\big(t, z(t)\big) = g\big(t, z(t) + \alpha + (\beta - \alpha)t\big) \\ &= g\big(t, y(t)\big) \end{aligned}$$

■

Example 3 Show that this problem has a unique solution:

$$\begin{cases} x'' = \big[5x - 10t + 35 + \sin(3x - 6t + 21)\big]e^t \\ x(0) = -7 \qquad x(1) = -5 \end{cases}$$

Solution Since the boundary values are not homogeneous in this problem, Theorem 1 does not apply immediately. To obtain an equivalent problem with 0 boundary values, we change the dependent variable as in Theorem 3. Let

$$z(t) = x(t) - \ell(t) \quad \text{and} \quad \ell(t) = -7 + 2t$$

Then

$$\begin{aligned} z'' = x'' &= \big[5x - 10t + 35 + \sin(3x - 6t + 21)\big]e^t \\ &= \big\{5(z + \ell) - 10t + 35 + \sin\big[3(z + \ell) - 6t + 21\big]\big\}e^t \\ &= \{5z + \sin 3z\}e^t \end{aligned}$$

The boundary values for the new variable z are

$$\begin{aligned} z(0) &= x(0) - \ell(0) = -7 + 7 = 0 \\ z(1) &= x(1) - \ell(1) = -5 - (-5) = 0 \end{aligned}$$

The boundary problem for z has a unique solution, as shown in Example 1. Hence the problem for x has a unique solution. ■

Example 4 Convert this two-point boundary-value problem to an equivalent one with 0 boundary values on the interval [0, 1]:

$$\begin{cases} x'' = x^2 + 3 - t^2 + xt \\ x(3) = 7 \qquad x(5) = 9 \end{cases} \tag{9}$$

Solution By Theorem 2, an equivalent problem is

$$\begin{cases} y'' = g(t, y) \\ y(0) = 7 \qquad y(1) = 9 \end{cases} \tag{10}$$

with

$$g(t, y) = 4f(3 + 2t, y) = 4\left[y^2 + 3 - (3 + 2t)^2 + y(3 + 2t)\right]$$

By Theorem 3, another equivalent problem is

$$\begin{cases} z'' = h(t, z) \\ z(0) = 0 \qquad z(1) = 0 \end{cases} \tag{11}$$

with $h(t, z) = g(t, z + 7 + 2t)$. Expressing h in detail, we have

$$h(t, z) = 4\left[(z + 7 + 2t)^2 + 3 - (3 + 2t)^2 + (z + 7 + 2t)(3 + 2t)\right]$$

We can solve boundary-value problem (11) for z and obtain the solution of boundary-value problem (10) from the equation

$$y(t) = z(t) + 7 + 2t$$

The solution of boundary-value problem (9) is then obtained from

$$x(t) = y\left(\frac{t - 3}{2}\right)$$

■

Example 5 Suppose that we compute the solution $z(0.5) = 8.2$ for the boundary-value problem (11). What is the corresponding solution to boundary-value problem (9)?

Solution Clearly, $y(0.5) = z(0.5) + 7 + 2(0.5) = 16.2$ and $0.5 = (t - 3)/2$ implies $t = 4$. The answer is

$$x(4) = y(0.5) = z(0.5) + 8 = 16.2$$

■

THEOREM 4 *Let f be a continuous function of (t, s), where $0 \leqq t \leqq 1$ and $-\infty < s < \infty$. Assume that on this domain*

$$|f(t, s_1) - f(t, s_2)| \leqq k|s_1 - s_1| \qquad (k < 8)$$

Then the two-point boundary-value problem

$$\begin{cases} x'' = f(t, x) \\ x(0) = x(1) = 0 \end{cases} \tag{12}$$

has a unique solution in $C[0, 1]$.

Proof (*In outline*) By appealing to Problem 10, we know that the boundary-value problem (12) is equivalent to the integral equation

$$x(t) = \int_0^1 G(t, s) f\big(s, x(s)\big)\, ds \tag{13}$$

in which G is the Green's function of Problem 6. The integral equation (13) has the form $x = F(x)$, where F is the operation defined by the integral. By using Banach's Contraction Mapping Theorem in the space $C[0, 1]$, we conclude that F has a unique fixed point and that Equations (12) and (13) have unique solutions. ■

Example 6 Show that this problem has a unique solution:

$$\begin{cases} x'' = 2\exp(t\cos x) \\ x(0) = x(1) = 0 \end{cases} \tag{14}$$

Solution Here $f(t, s) = 2\exp(t\cos s)$. By the Mean-Value Theorem,

$$|f(t, s_1) - f(t, s_2)| = \left|\frac{\partial f}{\partial s}(t, s_3)\right| \; |s_1 - s_2|$$

The derivative needed here satisfies the relation

$$\left|\frac{\partial f}{\partial s}\right| = |2\exp(t\cos s)(-t\sin s)| \leqq 2e < 5.437 < 8$$

By Theorem 4, the two-point boundary-value problem (14) has a unique solution. ■

PROBLEMS 8.7

1. Solve the two-point boundary-value problem $x'' = x$, $x(0) = 1$, $x(1) = 1$.
2. Determine all pairs (α, β) for which the problem $x'' = x$, $x(0) = \alpha$, $x(1) = \beta$ has a solution.
3. Repeat Problem 2 for $x'' = -x$, $x(0) = \alpha$, $x(\pi) = \beta$.
4. Solve the problem $x'' - 2x' + x = 0$, $x(0) = \alpha$, $x(1) = \beta$. Are there any pairs (α, β) for which no solution exists?
5. Give an example of a two-point boundary-value problem of type (3) in the text for which there is more than one solution. *Hint:* Consider Problem 3.

6. Show that the two-point boundary-value problem $x'' = f(t)$, $x(0) = x(1) = 0$ is solved by the formula

$$x(t) = \int_0^1 G(t, s) f(s)\, ds$$

in which

$$G(t, s) = \begin{cases} s(t-1) & 0 \leqq s \leqq t \leqq 1 \\ t(s-1) & 0 \leqq t \leqq s \leqq 1 \end{cases}$$

The function G is known as the **Green's function** for this problem.

7. Consider these two boundary-value problems:

$$\textbf{(i)} \quad \begin{cases} x'' = f(t, x, x') \\ x(a) = \alpha \qquad x(b) = \beta \end{cases}$$

$$\textbf{(ii)} \quad \begin{cases} x'' = h^2 f(a + th, x, h^{-1}x') \\ x(0) = \alpha \qquad x(1) = \beta \end{cases}$$

Show that if x is a solution of **(ii)**, then the function $y(t) = x((t-a)/h)$ solves **(i)**, where $h = b - a$.

8. Illustrate Theorem 2 with the boundary-value problem

$$\begin{cases} x'' = t + x^2 - 3x' \\ x(3) = \alpha \qquad x(7) = \beta \end{cases}$$

This is equivalent to

$$\begin{cases} x'' = 48 + 64t + 16x^2 - 12x' \\ x(0) = \alpha \qquad x(1) = \beta \end{cases}$$

9. Use Theorem 4 to establish that this two-point problem has a unique solution:

$$\begin{cases} x'' = \cos(tx) \\ x(0) = 1 \qquad x(1) = 4 \end{cases}$$

10. Show that the boundary-value problem

$$\begin{cases} x'' = f(t, x) \\ x(0) = 0 \qquad x(1) = 0 \end{cases}$$

is equivalent to the integral equation

$$x(t) = \int_0^1 G(t, s) f\big(s, x(s)\big)\, ds$$

in which G is the Green's function of Problem 6.

11. Prove that the following two-point boundary-value problem has a unique solution:

$$\begin{cases} x'' = (t^3 + 5)x + \sin t \\ x(0) = x(1) = 0 \end{cases}$$

12. Prove that this problem has a unique solution:

$$\begin{cases} x'' = \tan^{-1} x + 2x + \cos t \\ x(0) = x(1) = 0 \end{cases}$$

13. Solve the boundary-value problem

$$\begin{cases} x'' = x^2 \\ x(0) = 2/3 \qquad x(1) = 3/8 \end{cases}$$

14. (Continuation) Solve Problem 13 when the boundary conditions are simplified to $x(0) = 0$, $x(1) = 1$. Refer to Davis [1962].

15. Solve the boundary-value problem

$$\begin{cases} x'' = x^3 \\ x(0) = 1 \qquad x(1) = 2 - \sqrt{2} \end{cases}$$

16. Prove that if x solves the differential equation $x'' = f(x)$, then so does y, where $y(t) = x(t + c)$.

17. Prove that this problem has a unique solution:

$$\begin{cases} x'' = \dfrac{1}{2}\exp\left\{\dfrac{1}{2}(t+1)\cos(x + 7 - 3t)\right\} \qquad (-1 \leqq t \leqq 1) \\ x(-1) = -10 \qquad x(1) = -4 \end{cases}$$

18. Find a two-point boundary-value problem of the form

$$\begin{cases} z'' = h(s, z) \\ z(0) = z(1) = 0 \end{cases}$$

which is equivalent to this problem:

$$\begin{cases} x'' = \cos(x^2 t^2) \\ x(3) = 5 \qquad x(7) = 12 \end{cases}$$

19. Suppose that v and w are known functions of t having these properties:

$$\begin{cases} v'' = \cos t + ve^t + v't^2 \qquad v(1) = 5 \qquad v(3) = 42 \\ w'' = \cos t + we^t + w't^2 \qquad w(1) = 8 \qquad w(3) = 9 \end{cases}$$

What function solves this two-point boundary-value problem?

$$\begin{cases} x'' = \cos t + xe^t + x't^2 \\ x(1) = 7 \quad x(3) = 20 \end{cases}$$

20. (Multiple-choice question)

(a) The two-point boundary-value problem

$$\begin{cases} x'' = -4x \\ x(0) = 1 \qquad x(\pi/2) = -1 \end{cases}$$

is an example having **(i)** no solution; **(ii)** exactly two solutions; **(iii)** exactly one solution; **(iv)** no solution in elementary functions; **(v)** more than one solution.

(b) Repeat part **(a)** for the two-point boundary-value problem

$$\begin{cases} x'' = -4x \\ x(0) = 1 \qquad x(\pi/2) = 2 \end{cases}$$

21. Consider the two-point boundary-value problem

$$\begin{cases} x'' = f(t, x) \\ x(0) = x(1) = 0 \end{cases}$$

For which function f can we be sure that a unique solution exists?

(i) $f(t, x) = t^2(1 + x^2)^{-1}$

(ii) $f(t, x) = t \sin x$

(iii) $f(t, x) = (\tan t)(\tan x)$

(iv) $f(t, x) = t/x$

(v) $f(t, x) = x^{1/3}$

22. Consider the two-point boundary-value problem

$$\begin{cases} x'' = f(t, x) \\ x(0) = x(1) = 0 \end{cases}$$

For which function f can we be sure that a unique solution exists?

(i) $f(t, x) = t^2 x^3$

(ii) $f(t, x) = t(\tan^{-1} x)$

(iii) $f(t, x) = tx^{4/3}$

(iv) $f(t, x) = \log(1 + x^2)$

(v) $f(t, x) = |tx|$

23. Prove that if $x(t)$ is a solution of

$$\begin{cases} x''(t) = f\big(t, x(t)\big) \\ x(a) = \alpha \qquad x(b) = \beta \end{cases}$$

then $y(s) = x(a + (b - a)s)$ is a solution of

$$\begin{cases} y''(s) = (b-a)^2 f\big(a + (b-a)s, y(s)\big) \\ y(0) = \alpha \qquad y(1) = \beta \end{cases}$$

and consequently $z(s) = y(s) - [\alpha + (\beta - \alpha)s]$ is a solution of

$$\begin{cases} z''(s) = (b-a)^2 f\big(a + (b-a)s, z(s) + \alpha + (\beta - \alpha)s\big) \\ z(0) = 0 \qquad z(1) = 0 \end{cases}$$

8.8 Boundary-Value Problems: Shooting Methods

The two-point boundary-value problem being considered is

$$\begin{cases} x'' = f(t, x, x') \\ x(a) = \alpha \qquad x(b) = \beta \end{cases} \tag{1}$$

One natural way to attack this problem is to solve the related initial-value problem, with a *guess* as to the appropriate initial value $x'(a)$. Then we can integrate the equation to obtain an approximate solution, hoping that $x(b) = \beta$. If not, then the guessed value of $x'(a)$ can be altered and we can try again. This process is called **shooting**, and there are ways of doing it systematically.

Let us denote the guessed value of $x'(a)$ by z, so that the corresponding initial-value problem is

$$\begin{cases} x'' = f(t, x, x') \\ x(a) = \alpha \qquad x'(a) = z \end{cases} \tag{2}$$

The solution of this initial-value problem will be denoted by x_z. The objective is to select z so that $x_z(b) = \beta$. We put

$$\phi(z) \equiv x_z(b) - \beta$$

so that our objective is simply to solve the equation $\phi(z) = 0$ for z. The methods considered in Chapter 3 for solving a single nonlinear equation are applicable here. For example, the bisection method, the secant method, and Newton's method are all available, in addition to functional iteration. The function ϕ is an *expensive* one to compute since each value of $\phi(z)$ is obtained by numerically solving an initial-value problem.

Secant Method

Let us recall how the secant method is applied to solve an equation $\phi(z) = 0$. With two values of $\phi(z)$ at hand, say $\phi(z_1)$ and $\phi(z_2)$, we pretend that ϕ is *linear*. Using the equation of the straight line through the points $(z_2, \phi(z_2))$ and $(z_1, \phi(z_1))$, we have

$$\phi(z) - \phi(z_2) = \left(\frac{\phi(z_1) - \phi(z_2)}{z_1 - z_2} \right) (z - z_2)$$

If we select z_3 so that $\phi(z_3) = 0$, we obtain the formula

$$z_3 = z_2 - \left(\frac{z_2 - z_1}{\phi(z_2) - \phi(z_1)} \right) \phi(z_2) \tag{3}$$

This procedure can be repeated to obtain a sequence of values $z_1, z_2, \dots, z_n$ by

$$z_n = z_{n-1} - \left(\frac{z_{n-1} - z_{n-2}}{\phi(z_{n-1}) - \phi(z_{n-2})} \right) \phi(z_{n-1})$$

This equation defines the secant method. (See Section 3.3.) When several values of z have been obtained for which $\phi(z)$ is nearly 0, we can stop this procedure and use polynomial interpolation to estimate a better value. Here is how this should be done: Suppose that $\phi(z_1), \phi(z_2), \dots, \phi(z_n)$ are small. Find a polynomial p that interpolates the table

$\phi(z_1)$	$\phi(z_2)$	$\cdots$	$\phi(z_n)$
z_1	z_2	$\cdots$	z_n

Thus, the polynomial p has the property $p(\phi(z_i)) = z_i$ for $1 \leqq i \leqq n$. The next estimate, z_{n+1}, is determined by the equation $p(0) = z_{n+1}$. This procedure amounts to an approximation of the **inverse function** of ϕ by the polynomial p. Its success depends on ϕ having a differentiable inverse in a neighborhood of the root. This in turn requires the assumption that the root in question be *simple*.

Linear Function

The shooting method as outlined above can be quite costly in computational effort, and it is important to consider techniques for economizing. Obviously any partial information about the correct value of $x'(a)$ should be exploited. It is also possible to solve the initial-value problems with a large step size, since high precision is essentially wasted in the first stages of the shooting method. Only when $\phi(z)$ is nearly 0 should a small step size be used.

There is one class of problems in which the secant method yields the exact solution in one step. This is when ϕ is in fact a *linear* function. This occurs when the differential equation is linear. In the linear case, the two-point boundary-value problem will have the form

$$\begin{cases} x'' = u(t) + v(t)x + w(t)x' \\ x(a) = \alpha \qquad x(b) = \beta \end{cases} \tag{4}$$

We shall assume in what follows that the functions u, v, and w are continuous on the interval $[a, b]$. Suppose that we have solved the differential equation in (4) twice, with two different initial conditions, obtaining solutions x_1 and x_2—say

$$\begin{cases} x_1(a) = \alpha \qquad x_1'(a) = z_1 \\ x_2(a) = \alpha \qquad x_2'(a) = z_2 \end{cases} \tag{5}$$

Now we form a linear combination of x_1 and x_2:

$$y(t) = \lambda x_1(t) + (1 - \lambda)x_2(t) \tag{6}$$

where λ is a parameter. It is easily verified that y solves the differential equation and meets the first of the two boundary conditions; that is, $y(a) = \alpha$. We select λ so that $y(b) = \beta$. Thus,

$$\beta = y(b) = \lambda x_1(b) + (1 - \lambda)x_2(b)$$

and

$$\lambda = \frac{\beta - x_2(b)}{x_1(b) - x_2(b)} \tag{7}$$

In a computer realization of these ideas for the linear problem (4), we can obtain x_1 and x_2 at the same time. Thus, the two initial-value problems to be solved could be specified as

$$\begin{cases} x'' = f(t, x, x') \\ x(a) = \alpha \qquad x'(a) = 0 \end{cases} \qquad \begin{cases} x'' = f(t, x, x') \\ x(a) = \alpha \qquad x'(a) = 1 \end{cases}$$

where $f(t, x, x') = u(t) + v(t)x + w(t)x'$. The solution of the first is x_1 and the solution of the second is x_2. To create a system of first-order equations without t appearing explicitly, we define also $x_0 = t$, $x_3 = x_1'$, $x_4 = x_2'$. Then the system of differential equations with initial values is

$$\begin{cases} x_0 = 1 & x_0(a) = a \\ x_1' = x_3 & x_1(a) = \alpha \\ x_2' = x_4 & x_2(a) = \alpha \\ x_3' = f(x_0, x_1, x_3) & x_4(a) = 0 \\ x_4' = f(x_0, x_2, x_4) & x_5(a) = 1 \end{cases} \tag{8}$$

This system should be given as input to a computer program for solving the initial-value problem. The approximate values of the discrete functions $x_1(t_i)$ and $x_2(t_i)$ for $a = t_0 \leqq t_i \leqq t_m = b$ should be stored in the computer memory as one-dimensional arrays. Next, the value of λ should be computed by Equation (7). Finally, the solution y should be computed at each desired value of t from Equation (6).

THEOREM 1 *If the linear two-point boundary-value problem* (4) *has a solution, then either* x_1 *itself is a solution or* $x_1(b) - x_2(b) \neq 0$ *(and* y *is a solution).*

Proof Let y_0, y_1, and y_2 be solutions of these initial-value problems:

$$\begin{array}{lll} y_0'' = u + vy_0 + wy_0' & y_0(a) = \alpha & y_0'(a) = 0 \\ y_1'' = vy_1 + wy_1' & y_1(a) = 1 & y_1'(a) = 0 \\ y_2'' = vy_2 + wy_2' & y_2(a) = 0 & y_2'(a) = 1 \end{array}$$

By the theory of second-order linear differential equations (in particular, Theorem 3 below), the general solution of the differential equation in (4) is

$$y_0 + c_1 y_1 + c_2 y_2$$

in which c_1 and c_2 are arbitrary constants. We see immediately that the functions x_1 and x_2 in (5) are special cases of the general solution given by

$$x_1 = y_0 + z_1 y_2 \qquad x_2 = y_0 + z_2 y_2 \tag{9}$$

Since it has been assumed that the problem in (4) has a solution, there exist c_1 and c_2 such that

$$\alpha = y_0(a) + c_1 y_1(a) + c_2 y_2(a)$$

$$\beta = y_0(b) + c_1 y_1(b) + c_2 y_2(b)$$

The first of these equations reduces to $c_1 = 0$, and thus we infer the existence of c_2 such that

$$\beta = y_0(b) + c_2 y_2(b) \tag{10}$$

If $x_1(b) - x_2(b) \neq 0$, then the function y defined by Equations (6) and (7) solves (4). If $x_1(b) - x_2(b) = 0$, then from Equations (9), we have $y_2(b) = 0$. Equation (10) tells us that $y_0(b) = \beta$, and Equations (9) tell us that x_1 is a solution. ■

Newton's Method

Let us return to the more general (nonlinear) two-point boundary-value problem of Equation (1), and consider how to apply Newton's method. Recall that x_z is defined as the solution of the problem

$$\begin{cases} x_z'' = f(t, x_z, x_z') \\ x_z(a) = \alpha \qquad x_z'(a) = z \end{cases} \tag{11}$$

We want to select z so that

$$\phi(z) \equiv x_z(b) - \beta = 0$$

Newton's formula for the function ϕ is

$$z_{n+1} = z_n - \frac{\phi(z_n)}{\phi'(z_n)} \tag{12}$$

To determine ϕ', we differentiate partially with respect to z all the equations in (11):

$$\begin{cases} \dfrac{\partial x_z''}{\partial z} = \dfrac{\partial f}{\partial t}\dfrac{\partial t}{\partial z} + \dfrac{\partial f}{\partial x_z}\dfrac{\partial x_z}{\partial z} + \dfrac{\partial f}{\partial x_z'}\dfrac{\partial x_z'}{\partial z} \\ \dfrac{\partial}{\partial z} x_z(a) = 0 \qquad \dfrac{\partial}{\partial z} x_z'(a) = 1 \end{cases} \tag{13}$$

With some simplification and the introduction of the new variable $v = \partial x_z / \partial z$, this becomes

$$\begin{cases} v'' = f_{x_z}(t, x_z, x_z')v + f_{x_z'}(t, x_z, x_z')v' \\ v(a) = 0 \qquad v'(a) = 1 \end{cases} \tag{14}$$

We recognize this set of equations as an initial-value problem. The differential equation in (14) is called the **first variational equation**. It can be solved step by step along with (11). At the end, $v(b)$ will be available, and we have

$$v(b) = \frac{\partial x_z(b)}{\partial z} = \phi'(z)$$

This enables us to use Newton's method, Equation (12), to find a root of ϕ.

Multiple Shooting

An elaboration of the shooting method is called **multiple shooting**. The basic strategy here is to divide the given interval $[a, b]$ into subintervals and attempt to solve the global problem in pieces. Let us describe what would be done if the interval were subdivided into just two parts, $[a, c]$ and $[c, b]$.

The original problem is as before; namely,

$$\begin{cases} x'' = f(t, x, x') \\ x(a) = \alpha \qquad x(b) = \beta \end{cases}$$

On the two subintervals, we solve initial-value problems to get two functions x_1 and x_2:

$$\begin{cases} x_1'' = f(t, x_1, x_1') & x_1(a) = \alpha \quad x_1'(a) = z_1 \quad (a \leqq t \leqq c) \\ x_2'' = f(t, x_2, x_2') & x_2(b) = \beta \quad x_2'(b) = z_2 \quad (c \leqq t \leqq b) \end{cases}$$

Notice that z_1 and z_2 are parameters at our disposal. The function x_1 is required only on the interval $[a, c]$, and x_2 is required only on $[c, b]$. The numerical solution of x_2 will proceed in the direction of *decreasing* t.

The parameters z_1 and z_2 are now to be adjusted until the piecewise function

$$x(t) = \begin{cases} x_1(t) & a \leqq t \leqq c \\ x_2(t) & c \leqq t \leqq b \end{cases}$$

becomes a solution to the problem. Thus, we require continuity in x and x' at the point c: $x_1(c) - x_2(c) = 0$ and $x_1'(c) - x_2'(c) = 0$. These two conditions are to be fulfilled by an adroit choice of z_1 and z_2. Typically this would be done by Newton's method in dimension 2, as in Section 3.5.

Multiple shooting with k subintervals will involve k **subfunctions**, each of them being obtained by numerically solving an initial-value problem. The initial values of these k subfunctions will form a set of $2k$ parameters. At each of the $k - 1$ interior division points of the interval, the continuity of the global function and its first derivative must be imposed. This provides $2k - 2$ conditions. There are two endpoint conditions, and so the number of conditions matches the number of parameters. The resulting system of nonlinear equations is solved iteratively—for example, by the higher-dimensional Newton method.

Much of the software that is available for two-point boundary-value problems is written for a system of first-order ordinary differential equations

$$X' = F(t, X)$$

where $X = (x_1, x_2, \ldots, x_n)^T$ and $F = (f_1, f_2, \ldots, f_n)^T$. The boundary conditions are often permitted to be quite general. For example, some codes allow the boundary conditions to be in the form

$$G\big(X(a), X(b)\big) = 0$$

where $G = (g_1, g_2, \ldots, g_n)^T$. Some software requires the user to furnish the Jacobian matrix J of F, which is the $n \times n$ matrix with entries $J_{ij} = \partial f_i / \partial x_j$.

Example 1 Suppose that software of the type described above is to be used on the problem

$$\begin{cases} x'' = tx + \cos x' \\ x(3) + x'(5) = 7 \qquad x'(3)^2 x(5) = 10 \end{cases}$$

What are F, G, and J?

Solution Set $x_1 = x$ and $x_2 = x'$. The problem can now be expressed as

$$\begin{cases} x_1' = x_2 & x_1(3) + x_2(5) - 7 = 0 \\ x_2' = tx_1 + \cos x_2 & x_2(3)^2 x_1(5) - 10 = 0 \end{cases}$$

The functions F and G can be read off from these equations. The Jacobian function is given by

$$J(t, X) = \begin{bmatrix} 0 & 1 \\ t & -\sin x_2 \end{bmatrix}$$

■

Second-Order Linear Equations

Two important theorems from the general theory of second-order linear differential equations are cited here. The standard reference for such matters is Coddington and Levinson [1955].

THEOREM 2 *If u, v, and w are continuous functions on the closed interval $[a, b]$, then for any pair of real numbers α and α' the initial-value problem*

$$\begin{cases} x'' = u + vx + wx' \\ x(a) = \alpha \qquad x'(a) = \alpha' \end{cases}$$

has a unique solution on $[a, b]$.

THEOREM 3 *Every solution of the (nonhomogeneous) equation*

$$x'' - vx - wx' = u \tag{15}$$

can be expressed in the form $x_0 + c_1x_1 + c_2x_2$, where x_0 is any particular solution of Equation (15), and x_1 and x_2 form a linearly independent set of solutions to the (homogeneous) equation

$$x'' - vx - wx' = 0$$

PROBLEMS 8.8

1. Using the true solution, find the function ϕ explicitly in this case:

$$\begin{cases} x'' = -x \\ x(0) = 1 \qquad x\left(\frac{\pi}{2}\right) = 3 \end{cases}$$

2. Determine the true solution and find the function ϕ explicitly in this case:

$$\begin{cases} x'' = -(x')^2 x^{-1} \\ x(1) = 3 \qquad x(2) = 5 \end{cases}$$

Solve the boundary-value problem using ϕ.

3. Solve analytically the three-point boundary-value problem:

$$\begin{cases} x''' = -e^t + 4(t+1)^{-3} \\ x(0) = -1 \qquad x(1) = 3 - e + 2\ln 2 \qquad x(2) = 6 - e^2 + 2\ln 3 \end{cases}$$

4. Show how the shooting method can be used to solve a two-point boundary-value problem of the following type, in which the constants α, β, and c_{ij} are all given:

$$\begin{cases} x'' = u(t) + v(t)x + w(t)x' \\ c_{11}x(a) + c_{12}x'(a) = \alpha \\ c_{21}x(b) + c_{22}x'(b) = \beta \end{cases}$$

Hint: Let x_1 solve the differential equation with initial conditions specified in such a way that $c_{11}x_x(a) + c_{12}x_1(a) = \alpha$. Let x_2 solve the differential equation with initial conditions $x_2(a) = -c_{12}$ and $x_2'(a) = c_{11}$. Consider $x_1 + \lambda x_2$.

5. What is the first variational equation corresponding to this linear differential equation?

$$x'' = a(t) + b(t)x + c(t)x'$$

6. What is the first variational equation for this differential equation?

$$x'' = \cos(tx) + \sin(t^2 x')$$

Can the first variational equation be solved numerically by itself, or must it be solved simultaneously with the equation for x''?

7. Show that if x_1 and x_2 solve $x'' = u(t) + v(t)x + w(t)x'$ with the initial conditions $x_1(a) = x_2(a) = \alpha$, $x_1'(a) = 0$, and $x_2'(a) = 1$, then $x_2 - x_1$ is the solution of the first variational problem in Equation (12).

8. Show that if we solve a linear two-point boundary-value problem by using Newton's method on ϕ, and if we compute ϕ' by using the first variational equation, the result is the same as the one given by Equations (6) and (7).

9. Prove that the function ϕ is linear in the case of a linear differential equation.

10. Solve the problem

$$\begin{cases} x'' = -9x \\ x(0) = 1 \qquad x(\frac{\pi}{6}) = 5 \end{cases}$$

by first finding the solution x_z to the problem

$$\begin{cases} x'' = -9x \\ x(0) = 1 \qquad x'(0) = z \end{cases}$$

Then adjust z so that $x_z(\frac{\pi}{6}) = 5$. Describe how the result would be altered if $x(\frac{\pi}{3}) = 5$.

11. Find the function ϕ explicitly in this case:

$$\begin{cases} x'' = -2t(x')^2 \\ x(0) = 1 \qquad x(1) = 1 + \frac{\pi}{4} \end{cases}$$

Using ϕ, solve the boundary-value problem.

12. If x_z is the solution of the initial-value problem $x'' = x$, $x(0) = 0$, $x'(0) = z$, what is $x_z(1)$? *Hint*: Multiply the differential equation by x and integrate.

13. For the two-point boundary-value problem $x'' = x$, with $x(0) = 0$ and $x(1) = 17$, what is the function $\phi(z)$?

14. If x_z is the solution of the initial-value problem $x'' = -x$, $x(0) = 5$, $x'(0) = z$, and if $\phi(z) = x_z(\pi/2) - 3$, what is $\phi'(z)$?

15. In solving the two-point boundary-value problem

$$\begin{cases} x'' - 37t^2x' = 95 \\ x(6) = 1 \qquad x(12) = 2 \end{cases}$$

using the shooting method based on the secant method, we obtain two pairs of numbers $(z_i, x_{z_i}(b))$ for $i = 1, 2$—say, $(4, 5)$ and $(2, 9)$. What initial-value problem is solved for the next iteration in this procedure?

16. Refer to the discussion of the linear two-point boundary-value problem and show that $c_1 = 0$ and $c_2 = \left[\beta - y_0(b)\right]/y_2(b)$.

17. (Continuation) Prove that if $y_2(b) = 0$, then the boundary-value problem will not be solvable for all (α, β).

18. Prove that the following procedure will solve the two-point boundary-value problem (4), provided that a solution exists. Solve two initial-value problems

$$\begin{cases} x_1'' = u + vx_1 + wx_1' \qquad x_1(a) = \alpha \qquad x_1'(a) = 0 \\ x_2'' = vx_2 + wx_2' \qquad x_2(a) = 0 \qquad x_2' = 1 \end{cases}$$

Then either x_1 itself is the solution or the solution is $x_1 + cx_2$ where c is the constant $[\beta - x_1(b)]/x_2(b)$.

19. Prove that the problem $x'' = x$, $x(a) = \alpha$, $x(b) = \beta$ always has a unique solution. (Assume that $a \neq b$.)

COMPUTER PROBLEMS 8.8

1. Write out the algorithm for the shooting method. Solve the two-point boundary-value problem:

$$\begin{cases} x'' = e^t + x\cos t - (t + 1)x' \\ x(0) = 1 \qquad x(1) = 3 \end{cases}$$

by the shooting method. Notice that the problem is linear. Use the Runge-Kutta method of order 4 with step size $h = 0.01$.

2. Write an algorithm for solving the linear two-point problem as outlined in the text. Refer to Equations (4)–(8). Write a general-purpose code based on this algorithm. The functions u, v, and w should be furnished by the user of the code in the form of subroutines.

3. (Continuation) Test the code in Computer Problem 2 on this example:

$$\begin{cases} u(t) = e^{t-3},\ v(t) = t^2 + 2,\ w(t) = \sin t \\ a = 2.6,\ b = 5.1,\ \alpha = 7.0,\ \beta = -3 \end{cases}$$

8.9 Boundary-Value Problems: Finite-Difference Methods

Another approach to the two-point boundary-value problem consists of an initial discretization of the t-interval followed by the use of approximate formulas for the derivatives. These two formulas are especially useful:

$$x'(t) = (2h)^{-1}\left[x(t+h) - x(t-h)\right] - \frac{1}{6}h^2 x'''(\xi) \tag{1}$$

$$x''(t) = h^{-2}\left[x(t+h) - 2x(t) + x(t-h)\right] - \frac{1}{12}h^2 x^{(4)}(\tau) \tag{2}$$

Second-Order Differential Equations

Suppose that the problem to be solved is

$$\begin{cases} x'' = f(t, x, x') \\ x(a) = \alpha \qquad x(b) = \beta \end{cases} \tag{3}$$

Let the interval $[a, b]$ be partitioned by points $a = t_0, t_1, t_2, \ldots, t_{n+1} = b$. These need not be equally spaced, but in practice they usually are. Indeed, if the points are *not* uniformly distributed, then more complicated versions of (1) and (2) will have to be introduced. So, for simplicity, we assume

$$t_i = a + ih \qquad 0 \leqq i \leqq n+1 \qquad h = (b-a)/(n+1) \tag{4}$$

The approximate value of $x(t_i)$ is denoted by y_i. The discrete version of (3) is then

$$\begin{cases} y_0 = \alpha \\ h^{-2}\left(y_{i-1} - 2y_i + y_{i+1}\right) = f\left(t_i, y_i, (2h)^{-1}(y_{i+1} - y_{i-1})\right) \qquad (1 \leqq i \leqq n) \\ y_{n+1} = \beta \end{cases} \tag{5}$$

The Linear Case

In Equation (5), the unknowns are $y_1, y_2, \ldots, y_n$, and there are n equations to be solved. If f involves y_i in a nonlinear way, these equations will be nonlinear and, in general, difficult to solve. However, let us assume that f is linear in x and x'. Then it has the form

$$f(t, x, x') = u(t) + v(t)x + w(t)x' \tag{6}$$

System (5) is now a linear system of equations, which can be written in the form

$$\begin{cases} & y_0 = \alpha \\ (-1-\frac{1}{2}hw_i)y_{i-1} + (2+h^2v_i)y_i + (-1+\frac{1}{2}hw_i)y_{i+1} & = -h^2u_i \quad (1 \leqq i \leqq n) \\ & y_{n+1} = \beta \end{cases} \tag{7}$$

We have written $u_i = u(t_i)$, $v_i = v(t_i)$, and so on.

Next, we introduce the abbreviations

$$\begin{aligned} a_i &= -1 - \frac{1}{2}hw_{i+1} \\ d_i &= 2 + h^2 v_i \\ c_i &= -1 + \frac{1}{2}hw_i \\ b_i &= -h^2 u_i \end{aligned}$$

so that the system of equations looks like this:

$$\begin{bmatrix} d_1 & c_1 & & & & \\ a_1 & d_2 & c_2 & & & \\ & a_2 & d_3 & c_3 & & \\ & & \ddots & \ddots & \ddots & \\ & & & a_{n-2} & d_{n-1} & c_{n-1} \\ & & & & a_{n-1} & d_n \end{bmatrix} \begin{bmatrix} y_1 \\ y_2 \\ y_3 \\ \vdots \\ y_{n-1} \\ y_n \end{bmatrix} = \begin{bmatrix} b_1 - a_0\alpha \\ b_2 \\ b_3 \\ \vdots \\ b_{n-1} \\ b_n - c_n\beta \end{bmatrix} \tag{8}$$

Since the elements not shown are 0, this system is tridiagonal and can be solved by a special Gaussian algorithm. Note that if h is small and $v_i > 0$, the matrix of the system is diagonally dominant since

$$|2 + h^2 v_i| > \left|1 + \frac{1}{2}hw_i\right| + \left|1 - \frac{1}{2}hw_i\right| = 2$$

Here we must assume that $|\frac{1}{2}hw_i| \leqq 1$ because then the terms $1 \pm \frac{1}{2}hw_i$ are both nonnegative. Henceforth, we assume that $v_i > 0$ and h is small enough so that $|\frac{1}{2}hw_i| < 1$.

The following inequality will be needed later:

$$
\begin{aligned}
|d_i| - |c_i| - |a_{i-1}| &= 2 + h^2 v_i - \left(1 - \frac{1}{2}hw_i\right) - \left(1 + \frac{1}{2}hw_i\right) \\
&= h^2 v_i
\end{aligned}
\tag{9}
$$

Convergence

We shall undertake to show that when h converges to 0, the discrete solution converges to the solution of the boundary-value problem. To know that the boundary-value problem

$$
\begin{cases}
x'' = u + vx + wx' \\
x(a) = \alpha \qquad x(b) = \beta
\end{cases}
\tag{10}
$$

has a unique solution, we invoke a theorem from Keller [1968, p. 9] that yields this conclusion under the hypotheses that u, v, and w belong to $C[a, b]$ and $v > 0$. These assumptions are therefore adopted. The theorem of Keller follows.

THEOREM 1 *The boundary-value problem*

$$
\begin{cases}
x'' = f(t, x, x') \\
c_{11}x(a) + c_{12}x'(a) = c_{13} \\
c_{21}x(b) + c_{22}x'(b) = c_{23}
\end{cases}
$$

has a unique solution on the interval $[a, b]$ provided that:

(i) *f and its first partial derivatives f_t, f_x, $f_{x'}$ are continuous on the domain $D = [a, b] \times \mathbb{R} \times \mathbb{R}$*

(ii) *$f_x > 0$, $|f_x| \leqq M$, $|f_{x'}| \leqq M$ on D*

(iii) $|c_{11}| + |c_{12}| > 0$, $|c_{21}| + |c_{22}| > 0$, $|c_{11}| + |c_{21}| > 0$, $c_{11}c_{12} \leqq 0 \leqq c_{21}c_{22}$

We denote by $x(t)$ the true solution of the problem, and by y_i the solution of the discrete problem. Notice that y_i depends on h. We shall estimate $|x_i - y_i|$ and show that it converges to 0 when $h \to 0$. Here, of course, x_i denotes $x(t_i)$.

With the aid of Formulas (1) and (2), we see that $x(t)$ satisfies the following system of equations, with $1 \leqq i \leqq n$:

$$
\begin{aligned}
&h^{-2}(x_{i-1} - 2x_i + x_{i+1}) - \frac{1}{12}h^2 x^{(4)}(\tau_i) \\
&\qquad = u_i + v_i x_i + w_i \left[(2h)^{-1}(x_{i+1} - x_{i-1}) - \frac{1}{6}h^2 x'''(\xi_i)\right]
\end{aligned}
\tag{11}
$$

On the other hand, the discrete solution satisfies the equations

$$
h^{-2}(y_{i-1} - 2y_i + y_{i+1}) = u_i + v_i y_i + w_i(2h)^{-1}(y_{i+1} - y_{i-1})
\tag{12}
$$

If we subtract Equation (12) from Equation (11) and write $e_i \equiv x_i - y_i$, the result is

$$h^{-2}(e_{i-1} - 2e_i + e_{i+1}) = v_i e_i + w_i(2h)^{-1}(e_{i+1} - e_{i-1}) + h^2 g_i \tag{13}$$

where

$$g_i = \frac{1}{12}x^{(4)}(\tau_i) - \frac{1}{6}x'''(\xi_i) \tag{14}$$

After collecting terms and multiplying by $-h^2$, we have an equation similar to Equation (7); namely,

$$\left(-1 - \frac{1}{2}hw_i\right) e_{i-1} + (2 + h^2 v_i)e_i + \left(-1 + \frac{1}{2}hw_i\right) e_{i+1} = -h^4 g_i \tag{15}$$

Using the coefficients introduced earlier, we write this in the form

$$a_{i-1}e_{i-1} + d_i e_i + c_i e_{i+1} = -h^4 g_i \tag{16}$$

Let $\lambda = \|e\|_\infty$ and select an index i for which

$$|e_i| = \|e\|_\infty = \lambda$$

Here e is the vector $e = (e_1, e_2, \dots, e_n)$. Then from Equation (16) we get

$$|d_i|\,|e_i| \leqq h^4|g_i| + |c_i|\,|e_{i+1}| + |a_{i-1}|\,|e_{i-1}|$$

and using Equation (9),

$$\begin{aligned} |d_i|\lambda &\leqq h^4\|g\|_\infty + |c_i|\lambda + |a_{i-1}|\lambda \\ \lambda\left(|d_i| - |c_i| - |a_{i-1}|\right) &\leqq h^4\|g\|_\infty \\ h^2 v_i \lambda &\leqq h^4\|g\|_\infty \\ \|e\|_\infty &\leqq h^2\left[\|g\|_\infty / \inf v(t)\right] \end{aligned}$$

By Equation (14), $\|g\|_\infty \leqq \|x^{(4)}\|_\infty/12 + \|x'''\|_\infty/6$. The expression in the brackets is a bound independent of h. Thus, we see that $\|e\|_\infty$ is $\mathcal{O}(h^2)$ as $h \to 0$.

PROBLEMS 8.9

1. Solve the two-point boundary-value problem

$$\begin{cases} x'' + 2x' + 10t = 0 \\ x(0) = 1 \qquad x(1) = 2 \end{cases}$$

for $x(\frac{1}{2})$ using the finite-difference method with $h = \frac{1}{2}$.

2. Show what modifications must be made in the linear equation (7) to cope with boundary conditions of the form

$$c_{11}x(a) + c_{12}x'(a) + c_{13} = c_{21}x(b) + c_{22}x'(b) + c_{23} = 0$$

3. Use Keller's theorem to prove that the boundary-value problem (10) has a unique solution, provided that the coefficient functions are continuous and $v > 0$.

COMPUTER PROBLEMS 8.9

1. Write a general-purpose computer program to solve linear two-point boundary-value problems by the finite-difference method as described in the text. Allow the user to furnish a, α, b, β, n and the functions u, v, w as in Equations (3), (4), and (6).
2. (Continuation) Test the program written in Computer Problem 1 on the following examples:

   (i) $\begin{cases} x'' = -x \\ x(0) = 3 \qquad x(\frac{\pi}{2}) = 7 \end{cases}$

   (ii) $\begin{cases} x'' = 2e^t - x \\ x(0) = 2 \qquad x(1) = e + \cos 1 \end{cases}$

   Also, compute the errors in the numerical solutions of these two test cases. The solutions are: **(i)** $x(t) = 7\sin t + 3\cos t$ **(ii)** $x(t) = e^t + \cos t$

---

*8.10 Boundary-Value Problems: Collocation

The method of collocation provides a strategy by which we can attack many problems in applied mathematics. First, we give a general description. Suppose that we have a linear operator L (such as an integral operator or differential operator) and we wish to solve the equation

$$Lu = w \tag{1}$$

In this equation, w is given and u is sought. Several approximate methods for solving Equation (1) begin by selecting some set of basic vectors $\{v_1, v_2, \ldots, v_n\}$ and then trying to solve Equation (1) with a vector u of the form

$$u = c_1v_1 + c_2v_2 + \cdots + c_nv_n \tag{2}$$

Since L is a linear operator, we have

$$Lu = \sum_{j=1}^{n} c_j L v_j$$

and thus Equation (1) leads to

$$\sum_{j=1}^{n} c_j L v_j = w \tag{3}$$

In general, we shall not be able to solve for the coefficients $c_1, c_2, \dots, c_n$ in System (3). But perhaps we can make Equation (3) *almost true*. In the method of **collocation**, the vectors u, w, and v_j are all functions on a common domain. We then require that the functions w and $\sum_{j=1}^{n} c_j L v_j$ agree in value at n prescribed points:

$$\sum_{j=1}^{n} c_j (L v_j)(t_i) = w(t_i) \qquad (1 \leqq i \leqq n) \tag{4}$$

This is a system of n linear equations from which we can compute values for the n unknown coefficients c_j. Of course, the functions v_j and the points t_i should be chosen so that the matrix with entries $(L v_j)(t_i)$ is nonsingular.

Sturm-Liouville Boundary-Value Problems

Now let us consider how this method would work on a Sturm-Liouville two-point boundary-value problem:

$$\begin{cases} u'' + pu' + qu = w \\ u(0) = 0 \qquad u(1) = 0 \end{cases} \tag{5}$$

Here the functions p, q, and w are all given and are assumed to be continuous on the interval $[0, 1]$. The function u is unknown; it, too, is defined on $[0, 1]$, but we expect it to be twice continuously differentiable. This problem becomes an example of the scheme outlined previously if we define the operator L by setting

$$Lu \equiv u'' + pu' + qu \tag{6}$$

We look for a solution in the vector space

$$V = \left\{ u \in C^2[0, 1] : u(0) = u(1) = 0 \right\} \tag{7}$$

Accordingly, if we select our set of basic functions $\{v_1, v_2, \dots v_n\}$ from V, then the homogeneous boundary conditions will be satisfied automatically. One set of functions that suggests itself is the (doubly indexed) set

$$v_{jk}(t) = t^j (1 - t)^k \qquad (j \geqq 1, \ k \geqq 1) \tag{8}$$

We verify readily that

$$v'_{jk} = j v_{j-1,k} - k v_{j,k-1} \tag{9}$$

$$v''_{jk} = j(j-1) v_{j-2,k} - 2jk v_{j-1,k-1} + k(k-1) v_{j,k-2} \tag{10}$$

From Equations (9) and (10), it is a simple matter to write down the function Lv_{jk}. If n functions are chosen from the set described in Equation (8), and if n points t_i are chosen in [0, 1], we can attempt to solve the collocation Equations (4) and obtain an approximate solution of the problem (5).

Cubic B-Splines

Perhaps a better choice of basic functions for such problems would be a set of B-splines. In describing how the B-splines can be used to advantage, let us take as our *model problem* a slightly more general one; namely,

$$\begin{cases} u'' + pu' + qu = w \\ u(a) = \alpha \qquad u(b) = \beta \end{cases} \tag{11}$$

In order that our basic functions possess two continuous derivatives, we consider only the B-splines B_i^k with $k \geqq 3$. For simplicity, we take $k = 3$. Also, we shall take the knots to be equally spaced: $t_{i+1} - t_i = h$. Finally, we shall use the knots as the collocation points. Let n be the number of basic functions to be used (and the number of coefficients to be determined). There should be a total of n conditions for a determination of the n coefficients. Since there will be two end conditions, namely,

$$\sum_{j=1}^{n} c_j v_j(a) = \alpha \quad \text{and} \quad \sum_{j=1}^{n} c_j v_j(b) = \beta \tag{12}$$

we see that there should be $n - 2$ collocation conditions:

$$\sum_{j=1}^{n} c_j (Lv_j)(t_i) = w(t_i) \qquad (1 \leqq i \leqq n - 2) \tag{13}$$

These considerations lead us to define

$$h = (b - a)/(n - 3) \tag{14}$$

$$t_i = a + (i - 1)h \qquad (i = 0, \pm 1, \pm 2, \ldots) \tag{15}$$

The knots t_i that lie in $[a, b]$ are $a = t_1, t_2, \ldots, t_{n-2} = b$ (as is easily verified). These knots are the collocation points. Some knots outside the interval $[a, b]$ will be needed for defining the B-splines B_j^3. The arrangement of knots is shown in Figure 8.4.

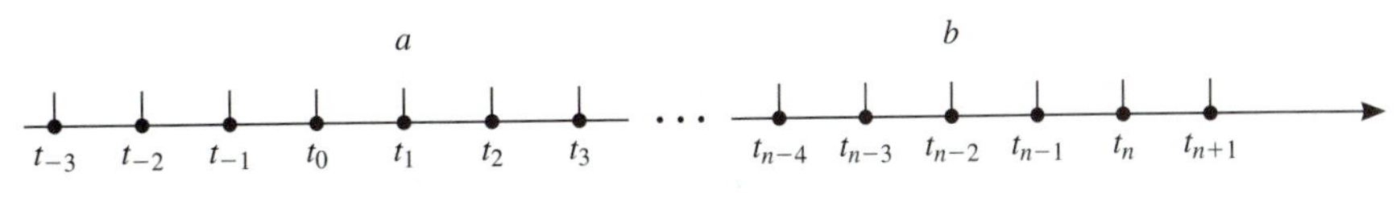

FIGURE 8.4 The configuration of knots

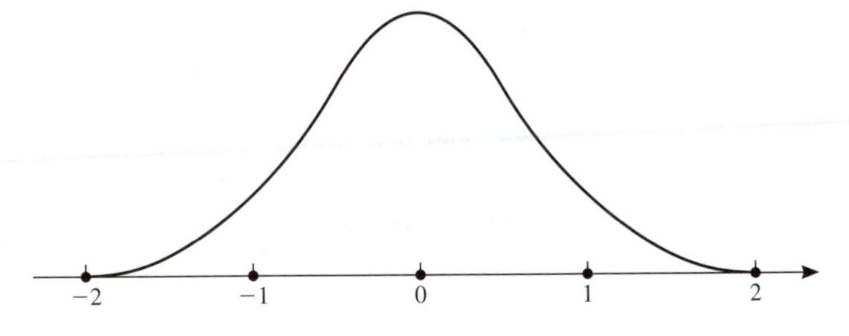

FIGURE 8.5 Cubic B-spline

The cubic B-splines of interest to us are the ones that are not identically 0 on $[a, b]$. These are $B^3_{-2}, B^3_{-1}, B^3_0, B^3_1, \ldots, B^3_{n-3}$. Accordingly, we put $v_j = B^3_{j-3}$ for $1 \leqq j \leqq n$.

Since the knots are equally spaced, the functions v_j can be obtained from a single B-spline denoted by B and defined by

$$B(t) = \begin{cases} (t+2)^3/6 & \text{on } [-2, -1] \\ \left[1 + 3(t+1) + 3(t+1)^2 - 3(t+1)^3\right]/6 & \text{on } [-1, 0] \\ B(-t) & \text{on } [0, 2] \\ 0 & \text{elsewhere} \end{cases} \tag{16}$$

Then, as we easily verify,

$$v_j(t) = B\left(\frac{t-a}{h} - j + 2\right) \tag{17}$$

The graph of the function B is shown in Figure 8.5, and the reader should verify that B is indeed a cubic spline.

The first and second derivatives of v_j are needed in computing $(Lv_j)(t_i)$. These derivatives are readily obtained from Equations (16) and (17). In a computer program to carry out the collocation method using the cubic B-splines, we would have to write subprograms to compute $B(t)$, $B'(t)$, and $B''(t)$. Then the matrix elements $(Lv_j)(t_i)$ could be computed efficiently. This will be a **banded matrix** because each function v_j has a small support. A general-purpose routine intended for *production* usage should exploit the sparse nature of the coefficient matrix.

COMPUTER PROBLEM 8.10

1. **(a)** Program a general-purpose code for solving a two-point boundary-value problem of the form

$$\begin{cases} u''(t) + p(t)u' + q(t)u(t) = w(t) \\ u(a) = \alpha \qquad u(b) = \beta \end{cases}$$

by the method of collocation. The user will specify: **(i)** functions p, q, w; **(ii)** real numbers a, α, b, β, with $a < b$; **(iii)** n, the number of *basic* functions to be used; and **(iv)** the *basic* functions v_i and their derivatives v_i', v_i''. The computer program will then generate $n-2$ collocation points t_i for $1 \leqq i \leqq n-2$ equally spaced in the interval $[a, b]$, including the endpoints. The program next sets up a linear system of n equations with n unknowns $c_1, c_2, \dots, c_n$ as follows:

$$\begin{cases} \displaystyle\sum_{j=1}^{n} c_j (Lv_j)(t_i) = w(t_i) & (1 \leqq i \leqq n-2) \\ \displaystyle\sum_{j=1}^{n} c_j v_j(a) = \alpha & \displaystyle\sum_{j=1}^{n} c_j v_j(b) = \beta \end{cases}$$

The program should then call a *linear equation solver* to compute the coefficients c_j. Finally, the approximate solution $u = \sum_{j=1}^{n} c_j v_j$ is tested by evaluating u and the *residual* $Lu - w$ at $2n-5$ points equally spaced in $[a, b]$. These points will include the collocation points (where the residual should be 0) as well as the midpoints between collocation points.

(b) Test the program written in part (a) by solving

$$\begin{cases} u'' + (\sin t)u' + (t^2+2)u = e^{t-3} \\ u(2.6) = 7 \qquad u(5.1) = -3 \end{cases}$$

Use cubic B-splines with equally spaced knots. The collocation points should be the knots on $[a, b]$.

*8.11 Linear Differential Equations

Here we consider systems of n linear differential equations with constant coefficients. The systems are assumed to be **autonomous**, which means that the independent variable t is not present explicitly. Such a system has the appearance

$$\begin{cases} x_1' = a_{11}x_1 + a_{12}x_2 + \cdots + a_{1n}x_n \\ x_2' = a_{21}x_1 + a_{22}x_2 + \cdots + a_{2n}x_n \\ \quad\vdots \\ x_n' = a_{n1}x_1 + a_{n2}x_2 + \cdots + a_{nn}x_n \end{cases} \tag{1}$$

In vector-matrix notation, this is simply

$$X' = AX \tag{2}$$

where $X = (x_1, x_2, \dots, x_n)^T$ and $X' = (x_1', x_2', \dots, x_n')^T$.

Eigenvalues and Eigenvectors

Let us attempt to solve this with a vector of the form $X(t) = e^{\lambda t}V$, where V is a constant vector. On substituting this *trial* solution in Equation (2), we obtain

$$\lambda e^{\lambda t} V = e^{\lambda t} AV \tag{3}$$

Thus, if

$$AV = \lambda V$$

then the vector function $e^{\lambda t}V$ is indeed a solution of Equation (2). We have proved the following theorem.

THEOREM 1 *If λ is an eigenvalue of the matrix A and if V is a corresponding eigenvector, then $X(t) = e^{\lambda t}V$ is a solution of the equation $X' = AX$.*

Theorem 1 indicates that solving the differential equation $X' = AX$ will involve some knowledge of the eigenvalues and eigenvectors of A. Furthermore, the theory of similar matrices will show us how to simplify a system of linear differential equations by changing variables. These matters will be taken up presently.

The most pleasant state of affairs for the equation $X' = AX$ is described in the next theorem.

THEOREM 2 *If the $n \times n$ matrix A has a linearly independent set of eigenvectors $V_1, V_2, \ldots, V_n$ with $AV_i = \lambda_i V_i$, then the solution space of the equation $X' = AX$ has a basis $X_i = e^{\lambda_i t}V_i$ with $1 \leqq i \leqq n$.*

Proof The set $\{X_1, X_2, \ldots, X_n\}$ is linearly independent because $\sum_{i=1}^{n} c_i X_i = 0$ leads to $\sum_{i=1}^{n} c_i e^{\lambda_i t} V_i = 0$, $c_i e^{\lambda_i t} = 0$, and $c_i = 0$, for $1 \leqq i \leqq n$.

To prove that the set in question spans the solution space of $X' = AX$, let X be any solution. As an element of $\mathbb{R}^n$ or $\mathbb{C}^n$, the initial-value vector $X(0)$ is a linear combination of $V_1, V_2, \ldots, V_n$; say

$$X(0) = \sum_{i=1}^{n} c_i V_i$$

Define $Y = \sum_{i=1}^{n} c_i X_i$. Then

$$Y' = \sum_{i=1}^{n} c_i X_i' = \sum_{i=1}^{n} c_i \lambda_i e^{\lambda_i t} V_i = \sum_{i=1}^{n} c_i e^{\lambda_i t} AV_i = A\left(\sum_{i=1}^{n} c_i X_i\right) = AY$$

Thus Y and X are solutions of the differential equation, and they have the same initial value: $Y(0) = X(0)$. (Why?) By the uniqueness theorem for the initial-value problem, we conclude that $Y = X$, or in other words $X = \sum_{i=1}^{n} c_i X_i$. ■

If A has the property mentioned in Theorem 2, then there is a nonsingular matrix P whose columns are the vectors $V_1, V_2, \ldots, V_n$. The equation $AV_i = \lambda_i V_i$

translates into matrix notation as

$$AP = P\Lambda \tag{4}$$

where Λ is the diagonal matrix having $\lambda_1, \lambda_2, \ldots, \lambda_n$ on its diagonal. Consider the change of (dependent) variables described by $X = PY$. Since P is nonsingular, we can recover Y from X. Now Y has this property:

$$Y' = P^{-1}X' = P^{-1}AX = P^{-1}APY = \Lambda Y \tag{5}$$

Thus, the differential equation for Y is much simpler than the one for X, since Λ is a diagonal matrix. The individual equations in the system $Y' = \Lambda Y$ are **uncoupled** and can be solved separately.

Example 1 Solve the initial-value problem $X' = AX$ when

$$A = \begin{bmatrix} 1 & 0 & 1 \\ 0 & 0 & 0 \\ 0 & 0 & -1 \end{bmatrix} \qquad X(0) = \begin{bmatrix} 5 \\ 7 \\ 6 \end{bmatrix}$$

Solution The matrix $A - \lambda I$ is

$$\begin{bmatrix} 1-\lambda & 0 & 1 \\ 0 & -\lambda & 0 \\ 0 & 0 & -1-\lambda \end{bmatrix}$$

and its determinant is the **characteristic polynomial of A**:

$$(1-\lambda)(-\lambda)(-1-\lambda)$$

The roots of this polynomial are the *eigenvalues of* A and are $\lambda_1 = 1$, $\lambda_2 = 0$, $\lambda_3 = -1$. For each of these, we find an eigenvector by solving $AV_i = \lambda_i V_i$. Placing these as columns in a matrix P, we obtain

$$P = \begin{bmatrix} 1 & 0 & 1 \\ 0 & 1 & 0 \\ 0 & 0 & -2 \end{bmatrix}$$

Next, we find that

$$P^{-1} = \begin{bmatrix} 1 & 0 & \frac{1}{2} \\ 0 & 1 & 0 \\ 0 & 0 & -\frac{1}{2} \end{bmatrix}$$

If $Y = (y_1, y_2, y_3)^T$, then the initial-value problem for Y is $Y' = \Lambda Y$, where

$$\Lambda = P^{-1}AP = \begin{bmatrix} 1 & 0 & 0 \\ 0 & 0 & 0 \\ 0 & 0 & -1 \end{bmatrix}$$

Hence, we have

$$\begin{cases} y_1' = y_1 \\ y_2' = 0 \\ y_3' = -y_3 \end{cases} \qquad Y(0) = P^{-1}X(0) = \begin{bmatrix} 8 \\ 7 \\ -3 \end{bmatrix}$$

and its solution is

$$y_1 = 8e^t \qquad y_2 = 7 \qquad y_3 = -3e^{-t}$$

Since $X = PY$, the corresponding solution for $X = (x_1, x_2, x_3)^T$ is

$$x_1 = 8e^t - 3e^{-t} \qquad x_2 = 7 \qquad x_3 = 6e^{-t}$$

■

Matrix Exponential

There is an elegant formal method of solving a system $X' = AX$. This method makes it unnecessary to refer to the eigenvalues of A until we wish to compute solutions numerically. We begin by defining the matrix exponential.

DEFINITION 1 *If A is a square matrix, we put*

$$e^A = I + A + \frac{1}{2!}A^2 + \frac{1}{3!}A^3 + \cdots \tag{6}$$

This definition is derived from the standard series

$$e^z = 1 + z + \frac{1}{2!}z^2 + \frac{1}{3!}z^3 + \cdots \tag{7}$$

by substituting a matrix for the complex variable z. To see that the series for e^A converges, we take any norm on $\mathbb{C}^n$ and use the corresponding subordinate matrix norm for $n \times n$ matrices. The *tail* of our series can be estimated as follows:

$$\left\| \sum_{k=m}^{\infty} \frac{1}{k!}A^k \right\| \leqq \sum_{k=m}^{\infty} \frac{1}{k!}\|A^k\| \leqq \sum_{k=m}^{\infty} \frac{1}{k!}\|A\|^k \tag{8}$$

This last expression is the tail of the ordinary exponential series when $z = \|A\|$. Thus, the tail of the series for e^A converges to 0 as $m \to \infty$. (This argument assumes as known the completeness of the space of $n \times n$ matrices with the given norm.)

If t is a real variable, then $tA = At$, and our definition yields

$$e^{At} = \sum_{k=0}^{\infty} \frac{t^k}{k!}A^k \tag{9}$$

Differentiation with respect to t in the series and subsequent simplification give us

$$\frac{d}{dt}e^{At} = Ae^{At} \tag{10}$$

THEOREM 3 *The solution of the initial-value problem*

$$\begin{cases} X' = AX \\ X(0) \text{ prescribed} \end{cases}$$

is $X(t) = e^{At}X(0)$.

Proof From the formula $X = e^{At}W$, with $W = X(0)$, we have at once

$$\begin{cases} X' = Ae^{At}W = AX \\ X(0) = e^{A0}W = W \end{cases}$$ ■

Diagonal and Diagonalizable Matrices

To use the preceding result in practice, it will be necessary to compute the matrix exponential in an efficient manner. Let us start with the case in which A is a *diagonal* matrix. If $A = \operatorname{diag}(\lambda_1, \lambda_2, \dots, \lambda_n)$, then $A^k = \operatorname{diag}(\lambda_1^k, \lambda_2^k, \dots, \lambda_n^k)$, as is easily verified. Hence, for such an A,

$$\begin{aligned} e^{At} &= \sum_{k=0}^{\infty} \frac{t^k}{k!} \operatorname{diag}(\lambda_1^k, \lambda_2^k, \dots, \lambda_n^k) \\ &= \operatorname{diag}(e^{\lambda_1 t}, e^{\lambda_2 t}, \dots, e^{\lambda_n t}) \end{aligned} \tag{11}$$

In this special case, the solution of the differential equation $X' = AX$ has the components

$$x_i(t) = e^{\lambda_i t} x_i(0) \qquad (1 \leqq i \leqq n)$$

The analysis just presented carries over at once to the case in which A is not diagonal but is **diagonalizable**. This term means that A is similar to a diagonal matrix, or in other words that $P^{-1}AP = \Lambda$ for some diagonal matrix Λ and some nonsingular matrix P. If this is true, the change of variable $X = PY$ changes the differential equation $X' = AX$ into $Y' = \Lambda Y$, as shown in Equation (5). The initial condition $X(0)$ becomes $Y(0) = P^{-1}X(0)$, and the solution is

$$X = PY = P\left(e^{\Lambda t}P^{-1}X(0)\right) = P \operatorname{diag}(e^{\lambda_1 t}, e^{\lambda_2 t}, \dots, e^{\lambda_n t})P^{-1}X(0) \tag{12}$$

Jordan Blocks

We have postponed until now the discussion of the case when A is not diagonalizable. This means that $\mathbb{C}^n$ does not have a basis consisting of eigenvectors of A. Two simple examples of such behavior are given by

$$J(\lambda, 2) = \begin{bmatrix} \lambda & 1 \\ 0 & \lambda \end{bmatrix} \qquad J(\lambda, 3) = \begin{bmatrix} \lambda & 1 & 0 \\ 0 & \lambda & 1 \\ 0 & 0 & \lambda \end{bmatrix}$$

Let us consider $J(\lambda, 3)$ in detail. Since this matrix is upper triangular, its diagonal elements are its eigenvalues. Thus, $J(\lambda, 3)$ has the single eigenvalue λ, and if we write out the equation $J(\lambda, 3)X = \lambda X$, we have

$$\begin{cases} \lambda x_1 + x_2 = \lambda x_1 \\ \lambda x_2 + x_3 = \lambda x_2 \\ \qquad\quad \lambda x_3 = \lambda x_3 \end{cases}$$

This obviously implies that $x_2 = x_3 = 0$, and therefore, the only solutions have the form $X = (\beta, 0, 0)^T$. The solutions form a one-dimensional space. In other words, the eigenvectors of $J(\lambda, 3)$ span only a one-dimensional subspace in $\mathbb{R}^3$ or $\mathbb{C}^3$. Matrices of the form $J(\lambda, k)$ exist for each $k \geqq 2$, and they all have the same property that we just observed for $J(\lambda, 3)$. These matrices are called **Jordan blocks**. The principal theorem in the subject is as follows.

THEOREM 4 *Every square matrix is similar to a block diagonal matrix having Jordan blocks on its diagonal.*

The special form referred to in the theorem is called the **Jordan canonical form** of the given matrix. Here are some examples of 3×3 matrices in Jordan canonical form:

$$\begin{bmatrix} 7 & 0 & 0 \\ 0 & 5 & 0 \\ 0 & 0 & 3 \end{bmatrix} \quad \begin{bmatrix} 5 & 1 & 0 \\ 0 & 5 & 0 \\ 0 & 0 & 3 \end{bmatrix} \quad \begin{bmatrix} 5 & 0 & 0 \\ 0 & 5 & 1 \\ 0 & 0 & 5 \end{bmatrix} \quad \begin{bmatrix} 0 & 1 & 0 \\ 0 & 0 & 0 \\ 0 & 0 & 0 \end{bmatrix} \quad \begin{bmatrix} 5 & 1 & 0 \\ 0 & 5 & 1 \\ 0 & 0 & 5 \end{bmatrix}$$

The first of these contains three Jordan blocks, whereas the second, third, and fourth contain two. The fifth matrix is itself a Jordan block.

A Jordan block can be written in the form

$$J(\lambda, k) = \lambda I_k + H_k \tag{13}$$

where I_k denotes the $k \times k$ identity matrix, and H_k denotes a $k \times k$ matrix of the form

$$H_k = \begin{bmatrix} 0 & 1 & 0 & \cdots & 0 & 0 & 0 \\ 0 & 0 & 1 & \cdots & 0 & 0 & 0 \\ 0 & 0 & 0 & \ddots & 0 & 0 & 0 \\ \vdots & \vdots & \vdots & \ddots & \ddots & \vdots & \vdots \\ 0 & 0 & 0 & \cdots & 0 & 1 & 0 \\ 0 & 0 & 0 & \cdots & 0 & 0 & 1 \\ 0 & 0 & 0 & \cdots & 0 & 0 & 0 \end{bmatrix} \tag{14}$$

The effect of multiplying a vector by H_k is easily seen. We have

$$\begin{bmatrix} 0 & 1 & 0 & \cdots & 0 & 0 \\ 0 & 0 & 1 & \cdots & 0 & 0 \\ 0 & 0 & 0 & \ddots & 0 & 0 \\ \vdots & \vdots & \vdots & \ddots & \ddots & \vdots \\ 0 & 0 & 0 & \cdots & 0 & 1 \\ 0 & 0 & 0 & \cdots & 0 & 0 \end{bmatrix} \begin{bmatrix} \xi_1 \\ \xi_2 \\ \xi_3 \\ \vdots \\ \xi_{k-1} \\ \xi_k \end{bmatrix} = \begin{bmatrix} \xi_2 \\ \xi_3 \\ \xi_4 \\ \vdots \\ \xi_k \\ 0 \end{bmatrix} \quad \textbf{(15)}$$

It now is evident that $H_k^k V = 0$ because each application of H_k to V removes one component. Thus, H_k is **nilpotent**; indeed, $H_k^k = 0$. This fact will be useful in computing e^{At} when A is a Jordan block. We have

$$\begin{aligned} e^{(\lambda I_k + H_k)t} &= e^{t\lambda I_k} e^{H_k t} \\ &= \sum_{j=0}^{\infty} \frac{(\lambda t I_k)^j}{j!} \sum_{j=0}^{\infty} \frac{(tH_k)^j}{j!} \\ &= e^{\lambda t}\left[I_k + tH_k + \frac{t^2}{2!}H_k^2 + \cdots + \frac{t^{k-1}}{(k-1)!}H_k^{k-1}\right] \end{aligned} \quad \textbf{(16)}$$

The series terminates as shown because the kth power (and all subsequent powers) of H_k is 0. Notice that in the preceding calculation we have used a valid case of the formula $e^{A+B} = e^A e^B$. (See Problem 13.)

Now let us solve the differential equation $X' = AX$ when A is a Jordan block, say $A = \lambda I_k + H_k$. The solution is obtained by applying Theorem 3 and Equation (16); it is

$$X(t) = e^{At}X(0) = e^{\lambda t}\left[I_k + tH_k + \frac{t^2}{2!}H_k + \cdots + \frac{t^{k-1}}{(k-1)!}H_k^{k-1}\right]X(0) \quad \textbf{(17)}$$

Example 2 Solve the initial-value problem $X' = AX$ when

$$A = \begin{bmatrix} 3 & 1 & 0 & 0 \\ 0 & 3 & 1 & 0 \\ 0 & 0 & 3 & 1 \\ 0 & 0 & 0 & 3 \end{bmatrix} \qquad X(0) = \begin{bmatrix} 7 \\ 5 \\ 3 \\ 9 \end{bmatrix}$$

Solution The matrix A is of the form $3I_4 + H_4$. The solution, given by Equation (17), is

$$X(t) = e^{3t}\left(I + tH_4 + \frac{1}{2}t^2H_4^2 + \frac{1}{6}t^3H_4^3\right)X(0)$$

$$= e^{3t}\begin{bmatrix}7\\5\\3\\9\end{bmatrix} + te^{3t}\begin{bmatrix}5\\3\\9\\0\end{bmatrix} + \frac{1}{2}t^2e^{3t}\begin{bmatrix}3\\9\\0\\0\end{bmatrix} + \frac{1}{6}t^3e^{3t}\begin{bmatrix}9\\0\\0\\0\end{bmatrix}$$

$$= \begin{bmatrix}7 + 5t + 1.5t^2 + 1.5t^3\\ 5 + 3t + 4.5t^2\\ 3 + 9t\\ 9\end{bmatrix} e^{3t}$$

■

Solution in Complete Generality

It is now possible to describe the solution of the system of differential equations $X' = AX$ in complete generality, starting with the Jordan canonical form of A and the similarity transformation that brings it about. Suppose that

$$P^{-1}AP = C$$

in which C is the Jordan canonical form of A. We know that the change of variables $X = PY$ will lead to the differential equation $Y' = CY$. (In this connection see Equation (5).) This differential equation, $Y' = CY$, can be divided into *uncoupled* blocks. To see this, consider an example such as

$$C = \begin{bmatrix}5 & 1 & 0 & 0 & 0\\ 0 & 5 & 1 & 0 & 0\\ 0 & 0 & 5 & 0 & 0\\ 0 & 0 & 0 & 7 & 1\\ 0 & 0 & 0 & 0 & 7\end{bmatrix}$$

The differential equations are

$$\begin{cases} y_1' = 5y_1 + y_2\\ y_2' = 5y_2 + y_3\\ y_3' = 5y_3\\ y_4' = 7y_4 + y_5\\ y_5' = 7y_5 \end{cases}$$

It is clear that the first group of three equations can be solved by itself, and the last pair can be solved by itself. The general case is quite analogous, and we see that each Jordan block in C gives rise to a set of differential equations that is uncoupled from the remaining equations. With each Jordan block, a piece of the solution is obtained as in Equation (17).

Example 3 Using the matrix C above, solve the initial-value problem $Y' = CY$, with initial conditions $Y(0) = (3, 2, 8, 4, 1)^T$.

Solution The two uncoupled systems are

$$\begin{bmatrix} y_1' \\ y_2' \\ y_3' \end{bmatrix} = \begin{bmatrix} 5 & 1 & 0 \\ 0 & 5 & 1 \\ 0 & 0 & 5 \end{bmatrix} \begin{bmatrix} y_1 \\ y_2 \\ y_3 \end{bmatrix} \qquad \text{initial value} \quad \begin{bmatrix} 3 \\ 2 \\ 8 \end{bmatrix}$$

$$\begin{bmatrix} y_4' \\ y_5' \end{bmatrix} = \begin{bmatrix} 7 & 1 \\ 0 & 7 \end{bmatrix} \begin{bmatrix} y_4 \\ y_5 \end{bmatrix} \qquad \text{initial value} \quad \begin{bmatrix} 4 \\ 1 \end{bmatrix}$$

By using the method illustrated in Example 2, we obtain these solutions:

$$\begin{bmatrix} y_1 \\ y_2 \\ y_3 \end{bmatrix} = e^{5t}\left(I + tH_3 + \frac{1}{2}t^2H_3^2\right)\begin{bmatrix} 3 \\ 2 \\ 8 \end{bmatrix} = e^{5t}\begin{bmatrix} 3 \\ 2 \\ 8 \end{bmatrix} + te^{5t}\begin{bmatrix} 2 \\ 8 \\ 0 \end{bmatrix} + \frac{1}{2}t^2e^{5t}\begin{bmatrix} 8 \\ 0 \\ 0 \end{bmatrix}$$

$$= \begin{bmatrix} 3 + 2t + 4t^2 \\ 2 + 8t \\ 8 \end{bmatrix} e^{5t}$$

$$\begin{bmatrix} y_4 \\ y_5 \end{bmatrix} = e^{7t}(I + tH_2)\begin{bmatrix} 4 \\ 1 \end{bmatrix} = e^{7t}\begin{bmatrix} 4 \\ 1 \end{bmatrix} + te^{7t}\begin{bmatrix} 1 \\ 0 \end{bmatrix} = \begin{bmatrix} 4 + t \\ 1 \end{bmatrix} e^{7t}$$ ■

In the theory of the linear differential equation $X' = AX$, the exponential e^{At} is called the **fundamental matrix** of the equation. We have seen that it is the key to solving the initial-value problem associated with the differential equation. If we possess the Jordan canonical form C of A and know the similarity transformation

$$P^{-1}AP = C$$

then we can change variables by $X = PY$, solve the equation $Y' = CY$, and return to X, obtaining ultimately

$$X = PY = Pe^{Ct}Y(0) = Pe^{Ct}P^{-1}X(0) \tag{18}$$

On the other hand, we know another form for the solution; namely, $X = e^{At}X(0)$. Comparing this to Equation (18), we conclude that

$$e^{At} = Pe^{Ct}P^{-1}$$

This can be regarded as one way of computing the fundamental matrix.

Nonhomogeneous Problem

The principles that have been established can now be applied to the **nonhomogeneous problem**

$$X' = AX + W \tag{19}$$

where W can be a vector function of t. Let us consider first the case when A is diagonalizable. Then a similarity transform $P^{-1}AP = \Lambda$ produces a diagonal matrix Λ. The change of variable $X = PY$ transforms Equation (19) to

$$Y' = \Lambda Y + P^{-1}W \tag{20}$$

This is a completely decoupled set of n equations, of which a typical one would be of the form

$$\eta'(t) = \lambda\eta(t) + g(t) \tag{21}$$

The solution of this is

$$\eta(t) = e^{\lambda t}\left[\eta(0) + \int_0^t e^{-\lambda s} g(s)\, ds\right] \tag{22}$$

Example 4 Solve the equation $X' = AX + W$ when

$$A = \begin{bmatrix} 0 & 0 & 0 \\ 0 & 1 & 0 \\ 0 & 0 & 2 \end{bmatrix} \qquad W = \begin{bmatrix} t^2 \\ t \\ \sin t \end{bmatrix} \qquad X(0) = \begin{bmatrix} 5 \\ 7 \\ 9 \end{bmatrix}$$

Solution The system comprises the following individual uncoupled equations:

$$\begin{cases} x_1' = t^2 \\ x_2' = x_2 + t \\ x_3' = 2x_3 + \sin t \end{cases}$$

The solutions, obtained by using Equation (22), are

$$x_1(t) = 5 + \int_0^t s^2\, ds = 5 + \frac{1}{3}t^3$$

$$x_2(t) = e^t\left[7 + \int_0^t e^{-s} s\, ds\right] = 8e^t - t - 1$$

$$x_3(t) = e^{2t}\left[9 + \int_0^t e^{-2s}\sin s\, ds\right] = \frac{46}{5}e^{2t} - \frac{2}{5}\sin t - \frac{1}{5}\cos t \qquad \blacksquare$$

If the matrix A in the equation $X' = AX + W$ is not diagonalizable, we can use the Jordan canonical form and a change of variables to split the problem into uncoupled subsystems. Thus, it suffices to illustrate the procedure with a single Jordan block.

Example 5 Solve the equation $X' = AX + W$ when

$$A = \begin{bmatrix} 5 & 1 & 0 \\ 0 & 5 & 1 \\ 0 & 0 & 5 \end{bmatrix} \qquad W = \begin{bmatrix} t^2 \\ t \\ \sin t \end{bmatrix} \qquad X(0) = \begin{bmatrix} 5 \\ 7 \\ 9 \end{bmatrix}$$

Solution The individual equations are

$$\begin{cases} x_1' = 5x_1 + x_2 + t^2 \\ x_2' = 5x_2 + x_3 + t \\ x_3' = 5x_3 + \sin t \end{cases}$$

The recommended procedure here is to begin at the bottom and solve the equations in reverse order. This leads to

$$\begin{aligned} x_3(t) &= e^{5t}\left\{9 + \int_0^t e^{-5s}\sin s\,ds\right\} \\ &= \left(9 + \frac{1}{26}\right)e^{5t} - \frac{5}{26}\sin t - \frac{1}{26}\cos t \\ x_2(t) &= e^{5t}\left\{7 + \int_0^t e^{-5s}\left[x_3(s) + s\right]ds\right\} \\ &= \left(7 + \frac{1}{25} - \frac{10}{26^2}\right)e^{5t} + \frac{1}{2}\left(9 + \frac{1}{26}\right)te^{5t} \\ &\quad + \frac{2}{26^2}(12\sin t + t\cos t) - \frac{1}{5}t - \frac{1}{25} \\ x_1(t) &= e^{5t}\left\{5 + \int_0^t e^{-5s}\left[x_2(s) + s^2\right]ds\right\} \\ &= \left(5 + \frac{74}{26^3}\right)e^{5t} + \left(7 + \frac{1}{25} - \frac{10}{26^2}\right)te^{5t} + \frac{1}{2}\left(9 + \frac{1}{26}\right)t^2e^{5t} \\ &\quad - \frac{2}{26^3}(55\sin t + 37\cos t) - \frac{1}{5}t^2 - \frac{1}{25}t \end{aligned}$$

■

The computation of a matrix exponential e^A is a task that should be undertaken with full cognizance of the possible pitfalls. We recommend to the reader the article of Moler and Van Loan [1978]. Their investigation indicates that the following four-step procedure should work well in most cases:

(i) Let j be the first positive integer such that $\|A\|/2^j \leqq 1/2$.
(ii) If a relative error tolerance ε is given, select p to be the first positive integer for which $2^{p-3}(p+1) \geqq 1/\varepsilon$.
(iii) Compute $e^{A/2^j}$ from the truncated Taylor series $e^z = \sum_{k=0}^{p} z^k/k!$
(iv) Square the computed $e^{A/2^j}$ from step **(iii)** j times to obtain $e^A = (e^{A/2^j})^{2^j}$.

The computed value of e^A from step **(iv)** is e^{A+E}, where E is a matrix satisfying $\|E\|/\|A\| \leqq \varepsilon$. This conclusion is Corollary 1 in the cited reference.

PROBLEMS 8.11

1. Find the general solution of the system $X' = AX$ when

$$A = \begin{bmatrix} 1 & 0 & 3 \\ -1 & 1 & -1 \\ 3 & 0 & 1 \end{bmatrix}$$

2. (Continuation) Find the solution of the equation in Problem 1 subject to the initial condition $X_0 = (-1, 4, 7)^T$.

3. Find the general solution of the system

$$\begin{cases} x_1' = 3x_1 - 5x_2 \\ x_2' = 2x_1 + x_2 \end{cases}$$

4. Prove that for any $n \times n$ matrix A, e^A is nonsingular. *Caution*: Do not suppose that $e^A e^B = e^{A+B}$ for all $n \times n$ matrices A and B. (See Problem 5.)

5. Show by example that the equation $e^A e^B = e^{A+B}$ is not always true. *Hint:* Consider e^{At}, e^{Bt}, and $e^{(A+B)t}$.

6. Prove that if $A = P^{-1}BP$, then $e^A = P^{-1}e^B P$.

7. Show in detail how the estimate in Equation (8) and a completeness argument are used to establish convergence in Equation (6).

8. Use induction on k to prove that if $V_1, V_2, \ldots, V_k$ are eigenvectors of a matrix corresponding to k distinct eigenvalues, then $\{V_1, V_2, \ldots, V_k\}$ is linearly independent.

9. Prove Equation (10) by differentiation in the series for e^{At}.

10. Use the result in Problem 6 to establish the result in Equation (12) directly—that is, without the change of variable.

11. Let B be an $n \times n$ matrix, and let $V_1, V_2, \ldots, V_k$ be vectors in $\mathbb{R}^n$ such that $V_1 \neq 0$, $BV_1 = 0$, $BV_2 = V_1, \ldots, BV_k = V_{k-1}$. Use induction on k to prove that $\{V_1, V_2, \ldots, V_k\}$ is linearly independent. How large can k be?

12. Find the exact conditions on a matrix of order 2 so that its Jordan canonical form will contain only one Jordan block.

13. Prove that $e^{A+B} = e^A e^B$ if and only if $AB = BA$. (See Problem 5.)

14. Prove that if A and B are diagonalized by the same similarity transformation (that is, PAP^{-1} and PBP^{-1} are diagonal), then $AB = BA$.

15. Prove that the solution of the linear system

$$X' = AX + V(t) \qquad X(0) = W$$

is given by

$$X(t) = e^{At}W + e^{At}\int_0^t e^{-As}V(s)\,ds$$

Explain what is meant by the indefinite integration appearing here.

16. What is the solution of the initial-value problem $X' = AX$ when X is specified at a point t_0 other than 0?

17. Show that the fundamental matrix for the system

$$X' = AX \quad \text{with} \quad A = \begin{bmatrix} -1 & 6 \\ 1 & -2 \end{bmatrix}$$

is

$$\frac{1}{5}\begin{bmatrix} 2e^{-4t} + 3e^t & -6e^{-4t} + 6e^t \\ -e^{-4t} + e^t & 3e^{-4t} + 2e^t \end{bmatrix}$$

18. Prove that the jth column in the fundamental matrix is a solution of the initial-value problem $X' = AX$, $X(0) = U_j$, where U_j is the jth standard unit vector.

19. Make a guess as to the matrix inverse of e^A and then prove that you are right. *Caution:* In general, $e^{A+B} \neq e^A e^B$.

20. Solve $X' = AX$ when

$$A = \begin{bmatrix} 5 & 4 & 3 \\ -1 & 0 & -3 \\ 1 & -2 & 1 \end{bmatrix} \quad \text{and} \quad X(0) = \begin{bmatrix} 1 \\ 5 \\ 3 \end{bmatrix}$$

21. Complete Example 4.

22. Prove that a matrix which is not diagonalizable is the limit of a sequence of diagonalizable matrices.

23. (Continuation) Consider the linear system $X' = AX$, $X(0) = V$ in which A is not diagonalizable. If B is a diagonalizable matrix close to A, what can be said about the solution of $Y' = BY$, $Y(0) = V$?

24. Prove that $\det e^A = e^{\operatorname{tr}(A)}$, where A is any $n \times n$ matrix and $\operatorname{tr}(A)$ is the **trace** of A—that is, the sum of the diagonal elements in A.

25. For an $n \times n$ matrix A, let $|A| = \sum_{i=1}^n \sum_{j=1}^n |a_{ij}|$. Prove that $|e^A| \leqq n - 1 + e^{|A|}$.

26. Although in general $e^A e^B \neq e^{A+B}$, prove that $e^{A+B} = \lim_{k\to\infty}[e^{A/k}e^{B/k}]^k$.

8.12 Stiff Equations

Stiffness, in a system of ordinary differential equations, refers to a wide disparity in the time scales of the components in the vector solution. Some numerical procedures that are quite satisfactory in general will perform poorly on stiff equations. This happens when stability in the numerical solution can be achieved only with very small step size.

Stiff differential equations arise in a number of applications. For example, in the control of a space vehicle the path is expected to be quite smooth, but very rapid corrections can be made in the course if any deviation from the programmed flight path is detected. Another source of such problems is in the monitoring of chemical processes, since there can be a great diversity of time scales for the physical and chemical changes occurring. In electrical circuit theory, stiff problems arise because transients with time scales of the order of a microsecond can be imposed on the generally smooth behavior of the circuit.

Euler's Method

The numerical difficulties can be illustrated by using Euler's method on a simple model problem. Euler's method is the Taylor method of order 1, and for the initial-value problem

$$\begin{cases} x' = f(t, x) \\ x(t_0) = x_0 \end{cases} \tag{1}$$

it proceeds by using the equation

$$x_{n+1} = x_n + hf(t_n, x_n) \qquad (n \geqq 0) \tag{2}$$

Consider now the result of using Euler's method on this simple test problem:

$$\begin{cases} x' = \lambda x \\ x(0) = 1 \end{cases} \tag{3}$$

Euler's method produces the numerical solution

$$\begin{cases} x_0 = 1 \\ x_{n+1} = x_n + h\lambda x_n = (1 + h\lambda)x_n \end{cases} \tag{4}$$

Thus, at the nth step, the approximate solution is

$$x_n = (1 + h\lambda)^n \tag{5}$$

On the other hand, the actual solution of Equation (3) is

$$x(t) = e^{\lambda t} \tag{6}$$

If $\lambda < 0$, the solution in Equation (6) is exponentially decaying. It is tending to the *steady state* of 0 as t goes to infinity. The numerical solution in Equation (5) will tend to 0 if and only if $|1 + h\lambda| < 1$. This forces us to chose $h > 0$ so that $1 + h\lambda > -1$, and since $\lambda < 0$, we must impose the condition $h < -2/\lambda$.

For example, if $\lambda = -20$, we shall have to take $h < 0.1$, although the solution that we are trying to track is extremely flat (and practically 0) shortly after the initial instant, $t = 0$, where $x = 1$. We note that $x(t) = e^{-20t} \leqq 2.1 \times 10^{-9}$ for $t \geqq 1$. Thus, the numerical solution must proceed with *small* steps in a region where the nature of the solution indicates that *large* steps may be taken. This phenomenon is one aspect of stiffness. A function such as e^{-20t} that decays almost immediately to 0 (that is, with very large negative slope) is said to be a **transient** because its physical effect is of brief duration. We expect that the numerical tracking of a transient function will require a small step size until the transient effect has become negligible; after that, a good numerical method should permit a large step size. Euler's method does not fulfill this desideratum.

Modified Euler's Method

By contrast, the implicit Euler's method *will* meet the criterion just mentioned. The implicit Euler's method is defined by the equation

$$x_{n+1} = x_n + hf(t_{n+1}, x_{n+1}) \qquad (n \geqq 0) \tag{7}$$

When this method is applied to the test problem in Equation (3), we obtain

$$\begin{cases} x_0 = 1 \\ x_{n+1} = x_n + h\lambda x_{n+1} \end{cases} \tag{8}$$

This gives

$$x_{n+1} = (1 - h\lambda)^{-1} x_n \tag{9}$$

Hence at the nth step, we shall have

$$x_n = (1 - h\lambda)^{-n} \tag{10}$$

For negative λ, our requirement that the numerical solution mimic the actual solution becomes

$$|1 - h\lambda|^{-1} < 1 \tag{11}$$

This is obviously true for *all* (positive) step sizes h.

Systems of Differential Equations

Similar considerations apply to systems of differential equations. Again we appeal to the principle that a good numerical method should perform well on simple linear test cases. (Recall that this principle was used in Section 8.5 to deduce *stability* and *consistency* as essential attributes of acceptable linear multistep methods.) Here is a simple test case involving a system of two differential equations:

$$\begin{cases} x' = \alpha x + \beta y & x(0) = 2 \\ y' = \beta x + \alpha y & y(0) = 0 \end{cases} \tag{12}$$

The solution of this system is

$$\begin{cases} x(t) = e^{(\alpha+\beta)t} + e^{(\alpha-\beta)t} \\ y(t) = e^{(\alpha+\beta)t} - e^{(\alpha-\beta)t} \end{cases} \tag{13}$$

If Euler's method is used to compute a numerical solution to Equations (12), the formulas for advancing the solution are

$$\begin{cases} x_{n+1} = x_n + h(\alpha x_n + \beta y_n) & x_0 = 2 \\ y_{n+1} = y_n + h(\beta x_n + \alpha y_n) & y_0 = 0 \end{cases} \tag{14}$$

The solutions of these difference equations are

$$\begin{cases} x_n = (1 + \alpha h + \beta h)^n + (1 + \alpha h - \beta h)^n \\ y_n = (1 + \alpha h + \beta h)^n - (1 + \alpha h - \beta h)^n \end{cases} \tag{15}$$

The case in which we are interested is defined by $\alpha < \beta < 0$. With this supposition, the solutions in Equations (13) are decaying exponentially to 0. In order that the numerical solutions in Equations (15) will mimic this behavior, we want

$$|1 + \alpha h + \beta h| < 1 \qquad |1 + \alpha h - \beta h| < 1 \tag{16}$$

The inequalities in (16) are equivalent to the single condition

$$0 < h < -2/(\alpha + \beta)$$

since $\alpha < \beta < 0$. To see what can happen, suppose that $\alpha = -20$ and $\beta = -19$. Then our solutions are combinations of e^{-39t} and e^{-t}. The first of these is a *transient* function, which after a short time interval is negligible compared to e^{-t}. Yet the transient will actually control the allowable step size throughout the entire numerical solution.

General Linear Multistep Methods

Let us examine the general linear multistep method of Section 8.5 to see what additional properties it should have in order to perform satisfactorily on our single test equation (3). The general method has the form

$$a_k x_n + a_{k-1} x_{n-1} + \cdots + a_0 x_{n-k} = h[b_k f_n + b_{k-1} f_{n-1} + \cdots + b_0 f_{n-k}] \tag{17}$$

When this is applied to the test problem, we obtain

$$a_k x_n + a_{k-1} x_{n-1} + \cdots + a_0 x_{n-k} = h\lambda(b_k x_n + b_{k-1} x_{n-1} + \cdots + b_0 x_{n-k}) \quad \textbf{(18)}$$

Thus, our numerical solution will solve the homogeneous linear difference equation

$$(a_k - h\lambda b_k)x_n + (a_{k-1} - h\lambda b_{k-1})x_{n-1} + \cdots + (a_0 - h\lambda b_0)x_{n-k} = 0 \quad \textbf{(19)}$$

The solutions of this equation will be combinations of certain *basic* solutions of the type $x_n = r^n$, where r is a root of the polynomial

$$\phi(z) = (a_k - h\lambda b_k)z^k + (a_{k-1} - h\lambda b_{k-1})z^{k-1} + \cdots + (a_0 - h\lambda b_0) \quad \textbf{(20)}$$

Notice that the polynomial defined in Equation (20) is of the form $\phi = p - \lambda hq$, where p and q are the polynomials that were useful in Section 8.5; namely,

$$p(z) = a_k z^k + a_{k-1} z^{k-1} + \cdots + a_1 z + a_0 \quad \textbf{(21)}$$

$$q(z) = b_k z^k + b_{k-1} z^{k-1} + \cdots + b_1 z + b_0 \quad \textbf{(22)}$$

A-Stability

If $\lambda < 0$ in the test problem (3), then the solution is exponentially decaying. For the numerical solution obtained from the multistep method in Equation (17) to reflect this behavior, it is necessary that all roots of the polynomial ϕ in Equation (20) lie in the disk $|z| < 1$. For complex values of λ, say $\lambda = \mu + i\nu$, the solution of the test problem is

$$x(t) = e^{\lambda t} = e^{\mu t} e^{i\nu t} = e^{\mu t}(\cos \nu t + i \sin \nu t)$$

Exponential decay in this case corresponds to $\mu < 0$. In order that the multistep method perform well on such problems, we would like the roots of ϕ to be interior to the unit disk whenever $h > 0$ and $\mathrm{Re}(\lambda) < 0$. This property is called ***A*-stability**.

The implicit Euler's method in Equation (7) is *A*-stable, as shown in Equation (11). The implicit trapezoid method, defined by

$$x_n - x_{n-1} = \frac{1}{2}h[f_n + f_{n-1}] \quad \textbf{(23)}$$

is *A*-stable because the polynomial ϕ is

$$\phi(z) = z - 1 - \lambda h \left(\frac{1}{2}z + \frac{1}{2}\right)$$

and its root is $z = (2 + \lambda h)/(2 - \lambda h)$; we see easily that this root is interior to the unit disk when $h > 0$ and $\mathrm{Re}(\lambda) < 0$.

An important theorem, due to Dahlquist [1963], states that an *A*-stable linear multistep method must be an implicit method, and its order cannot exceed 2. This result places a severe restriction on *A*-stable methods. The implicit trapezoid rule is often used on stiff equations because it has the least truncation error among all A-stable linear multistep methods.

Region of Absolute Stability

Every multistep method possesses a **region of absolute stability**. This is the set of complex numbers ω such that the roots of $p - \omega q$ lie in the interior of the unit disk. From the preceding discussion it follows that a method will work well on the test problem $x' = \lambda x$ if λh lies in the region of absolute stability. We notice also that a method is A-stable if and only if its region of absolute stability contains the left half-plane.

Example 1 What is the region of absolute stability for the Euler method?

Solution The Euler method is defined by the equation

$$x_n - x_{n-1} = hf_{n-1}$$

Hence, $p(z) = z - 1$, $q(z) = 1$, and $\phi(z) = z - 1 - \omega$, where $\omega = \lambda h$. Now the root of ϕ is $z = 1 + \omega$. For *all* roots of ϕ to be interior to the unit disk, we require $|1 + \omega| < 1$. This is a disk of radius 1 in the complex plane with center at -1. ■

Since A-stability is such a severe restriction, methods lacking the property are nevertheless used on stiff problems. When this is done, we hope that $\omega = \lambda h$ will lie in the region of absolute stability for the method. The quantity λ pertains to the differential equation being solved. For a single linear equation, λ is the coefficient of x in the equation. For a single nonlinear equation, λ can be a constant that arises by locally approximating the equation by a linear equation. For a system of linear equations, say $X' = AX$, λ can be any one of the eigenvalues of A. Thus, h times each eigenvalue of A should lie in the region of absolute stability for the method being used. For a system of nonlinear equations, we can contemplate local approximation of the given system by linear systems and application of the preceding criterion. This idea is usually impossible to implement, and accordingly the only safe strategy in the case of difficult stiff problems may be to revert to the trapezoid rule. Some of the best current codes for the initial-value problem make an effort to detect stiffness in the course of the step-by-step solution and to take appropriate defensive action when stiffness occurs. The reader is referred to Gear [1971], Shampine and Allen [1973], and Shampine and Gordon [1975] for details. In general, the higher-order methods have smaller regions of absolute stability, and in the presence of stiffness these adaptive codes shift to methods of lower order having more favorable regions of absolute stability. (See Byrne and Hindmarsh [1987] or Shampine and Gear [1979].)

Nonlinear Equations

To see the relevance of the foregoing considerations to systems of *nonlinear* equations, let us turn to a typical system, assumed to be autonomous:

$$X' = F(X) \tag{24}$$

Here X is a vector function of t having n component functions $x_1, x_2, \ldots, x_n$. On the right side there is a function $F : \mathbb{R}^n \rightarrow \mathbb{R}^n$. This function has component functions f_i. The **Jacobian matrix** of F is the matrix $J = (J_{ij})$ defined by $J_{ij} = \partial f_i / \partial x_j$. At a point X_0, the linearized form of Equation (24) is

$$X' = F(X_0) + J(X_0)(X - X_0) \tag{25}$$

where $J(X_0)$ indicates the evaluation of J at X_0. Equation (25) is a linear differential equation, and the eigenvalues of the matrix $J(X_0)$ figure in the theory. If Equation (24) is stiff in the vicinity of X_0, these eigenvalues satisfy inequalities

$$\text{Re}(\lambda_1) \leqq \text{Re}(\lambda_2) \leqq \cdots \leqq \text{Re}(\lambda_n) < 0$$

and $\text{Re}(\lambda_1)$ is much smaller than $\text{Re}(\lambda_n)$.

If a multistep method is to perform well on such a problem, then $h\lambda_i$ should lie in the region of absolute stability for that method. If the λ_i were known, this information could be used to select h appropriately.

PROBLEMS 8.12

1. Find the region of absolute stability for the implicit Euler method defined by Equation (7).
2. Determine whether the problem $x'' = (57 + \sin t)x$ is or is not stiff.
3. Determine whether this problem is stiff:

$$\begin{cases} x'' = -20x' - 19x \\ x(0) = 2 \qquad x'(0) = -20 \end{cases}$$

4. Consider the multistep method

$$x_n + \alpha x_{n-1} - (1 + \alpha)x_{n-2} = \frac{1}{2}h\big[-\alpha f_n + (4 + 3\alpha)f_{n-1}\big]$$

   Determine α so that the method is stable, consistent, convergent, A-stable, and of second order. (See Sections 8.5 and 8.4.)
5. Verify that Equations (15) provide the solution of Equations (14).
6. Find the region of absolute stability of the implicit trapezoid rule, Equation (23).
7. Show that the two inequality relations (16) are equivalent to $0 < h < -2/(\alpha + \beta)$ when $\alpha < \beta < 0$.

COMPUTER PROBLEMS 8.12

1. Solve the differential equation in Problem 2 with initial conditions $x(0) = 1$, $x'(0) = -\sqrt{57}$. (The **Riccati transformation** is $y = x'/x$. It can be used to obtain an equivalent pair of equations $y' = 57 + \sin(t) - y^2, x' = xy$ with $y(0) = -\sqrt{57}$ and $x(0) = 1$ so that available software can be applied.)

2. Test the **implicit midpoint method**

$$x_n - x_{n-1} = hf\left(t_{n-1} + \frac{1}{2}h, \frac{1}{2}(x_n + x_{n-1})\right)$$

on the equation $x' = \lambda x$, with $\lambda < 0$, to determine whether its performance will be good on stiff problems.

3. A certain chemical reaction is described by a stiff system

$$\begin{cases} x_1' = -1000x_1 + x_2 & x_1(0) = 1 \\ x_2' = 999x_1 - 2x_2 & x_2(0) = 0 \end{cases}$$

Show that x_1 decays rapidly, whereas x_2 decays slowly. Such a system is difficult to solve since a very small step size will be necessary.

CHAPTER NINE
Numerical Solution of Partial Differential Equations

This chapter introduces the subject of the numerical solution of partial differential equations. The numerical calculations that arise in this area can easily tax the resources of the largest and fastest computers. Because of the immense computing burden usually involved in solving partial differential equations, this branch of numerical analysis is one in which there is widespread current research activity. As we consider several representative problems and the procedures available to solve them, it will become clear why storage and run-time demands (even on supercomputers) are so enormous for these problems.

9.1 Parabolic Equations: Explicit Methods

We begin with a representative partial differential equation of **parabolic type**.

Heat Equation

Such an equation is the **heat equation**, which is also known as the **diffusion equation**. If units for physical quantities are suitably chosen, the heat equation will have this form:

$$u_{xx} + u_{yy} + u_{zz} = u_t \tag{1}$$

In this equation, u is a function of x, y, z, and t. The notation u_t denotes $\partial u/\partial t$, u_{xx} denotes $\partial^2 u/\partial x^2$, and so on. Equation (1) governs the temperature u at time t at position (x, y, z) in a three-dimensional body. We write $u(x, y, z, t)$ for the value of u at the *point* (x, y, z, t). Thus, u is a real-valued function of four real variables.

As in the theory of *ordinary* differential equations, a properly posed physical problem never consists of a partial differential equation alone; there must be additional boundary conditions in sufficient number to specify a *unique* solution to the problem. We now take as a model problem the heat equation (1) in its one-dimensional version together with subsidiary conditions as follows:

$$\begin{cases} u_{xx} = u_t & (t \geqq 0, \quad 0 \leqq x \leqq 1) \\ u(x, 0) = g(x) & (0 \leqq x \leqq 1) \\ u(0, t) = a(t) & (t \geqq 0) \\ u(1, t) = b(t) & (t \geqq 0) \end{cases} \tag{2}$$

System (2) models the temperature distribution in a rod of length 1, whose ends are maintained at temperatures $a(t)$ and $b(t)$, respectively. (See Figure 9.1.) We assume that the functions g, a, and b are to be given. Furthermore, an *initial* temperature profile is prescribed by the function g. The function u is a function of (x, t); there is no dependence on y or z. Shown in Figure 9.2 is the domain in which $u(x, t)$ is sought; it is a subset of the plane determined by the variables x and t.

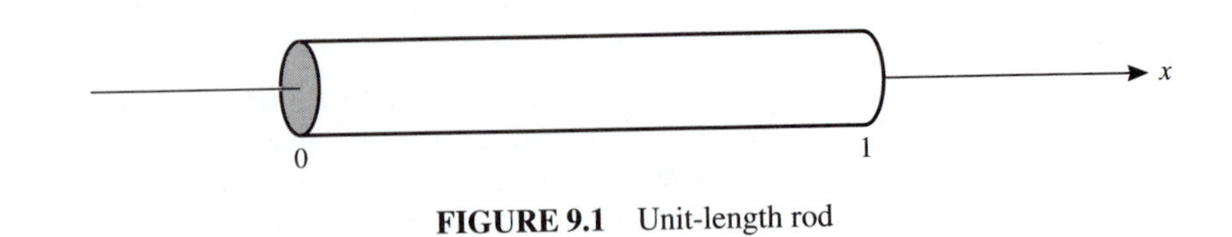

FIGURE 9.1 Unit-length rod

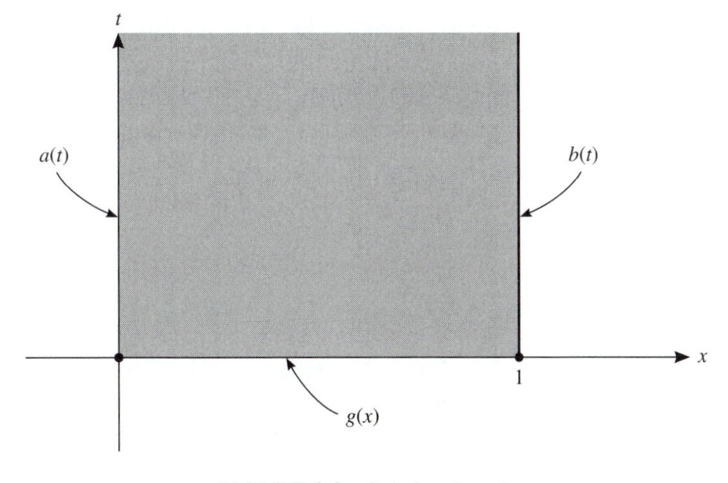

FIGURE 9.2 Solution domain

Finite-Difference Method

One of the principal strategies used in solving problems such as (2) numerically is the **finite-difference method**. It involves an initial **discretization** of the domain as follows:

$$\begin{cases} t_j = jk & (j \geqq 0) \\ x_i = ih & (0 \leqq i \leqq n+1) \end{cases} \tag{3}$$

The variables t and x have different step sizes, denoted by k and h, respectively. Since x runs over the interval [0, 1], we have $h = 1/(n+1)$. Our objective is to compute approximate values of the solution function u at the so-called **mesh points** (x_i, t_j).

The next step in the process is to select some simple formulas for approximating the derivatives appearing in the differential equation. Some familiar basic formulas that can be used are these:

$$f'(x) \approx \frac{1}{h}[f(x+h) - f(x)] \tag{4}$$

$$f'(x) \approx \frac{1}{2h}[f(x+h) - f(x-h)] \tag{5}$$

$$f''(x) \approx \frac{1}{h^2}[f(x+h) - 2f(x) + f(x-h)] \tag{6}$$

Of course, there exist many other such formulas, affording various degrees of precision.

Now, we replace the differential problem (2) by a discretized version of it using Equations (4) and (6). Since these two problems will have different solutions, we use another letter, v, for the discrete problem:

$$\frac{1}{h^2}[v(x+h,t) - 2v(x,t) + v(x-h,t)] = \frac{1}{k}[v(x,t+k) - v(x,t)] \tag{7}$$

To simplify the notation, we let $x = x_i$ and $t = t_j$ in Equation (7) and then abbreviate $v(x_i, t_j)$ by v_{ij}. The result is

$$\frac{1}{h^2}\left(v_{i+1,j} - 2v_{ij} + v_{i-1,j}\right) = \frac{1}{k}\left(v_{i,j+1} - v_{ij}\right) \tag{8}$$

When $j = 0$ in Equation (8), all the terms appearing there are known, with the exception of v_{i1}. Indeed, the initial temperature distribution g gives us

$$g(x_i) = u(x_i, 0) = v_{i0}$$

In other words, we know the correct value of u (and of v) at the t-level corresponding to $t = 0$. The values of v_{i1} can therefore be *calculated* from Equation (8). For this purpose, the equation is written in the equivalent form

$$v_{i,j+1} = \frac{k}{h^2}(v_{i+1,j} - 2v_{i,j} + v_{i-1,j}) + v_{i,j}$$

or, with the abbreviation $s = k/h^2$,

$$v_{i,j+1} = sv_{i-1,j} + (1-2s)v_{ij} + sv_{i+1,j} \tag{9}$$

By means of Equation (9), the numerical solution can be advanced step by step in the t-direction. The sketch in Figure 9.3 shows a typical set of four mesh points

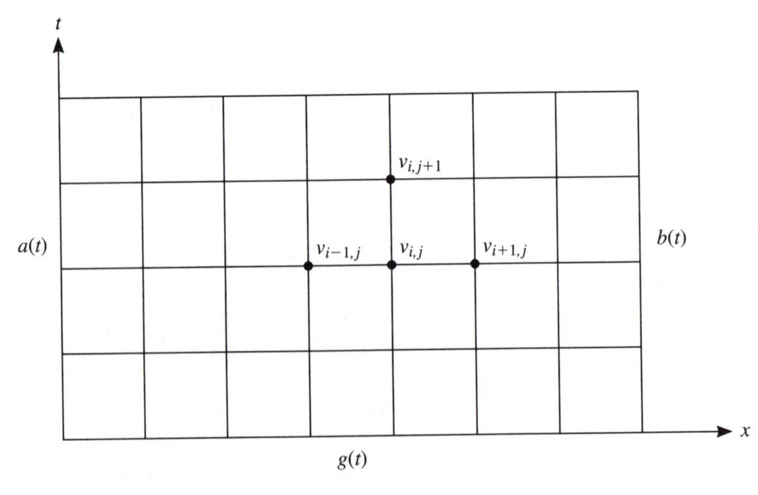

FIGURE 9.3 Typical grid

involved in Equation (9). Since Equation (9) gives the *new* values $v_{i,j+1}$ explicitly in terms of *previous* values $v_{i-1,j}, v_{ij}, v_{i+1,j}$, the method based on this equation is called an **explicit method**.

It is to be emphasized that even if the computation of successive v_{ij} is carried out with complete precision, the numerical solution will differ from the solution of the partial differential equation. This is because the function v is a solution of a different problem—namely, the finite-difference *analogue* of the differential equation. In fact, we must expect a rather large *difference* because of the low precision in the approximate differentiation formulas being used.

Algorithm

An algorithm for carrying out the computation for the preceding explicit method involves two phases: initialization and solution. Here the number of mesh points x_i in $[0, 1]$ is $n + 2$, the step size in the t-variable is k, and the number of steps in t to be computed is M.

input n, k, M
$h \leftarrow 1/(n+1)$
$s \leftarrow k/h^2$
$w_i \leftarrow g(ih) \qquad (0 \leqq i \leqq n+1)$
$t \leftarrow 0$
output $0, t, (w_0, w_1, \dots, w_{n+1})$
for $j = 1$ **to** M **do**
  $v_0 \leftarrow a(jk)$
  $v_{n+1} \leftarrow b(jk)$
  for $i = 1$ **to** n **do**
    $v_i \leftarrow sw_{i-1} + (1-2s)w_i + sw_{i+1}$
  end do
  $t \leftarrow jk$
  output $j, t, (v_0, v_1, \dots, v_{n+1})$
  $(w_1, w_2, \dots, w_n) \leftarrow (v_1, v_2, \dots, v_n)$
end do

The reader is urged to program the algorithm and perform the numerical experiments described in Computer Problem 1. Experiments of this type lead to the conclusion that not all pairs of step sizes (h, k) will be satisfactory. An analysis following the algorithm will show why this is true. We shall analyze the special case when $a(t) = b(t) = 0$.

Stability Analysis

Equation (9), which defines the numerical process, can be interpreted using matrix-vector notation. Let the vector of values at time $t = jk$ be denoted by V_j. Thus,

$$V_j = \begin{bmatrix} v_{1j} \\ v_{2j} \\ \vdots \\ v_{nj} \end{bmatrix} \tag{10}$$

Equation (9) can be written as

$$V_{j+1} = AV_j \tag{11}$$

in which A is the $n \times n$ matrix given by

$$A = \begin{bmatrix} 1-2s & s & 0 & \cdots & 0 \\ s & 1-2s & s & \cdots & 0 \\ 0 & s & 1-2s & \cdots & 0 \\ \vdots & \vdots & \vdots & \ddots & \vdots \\ 0 & 0 & 0 & \cdots & 1-2s \end{bmatrix} \tag{12}$$

Notice that $v_{0j} = v_{n+1,j} = 0$ because $a(t) = b(t) = 0$. Hence, on each horizontal line only n unknown values are present.

At this stage, the argument can proceed in two different ways. The first argument depends on the physical fact that the temperature in the bar will tend to 0, as $t \to \infty$, because the ends are being maintained at temperature 0. We therefore must insist that our numerical solution tend to 0 also as $t \to \infty$. Since $V_{j+1} = AV_j$, we have

$$V_j = AV_{j-1} = A^2V_{j-2} = \cdots = A^jV_0$$

By a theorem in Section 4.6, the following two conditions are equivalent:

(i) $\lim_{j\to\infty} A^jV = 0$ for all vectors V

(ii) $\rho(A) < 1$

Recall that $\rho(A)$ is the special radius of the matrix A. Hence, the parameter $s = k/h^2$ should be chosen so that $\rho(A) < 1$.

The same conclusion can be reached without appeal to the physical problem of temperature distribution by analyzing instead the effect of roundoff errors in the numerical computation. Suppose that at a certain step (which we may assume to be the first step), an error is introduced. Then instead of the vector V_0, we shall have a perturbation of it, say $\widetilde{V}_0$. The explicit method will produce vectors $\widetilde{V}_j = A^j\widetilde{V}_0$, and the error at the jth step is

$$V_j - \widetilde{V}_j = A^jV_0 - A^j\widetilde{V}_0 = A^j(V_0 - \widetilde{V}_0)$$

To ensure that this error will die away as $j \to \infty$, we again must require that $\rho(A) < 1$.

A determination of the eigenvalues will be given later; the result of this work is that the eigenvalues of A are

$$\lambda_j = 1 - 2s(1 - \cos\theta_j) \qquad \theta_j = \frac{j\pi}{n+1} \qquad (1 \leqq j \leqq n) \tag{13}$$

For $\rho(A)$ to be less than 1, we require

$$-1 < 1 - 2s(1 - \cos\theta_j) < 1$$

This is true if and only if $s < (1 - \cos\theta_j)^{-1}$. Since s is positive, the greatest restriction on s occurs when $\cos\theta_j = -1$. Since θ_j will be very close to π when $j = n$, we must require $s \leqq 1/2$.

To summarize: For the preceding explicit algorithm to be **stable**, it is necessary to assume that $s = k/h^2 \leqq 1/2$.

The severe restriction $k \leqq h^2/2$ forces this method to be very slow. For example, if $h = 0.01$, then the largest permissible value of k is 5×10^{-5}. If we want to compute a solution for $0 \leqq t \leqq 10$, then the number of time steps must be $\frac{1}{2} \times 10^6$ and the number of mesh points must be over 20 million! So the elegance and simplicity of the explicit method are accompanied by an unacceptable inefficiency.

To complete the foregoing analysis, it is necessary to prove that the eigenvalues of A are as asserted in Equation (13). We begin by observing that A in Equation (11) can be written as

$$A = I - sB$$

in which B is the $n \times n$ matrix

$$B = \begin{bmatrix} 2 & -1 & 0 & \cdots & 0 \\ -1 & 2 & -1 & \cdots & 0 \\ 0 & -1 & 2 & \cdots & 0 \\ \vdots & \vdots & \vdots & \ddots & \vdots \\ 0 & 0 & 0 & \cdots & 2 \end{bmatrix} \tag{14}$$

The eigenvalues λ_i of A are related to the eigenvalues μ_i of B via the equation

$$\lambda_i = 1 - s\mu_i$$

Thus, it suffices to determine the eigenvalues of B.

LEMMA 1 *Let $x = (\sin\theta, \sin 2\theta, \ldots, \sin n\theta)^T$. If $\theta = j\pi/(n+1)$, then x is an eigenvector of B corresponding to the eigenvalue $2 - 2\cos\theta$.*

Proof Let $B_{ij} = 2\delta_{ij} - \delta_{i+1,j} - \delta_{i-1,j}$, where $B = (B_{ij})$ and δ_{ij} is the Kronecker delta. Hence, if $2 \leqq i \leqq n - 1$, then

$$
\begin{aligned}
(Bx)_i &= \sum_{j=1}^{n} (2\delta_{ij} - \delta_{i+1,j} - \delta_{i-1,j})x_j \\
&= 2x_i - x_{i+1} - x_{i-1} \\
&= 2\sin i\theta - [\sin(i+1)\theta + \sin(i-1)\theta] \\
&= 2\sin i\theta - 2\sin i\theta \cos\theta \\
&= (2 - 2\cos\theta)\sin i\theta \\
&= (2 - 2\cos\theta)x_i
\end{aligned}
$$

Here we used the standard relation $\sin(\alpha + \beta) + \sin(\alpha - \beta) = 2\sin\alpha\cos\beta$. If $i = 1$ or $i = n$, we can use the same calculation provided that $x_0 = 0$ and $x_{n+1} = 0$. The formula $x_i = \sin i\theta$ automatically gives $x_0 = 0$, and the equation $x_{n+1} = 0$ will be true if $\sin(n+1)\theta = 0$ or $(n+1)\theta = j\pi$ for some integer j. The two special cases ($i = 1$ and $i = n$) are then as follows:

$$
\begin{aligned}
(Bx)_1 &= 2x_1 - x_2 = 2x_1 - x_2 - x_0 = (2 - 2\cos\theta)x_1 \\
(Bx)_n &= 2x_n - x_{n-1} = 2x_n - x_{n+1} - x_{n-1} = (2 - 2\cos\theta)x_n
\end{aligned}
$$

Putting all n of these equations together, we have

$$
Bx = (2 - 2\cos\theta)x
$$

■

Stability Analysis: Fourier Method

The question of *stability* in numerical procedures for partial differential equations arises in almost all problems involving time as an independent variable. This is natural, since solutions over long time intervals may be of interest. The method used above to analyze stability in the explicit method is called the **matrix method**. Another method, ascribed to von Neumann, can be termed the **Fourier method**. In this method, we try to find a solution of the finite-difference equations having the form

$$
v_{jn} = e^{ij\beta h} e^{n\lambda k} \qquad \left(i = \sqrt{-1}\right) \tag{15}
$$

(Here we use j for the first subscript instead of i so that it will not be confused with the complex number $i = \sqrt{-1}$.) Once this has been done, by a suitable choice of λ, the behavior of this solution is examined as $t \to \infty$ or $n \to \infty$. Obviously, this depends on the factor $e^{\lambda nk} = (e^{\lambda k})^n$ in Equation (15). If $|e^{\lambda k}| > 1$, this solution becomes unbounded. Since any numerical solution will eventually be contaminated by all the unwanted extraneous solutions, this unbounded solution will dominate the true solution, which is decaying exponentially.

Why do we consider a solution of the form of Equation (15)? The solution of the finite-difference equation should have the same form as the solution of the basic partial differential equation. The heat equation (2) has a solution of the form $u(x, t) = e^{-\pi^2 t}\sin(\pi x)$. This explains why we seek a solution of the difference equation of a similar form.

Let us carry out this analysis on the explicit method given by Equation (9). (We implicitly assume that $k > 0$ here.) On substituting the trial solution (15) in Equation (9), we obtain

$$e^{ij\beta h}e^{(n+1)\lambda k} = se^{i(j-1)\beta h}e^{n\lambda k} + (1-2s)e^{ij\beta h}e^{n\lambda k} + se^{i(j+1)\beta h}e^{n\lambda k}$$

After removing the factor $e^{ij\beta h}e^{n\lambda k}$, the result is

$$\begin{aligned} e^{\lambda k} &= se^{-i\beta h} + 1 - 2s + se^{i\beta h} \\ &= 2s\cos\beta h + 1 - 2s \\ &= 1 - 2s(1 - \cos\beta h) \\ &= 1 - 4s\sin^2(\beta h/2) \end{aligned} \tag{16}$$

Here we used the familiar equations

$$e^{i\theta} = \cos\theta + i\sin\theta \qquad 1 - \cos\theta = 2\sin^2(\theta/2)$$

Recall that $s = k/h^2$, where k and h are step sizes in t and x, respectively. It is clear from Equation (16) that $e^{\lambda k} \leqq 1$ (as long as β is real). To have stability, we require also that $e^{\lambda k} \geqq -1$, and this leads to the restriction

$$s\sin^2(\beta h/2) \leqq 1/2$$

Since $\sin^2(\beta h/2)$ can be close to 1, we must have $s \leqq 1/2$ for stability.

PROBLEMS 9.1

1. Prove that for fixed n, the stability requirement in the explicit algorithm is

$$s < \left(1 + \cos\frac{\pi}{n+1}\right)^{-1}$$

For $n = 10$, what values of k are satisfactory?

2. Show that the function

$$u(x, t) = \sum_{n=1}^{N} c_n \exp(-n^2\pi^2 t)\sin n\pi x$$

solves the heat conduction problem $u_{xx} = u_t$ with boundary conditions

$$\begin{cases} u(x, 0) = \displaystyle\sum_{n=1}^{N} c_n \sin n\pi x \\ u(0, t) = u(1, t) = 0 \end{cases}$$

3. Establish the claim that approximately 10^7 mesh points are required for stability when $h = 10^{-2}$ and $0 \leqq t \leqq 10$. What are the corresponding numbers when $h = 10^{-4}$?

COMPUTER PROBLEM 9.1

1. Using the algorithm for the *explicit* method, find a numerical solution to this heat conduction problem:

$$\begin{cases} u_{xx} = u_t \\ u(x,0) = \sin \pi x \\ u(0,t) = u(1,t) = 0 \end{cases}$$

Use $h = 0.1$, $k = 0.005125$, and $M = 200$ in the first experiment. Compare the computed solution to the exact solution $u(x,t) = \exp(-\pi^2 t)\sin(\pi x)$. Then change k to 0.006 and M to 171 in the second experiment.

9.2 Parabolic Equations: Implicit Methods

We continue to study the model problem of heat conduction. Recall that

$$\begin{cases} u_{xx} = u_t & (t \geqq 0, \quad 0 \leqq x \leqq 1) \\ u(x,0) = g(x) & (0 \leqq x \leqq 1) \\ u(0,t) = u(1,t) = 0 & (t \geqq 0) \end{cases} \tag{1}$$

As in Section 9.1, the derivatives are replaced by approximations, and we use $v(x,t)$ to denote the solution of the discretized problem. The finite-difference equation to be considered now is

$$\frac{1}{h^2}\left[v(x+h,t) - 2v(x,t) + v(x-h,t)\right] = \frac{1}{k}\left[v(x,t) - v(x,t-k)\right] \tag{2}$$

Using notation established in Section 9.1, we write this as

$$\frac{1}{h^2}\left[v_{i+1,j} - 2v_{ij} + v_{i-1,j}\right] = \frac{1}{k}\left[v_{ij} - v_{i,j-1}\right] \tag{3}$$

This equation seems to differ only superficially from Equation (7) in Section 9.1, but it will require a different type of algorithm for its solution. Observe that three terms in Equation (3) refer to v at the j-level (t) and only one term refers to v at the $(j-1)$-level ($t-k$). If v is known at the mesh points on the $(j-1)$-level, then the values at the j-level can be computed from Equation (3), but only by solving a *system* of equations. We therefore rewrite Equation (3) in the following form, using $s = k/h^2$:

$$-sv_{i-1,j} + (1+2s)v_{ij} - sv_{i+1,j} = v_{i,j-1} \qquad (1 \leqq i \leqq n) \tag{4}$$

As before, let V_j be the vector

$$V_j = \begin{bmatrix} v_{1j} \\ v_{2j} \\ \vdots \\ v_{nj} \end{bmatrix}$$

Then Equation (4) represents a system of n equations for determining the vector V_j, assuming of course that V_{j-1} is already known. This system of equations has the form

$$AV_j = V_{j-1} \tag{5}$$

in which A is the $n \times n$ matrix

$$A = \begin{bmatrix} 1+2s & -s & 0 & \cdots & 0 \\ -s & 1+2s & -s & \cdots & 0 \\ 0 & -s & 1+2s & \cdots & 0 \\ \vdots & \vdots & \vdots & \ddots & \vdots \\ 0 & 0 & 0 & \cdots & 1+2s \end{bmatrix} \tag{6}$$

Formally, the solution of Equation (5) is given by

$$V_j = A^{-1}V_{j-1}$$

and from this we obtain

$$V_j = A^{-1}(A^{-1}V_{j-2}) = A^{-1}\big(A^{-1}(A^{-1}V_{j-3})\big) = \cdots = A^{-j}V_0$$

The vector V_0 is known since it contains initial values $u(ih, 0)$. By the same reasoning that was used in Section 9.1, we must require $\rho(A^{-1}) < 1$ for the stability of the algorithm. Using the matrix B from Equation (14) in Section 9.1, we see that

$$A = I + sB$$

From Lemma 1 in Section 9.1, the eigenvalues of A are

$$\lambda_i = 1 + 2s(1 - \cos\theta_i) \qquad \theta_i = \frac{i\pi}{n+1} \qquad (1 \leqq i \leqq n) \tag{7}$$

Since these obviously satisfy $\lambda_i > 1$, the eigenvalues of A^{-1} lie in the interval $(0, 1)$, and we conclude that the proposed method is **stable** for all values of h and k. This method is sometimes called the **fully implicit method**.

Algorithm

A pseudocode to carry out the fully implicit method can be formulated now. This program will refer to two other routines, one of which furnishes initial values $g(x)$. The other routine is named `tri`; its purpose is to solve a tridiagonal system of linear equations. The pseudocode given at the end of Section 4.3 is suitable here. In that routine, the diagonal elements are labeled $d_1, d_2, \ldots, d_n$. The superdiagonal elements are $c_1, c_2, \ldots, c_{n-1}$, and the subdiagonal elements are $a_1, a_2, \ldots, a_{n-1}$. The numbers on the right-hand side are $b_1, b_2, \ldots, b_n$, and the solution was given the name $x_1, x_2, \ldots, x_n$. Hence,

$$\texttt{tri}(n, a, d, c, b; x)$$

takes input n, a, d, c, b and provides output x. In the present application, a, d, and c will be assigned constant values. We shall use the initial values $v_i = g(ih)$ in place of b. The routine `tri` provides the next value of the vector v, which overwrites the previous value. The diagonal entries d_i are altered in `tri` and therefore must be reinitialized at each time step.

```
input n, k, M
h ← 1/(n + 1)
s ← k/h^2
for i = 1 to n do
    v_i ← g(ih)
end do
t ← 0
output 0, t, (v_1, v_2, ..., v_n)
for i = 1 to n − 1 do
    c_i ← −s
    a_i ← −s
end do
for j = 1 to M do
    for i = 1 to n do
        d_i ← 1 + 2s
    end do
    call tri(n, a, d, c, v; v)
    t ← jk
    output j, t, (v_1, v_2, ..., v_n)
end do
```

Crank-Nicolson Method

It is possible to combine the implicit and explicit methods into a more general formula containing a parameter θ. This formula is

$$\frac{\theta}{h^2}\left(v_{i+1,j} - 2v_{ij} + v_{i-1,j}\right) + \frac{1-\theta}{h^2}\left(v_{i+1,j-1} - 2v_{i,j-1} + v_{i-1,j-1}\right) = \frac{1}{k}\left(v_{ij} - v_{i,j-1}\right) \tag{8}$$

We see at once that when $\theta = 0$, this formula yields the explicit scheme discussed in the previous section. (See Equation (8) in Section 9.1.) When $\theta = 1$, the formula reduces to the implicit scheme discussed above. The special case $\theta = \frac{1}{2}$ leads to a numerical procedure known by the names of its inventors, John Crank and Phyllis Nicolson.

We now investigate the Crank-Nicolson method in more detail. The pertinent formula is written as follows, where as usual all the *new* points (corresponding to j) have been put on the left and the *old* points (corresponding to $j - 1$) have been put on the right:

$$-sv_{i-1,j} + (2 + 2s)v_{ij} - sv_{i+1,j} = sv_{i-1,j-1} + (2 - 2s)v_{i,j-1} + sv_{i+1,j-1} \tag{9}$$

Here $s = k/h^2$. As before, a vector V_j containing entries v_{ij} for $1 \leqq i \leqq n$ is introduced. Equation (9) has the vector form

$$(2I + sB)V_j = (2I - sB)V_{j-1} \tag{10}$$

where B is as before; see Equation (14) in Section 9.1. By familiar reasoning, the stability of the method is assured if

$$\rho\left[(2I + sB)^{-1}(2I - sB)\right] < 1 \tag{11}$$

If $\mu_1, \mu_2, \ldots, \mu_n$ are the eigenvalues of B, then the requirement (11) becomes

$$\left|(2 + s\mu_i)^{-1}(2 - s\mu_i)\right| < 1 \tag{12}$$

Since $\mu_i = 2(1 - \cos\theta_i)$, we see that $0 < \mu_i < 4$. A little algebra then shows that Inequality (12) is true. The Crank-Nicolson method is therefore stable for all values of the ratio $s = k/h^2$.

Stability is, of course, not the only criterion used in selecting the step sizes h and k in these methods. In general, the smaller we take h and k, the more nearly the discretized problem will mimic the original differential equation. What is required is a theorem to guarantee that the solution $v(x, t)$ of the discrete problem converges to the solution $u(x, t)$ of the original problem when $h \to 0$ and $k \to 0$. Such a result is our next concern.

Analysis

The *errors* at the mesh points are defined by the equation

$$e_{ij} = u(x_i, t_j) - v(x_i, t_j) \tag{13}$$

Now we abbreviate $u(x_i, t_j)$ and $v(x_i, t_j)$ by u_{ij} and v_{ij}. Thus, $v_{ij} = u_{ij} - e_{ij}$. Let us analyze the explicit method for the one-dimensional heat equation (from the preceding section) for its convergence properties. The equation that defines this method is

$$v_{i,j+1} = s(v_{i-1,j} - 2v_{ij} + v_{i+1,j}) + v_{i,j} \tag{14}$$

This is Equation (9) from Section 9.1, and as usual $s = k/h^2$. In Equation (14), we replace v by $u - e$, getting

$$\begin{aligned} u_{i,j+1} - e_{i,j+1} &= s(u_{i-1,j} - 2u_{ij} + u_{i+1,j}) + u_{ij} \\ &\quad - s(e_{i-1,j} - 2e_{ij} + e_{i+1,j}) - e_{ij} \end{aligned}$$

This can be rearranged in the form

$$\begin{aligned} e_{i,j+1} &= se_{i-1,j} + (1-2s)e_{ij} + se_{i+1,j} \\ &\quad - s\left[u_{i-1,j} - 2u_{ij} + u_{i+1,j}\right] + \left[u_{i,j+1} - u_{ij}\right] \end{aligned} \tag{15}$$

To simplify the bracketed terms in this equation, we refer to previously established formulas for numerical differentiation; namely,

$$f''(x) = \frac{1}{h^2}\left[f(x+h) - 2f(x) + f(x-h)\right] - \frac{h^2}{12}f^{(4)}(\xi) \tag{16}$$

$$g'(t) = \frac{1}{k}\left[g(t+k) - g(t)\right] - \frac{k}{2}g''(\tau) \tag{17}$$

Using these formulas, we obtain from Equation (15)

$$\begin{aligned} e_{i,j+1} &= se_{i-1,j} + (1-2s)e_{i,j} + se_{i+1,j} \\ &\quad - s\left[h^2 u_{xx}(x_i, t_j) + \frac{h^4}{12}u_{xxxx}(\xi_i, t_j)\right] \\ &\quad + \left[ku_t(x_i, t_j) + \frac{k^2}{2}u_{tt}(x_i, \tau_j)\right] \end{aligned} \tag{18}$$

Now use the facts that $sh^2 = k$ and $u_{xx} = u_t$. Equation (18) can then be written in the form

$$\begin{aligned} e_{i,j+1} &= se_{i-1,j} + (1-2s)e_{ij} + se_{i+1,j} \\ &\quad - kh^2\left[\frac{1}{12}u_{xxxx}(\xi_i, t_j) - \frac{s}{2}u_{tt}(x_i, \tau_j)\right] \end{aligned} \tag{19}$$

Let us confine (x, t) to the compact set $S = \{(x, t) : 0 \leqq x \leqq 1 \text{ and } 0 \leqq t \leqq T\}$. Then put

$$M = \frac{1}{12} \max |u_{xxxx}(x, t)| + \frac{s}{2} \max |u_{tt}(x, t)| \tag{20}$$

where the maxima are over all $(x, t) \in S$. We also define an error vector

$$E_j = \begin{bmatrix} e_{1j} \\ e_{2j} \\ \vdots \\ e_{nj} \end{bmatrix}$$

and put

$$\|E_j\|_\infty = \max_{1 \leqq i \leqq n} |e_{ij}|$$

Finally, we assume that $1 - 2s \geqq 0$. Then from Equation (19), we infer that

$$\begin{aligned} |e_{i,j+1}| &\leqq s|e_{i-1,j}| + (1 - 2s)|e_{ij}| + s|e_{i+1,j}| + kh^2M \\ &\leqq s\|E_j\|_\infty + (1 - 2s)\|E_j\|_\infty + s\|E_j\|_\infty + kh^2M \\ &= \|E_j\|_\infty + kh^2M \end{aligned} \tag{21}$$

Since the right side of Equation (21) does not involve i, we obtain

$$\begin{aligned} \|E_{j+1}\|_\infty &\leqq \|E_j\|_\infty + kh^2M \leqq \|E_{j-1}\|_\infty + 2kh^2M \leqq \cdots \\ &\leqq \|E_0\|_\infty + (j + 1)kh^2M \end{aligned}$$

Since the solution $v(x, t)$ is started with the correct initial values, we have $E_0 = 0$, and therefore

$$\|E_j\|_\infty \leqq jkh^2M$$

Now let $jk = t$, so that $\|E_j\|_\infty$ represents the maximum error at mesh points on the t-level. Then, because $t \leqq T$,

$$\|E_j\|_\infty \leqq Th^2M = \mathcal{O}(h^2)$$

Thus, as $h \to 0$, the maximum errors on any fixed t-level converge to 0 with the rapidity of h^2, provided that $s = k/h^2 \leqq 1/2$ and that the functions u_{xxxx} and u_{tt} are continuous.

Summary

We have discussed three methods for obtaining the numerical solution of the one-dimensional heat conduction problem:

Method	Matrix Equation
Explicit	$V_{j+1} = (I - sB)V_j$
Implicit	$(I + sB)V_j = V_{j-1}$
Crank-Nicolson	$(2I + sB)V_j = (2I - sB)V_{j-1}$

One can easily implement these methods with only two routines involving tridiagonal systems:

(i) Compute $y = Tx$.
(ii) Solve $Tx = b$ for x.

PROBLEMS 9.2

1. Show that if $r > 0$, then the largest eigenvalue of the matrix $(I + rB)^{-1}(I - rB)$ is $(1 - q)/(1 + q)$, where $q = 4r\sin^2\left[\pi/(2n + 2)\right]$.

2. Prove that the stability condition for the method defined by Equation (8) is the inequality $s \leqq (2 - 4\theta)^{-1}$ if $0 \leqq \theta < \frac{1}{2}$, but if $\frac{1}{2} \leqq \theta \leqq 1$, there is no restriction on s.

3. Carry out a convergence analysis for the fully implicit method. *Hint:* At some stage in your analysis you should have a vector equation of the form

$$(I + sB)E_j = E_{j-1} - C_j$$

where C_j is a vector containing components of the form

$$\frac{k^2}{2}u_{tt}(x_i, t_j + \theta_2 k) + \frac{sh^4}{12}u_{xxxx}(x_i + \theta_1 h, t_j)$$

4. Generalize the algorithm in this section to handle end conditions such as in the problem

$$\begin{cases} u_{xx} = u_t & (t \geqq 0, \quad 0 \leqq x \leqq 1) \\ u(x, 0) = g(x) & (0 \leqq x \leqq 1) \\ u(0, t) = a(t) & (t \geqq 0) \\ u(1, t) = b(t) & (t \geqq 0) \end{cases}$$

COMPUTER PROBLEMS 9.2

1. Use the fully implicit method to solve this heat conduction problem on the unit square:

$$\begin{cases} u_{xx} = u_t \\ u(x, 0) = (x - x^2)e^x \\ u(0, t) = u(1, t) = 0 \end{cases}$$

Suggested values are $n = 20$, $M = 50$, and $k = 0.05$.

2. Write and test a program to execute the Crank-Nicolson procedure on the following problem:

$$\begin{cases} u_{xx} - \alpha u_t = f(x) & (0 < x < L, \quad 0 < t) \\ u(x, 0) = g(x) & (0 < x < L) \\ u(0, t) = u(L, t) = 0 & (0 < t) \end{cases}$$

3. Compare the explicit, implicit, and Crank-Nicolson methods using two routines involving operations on tridiagonal systems and test them on Computer Problem 1 of Section 9.1.

9.3 Problems Without Time Dependence: Finite-Difference Methods

Laplace's equation in two variables is a typical partial differential equation in which time is *not* one of the variables:

$$u_{xx} + u_{yy} = 0 \tag{1}$$

In this equation, u is a function of (x, y). A concrete physical problem involving Equation (1) would contain also the specification of a region Ω in the xy-plane where the solution is sought, and there would be boundary conditions placed on u (such as values that u or its normal derivative were to assume on the boundary of Ω). In dealing with a region Ω in $\mathbb{R}^2$, we assume that Ω is an open set. Its boundary is denoted by $\partial\Omega$, and its closure is denoted by $\overline{\Omega}$. Thus, $\overline{\Omega} = \Omega \cup \partial\Omega$.

Dirichlet Problem

A problem that occurs in the study of heat, electricity, and many other branches of physics is the **Dirichlet problem**. In the two-dimensional version, an open region Ω is prescribed in $\mathbb{R}^2$. A function g, defined on the boundary of Ω, is also given. We seek then a function u that is continuous on the closure $\overline{\Omega}$, satisfies Laplace's equation in Ω, and equals g on the boundary. This is summarized by

$$\begin{cases} u_{xx} + u_{yy} = 0 & \text{in } \Omega \\ u(x, y) = g(x, y) & \text{on } \partial\Omega \\ u \text{ continuous on } \overline{\Omega} \end{cases} \tag{2}$$

If Ω is subject to some mild restrictions and if g is continuous, then it can be proved that the Dirichlet problem has a unique solution.

To illustrate some numerical techniques, we shall consider the Dirichlet problem on a square. This problem can be solved by the analytic techniques of separation of variables and Fourier series. But for other regions, the numerical methods discussed here and in the next section are often needed. Also, it should be emphasized that even if a solution in the form of an infinite series can be given, a numerical solution involving a discretization of the problem may be preferable.

In the illustrative problem, the region is the open unit square

$$\Omega = \{(x, y) : 0 < x < 1,\ 0 < y < 1\}$$

The boundary function g is arbitrary for the moment. Its values are to be furnished by a suitable subroutine in the computer program.

Finite Difference

One approach to the numerical solution of problem (2) employs finite-difference approximations to the derivatives. A formula familiar from Section 9.2 can be used; namely,

$$f''(x) = \frac{1}{h^2}\left[f(x+h) - 2f(x) + f(x-h)\right] + \mathcal{O}(h^2) \tag{3}$$

First, a network of grid points is established in the square $\overline{\Omega}$:

$$(x_i, y_j) = (ih, jh) \qquad (0 \leqq i, j \leqq n+1) \qquad h = \frac{1}{n+1} \tag{4}$$

Notice that the same step size is being used for both variables. Next, the differential equation (1) at the mesh point (x_i, y_j) is replaced by its finite-difference analogue at that point, which is

$$\frac{1}{h^2}[v_{i-1,j} - 2v_{ij} + v_{i+1,j}] + \frac{1}{h^2}[v_{i,j-1} - 2v_{ij} + v_{i,j+1}] = 0$$

or

$$4v_{ij} - v_{i-1,j} - v_{i+1,j} - v_{i,j-1} - v_{i,j+1} = 0 \tag{5}$$

Here v_{ij} is intended to approximate $u(x_i, y_j)$. A different variable is used because we want to distinguish between the solutions of two different problems: first, the **discrete problem** (that is, the finite-difference equation) and second, the **continuous problem** (that is, the original partial differential equation).

The values of v_{ij} are *known* when $i = 0$ or $n+1$ and when $j = 0$ or $n+1$, since these are the prescribed boundary values in the problem (as given by the function g). Thus in Equation (5), some terms v_{ij} may be known while others are unknown. This means, in effect, that we shall be solving a nonhomogeneous system of linear equations because all *known* values will have been transferred to the right. To take a simple case, let $n = 3$. There is one equation of type (5) for each *interior* grid point. These are depicted in Figure 9.4 by large dots.

The unknown quantities in this problem can be ordered in many ways (corresponding to the different orderings of the mesh points). We select the one known

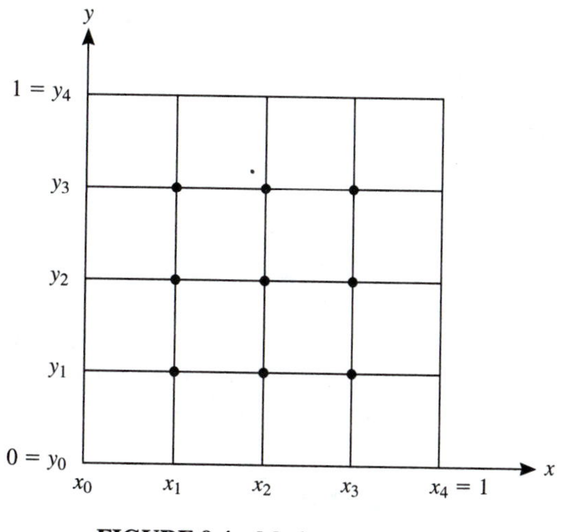

FIGURE 9.4 Mesh on unit square

as the **natural ordering**:

$$v = \left[v_{11}, v_{21}, v_{31}, v_{12}, v_{22}, v_{32}, v_{13}, v_{23}, v_{33}\right]$$

Likewise, the nine linear equations can be ordered in many ways. We decide to order them by associating Equation (5) with its *central point* (x_i, y_j) and then ordering the equations as we ordered the points. The result is as follows, where all known quantities have been placed on the right.

$$\begin{array}{ll}
4v_{11} - v_{21} - v_{12} & = v_{10} + v_{01} \\
4v_{21} - v_{11} - v_{31} - v_{22} & = v_{20} \\
4v_{31} - v_{21} - v_{32} & = v_{30} + v_{41} \\
4v_{12} - v_{11} - v_{22} - v_{13} & = v_{02} \\
4v_{22} - v_{21} - v_{12} - v_{32} - v_{23} & = 0 \\
4v_{32} - v_{31} - v_{22} - v_{33} & = v_{42} \\
4v_{13} - v_{12} - v_{23} & = v_{03} + v_{14} \\
4v_{23} - v_{22} - v_{13} - v_{33} & = v_{24} \\
4v_{33} - v_{32} - v_{23} & = v_{43} + v_{34}
\end{array}$$

This system has the form $Av = b$, in which the matrix A is of order 9×9. Out of 81 elements, only 33 are nonzero. The coefficient matrix follows.

$$A = \begin{bmatrix}
\begin{bmatrix} 4 & -1 & 0 \\ -1 & 4 & -1 \\ 0 & -1 & -4 \end{bmatrix} & \begin{bmatrix} -1 & 0 & 0 \\ 0 & -1 & 0 \\ 0 & 0 & -1 \end{bmatrix} & \begin{bmatrix} 0 & 0 & 0 \\ 0 & 0 & 0 \\ 0 & 0 & 0 \end{bmatrix} \\
\begin{bmatrix} -1 & 0 & 0 \\ 0 & -1 & 0 \\ 0 & 0 & -1 \end{bmatrix} & \begin{bmatrix} 4 & -1 & 0 \\ -1 & 4 & -1 \\ 0 & -1 & 4 \end{bmatrix} & \begin{bmatrix} -1 & 0 & 0 \\ 0 & -1 & 0 \\ 0 & 0 & -1 \end{bmatrix} \\
\begin{bmatrix} 0 & 0 & 0 \\ 0 & 0 & 0 \\ 0 & 0 & 0 \end{bmatrix} & \begin{bmatrix} -1 & 0 & 0 \\ 0 & -1 & 0 \\ 0 & 0 & -1 \end{bmatrix} & \begin{bmatrix} 4 & -1 & 0 \\ -1 & 4 & -1 \\ 0 & -1 & 4 \end{bmatrix}
\end{bmatrix}$$

In this matrix, the essential block structure has been indicated. We can write, with obvious notation,

$$A = \begin{bmatrix} T & -I & 0 \\ -I & T & -I \\ 0 & -I & T \end{bmatrix}$$

With the preceding special case as a model, let us proceed to the general case, where n can be any integer. The number of interior grid points is n^2, and with each of these points we have associated an unknown value $u_{ij} = u(x_i, y_j)$. The function u is to be a solution of Equation (2). After the discretization, we have a linear system of equations governing the n^2 unknowns v_{ij}, where $1 \leqq i \leqq n$ and $1 \leqq j \leqq n$.

Algorithm

The system of equations will be *sparse* because each equation will contain at most five unknowns (there are at most $5n$ nonzeros out of the total set of n^2). Iterative procedures such as those discussed in Section 4.6 can be quite effective in this situation. The known boundary values are

$$v_{ij} = g(x_i, y_j) \text{ when } i \text{ or } j \text{ equals } 0 \text{ or } n + 1 \tag{6}$$

Thus in the algorithm, the full array of elements v_{ij} allows values $0 \leqq i, j \leqq n + 1$ and contains $(n + 2)^2$ elements. For an illustration, we propose the Gauss-Seidel iterative method. For this procedure, Equation (5) associated with the point (x_i, y_j) is written in the form

$$v_{ij} = (v_{i-1,j} + v_{i+1,j} + v_{i,j-1} + v_{i,j+1})/4 \tag{7}$$

This equation is used for *updating* v_{ij}. When this equation is used, the value obtained from the right-hand side replaces the *old* value of v_{ij}. If we are careful to use Equation (7) only when $1 \leqq i, j \leqq n$, then only interior grid points will be involved in this updating. Boundary grid points will occur only on the right in Equation (7). The broad outline of the computation is now as follows:

(i) Place the boundary values into the appropriate v_{ij}.
(ii) Provide suitable starting values for interior nodes.
(iii) Carry out M steps of the Gauss-Seidel iteration.

A fragment of pseudocode for the first task is

```
for i = 0 to n + 1 do
    v_i0 = g(x_i, 0);  v_i,n+1 = g(x_i, 1)
    v_n+1,i = g(1, y_i);  v_0i = g(0, y_i)
end do
```

Pseudocode for the third task is

```
for k = 1 to M do
    for j = 1 to n do
        for i = 1 to n do
            v_ij = (v_{i-1,j} + v_{i+1,j} + v_{i,j-1} + v_{i,j+1})/4.0
        end do
    end do
end do
```

Notice that the nested loops in this algorithm carry out the updating of the vector v in the order selected (the *natural* order).

For a concrete case, we took the function g to be given by the formula

$$g(x, y) = 10^{-4} \sinh(3\pi x) \sin(3\pi y)$$

In the computer program, this is used only for boundary points of the square. But g is harmonic, and so g is also the solution to the problem. It is thus possible to compute the approximate solution v_{ij} and compare to the true solution $u(x_i, y_j) = g(x_i, y_j)$. We let $n = 18$, so that the matrix A is 324×324. This matrix is not stored in the computer because the Gauss-Seidel method does not require this. Instead, we use Equation (7) and need only the vector v, which has 324 components, v_{ij} ($0 \leqq i \leqq 18$ and $0 \leqq j \leqq 18$). After 200 iterations, the error was found to satisfy the inequality

$$|v_{ij} - u(x_i, y_j)| < 0.345 \times 10^{-7}$$

PROBLEMS 9.3

1. Let ∇^2 denote the Laplacian operator: $\nabla^2 u = u_{xx} + v_{yy}$. Prove that a problem of the form

$$\begin{cases} \nabla^2 u = f & \text{in } \Omega \\ u = 0 & \text{on } \partial\Omega \end{cases}$$

can be solved by the following three steps: **(i)** Find g such that $\nabla^2 g = f$. **(ii)** Solve the Dirichlet problem in Ω, using $-g$ for boundary values. **(iii)** Add g to the function obtained in step **(ii)** to obtain u.

2. (Continuation) Describe a Dirichlet problem equivalent to

$$\begin{cases} \nabla^2 u = (6 - 3x^2) \sin y - y \cos x & \text{in } \Omega \\ u = 0 & \text{on } \partial\Omega \end{cases}$$

3. One way of ordering the points in the plane is by **lexicographic ordering**. This refers to the ordering used in a dictionary. In a dictionary, we place first all words beginning with "*a*." Among these, we order the words according to the second

letter. In case of ties, we look at the third letter, and so on. Thus, in the plane, lexicographic ordering is defined by

$$(x, y) \leqq (u, v) \text{ if } (x < u) \text{ or } (x = u \text{ and } y \leqq v)$$

What matrix arises in the model problem if the mesh points and their equations are ordered lexicographically?

4. In the model problem, with n^2 interior grid points, how many equations will be homogeneous in general? How many equations will have five nonzero coefficients on the left? How many equations will have four nonzero coefficients on the left? How many will have three nonzero coefficients? What is the precise number of nonzero coefficients in the matrix?

5. Find the *analytic* solution of the Dirichlet problem on the unit square when the boundary values are 0 except for $u(x, 0) = \sin 4\pi x$. Use the method of separation of variables and Fourier series.

6. Let ∇^2 be the Laplacian in two dimensions, and let δ be the discrete Laplacian, defined by

$$\begin{aligned}(\delta f)(x, y) &= [f(x+1, y) - 2f(x, y) + f(x-1, y)] \\ &= [f(x, y+1) - 2f(x, y) + f(x, y-1)]\end{aligned}$$

Prove or disprove that for each (x, y) there is a (ξ, η) such that

$$(\delta f) = (\nabla^2 f)(\xi, \eta)$$

What about the one-dimensional version of this?

COMPUTER PROBLEMS 9.3

1. Program the procedure outlined in the text for solving the Dirichlet problem on a square. The number of points, the number of iterations, and the frequency of printing should be controlled by parameters that are easily changed. The boundary values should be provided by a subprogram. Test your program using

$$g(x, y) = 4xy(x - y)(x + y)$$

Compare the solutions of the continuous and discrete problems.

2. Program the example in the text and see whether your computer produces results similar to those cited.

3. Modify the programs given to use the SOR (successive overrelaxation) method instead of the Gauss-Seidel method.

*9.4 Problems Without Time Dependence: Galerkin and Ritz Methods

Galerkin methods are used widely on problems in which it is required to determine an unknown *function*. Of course, differential equations and integral equations are in this category. We begin with a broad abstract statement of the principles as they apply to *any* linear problem. Then we illustrate them with a numerical example in which we solve the Dirichlet problem on a rectangle.

Galerkin Method

Suppose we are confronted with a problem of the form

$$Lu = f \tag{1}$$

in which L is a linear operator, f is a given function, and u is a function to be determined from the equation. In the **Galerkin method**, we select a set of **basic functions** or **test functions** $u_1, u_2, \dots, u_n$. We then attempt to solve Equation (1) with a suitable linear combination of these basic functions. Setting

$$u = \sum_{j=1}^{n} c_j u_j$$

and using the linearity of L, we obtain

$$\sum_{j=1}^{n} c_j L u_j = f \tag{2}$$

In a typical situation, this equation is **inconsistent** because f will not generally lie in the vector space spanned by the functions Lu_j. We therefore solve Equation (2) *approximately* and obtain thereby an approximate solution of Equation (1). The approximate solving of Equation (2) can be carried out according to many different criteria, each of which leads to a different approximate solution u. The most natural approach is to select $c_1, c_2, \dots, c_n$ so that some norm is minimized:

$$\left\| \sum_{j=1}^{n} c_j L u_j - f \right\| = \min \tag{3}$$

This is a problem in best approximation. We are approximating f by the nearest element in the subspace generated by the functions Lu_j. This is relatively easy in an inner-product space because the technique of orthogonal projection is available.

A very general way of obtaining an approximate solution of Equation (2) is to select a set of linear functionals $\phi_1, \phi_2, \dots, \phi_n$ and to impose the condition

$$\phi_i\left(\sum_{j=1}^{n} c_j L u_j - f\right) = 0 \qquad (1 \leqq i \leqq n) \tag{4}$$

By the linearity of the functionals, this becomes

$$\sum_{j=1}^{n} \phi_i(Lu_j)c_j = \phi_i(f) \qquad (1 \leqq i \leqq n) \tag{5}$$

Equation (5) is a system of n linear equations in the n unknowns c_j. If the functionals are point-evaluation functionals, defined by

$$\phi_i(v) = v(x_i) \qquad (1 \leqq i \leqq n) \tag{6}$$

then the method outlined above is called **collocation**. We shall refer to all manifestations of the preceding strategy as **Galerkin methods**. The classical Galerkin method is a particular case of Equation (5) in a Hilbert space, where $\phi_i(v) = \langle u_i, v\rangle$. Thus, the equations to be solved in this case are

$$\sum_{j=1}^{n} c_j\langle u_i, Lu_j\rangle = \langle u_i, f\rangle \qquad (1 \leqq i \leqq n)$$

In Section 8.10, the collocation method was used to solve two-point boundary-value problems involving differential equations. Here we shall illustrate the Galerkin method for the **Dirichlet problem**. The example chosen is one that can be solved also with separation of variables and Fourier series.

Dirichlet Problem

Consider an open rectangle in the plane, described by

$$\Omega = \{(x, y) : |x| < 1,\ |y| < 2\} \tag{7}$$

We seek a continuous function u, defined on $\overline{\Omega}$, such that

$$\begin{cases} \nabla^2 u = 0 & \text{in } \Omega \\ u(x, y) = x^2 + y^2 & \text{on } \partial\Omega \end{cases} \tag{8}$$

This problem can be interpreted as one of the form $Lu = f$; we simply define L and f by the equations

$$Lu = \begin{bmatrix} \nabla^2 u \\ u \mid \partial\Omega \end{bmatrix} \qquad f(x, y) = \begin{bmatrix} 0 \\ x^2 + y^2 \end{bmatrix} \tag{9}$$

The notation $u \mid S$ denotes the restriction of the function u to a set S. It is a linear operation because

$$(\alpha u + \beta v) \mid S = \alpha(u \mid S) + \beta(v \mid S)$$

To use a Galerkin method, it is necessary to select a suitable set of basic functions. For the problem under discussion, we choose functions $u_1, u_2, \ldots, u_n$ that satisfy the homogeneous part of the problem. That is, we select u_i such that $\nabla^2 u_i = 0$ in Ω. Such functions are said to be **harmonic** in Ω. A rich supply of harmonic functions is available by virtue of the principle that the real and imaginary parts of any analytic function of a complex variable are harmonic. The first few power functions, z^k, provide a convenient set of harmonic polynomials:

$$\begin{aligned}
z^0 &= 1\\
z^1 &= x + iy\\
z^2 &= (x^2 - y^2) + (2xy)i\\
z^3 &= (x^3 - 3xy^2) + (3x^2y - y^3)i\\
z^4 &= (x^4 - 6x^2y^2 + y^4) + (4x^3y - 4xy^3)i\\
z^5 &= (x^5 - 10x^3y^2 + 5xy^4) + (5x^4y - 10x^2y^3 + y^5)i\\
z^6 &= (x^6 - 15x^4y^2 + 15x^2y^4 - y^6) + (6x^5y - b20x^3y^3 + 6xy^5)i
\end{aligned}$$

Since our problem possesses symmetry with respect to the coordinate axes and the origin, we select real and imaginary parts of z^k that have the same symmetries. This limits us to the real part of z^k for k even. Thus, if we let $n = 4$, we could use as the basic functions

$$\begin{aligned}
u_1(x, y) &= 1\\
u_2(x, y) &= x^2 - y^2\\
u_3(x, y) &= x^4 - 6x^2y^2 + y^4\\
u_4(x, y) &= x^6 - 15x^4y^2 + 15x^2y^4 - y^6
\end{aligned}$$

For this choice of the basic functions, it is clear that the function $u = \sum_{j=1}^{4} c_j u_j$ will automatically satisfy $\nabla^2 u = 0$, and that the coefficients c_j should be chosen so that the boundary conditions are approximately satisfied. The equation that must be solved approximately is

$$\sum_{j=1}^{4} c_j u_j(x, y) = x^2 + y^2 \quad \text{on } \partial\Omega \tag{10}$$

Because of the symmetries, it suffices to consider only the part of $\partial\Omega$ that lies in the first quadrant.

To illustrate the collocation method, we select four points on $\partial\Omega$—namely, $(0, 2)$, $(1, 0)$, $(1, 1)$, and $(1, 2)$. We insist that Equation (10) be valid just at these four points. We are led then to the following system of linear equations:

$$\begin{bmatrix} 1 & -4 & 16 & -64 \\ 1 & 1 & 1 & 1 \\ 1 & 0 & -4 & 0 \\ 1 & -3 & -7 & 117 \end{bmatrix} \begin{bmatrix} c_1 \\ c_2 \\ c_3 \\ c_4 \end{bmatrix} = \begin{bmatrix} 4 \\ 1 \\ 2 \\ 5 \end{bmatrix}$$

The approximate solution of this is

$$c = (1.8261,\ -0.7870,\ -0.04348,\ 0.004348)^T$$

The largest discrepancy between $\sum_{j=1}^{4} c_j u_j$ and the function $x^2 + y^2$ on the boundary of the rectangle is then approximately 0.074.

In the preceding example, let us retain the set of base functions but follow another approach in "solving" Equation (10). We select a set of m points (x_i, y_i) on the boundary of Ω and attempt to minimize the expression

$$\sum_{i=1}^{m} \left[\sum_{j=1}^{4} c_j u_j(x_i, y_i) - (x_i^2 + y_i^2) \right]^2 \tag{11}$$

with respect to c_j. This procedure includes collocation as a special case because we could use $m = 4$. But by taking more points, we can simulate the minimization of the squared ℓ^2-norm

$$\int_{\partial\Omega} \left[\sum_{j=1}^{4} c_j u_j(x, y) - (x^2 + y^2) \right]^2 dx\,dy \tag{12}$$

As a practical test of this approach, we selected 76 equally spaced points in the first-quadrant portion of $\partial\Omega$. The minimization of expression (11) becomes a least-squares matrix problem having 76 equations and 4 unknowns. The result of this computation is

$$c = (1.8216,\ -0.7811,\ -0.04458,\ 0.004052)^T$$

The maximum deviation among the 76 points is 0.049. The 76 points used were $(1, i/25)$ for $0 \leqq i \leqq 50$ and $(i/25, 2)$ for $0 \leqq i \leqq 24$.

Finally, one can attempt the minimization of the quantity

$$\max_{1 \leqq i \leqq m} \left| \sum_{j=1}^{4} c_j u_j(x_i, y_i) - (x_i^2 + y_i^2) \right| \tag{13}$$

This calls for the **minimax solution** of an overdetermined system of m linear equations in four unknowns. For large m, this procedure simulates the minimization of the uniform norm

$$\max_{(x,y)\in\partial\Omega}\left|\sum_{j=1}^{4} c_j u_j(x, y) - (x^2 + y^2)\right| \tag{14}$$

This problem, or its discrete analogue (13), can be solved by the Remez algorithm. (See Section 6.9.) When the expression (13) is minimized using the same set of 76 points as in the least-squares solution, the value of c turns out to be

$$c = (1.8072,\ -0.7950,\ -0.0400,\ 0.003692)^T$$

and the maximum deviation among the 76 points is 0.033. The additional effort expended in obtaining the solution by this more elaborate method is hardly justified by the marginal improvement in accuracy. It would be better to increase the number of basic functions.

Poisson's Equation

Boundary-value problems involving **Poisson's equation** are often encountered. A typical one might be like this:

$$\begin{cases} \nabla^2 w = f & \text{in } \Omega \\ w = g & \text{on } \partial\Omega \end{cases} \tag{15}$$

One way to attack this problem is to split it into two simpler problems; namely,

$$\begin{cases} \nabla^2 v = 0 & \text{in } \Omega \\ v = g & \text{on } \partial\Omega \end{cases} \qquad \begin{cases} \nabla^2 u = f & \text{in } \Omega \\ u = 0 & \text{on } \partial\Omega \end{cases}$$

After u and v have been obtained, the solution to the original problem is $w = u + v$. These two simpler problems have the advantage that each has a homogeneous part. This feature is exploited in the Galerkin method. Thus, in solving the problem involving u, one could select base functions that are 0 on the boundary of Ω. If $u_1, u_2, \ldots, u_n$ have this property, then the same is true for an arbitrary linear combination. Next, we set $u = \sum_{j=1}^{n} c_j u_j$ and try to make the equation $\nabla^2 u = f$ true by choosing the coefficients c_j. This leads to the equation

$$\sum_{j=1}^{n} c_j \nabla^2 u_j = f \tag{16}$$

As is usual in these procedures, no choice of coefficients will make this equation true, and an approximate solution must be sought.

Rayleigh-Ritz Method

Another strategy for solving the problem

$$\begin{cases} \nabla^2 u = f & \text{in } \Omega \\ u = 0 & \text{on } \partial\Omega \end{cases} \tag{17}$$

is the **Rayleigh-Ritz method**. The region Ω is held fixed, and we work in an inner-product space V whose elements are functions u such that u_{xx} and u_{yy} are continuous in Ω and such that $u(x, y) = 0$ on the boundary of Ω. It is assumed that f is continuous in Ω. Therefore, the solution to the problem should be in the space V. The inner product in V is defined by the equation

$$\langle u, v\rangle = \iint_\Omega u(x, y)\, v(x, y)\, dx\, dy \tag{18}$$

THEOREM 1 *The operator $-\nabla^2$ is self-adjoint and positive definite on the inner-product space V.*

Proof The property of being self-adjoint (or *symmetric*) is expressed by the equation

$$\langle -\nabla^2 u, v\rangle = \langle u, -\nabla^2 v\rangle \tag{19}$$

To prove Equation (19), we shall require Green's Theorem in the plane, which states that for decent functions and decent regions,

$$\iint_\Omega (P_x + Q_y)\, dx\, dy = \int_{\partial\Omega} (P\, dy - Q\, dx)$$

Using Green's Theorem, we can compute as follows, exploiting the fact that u and v vanish on $\partial\Omega$:

$$\begin{aligned} \langle \nabla^2 u, v\rangle &= \iint_\Omega (u_{xx} + u_{yy}) v\, dx\, dy \\ &= \iint_\Omega \left[(u_x v)_x + (u_y v)_y - u_x v_x - u_y v_y\right] dx\, dy \\ &= \int_{\partial\Omega} (u_x v\, dy - u_y v\, dx) - \iint_\Omega (u_x v_x + u_y v_y)\, dx\, dy \\ &= -\iint_\Omega (u_x v_x + u_y v_y)\, dx\, dy \end{aligned}$$

The final expression in this calculation involves u and v in a symmetric manner, and therefore

$$\langle \nabla^2 u, v\rangle = \langle u, \nabla^2 v\rangle \tag{20}$$

For the positive definiteness of $-\nabla^2$, we must prove that if $u \in V$ and $u \neq 0$, then

$$\langle -\nabla^2 u, u\rangle > 0$$

The preceding calculation indicates that

$$\langle -\nabla^2 u, u\rangle = \iint_\Omega \left[(u_x)^2 + (u_y)^2\right] dx\, dy \tag{21}$$

The value of this integral is certainly nonnegative, and the only question is whether it can be 0 for a function u that is *not* 0. If the value of the integral is 0, then $u_x = u_y = 0$ in Ω. Hence, u is at the same time a function of y alone and a function of x alone. Thus, u must be constant in Ω. But as a member of V, u must vanish on $\partial\Omega$. Hence, $u = 0$. ■

Since the operator $-\nabla^2$ is positive definite and symmetric on V, a new inner product can be defined by

$$[u, v] \equiv \langle -\nabla^2 u, v\rangle = \iint_\Omega (u_x v_x + u_y v_y)\, dx\, dy \tag{22}$$

This new inner product induces a new norm in the space V via the equation

$$\|u\| = [u, u]^{1/2}$$

In the Rayleigh-Ritz method, we select basic functions $u_1, u_2, \dots, u_n$ in V and try to determine $c_1, c_2, \dots, c_n$ so that $\sum_{j=1}^n c_j u_j$ is *as close as possible to the true solution* in the norm $\|\cdot\|$. Since the machinery of inner-product spaces is available, we know that the *normal* equations determine the coefficients c_j. Thus, if u is the true solution, the orthogonality condition must hold:

$$u - \sum_{j=1}^n c_j u_j \perp u_i \qquad (1 \leqq i \leqq n)$$

This leads at once to the equations

$$\sum_{j=1}^n c_j [u_j, u_i] = [u, u_i] \qquad (1 \leqq i \leqq n)$$

These equations can be simplified by the use of definition (22) and the relation

$$[u, u_i] = \langle -\nabla^2 u, u_i\rangle = \langle -f, u_i\rangle \qquad (1 \leqq i \leqq n)$$

The result is the set of normal equations:

$$\sum_{j=1}^n c_j [u_j, u_i] = -\langle f, u_i\rangle \qquad (1 \leqq i \leqq n) \tag{23}$$

Observe that the unknown function u is not present in the system (23).

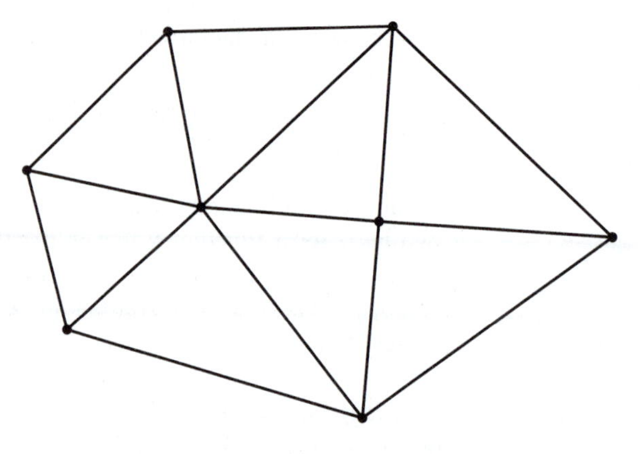

FIGURE 9.5 A proper triangulation

Finite-Element Method

When the Galerkin method employs basic functions that are piecewise polynomials, the procedure is called the **finite-element method**. In one manifestation of this strategy, the region Ω is assumed to be polygonal and is subjected to **triangulation**, as illustrated in Figure 9.5.

A linear function of (x, y) is of the form $ax + by + c$. One such function can be defined on each triangle by assigning the values of the function at the three vertices. The piecewise linear function obtained in this way is continuous because the triangles share sides in any proper triangulation. Thus, we do not permit triangulations such as shown in Figure 9.6 since the piecewise linear functions obtained by assigning arbitrary values at the vertices of each triangle may not be continuous.

The rule for proper triangulations is that any point that is a vertex of some triangle must be a vertex of each triangle to which it belongs. In Figure 9.6, point e is a vertex of $\triangle abe$ and belongs to $\triangle dbc$ but is not a vertex of the latter.

The finite-element method has developed into such a vast subject that we advise readers to consult the specialized literature devoted to it. See, for example, Becker, Carey, and Oden [1981], Mitchell and Wait [1977], Oden [1972], Oden and Reddy [1976], Strang and Fix [1973], Vichnevetsky [1981], Wait and Mitchell [1986], and Zienkiewicz and Morgan [1983].

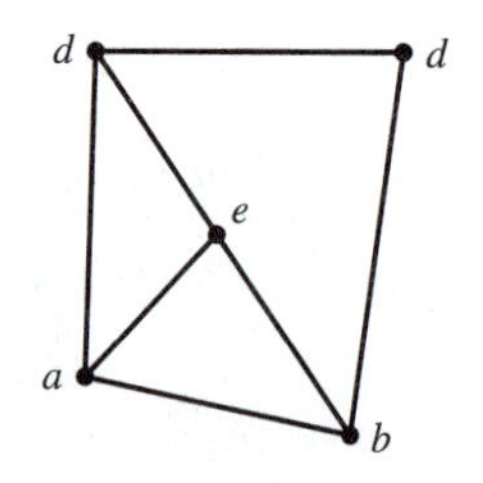

FIGURE 9.6 An improper triangulation

PROBLEMS 9.4

1. Prove that if w is an analytic function of z (with $w = u + iv$ and $z = x + iy$), then u and v are harmonic. *Hint:* Use the Cauchy-Riemann equations.
2. Let $z^n = u_n + iv_n$. Prove that u_n and v_n can be generated recursively by the formulas

$$\begin{cases} u_0 = 1 \quad v_0 = 0 \\ u_{n+1} = xu_n - yv_n \quad v_{n+1} = xv_n + yu_n \end{cases}$$

3. (Continuation) Prove that for even n, u_n and v_n have the properties

$$u_n(-x, y) = u_n(x, -y) = u_n(x, y) \qquad v_n(-x, y) = v_n(x, -y) = -v_n(x, y)$$

4. Prove that the Dirichlet problem defined by Equations (7) and (8) has a solution that satisfies the symmetry condition of Problem 3.
5. The Dirichlet problem for the wave equation is usually not solvable. Show, for example, that there is no function satisfying $u_{xy} = 0$ in the unit square and taking boundary values $u(x, 0) = x$, $u(0, y) = y$, and $u(1, y) = 1$.
6. Prove that if $u_0 + iv_0$ is analytic, then the same is true of $u_n + iv_n$, where these functions are generated by the formulas

$$u_{n+1} = xu_n - yv_n \qquad v_{n+1} = xv_n + yu_n$$

7. Derive the equations for the basic functions $u_4(x, y)$ and $u_5(x, y)$.

COMPUTER PROBLEM 9.4

1. Program the examples given in the text to verify their correctness. (Of course, differences will arise from using computers with different word lengths.)

*9.5 First-Order Partial Differential Equations: Characteristic Curves

As in the theory of ordinary differential equations, higher-order partial differential equations can be replaced by systems of first-order equations. The following examples will show how this can be done.

Systems of First Order

Example 1 Turn the one-dimensional heat equation considered in Section 9.1

$$u_{xx} = u_t \tag{1}$$

into a system of first-order equations.

Solution By introducing the variable $v = u_x$, we obtain an equivalent system:

$$\begin{cases} v_x - u_t = 0 \\ u_x - v = 0 \end{cases} \tag{2}$$

■

Example 2 Show that the three-dimensional heat equation

$$u_{xx} + u_{yy} + u_{zz} = u_t \tag{3}$$

can be treated in the same way as in Example 1.

Solution By introducing variables $u^{(1)} = u$, $u^{(2)} = u_x$, $u^{(3)} = u_y$, and $u^{(4)} = u_z$, we arrive at an equivalent system of first-order equations

$$\begin{cases} u_x^{(2)} + u_y^{(3)} + u_z^{(4)} - u_t^{(1)} = 0 \\ u_x^{(1)} = u^{(2)} \\ u_y^{(1)} = u^{(3)} \\ u_z^{(1)} = u^{(4)} \end{cases} \tag{4}$$

■

Example 3 Repeat Example 1 for the general second-order partial differential equation with two independent variables written as

$$F(x, y, u, u_x, u_y, u_{xx}, u_{xy}, u_{yy}) = 0 \tag{5}$$

Solution This is equivalent to the following system of first-order equations:

$$\begin{cases} F(x, y, u, v, w, v_x, v_y, w_y) = 0 \\ u_x = v \\ u_y = w \end{cases} \tag{6}$$

■

Characteristic Curves

We now take up the concept of **characteristic curves**, or simply **characteristics**, for partial differential equations. One type of characteristic curve for a given equation is a curve on which the solution is constant. To illustrate this possibility, consider the following first-order equation (which is closely related to the second-order wave equation):

$$u_x + cu_y = 0 \tag{7}$$

Let a curve in the xy-plane be given as the graph of the function $y = y(x)$. Along this curve, the values of u are functions of x alone—namely, $u(x, y(x))$. Suppose that the curve is one on which this expression is constant. Then the following condition must be fulfilled:

$$0 = \frac{d}{dx}u\big(x, y(x)\big) = u_x + u_y\frac{dy}{dx} \tag{8}$$

The curve that we are seeking is then a solution of the ordinary differential equation

$$\frac{dy}{dx} = -\frac{u_x}{u_y} \tag{9}$$

In this example, Equation (9) can be simplified by using Equation (7). The result is

$$\frac{dy}{dx} = c \tag{10}$$

The solutions of Equation (10) are straight lines of slope c in the xy-plane. Through any given point (x_0, y_0) in the plane there passes exactly one of these characteristic lines, and its equation is

$$y - y_0 = c(x - x_0) \tag{11}$$

What is the usefulness of such characteristic curves? Suppose that the problem to be solved is not the differential equation (7) by itself but rather a problem with an auxiliary condition; for example,

$$\begin{cases} u_x + cu_y = 0 \\ u(x, 0) = f(x) \end{cases} \tag{12}$$

where f is a given function. To compute the solution at any point (x_0, y_0), we first determine the characteristic curve through this point and then follow this curve to a point of the form $(x, 0)$. At such a point, the solution is known, namely, $u(x, 0) = f(x)$, and along the curve, the solution is constant. Hence, $u(x_0, y_0)$ equals the value of $f(x)$, provided that $(x, 0)$ is on the characteristic curve passing through (x_0, y_0). In our example, these curves are straight lines. Figure 9.7 depicts the situation.

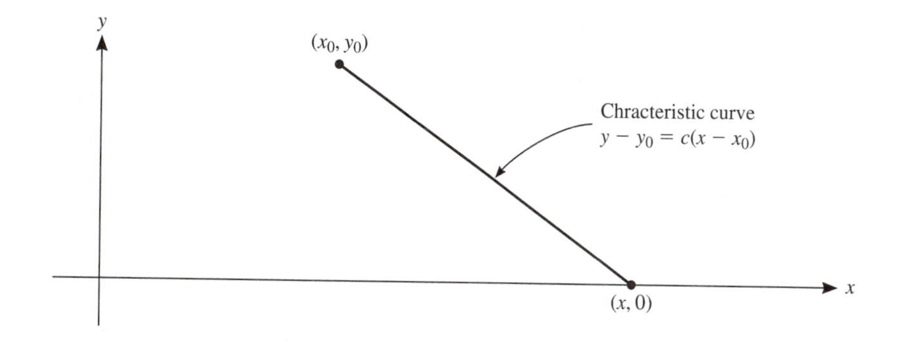

FIGURE 9.7 Characteristic curves: straight lines

The condition that $(x, 0)$ lie on the characteristic curve given by Equation (11) is simply

$$0 - y_0 = c(x - x_0) \tag{13}$$

Thus, $x = x_0 - c^{-1}y_0$ and

$$u(x_0, y_0) = f(x_0 - c^{-1}y_0) \tag{14}$$

Equation (14) gives the solution of the original problem (12). Of course, we can write it simply as $u(x, y) = f(x - c^{-1}y)$. It is readily verified directly that this function satisfies Equation (12).

Here is another example to illustrate the use of characteristic curves.

Example 4 Solve this problem by the method of characteristics:

$$\begin{cases} u_x + yu_y = 0 \\ u(0, y) = f(y) \end{cases} \tag{15}$$

Solution This is a partial differential equation to be solved at all points in the xy-plane, and the values of the solution function are prescribed on the y-axis. If we proceed as before, the ordinary differential equation that describes the characteristic curves will be

$$\frac{dy}{dx} = -\frac{u_x}{u_y} = y \tag{16}$$

The solution passing through (x_0, y_0) is

$$y = y_0 e^{x - x_0} \tag{17}$$

The intersection of this characteristic curve with the y-axis occurs at $y = y_0 e^{-x_0}$. Then we have

$$u(x_0, y_0) = f(y_0 e^{-x_0}) \tag{18}$$

as the solution. Of course, we would usually write this as

$$u(x, y) = f(ye^{-x}) \tag{19}$$

It is easily verified directly that this is the solution of the problem in Equation (15). ■

The reader is invited to try some of the problems—for example, Problem 1, which is elementary.

General Theory of Characteristic Curves

The general theory of characteristic curves should encompass more general equations and not treat the variables x and y in an unsymmetrical way. Furthermore, we do not expect the solution to be *constant* along a characteristic curve; we ask only that it satisfy some *ordinary* differential equation along a characteristic.

Consider the equation

$$au_x + bu_y = c \tag{20}$$

in which we allow a, b, and c to be functions of x, y, and u. Suppose that the value of the solution is known at one point (x_0, y_0). This could be the result of prior calculations or a *boundary value* that had been prescribed. We shall indicate now how additional solution values of Equation (20) can be obtained by integrating some *ordinary* differential equations. Consider this system of three ordinary differential equations with initial condition:

$$\begin{cases} \dfrac{dx}{ds} = a & \dfrac{dy}{ds} = b & \dfrac{du}{ds} = c \\ x(0) = x_0 & y(0) = y_0 & u(0) = u_0 = u(x_0, y_0) \end{cases} \tag{21}$$

To produce a table of $x(s)$, $y(s)$, $u(s)$ starting at $s = 0$, one can integrate this system numerically. It is not hard to verify that this table will provide solution points of the *partial* differential equation (20). Indeed,

$$\begin{aligned} a\big(x(s), y(s), u(s)\big)u_x\big(x(s), y(s)\big) &+ b\big(x(s), y(s), u(s)\big)u_y\big(x(s), y(s)\big) \\ &= x'(s)u_x\big(x(s), y(s)\big) + y'(s)u_y\big(x(s), y(s)\big) \\ &= \frac{d}{ds}u\big(x(s), y(s)\big) = c\big(x(s), y(s), u(s)\big) \end{aligned}$$

(In this calculation, primes indicate differentiation with respect to s.)

If the values of $u(x, y)$ have been prescribed along some curve Γ (not a characteristic curve!), then in principle one can start at any point of Γ, integrate System (21), and obtain additional values of $u(x, y)$ along arcs. (See Figure 9.8 for help in understanding this idea.) The arcs obtained in this process are characteristic curves for Equation (20).

Example 5 Determine the ordinary differential equations that define the characteristic curves belonging to this equation:

$$\sin(x^2 + y^2)u_x + (3x + y^2)u_y = e^{xy}$$

Solution According to the general theory, the system should be

$$\begin{cases} x' = \sin(x^2 + y^2) \\ y' = 3x + y^2 \\ u' = e^{xy} \end{cases} \tag{22}$$

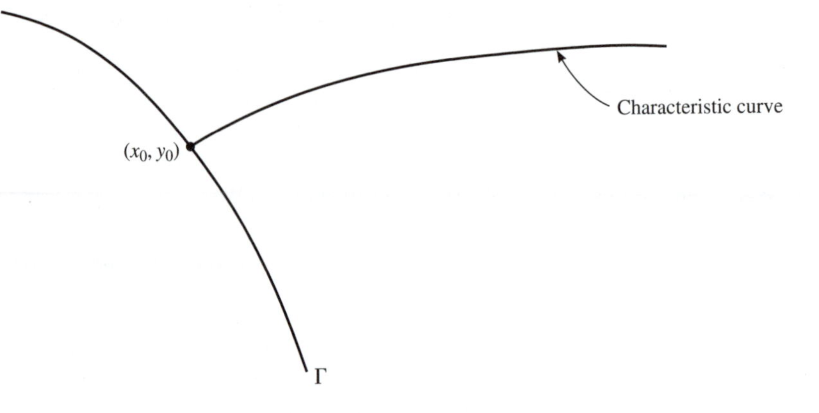

FIGURE 9.8 Characteristic curves: arcs

Since a and b do not involve u in this example, the characteristic curves (in the xy-plane) can be obtained by integrating only the first two equations in (22). ■

Example 6 Consider the partial differential equation

$$6u_x + xu_y = y$$

Given that $u(3, 3) = 4$, what value of $u(15, 21)$ is obtained by following a characteristic curve?

Solution The characteristic curve through the initial point is governed by the initial-value problem

$$\begin{cases} x' = 6 & y' = x & u' = y \\ x(0) = 3 & y(0) = 3 & u(0) = 4 \end{cases}$$

Integrating these equations, we obtain

$$\begin{cases} x = 6s + 3 \\ y = 3s^2 + 3s + 3 \\ u = s^3 + \frac{3}{2}s^2 + 3s + 4 \end{cases}$$

Letting $s = 2$, we arrive at $x = 15$, $y = 21$, $u = 24$. ■

Example 7 Use the method of characteristics to find a solution of the boundary-value problem

$$\begin{cases} 6u_x + xu_y = y \\ u(x, y) = 4 \text{ when } x = y \end{cases}$$

Solution Let us find the characteristic curves passing through the point $(x, y) = (r, r)$, $r \leqq 6$. We solve the system

$$\begin{cases} x' = 6 \\ x(0) = r \end{cases} \quad \begin{matrix} y' = x \\ y(0) = r \end{matrix} \quad \begin{matrix} u' = y \\ u(0) = 4 \end{matrix}$$

and obtain

$$\begin{cases} x = 6s + r \\ y = 3s^2 + rs + r \\ u = s^3 + \frac{1}{2}rs^2 + rs + 4 \end{cases} \tag{23}$$

If (x, y) is a given point in the plane, we try to determine a corresponding (r, s) using the first two equations in (23). The result of this calculation is

$$\begin{cases} r = 6 - \sqrt{(x-6)^2 + 12(x-y)} \\ s = (x-r)/6 \end{cases}$$

With the values of r and s at hand, the value of u at (x, y) can be computed from the third equation in (23). Thus, for example, if $(x, y) = (15, 21)$, then $(r, s) = (3, 2)$ and $u = 24$, as in Example 6. ■

Example 8 Solve the boundary-value problem

$$\begin{cases} xu_x + yuu_y = xy \\ u(x, y) = 2xy \text{ when } xy = 3 \end{cases} \tag{24}$$

by the method of characteristics.

Solution The equations for the characteristics are

$$\begin{cases} x' = x \\ x(0) = x_0 \end{cases} \quad \begin{cases} y' = uy \\ y(0) = 3/x_0 \end{cases} \quad \begin{cases} u' = xy \\ u(0) = 6 \end{cases}$$

Since

$$(xy)' = x'y + xy' = xy + xuy = xy(1+u) = u'(1+u)$$

an integration with respect to s can be carried out, yielding

$$xy = u + \frac{1}{2}u^2 + c \tag{25}$$

By imposing the initial conditions, we find that $c = -21$. Equation (25) can be solved for u, with the result

$$u = -1 + \sqrt{43 + 2xy}$$

■

In the foregoing examples, there is another way of interpreting our work. Let us use r as a parameter to describe points on the data curve Γ. As before, s is the parameter that describes points on any one of the characteristic curves. We assume that $s = 0$ for points on Γ. (See Figure 9.9.) The point marked $\otimes$ is described by $r = 4$ and $s = 2$.

In the same way, other points in the plane can be described in coordinates (r, s). It is not necessarily true that *every* point in the plane can be assigned values of r and s. Furthermore, it can happen that characteristic curves corresponding to two different values of r can meet. In that case, the intersection point will not have a unique (r, s)-description. If the data curve Γ is not a characteristic curve, and if the coefficient functions a, b, and c are smooth, then points near Γ will have unique coordinates (r, s).

Using r and s in the way just outlined, we can describe our work as follows: The functions $(r, s) \longmapsto x(r, s)$, $y(r, s)$, and $u(r, s)$ are determined by the differential equations

$$\begin{cases} \dfrac{\partial x}{\partial s} = a & \dfrac{\partial y}{\partial s} = b & \dfrac{\partial u}{\partial s} = c \\ x(r, 0) = f(r) & y(r, 0) = g(r) & u(r, 0) = h(r) \end{cases}$$

It is assumed that Γ is given parametrically by

$$x = f(r) \qquad y = g(r)$$

The integration of the three differential equations produces three functions that represent a surface parametrically:

$$x = x(r, s) \qquad y = y(r, s) \qquad u = u(r, s)$$

The surface obviously contains the space curve

$$\Gamma^* \;:\; x = f(r) \qquad y = g(r) \qquad u = h(r)$$

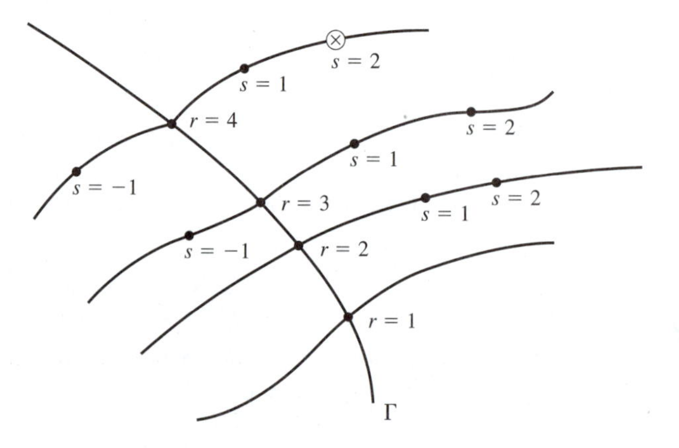

FIGURE 9.9 Data curve Γ

Thus, we have found a surface that solves the original partial differential equation and contains a given space curve, Γ^*. Notice that Γ is the projection of Γ^* onto the xy-plane.

PROBLEMS 9.5

1. Use the method of characteristics to solve these problems:

(a) $\begin{cases} u_x + xu_y = 0 \\ u(0, y) = f(y) \end{cases}$

(b) $\begin{cases} u_x + 2uu_y = 0 \\ u(0, y) = f(y) \end{cases}$

(c) $\begin{cases} xu_x + 2yu_y = 0 \\ u(1, y) = f(y) \end{cases}$

2. Find the value $u(17, 3)$ given that

$$\begin{cases} u_x + yu_y = 0 \\ u(18, 3e) = k\pi/2 \end{cases}$$

Hint: The example in the text will be useful.

3. Verify that the function obtained in the solution of Example 8 solves the given problem.

4. Find the solution of the differential equation in Example 6, given that $u = e^x \sin y$ on the curve $y = x^3$. Explain how the numerical difficulties can be surmounted. Find $u(7, 5)$ by making use of the fact that the point $(7, 5)$ is on a characteristic curve through $(1, 1)$.

5. Explain, in the light of the general theory, why the solution functions for Equations (7) and (15) are *constant* along the characteristic curves.

6. Show that the solution found in Example 7 is not unique. *Hint:* In solving for (r, s) when (x, y) is given, there are two solutions of the quadratic equations.

7. Solve by the method of characteristics:

$$\begin{cases} u_x + u_y = u^2 \\ u(x, y) = y \text{ on line } x + y = 0 \end{cases}$$

8. Solve this boundary-value problem by the method of characteristics, and account for any difficulties:

$$\begin{cases} u_x + 2u_y = u \\ u = 1 \text{ when } y = 2x \end{cases}$$

9. Solve by the method of characteristics:

$$\begin{cases} uu_x + u_y = 1 \\ u = r \text{ on the curve } x = r^2, \quad y = 2r \end{cases}$$

10. Solve by the method of characteristics:

$$\begin{cases} u_x + 2u_y = y \\ u = r^2 \text{ on circle } x = \cos r, \quad y = \sin r \end{cases}$$

11. Consider the partial differential equation $au_x + bu_y = 0$, in which a and b are functions of x only. Find the equation of the curve passing through (x_0, y_0) on which the values of u remain constant.

---

*9.6 Quasilinear Second-Order Equations: Characteristics

In this section, we discuss the equation

$$au_{xx} + bu_{xy} + cu_{yy} + e = 0 \tag{1}$$

We permit a, b, c, and e to be functions of x, y, u, u_x, and u_y. Such an equation is said to be **quasilinear**. As in our study of characteristic curves for first-order equations, we ask here how the solution function behaves along a curve in the xy-plane.

Characteristic Curves

Let C be a curve in the xy-plane given parametrically by

$$x = x(s) \qquad y = y(s) \qquad (s \in \mathbb{R})$$

The traditional notation $p = u_x$ and $q = u_y$ will simplify our work. We regard x, y, p, q, and u indirectly as functions of s. By direct differentiation, we have

$$\frac{dp}{ds} = \frac{\partial p}{\partial x}\frac{dx}{ds} + \frac{\partial p}{\partial y}\frac{dy}{ds} = u_{xx}x' + u_{xy}y' \tag{2}$$

$$\frac{dq}{ds} = \frac{\partial q}{\partial x}\frac{dx}{ds} + \frac{\partial q}{\partial y}\frac{dy}{ds} = u_{xy}x' + u_{yy}y' \tag{3}$$

From Equation (2), we can solve for u_{xx}, getting

$$u_{xx} = \left(\frac{dp}{ds} - u_{xy}y'\right) \Big/ x' = \frac{dp}{ds}\frac{ds}{dx} - u_{xy}\frac{dy}{ds}\frac{ds}{dx} = \frac{dp}{dx} - u_{xy}\frac{dy}{dx} \tag{4}$$

Similarly from Equation (3), we have

$$u_{yy} = \left(\frac{dq}{ds} - u_{xy}x'\right) \Big/ y' = \frac{dq}{ds}\frac{ds}{dy} - u_{xy}\frac{dx}{ds}\frac{ds}{dy} = \frac{dq}{dy} - u_{xy}\frac{dx}{dy} \tag{5}$$

These equations, (4) and (5), are valid along the curve C. If the expressions just derived are substituted in Equation (1), the result is

$$a\left(\frac{dp}{dx} - u_{xy}\frac{dy}{dx}\right) + bu_{xy} + c\left(\frac{dq}{dy} - u_{xy}\frac{dx}{dy}\right) + e = 0 \qquad \textbf{(6)}$$

When Equation (6) is multiplied by dy/dx, the result is

$$a\left[\frac{dp}{dx}\frac{dy}{dx} - u_{xy}\left(\frac{dy}{dx}\right)^2\right] + bu_{xy}\frac{dy}{dx} + c\left[\frac{dq}{dx} - u_{xy}\right] + e\frac{dy}{dx} = 0 \qquad \textbf{(7)}$$

Collecting terms in (7), we have

$$-u_{xy}\left[a\left(\frac{dy}{dx}\right)^2 - b\frac{dy}{dx} + c\right] + a\frac{dp}{dx}\frac{dy}{dx} + c\frac{dq}{dx} + e\frac{dy}{dx} = 0 \qquad \textbf{(8)}$$

The curve C, unspecified until now, is to be chosen so that the term u_{xy} disappears from Equation (8). Thus, C is described by the differential equation

$$a\left(\frac{dy}{dx}\right)^2 - b\frac{dy}{dx} + c = 0 \qquad \textbf{(9)}$$

Such a curve C is called a **characteristic curve** of the differential Equation (1).

Classification

Since Equation (9) is a quadratic equation in dy/dx, the nature of the characteristic curves is determined by the **discriminant**

$$\Delta \equiv b^2 - 4ac$$

If $\Delta > 0$ at a certain value (x, y, u), then the differential equation is said to be **hyperbolic** there. If $\Delta = 0$, the equation is said to be **parabolic**, and if $\Delta < 0$, it is called **elliptic**. This classification can vary from point to point in the xy-plane, and it can depend also on the solution u, since a, b, and c are permitted to depend on x, y, u, u_x, and u_y. In the **linear case**, the coefficient functions a, b, and c depend only on x and y, and the classification becomes simpler.

This classification of equations is nicely illustrated by three familiar equations that are prototypical:

Name	**Form**	a	b	c	Δ	**Type**
Heat	$u_{xx} - u_y = 0$	1	0	0	0	Parabolic
Wave	$u_{xx} - u_{yy} = 0$	1	0	−1	4	Hyperbolic
Laplace	$u_{xx} + u_{yy} = 0$	1	0	1	−4	Elliptic

Example 1 Classify the following equation as to its type:

$$(x + y)u_{xx} + (1 + x^2)u_{yy} = 0$$

Solution The discriminant of this equation is

$$\Delta(x, y) = -4(x + y)(1 + x^2)$$

On the domain where $x + y > 0$, the equation is elliptic. On the domain where $x + y < 0$, the equation is hyperbolic, and on the line $x + y = 0$, it is parabolic. ■

Example 2 Find the characteristic curves belonging to the equation

$$yu_{xx} + (x + y^2)u_{xy} + xyu_{yy} = 0 \tag{10}$$

Solution The differential equation describing the characteristic curves is Equation (9). In this example, it is

$$y\left(\frac{dy}{dx}\right)^2 - (x + y^2)\frac{dy}{dx} + xy = 0 \tag{11}$$

The discriminant is

$$\Delta = b^2 - 4ac = (x + y^2)^2 - 4xy^2 = (x - y^2)^2$$

Since $\Delta > 0$ (except on the curve $x = y^2$), the partial differential equation (10) is hyperbolic, and there are two ordinary differential equations to solve for the characteristic curves; namely,

$$\frac{dy}{dx} = \frac{b \pm \sqrt{\Delta}}{2a} = \frac{(x + y^2) \pm |x - y^2|}{2y} \tag{12}$$

Suppose $x > y^2$. The first equation, corresponding to the $+$ sign, is

$$\frac{dy}{dx} = \frac{x}{y}$$

whose solution is the family of hyperbolas $y^2 - x^2 = \alpha$. The second equation, corresponding to the $-$ sign, is

$$\frac{dy}{dx} = y$$

whose solution is the family of exponential curves $y = \beta e^x$. When $x < y^2$, these two cases are reversed. When $x = y^2$, Equation (10) is parabolic and the characteristic curve is again the solution of the ordinary differential equation $dy/dx = y$. ■

Algorithm

Now let us return to Equation (8), and assume that the curve C is a characteristic curve. On C, the slope function dy/dx obeys Equation (9), and so Equation (8) simplifies to

$$a\frac{dp}{dx}\frac{dy}{dx} + c\frac{dq}{dx} + e\frac{dy}{dx} = 0 \tag{13}$$

We shall now indicate how, in the case of a hyperbolic equation, Equation (13) can be used locally to compute a solution numerically. In this case, there will be two characteristic curves through a given point (x_0, y_0). Consider the situation shown in Figure 9.10, where two characteristic curves pass through point A.

To simplify the notation, we introduce functions v_1 and v_2 that arise from solving the quadratic equation (9):

$$\begin{cases} v_1 = (b + \sqrt{\Delta}\,)/(2a) \\ v_2 = (b - \sqrt{\Delta}\,)/(2a) \end{cases} \tag{14}$$

The two characteristic curves are then given by the differential equations

$$\frac{dy}{dx} = v_1 \qquad \frac{dy}{dx} = v_2 \tag{15}$$

Equation (13) can be simplified by using the fact that the functions v_1 and v_2 of Equation (14) satisfy these relationships:

$$v_1 + v_2 = b/a \qquad v_1 v_2 = c/a \tag{16}$$

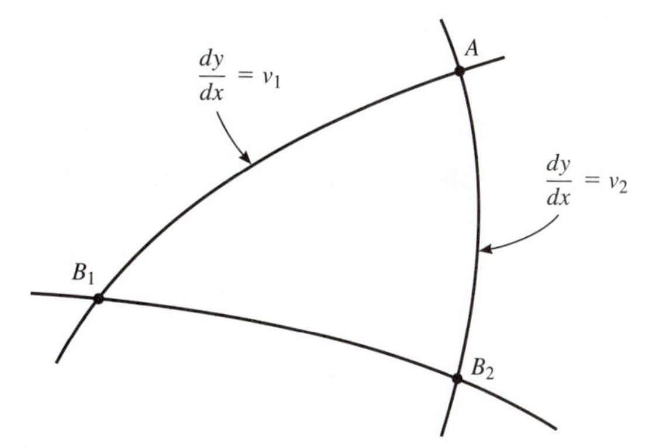

FIGURE 9.10 Characteristic curves AB_1 and AB_2

The simpler version of Equation (13), written for both characteristic curves, is

$$\begin{cases} \dfrac{dp}{dx} + v_1 \dfrac{dq}{dx} = -e/a & \text{when} \quad \dfrac{dy}{dx} = v_2 \\ \dfrac{dp}{dx} + v_2 \dfrac{dq}{dx} = -e/a & \text{when} \quad \dfrac{dy}{dx} = v_1 \end{cases} \tag{17}$$

To solve Equations (17), a finite-difference method can be used. The equations to be solved are discretized versions of Equations (15):

$$\frac{y(A) - y(B_1)}{x(A) - x(B_1)} = \frac{v_1(A) + v_1(B_1)}{2} \tag{18}$$

$$\frac{y(A) - y(B_2)}{x(A) - x(B_2)} = \frac{v_2(A) + v_2(B_2)}{2} \tag{19}$$

and discretized versions of Equations (17):

$$\frac{p(A) - p(B_2)}{x(A) - x(B_2)} + \left[\frac{v_1(A) + v_1(B_2)}{2}\right]\left[\frac{q(A) - q(B_2)}{x(A) - x(B_2)}\right] = -\frac{1}{2}\left[(e/a)(A) + (e/a)(B_2)\right] \tag{20}$$

$$\frac{p(A) - p(B_1)}{x(A) - x(B_1)} + \left[\frac{v_2(A) + v_2(B_1)}{2}\right]\left[\frac{q(A) - q(B_1)}{x(A) - x(B_1)}\right] = -\frac{1}{2}\left[(e/a)(A) + (e/a)(B_1)\right] \tag{21}$$

Here A and B_1 are points on the characteristic curve $dy/dx = v_1$, and A and B_2 are points on the characteristic curve $dy/dx = v_2$. The formula for computing $u(A)$ is

$$u(A) = u(B_1) + \left[\frac{p(A) + p(B_1)}{2}\right][x(A) - x(B_1)] + \left[\frac{q(A) + q(B_1)}{2}\right][y(A) - y(B_1)] \tag{22}$$

This is the finite-difference analogue of the equation

$$du = u_x\,dx + u_y\,dy = p\,dx + q\,dy$$

since $p = u_x$ and $q = u_y$. Observe that in Equations (18)–(22), average values of functions along various arcs are systematically used.

Suppose now that we know x, y, u, p, and q at the two points B_1 and B_2. Then Equations (18)–(22) can be interpreted as equations from which the *new* values $x(A)$, $y(A)$, $u(A)$, $p(A)$, and $q(A)$ are to be determined. This itself could be a source of difficulty because the equations are nonlinear. In numerical practice, these would usually be solved by iteration. Here is the outline of such a procedure:

(i) Start with a guess as to the correct values of $x(A)$, $y(A)$, $u(A)$, $p(A)$, $q(A)$. These guesses can be simple perturbations of the most recent values of these functions, say at B_1.
(ii) Compute $v_1(A)$, $v_2(A)$, and $(e/a)(A)$ using the most recent values of $x(A)$, $y(A)$, $p(A)$, $q(A)$, $u(A)$.
(iii) Use Equations (18) and (19) to recompute $y(A)$ and $x(A)$. Then use Equations (20) and (21) to recompute $p(A)$ and $q(A)$. Use Equation (22) to recompute $u(A)$. Return to step **(ii)** if the new values differ substantially from the old ones.

In step **(i)**, initial guesses for the quantities being computed can be easily obtained by using less accurate forms of Equations (18)–(21). Thus, we could replace the average values appearing in these equations by values at B_1 and B_2. When this is done, Equations (18) and (19) are linear and quickly solved to produce initial values for the iteration. This work results in the following formulas:

$$\begin{aligned}
x(A) &= \frac{y(B_2) - y(B_1) + x(B_1)v_1(B_1) - x(B_2)v_2(B_2)}{v_1(B_1) - v_2(B_2)} \\
y(A) &= y(B_1) + v_1(B_1)\left[x(A) - x(B_1)\right] \\
R &= p(B_2) - p(B_1) + v_1(B_2)q(B_2) - v_2(B_1)q(B_1) \\
S &= (e/a)(B_2)\left[x(A) - x(B_2)\right] - (e/a)(B_1)\left[x(A) - x(B_1)\right] \\
q(A) &= (R - S)/\left[v_1(B_2) - v_2(B_1)\right] \\
p(A) &= p(B_2) - (e/a)(B_2)\left[x(A) - x(B_2)\right] - v_1(B_2)\left[q(A) - q(B_2)\right]
\end{aligned}$$

If the quasilinear equation (1) is in fact *linear*, then a, b, c, Δ, v_1, and v_2 are functions of x and y only. The function u and its derivatives will *not* appear in the preceding six functions. In this case, Equations (18) and (19) can be solved together for $x(A)$ and $y(A)$. After that, Equations (20) and (21) can be solved together for $p(A)$ and $q(A)$.

Example 3 Write an informal code to solve this boundary-value problem by the method of characteristics:

$$\begin{cases} u_{xx} - 4u_{yy} - u_y = 0 \\ u(x,0) = f(x) \quad u_y(x,0) = g(x) \qquad (0 \leqq x \leqq 1) \end{cases}$$

Solution We select n equally spaced points x_j in the interval $[0, 1]$, and compute the numerical solution at intersection points of characteristic curves originating at the x_j. Equations (14) give us $v_1 = 2$ and $v_2 = -2$, since $\Delta = 16$. The differential equations of the characteristic curves are $dy/dx = 2$ and $dy/dx = -2$. These *curves* are straight lines, as shown in Figure 9.11, where $n = 8$ is used. In this example, Equations

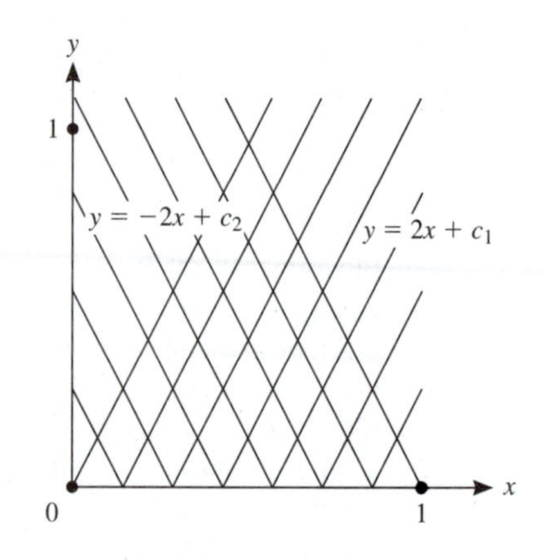

FIGURE 9.11 Characteristic curves in Example 3

(17) take the following form:

$$\begin{cases} \dfrac{dp}{dx} + 2\dfrac{dq}{dx} - q = 0 & \text{when } \dfrac{dy}{dx} = -2 \\ \dfrac{dp}{dx} - 2\dfrac{dq}{dx} - q = 0 & \text{when } \dfrac{dy}{dx} = 2 \end{cases}$$

since $q = u_y = -e$. Now let $h = 1/(n-1)$. This is the spacing between the discrete points x_j on the interval [0, 1]. Referring to Figure 9.12, we see that if $B_1 = (x, y)$, then $B_2 = (x + h, y)$ and $A = (x + \frac{h}{2}, y + h)$. Thus, we can dispense with Equations (18) and (19). We use $e = -u_y = -q$ in Equations (20) and (21). After simplification, these become

$$\begin{cases} p(A) + \left(2 + \dfrac{h}{4}\right) q(A) = p(B_2) + \left(2 - \dfrac{h}{4}\right) q(B_2) \\ p(A) - \left(2 + \dfrac{h}{4}\right) q(A) = p(B_1) - \left(2 - \dfrac{h}{4}\right) q(B_1) \end{cases}$$

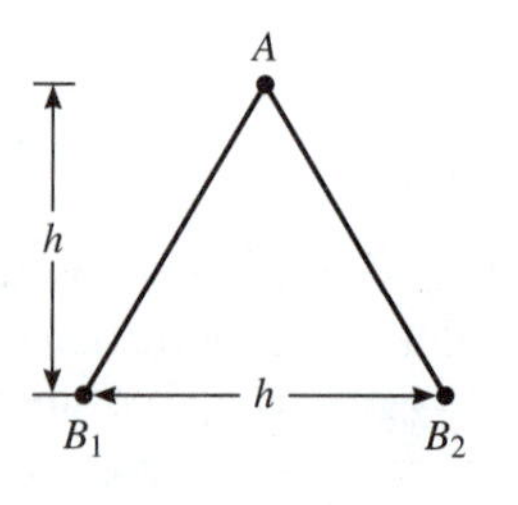

FIGURE 9.12

The solution of this pair of linear equations is

$$p(A) = \frac{1}{2}\left[p(B_1) + p(B_2)\right] + \left(1 - \frac{h}{8}\right)\left[q(B_2) - q(B_1)\right]$$

$$q(A) = \left\{p(B_2) - p(B_1) + \left(2 - \frac{h}{4}\right)\left[q(B_2) + q(B_1)\right]\right\} \Big/ \left(4 + \frac{h}{2}\right)$$

Formula (22), suitably simplified, reads

$$u(A) = u(B_1) + \frac{h}{4}\left[p(A) + p(B_1)\right] + \frac{h}{2}\left[q(A) + q(B_1)\right]$$

An algorithm for this procedure follows:

input n
$h \leftarrow 1/(n-1)$
for $j = 1$ **to** n **do**
  $x_j \leftarrow (j-1)h$
  $y_j \leftarrow 0$
  $u_j \leftarrow f(x_j)$
  $p_j \leftarrow f'(x_j)$
  $q_j \leftarrow g(x_j)$
  output x_j, y_j, p_j, q_j, u_j
end do
for $i = 1$ **to** n **do**
  $y_i \leftarrow ih$
  for $j = 1$ **to** $n - i$ **do**
    $x_j \leftarrow x_j + h/2$
    $\tilde{p} \leftarrow [p_j + p_{j+1}]/2 + (1 - h/8)[q_{j+1} - q_j]$
    $\tilde{q} \leftarrow \left[p_{j+1} - p_j + (2 - h/4)(q_{j+1} + q_j)\right]/(4 + h/2)$
    $u_j \leftarrow u_j + h(\tilde{p} + p_j)/4 + h(\tilde{q} + q_j)/2$
    $p_j \leftarrow \tilde{p}$
    $q_j \leftarrow \tilde{q}$
    output x_j, y_j, p_j, q_j, u_j
  end do
end do

■

An Alternative Approach to Characteristics

There is another approach to the theory of characteristics; its starting point is the question of whether a change of variables from (x, y) to (ξ, η), say, can lead to any simplification in the differential equation (1). Suppose, then, that new variables are introduced with equations

$$\xi = \xi(x, y) \qquad \eta = \eta(x, y) \tag{23}$$

The effect of this change in the differential equation

$$au_{xx} + bu_{xy} + cu_{yy} + e = 0 \tag{24}$$

can be determined by first computing

$$\begin{aligned}
u_x &= u_\xi \xi_x + u_\eta \eta_x \\
u_y &= u_\xi \xi_y + u_\eta \eta_y \\
u_{xx} &= u_\xi \xi_{xx} + u_{\xi\xi} \xi_x^2 + u_{\xi\eta} \eta_x \xi_x + u_\eta \eta_{xx} + u_{\eta\xi} \xi_x \eta_x + u_{\eta\eta} \eta_x^2 \\
u_{xy} &= u_\xi \xi_{xy} + u_{\xi\xi} \xi_y \xi_x + u_{\xi\eta} \eta_y \xi_x + u_\eta \eta_{xy} + u_{\eta\xi} \xi_y \eta_x + u_{\eta\eta} \eta_y \eta_x \\
u_{yy} &= u_\xi \xi_{yy} + u_{\xi\xi} \xi_y^2 + u_{\xi\eta} \eta_y \xi_y + u_\eta \eta_{yy} + u_{\eta\xi} \xi_y \eta_y + u_{\eta\eta} \eta_y^2
\end{aligned}$$

Substituting these expressions in Equation (24), we obtain

$$\begin{aligned}
u_{\xi\xi}[a\xi_x^2 + b\xi_x\xi_y + c\xi_y^2] &+ u_{\xi\eta}[2a\xi_x\eta_x + b\xi_x\eta_y + b\xi_y\eta_x + 2c\xi_y\eta_y] \\
&+ u_{\eta\eta}[a\eta_x^2 + b\eta_y\eta_x + c\eta_y^2] + f = 0
\end{aligned} \tag{25}$$

In a, b, c, and e, we must also make the appropriate substitutions. All the second derivative terms are shown, and f contains all other terms. (See Problem 8.) Since we are interested here in the hyperbolic case, the quadratic equation

$$a\lambda^2 + b\lambda + c = 0$$

will be assumed to have two real roots—namely, $\left(-b \pm \sqrt{\Delta}\,\right)/(2a)$, where the discriminant, $\Delta = b^2 - 4ac$, is positive. If we choose our coordinate transformations so that

$$\begin{aligned}
\xi_x/\xi_y &= \left(-b + \sqrt{\Delta}\,\right)/(2a) \\
\eta_x/\eta_y &= \left(-b - \sqrt{\Delta}\,\right)/(2a)
\end{aligned}$$

then the differential equation (25) simplifies to

$$u_{\xi\eta}\left[2a\xi_x\eta_x + b(\xi_x\eta_y + \xi_y\eta_x) + 2c\xi_y\eta_y\right] + f = 0 \tag{26}$$

This equation is the **canonical form** of a hyperbolic equation. The curves described by setting

$$\xi(x, y) = \text{const.} \qquad \eta(x, y) = \text{const.} \tag{27}$$

are the **characteristic curves**. These curves solve the same differential equations that were obtained previously in Equation (9), as is easily verified.

Example 4 Find the region in which the equation

$$u_{xx} - yu_{yy} = 0 \tag{28}$$

is hyperbolic, and determine a change of variables that reduces the equation to canonical form.

Solution The discriminant is $b^2 - 4ac = 4y$, and Equation (28) is therefore hyperbolic in the upper half-plane. To find the change of variables, we solve the differential equations

$$\xi_x/\xi_y = \left(-b + \sqrt{b^2 - 4ac}\right)/(2a) = y^{1/2}$$

$$\eta_x/\eta_y = \left(-b - \sqrt{b^2 - 4ac}\right)/(2a) = -y^{1/2}$$

As solutions, we can take

$$\xi = x + 2y^{1/2} \qquad \eta = x - 2y^{1/2}$$

In terms of the new variables, Equation (28) becomes

$$4u_{\xi\eta} + \frac{2}{\xi - \eta}(u_\xi - u_\eta) = 0$$

■

PROBLEMS 9.6

1. Show that if $b = 0$ and $ac < 0$, then the differential equation (1) is hyperbolic. Show that its characteristic curves are solutions of $dy/dx = \sqrt{-c/a}$.
2. Prove that if $c = -a$ and $b^2 > 4a^2$, then the differential equation (1) is hyperbolic, and at each point of the xy-plane the two characteristic curves through the point are perpendicular to each other.
3. Find a second-order partial differential equation whose characteristic curves are $x \cos \alpha + y \sin \alpha = 0$ and $x^2 + y^2 = \beta$.
4. Classify these equations as hyperbolic, parabolic, or elliptic:
   (a) $yu_{xx} + xu_{xy} + u_{yy} + u + u_x = 0$
   (b) $xyu_{xy} + e^x u_x + yu_y = 0$
   (c) $3u_{xx} + u_{xy} + u_{yy} + 2yu + 7 = 0$
5. Find the region where the equation $xy^2 u_{xx} = u_{yy}$ is hyperbolic. Determine its canonical form and the simplifying change of variables.
6. Verify that the curves along which $\xi(x, y)$ or $\eta(x, y)$ remains constant are characteristic curves as defined in Equation (9).
7. After making the change of variables in Equation (23), Equation (24) becomes Equation (25). The latter has the form

$$\alpha u_{\xi\xi} + \beta u_{\xi\eta} + \gamma u_{\eta\eta} + e = 0$$

Prove that

$$\beta^2 - 4\alpha\gamma = (b^2 - 4ac)J^2$$

where J is the Jacobian of the transformation:

$$J = \begin{vmatrix} \xi_x & \xi_y \\ \eta_x & \eta_y \end{vmatrix}$$

Finally, conclude that if the Jacobian does not vanish, then the type of our differential equation does not change.

8. In Equation (25) show that

$$f = e + au_\xi \xi_{xx} + au_\eta \eta_{xx} + bu_\xi \xi_{xy} + bu_\eta \eta_{xy} + cu_\xi \xi_{yy} + cu_\eta \eta_{yy}$$

9. Show that the canonical form of the differential equation in Example 4 is

$$16u_{\xi\eta} + u_\xi + u_\eta = 0$$

10. Fill in the details of the derivation of the formulas just before Example 3.

11. Work out the solutions given in Example 3.

12. Verify the last equation in Example 4.

COMPUTER PROBLEM 9.6

1. Write a general code to solve this hyperbolic problem, using the method of characteristics:

$$\begin{cases} au_{xx} + bu_{yy} + cu_x + du_y = 0 \\ u(0, y) = f(y) \quad u_x(0, y) = g(y) \qquad (0 \leqq y \leqq 1) \end{cases}$$

Assume that a, b, c, and d are constants satisfying $ab < 0$. Take equally spaced points on the y-axis in the interval $0 \leqq y \leqq 1$.

---

*9.7 Other Methods for Hyperbolic Problems

We shall develop first some finite-difference methods for hyperbolic systems of first-order linear partial differential equations. We start with a single equation accompanied by an initial condition:

$$\begin{cases} u_t = \alpha u_x \\ u(x, 0) = f(x) \qquad (-\infty < x < \infty) \end{cases} \tag{1}$$

Here α is a (real) constant. Since the solution is known on the line $t = 0$, it is natural to attempt a **marching method** to extend the solution to additional lines $t = k$, $t = 2k$, and so on, where k is the step size adopted for the t-variable.

Lax-Wendroff Method

Assume that the solution u is continuously differentiable any number of times in both variables x and t. Then by Taylor's Theorem,

$$u(x, t + k) = u + ku_t + \frac{k^2}{2!} u_{tt} + \frac{k^3}{3!} u_{ttt} + \cdots \tag{2}$$

On the right side of this equation, it is understood that u and its derivatives will be evaluated at the *base point* (x, t). Also, the series should be terminated at some stage, and an appropriate error term supplied.

Since the function u is to solve (1), we shall have $u_t = \alpha u_x$, $u_{tt} = \alpha^2 u_{xx}$, and so on. (Problem 1 asks for a formal proof of this.) Thus in Equation (2), we can write

$$u(x, t + k) = u + k\alpha u_x + \frac{(k\alpha)^2}{2!} u_{xx} + \frac{(k\alpha)^3}{3!} u_{xxx} + \cdots \tag{3}$$

If we wish to design a numerical procedure having second-order accuracy in the step size, the series in (3) can be truncated with the term u_{xx}. Then the derivatives on the right in Equation (3) can be replaced by second-order finite-difference approximations. Let h be the step size adopted for the x-variable. Then the formula for advancing the solution is

$$v(x, t + k) = v(x, t) + k\alpha \left[\frac{v(x + h, t) - v(x - h, t)}{2h} \right] + \frac{k^2\alpha^2}{2} \left[\frac{v(x + h, t) - 2v(x, t) + v(x - h, t)}{h^2} \right]$$

so that

$$v(x, t + k) = (2s^2 + s)v(x + h, t) + (1 - 4s^2)v(x, t) + (2s^2 - s)v(x - h, t) \tag{4}$$

where $s = (\alpha k)/(2h)$. The method embodied in Equation (4) is the **Lax-Wendroff method**. We have changed to the variable v because the solution of Equation (4) is not generally a solution of Equation (3).

Obviously there is no obstruction to deriving higher-order methods from Equation (3). Problem 2 asks for the derivation of a third-order method using these principles.

Stability Analysis

In the simple model problem given by (1), the solution can be written down at once. It is

$$u(x, t) = f(x + \alpha t)$$

If grid points (jh, nk) are introduced, and if we set $v_{jn} = v(jh, nk)$, then Equation (4) will take the following form:

$$v_{j,n+1} = (2s^2 + s)v_{j+1,n} + (1 - 4s^2)v_{jn} + (2s^2 - s)v_{j-1,n} \tag{5}$$

in which $s = (k\alpha)/(2h)$. A stability analysis can be carried out using the *Fourier method* of Section 9.1. To do this, one searches for a solution of Equation (5) having the form

$$v_{jn} = e^{ij\beta h}e^{n\lambda k} \qquad (i = \sqrt{-1})$$

On substituting this trial solution in Equation (5) and simplifying, we obtain

$$\begin{aligned} e^{\lambda k} &= 1 - 4s^2 + e^{i\beta h}(2s^2 + s) + e^{-i\beta h}(2s^2 - s) \\ &= 1 - 4s^2 + s(e^{i\beta h} - e^{-i\beta h}) + 2s^2(e^{i\beta h} + e^{-i\beta h}) \\ &= 1 - 4s^2 + 2is \sin \beta h + 4s^2 \cos \beta h \end{aligned}$$

Here we have used Euler's relation

$$e^{i\beta h} = \cos \beta h + i \sin \beta h$$

For stability, we require $|e^{\lambda h}| \leqq 1$. A computation yields

$$|e^{\lambda h}|^2 = 1 - 16s^2 \sin^2 \theta \,(1 - 4s^2 \sin^2 \theta + \cos^2 \theta) \tag{6}$$

where $\theta = \beta h/2$. The stability condition is now

$$1 - 4s^2 \sin^2 \theta + \cos^2 \theta \geqq 0 \tag{7}$$

The minimum value of the expression on the left occurs when $\theta = \pi/4$, and the minimum is nonnegative if $|s| \leqq 1/2$. Thus, the stability requirement (a *necessary* condition) is that $k|\alpha| \leqq h$.

Systems of Equations

Next, consider a hyperbolic system of equations, which is taken to be of the form

$$U_t = AU_x \tag{8}$$

Here U is a vector of functions, say $u^{(1)}, u^{(2)}, \ldots, u^{(m)}$. Each component is a function of (x, t). The matrix A is $m \times m$, and it must have m distinct real eigenvalues if the problem is classified as hyperbolic. The vector form of the Lax-Wendroff method is the analogue of Equation (4):

$$V(x, t + k) = V(x, t) + \left(\frac{k}{2h}\right) A\big[V(x + h, t) - V(x - h, t)\big]$$
$$+ \left(\frac{k^2}{2h^2}\right) A^2\big[V(x + h, t) - 2V(x, t) + V(x - h, t)\big] \qquad \textbf{(9)}$$

so that

$$V(x, t + h) = (\tau A)(2\tau A + I)V(x + h, t) + \big(I - (2\tau A)^2\big)V(x, t)$$
$$+ (\tau A)(2\tau A - I)V(x - h, t)$$

where $\tau = k/(2h)$. Without going into details, we mention only that this numerical method is stable if and only if each eigenvalue λ of A satisfies $k|\lambda| \leqq h$; that is, the spectral radius $\rho(A)$ of A should satisfy the inequality $k\rho(A) \leqq h$.

Wendroff's Implicit Method

The preceding discussion applies to a differential equation with only initial values supplied. If the variable x is confined to an interval, which we can take to be $[0, 1]$, then a properly posed problem will assign values on the entire boundary of the region described by $0 \leqq x \leqq 1$ and $t \geqq 0$. In this case, *implicit* numerical methods can be used on the differential equation in (1). As might be expected from the developments in Section 9.2, better stability properties can be obtained from implicit methods. One such method is known by Wendroff's name. We write it first in a form that suggests its origin:

$$\frac{1}{2}\left(\frac{v_{j,n+1} - v_{j,n}}{k} + \frac{v_{j+1,n+1} - v_{j+1,n}}{k}\right)$$
$$= \frac{\alpha}{2}\left(\frac{v_{j+1,n} - v_{jn}}{h} + \frac{v_{j+1,n+1} - v_{j,n+1}}{h}\right) \qquad \textbf{(10)}$$

Here one sees *averages* of first-order differences being used to represent derivatives. The truncation error of this method is $\mathcal{O}(h^2 + k^2)$.

Stability Analysis

Before undertaking a stability analysis, we write Equation (10) in this form:

$$v_{j,n+1}(1 + r) + v_{j+1,n+1}(1 - r) = v_{jn}(1 - r) + v_{j+1,n}(1 + r) \qquad \textbf{(11)}$$

in which $r = \alpha k/h$. Next, we seek a solution of Equation (11) in the form

$$v_{jn} = e^{ij\beta h}e^{n\lambda k}$$

Substitution of this into Equation (11) and subsequent simplification lead to

$$e^{\lambda k}(1 + r) + e^{\lambda k} e^{i\beta h}(1 - r) = 1 - r + e^{i\beta h}(1 + r)$$

Hence,

$$e^{\lambda k} = \left[1 + e^{i\beta h} + r(e^{i\beta h} - 1)\right] / \left[1 + e^{i\beta h} - r(e^{i\beta h} - 1)\right] \tag{12}$$

In order that $|e^{\lambda k}| \leqq 1$, it is necessary and sufficient that

$$|1 + e^{i\beta h} + r(e^{i\beta h} - 1)| \leqq |1 + e^{i\beta h} - r(e^{i\beta h} - 1)| \tag{13}$$

Appealing to Problem 6, we find as an equivalent condition

$$(1 + \cos \beta h)(r \cos \beta h - r) + (\sin \beta h)(r \sin \beta h) \leqq 0 \tag{14}$$

The left side of this equation is 0, and the method is stable for all α, k, and h.

Error Analysis

Now we undertake a proof of the assertion made previously that the truncation error in Wendroff's method is $\mathcal{O}(h^2 + k^2)$. The exact meaning of this requires some additional explanation. Our differential equation is of the form $Lu = 0$, with L defined by $Lu \equiv u_t - \alpha u_x$. From Equation (10), the finite-difference operator that replaces L in the numerical process is of the form

$$\frac{1}{2}(A + B) - \frac{\alpha}{2}(C + D)$$

where A, B, C, and D are given by

$$(Au)(x, t) = k^{-1}\left[u(x, t + k) - u(x, t)\right]$$

$$(Bu)(x, t) = k^{-1}\left[u(x + h, t + k) - u(x + h, t)\right]$$

$$(Cu)(x, t) = h^{-1}\left[u(x + h, t) - u(x, t)\right]$$

$$(Du)(x, t) = h^{-1}\left[u(x + h, t + k) - u(x, t + k)\right]$$

Using central differences, we have

$$(Au)(x, t) = u_t\left(x, t + \frac{1}{2}k\right) + \mathcal{O}(k^2)$$

$$(Bu)(x, t) = u_t\left(x + h, t + \frac{1}{2}k\right) + \mathcal{O}(k^2)$$

$$(Cu)(x, t) = u_x\left(x + \frac{1}{2}h, t\right) + \mathcal{O}(h^2)$$

$$(Du)(x, t) = u_x\left(x + \frac{1}{2}h, t_k\right) + \mathcal{O}(h^2)$$

We are *not* asserting that for any u that is sufficiently smooth,

$$\left\{L - \frac{1}{2}(A + B) + \frac{\alpha}{2}(C + D)\right\} u = \mathcal{O}(h^2 + k^2)$$

This equation will be established only for functions that *satisfy the differential equation*. Thus in the proof, it is assumed that $Lu = 0$, and consequently, we only have to prove

$$(A + B)u - \alpha(C + D)u = \mathcal{O}(h^2 + k^2)$$

LEMMA 1 *The discretization error in the Wendroff method is* $\mathcal{O}(h^2 + k^2)$.

Proof From Taylor's Theorem and the differential equation, we have

$$\begin{aligned} A &= u_t(x, t) + \frac{1}{2}ku_{tt}(x, t) + \mathcal{O}(k^2) \\ &= \alpha u_x(x, t) + \frac{1}{2}\alpha^2 k u_{xx}(x, t) + \mathcal{O}(k^2) \end{aligned}$$

Replacing x by $x + h$ in this last equation yields

$$\begin{aligned} B &= \alpha u_x(x + h, t) + \frac{1}{2}\alpha^2 k u_{xx}(x + h, t) + \mathcal{O}(k^2) \\ &= \alpha u_x(x, t) + \alpha h u_{xx}(x, t) + \mathcal{O}(h^2) + \frac{1}{2}\alpha^2 k u_{xx}(x, t) + \mathcal{O}(hk) + \mathcal{O}(k^2) \end{aligned}$$

Similarly, we get

$$\begin{aligned} C &= u_x(x, t) + \frac{1}{2}hu_{xx}(x, t) + \mathcal{O}(h^2) \\ D &= u_x(x, t + k) + \frac{1}{2}hu_{xx}(x, t + k) + \mathcal{O}(h^2) \\ &= u_x(x, t) + ku_{xt}(x, t) + \mathcal{O}(k^2) + \frac{1}{2}hu_{xx}(x, t) + \mathcal{O}(hk) + \mathcal{O}(h^2) \\ &= u_x(x, t) + \alpha k u_{xx}(x, t) + \mathcal{O}(k^2) + \frac{1}{2}hu_{xx}(x, t) + \mathcal{O}(hk) + \mathcal{O}(h^2) \end{aligned}$$

Thus, we have

$$\begin{aligned} A + B - \alpha(C + D) &= \mathcal{O}(k^2) + \mathcal{O}(hk) + \mathcal{O}(h^2) \\ &= \mathcal{O}(h^2 + k^2) \end{aligned}$$

■

Galerkin Methods

Galerkin methods can also be applied to hyperbolic problems. One way of doing so will be illustrated with a single first-order equation accompanied by initial and boundary conditions:

$$\begin{cases} u_t = \alpha u_x & \\ u(x, 0) = g(x) & (0 \leqq x \leqq 1) \\ u(0, t) = u(1, t) = 0 & (t > 0) \end{cases} \tag{15}$$

One begins by selecting some basic functions of x, say $w_1, w_2, \ldots, w_n$. We try to solve our problem with a function of the form

$$u(x, t) = \sum_{j=1}^{n} v_j(t) w_j(x) \tag{16}$$

The functions $v_1, v_2, \ldots, v_n$ are available to assist in this task. When the trial function (16) is substituted in the differential equation, the result is

$$\sum_{j=1}^{n} \left[v_j'(t) w_j(x) - \alpha v_j(t) w_j'(x) \right] = 0 \tag{17}$$

As is usual in following this strategy, we do not expect Equation (17) to be consistent. That is, we do not expect to find functions v_j that make the equation true. The reason for this, of course, is that the solution of the boundary-value problem (15) will probably not be expressible in terms of the chosen functions w_j as in Equation (16). In the Galerkin method (and in other similar methods), an approximate solution of Equation (17) is sought. At the same time, we must take into account the boundary and initial conditions in (15). For simplicity, let us assume that each basic function w_j vanishes at 0 and at 1. Then the homogeneous boundary conditions will be automatically satisfied by the trial function in (16). Next, we take the inner product of both sides of Equation (17) with w_i for $1 \leqq i \leqq n$. The result is

$$\sum_{j=1}^{n} \left[v_j'(t) \langle w_j, w_i \rangle - \alpha v_j(t) \langle w_j', w_i \rangle \right] = 0 \qquad (1 \leqq i \leqq n) \tag{18}$$

Here the notation is $\langle f, g \rangle = \int_0^1 f(x) g(x)\, dx$. Equation (18) is a system of n linear homogeneous differential equations for the n unknown functions v_j. We observe that there is a great advantage in choosing $\{w_1, w_2, \ldots, w_n\}$ to be an **orthonormal system** on the interval [0, 1]. Assuming that this is the case, Equation (18) takes the form

$$V' = AV \tag{19}$$

where $V = (v_1, v_2, \ldots, v_n)^T$ and A is an $n \times n$ matrix whose elements are

$$a_{ij} = \alpha\langle w_j', w_i\rangle$$

The initial condition in (15) now becomes

$$\sum_{j=1}^{n} v_j(0)w_j(x) = g(x) \tag{20}$$

Again, this may be impossible to satisfy exactly, and if so, one can choose $v_j(0)$ so that Equation (20) is as nearly satisfied as possible, say in the L^2-norm. Since the functions w_j have been assumed to form an orthonormal system, this means that

$$v_j(0) = \langle g, w_j\rangle \tag{21}$$

Equation (20) provides the initial conditions for the solution of the system (19). As we know from Section 8.11, the solution of this system is

$$V(t) = e^{tA}V(0)$$

A convenient orthonormal system consisting of functions that vanish at 0 and 1 is

$$w_j(x) = 2^{-1/2}\sin \pi jx$$

Another system can be constructed by using the Gram-Schmidt procedure on the sequence of functions $x \longmapsto (x^2 - x)x^j$, where $j = 1, 2, \ldots, n$.

PROBLEMS 9.7

1. Prove formally that if $u_t = \alpha u_x$, then for $n = 0, 1, 2, \ldots,$

$$\frac{\partial^n u}{\partial t^n} = \alpha^n \frac{\partial^n u}{\partial x^n}$$

   Prove also the vector case.

2. Derive a third-order approximation method from Equation (3).
3. Provide all the details in the argument leading to Equations (6) and (7) and to the stability condition $k|\alpha| \leqq h$.
4. Carry out a stability analysis on this numerical procedure for Equation (1):

$$\frac{1}{2k}(v_{j,n+1} - v_{j,n-1}) = \frac{\alpha}{2h}(v_{j+1,n} - v_{j-1,n})$$

5. Investigate the stability of Euler's method for Equation (1):

$$v_{j,n+1} = v_{j,n} + \frac{\alpha k}{2h}(v_{j+1,n} - v_{j-1,n})$$

6. Let u and v be two complex numbers, say $u = x + iy$ and $v = a + ib$. Show that these inequalities are equivalent:

   (i) $|u + v| \leqq |u - v|$

   (ii) $xa + yb \leqq 0$

   Show that Equation (14) follows.

7. Refer to Wendroff's method and show that the leading term in the truncation error is

$$\frac{1}{12}(\alpha h^2 - \alpha^3 k^2)u_{xxx}(x, t)$$

COMPUTER PROBLEMS 9.7

1. Program the Lax-Wendroff method in Equation (4). Assume that the numerical solution is to be computed on the line segment defined by $a \leqq x \leqq b$ and $t = T$, where a, b, and T are prescribed. Of course, f and α are also prescribed. The user will wish to assign the step sizes.

2. (Continuation) Test the program written in Computer Problem 1 on the problem $u_t = 2u_x$, with $u(x, 0) = (1 - x)^2$. Take $h = 0.02$, $k = 0.01$, $a = 1$, $b = 2$, and $T = 1$. Compare the numerical solution to the true solution.

9.8 Multigrid Method

The **multigrid method** for the numerical solution of differential equations is based on discretization and the subsequent approximation of derivatives by finite-difference formulas. The distinguishing feature of this method is that a number of different grids are used on the domain, ranging from **coarse** to **fine**. A numerical solution on a coarse grid can be computed quickly, but it will have low accuracy. Still, it might be useful as a starting point for an iterative solution on a finer grid. This is only one aspect of the multigrid strategy, but let us start by showing how it would work on a simple example.

Illustrative Example

Consider the following two-point boundary-value problem:

$$\begin{cases} u''(x) = f(x) \\ u(0) = u(1) = 0 \end{cases} \tag{1}$$

This example is only illustrative; the real power of the multigrid strategy will emerge only when we begin to apply it to partial differential equations.

If a step size h is chosen, we can discretize problem (1) in this standard way:

$$h^{-2}(v_{j-1} - 2v_j + v_{j+1}) = f_j \qquad (1 \leqq j \leqq n) \tag{2}$$

Here we have adopted these definitions:

$$h = \frac{1}{n+1} \qquad x_j = jh \qquad v_j \approx u(x_j) \qquad f_j = f(x_j) \qquad v_0 = v_{n+1} = 0$$

The $n \times n$ system of Equations (2) is not difficult to solve directly, but in view of what will be necessary in solving partial differential equations, we elect to treat (2) with an *iterative* method and select for this purpose the Gauss-Seidel procedure. Of course, an iterative method will produce a satisfactory solution sooner if it is provided with a good starting vector. One way to obtain a starting point for the problem at hand is to solve the same problem on a *coarser* grid. Thus, we can imagine System (2) written down with a *larger* step size h. This system will have fewer equations, but the equations will have the same structure. The system of equations is of the form

$$\begin{bmatrix} 2 & -1 & 0 & \cdots & 0 \\ -1 & 2 & -1 & \cdots & 0 \\ 0 & -1 & 2 & \cdots & 0 \\ \vdots & \vdots & \vdots & \ddots & \vdots \\ 0 & 0 & 0 & \cdots & 2 \end{bmatrix} \begin{bmatrix} v_1 \\ v_2 \\ v_3 \\ \vdots \\ v_n \end{bmatrix} = \begin{bmatrix} -h^2 f_1 \\ -h^2 f_2 \\ -h^2 f_3 \\ \vdots \\ -h^2 f_n \end{bmatrix}$$

Hence, this too can be solved with the Gauss-Seidel iteration. If we elect to do so, then we shall wish to begin with a good starting vector, and such a starting point can be obtained by going to a still coarser grid, and so on. Thus, the logical place to begin is with the coarsest grid, which is System (2) in the simplest case; namely, $n = 1$. When $n = 1$, $h = 1/2$ and there is only one equation in the system. Its solution is

$$v_0 = 0 \qquad v_1 = -\frac{1}{8} f\left(\frac{1}{2}\right) \qquad v_2 = 0$$

Now we are ready to pass to a finer grid. We divide h by 2, replace n by $2n + 1$, and set up the new system of equations. Thus, $n = 3$ and $h = 1/4$. Let us temporarily use w as the vector involved in this new system. We shall use v to assign suitable starting values to the components w_0, w_1, w_2, w_3, w_4. The sketch in Figure 9.13 shows the relation between v and w. The assignment of values to the vector w, based on information carried by the vector v, can be accomplished in many ways. A simple interpolation scheme is usually used. For the illustration here, we shall copy values of v to w where appropriate, and use averages for the other components. Thus, these five equations are to be used:

$$w_0 = v_0 \qquad w_2 = v_1 \qquad w_4 = v_2 \qquad w_1 = (v_0 + v_1)/2 \qquad w_3 = (v_1 + v_2)/2$$

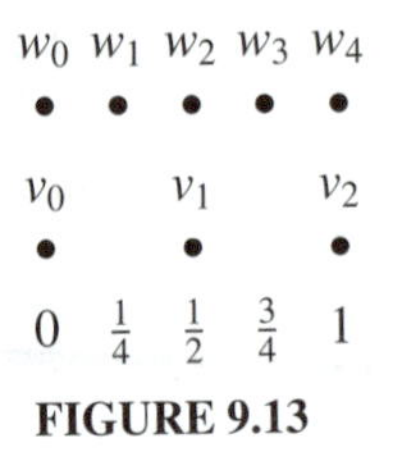

FIGURE 9.13

Having done this, we discard the old v-vector and replace it by w. Next, with the starting vector v just constructed, we carry out a few iterations of the Gauss-Seidel method. Then a new grid is formed by halving h and changing n accordingly. After that, the process is repeated.

```
input m, k
h ← 1/2 ; n ← 1
v_0 ← 0 ; v_2 ← 0 ; v_1 ← −(1/8) f(h)
for i = 2 to m do
    h ← h/2 ; n ← 2n + 1
    for j = 0 to (n + 1)/2 do
        w_{2j} ← v_j
    end do
    for j = 1 to (n + 1)/2 do
        w_{2j−1} ← (v_{j−1} + v_j)/2
    end do
    for j = 0 to n + 1 do
        v_j ← w_j
    end do
    for p = 1 to k do
        for j = 1 to n do
            v_j ← [v_{j−1} + v_{j+1} − h^2 f(jh)]/2
        end do
    end do
    output (v_i)
end do
```

In the pseudocode, f is the function appearing in the original problem (1). The number of grids used (including the first) is m, and k is the number of Gauss-Seidel iterations to be carried out on each grid. The transfer of information from v to w is called the **interpolation phase**. For the general step of the algorithm, we use formulas

$$w_{2j} = v_j \qquad w_{2j-1} = (v_{j-1} + v_j)/2 \tag{3}$$

In the pseudocode, the loop starting at "**for** $p = 1$ **to** k **do**" is executing k steps of the Gauss-Seidel iteration. Thus, Equation (2) is solved for the jth unknown, leading to

$$v_j = (v_{j-1} + v_{j+1} - h^2 f_j)/2 \tag{4}$$

This formula is used to update each v_j, and the whole iterative process is carried out k times.

The interpolation process that produces w from v and then replaces v by w can be made much simpler, as suggested in Problem 1. The code serves only a didactic purpose, but it can be used for experiments—especially on problems for which the solution is known because then the difference between the actual solution and the numerical solution can be examined. This was done in the case $f(x) = \cos x$, for which the solution of the two-point boundary-value problem is

$$u(x) = -\cos x + x(\cos 1 - 1) + 1 \tag{5}$$

Letting m (the number of grid refinements) be 6 and letting k (the number of Gauss-Seidel iterations) be 3, we find a discrepancy of about 10^{-3} between the computed and the true solutions. The value of h on the last grid is $1/64$, and $h^2 \approx 2 \times 10^{-4}$.

Damping of Errors

Another major component of the multigrid method is a systematic progression from fine grids to coarser grids. This direction is just opposite to the one previously explained, in which information from a coarse grid was passed *upward* to provide a starting point for the iterations on a finer grid. The reason for incorporating coarser grids at a later point in the computation is that low-frequency errors tend to be efficiently damped in the iterations associated with a coarse grid. Conversely, high-frequency errors tend to be efficiently damped in the iterations on a fine grid. This important characteristic is exploited in the multigrid strategy, and to a large extent it accounts for the success of the method.

A simple numerical experiment using the differential equation (1) will show the damping of errors of different frequencies. Consider the homogeneous problem

$$\begin{cases} u'' = 0 \\ u(0) = u(1) = 0 \end{cases} \tag{6}$$

whose solution is obviously $u(x) \equiv 0$. Let us solve this problem with the Gauss-Seidel iteration described by Equation (4), taking $f = 0$ in that equation and starting with a sinusoidal function

$$v_j = \sin\left(\frac{jp\pi}{n+1}\right) \qquad (0 \leqq j \leqq n+1)$$

Here n (as previously used) is the number of interior nodes on the interval (0, 1). It controls the fineness of the grid. In this experiment, the vector v can be considered as the error, since the solution of the problem is $u \equiv 0$. The parameter p controls

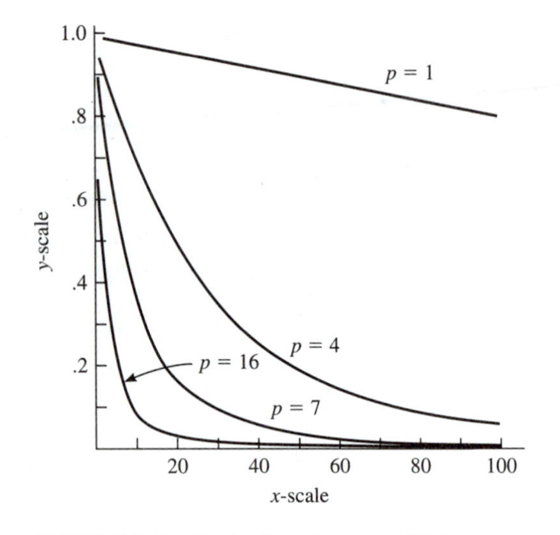

FIGURE 9.14 Reduction of error in 100 iterations

the *frequency* of the initial error. A pseudocode to execute k iterations using four prescribed frequencies is given. The norm of the vector is computed at the end of each Gauss-Seidel iteration.

When this program was run with $n = 63$, $k = 100$, and $p = 1, 4, 7, 16$, the results in Figure 9.14 were obtained. It is obvious that the errors of high frequency are rapidly damped by the iteration, while low-frequency errors are damped only slightly.

```
input n, k, p1, p2, p3, p4
for p = p1, p2, p3, p4 do
    for j = 0 to n + 1 do
        v_j = sin((jpπ)/(n + 1))
    end do
    for i = 1 to k do
        for j = 1 to n do
            v_j = (v_{j-1} + v_{j+1})/2
        end do
        ρ_i = ||v||_∞ = max_{1≦j≦n} |v_j|
        output p, i, ρ_i
    end do
end do
```

Analysis

To analyze the damping effect of iteration on errors, we choose the Jacobi method as an illustration. When the Jacobi iteration is used on a system $Ax = b$, the iteration formula is

$$x^{(k+1)} = (I - D^{-1}A)x^{(k)} + D^{-1}b \tag{7}$$

where D is the diagonal of A. In the example of this section, the matrix A involved in Equation (2) is

$$A = \begin{bmatrix} 2 & -1 & 0 & \cdots & 0 \\ -1 & 2 & -1 & \cdots & 0 \\ 0 & -1 & 2 & \cdots & 0 \\ \vdots & \vdots & \vdots & \ddots & \vdots \\ 0 & 0 & 0 & \cdots & 2 \end{bmatrix}$$

and the right-hand side is $b = h^2 f_j$. The iteration matrix is therefore

$$G \equiv I - D^{-1}A = \begin{bmatrix} 0 & \frac{1}{2} & 0 & \cdots & 0 \\ \frac{1}{2} & 0 & \frac{1}{2} & \cdots & 0 \\ 0 & \frac{1}{2} & 0 & \cdots & 0 \\ \vdots & \vdots & \vdots & \ddots & \vdots \\ 0 & 0 & 0 & \cdots & 0 \end{bmatrix}$$

It is clear that

$$G = I - \frac{1}{2}A \tag{8}$$

By Lemma 1 in Section 9.1, the eigenvalues of A are

$$\mu_j = 2 - 2\cos\frac{j\pi}{n+1} \qquad (1 \leqq j \leqq n) \tag{9}$$

The eigenvector corresponding to μ_j is

$$v^{(j)} = \left(\sin\frac{j\pi}{n+1},\ \sin\frac{2j\pi}{n+1}, \ldots,\ \sin\frac{nj\pi}{n+1}\right) \tag{10}$$

By Equation (8), the eigenvalues λ_j of the iteration matrix G are

$$\lambda_j = 1 - \frac{1}{2}\mu_j = \cos\frac{j\pi}{n+1} \qquad (1 \leqq j \leqq n)$$

The spectral radius of the iteration matrix is therefore

$$\rho(G) = \cos\frac{\pi}{n+1} \approx 1 - \frac{1}{2}\left(\frac{\pi}{n+1}\right)^2 = 1 - \frac{\pi^2}{2}h^2$$

Now we can begin to draw some conclusions. First, since $\rho(G) < 1$, the Jacobi iteration will converge, by Theorem 5 in Section 4.6. Second, we notice that for small values of h, the spectral radius is near 1, and this implies that on the finer grids convergence will be very slow. Third, we can use the eigenvectors of

A in Equation (10) to understand the damping of errors. The vectors $v^{(j)}$ are also eigenvectors of G, and they form a basis for $\mathbb{R}^n$. Hence, any error that is present in the numerical solution must be a linear combination of $v^{(1)}, v^{(2)}, \ldots, v^{(n)}$. It is sufficient to consider a single component of the error, say $v^{(j)}$. Obviously, iteration k with G on $v^{(j)}$ will produce $\lambda_j^k v^{(j)}$, and this will converge to 0 as $k \to \infty$ because $|\lambda_j| < 1$. But the strength of this **damping effect** will be greater when $|\lambda_j|$ is least. The small values of $|\lambda_j|$ are associated with j in the middle of the range $1 \leqq j \leqq n$, whereas for $j \approx 1$ or $j \approx n$, we have $|\lambda_j| \approx 1$.

Restriction and Grid Correction

The entire multigrid process will contain many steps in which results obtained on one grid are passed to another grid for the next set of calculations. We have already indicated how the calculations proceed *upward* from a coarse grid to a fine grid. In proceeding *downward*, let us assume that an approximate solution vector to System (2) is available and is denoted by v^i. Instead of iterating in System (2) with the current value of h, we decide to descend to a *coarser* grid for the iterations because this will proceed faster. Our objective is to refine v^i by adding a suitable correction to it. If e^i is the correction and if the matrix in System (2) is denoted by A^i, then we want e^i to solve the equation

$$A^i(v^i + e^i) = f^i \tag{11}$$

Here f^i represents the function f on the current grid. We write the former equation as

$$A^i e^i = f^i - A^i v^i = r^i$$

where r^i is the residual vector engendered by the approximate solution v^i. The system $A^i e^i = r^i$ is to be solved by passing to a coarser grid and using there a few iterations of the Gauss-Seidel method. On the coarser grid, we shall have a system of equations

$$A^{i-1} e^{i-1} = r^{i-1} \tag{12}$$

The vector r^{i-1} is obtained from r^i by some elementary transfer of information from a fine grid to a coarser grid. One way of doing this is to write

$$r_j^{i-1} = r_{2j}^i \tag{13}$$

A more refined method, using all the information in r^i, is given by a weighted average

$$r_j^{i-1} = \frac{1}{4} r_{2j-1}^i + \frac{1}{2} r_{2j}^i + \frac{1}{4} r_{2j+1}^i \tag{14}$$

No matter what formulas are used, this process is called **restriction**.

After a few iterations of the Gauss-Seidel method on Equation (12), we shall have a vector e^{i-1}. By means of the interpolation process previously discussed, we obtain a vector e^i and then update v^i by the operation

$$v^i \leftarrow v^i + e^i$$

This procedure is called the **coarse grid correction scheme**.

V-Cycle Algorithm

Our final objective is to describe what is termed a ***V*-cycle** in the multigrid algorithm. The *V*-cycle receives its name from the diagram in Figure 9.15 that describes the course of the computations. Each circle in the diagram denotes a block of computing performed on one grid. The computation begins on the finest grid. After a small number of iterations, the residual, r, is computed, and control passes to a coarser grid, where an equation of the form $Az = r$ is processed. The process is repeated, passing "downward" through a succession of grids until a system of equations on the coarsest grid is obtained. This is usually solved exactly because it consists of a small number of equations, possibly just one. In the upward part of the diagram, information is passed to finer grids by interpolation processes and additional iterations are performed.

To give a formal description of the *V*-cycle, we first number the grids $\mathcal{G}^1, \mathcal{G}^2, \dots, \mathcal{G}^m$, where $\mathcal{G}^1$ denotes the coarsest grid and $\mathcal{G}^m$ the finest. On grid $\mathcal{G}^i$ a matrix A^i is prescribed. The problem that we wish to solve is $A^m v^m = f^m$; this is the discretized version of the given boundary-value problem on the finest grid. Thus f^m is an *input* vector. Successive f^i for $i = m-1, m-2, \dots, 1$ will, however, be computed. The steps in the *V*-cycle are then as follows:

(i) Set i equal to m, and put the data from the boundary-value problem into f^m. Put the best available guess into v^m.
(ii) Apply an iterative method k times on the system $A^i v^i = f^i$, starting with the current v^i. Compute the residual: $r^i \leftarrow f^i - A^i v^i$. Apply a restriction operator $R_i : f^{i-1} \leftarrow R_i r^i$. Subtract 1 from i.
(iii) If $i = 1$, go to step **(iv)**. If $i > 1$, go to step **(ii)**.
(iv) Solve $A^1 v^1 = f^1$ exactly.
(v) Add 1 to i. Apply an extension operator E_i and then add the correction term: $v^i \leftarrow v^i + E_i v^{i-1}$. Apply an iterative method k times to the system $A^i v^i = f^i$, starting with the current v^i.
(vi) If $i = m$, send v^m to output and stop. If $i < m$, go to step **(v)**.

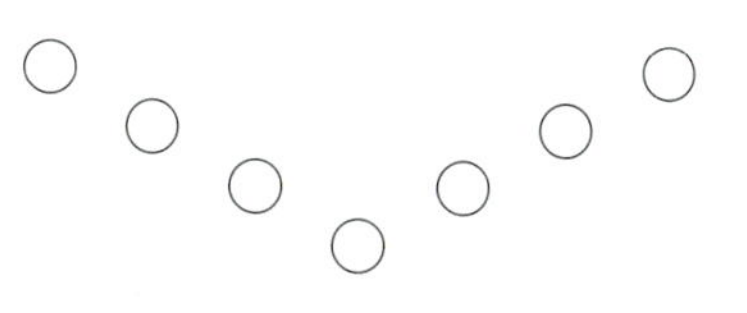

FIGURE 9.15

A pseudocode for applying the V-cycle in the multigrid method to our model problem is given next. (To save space, an abbreviated notation for loops has been used.) In grid G^i, the step size is $h = 2^{-i}$. A temporary work area, w, is used to store intermediate quantities. The restriction operator used in the code is the simplest one, whereby even-indexed components are simply copied into the vector on the next coarser grid as in Equation (13). Everything else in the code conforms to the algorithm outlined previously. The Gauss-Seidel iterative method is used. In this algorithm, elements of the vector v^i_j could be stored in a two-dimensional array such as $V(J, I) \leftarrow v^i_j$ so that the columns of the array correspond to grid values at the various levels, if this is a concern.

The pseudocode was used for various numerical experiments on the model problem mentioned previously:

$$\begin{cases} u'' = \cos x \\ u(0) = u(1) = 0 \end{cases}$$

The solution of this is given in Equation (5). For example, we set $m = 7$, which corresponds to $h = 1/128$ on the finest grid. Then various values of k were used to determine the effect of this parameter on the error. Here *error* refers to $\max_{0 \leqq j \leqq n+1} |u(x_j) - v^m_j|$, where u is the true solution, $x_j = jh$, and v^m is the approximate solution on the finest grid produced by the algorithm with $h = 2^{-m}$. Each time the number of Gauss-Seidel iterations was increased by 1, the new error was approximately $2/5$ of the preceding error. This held true for $3 \leqq k \leqq 8$. Further increases in k led to less dramatic reductions of the error, with reduction factors of approximately 0.6, 0.7, 0.8, and 0.9.

input m, k
$n \leftarrow 2^m - 1$
$h \leftarrow 1/(n+1)$
$v^i_j \leftarrow 0;\ f^i_j \leftarrow 0 \qquad (1 \leqq i \leqq m,\ 0 \leqq j \leqq n+1)$
$f^m_j \leftarrow f(jh) \qquad (1 \leqq j \leqq n)$
for $i = m$ **to** 2 **step** -1 **do**
  for $p = 1$ **to** k **do**
    $v^i_j \leftarrow [v^i_{j-1} + v^i_{j+1} - h^2 f^i_j]/2 \qquad (1 \leqq j \leqq n)$
  end do
  $w_j \leftarrow f^i_j - [v^i_{j-1} - 2v^i_j + v^i_{j+1}]/h^2 \qquad (1 \leqq j \leqq n)$
  $f^{i-1}_j \leftarrow w_{2j} \qquad (1 \leqq j \leqq \frac{n-1}{2})$
  $h \leftarrow 2h$
  $n \leftarrow (n-1)/2$
end do
$v^1_1 = -f(1/2)/8$
for $i = 2$ **to** m **do**
  $h \leftarrow h/2$
  $n \leftarrow 2n + 1$
  $w_{2j} \leftarrow v^{i-1}_j \qquad (0 \leqq j \leqq \frac{n+1}{2})$
  $w_{2j-1} \leftarrow [v^{i-1}_{j-1} + v^{i-1}_j]/2 \qquad (1 \leqq j \leqq \frac{n+1}{2})$
  $v^i_j \leftarrow v^i_j + w_j \qquad (0 \leqq j \leqq n+1)$

```
        for p = 1 to k do
            v_j^i ← [v_{j-1}^i + v_{j+1}^i − h^2 f_j^i]/2      (1 ≦ j ≦ n)
        end do
    end do
    output v_j^m      (0 ≦ j ≦ n + 1)
```

Operation Count

To estimate the computing costs of the multigrid algorithm, let us examine the V-cycle as given in the code, and count the arithmetic operations in it.

In the downward part of the process, m different grids are used. On grid G^i there are 2^i points, 2^i unknowns, and 2^i equations. Each updating of a variable requires four operations in the Gauss-Seidel method. Thus on each grid, we shall expend $4k2^i$ operations in the iteration. The residual calculation adds $5 \cdot 2^i$ and the restriction operator adds none. Hence on the ith grid, $(4k+5)2^i$ operations are used. The total for all m grids is

$$\sum_{i=1}^{m}(4k+5)2^i \approx (4k+5)2^{m+1} = (8k+10)2^m$$

A similar count on the upward part of the V-cycle produces a total of about $(8k+4)2^m$. Thus for the entire V-cycle, about $16(k+1)2^m$ operations are involved.

What would the corresponding numbers be for a two-dimensional problem such as $\nabla^2 u = f(x, y)$ on the unit square? Each equation in the familiar discretization involves five unknowns, and the updating of each variable by the Gauss-Seidel procedure will have about six operations. The number of variables on the ith grid is now $(2^i)^2$ because there are two variables. Thus, we see that the principal effect on our counting is to change 2^m to 4^m. The factor multiplying 4^m will be somewhat larger than $16(k+1)$, but it will still be linear in k.

The preceding remarks show that the computational effort is an exponential function of m (the number of grids) and a linear function of k (the number of iterations prescribed). This conclusion is valid for any number of dimensions.

PROBLEM 9.8

1. The interpolation phase computes w from v and then replaces v by w. Find an efficient code to do this, using only the v-array.

COMPUTER PROBLEMS 9.8

1. Repeat the numerical experiment reported in the text that is designed to show the reduction of errors of different frequencies.

2. Repeat the numerical experiment in the text to solve the model problem on a succession of finer grids.

3. Repeat the numerical experiment in the text using the V-cycle algorithm.

4. Program the V-cycle algorithm for the following two-dimensional problem:

$$\begin{cases} u_{xx} + u_{yy} = f(x, y) & (0 < x < 1, \quad 0 < y < 1) \\ u(x, y) = 0 \text{ on boundary} \end{cases}$$

Test your program when $f(x, y) = 2x(x - 1) + 2y(y - 1)$. The true solution is $u(x, y) = xy(1 - x)(1 - y)$.

5. In the first code of this section, adjust the loops so that v_0 and v_{n+1} (which remain constant) are not being copied back and forth.

6. Carry out some numerical experiments using the first code in this section to determine whether there is an optimal value for k, the number of Gauss-Seidel iterations.

7. Generalize the codes in this section to accommodate the problem $u''(x) = f(x)$, $u(a) = \alpha$, $u(b) = \beta$.

*9.9 Fast Methods for Poisson's Equation

Poisson's equation in two variables is

$$u_{xx} + u_{yy} = f(x, y) \tag{1}$$

In a typical physical problem involving this equation, we would seek a function u that satisfies Equation (1) in some prescribed open region Ω and fulfills some prescribed conditions on the boundary of Ω (denoted by $\partial\Omega$). The function f is defined on Ω.

In recent years, Fourier analysis has been applied to obtain fast algorithms for solving such boundary-value problems. These new methods take advantage of the fast Fourier transform. Here a simple model problem will be used to illustrate the innovations that lead to these new algorithms.

Model Problem

The model problem is described thus:

$$\begin{cases} \Omega = \{(x, y) : 0 < x < 1,\ 0 < y < 1\} \\ u_{xx} + u_{yy} = f(x, y) \text{ in } \Omega \\ u(x, y) = 0 \text{ on } \partial\Omega \end{cases} \tag{2}$$

We embark on a discretization by putting

$$h = \frac{1}{n + 1} \qquad x_i = ih \qquad y_j = jh \qquad (0 \leqq i, j \leqq n + 1)$$

A familiar approach to problem (2) then introduces

$$v_{ij} \approx u(x_i, y_j) \qquad f_{ij} = f(x_i, y_j)$$

One discretized version of our problem is then

$$h^{-2}(v_{i+1,j} - 2v_{ij} + v_{i-1,j}) + h^{-2}(v_{i,j+1} - 2v_{ij} + v_{i,j-1}) = f_{ij} \tag{3}$$

In Equation (3), the range of i and j is $\{1, 2, \dots, n\}$. Most of the terms in Equation (3) are unknown, but because of the boundary conditions, we require

$$v_{0j} = v_{n+1,j} = v_{i0} = v_{i,n+1} = 0 \tag{4}$$

The traditional way of proceeding at this juncture is to solve system (3) by an iterative method. Here there are n^2 equations and n^2 unknowns. The computational effort to solve this system using, say, successive overrelaxation is $\mathcal{O}(n^3 \log n)$. The alternative approach involving fast transforms will bring this effort down to $\mathcal{O}(n^2 \log n)$.

Fast Fourier Sine Transform

A solution of system (3) will be sought in the following form:

$$v_{ij} = \sum_{k=1}^{n} \widehat{v}_{kj} \sin ik\phi \qquad (0 \leqq i, j \leqq n+1) \tag{5}$$

where $\phi = \pi/(n+1)$. Here the numbers $\widehat{v}_{kj}$ are unknowns that we wish to determine. They represent the Fourier sine transform of the function v. Once the $\widehat{v}_{kj}$ have been determined, the fast Fourier sine transform can be used to compute v_{ij} efficiently.

If the v_{ij} from Equation (5) are substituted into Equation (3), the result is

$$\begin{aligned} &\sum_{k=1}^{n} \widehat{v}_{kj}\left[\sin(i+1)k\phi - 2\sin ik\phi + \sin(i-1)k\phi\right] \\ &\quad + \sum_{k=1}^{n} \sin ik\phi \left[\widehat{v}_{k,j+1} - 2\widehat{v}_{kj} + \widehat{v}_{k,j-1}\right] = h^2 f_{ij} \end{aligned} \tag{6}$$

Now we use a trigonometric identity in Lemma 2 (below) to simplify the first summation. At the same time, we introduce the sine transform of f_{ij}:

$$f_{ij} = \sum_{k=1}^{n} \widehat{f}_{kj} \sin ik\phi$$

The result is

$$\sum_{k=1}^{n} \widehat{v}_{kj}(-4 \sin ik\phi) \left(\sin^2 \frac{k}{2}\phi \right) + \sum_{k=1}^{n} \sin ik\phi (\widehat{v}_{k,j+1} - 2\widehat{v}_{kj} + \widehat{v}_{k,j-1}) = h^2 \sum_{k=1}^{n} \widehat{f}_{kj} \sin ik\phi \tag{7}$$

By Lemma 1 (below), the matrix having elements $\sin ij\phi$ is nonsingular. Therefore, we can deduce from Equation (7) that

$$\widehat{v}_{kj} \left(-4 \sin^2 \frac{k\phi}{2} \right) + \widehat{v}_{k,j+1} - 2\widehat{v}_{kj} + \widehat{v}_{k,j-1} = h^2 \widehat{f}_{kj} \tag{8}$$

Equation (8) appears at first glance to be another system of n^2 equations in n^2 unknowns, which is only slightly different from the original system (3). But closer inspection reveals that in Equation (8), k can be held fixed, and the resulting system of n equations can be easily and *directly* solved since it is tridiagonal. Thus for fixed k, the unknowns in (8) form a vector

$$(\widehat{v}_{k1}, \widehat{v}_{k2}, \ldots, \widehat{v}_{kn})$$

in $\mathbb{R}^n$. The procedure used above has **decoupled** the original system of n^2 equations into n systems of n equations each. A tridiagonal system of n equations can be solved in $\mathcal{O}(n)$ operations (in fact, fewer than $10n$ operations are needed). Thus, we can solve n tridiagonal systems at a cost of $10n^2$. The fast Fourier sine transform uses $\mathcal{O}(n \log n)$ operations on a vector with n components. Thus, the total computational burden in the fast Poisson method is $\mathcal{O}(n^2 \log n)$.

Additional Details

Some details in the preceding discussion need to be addressed. First, observe that from Equation (5), the boundary conditions

$$v_{0j} = v_{n+1,j} = 0 \qquad (0 \leqq j \leqq n+1)$$

will be automatically met without placing any restrictions on the coefficients $\widehat{v}_{kj}$. However, the remaining boundary conditions

$$v_{i0} = v_{i,n+1} = 0$$

will not be fulfilled unless we add two equations:

$$\sum_{k=1}^{n} \widehat{v}_{k0} \sin ik\phi = \sum_{k=1}^{n} \widehat{v}_{k,n+1} \sin ik\phi = 0 \qquad (0 \leqq i \leqq n+1) \tag{9}$$

Now Equation (9) states that the two vectors

$$(\widehat{v}_{10}, \widehat{v}_{20}, \ldots, \widehat{v}_{n,0}) \qquad \text{and} \qquad (\widehat{v}_{1,n+1}, \widehat{v}_{2,n+1}, \ldots, \widehat{v}_{n,n+1})$$

must be orthogonal to the rows of a matrix that is (by Lemma 1) nonsingular. Hence, we must define

$$\widehat{v}_{k0} = \widehat{v}_{k,n+1} = 0 \qquad (1 \leqq k \leqq n)$$

LEMMA 1 *The $(n-1) \times (n-1)$ matrix A having elements*

$$a_{kj} = (2/n)^{1/2} \sin \frac{kj\pi}{n} \qquad (1 \leqq k \leqq n-1,\ 1 \leqq j \leqq n-1)$$

is symmetric and orthogonal. Hence, $A^2 = I$.

Proof We compute the generic element of A^2:

$$\begin{aligned} A^2_{kj} &= \sum_{\nu=1}^{n-1} a_{k\nu} a_{\nu j} = \frac{2}{n} \sum_{\nu=1}^{n-1} \sin \frac{k\nu\pi}{n} \sin \frac{j\nu\pi}{n} \\ &= \frac{1}{n} \sum_{\nu=1}^{n-1} \left[\cos \frac{\nu(k-j)\pi}{n} - \cos \frac{\nu(k+j)\pi}{n} \right] \\ &= \frac{1}{n} \operatorname{Re} \left[\sum_{\nu=0}^{n-1} (e^{i\nu\phi} - e^{i\nu\theta}) \right] \end{aligned} \tag{10}$$

where $\phi = (k-j)\pi/n$ and $\theta = (k+j)\pi/n$.

If $k = j$, then $\phi = 0$ and $\theta = 2k\pi/n$. Moreover, θ is not a multiple of 2π because $1 \leqq k \leqq n-1$. Hence in this case

$$A^2_{kk} = \frac{1}{n} \operatorname{Re} \left[n - \frac{e^{in\theta} - 1}{e^{i\theta} - 1} \right] = 1$$

Here we noted that $e^{in\theta} = e^{i2k\pi} = 1$.

If $k \neq j$, then neither ϕ nor θ is a multiple of 2π, and the geometric series in Equation (10) can be summed with the usual formula. The result is

$$A^2_{kj} = \frac{1}{n} \operatorname{Re} \left[\frac{e^{in\phi} - 1}{e^{i\phi} - 1} - \frac{e^{in\theta} - 1}{e^{i\theta} - 1} \right]$$

If $k - j$ is even, then so is $k + j$, and in this case $e^{in\phi} = e^{in\theta} = 1$. Hence, $A^2_{kj} = 0$. On the other hand, if $k - j$ is odd, then so is $k + j$. In this case,

$e^{in\phi} = e^{in\theta} = -1$. Thus, we will want to prove that

$$\text{Re}\left[\frac{-2}{e^{i\phi} - 1} - \frac{-2}{e^{i\theta} - 1}\right] = 0 \tag{11}$$

To establish this, first note that for $z \neq 0$,

$$\text{Re}\left[\frac{1}{z}\right] = \text{Re}\left[\frac{\bar{z}}{z\bar{z}}\right] = \frac{\text{Re}[z]}{|z|^2}$$

Next apply this to $z = e^{i\phi} - 1$, getting

$$\text{Re}\left[\frac{1}{e^{i\phi} - 1}\right] = \frac{\cos\phi - 1}{(\cos\phi - 1)^2 + \sin^2\phi} = \frac{\cos\phi - 1}{2 - 2\cos\phi} = -\frac{1}{2}$$

Of course, this is true when $\phi = \theta$, and thus, the two terms on the left in Equation (11) are identical. ■

LEMMA 2

$$\sin(A + B) - 2\sin A + \sin(A - B) = -4\sin A \sin^2\frac{B}{2}$$

Proof Use these familiar identities:

$$\sin(A \pm B) = \sin A \cos B \pm \cos A \sin B$$
$$1 - \cos 2A = 2\sin^2 A$$

■

COMPUTER PROBLEM 9.9

1. Solve model problem (2) with various values of n and a computer routine from your program library that carries out the fast Poisson solving procedure.

Chapter Ten

Linear Programming and Related Topics

*10.1 Convexity and Linear Inequalities

The theory of linear *inequalities* closely parallels the more familiar theory of linear *equations*. We shall develop the basic parts of this subject here and indicate some of its applications.

Basic Concepts

All of the vectors and matrices that are employed in this subject are real (not complex), since the theory utilizes in an essential way the *order* structure of the real line. For two points (vectors) x and y in $\mathbb{R}^n$, we write

$$x \geqq y \qquad \text{if and only if} \qquad x_i \geqq y_i \qquad (1 \leqq i \leqq n)$$

Similarly, we define $x \leqq y$, $x > y$, or $x < y$ by component-wise inequalities. In particular, it should be noted that $x > y$ is *not* the same as ($x \geqq y$ and $x \neq y$). (The reason is that $x \neq y$ does not mean $x_i \neq y_i$ for *all* i; rather, it means that $x_i \neq y_i$ for *some* i.)

A system of m weak linear inequalities in n variables can be written as

$$Ax \geqq b \tag{1}$$

where A is $m \times n$, x is $n \times 1$, and b is $m \times 1$. The individual inequalities in this system are

$$a_{i1}x_1 + a_{i2}x_2 + \cdots + a_{in}x_n \geqq b_i \qquad (1 \leqq i \leqq m)$$

or

$$A_{\text{row}_i} x \geqq b_i$$

where A_{row_i} is the ith row vector of the matrix A. A fundamental question concerning such a system is whether it is **consistent**. In other words, does there exist an x such that $Ax \geqq b$? If it is consistent, we shall want to have algorithms for obtaining solutions. Linear inequalities with more general systems

$$\begin{cases} \sum_{j=1}^{n} a_{ij}x_j \geqq b_i & (1 \leqq i \leqq m_1) \\ \sum_{j=1}^{n} a_{ij}x_j > b_i & (m_1 + 1 \leqq i \leqq m) \end{cases}$$

are possible, but for simplicity we consider the System (1).

To see what to expect, let us consider a simple example in which $n = 2$ and $m = 3$:

$$\begin{cases} x_1 + x_2 \geqq 2 \\ x_1 - x_2 \geqq -1 \\ -3x_1 + x_2 \geqq -6 \end{cases} \tag{2}$$

or

$$\begin{bmatrix} 1 & 1 \\ 1 & -1 \\ -3 & 1 \end{bmatrix} \begin{bmatrix} x_1 \\ x_2 \end{bmatrix} \geqq \begin{bmatrix} 2 \\ -1 \\ -6 \end{bmatrix}$$

The points that satisfy $x_1 + x_2 \geqq 2$ lie on one side of the line given by $x_1 + x_2 = 2$. By graphing this line and the two others that arise from System (2), we can determine the set of all points that solve System (2). It is the triangle shown in Figure 10.1. It is easily seen from the figure that if we reverse all the inequalities in System (2), then the resulting system is **inconsistent**. Also, by making some minor changes in the coefficient matrix, we can produce a system whose set of solutions is **unbounded**. (For example, change -3 to -1 in the third inequality.) The sketch also shows that the solution set is **convex**, a fact to be elaborated now. (For the basic theory of convex sets, the reader should consult Section 6.9.)

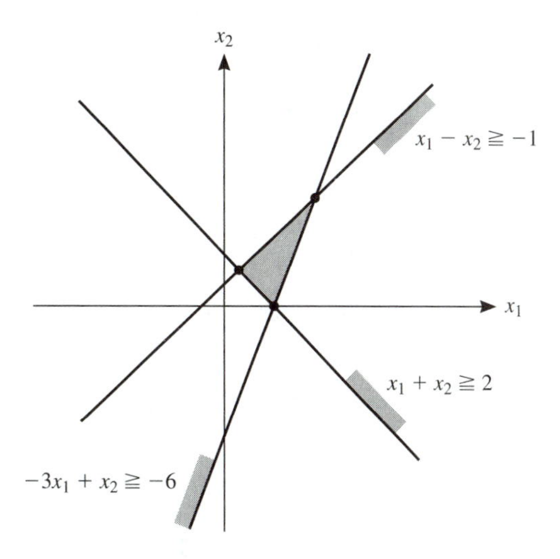

FIGURE 10.1 Solution set for system (2)

Convex Sets and Convex Hull

DEFINITION 1 *A subset K in a linear space is said to be* **convex** *if the line segment joining any two points of K lies wholly within K.*

The algebraic expression of this property is

$$\left.\begin{matrix} x, y \in K \\ 0 \leqq \theta \leqq 1 \end{matrix}\right\} \Rightarrow \theta x + (1 - \theta)y \in K$$

Some elementary facts about convex sets are contained in the next few theorems. The term **convex combination** refers to a linear combination of points in which the coefficients are nonnegative and sum to 1.

THEOREM 1 *If K is convex, then any convex combination of points in K also belongs to K.*

Proof Theorem 1 is established by induction. The case $m = 1$ is trivial, and the case $m = 2$ is true by the definition of convexity. To prove the case m from the case $m - 1$, we write

$$\sum_{i=1}^{m} \lambda_i x^{(i)} = \lambda_m x^{(m)} + (1 - \lambda_m) \sum_{i=1}^{m-1} \frac{\lambda_i}{1 - \lambda_m} x^{(i)}$$

and observe that $x^{(i)}$ are points in K, $0 \leqq \lambda_m \leqq 1$, and $1 - \lambda_m = \sum_{i=1}^{m-1} \lambda_i$. Then the coefficients $\lambda_i/(1 - \lambda_m)$ are nonnegative and sum to 1. ■

THEOREM 2 *The intersection of a family of convex sets is also convex.*

Proof Suppose K_α is a convex set for each α in an index set. If $x, y \in \bigcap K_\alpha$ and $0 \leqq \theta \leqq 1$, then $\theta x + (1-\theta)y \in K_\alpha$ for each α because K_α is convex. Hence, $\theta x + (1-\theta)y \in \bigcap K_\alpha$. ■

Here are some theorems concerning the solution sets of System (2).

THEOREM 3 *The solution set of a system of linear inequalities is a convex set.*

Proof The solution set for a single linear inequality $a^T x \geqq \beta$ is convex, as is shown by the calculation

$$a^T\left(\theta x + (1-\theta)y\right) = \theta a^T x + (1-\theta)a^T y \geqq \theta\beta + (1-\theta)\beta = \beta$$

where x and y are members of this solution set. The solution set of a *system* of linear inequalities is the intersection of the solution sets of the individual inequalities that make up the system. This set is also convex because the intersection of any family of convex sets is convex. ■

Recall from Section 6.9 that the **convex hull** of a set S is the set of all convex linear combinations of points in S and is often denoted by co(S).

THEOREM 4 *The convex hull of a set S is the smallest convex set containing S.*

Proof Let K be the convex hull of S. Let T be any other convex set containing S. We want to prove that $K \subseteqq T$. A typical element of K is of the form $x = \sum_{i=1}^n \lambda_i x_i$, where $x_i \in S$, $\lambda_i \geqq 0$, and $\sum_{i=1}^n \lambda_i = 1$. Obviously $x_i \in T$ because T contains S. Since T is convex, $x \in T$. Since x was an arbitrary element of K, $K \subseteqq T$. ■

THEOREM 5 **Separation Theorem I** *Let X be a closed convex set in $\mathbb{R}^n$. If p is a point not in X, then for some $v \neq 0$ we have*

$$\langle v, p\rangle < \inf_{x\in X}\langle v, x\rangle$$

Proof Let S be a closed ball centered at p, with radius large enough to ensure that S intersects X. Then $S \bigcap X$ is compact. The function $x \longmapsto \|x - p\|$, being continuous, assumes its minimum value on $S \bigcap X$, say at ξ. If $x \in X$ and $0 < \theta < 1$, then $y \equiv \theta x + (1-\theta)\xi \in K$, and consequently $\|y - p\| \geqq \|\xi - p\|$, as is easily verified. Thus,

$$\begin{aligned}
\|\xi - p\|^2 &\leqq \|\theta x + (1-\theta)\xi - p\|^2 \\
&= \|\xi - p + \theta(x - \xi)\|^2 \\
&= \|\xi - p\|^2 + 2\theta\langle \xi - p, x - \xi\rangle + \theta^2\|x - \xi\|^2
\end{aligned}$$

Hence,

$$0 \leqq 2\langle \xi - p, x - \xi\rangle + \theta\|x - \xi\|^2$$

By letting θ converge to 0, we obtain

$$0 \leqq \langle \xi - p, x - \xi \rangle$$

By letting $v = \xi - p$, we can write this last inequality as

$$0 \leqq \langle v, x - p + p - \xi \rangle = \langle v, x - p - v \rangle = \langle v, x - p \rangle - \|v\|^2$$

Hence, we obtain an inequality stronger than the one asserted:

$$0 < \|v\|^2 \leqq \langle v, x \rangle - \langle v, p \rangle \qquad \blacksquare$$

Proof (*Alternative proof*) Since p is not in X, 0 is not in the translated set

$$X - p = \{x - p : x \in X\}$$

Now we apply Lemma 2 of Section 6.9 to the set $X - p$. By examining the proof of that lemma, we conclude that there is a vector $v \neq 0$ with the property

$$\langle v, x - p \rangle \geqq \langle v, v \rangle \qquad (x \in X)$$

This yields

$$\langle v, p \rangle \leqq \langle v, x \rangle - \|v\|^2 \qquad (x \in X) \qquad \blacksquare$$

The theorem, as stated, is true in Hilbert space. The proof must be modified, however. The existence of the point ξ now cannot be based on the compactness of $S \bigcap X$ because closed and bounded sets in Hilbert space are not necessarily compact. One can proceed as follows: Let $d = \inf\{\|x - p\| : x \in X\}$, and select a sequence $x_i \in X$ such that $\|x_i - p\| \to d$. By the Parallelogram Law,

$$\begin{aligned} \|x_i - x_j\|^2 &= \|(p - x_j) - (p - x_i)\|^2 \\ &= 2\|p - x_j\|^2 + 2\|p - x_i\|^2 - 4\|p - (x_i + x_j)/2\|^2 \\ &\leqq 2\|p - x_j\|^2 + 2\|p - x_i\|^2 - 4d^2 \to 0 \end{aligned}$$

This shows that the sequence x_i has the Cauchy property. Hence, it converges to a point ξ, and it is easily seen that $\xi \in X$ (because X is closed) and that $\|x - \xi\| = d$ (because of continuity). The remainder of the proof need not be changed for Hilbert space.

Suitable versions of the Separation Theorem are true in more general spaces. One geometric application will be cited now. A **closed half-space** in $\mathbb{R}^n$ is defined to be a set of the form

$$\{x : \langle a, x \rangle \geqq \lambda\}$$

where $a \in \mathbb{R}^n$, $\lambda \in \mathbb{R}$, and $a \neq 0$.

THEOREM 6 *Every closed convex set in $\mathbb{R}^n$ is the intersection of all the closed half-spaces that contain it.*

Proof Let X be a closed and convex set. It is clear that X is contained in the intersection of the closed half-spaces that contain X. For the reverse inclusion, suppose that p is a point not in X. By the Separation Theorem, there is a vector $v \neq 0$ such that

$$\langle v, p \rangle < \inf_{x \in X} \langle v, x \rangle$$

Putting

$$\lambda = \inf_{x \in X} \langle v, x \rangle$$

we see that

$$X \subseteqq \{x : \langle v, x \rangle \geqq \lambda\}$$

but this half-space does not contain p. ■

THEOREM 7 **Separation Theorem II** *Let X be closed convex set and Y a compact convex set (in $\mathbb{R}^n$). If X and Y do not intersect, then there exists a $v \in \mathbb{R}^n$ such that*

$$\inf_{x \in X} \langle v, x \rangle > \sup_{y \in Y} \langle v, y \rangle$$

Proof First, we show that the set $Z = X - Y$ is closed. Let $z_k \in Z$ and suppose that $z_k \to z$. Is $z \in Z$? We write $z_k = x_k - y_k$ with $x_k \in X$ and $y_k \in Y$. By the compactness of Y, there is a convergent subsequence $y_{k_i} \to y \in Y$. Then $z_{k_i} \to z$ and $x_{k_i} \to z + y$. Since X is closed, $z + y \in X$. Thus, $z \in X - Y$. Next we note that $0 \notin X - Y$ because the equation $0 = x - y$ (with $x \in X, y \in Y$) would imply that X and Y contained a common point. Also, a quick calculation shows that Z is convex. We can therefore apply the Separation Theorem I to Z, concluding that a vector v exists such that

$$\langle v, 0 \rangle < \inf_{x \in X,\, y \in Y} \langle v, x - y \rangle$$

If ε denotes the infimum in the inequality, we have $\langle v, x - y \rangle \geqq \varepsilon$ for $x \in X$ and $y \in Y$. This leads at once to the inequality asserted in the statement of the theorem. ■

Extreme Points

An **extreme point** of a convex set K is a point $x \in K$ that cannot be written as $x = \theta y + (1 - \theta)z$ with $0 < \theta < 1$, $y \in K$, $z \in K$, and $y \neq z$. In other words, it is *not* an **interior point** of any line segment belonging to k. An equivalent definition is that x is an extreme point of the convex set K if $K \setminus \{x\}$ is convex. As an example, the vertices of a cube are its only extreme points. The extreme points of a solid sphere are all its boundary points.

The next theorem is the **Finite-Dimensional Version of the Krein-Milman Theorem**. (For that theorem, see, for example, Royden [1968].)

THEOREM 8 *Every compact convex set in n-space is the closure of the convex hull of the set of its extreme points.*

Proof The proof is by induction on n. If $n = 1$, the convex set is a closed and bounded interval. The extreme points are the endpoints of the interval, and so the theorem is obviously true in this case. Now assume the validity of the theorem for dimensions less than n, and let K be a compact convex set in n-space. Let E be the set of extreme points of K, and let H denote the convex hull of E. It is to be proved that $\overline{H} = K$ (where $\overline{H}$ denotes the closure of H). Since K is convex and $E \subseteqq K$, we have $H \subseteqq K$. Since K is closed, $\overline{H} \subseteqq K$. We must show that this latter inclusion is not a proper inclusion. Suppose that p is a point of $K \backslash \overline{H}$. By making a translation ($x \longmapsto x - p$), we can assume that $p = 0$. Since $0 \notin \overline{H}$, the Theorem on Homogeneous Inequalities (Theorem 3 in Section 6.9) implies the existence of a vector v such that $\langle v, u \rangle < 0$ for all $u \in \overline{H}$. We put

$$c = \sup\{\langle v, x \rangle : x \in K\}$$

Since $0 \in K$, we have $c \geqq 0$. Since K is compact, this supremum is attained. That means that the set

$$K' = \{x \in K : \langle v, x \rangle = c\}$$

is nonempty. The set K' is also compact, convex, and of dimension $n - 1$, since it lies on a hyperplane. By the induction hypothesis, K' has at least one extreme point, z. Now z is in fact an extreme point of K, as we shall prove. Suppose $z = \theta z_1 + (1 - \theta) z_2$ with $0 < \theta < 1$ and $z_i \in K$. Then

$$c = \langle v, z \rangle = \theta \langle v, z_1 \rangle + (1 - \theta) \langle v, z_2 \rangle \leqq \theta c + (1 - \theta) c = c$$

Hence, $\langle v, z_1 \rangle = \langle v, z_2 \rangle = c$, and $z_i \in K'$. But z is an extreme point of K', and therefore $z_1 = z_2$. This proves that $z \in E$. Hence, $\langle v, z \rangle < 0 \leqq c$: a contradiction. ■

The importance of extreme points in optimization problems stems from the fact that in finding the minimum of a linear function on a compact convex set, we can confine the search to the extreme points. Here is the formal result.

THEOREM 9 *Let K be a compact convex set in $\mathbb{R}^n$, and let f be a linear functional on $\mathbb{R}^n$. The maximum and minimum of f on K are attained at extreme points of K.*

Proof Let

$$c = \sup\{f(x) : x \in K\}$$

Since f is continuous and K is compact, the set

$$K' = \{x \in K : f(x) = c\}$$

is nonempty. It is also convex and compact. Hence, by the Krein-Milman Theorem, K' possesses an extreme point, z. By reasoning familiar from the proof of the Krein-Milman Theorem, z is an extreme point of K. The proof for the minimum of f on K follows by considering the maximum of $-f$. ■

Theorem 9 is valid for a compact convex set in any locally convex linear topological space. It is also true for any continuous convex functional. The Krein-Milman Theorem is true as stated in any locally convex linear topological space. In the finite-dimensional case, it is not necessary to take the closure. See, for example, Holmes [1972, p. 82].

PROBLEMS 10.1

1. Prove that every closed convex set is the intersection of all the open half-spaces that contain it. An open half-space is a set of the form $\{x : \langle a, x\rangle > \lambda\}$.
2. Show that $p \in \text{co}(X)$ if and only if $0 \in \text{co}(X - p)$.
3. Show that if S and T are convex sets, then λS, $S + T$, and $S - T$ are convex. (The set $S + T$ is defined as the set of all sums $s + t$, where $s \in S$ and $t \in T$.)
4. Prove that if L is a linear map and K is a convex set, then $L(K)$ is convex. (The set $L(K)$ is defined as the set of all points $L(x)$, where $x \in K$.)
5. Prove that

$$\text{co}(\lambda S) = \lambda\, \text{co}(S) \qquad \text{co}(S + T) = \text{co}(S) + \text{co}(T)$$

6. Let X be a closed convex set in a Hilbert space. Prove that if $p \notin X$, $\xi \in X$, and $\|p - \xi\| = \text{dist}(p, X)$, then $\langle p - \xi, x - \xi\rangle \leqq 0$ for all $x \in X$.
7. Let U be a compact set in $\mathbb{R}^n$. Prove that if the system of linear inequalities $\langle u, x\rangle > 0$ for $u \in U$ is inconsistent, then it contains an inconsistent subsystem of at most $n + 1$ inequalities.
8. Prove that if K is a closed convex set in Hilbert space, then for each point $p \notin K$ there is a unique closest point k in K.
9. Prove that the mapping $p \longmapsto k$ defined in Problem 8 is nonexpansive. Thus, if (p_1, k_1) and (p_2, k_2) are mapping pairs, then $\|k_1 - k_2\| \leqq \|p_1 - p_2\|$.
10. Can a bounded set have a convex complement?
11. Prove that if the sets X_i are convex, then $\text{co}(X_1 \bigcup \cdots \bigcup X_k)$ is the set of all convex combinations $\sum_{i=1}^{k} \theta_i x_i$, where $x_i \in X_i$.
12. Prove that the closure of a convex set is convex.
13. Let K be a convex set, p an interior point of K, and q any point of K. Show that if $0 < \theta < 1$, then $\theta p + (1 - \theta)q$ is an interior point of K.

14. Prove that if the system of inequalities $\langle u, x\rangle > 0$ for $u \in U$ is consistent, then so is the system $\langle u, x\rangle > 0$ for $u \in \text{co}(U)$.

15. Show that if S and T are compact and convex, then so is $\text{co}(S \bigcup T)$.

16. Show that if X is a bounded set in $\mathbb{R}^n$, then, for any p,

$$\sup\{\|p - x\| : x \in X\} = \sup\{\|p - x\| : x \in \text{co}(X)\}$$

17. Prove that if U is a compact convex set and the system $\langle u, x\rangle > 0$ is inconsistent ($u \in U$), then the same is true for the system $\langle u, x\rangle > 0$, with u in the set of extreme points of U.

18. Show that any convex set in the plane is the union of all triangles whose vertices lie in the given set.

19. Let X be a finite set containing at least $n + 2$ points in $\mathbb{R}^n$. Prove that it is possible to write $X = X_1 \bigcup X_2$ where $\text{co}(X_1) \bigcap \text{co}(X_2) = \varnothing$.

20. Prove that the convex hull of an open set in $\mathbb{R}^n$ is open.

21. Show by an example that the convex hull of a closed set need not be closed.

22. For a set X in linear space of possibly infinite dimension, let H_n be the set of all points that can be written as $\sum_{i=1}^n \theta_i x_i$, with $\theta_i \geqq 0$, $\sum_{i=1}^n \theta_i = 1$ and $x_i \in X$. Prove that $\bigcup_{n=1}^\infty H_n$ is convex. Prove that this set is $\text{co}(X)$.

23. In $\mathbb{R}^n$, let the line segment joining x and y be denoted by $\overline{xy}$. For a given set X_0, define $X_1, X_2, \ldots, X_k$ by putting

$$X_{k+1} = \bigcup\{\overline{xy} : x \in X_k,\ y \in X_k\}$$

Prove that $X_{2^n+1} = \text{co}(X_0)$.

*10.2 Linear Inequalities

Let X be a vector space over the real field. A **linear functional** is a linear map of X into $\mathbb{R}$. A **linear inequality** is a statement $f(x) \geqq \alpha$ in which f is a linear functional. A single inequality of this type has for its set of solutions a **half-space** $\{x : f(x) \geqq \alpha\}$. Of much more interest are **systems of linear inequalities** $f_i(x) \geqq \alpha_i$, for $i \in I$, where I is some index set, not necessarily finite.

Systems of Homogeneous Equations

We begin with a theorem from linear algebra concerning systems of homogeneous equations and then proceed to an analogue concerning inequalities. We use the notation f^0 for the **null space of a functional** f:

$$f^0 = \{x \in X : f(x) = 0\}$$

Also, $\mathcal{L}(f_1, f_2, \ldots, f_m)$ denotes the **linear span** of the set $\{f_1, f_2, \ldots, f_m\}$.

THEOREM 1 Linear Dependence Theorem *These are equivalent properties for a set of linear functionals:*

(i) $f^0 \supseteqq \bigcap_{i=1}^{m} f_i^0$

(ii) $f \in \mathcal{L}(f_1, f_2, \ldots, f_m)$

Proof The implication **(ii)** $\Rightarrow$ **(i)** is easy. Indeed, if $f = \sum_{i=1}^{m} \lambda_i f_i$, and if $x \in \bigcap_{i=1}^{m} f_i^0$, then $f_i(x) = 0$ for all i and obviously $f(x) = 0$.

For the converse, we proceed by induction on m. Let $m = 1$ and suppose that $f^0 \supseteqq f_1^0$. If $f_1 = 0$, then $f_1^0 = X$. Consequently, $f^0 = X$ and $f = 0$. Hence, $f \in \mathcal{L}(f_1)$ in this case. If $f_1 \neq 0$, take a point y such that $f_1(y) = 1$, and let x be any point in X. We have $f_1[x - f_1(x)y] = f_1(x) - f_1(x)f_1(y) = 0$, so that $x - f_1(x)y \in f_1^0$. By hypothesis, $x - f_1(x)y \in f^0$. Hence, $f(x) - f_1(x)f(y) = 0$ or $f = f(y)f_1$. Thus, $f \in \mathcal{L}(f_1)$.

For the inductive step, suppose that the theorem has been proved for the integer m. Suppose

$$f^0 \supseteqq \bigcap_{i=1}^{m+1} f_i^0$$

Let $Y = f_{m+1}^0$. This is a subspace of X. Denoting restriction to Y by the notation $f \mid Y$, we have

$$(f \mid Y)^0 \supseteqq \bigcap_{i=1}^{m} (f_i \mid Y)$$

By the induction hypothesis,

$$f \mid Y = \sum_{i=1}^{m} \lambda_i f_i \mid Y$$

for appropriate λ_i. Now two equivalent equations are $(f - \sum_{i=1}^{m} \lambda_i f_i) \mid Y = 0$ and $(f - \sum_{i=1}^{m} \lambda_i f_i)^0 \supseteqq f_{m+1}^0$. Using the case $m = 1$ of the present theorem, we conclude that

$$f - \sum_{i=1}^{m} \lambda_i f_i = \lambda_{m+1} f_{m+1}$$

for some λ_{m+1}. ■

Linear Inequalities

Now we come to an exact analogue of this theorem for linear inequalities. Instead of the null spaces of the functionals, we use the **half-spaces**:

$$f^+ = \{x \in X : f(x) \geqq 0\}$$

Instead of the ordinary linear span of $f_1, f_2, \ldots, f_m$, we consider the **cone** that they generate:

$$C\{f_1, f_2, \ldots, f_m\} = \left\{ \sum_{i=1}^{m} \lambda_i f_i : \lambda_i \geqq 0 \right\}$$

THEOREM 2 **Farkas's Theorem, 1902** *These are equivalent for linear functionals f and f_i:*

(i) $f^+ \supseteq \bigcap_{i=1}^{m} f_i^+$

(ii) $f \in C(f_1, f_2, \ldots, f_m)$

Proof If **(ii)** is true then **(i)** follows easily. Indeed suppose that

$$f = \sum_{i=1}^{m} \lambda_i f_i$$

with $\lambda_i \geqq 0$. If $x \in \bigcap_{i=1}^{m} f_i^+$, then $f_i(x) \geqq 0$ for all i, and it follows obviously that $f(x) \geqq 0$ also.

For the converse, we give the proof under the assumption that $X = \mathbb{R}^n$. Every linear functional is of the form $f(x) = \langle v, x \rangle$ for an appropriate v. Assume now that $f \notin C$, where C is the cone $C(f_1, f_2, \ldots, f_m)$. By the Separation Theorem, there is a vector v such that

$$\langle v, f \rangle < \inf_{u \in C} \langle v, u \rangle$$

Let

$$k = \inf_{u \in C} \langle v, u \rangle$$

Since $0 \in C$, we have $k \leqq 0$. If $k < 0$, we select $u \in C$ such that $k \leqq \langle v, u \rangle < 0$. For all positive t, we have $tu \in C$. If t is sufficiently large, we shall have $\langle v, tu \rangle < k$, which is a contradiction. Hence, $k = 0$ and

$$\langle v, f \rangle < 0 \leqq \langle v, u \rangle$$

for $u \in C$. It follows that $f(v) < 0 \leqq f_i(v)$, showing that **(i)** is false. ■

Farkas's Theorem has the following matrix-vector formulation.

THEOREM 3 *The following properties of a matrix A and a vector c are equivalent:*

(i) *For all x, if $Ax \geqq 0$ then $c^T x \geqq 0$.*

(ii) *For some y, $y \geqq 0$ and $c = A^T y$.*

Consistent and Inconsistent Systems

Next we present a nonhomogeneous analogue to Theorem 3. In it we use the following terminology: One system of inequalities, $f_i(x) \geqq \alpha_i$, is said to be a *consequence* of another system, $g_i(x) \geqq \beta_i$, if each solution of the second satisfies the first. Thus,

$$\{x : g_i(x) \geqq \beta_i \text{ for all } i\} \subseteqq \{x : f_i(x) \geqq \alpha_i \text{ for all } i\}$$

THEOREM 4 **Nonhomogeneous Farkas's Theorem** *If a linear inequality $f(x) \geqq \alpha$ is a consequence of a consistent system of linear inequalities*

$$g_i(x) \geqq \beta_i \qquad (1 \leqq i \leqq n)$$

then, for suitable $\theta_i \geqq 0$, we have

$$f = \sum_{i=1}^{n} \theta_i g_i \qquad \textit{and} \qquad \sum_{i=1}^{n} \theta_i \beta_i \geqq \alpha$$

Proof Consider the system

$$g_i(x) - \lambda\beta_i \geqq 0 \qquad (\lambda > 0) \tag{1}$$

If the pair (x, λ) solves System (1), then $g_i(x) \geqq \lambda\beta_i$ and $g_i(x/\lambda) \geqq \beta_i$. By the hypothesis in the theorem, $f(x/\lambda) \geqq \alpha$. Hence,

$$f(x) - \lambda\alpha \geqq 0 \tag{2}$$

This shows that Inequality (2) is a consequence of System (1). Now consider the system

$$g_i(x) - \lambda\beta_i \geqq 0 \qquad (\lambda \geqq 0) \tag{3}$$

If (2) is a consequence of (3), then the homogeneous form of Farkas's Theorem applies, and we conclude that

$$(f; -\alpha) = \sum_{i=1}^{n} \theta_i(g_i; -\beta_i) + \theta_0(0; 1) \qquad (\theta_i \geqq 0)$$

where $(f; -\alpha)$ denotes simply the pair, considered as a vector. This conclusion can be written in the form

$$f = \sum_{i=1}^{n} \theta_i g_i \qquad \alpha = \sum_{i=1}^{n} \theta_i\beta_i - \theta_0 \leqq \sum_{i=1}^{n} \theta_i\beta_i$$

This is the assertion to be proved. We are not finished, however, because it may happen that (2) is not a consequence of (3), although, as we showed, it *is* a consequence of (1). In this case, there is a solution of (3) that solves neither (1) nor (2). Such a solution pair $(u; \lambda)$ must have $\lambda = 0$ because otherwise it solves (1) and hence (2). Thus, $g_i(u) \geqq 0 > f(u)$. By the hypotheses of the theorem, there is a vector v such that $g_i(v) \geqq \beta_i$. Select a positive number λ such that $f(u + \lambda v) < \lambda\alpha$. This is possible because as $\lambda \downarrow 0$, the right side approaches 0 while the left side approaches the negative number $f(u)$. We have a contradiction of hypotheses, since $f(u/\lambda + v) < \alpha$ while $g_i(u/\lambda + v) \geqq g_i(v) \geqq \beta_i$. ■

Matrix-Vector Forms

The matrix-vector version of Theorem 4 follows.

THEOREM 5 *If the system*

$$Ax \geqq b$$

is consistent, and the system

$$Ax \geqq b \qquad c^T x < \alpha$$

is inconsistent, then the system

$$A^T y = c \qquad y^T b \geqq \alpha \qquad y \geqq 0$$

is consistent.

THEOREM 6 *If the system*

$$Ax = b \qquad x \geqq 0 \tag{4}$$

is inconsistent, then the system

$$A^T y \geqq 0 \qquad b^T y \leqq 0 \tag{5}$$

is consistent.

Proof If (4) is inconsistent, then

$$b \notin K \equiv \{Ax : x \geqq 0\}$$

Since K is closed and convex, the Separation Theorem of Section 10.1 applies, and there exists a vector y such that

$$\langle y, b\rangle < \inf_{x \geqq 0} \langle y, Ax\rangle$$

Since x can be 0 in this calculation, we have $\langle y, b\rangle < 0$. It remains to be verified that $A^T y \geqq 0$. If this is not true, then for some index α, $(A^T y)_\alpha < 0$. Let x be the vector with coordinates $x_j = \lambda\delta_{\alpha j}$. Then

$$\langle y, Ax\rangle = \sum_{j=1}^{n}(A^T y)_j x_j = \lambda(A^T y)_\alpha \longrightarrow -\infty \qquad (\text{ as } \lambda \longrightarrow +\infty)$$

For suitable λ, we then must have $\langle y, Ax\rangle < \langle y, b\rangle$, which is a contradiction. Thus, y solves System (5). ■

THEOREM 7 *If the system*

$$Ax \leqq b \qquad x \geqq 0 \tag{6}$$

is inconsistent, then the system

$$A^T y \geqq 0 \qquad b^T y < 0 \qquad y \geqq 0 \tag{7}$$

is consistent.

Proof If (6) is inconsistent, then so is

$$Ax + z = b \qquad x \geqq 0 \qquad z \geqq 0 \tag{8}$$

We write (8) in the following form:

$$\begin{bmatrix} A & I \end{bmatrix}\begin{bmatrix} x \\ z \end{bmatrix} = b \qquad \begin{bmatrix} x \\ z \end{bmatrix} \geqq 0 \tag{9}$$

By Theorem 6, the system

$$\begin{bmatrix} A^T \\ I \end{bmatrix} y \geqq 0 \qquad b^T y < 0 \tag{10}$$

is consistent. This is System (7) in disguise. ■

PROBLEMS 10.2

1. Prove that for an $m \times n$ matrix A, one or the other of these systems is consistent, but not both:

   (i) $Ax = 0 \qquad x \geqq 0 \qquad x \neq 0$

   (ii) $A^T y > 0$

2. Prove that for an $m \times n$ matrix A, one or the other of these systems is consistent, but not both:

   (i) $Ax \leqq 0 \qquad x \geqq 0 \qquad x \neq 0$

   (ii) $A^T y > 0 \qquad y > 0$

   (*Note:* This is known as **Ville's Theorem**, 1938.)

3. Define $P_n = \{x \in \mathbb{R}^n : x \geqq 0 \text{ and } \sum_{i=1}^n x_i = 1\}$. Prove that for any $m \times n$ matrix A, either $Ax \geqq 0$ for some $x \in P_n$ or $A^T y \leqq 0$ for some $y \in P_m$.

4. Using the notation of Problem 3, prove that for any $m \times n$ matrix A,

$$\max_{x \in P_n} \min_{y \in P_m} y^T Ax \leqq \min_{y \in P_m} \max_{x \in P_n} y^T Ax$$

Hint: Start with $\min_y y^T Ax \leqq \max_x y^T Ax$.

5. Prove that if U is the $m \times n$ matrix consisting entirely of 1's, then for all $x \in P_n$ and all $y \in P_m$,

$$y^T(A - \lambda U)x = y^T Ax - \lambda$$

6. (**Min-Max Theorem of Game Theory**) Prove that

$$\max_{x \in P_n} \min_{y \in P_m} y^T Ax = \min_{y \in P_m} \max_{x \in P_n} y^T Ax$$

Hints: If the inequality in Problem 4 is strict, then let λ be a number between the two quantities. Use Problem 5. Apply Problem 3 to the matrix $A - \lambda U$.

7. Prove that if the inequalities $Ax \geqq b$ have no solution, then for some y the following hold: $y \geqq 0$, $A^T y = 0$, and $b^T y = 1$.

8. Prove that if the inequality $Ax \geqq b$ has no nonnegative solution, then the inequalities $A^T y \leqq 0$, $b^T y > 0$ have a nonnegative solution.

9. Prove that if the system $Ax = 0$, $x \geqq 0$, $x \neq 0$ is inconsistent, then the system $A^T y < 0$ is consistent.

10. Prove that if the system $Ax \geqq 0$, $x \geqq 0$, $x \neq 0$ is inconsistent, then the system $A^T y < 0$, $y \geqq 0$ is consistent.

11. Prove that if the system $Ax = 0$, $x > 0$ is inconsistent, then the system $Ax \leqq 0$ is consistent.

12. Prove that if A is an $n \times (n+1)$ matrix, then the system $Ax \geqq 0$, $x \neq 0$ is consistent.

13. Prove that for any $m \times n$ matrix A, the system $Ax > 0$ is consistent if and only if the system $A^T y = 0$, $y > 0$ is inconsistent.

14. Find a necessary and sufficient condition that the set

$$K = \{x \in \mathbb{R}^n : Ax \leqq b\}$$

be bounded.

15. Prove that one or the other of these systems is consistent, but not both:

(i) $Ax = b$

(ii) $A^T y = 0 \qquad b^T y = 1$

10.3 Linear Programming

The term **linear programming** does not refer to the programming of a computer but to the programming of business or economic enterprises. An explicit and technical meaning has been assigned to the term: It means finding the maximum value of a linear function of n real variables over a convex polyhedral set in $\mathbb{R}^n$. We adopt the following standard form for such a problem.

LP PROBLEM 1 **First Standard Form** *Let $c \in \mathbb{R}^n$, $b \in \mathbb{R}^m$, and $A \in \mathbb{R}^{m \times n}$. Find the maximum of $c^T x$ subject to the constraints $x \in \mathbb{R}^n$, $Ax \leqq b$, and $x \geqq 0$.*

We remind the reader that if $x = (x_1, x_2, \dots, x_n)^T$, then the vector inequality $x \geqq 0$ means that $x_i \geqq 0$ for all $i \in \{1, 2, \dots, n\}$. Likewise, the inequality

$$Ax \leqq b$$

means that

$$\sum_{j=1}^{n} a_{ij} x_j \leqq b_i \qquad \text{for all} \quad i \in \{1, 2, \dots, m\}$$

Next we describe some terminology that is commonly used. The **feasible set** in our problem is the set

$$K = \{x \in \mathbb{R}^n : Ax \leqq b,\ x \geqq 0\}$$

The *value of the problem* is the number

$$v = \sup\{c^T x : x \in K\}$$

A **feasible point** is any element of K. A **solution** or **optimal feasible point** is any $x \in K$ such that $c^T x = v$. The function $x \longmapsto c^T x = \sum_{j=1}^{n} c_j x_j$ is the **objective function**. Since the problem is completely determined by the data A, b, c, we refer to it as linear programming problem (A, b, c).

Techniques for Converting Problems

Almost any problem involving the optimization of a linear function with variables subject to linear inequalities can be put into the linear programming format. Doing this often requires one or more of the following ideas:

(i) If it is desired to minimize $c^T x$, that is the same as maximizing $-c^T x$.
(ii) Any constraint of the form $a^T x \geqq \beta$ is equivalent to $-a^T x \leqq -\beta$.
(iii) Any constraint of the form $a^T x = \beta$ is equivalent to $a^T x \leqq \beta$, $-a^T x \leqq -\beta$.
(iv) Any constraint of the form $|a^T x| \leqq \beta$ is equivalent to $a^T x \leqq \beta$, $-a^T x \leqq \beta$.

(v) If the objective function includes an additive constant, it has no effect on the solution. Thus, the maximum of $c^T x + \beta$ occurs at the same points as the maximum of $c^T x$.

(vi) If a given problem does *not* require a variable x_j to be nonnegative, we can replace x_j by the difference of two variables that *are* required to be nonnegative: $x_j = u_j - v_j$, say.

Example 1 Convert this problem to a linear programming problem in standard form:

$$\text{Minimize:} \quad 7x_1 - x_2 + x_3 - 4$$

$$\text{Constraints:} \begin{cases} x_1 + x_2 - x_3 \geqq 2 \\ 3x_1 + 4x_2 + x_3 = 6 \\ |x_1 - 2x_2 + 3x_3| \leqq 5 \\ x_1 \geqq 0, x_2 \leqq 0 \end{cases}$$

Solution We let $u_1 = x_1$, $u_2 = -x_2$, and $u_3 - u_4 = x_3$. Now we have

$$\text{Maximize:} \quad -7u_1 - u_2 - u_3 + u_4$$

$$\text{Constraints:} \begin{cases} -u_1 + u_2 + u_3 - u_4 \leqq -2 \\ 3u_1 - 4u_2 + u_3 - u_4 \leqq 6 \\ -3u_1 + 4u_2 - u_3 + u_4 \leqq -6 \\ u_1 + 2u_2 + 3u_3 - 3u_4 \leqq 5 \\ -u_1 - 2u_2 - 3u_3 + 3u_4 \leqq 5 \\ u_1 \geqq 0, u_2 \geqq 0, u_3 \geqq 0, u_4 \geqq 0 \end{cases}$$

■

A given linear programming problem (A, b, c) may or may not have a solution. To begin with, the feasible set K may be empty, and then no solution exists. If the feasible set is nonempty and unbounded, it can happen that the objective function has no upper bound on K. In that case $v = +\infty$, and there is no solution. If K is empty, $v = -\infty$, and there is no solution. If K is nonempty and bounded, then there exists at least one solution. This is a consequence of the fact that K is then compact (closed and bounded in $\mathbb{R}^n$), and thus the objective function (which is continuous) *attains* its supremum on K.

Dual Problem

With any linear programming problem (A, b, c), we can associate another problem $(-A^T, -c, -b)$. This problem is called the **dual** of the original. For example, the problem

$$\text{Maximize:} \quad 3x_1 - 2x_2$$

$$\text{Constraints:} \begin{cases} 7x_1 + x_2 \leqq 18 \\ -3x_1 + 5x_2 \leqq 25 \\ 6x_1 - x_2 \leqq 13 \\ x_1 \geqq 0, x_2 \geqq 0 \end{cases}$$

has the following dual problem:

$$\text{Maximize:} \quad -18y_1 - 25y_2 - 13y_3$$

$$\text{Constraints:} \begin{cases} -7y_1 + 3y_2 - 6y_3 \leqq -3 \\ -y_1 - 5y_2 + y_3 \leqq 2 \\ y_1 \geqq 0,\ y_2 \geqq 0,\ y_3 \geqq 0 \end{cases}$$

The relationship between a linear programming problem and its dual is the subject of **duality theory**. Some of its salient results will now be developed.

THEOREM 1 *If x is a feasible point for a linear programming problem (A, b, c), and if y is a feasible point for the dual problem $(-A^T, -c, -b)$, then*

$$c^T x \leqq y^T A x \leqq b^T y$$

If equality occurs here, then x and y are solutions of their respective problems.

Proof The points x and y satisfy

$$x \geqq 0 \qquad Ax \leqq b \qquad y \geqq 0 \qquad -A^T y \leqq -c$$

From this, we have

$$c^T x \leqq (A^T y)^T x = y^T A x \leqq y^T b = b^T y$$

The values v_1 and v_2 of the two problems must therefore satisfy

$$c^T x \leqq v_1 \leqq b^T y$$
$$-b^T y \leqq v_2 \leqq -c^T x$$

If $c^T x = b^T y$, then clearly $c^T x = v_1 = b^T y = -v_2$. ■

Theorem 1 can often be used to estimate the value, v_1, of a linear programming problem. If a feasible point x is known, and if a feasible point y for the dual problem is known, then the inequality $c^T x \leqq v_1 \leqq b^T y$ defines an interval containing v_1.

THEOREM 2 *If a linear programming problem and its dual both have feasible points, then both problems have solutions, and their values are the negatives of each other.*

Proof By Theorem 1, it suffices to prove that an x and y exist such that

$$x \geqq 0 \qquad Ax \leqq b \qquad y \geqq 0 \qquad -A^T y \leqq -c \qquad c^T x \geqq b^T y$$

Indeed, such a pair (x, y) gives a solution, x, for the original problem and a solution, y, for its dual. Our task is to prove that the following system of linear inequalities is consistent:

$$\begin{bmatrix} A & 0 \\ 0 & -A^T \\ -c^T & b^T \end{bmatrix} \begin{bmatrix} x \\ y \end{bmatrix} \leqq \begin{bmatrix} b \\ -c \\ 0 \end{bmatrix} \qquad \begin{bmatrix} x \\ y \end{bmatrix} \geqq \begin{bmatrix} 0 \\ 0 \end{bmatrix}$$

Let us assume that this system is *inconsistent* and try to deduce a contradiction. By Theorem 7 in Section 10.2, the following system is *consistent*:

$$\begin{cases} \begin{bmatrix} A^T & 0 & -c \\ 0 & -A & b \end{bmatrix} \begin{bmatrix} u \\ v \\ \lambda \end{bmatrix} \geqq \begin{bmatrix} 0 \\ 0 \end{bmatrix} \\ \begin{bmatrix} b^T & -c^T & 0 \end{bmatrix} \begin{bmatrix} u \\ v \\ \lambda \end{bmatrix} < 0 \\ \begin{bmatrix} u \\ v \\ \lambda \end{bmatrix} \geqq 0 \end{cases}$$

Here u and v are vectors and λ is a constant. Let (u, v, λ) satisfy this system. Then

$$\begin{cases} A^T u - \lambda c \geqq 0 & -Av + \lambda b \geqq 0 & b^T u - c^T v < 0 \\ u \geqq 0 & v \geqq 0 & \lambda \geqq 0 \end{cases}$$

Suppose first that $\lambda > 0$. Then $\lambda^{-1} v$ is feasible for the problem (A, b, c), and $\lambda^{-1} u$ is feasible for the dual problem $(-A^T, -c, -b)$. Hence by Theorem 1, we have $c^T(\lambda^{-1} v) \leqq b^T(\lambda^{-1} u)$ and $b^T u - c^T v \geqq 0$, a contradiction.

If $\lambda = 0$, then $A^T u \geqq 0 \geqq Av$. Taking x feasible for the original problem and y feasible for the dual problem, we arrive at contradiction of a previous inequality:

$$c^T v \leqq (A^T y)^T v = y^T A v \leqq 0 \leqq (A^T u)^T x = u^T (Ax) \leqq u^T b = b^T u \qquad \blacksquare$$

THEOREM 3 *If either a linear programming problem or its dual has a solution, then so has the other.*

Proof Since the dual of the dual problem is the original problem, it is only necessary to prove one case of this theorem. Suppose that the dual problem $(-A^T, -c, -b)$ has a solution y_0. Then the system of inequalities

$$-A^T y \leqq -c \qquad y \geqq 0 \qquad -b^T y \geqq -b^T y_0$$

is inconsistent. The corresponding system in which we omit the third inequality is, of course, consistent. We write the inconsistent system in the form

$$\begin{bmatrix} A^T \\ I \end{bmatrix} y \geqq \begin{bmatrix} c \\ 0 \end{bmatrix} \qquad b^T y \leqq b^T y_0$$

Now use the Nonhomogeneous Farkas's Theorem from Section 10.2. It implies the consistency of the system

$$\begin{bmatrix} A & I \end{bmatrix} \begin{bmatrix} x \\ u \end{bmatrix} = b \qquad \begin{bmatrix} x^T & u^T \end{bmatrix} \begin{bmatrix} c \\ 0 \end{bmatrix} \geqq b^T y_0 \qquad \begin{bmatrix} x \\ u \end{bmatrix} \geqq \begin{bmatrix} 0 \\ 0 \end{bmatrix}$$

Thus,

$$Ax + u = b \qquad x^T c \geqq b^T y_0 \qquad x \geqq 0 \qquad u \geqq 0$$

It follows that

$$Ax \leqq b \qquad c^T x \geqq b^T y_0 \qquad x \geqq 0$$

Hence, x is a solution of the original problem, by Theorem 1. ■

THEOREM 4 *Let x and y be feasible points for a linear programming problem and its dual, respectively. These points are solutions of their respective problems if and only if $(Ax)_i = b_i$ for each index i such that $y_i > 0$, and $(A^T y)_i = c_i$ for each index i such that $x_i > 0$.*

Proof If x and y are solutions, then by Theorems 1 and 2,

$$y^T b = b^T y = y^T Ax = c^T x = x^T c$$

This yields the equation $y^T(b - Ax) = 0$. Since $y \geqq 0$ and $b - Ax \geqq 0$, we conclude that $y_i(b_i - (Ax)_i) = 0$ for each i. Thus, $(Ax)_i = b_i$ whenever i is an index for which $y_i > 0$. The other condition follows by a symmetry argument. For the converse, suppose that $y_i(b_i - (Ax)_i) = 0$ and $x_i(c_i - (A^T y)_i) = 0$ for all i. Then

$$b^T y = y^T b = y^T Ax = x^T A^T y = x^T c = c^T x$$

By Theorem 1, x and y are solutions of their respective problems. ■

PROBLEMS 10.3

1. Convert these problems to the standard and dual linear programming form.

(a) Minimize: $3x_1 + x_2 - 5x_3 + 2$

$$\text{Constraints: } \begin{cases} x_1 \geqq x_2 \\ x_2 \leqq 0 \\ -x_1 + 4x_3 \geqq 0 \\ x_1 + x_2 + x_3 = 0 \end{cases}$$

(b) Minimize: $|x_1 + x_2 + x_3|$

$$\text{Constraints: } \begin{cases} x_1 - x_2 = 5 \\ x_2 - x_3 = 7 \\ x_1 \leqq 0, x_3 \geqq 2 \end{cases}$$

(c) Minimize: $|x_1| - |x_2|$

$$\text{Constraints: } \begin{cases} x_1 + x_2 = 5 \\ 2x_1 + 3x_2 - x_3 \leqq 0 \\ x_3 \geqq 4 \end{cases}$$

2. What can you prove about the linear programming problem (A, b, c) if every feasible point is a solution?

10.4 The Simplex Algorithm

By means of standard techniques, a linear programming problem can be put into the following form:

LP PROBLEM 2 **Second Standard Form** *Maximize $c^T x$ subject to constraints $Ax = b$ and $x \geqq 0$. Here $x \in \mathbb{R}^n$, $c \in \mathbb{R}^n$, $b \in \mathbb{R}^m$, and $A \in \mathbb{R}^{m \times n}$.*

Basic Concepts

Recall that the expression $c^T x$ defines the **objective function**. We note that if a linear programming problem has constraints of the form

$$Ax \leqq b$$

then, by introducing a vector $u \geqq 0$ (whose components are called **slack variables**), we can write

$$Ax + u = b$$

thus obtaining the second standard form given above.

For a problem in the second standard form, the **feasible set** is

$$K = \{x \in \mathbb{R}^n : Ax = b,\ x \geqq 0\}$$

For any $x \in K$, we define the set of indices for which the components of x are positive by

$$I(x) = \{i : 1 \leqq i \leqq n \text{ and } x_i > 0\}$$

The columns of the matrix A are denoted by $A_1, A_2, \ldots, A_n$. The equation

$$Ax = b$$

becomes

$$\sum_{i=1}^{n} x_i A_i = b$$

Now we can state an important theorem.

THEOREM 1 *Let $x \in K$. The following properties of x are equivalent:*

(i) *x is an extreme point of K.*
(ii) *$\{A_i : i \in I(x)\}$ is linearly independent.*

Proof Suppose that **(ii)** is true. We shall prove that **(i)** is true. Let $x = \theta u + (1 - \theta)v$, where $u \in K$, $v \in K$, and $0 < \theta < 1$. For each $i \notin I(x)$, we have

$$0 = x_i = \theta u_i + (1 - \theta)v_i$$

Recalling that $u_i \geqq 0$ and $v_i \geqq 0$, we conclude that $u_i = v_i = 0$ for $i \notin I(x)$. Then we have

$$0 = Au - Av = \sum_{i=1}^{n}(u_i - v_i)A_i = \sum_{i \in I(x)}(u_i - v_i)A_i$$

By **(ii)**, we conclude that $u_i = v_i$. Hence, x is an extreme point of K.

Now suppose that **(i)** is true. If

$$\sum_{i \in I(x)} w_i A_i = 0$$

then put $w_i = 0$ for $i \notin I(x)$. Obviously

$$\sum_{i \in I(x)} (x_i \pm \lambda w_i)A_i = b$$

Since $x_i > 0$ for $i \in I(x)$, we can take $\lambda \neq 0$ and so small that $x_i + \lambda w_i > 0$ and $x_i - \lambda w_i > 0$ for $i \in I(x)$. Then $u = x + \lambda w$ and $v = x - \lambda w$ are feasible points. Since $x = \frac{1}{2}(u + v)$ and x is an extreme point of K, we conclude that $u = v$ and that $w = 0$. This proves **(ii)**. ∎

COROLLARY 1 *The feasible set K can have only a finite number of extreme points.*

Proof Let E be the set of extreme points of K. For each $x \in E$, $I(x) \subseteqq \{1, 2, \dots, n\}$, so $I : E \longrightarrow 2^{\{1,2,\dots,n\}}$. (The notation 2^S denotes the **family of all subsets of a set S**.) The mapping I thus defined (on E) is one to one. To verify this, let $x, y \in E$ and $x \neq y$.

Then

$$b = Ax = \sum_{i=1}^{n} x_i A_i = \sum_{i \in I(x)} x_i A_i$$

$$b = Ay = \sum_{i=1}^{n} y_i A_i = \sum_{i \in I(y)} y_i A_i$$

If $I(x) = I(y)$, we have here a contradiction of the linear independence of the set $\{A_i : i \in I(x)\}$. This mapping of E injectively into $2^{\{1,2,\ldots,n\}}$ shows that the number of elements of E cannot exceed 2^n. ■

The simplex algorithm of Dantzig [1948] consists of two parts. In the first part, an initial extreme point, $x^{(1)}$, of K is found. In the second part, a finite sequence of extreme points is generated, starting with $x^{(1)}$, such that the objective function increases with each point generated. Let this sequence be denoted by $x^{(1)}, x^{(2)}, \ldots$. It follows that $c^T x^{(1)} < c^T x^{(2)} < \cdots$. If the problem has no solution, then this fact is discovered in the course of the algorithm. If the linear programming problem has a solution, then at a finite stage in the algorithm, an extreme point $x^{(k)}$ is produced that is a solution; that is, it maximizes the objective function.

Abstract Form

Here we describe an abstract form of the second part of the algorithm. We adopt the following hypothesis.

ASSUMPTION 1 **Nondegeneracy Assumption** *Each extreme point x of the feasible set K has exactly m positive components.*

Suppose that x is an extreme point of K. By Theorem 1, the set $\{A_i : i \in I(x)\}$ is linearly independent. Hence by the Nondegeneracy Assumption, it is a basis for $\mathbb{R}^m$. (For this reason, we find such an x referred to in the literature as a **basic feasible point**.) There must exist coefficients D_{ij} such that

$$A_j = \sum_{i \in I(x)} D_{ij} A_i \qquad (1 \leqq j \leqq n)$$

If $i \notin I(x)$, we put $D_{ij} = 0$. Then

$$A_j = \sum_{i=1}^{n} D_{ij} A_i$$

or

$$A = AD^T$$

We define

$$d = D^T c$$

where c is the vector appearing in the objective function. Note that D and d depend on x and will change in the course of the algorithm.

For any index $q \notin I(x)$ and for any $\lambda \in \mathbb{R}$,

$$\begin{aligned} b = Ax &= \sum_{i=1}^{n} x_i A_i + \lambda A_q - \lambda \sum_{i=1}^{n} D_{iq} A_i \\ &= \sum_{i=1}^{n} (x_i - \lambda D_{iq} + \lambda \delta_{iq}) A_i \\ &\equiv \sum_{i=1}^{n} y_i A_i \end{aligned}$$

Hence,

$$Ay = b$$

where $y = x - \lambda D_q + \lambda d_q$. Here D_q denotes the qth column of D. Our objective now is to select q and λ so that y is an extreme point of K (that is, $y \geqq 0$) and $c^T y > c^T x$.

Since $q \notin I(x)$, $D_{qq} = 0$ and $x_q = 0$. Hence, $y_q = \lambda$. Because y is to be a feasible point, we require $\lambda \geqq 0$. Now compute:

$$\begin{aligned} c^T y &= \sum_{i=1}^{n} c_i x_i - \lambda \sum_{i=1}^{n} c_i D_{iq} + \lambda \sum_{i=1}^{n} c_i \delta_{iq} \\ &= c^T x - \lambda c^T D_q + \lambda c_q \\ &= c^T x + \lambda (c_q - d_q) \end{aligned}$$

To increase the value of the objective function, we shall select q so that $c_q > d_q$. If $c \leqq d$, no q exists, and the computations stop; x is a solution. Otherwise, it is usual to select q so that $c_q - d_q$ is as large as possible. From now on q is fixed.

The choice of λ is made so that $\lambda > 0$ and so that $I(y)$ will have at most m elements. From the definition of y, we see that $I(y) \subseteqq I(x) \bigcup \{q\}$. Since $I(x)$ has exactly m elements, we select λ so that one of the terms $x_i - \lambda D_{iq}$ is 0 while the others are $\geqq 0$. Observe first, however, that if $D_{iq} \leqq 0$ for $1 \leqq i \leqq n$, then $y \in K$ for all $\lambda > 0$. By the preceding formula (for $c^T y$), we see that in this case, $\lim_{\lambda \to \infty} c^T y = +\infty$. Hence, there is no solution because the objective function is unbounded on K. If $D_{iq} > 0$ for one or more values of i, we think of λ increasing from the value 0. At the beginning, $x_i - \lambda D_{iq} > 0$ for $i \in I(x)$. Some of these terms are decreasing, and when one of them becomes 0, we take the corresponding λ. Formally,

$$\lambda = \min\left\{\frac{x_i}{D_{iq}} : x_i > 0,\ D_{iq} > 0\right\}$$

Having thus fully determined y, we want to verify that it is an extreme point. First let $\lambda = x_p/D_{pq}$, where p is such that $x_p > 0$ and $D_{pq} > 0$. Then

$$I(y) \subseteqq I(x) \bigcup \{q\} \setminus \{p\}$$

By Theorem 1, it suffices to prove that the set

$$\left\{A_i : i \in I(x) \bigcup \{q\} \setminus \{p\}\right\}$$

is linearly independent. Suppose that

$$\sum_{i \in I(x)} \beta_i A_i + \beta_q A_q = 0 \qquad \text{with} \qquad \beta_p = 0$$

If $\beta_q = 0$, then this equation reduces to

$$\sum_{i \in I(x)} \beta_i A_i = 0$$

The independence of $\{A_i : i \in I(x)\}$ then implies that $\beta_i = 0$ for $i \in I(x)$. Hence, *all* the β_i's are 0 in this case. We can therefore proceed to the case when $\beta_q \neq 0$. By homogeneity, we can assume $\beta_q = -1$. Our equation now reads

$$A_q = \sum_{i \in I(x)} \beta_i A_i \qquad \text{with} \qquad \beta_p = 0$$

We also have

$$A_q = \sum_{i=1}^{n} D_{iq} A_i = \sum_{i \in I(x)} D_{iq} A_i$$

By the linear independence of $\{A_i : i \in I(x)\}$, we conclude that $\beta_i = D_{iq}$. But this cannot be true, since $\beta_p = 0$ and $D_{pq} > 0$ as noted previously. Hence, β_q must be 0. We have established that $y \in K$ and is an extreme point.

One detail remains to be proved; namely, that if $c \leqq d$, then x is a solution. Let u be any feasible point. Then $Ax = b = Au = A(D^T u)$. It follows that $x = D^T u$ since Ax and $AD^T u$ are linear combinations of $\{A_i : i \in I(x)\}$. Then we have the condition $c^T u \leqq d^T u = c^T Du = c^T x$, as we wished to prove.

Example 1 We shall illustrate how the simplex algorithm works in a concrete example. This is the problem to be considered:

$$\text{Maximize:} \quad F(x) = x_1 + 2x_2 + x_3$$

$$\text{Constraints:} \begin{cases} x_1 + x_2 + x_5 = 1 \\ x_1 + x_3 + x_4 + x_5 = 1 \\ x_i \geqq 0 \quad (1 \leqq i \leqq 5) \end{cases}$$

Solution Thus, the data are

$$A = \begin{bmatrix} 1 & 1 & 0 & 0 & 1 \\ 1 & 0 & 1 & 1 & 1 \end{bmatrix} \qquad b = \begin{bmatrix} 1 \\ 1 \end{bmatrix} \qquad c^T = \begin{bmatrix} 1\ 2\ 1\ 0\ 0 \end{bmatrix}$$

We start with $x = (0, 1, 0, 1, 0)^T$ and $I(x) = \{2, 4\}$. Observe that $\{A_2, A_4\}$ is a basis for $\mathbb{R}^2$. Hence, by Theorem 1, x is an extreme point of the feasible set, or a *basic feasible point*.

Each column of A is a linear combination of A_2 and A_4. In fact, we have, of course, $A_1 = A_2 + A_4$, $A_2 = A_2$, $A_3 = A_4$, $A_4 = A_4$, and $A_5 = A_2 + A_4$. The D-matrix is therefore

$$D = \begin{bmatrix} 0 & 0 & 0 & 0 & 0 \\ 1 & 1 & 0 & 0 & 1 \\ 0 & 0 & 0 & 0 & 0 \\ 1 & 0 & 1 & 1 & 1 \\ 0 & 0 & 0 & 0 & 0 \end{bmatrix}$$

The vector d^T is a linear combination of the rows D^i in D:

$$d^T = c^T D = \sum_{i=1}^{n} c_i D^i = D^1 + 2D^2 + D^3 = \begin{bmatrix} 2 & 2 & 0 & 0 & 2 \end{bmatrix}^T$$

The vector $c - d$ is given by

$$c - d = [-1 \quad 0 \quad 1 \quad 0 \quad -2]^T$$

Only one component is positive, and $q = 3$. The vector $x = \lambda D_q$ (with D_q denoting the qth column in D) is

$$x - \lambda D_q = [0 \quad 1 \quad 0 \quad 1 \quad 0]^T - \lambda[0 \quad 0 \quad 0 \quad 1 \quad 0]^T = [0 \quad 1 \quad 0 \quad 1 - \lambda \quad 0]^T$$

We take $\lambda = 1$; then the y-vector is

$$y = x - \lambda D_q + \lambda d_3 = [0 \quad 1 \quad 1 \quad 0 \quad 0]^T$$

The process is now repeated with y in place of x. Without giving the details, we find in the next step that $d = (3, 2, 1, 1, 3)^T$. Since $c \leqq d$, y is a solution and $F(y) = 3$. ■

Tableau Method

A practical realization of the simplex algorithm is often accomplished by exhibiting the data in a **tableau**, which is then modified in successive steps according to certain rules. We shall illustrate this organization of the algorithm with an example of modest size:

$$\text{Maximize:} \quad F(x) = 6x_1 + 14x_2$$
$$\text{Constraints:} \begin{cases} 2x_1 + x_2 \leqq 12 \\ 2x_1 + 3x_2 \leqq 15 \\ x_1 + 7x_2 \leqq 21 \\ x_1 \geqq 0, x_2 \geqq 0 \end{cases}$$

In preparation for the simplex algorithm, we introduce slack variables and rewrite the problem like this:

$$\text{Maximize:} \quad F(x) = 6x_1 + 14x_2 + 0x_3 + 0x_4 + 0x_5$$
$$\text{Constraints:} \begin{cases} 2x_1 + x_2 + x_3 = 12 \\ 2x_1 + 3x_2 + x_4 = 15 \\ x_1 + 7x_2 + x_5 = 21 \\ x_1 \geqq 0,\ x_2 \geqq 0,\ x_3 \geqq 0,\ x_4 \geqq 0,\ x_5 \geqq 0 \end{cases}$$

Hence, we have

$$\text{Maximize:} \quad F(x) = (6,\ 14,\ 0,\ 0,\ 0)^T x$$
$$\text{Constraints:} \begin{cases} \begin{bmatrix} 2 & 1 & 1 & 0 & 0 \\ 2 & 3 & 0 & 1 & 0 \\ 1 & 7 & 0 & 0 & 1 \end{bmatrix} x = \begin{bmatrix} 12 \\ 15 \\ 21 \end{bmatrix} \\ x = (x_1,\ x_2,\ x_3,\ x_4,\ x_5)^T \geqq 0 \end{cases}$$

Our first vector will be $x = (0, 0, 12, 15, 21)^T$. All of these data are summarized in the first tableau:

6	14	0	0	0	0
2	1	1	0	0	12
2	3	0	1	0	15
1	7	0	0	1	21
0	0	12	15	21	

Every step of the simplex method begins with a tableau. The top row contains coefficients that pertain to the objective function F. The current value of $F(x) = c^T x$ is displayed in the top right corner. The next m rows in the tableau represent a system of equations embodying the equality constraints. Remember that elementary row operations can be performed on this system of equations without altering the set of solutions. The last row of the tableau contains the current x-vector. Notice that

$F(x) = c^T x$ is easily computed using the top row and the bottom row. The preceding tableau is of the general form

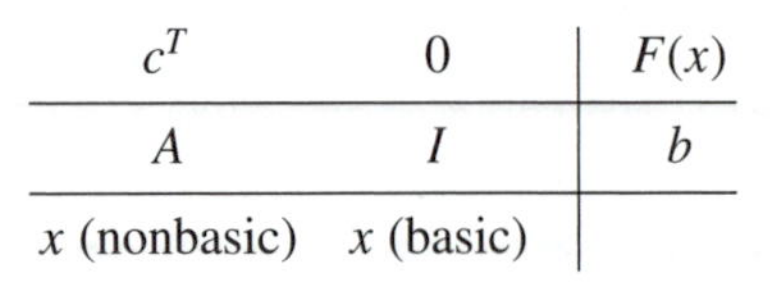

c^T	0	$F(x)$
A	I	b
x (nonbasic)	x (basic)	

Tableau Rules

Each tableau that occurs in the simplex method must satisfy these five rules:

(i) The x-vector must satisfy the equality constraints $Ax = b$.
(ii) The x-vector must satisfy the inequality $x \geqq 0$.
(iii) There are n components of x (designated **nonbasic variables**) that are 0. The remaining m components are usually nonzero and are designated **basic variables**. (Here n and m correspond to the values associated with the original problem before slack variables are introduced.)
(iv) In the matrix that defines the constraints, each basic variable occurs in only one row.
(v) The objective function F must be expressed only in terms of nonbasic variables.

Illustration Continued

In the first tableau for the preceding example, the basic variables are x_3, x_4, and x_5. The nonbasic variables are x_1 and x_2. We see at once that all five rules are true for this tableau.

In each step, we examine the current tableau to see whether the value of $F(x)$ can be increased by allowing a nonbasic variable to become a basic variable. In our example, we see that if we allow x_1 or x_2 to increase (and compensate by adjusting x_3, x_4, x_5), then the value of $F(x)$ will indeed increase. Since the coefficient 14 in F is greater than the coefficient 6, a unit increase in x_2 will increase $F(x)$ faster than a unit increase in x_1. Hence, we hold x_1 fixed at 0 and allow x_2 to increase as much as possible. These constraints apply:

$$
\begin{aligned}
0 &\leqq x_3 = 12 - x_2 \\
0 &\leqq x_4 = 15 - 3x_2 \\
0 &\leqq x_5 = 21 - 7x_2
\end{aligned}
$$

These constraints tell us that

$$
x_2 \leqq 12 \qquad x_2 \leqq 5 \qquad x_2 \leqq 3
$$

The most stringent of these is the inequality $x_2 \leqq 3$, and therefore x_2 is allowed to increase to 3. The resulting values of x_3, x_4, and x_5 are obtained by the three given constraints. Hence, our new x-vector is

$$x = \begin{bmatrix} 0 & 3 & 9 & 6 & 0 \end{bmatrix}^T$$

The new basic variables are x_2, x_3, and x_4, and we must now determine the next tableau in accordance with the preceding five rules. In order to satisfy rule **(v)**, we note that $x_2 = (21 - x_5)/7$. When this is substituted in F, we find a new form for the objective function:

$$\begin{aligned} F(x) &= 6x_1 + 14x_2 \\ &= 6x_1 + 14(21 - x_5)/7 = 6x_1 - 2x_5 + 42 \end{aligned}$$

To satisfy rule **(iv)**, Gaussian elimination steps (elementary row operations) are applied, using 7 as the pivot element. The purpose of this is to eliminate x_2 from all but one equation. After all this work has been carried out, step 1 is finished and step 2 begins with the second tableau, which is

6	0	0	0	−2	42
13/7	0	1	0	−1/7	9
11/7	0	0	1	−3/7	6
1	7	0	0	1	21
0	3	9	6	0	

The situation presented now is similar to that at the beginning. The nonbasic variables are x_1 and x_5. Any increase in x_5 will decrease $F(x)$, and so it is x_1 that is now allowed to become a basic variable. Hence, we hold x_5 fixed at 0 and allow x_1 to increase as much as possible. These constraints apply:

$$\begin{aligned} 0 &\leqq x_3 = 9 - (13/7)x_1 \\ 0 &\leqq x_4 = 6 - (11/7)x_1 \\ 0 &\leqq 7x_2 = 21 - x_1 \end{aligned}$$

These lead to

$$x_1 \leqq 63/13 \qquad x_1 \leqq 42/11 \qquad x_1 \leqq 21$$

The new basic variable x_1 is allowed to increase to only 42/11, and new values of x_3, x_4, and x_2 are computed from the tableau or from the constraint equations immediately above. The new x-vector is

$$x = \begin{bmatrix} 42/11 & 27/11 & 21/2 & 0 & 0 \end{bmatrix}^T$$

The nonbasic variables are now x_4 and x_5. To satisfy rule **(v)**, we use the substitution $x_1 = (7/11)(6 - x_4)$. Then

$$\begin{aligned} F(x) &= 6x_1 - 2x_5 + 42 \\ &= (42/11)(6 - x_4) - 2x_5 + 42 \\ &= -(42/11)x_4 - 2x_5 + 714/11 \end{aligned}$$

It is not necessary to complete the third tableau since both coefficients in F are negative. This signifies that the current x is a solution because neither of the nonbasic variables x_4 and x_5 can become basic variables without decreasing $F(x)$. Thus, in the original problem, the maximum value is $F(42/11,\ 27/11) = 630/11$.

Summary

On the basis of this example and the explanation given, we can summarize the work to be done on any given tableau as follows:

(i) If all coefficients in F (that is, the top row in the tableau) are $\leqq 0$, then the current x is the solution.
(ii) Select the nonbasic variable whose coefficient in F is positive and as large as possible. This variable becomes a new basic variable. Call it x_j.
(iii) Divide each b_i by the coefficient of the new basic variable in that row, a_{ij}. The value assigned to the new basic variable is the least of these ratios. Thus, if b_k/a_{kj} is the least, we set $x_j = b_k/a_{kj}$.
(iv) Using pivot element a_{kj}, create 0's in column j of A with Gaussian elimination steps.

Cost Estimates

In the practical applications of the simplex algorithm, there is a wide gulf between theoretical bounds on the cost and the actual cost. The upper bound on the number of simplex steps is the binomial coefficient

$$\binom{n}{m}$$

whereas in practice it is usual for the number of steps to be not more than $2m$. Even in a problem of *modest* size, say when $n = 300$ and $m = 100$, the value of the preceding binomial coefficient is astronomical. In fact, by using Stirling's formula (see Problem 1), we have

$$\binom{300}{100} = \frac{300!}{200!\,100!} \approx 4 \times 10^{81}$$

Other Algorithms

A new algorithm for linear programming was announced by Karmarkar [1984]. The new algorithm is claimed to be superior to the simplex method when the number of variables reaches 15,000 or more.

The cost of the Karmarkar algorithm is a polynomial function of the size of the problem, whereas the cost of the simplex algorithm is an exponential function of size. This in itself does not mean that the new algorithm is better, but it is an encouraging sign. A previous algorithm, due to Khachian, also has a cost function that is polynomial in size, but that algorithm never was competitive with the simplex method because successive steps required greater and greater precision.

PROBLEMS 10.4

1. Everyone should know **Stirling's formula**, which gives an estimate of $n!$; namely,

$$n! \approx \sqrt{2\pi n}\left[\frac{n}{e}\right]^n$$

Use this to derive the approximate formula:

$$\binom{n}{m} = \frac{n!}{m!\,(n-m)!} \approx \sqrt{\frac{n}{2\pi m(n-m)}}\left[\frac{n}{n-m}\right]^n\left[\frac{n-m}{m}\right]^m$$

2. Use Problem 1 to verify that

$$\binom{300}{100} \approx 4 \times 10^{81}$$

3. Solve this problem using the outline of the simplex method and a tableau as presented in this section:

$$\text{Maximize:} \quad F(x) = 2x_1 - 3x_2$$
$$\text{Constraints:} \begin{cases} 2x_1 + 5x_2 \geqq 10 \\ x_1 + 8x_2 \leqq 24 \\ x_1 \geqq 0,\ x_2 \geqq 0 \end{cases}$$

4. Consider the description of the simplex algorithm given in this section. Prove that if x is a solution, then $c \leqq d$.

5. **(a)** Repeat the solution of the first example in this section using a tableau.
   (b) Repeat the solution of the second example using the simplex algorithm.

6. Solve the following problem using the tableau method:

$$\text{Maximize:} \quad F(x) = 6x_1 + 14x_2$$
$$\text{Constraints:} \begin{cases} x_1 + x_2 \leqq 12 \\ 2x_1 + 3x_2 \leqq 15 \\ x_1 + 7x_2 \leqq 21 \\ x_1 \geqq 0,\ x_2 \geqq 0 \end{cases}$$

Now change the second inequality to $2x_1 + 3x_2 \geqq 15$ and repeat the solution.

APPENDIX A

Overview of Mathematical Software

A tremendous amount of mathematical software is available worldwide with more being developed each day. To find the most up-to-date information, a browser to the World Wide Web (www) on the Internet should be used. One can execute a search for available mathematical software in a particular application area of interest.

It is helpful to classify mathematical software into three categories: **(i)** public domain, **(ii)** freely accessible (some usage restrictions apply), and **(iii)** proprietary (license agreement required). Public domain has a specific legal meaning, implying that any use is permitted, including modification, resale, and so on. From the Internet, one can download either public domain or freely accessible software, and in many cases, one can obtain free demonstration copies of commercial software packages. Some of the software on the Internet has usage restrictions imposed by the authors, such as copyright, that allow unrestricted use for research and educational purposes only. (An example of this is code from the ACM Algorithms collection to be discussed below.) On the other hand, proprietary software must be purchased or leased from the developing company or from a computer store that sells software. Notice that whether money changes hands before the software can be used is not addressed in this breakdown of software into categories. For example, consider these seemingly contradictory examples: **(i)** paying for public domain software (Netlib on CD-ROM), **(ii)** paying for accessible software—you can download `netscape` freely, but you had better send a check if your use is non-educational, and **(iii)** software that is given away for free but is proprietary in the sense that a license agreement must be signed before you can get it.

In the following, we will give a quick overview of some available mathematical software with pointers to the www-addresses where additional information and some of the software can be found. Also mentioned are systems for helping in the search for mathematical software that solves specific problems. This is not a complete and comprehensive listing since developments are proceeding at such a rapid rate that any listing is soon out of date!

Since mathematical software is written in a variety of different programming languages and for a wide range of different computer architectures, it is difficult to know how to organize the discussion. We are guided by an excellent overview of mathematical special functions by Lozier and Olver [1994]. They organize software into the categories of *Software Packages*, *Intermediate Libraries*, *Comprehensive Libraries*, and *Interactive Systems*. First, *Software Packages* contain one or more subroutines for solving a particular problem in a subfield of numerical mathematics. Second, *Intermediate Libraries* are usually collections of subprograms for use on small computers or PCs. Some of these libraries contain a collection of useful mathematical functions in one or more computer languages. They may have been written by the manufacturer of the computer equipment or the developer of the compiler. Also, some intermediate libraries are available for PCs that are subsets of general-purpose mathematical libraries. Third, *Comprehensive Libraries* contain subroutines that have been assimilated into high-quality software containing many unifying features such as uniform documentation, uniform style of usage, and error-handling conditions. Finally, *Interactive Systems* are fully interactive computer environments with a powerful set of keyboard commands so that the user can avoid the compile-link-execute cycle. Usually there is an evolutionary process that mathematical software goes through from the introduction of the original idea in a journal or report to general acceptance by the scientific community of the algorithm and eventually the incorporation of the software into a large mathematical library or a sophisticated interactive computer system.

Rather than getting into a detailed discussion of particular software packages and intermediate libraries, we start by mentioning some searching systems for finding available mathematical software on the Internet and then giving general information related to mathematical software research and developments. Finally, we discuss a number of comprehensive libraries and interactive systems.

Searching Systems

Guide to Available Mathematical Software is an on-line, cross-indexed, and virtual repository of mathematical and statistical software for use in computational science and engineering. It was developed as a National Institute of Standards and Technology (NIST) project for providing scientists and engineers with improved access to reusable computer software. The user is guided through a problem decision tree to search for appropriate software for the particular problem to be solved or for the package/module name. Rather than providing a physical repository, the guide provides transparent access to multiple repositories maintained by NIST and others. One can obtain abstracts, documentation, and source code from the URL site.

`http://gams.nist.gov`

Netlib is a repository for mathematical software, documents (papers, reports, etc.), databases (address lists—e-mail and mail, conferences, performance data, etc.), and other useful mathematical information. There is a keyword searching capability for obtaining mathematical software as well as for searching back issues of the weekly NA-digest newsletter. This is a system developed to serve the community of numerical analysts and other researchers.

`http://www.netlib.org`

For example, one obtains a listing of all scientific computing and applied mathematics educational programs as follows.

`http://www.netlib.org/nse/cs_edu.html`

General Information

Homepages have been established by various research interest groups. A sampling of some of these are:

- Recent research developments on interval arithmetic:

  `http://cs.utep.edu/interval-comp/main.html.`

- A decision tree for optimization software:

  `http://plato.la.asu.edu/guide.html`

Mathematics Archive WWW Server provides an organized Internet access to a wide variety of mathematical resources. Primary emphasis is on material that is useful in teaching mathematics and educational software.

`http://archives.math.utk.edu`

Mathematics Information Servers is an extensive list of mathematics servers on the World Wide Web with references to many academic departments, electronic journals, sources for preprints of articles, and information on mathematical software.

`http://www.math.psu.edu/MathLists`

News Groups are used for a wide range of discussions and questions-and-answers postings in an unmoderated forum. Some USENET news groups for mathematical software are

```
sci.math.num-analysis
sci.math.research
sci.math.symbolic
```

Newsletters are available on the Internet with announcements and general information on mathematical software. Examples of these newsletters are

- NA-digest:

  `http://www.netlib.org/na-net`

- Approximation theory newsletter:

  e-mail: `at-net@leeor.technion.ac.il`

- Multigrid newsletter:

  `http://na.cs.yale.edu/mgnet/www/mgnet.html`

- Wavelet newsletter:

  `http://www.scarolina.edu/~wavelet`

Textbooks with associated mathematical software are widely available. Many numerical analysis and numerical methods textbooks come with or have associated software. In these books, one may find for each problem area a general discussion of analytical mathematics, the presentation of algorithms, and perhaps the actual implementation of them into computer routines written in one or more computer languages. The algorithms may be listed in the textbook, or the software may be available for purchase on a diskette or freely available so that the interested reader can download it from the Internet. For example, software supporting this textbook is available by anonymous-ftp as follows:

```
ftp ftp.ma.utexas.edu
user: anonymous
password: <user ID>
cd pub/papers/CNA/kincaid-cheney
get README
ls
```

or

```
ftp ftp.brookscole.com
user: anonymous
password: <user ID>
cd /brookscole/Mathematics/Texts_by_Authors
cd Kincaid_Cheney
get README
ls
```

Journals are published for the dissemination of recent research developments and associated algorithms and mathematical software. The bibliography to this book contains a listing of many of the primary numerical analysis journals. For some of these journals, tables of contents are available to readers over the Internet. Some of the organizations that publish research journals and disseminate mathematical software are:

- Association of Computing Machinery:

  `http://www.acm.org`

- American Mathematical Society:

  `http://www.ams.org`

- Society for Industrial and Applied Mathematics:

  `http://www.siam.org`

- Journal of Approximation Theory:

  `http://www.math.ohio-state.edu/JAT`

For example, the *ACM Transactions on Mathematical Software (TOMS)* publishes refereed articles and computer routines/packages. The ACM algorithm policy requires the software to be self-contained with adequate documentation, to have a test program with sample output, and to be reasonably portable over a variety of different computers. There is a TOMS homepage at

`http://www.acm.org/pubs/toms`

with a searchable table of contents to the algorithm papers and links to the software. The software is available from the ACM Algorithms Distribution Service and from Netlib. A classification system is used for indexing the algorithms, and a data base of them is maintained. You can obtain information from the ACM Algorithms Distribution Service at the URL site:

`http://acm.org/catalog/journals/120000.html`

Other journals exist for the exchange of software and related information in particular scientific disciplines. For example, the journal of *Computer Physics Communications* publishes papers on the computational aspects of physics and physical chemistry with refereed computer programs. The *Applied Statistics* journal publishes literature on statistical computing with refereed statistical software. Recently, entirely electronic journals have appeared with technical articles related to the development of numerical analysis algorithms. Articles from these journals are available over the Internet for local printing. One such journal is the *Electronic Transactions on Numerical Analysis* at

`http://etna.mcs.kent.edu`

Comprehensive Libraries

Comprehensive libraries are large collections of mathematical routines all written in a uniform style and with high standards of quality and robustness. Some of these libraries are listed here.

CMLIB is the Core Mathematics LIBrary of the National Institute of Science and Technology (NIST). It is a collection of approximately 750 high-quality public-domain Fortran subprograms that are easily transportable. The subroutines in this library solve many of the standard problems in mathematics and statistics. It contains mostly externally available software programs such as BLAS, EISPACK, FISHPACK, FCNPACK, FITPACK, LINPACK, and QUADPACK.

`http://gams.nist.gov`

CERN Library is maintained by the European Laboratory for Particle Physics. This library is primarily for the support of high-energy physics research, but it contains many routines for general mathematical use. With some restrictions, the library is distributed to outside organizations.

`http://consult.cern.ch`

ESSL is an Engineering and Scientific Subroutine Library for use on IBM computers. It is a state-of-the-art collection of over 450 mathematical routines for use on a wide range of IBM computers for solving scientific and engineering applications. The library has been tuned for specific IBM computer architectures such as workstations and parallel computers. It can be called from applications written in Fortran, C, or C++.

`http://www.ibm.com`

LibSci is a library of commonly used mathematical and scientific routines developed for use on Cray computer systems. For example, it includes routines for linear algebra, fast Fourier transforms, filtering, packing/unpacking, and vector gather/scatter.

`http://www.cray.com/craysoft`

IMSL Libraries are C-coded or Fortran-coded numerical and graphical libraries developed by Visual Numerics, Inc. These libraries contain a large collection of subroutines and function subprograms (over 500) that provide access to high-quality implementations of numerical methods in mathematics and statistics. They have evolved over approximately 25 years. These libraries are available for use on a wide range of computer platforms with subsets of them available for use on PCs. Other mathematical software products are available from Visual Numerics, such as PV-WAVE and Stanford Graphics.

`http://www.vni.com`

NAG Libraries are Fortran77/90-coded or C-coded numerical and statistical libraries for scientists, engineers, researchers, and software developers with applications involving mathematics, statistics, and optimization. Developed by the Numerical Algorithms Group, these software libraries have comprehensive numerical capabilities and were coded with the collaboration of numerical and statistical specialists. The largest version of the library has over 1000 routines. Some of this software has evolved over more than 20 years, producing state-of-the-art products with robust performance on over 80 computing platforms from PCs to supercomputers. Also, NAG developed the first fully standard Fortran 90 compiler, and the computer algebra system Axiom is available from NAG.

`http://www.nag.com`

SLATEC is a large collection of Fortran mathematical subprograms distributed by the Department of Energy (DOE) Energy Science and Technology Center:

`http://www.doe.gov/html/osti/estsc/estsc.html`

or from Netlib

`http://www.netlib.org/slatec`

It is characterized by portability, good numerical technology, good documentation, robustness, and quality assurance. The primary impetus for this library was to provide portable, nonproprietary, mathematical software for supercomputers at a consortium of government research laboratory. The original acronym stood for the national laboratories involved (Sandia, Los Alamos, Air Force Technical Exchange Committee). Subsequently, the library committee admitted three additional national laboratories (Lawrence Livermore, Oak Ridge, Sandia Livermore) plus the National Energy Supercomputer Center at Lawrence Livermore and the National Institute of Standards and Technology.

In the United States, national laboratories, national supercomputer centers, and various government agencies have large collections of mathematical software. For links to some of these research centers, see

`http://www.nsf.gov/nsf/homepage/links.html`

In addition to those listed above, some other general mathematical software libraries are BSCLIB, NSWC, NUMAL, NUMPAC, PORT, Scientific Desk, VECLIB, and more.

Interactive Systems

In general, a fully interactive mathematical software system contains a powerful set of commands that the user enters at a computer terminal or workstation by using the keyboard or the mouse to click on an icon. An immediate response is displayed on the screen. It may be a computation (numerical or symbolic) or a visual display such as a graph or figure. The programming burden is reduced since there is no compile-link-execute cycle. The capabilities of interactive systems can be extended by customizing the set of commands or icons. Interactive systems are able to integrate nonnumerical tasks with numerical computations. It seems that graphical and symbolic computing are best done in an interactive computing environment. For this reason, a recent trend is combining numerical computing with symbolic computing and graphical visualization into a totally interactive system.

Computer algebra systems have special capabilities useful in numerical mathematics. One such feature is arbitrary precision or multiple-precision floating-point arithmetic. In general, programming languages use the computer hardware for doing computer arithmetic so that the precision is fixed. The primary purpose of computer algebra systems is for exact mathematical calculations with floating-point computations a secondary capability. Nevertheless, one obtains a bonus of being able to carry out arbitrary precision floating-point computations in these systems. Unless the user specifies otherwise, symbolic systems generally avoid evaluations that introduce inexact results and evaluate expressions symbolically (with numbers rendered as rational fractions having arbitrarily long numerators and denominators or represented as symbols). The user can request that floating-point evaluations of numbers be carried out with arbitrarily long precision.

The following are a sampling of interactive mathematical software systems.

CPLEX is a software package used to solve linear programming problems, including integer, mixed integer, and network linear programming problems. CPLEX can be used as an interactive program or as a callable subroutine library. The CPLEX interactive problem solver allows users to enter, modify, and solve problems from computer terminals. This is particularly useful if one is solving a problem once or if one is prototyping a solution method. The callable library provides access to the CPLEX optimization, utility, problem modification, query, and file I/O routine directly from C or Fortran programs.

`http://www.cplex.com`

HiQ is an object-based numerical analysis and data visualization software package. HiQ solves math, science, and engineering problems using a methodology that combines a worksheet interface, interactive analysis, data visualization, an extensive mathematics library, and a script-programming language. HiQ is an interactive problem-solving environment for Macintosh and Power Macintosh computers.

`http://www.natinst.com`

Macsyma is a computer algebra system that supports symbolic, graphical, and numerical computing on a variety of different computer platforms—personal computers, scientific workstations, and mainframes. By programming in either Lisp or an Algol-like procedural language, one can extend the built-in capabilities of the system.

`http://www.macsyma.com`

Maple is an interactive symbolic computation system containing symbolic, numerical, graphical, and programming capabilities. Maple performs equation solving, linear algebra, calculus, complex analysis, and more with virtually unlimited precision. Maple was developed as an interactive system for computer algebraic manipulations associated with symbolic computing, but many more expanded capabilities have been added. It is available on a wide range of computers from personal computers and workstations to supercomputers.

`http://www.maplesoft.com`

Mathematica is an interactive software system for numerical, symbolic, and graphical computing as well as visualizations. The user can integrate Mathematica output (computations, graphics, animations, etc.) with ordinary text entirely with the Mathematica system for the preparations of complete electronic documents to be used as technical reports or presentations. A wide variety of Mathematica application libraries are available. MathLink is a communication protocol that allows the exchange of information between Mathematica and other computer packages such as Matlab or Excel. A programming language based on pattern matching can be used for extending the capabilities of the Mathematica system. Mathematica is available on over 20 platforms from PCs to scientific workstations to large-scale scientific mainframe computers.

`http://www.mathematica.com`

Matlab is a computing environment that provides computation, visualization, and application-specific tool boxes. Matrix notation is used to produce a *matrix laboratory* with a built-in set of commands for standard algorithms involving numerical computations. A matrix-oriented language may be used for large-scale computation and data analysis. Interactive 2-D and 3-D graphical capabilities are available for analysis, transformation, and visualization of data. Matlab can dynamically link with C or Fortran programs. The package is available on a wide range of computers such as PCs, workstations, and supercomputers. Matlab has more than 20 tool boxes available for specialized applications such as signal and image processing, control system design, frequency domain identification, robust control design, mathematics, statistics, and data analysis, neural networks and fuzzy logic, optimization, and splines. Also, symbolic computing is available with an tool box interface to Maple V.

`http://www.mathworks.com`

REDUCE is an interactive computer algebra system designed for general algebraic computations of interest to mathematicians, scientists, and engineers. While it is used by many as an algebraic calculator for problems that can be done by hand, the main aim of REDUCE is to support calculations that are feasible only by a computer solution.

`http://www.rrz.uni-koeln.de/REDUCE`

Some of the symbolic algebraic computer systems other than those listed here are Axiom, Derive, GANITH, Magma, Mathcad, Milo, MuPAD, Pari, Schur, and SymbMath.

BIBLIOGRAPHY

Abbreviations

ACMCOM	*Association for Computing Machinery (ACM) Communications*
ACMJ	*ACM Journal*
ACMTOMS	*ACM Transactions on Mathematical Software*
AMM	*American Mathematical Monthly*
AMS	*American Mathematical Society*
AN	*Acta Numerica*
ANM	*Applied Numerical Mathematics*
ANSI	*American National Standards Institute, Inc.*
CJ	*Computer Journal*
IEEE	*Institute of Electrical and Electronic Engineers*
IJNME	*International Journal Numerical Methods Engineering*
IMAJNA	*Institute for Mathematics and Its Applications, Journal of Numerical Analysis*
JAT	*Journal of Approximation Theory*
JCAM	*Journal Computing and Applied Mathematics*
JRNBS	*Journal Research National Bureau of Standards*
LAA	*Linear Algebra and Applications*
MAA	*Mathematical Association of America*
MC	*Mathematics of Computation*
MI	*Mathematics Intelligencer*
NM	*Numerische Mathematik*
SA	*Scientific American*
SIAM	*Society for Industrial and Applied Mathematics*
SIAMMAA	*SIAM Journal on Matrix Analysis and Applications*
SIAMNA	*SIAM Journal of Numerical Analysis*
SIAMREV	*SIAM Review*
SIAMSSC	*SIAM Journal on Scientific and Statistical Computing*
ZAMP	*Zeitschrift für angewandte Mathematik und Physik*

Abramowitz, M., and I. A. Stegun. 1956. "Abscissas and weights for Gaussian quadratures of high order." *JRNBS* **56**, 35–37.

Abramowitz, M., and I. A. Stegun (eds.). 1964. *Handbook of Mathematical Functions with Formulas, Graphs, and Mathematical Tables.* National Bureau of Standards. (Reprinted New York: Dover, 1965.)

Acton, F. S. 1959. *Analysis of Straight-Line Data.* New York: Wiley. (Reprinted New York: Dover, 1966.)

Ahlfors, L. V. 1966. *Complex Analysis.* New York: McGraw-Hill.

Aho, A., J. Hopcroft, and J. Ullman. 1974. *The Design and Analysis of Computer Algorithms.* Reading, MA: Addison-Wesley.

Aiken, R. C. (ed.). 1985. *Stiff Computation.* New York: Oxford University Press.

Alefeld, G., and R. Grigorieff (eds.). 1980. *Fundamentals of Numerical Computation.* Berlin: Springer.

Alefeld, G., and J. Herzberger. 1983. *Introduction to Interval Computations.* New York: Academic Press.

Alexander, J. C., and J. A. Yorke. 1978. "The homotopy continuation method: Numerically implemented topological procedures." *Transactions AMS* **242**, 271–284.

Allgower, E., and K. Georg. 1980. "Simplicial and continuation methods for approximating fixed points and solutions to systems of equations." *SIAMREV* **22**, 28–85.

Allgower, E., and K. Georg. 1990. *Numerical Continuation Methods.* New York: Springer-Verlag.

Allgower, E. L., K. Glasshoff, and H.-O. Peitgen (eds.). 1981. *Numerical Solution of Nonlinear Equations: Lecture Notes in Mathematical* **878**. New York: Springer-Verlag.

Ames, W. F. 1977. *Numerical Methods for Partial Differential Equations.* New York: Academic Press.

Anderson, E., Z. Bai, C. Bischof, J. Demmel, J. Dongarra, J. Du Croz, A. Greenbaum, S. Hammarling, A. McKenney, S. Ostrouchov, and D. Sorensen. 1995. *LAPACK Users' Guide - Release 2.0.* Philadelphia: SIAM. (To view HTML version, use the URL address: http://www.netlib.org/lapack/lug/lapack_lug.html.)

ANSI/IEEE. 1985. "IEEE standard for binary floating-point arithmetic." ANSI/IEEE Std. 754–1985. New York: IEEE.

ANSI/IEEE. 1987. "A radix-independent standard for floating-point arithmetic." IEEE Std. 854–1987. New York: IEEE.

Arbel, A. 1993. *Exploring Interior-Point Linear Programming Algorithms and Software.* Cambridge, MA: MIT Press.

Argyros, I. K., and F. Szidarovszky. 1993. *The Theory and Applications of Iteration Methods.* Boca Raton, FL: CRC Press.

Ascher, U. M., R. M. M. Mattheij, and R. D. Russell. 1995. *Numerical Solution of Boundary Value Problems for Ordinary Differential Equations.* Philadelphia: SIAM.

Atkinson, K. 1985. *Elementary Numerical Analysis.* New York: Wiley.

Axelsson, O. 1980. "A generalized conjugate direction method and its application on a singular perturbation problem." In *Numerical Analysis: Lecture Notes in Mathematics* **773**. New York: Springer-Verlag.

Axelsson, O. 1994. *Iterative Solution Methods.* New York: Cambridge University Press.

Ayoub, R. 1974. "Euler and the zeta function." *AMM* **81**, 1067–1086.

Aziz, A. K. (ed.). 1969. *Numerical Solution of Differential Equations.* New York: van Nostrand.

Aziz, A. K. (ed.). 1974. *Numerical Solutions of Boundary Value Problems for Ordinary Differential Equations.* New York: Academic Press.

Babuška, I., M. Prager, and E. Vitasék. 1966. *Numerical Processes in Differential Equations.* New York: Wiley-Interscience.

Backus, J. 1979. "The history of Fortran I, II, and III." *Annals of the History of Computing* **1**, 21–37.

Bailey, P. B., L. F. Shampine, and P. E. Waltman. 1968. *Nonlinear Two-Point Boundary-Value Problems.* New York: Academic Press.

Bak, J., and D. J. Newman. 1982. *Complex Analysis.* New York: Springer-Verlag.

Baker, C. T. A., C. A. H. Paul, and D. R. Willé. 1995. "Issues in the numerical solution of evolutionary delay differential equations." *Advances in Computational Mathematics* **3**, 171–196.

Barnes, E. R. 1986. "A variation on Karmarkar algorithm for solving linear programming problems." *Mathematical Programming* **36**, 174–182.

Barnhill, R., R. P. Dube, and F. F. Little. 1983. "Properties of Shepard's surfaces." *Rocky Mtn. J. Math.* **13**, 365–382.

Barnhill, R., and A. Riesenfeld. 1974. *Computer Aided Geometric Design.* New York: Academic Press.

Barnsley, M. 1988. *Fractals Everywhere.* New York: Academic Press.

Barnsley, M., and A. Sloan. 1988. "A better way to compress images." *Byte* **13**, 215–223.

Barrodale, I., and C. Phillips. 1975. "Solution of an overdetermined system of linear equations in the Chebyshev norm." *ACMTOMS* **1**, 264–270.

Barrodale, I., and F. D. K. Roberts. 1974. "Solution of an overdetermined system of equations in the ℓ_1 norm." *ACMCOM* **17**, 319–320.

Barrodale, I., F. D. K. Roberts, and B. L. Ehle. 1971. *Elementary Computer Applications.* New York: Wiley.

Bartels, R. H. 1971. "A stabilization of the simplex method." *NM* **16**, 414–434.

Bartels, R., J. Beatty, and B. Barsky. 1987. *An Introduction to Splines for Use in Computer Graphics and Geometric Modeling.* Los Altos, CA: Morgan Kaufmann.

Bartle, R. G. 1976. *The Elements of Real Analysis.* 2nd ed. New York: Wiley.

Becker, E. B., G. F. Carey, and J. T. Oden. 1981. *Finite Elements: An Introduction.* Vol. 1. Englewood Cliffs, NJ: Prentice-Hall.

Bell, E. T. 1975. *Men of Mathematics.* New York: Simon & Schuster.

Bell, G., and S. Glasstone. 1970. *Nuclear Reactor Theory.* New York: van Nostrand-Reinhold.

Bellman, R., and K. L. Cooke. 1963. *Differential-Difference Equations.* New York: Academic Press.

Belsley, D. A., E. Kuh, and R. Welsch. 1981. *Regression Diagnostics: Identifying Influential Data and Sources of Colinearity.* New York: Wiley.

Bender, C. M., and S. A. Orszag. 1978. *Advanced Mathematical Methods for Scientists and Engineers.* New York: McGraw-Hill.

Birkhoff, G., and R. E. Lynch. 1984. *Numerical Solution of Elliptic Problems.* Philadelphia: SIAM.

Bischof, C., A. Carle, P. Khademi, and A. Mauer. 1994. The ADIFOR 2.0 system for the automatic differentiation of Fortran 77 programs," Mathematics and Computer Sciences Report ANL/MCS-P481-1194. Argonne, IL: Argonne National Laboratory.

Björck, Å. 1967. "Solving linear least squares problems by Gram-Schmidt orthogonalization." *BIT* **7**, 1–21.

Björck, Å., and C. C. Paige. 1992. "Loss and recapture of orthogonality in the modified Gram-Schmidt algorithm." *SIAMMAA* **13**, 176–190.

Bloomfield, P. 1976. *Fourier Analysis of Time Series: An Introduction.* New York: Wiley-Interscience.

Blum, E. K. 1972. *Numerical Analysis and Computation: Theory and Practice.* Reading, MA: Addison-Wesley.

Bodewig, E. 1946. "Sur la méthode de Laguerre pour l'approximation des racines de certaines équations algébriques et sur la critique d'Hermite." *Nederl. Acad. Wetensch. Proc.* **49**, 911–921.

Boggs, P., R. H. Byrd, and R. B. Schnabel. 1985. *Numerical Optimization 1984.* Philadelphia: SIAM.

Bohman, H. 1952. "On approximation of continuous and analytic functions." *Arkiv för Matematik* **2**, 43–56.

Boisvert, R. F., S. E. Howe, D. K. Kahaner, and J. L. Springmann. 1990. "Guide to available mathematical software." Center for Computing and Applied Mathematics. Gaithersburg, MD: National Institute of Standards and Technology.

Boisvert, R. F., and R. A. Sweet. 1982. "Sources and development of mathematical software for elliptic boundary value problems." In *Sources and Development of Mathematical Software* (W. Cowell, ed.). Englewood Cliffs, NJ: Prentice-Hall.

de Boor, C. 1971. "CADRE: An algorithm for numerical quadrature." In *Mathematical Software* (J. R. Rice, ed.). New York: Academic Press.

de Boor, C. 1976. "Total positivity of the spline collocation matrix." *Indiana University Journal of Mathematics* **25**, 541–551.

de Boor, C. 1984. *A Practical Guide to Splines.* 2nd ed. New York: Springer-Verlag.

de Boor, C., and G. H. Golub (eds.). 1978. *Recent Advances in Numerical Analysis.* New York: Academic Press.

Borwein, J. M., and P. B. Borwein. 1984. "The arithmetic-geometric mean and fast computation of elementary functions." *SIAMREV* **26**, 351–366.

Botha, J. F., and G. F. Pinder. 1983. *Fundamental Concepts in the Numerical Solution of Differential Equations.* New York: Wiley.

Boyce, W. E., and R. C. DiPrima. 1977. *Elementary Differential Equations and Boundary Value Problems.* New York: Wiley.

Braess, D. 1984. *Nonlinear Approximation Theory.* New York: Springer-Verlag.

Bramble, J. H. (ed.). 1966. *Numerical Solution of Partial Differential Equations.* New York: Academic Press.

Bratley, P., B. L. Fox, and L. Schrage. 1987. *A Guide to Simulation.* New York: Springer-Verlag.

Brenan, K. E., S. L. Campbell, and L. R. Petzold. 1995. *Numerical Solution of Initial-Value Problems in Differential- Algebraic Equations.* Philadelphia: SIAM.

Brent, R. P. 1973. *Algorithms for Minimization Without Derivatives.* Englewood Cliffs, NJ: Prentice-Hall.

Brent, R. P. 1976. "Fast multiple precision evaluation of elementary functions." *ACMJ* **23**, 242–251.

Brezinski, C. 1994. "The generalizations of Newton's interpolation formula due to Mühlbach and Andoyer." *Electronic Transactions on Numerical Analysis* **2**, 130–137.

Briggs, W. T. 1987. *A Multigrid Tutorial.* Philadelphia: SIAM.

Briggs, W. T., and V. E. Henson. 1995. *The DFT: An Owner's Manual for the Discrete Fourier Transform.* Philadelphia: SIAM.

Brigham, E. O. 1974. *The Fast Fourier Transform.* Englewood Cliffs, NJ: Prentice-Hall.

Brophy, J. F., and P. W. Smith. 1988. "Prototyping Karmarkar's algorithm using MATH/PROTAN." *Directions* **5**, 2–3, Houston: IMSL Corp.

Brown, P. J. (ed.). 1977. *Software Portability.* New York: Cambridge University Press.

Brown, P. N., G. D. Byrne, and A. C. Hindmarsh. 1989. "VODE: a variable coefficient ODE solver." *SIAMSSC* **10**, 1039–1051.

Bunch, J. R., and D. J. Rose (eds.). 1976. *Sparse Matrix Computations.* New York: Academic Press.

Burden, R. L., and J. D. Faires. 1993. *Numerical Analysis.* 5th ed. Boston: PWS-Kent.

Burrage, K. 1978. "A special family of Runge-Kutta methods for solving stiff differential equations." *BIT* **18**, 22–41.

Burrage, K. 1995. *Parallel and Sequential Methods for Ordinary Differential Equations.* New York: Oxford University Press.

Butcher, J. C. 1987. *The Numerical Analysis of Ordinary Differential Equations: Runge-Kutta and General Linear Methods.* New York: Wiley.

Buzbee, B. L. 1984. "The SLATEC common mathematical library." In *Sources and Development of Mathematical Software* (W. R. Cowell, ed.). Englewood Cliffs, NJ: Prentice-Hall.

Byrne, G. D., and C. A. Hall (eds.). 1973. *Numerical Solution of Systems of Nonlinear Algebraic Equations.* New York: Academic Press.

Byrne, G., and A. Hindmarsh. 1987. "Stiff ODE solvers: A review of current and coming attractions." *Journal Computational Physics* **70**, 1–62.

Calvo, M., J. I. Montijano, and L. Rández. 1993. "On the change of stepsizes in multistep codes." *Num. Alg.* **4**, 283–304.

Carter, L. L., and E. D. Cashwell. 1975. "Particle-transport with the Monte Carlo method." *ERDA Critical Review Series* TID–26607. Springfield, VA: National Technical Information Service.

Cash, J. R. 1979. *Stable Recursions.* New York: Academic Press.

Cassels, J. W. S. 1981. *Economics for Mathematicians.* New York: Cambridge University Press.

Chaitlin, G. J. 1975. "Randomness and mathematical proof." *SA* May, 47–52.

Chambers, J. M. 1977. *Computational Methods for Data Analysis.* New York: Wiley.

Chatterjee, S., and B. Price. 1977. *Regression Analysis by Example.* New York: Wiley.

Cheney, E. W. 1982. *Introduction to Approximation Theory.* New York: Chelsea.

Cheney, W., and D. Kincaid. 1994. *Numerical Mathematics and Computing.* 3rd ed. Pacific Grove, CA: Brooks/Cole.

Cherkasova, M. P. 1972. *Collected Problems in Numerical Analysis.* New York: Walters-Noordhoff.

Childs, B., M. Scott, J. W. Daniel, E. Denman, and P. Nelson (eds.). 1979. *Codes for Boundary Value Problems in Ordinary Differential Equations*: *Lecture Notes in Computer Science* **76**. New York: Springer-Verlag.

Chow, S. N., J. Mallet-Paret, and J. A. Yorke. 1978. "Finding zeros of maps: Homotopy methods that are constructive with probability one." *MC* **32**, 887–899.

Chui, C. K. 1988. "Multivariate splines." *SIAM Regional Conference Series in Mathematics* **54**.

Chung, K. C., and T. H. Yao. 1977. "On lattices admitting unique Lagrange interpolations." *SIAMNA* **14**, 735–743.

Cline, A. K. 1974a. "Scalar and planar valued curve-fitting using splines under tension." *ACMCOM* **17**, 218–220.

Cline, A. K. 1974b. "Six subprograms for curve-fitting using splines under tension." *ACMCOM* **17**, 220–223.

Cline, A. K., C. B. Moler, G. W. Stewart, and J. H. Wilkinson. 1979. "An estimate for the condition number of a matrix." *SIAMNA* **16**, 368–375.

Coddington, E. A., and N. Levinson. 1955. *Theory of Ordinary Differential Equations.* New York: McGraw-Hill.

Cody, W. J. 1981. "Analysis of proposals for the floating-point standard." *Computer* **14**, 63–68.

Cody, W. J. 1988. "Floating-point standards—theory and practice." In *Reliability in Computing,* 99–107. New York: Academic Press.

Cody, W. J., J. T. Coonen, D. M. Gay, K. Hanson, D. Hough, W. Kahan, R. Karpiski, J. Palmer, F. N. Ris, and D. Stevenson. 1984. "A proposed radix- and wordlength-independent standard for floating-point arithmetic." *IEEE Micro.* **4**, 86–100.

Cody, W. J., and W. Waite. 1980. *Software Manual for the Elementary Functions.* Englewood Cliffs, NJ: Prentice-Hall.

Cohen, A. M. 1974. "A note on pivot size in Gaussian elimination." *LAA* **8**, 361–368.

Coleman, T. F., and C. Van Loan. 1988. *Handbook for Matrix Computations.* Philadelphia: SIAM.

Collatz, L. 1966a. *Functional Analysis and Numerical Mathematics.* 3rd ed. New York: Academic Press.

Collatz, L. 1966b. *The Numerical Treatment of Differential Equations.* New York: Springer-Verlag.

Concus, P., G. H. Golub, and D. P. O'Leary. 1976. "A generalized conjugate gradient method for the numerical solution of elliptical partial differential equations." In *Sparse Matrix Computations* (J. R. Bunch and D. J. Rose, eds.). New York: Academic Press.

Conte, S. D., and C. de Boor. 1980. *Elementary Numerical Analysis.* 3rd ed. New York: McGraw-Hill.

Cooley, J. W., P. A. Lewis, and P. P. Welch. 1967. "Historical notes on the fast Fourier transform." *Proceedings of the IEEE* **55**, 1675–1677.

Coonen, J. T. 1980. "An implementation guide to a proposed standard for floating-point arithmetic." *Computer* **13**, 68–79.

Coonen, J. T. 1981. "Underflow and the denormalized numbers." *Computer* **14**, 75–87.

Cowell, W. (ed.). 1977. "Portability of numerical software." In *Lecture Notes in Computer Science* **57**. New York: Springer-Verlag.

Crowder, H., R. S. Dembo, and J. M. Mulvey. 1979. "On reporting computational experiments with mathematical software." *ACMTOMS* **5**, 193–203.

Cryer, C. W. 1968. "Pivot size in Guassian elimination." *NM* **12**, 335–345.

Cullum, J., and R. A. Willoughby (eds.). 1986. *Large Scale Eigenvalue Problems.* Amsterdam: Elsevier.

Curry, J. H., L. Garnett, and D. Sullivan. 1983. "On the iteration of a rational function: Computer experiments with Newton's method." *Comm. Math. Physics* **91**, 267–277.

Dahlquist, G. 1956. "Convergence and stability in the numerical integration of ordinary differential equations." *Math. Scand.* **4**, 33–35.

Dahlquist, G. 1963. "A special stability problem for linear multistep methods." *BIT* **3**, 27–43.

Dahlquist, G., and A. Björck. 1974. *Numerical Methods.* Englewood Cliffs, NJ: Prentice-Hall.

Daniel, J. W., and R. E. Moore. 1970. *Computation and Theory in Ordinary Differential Equations.* San Francisco: Freeman.

Dano, S. 1974. *Linear Programming in Industry.* 4th ed. New York: Springer-Verlag.

Dantzig, G. B. 1948. "Programming in a linear structure." Washington, DC: U.S. Air Force, Comptroller's Office.

Dantzig, G. B. 1963. *Linear Programming and Extensions.* Princeton, NJ: Princeton University Press.

Datta, B. N. 1994. *Numerical Linear Algebra and Applications* Pacific Grove, CA: Brooks/Cole.

Davis, H. T. 1962. *Introduction to Nonlinear Differential and Integral Equations.* New York: Dover.

Davis, P. J. 1982. *Interpolation and Approximation.* New York: Dover.

Davis, P. J., and P. Rabinowitz. 1956. "Abscissas and weights for Gaussian quadratures of high order." *JRNBS* **56**, 35–37.

Davis, P. J., and P. Rabinowitz. 1984. *Methods of Numerical Integration.* 2nd ed. New York: Academic Press.

Day, J., and B. Peterson. 1988. "Growth in Gaussian elimination." *AMM* **95**, 489–513.

Dejon, B., and P. Henrici (eds.). 1969. *Constructive Aspects of the Fundamental Theory of Algebra.* New York: Wiley.

Dekker, K., and J. G. Verwer. 1984. *Stability of Runge-Kutta Methods for Stiff Nonlinear Differential Equations.* Amsterdam: Elsevier Science.

Dekker, T. J. 1969. "Finding a zero by means of successive linear interpolation." In *Constructive Aspects of the Fundamental Theorem of Algebra* (B. Dejon and P. Henrici, eds.). New York: Wiley-Interscience.

Delves, L. M., and J. Mohamed. 1985. *Computational Methods for Integral Equations.* New York: Cambridge University Press.

Demmel, J., and K. Veselić. 1992. "Jacobi's method is more accurate than QR." *SIAMMAA* **13**, 1204–1245.

Dennis, J. E., Jr., and J. Moré. 1974. "Quasi-Newton methods, motivation and theory." *SIAMREV* **19**, 46–89.

Dennis, J. E., Jr., and R. B. Schnabel. 1983. *Numerical Methods for Unconstrained Optimization and Nonlinear Equations.* Englewood Cliffs, NJ: Prentice-Hall.

Deuflhard, P., and G. Heindl. 1979. "Affine invariant convergence theorems for Newton's method and extensions to related methods." *SIAMNA* **16**, 1–10.

Dewdney, A. K. 1988. "Computer recreations: Random walks that lead to fractal crowds." *SA,* December.

Diekmann, O., S. A. Van Gils, S. M. Verduyn Lunel, and H. O. Walther. 1995. *Delay Equations Appl. Math. Sci.* **110**. New York: Springer-Verlag.

Dieudonné, J. 1960. *Foundations of Modern Analysis.* New York: Academic Press.

de Doncker, E., and I. Robinson. 1984. "An algorithm for automatic integration over a triangle using nonlinear extrapolation." *ACMTOMS* **10**, 1–16.

Dongarra, J. J., J. R. Bunch, C. B. Moler, and G. W. Stewart. 1979. `LINPACK` *Users Guide.* Philadelphia: SIAM.

Dongarra, J. J., and D. W. Walker. 1995. "Software libraries for linear algebra computations on high performance computers." *SIAMREV* **37**, 151–180.

Draper, N. R., and H. Smith. 1981. *Applied Regression Analysis.* New York: Wiley.

Driver, R. 1977. *Ordinary and Delay Differential Equations.* New York: Springer-Verlag.

Duff, I. S., A. M. Erisman, and J. K. Reid. 1986. *Direct Methods for Sparse Matrices.* New York: Oxford University Press.

Duffy, D. G. 1993. "On the numerical inversion of Laplace transforms: Comparison of three new methods on characteristic problems from applications." *ACMTOMS* **19**, 333–359.

Durand, E. 1960. "Solutions Numériques des Équations Algébriques." (2 vols.) Paris: Mason.

Eaves, B. C. 1976. "A short course in solving equations with PL homotopies." *SIAM—AMS Proceedings* **9**, 73–144.

Eaves, B. C., F. J. Gould, H.-O. Peitgen, and M. J. Todd (eds.). 1983. *Homotopy Methods and Global Convergence.* New York: Plenum.

Edelman, A. 1992. "The complete pivoting conjecture for Gaussian elimination is false." Department of Mathematics. Berkeley, CA: Lawrence Berkeley Laboratory and University of California Berkeley.

Edelman, A. 1994. "When is `x * (1/x)` $\neq$ `1` ?" Department of Mathematics. Cambridge, MA: Massachusetts Institute of Technology.

Eggermont, P. P. B. 1988. "Noncentral difference quotients and the derivative." *AMM* **95**, 551–553.

Elliott, D. F., and K. R. Rao. 1982. *Fast Transforms: Algorithms, Analyses, Applications.* New York: Academic Press.

Engels, H. 1980. *Numerical Quadrature and Cubature.* New York: Academic Press.

Epperson, J. F. 1987. "On the Runge example." *AMM* **4**, 329–341.

Farwig, R. 1986. "Rate of convergence of Shepard's global interpolation formula." *MC* **46**, 577–590.

Fatunla, S. O. 1988. *Numerical Methods for Initial Value Problems in Ordinary Differential Equations.* New York: Academic Press.

Fefferman, C. 1967. "An easy proof of the fundamental theorem of algebra." *AMM* **74**, 854–855.

Fehlberg, E. 1969. "Klassische Runge-Kutta Formeln fünfter und siebenter Ordnung mit Schrittweitenkontrolle." *Computing* **4**, 93–106.

Feldstein, A., and P. Turner. 1986. "Overflow, underflow, and severe loss of significance in floating-point addition and subtraction." *IMAJNA* **6**, 241–251.

Ficken, F. A. 1951. "The continuation method for functional equations." *Communications Pure & Applied Mathematics* **4**, 435–456.

Flehinger, B. J. 1966. "On the probability that a random integer has initial digit A." *AMM* **73**, 1056–1061.

Forsythe, G. E. 1957. "Generation and use of orthogonal polynomials for data-fitting with a digital computer." *SIAM Journal* **5**, 74–88.

Forsythe, G. E., M. A. Malcolm, and C. B. Moler. 1977. *Computer Methods for Mathematical Computations.* Englewood Cliffs, NJ: Prentice-Hall.

Forsythe, G. E., and C. B. Moler. 1967. *Computer Solution of Linear Algebraic Systems.* Englewood Cliffs, NJ: Prentice-Hall.

Forsythe, G. E., and W. R. Wasow. 1960. *Finite-Difference Methods for Partial Differential Equations.* New York: Wiley.

Fosdick, L. D. (ed.). 1979. *Performance Evaluation of Numerical Software.* Amsterdam: North-Holland.

Fosdick, L. 1993. "IEEE Arithmetic Short Reference." High Performance Scientific Computing, University of Colorado at Boulder.

Foster, L. V. 1981. "Generalizations of Laguerre's method: Higher order methods." *SIAMNA* **18**, 1004–1018.

Foster, L. V. 1994. "Gaussian elimination with partial pivoting can fail in practice." *SIAMMAA* **15**, 1354–1362.

Fournier, A., D. Fussell, and L. Carpenter. 1982. "Computer rendering of stochastic models." *ACMCOM* **25**, 371–384.

Fox, L. 1987. *Biographical Memoirs of Fellows of the Royal Society: James Hardy Wilkinson 1919–1986,* Vol. 33. London: Royal Society.

Fox, P. A., A. D. Hall, and N. L. Schryer. 1978. "Framework for a portable library." *ACMTOMS* **4**, 177–188.

Francis, J. G. F. 1961. "The QR transformation: A unitary analogue to the LR transformation." Parts 1 and 2. *Computing Journal* **4**.

Franke, R. 1982. "Scattered data interpolation: Tests of some methods." *MC* **38**, 181–200.

Fritsch, F. N., and R. E. Carlson. 1980. "Monotone piecewise cubic interpolation." *SIAMNA* **17**, 238–246.

Fröberg, C. E. 1969. *Introduction to Numerical Analysis.* 2nd ed. Reading, MA: Addison-Wesley.

Gaffney, P. 1987. "When things go wrong..." Report BSC87/1. Bergen, Norway: IBM Bergen Scientific Centre.

Galeone, L. 1977. "Generalizzazione del methodo di Laguerre." *Calcolo* **14**, 121–131.

Garbow, B. S., J. M. Boyle, J. J. Dongarra, and C. B. Moler. 1972. *Matrix Eigensystem Routines:* `EISPACK` *Guide Extension.* New York: Springer-Verlag.

Garcia, C. B., and F. J. Gould. 1980. "Relations between several path-following algorithms and local and global Newton methods." *SIAMREV* **22**, 263–274.

Garcia, C. B., and W. I. Zangwill. 1981. *Pathways to Solutions, Fixed Points, and Equilibria.* Englewood Cliffs, NJ: Prentice-Hall.

Gardner, M. 1961. *Mathematical Puzzles and Diversions.* New York: Simon & Schuster.

Gasca, M., and J. I. Maeztu. 1982. "On Lagrange and Hermite interpolation in $\mathbb{R}^k$." *NM* **39**, 1–14.

Gautschi, W. 1961. "Recursive computation of certain integrals." *ACMJ* **8**, 21–40.

Gautschi, W. 1967. "Computational aspects of three-term recurrence relations." *SIAMREV* **9**, 24–82.

Gautschi, W. 1975. "Computational methods in special functions." In *Theory and Applications of Special Functions* (R. Askey, ed.). New York: Academic Press, 1–98.

Gautschi, W. 1976. "Advances in Chebyshev quadrature." In *Numerical Analysis* (G. A. Watson, ed.). *Lecture Notes in Mathematics* **506**. New York: Springer-Verlag.

Gautschi, W. 1979. "Families of algebraic test equations." *Calcolo* **16**, 383–398.

Gautschi, W. 1983. "How and how not to check Gaussian quadrature formulae." *BIT* **23**, 209–216.

Gautschi, W. 1984. "Questions of numerical condition related to polynomials." In *Studies in Numerical Analysis* (G. H. Golub, ed.), 140–177. Washington, DC: MAA.

Gear, C. W. 1971. *Numerical Initial Value Problems in Ordinary Differential Equations.* Englewood Cliffs, NJ: Prentice-Hall.

Gekeler, E. 1984. *Discretization Methods for Stable Initial Value Problems: Lecture Notes in Mathematics* **1044**. New York: Springer-Verlag.

Gentleman, W. M. 1972. "Implementing Clenshaw-Curtis quadrature." *ACMJ* **15**, 337–342.

George, A., and J. W. Liu. 1981. *Computer Solution of Large Sparse Positive Definite Systems.* Englewood Cliffs, NJ: Prentice-Hall.

George, A., J. W. Liu, and E. Ng. 1980. "User guide for `SPARSPACK`: Waterloo sparse linear equations package." Computer Science Dept. Report CS-78-30 (revised 1980). Waterloo, Canada: University of Waterloo.

Gerald, C. F., and P. O. Wheatley. 1989. *Applied Numerical Analysis.* 4th ed. Reading, MA: Addison-Wesley.

Ghizetti, A., and A. Ossiccini. 1970. *Quadrature Formulæ.* New York: Academic Press.

Gill, P. E., G. H. Golub, W. Murray, and M. A. Saunders. 1974. "Methods for modifying matrix factorizations." *MC* **28**, 505–535.

Gill, P. E., and W. Murray. 1974. "Newton-type methods for unconstrained and linearly constrained optimization." *Mathematical Programming* **28**, 311–350.

Gill, P. E., W. Murray, and M. H. Wright. 1981. *Practical Optimization.* New York: Academic Press.

Gladwell, I., L. F. Shampine, and R. W. Brankin. 1987. "Automatic selection of the initial stepsize for an ODE solver." *JCAM* **18**, 175–192.

Gladwell, J., and R. Wait. 1979. *A Survey of Numerical Methods for Partial Differential Equations.* New York: Oxford University Press.

Glatz, G. 1978. "Stabile Deflationsalgorithmen bei der numerischen Berechnung von Polynomnullstellen." *Zeitschrift für Angewandte Mathematik und Mechanik* **58**, T416–T418.

Glieck, J. 1987. *Chaos.* New York: Viking Press.

Goldstein, A. A. 1966. *Constructive Real Analysis.* New York: Harper & Row.

Goldstine, H. H. 1977. *A History of Numerical Analysis from the 16th Through the 19th Century.* New York: Springer-Verlag.

Golub, G. H. (ed.). 1984. *Studies in Numerical Analysis.* Washington, DC: MAA.

Golub, G. H., and D. P. O'Leary. 1989. "Some history of the conjugate gradient and Lanczos methods." *SIAMREV* **31**, 50–102.

Golub, G. H., and J. M. Ortega. 1992. *Scientific Computing and Differential Equations.* New York: Academic Press.

Golub, G. H., and C. F. Van Loan. 1980. "An analysis of the total least squares problem." *SIAMNA* **17**, 883–893.

Golub, G. H., and C. F. Van Loan. 1989. *Matrix Computations.* 2nd ed. Baltimore, MD: Johns Hopkins University Press.

Gonzaga, C. C. 1992. "Path-following methods for linear programming." *SIAMREV* **34**, 167–224.

Good, I. J. 1972. "What is the most amazing approximate integer in the universe?" *Pi Mu Epsilon Journal* **5**, 314–315.

Gordon, W. J., and J. A. Wixom. 1978. "Shepard's method of 'metric interpolation' to bivariate and multivariate interpolation." *MC* **32**, 253–264.

Gould, N. 1991. "On growth in Gaussian elimination with complete pivoting." *SIAMMAA* **12**, 354–361.

Gourlay, A. R., and G. A. Watson. 1973. *Computational Methods for Matrix Eigenvalues.* New York: Wiley.

Greenspan, D. 1965. *Introductory Numerical Analysis of Elliptic Boundary Value Problems.* New York: Harper & Row.

Gregory, J. A. (ed.). 1986. *The Mathematics of Surfaces.* New York: Oxford University Press.

Gregory, R. T. 1980. *Error-Free Computation.* Huntington, NY: Krieger.

Gregory, R. T., and D. Karney. 1969. *A Collection of Matrices for Testing Computational Algorithms.* New York: Wiley.

Griewank, A., and G. F. Corliss. 1991. *Automatic Differentiation of Algorithms: Theory, Implementation, and Applications.* Philadelphia: SIAM.

Griffiths, P., and J. Harris. 1978. *Principles of Algebraic Geometry.* New York: Wiley.

Gustafson, B., and J. Oliger. 1995. *Time Dependent Problems and Difference Equations.* New York: Wiley.

Haar, A. 1918. "Die minkowskische Geometrie und die Annäherung an stetige Funktionen." *Mathematische Annalen* **78**, 294–311.

Haber, S. 1970. "Numerical Evaluation of Multiple Integrals." *SIAMREV* **12**, 481–526.

Haberman, R. 1977. *Mathematical Models.* Englewood Cliffs, NJ: Prentice-Hall.

Hackbusch, W. 1995. *Iterative Solution of Large Sparse Systems of Equations.* New York: Springer-Verlag.

Hackbusch, W., and U. Trottenberg (eds.). 1982. *Multigrid Methods: Lecture Notes in Mathematics* **960**. New York: Springer-Verlag.

Hageman, L. A., and D. M. Young. 1981. *Applied Iterative Methods.* New York: Academic Press.

Hairer, E., S. P. Nörsett, and G. Wanner. 1987. *Solving Ordinary Differential Equations I—Nonstiff Problems.* New York: Springer-Verlag.

Hairer, E., S. P. Nörsett, and G. Wanner. 1991. *Solving Ordinary Differential Equations II—Stiff and Differential-Algebraic Problems.* New York: Springer-Verlag.

Hammerlin, G. (ed.). 1982. *Numerical Integration.* New York: Birkhäuser-Verlag.

Hammersley, J. M., and DC Handscomb. 1964. *Monte Carlo Methods.* London: Methuen.

Hamming, R. W. 1973. *Numerical Methods for Scientists and Engineers.* New York: McGraw-Hill.

Hansen, E. R. 1969. *Topics in Interval Analysis.* New York: Oxford University Press.

Hardy, G. H. 1960. *A Course of Pure Mathematics.* 10th ed. New York: Cambridge University Press.

Hardy, R. L. 1971. "Multiquadric equations of topography and other irregular surfaces." *Journal Geophysical Research* **76**, 1905–1915.

Hart, J. F., E. W. Cheney, C. L. Lawson, H. J. Maehly, C. K. Mesztenyi, J. R. Rice, H. G. Thacher, Jr., and C. Witzgall. 1968. *Computer Approximations.* New York: Wiley. (Reprinted Huntington, NY: Krieger, 1978.)

Hartley, P. H. 1976. "Tensor product approximations to data defined on rectangle meshes in n-space." *CJ* **19**, 348–352.

Heller, D. 1978. "A survey of parallel algorithms in numerical linear algebra." *SIAMREV* **20**, 740–777.

Hennell, M. A., and L. M. Delves (eds.). 1980. *Production and Assessment of Numerical Software.* New York: Academic Press.

Henrici, P. 1962. *Discrete Variable Methods in Ordinary Differential Equations.* New York: Wiley.

Henrici, P. 1963. *Error Propagation for Difference Methods.* New York: Wiley.

Henrici, P. 1964. *Elements of Numerical Analysis.* New York: Wiley.

Henrici, P. 1974. *Applied and Computational Complex Analysis.* (3 volumes) New York: Wiley.

Hestenes, M. R. 1980. *Conjugate Direction Methods in Optimization.* New York: Springer-Verlag.

Hestenes, M. R., and E. Stiefel. 1952. "Methods of conjugate gradient for solving linear systems." *JRNBS* **45**, 409–436.

Hestenes, M. R., and J. Todd. 1991. *Mathematicians Learning to Use Computers.* Special Publication 730. Gaithersburg, MD: National Institute of Standards and Technology.

Hetzel, W. C. (ed.). 1973. *Program Test Methods.* Englewood Cliffs, NJ: Prentice-Hall.

Higham, N. J. 1996. *Accuracy and Stability of Numerical Algorithms.* Philadelphia: SIAM

Higham, N. J., and D. J. Higham. 1989. "Large growth factors in Gaussian elimination with pivoting." *SIAMMAA* **10**, 155–164.

Higham, N. J., and N. Trefethen. 1991. "Complete pivoting conjecture is disproved." *SIAM News* **24**, 9.

Hindmarsh, A. 1980. "`LSODE` and `LSODEI`: Two initial value ordinary differential equations solvers." *ACM Special Interest Group in Numerical Methods Newsletter* **15**, 10–11.

Hirsch, M. W., and S. Smale. 1979. "On algorithms for solving $f(x) = 0$." *Communications of Pure and Applied Mathematics* **32**, 281–312.

Holmes, R. B. 1972. *A Course on Optimization and Best Approximation.* New York: Springer-Verlag.

Horn, R. A., and C. R. Johnson. 1986. *Matrix Analysis.* New York: Cambridge University Press.

Hough, D. 1981. "Applications of the proposed IEEE 754 standard for floating-point arithmetic." *Computer* **14**, 70–74.

Householder, A. S. 1964. *The Theory of Matrices in Numerical Analysis.* New York: Blaisdell. (Reprinted New York: Dover, 1974.)

Householder, A. S. 1970. *The Numerical Treatment of a Single Nonlinear Equation.* New York: McGraw-Hill.

Hull, T. E., W. H. Enright, B. M. Fellen, and A. E. Sedgwick. 1972. "Comparing numerical methods for ordinary differential equations." *SIAMNA* **9**, 603–637.

IEEE. 1981. "A proposed standard for binary floating-point arithmetic: Draft 8.0 of IEEE Task P754." *Computer* **14**, March.

IEEE. 1985. "IEEE standard for binary floating point arithmetic." *ANSI/IEEE Standard* **P754**. New York: IEEE.

IEEE. 1987. "A radix-independent standard for floating-point arithmetic." IEEE Std. 754–1987. New York: IEEE.

IMSL. 1995. *IMSL Library Reference Manual.* Houston: Visual Numerics, Inc.

Isaacson, E., and H. B. Keller. 1966. *Analysis of Numerical Methods.* New York: Wiley.

Iserles, A. 1994. "Numerical analysis of delay differential equations with variable delays," *Annals of Numer. Math.* **1**, 133–152.

Jackson, K. R., W. H. Enright, and T. E. Hull. 1978. "A theoretical criterion for comparing Runge-Kutta formulas." *SIAMNA* **15**, 618–641.

Jacobs, D. (ed.). 1978. *Numerical Software—Needs and Availability.* New York: Academic Press.

Jain, M. K. 1984. *Numerical Solution of Differential Equations.* 2nd ed. New York: Wiley.

Jenkins, M. A., and J. F. Traub. 1970a. "A three-stage algorithm for real polynomials using quadratic iteration." *SIAMNA* **7**, 545–566.

Jenkins, M. A., and J. F. Traub. 1970b. "A three-stage variable-shift iteration for polynomial zeros." *NM* **14**, 252–263.

Jennings, A. 1977. *Matrix Computations for Engineers and Scientists.* New York: Wiley.

Jerome, J. W. 1985. "Approximate Newton methods and homotopy for stationary operator equations." *Constructive Approximation* **1**, 271–285.

Johnson, L. W., and R. D. Riess. 1982. *Numerical Analysis.* 2nd ed. Reading, MA: Addison-Wesley.

Joubert, W. D., G. F. Carey, N. A. Berner, A. Kalhan, H. Kohli, A. Lorber, R. T. Mclay, and Y. Shen. 1995. "PCG Reference Manual." Center for Numerical Analysis Report CNA–274. Austin, TX: University of Texas at Austin.

Joyce, D. 1971. "Survey of extrapolation processes in numerical analysis." *SIAMREV* **13**, 435–490.

Kahan, W. 1967. "Laguerre's method and a circle which contains at least one zero of a polynomial." *SIAMNA* **4**, 474–482.

Kahan, W. 1993. "A fear of constants and disdain for singularities." Berkeley, CA: University of California at Berkeley.

Kahaner, D. 1970. "Matrix description of the fast Fourier transform." *IEEE Transactions Audio and Electroacoustics* **AU-18**, 422–450.

Kahaner, D. 1978. "The fast Fourier transform by polynomial evaluation." *ZAMP* **29**, 387–394.

Kahaner, D., C. Moler, and S. Nash. 1989. *Numerical Methods and Software.* Englewood Cliffs, NJ: Prentice-Hall.

Kaps, P., and P. Rentrop. 1979. "Generalized Runge-Kutta methods of order four with step size control for stiff ordinary differential equations." *NM* **33**, 55–68.

Karmarkar, N. 1984. "A new polynomial-time algorithm for linear programming." *Combinatorica* **4**, 373–395.

Karon, J. M. 1978. "Computing improved Chebyshev approximations by the continuation method." *SIAMNA* **15**, 1269–1288.

Kearfott, R. B., M. Dawande, K. Du, and C. Hu. 1994. "A portable Fortran 77 interval standard function library." *ACMTOMS* **20**, 447–459.

Keller, H. B. 1968. *Numerical Methods for Two-Point Boundary-Value Problems.* Waltham, MA: Blaisdel.

Keller, H. B. 1976. *Numerical Solution of Two-Point Boundary Value Problems.* Philadelphia: SIAM.

Keller, H. B. 1977. "Numerical solution of bifurcation and nonlinear eigenvalue problems." In *Applications of Bifurcation Theory* (P. Rabinowitz, ed.), 359–384. New York: Academic Press.

Keller, H. B. 1978. "Global homotopies and Newton methods." In *Recent Advances in Numerical Analysis* (C. de Boor and G. H. Golub, eds.), 73–94. New York: Academic Press.

Kelley, C. T. 1995. *Iterative Methods for Linear and Nonlinear Equations.* Philadelphia: SIAM.

Kennedy, W. J., and J. E. Gentle. 1988. *Statistical Computing.* New York: Dekker.

Kernighan, B. W., and P. J. Plauger. 1974. *The Elements of Programming Style.* New York: McGraw-Hill.

Khovanskii, A. N. 1963. *The Application of Continued Fractions and Their Generalizations to Problems in Approximation Theory.* Groningen: Noordhoff.

Kincaid, D. R., and T. C. Oppe. 1988. "A parallel algorithm for the general *LU* factorization." *Communications Applied Numerical Methods* **4**, 349–359.

Kincaid, D. R., T. C. Oppe, and D. M. Young. 1989. "`ITPACKV 2D` user's guide." Center for Numerical Analysis Report CNA–232. Austin, TX: University of Texas at Austin.

Kincaid, D. R., J. R. Respess, D. M. Young, and R. G. Grimes. 1982. "`ITPACK 2C`: A Fortran package for solving large sparse linear systems by adaptive accelerated iterative methods." *ACMTOMS* **8**, 302–322.

Kincaid, D. R., and D. M. Young. 1979. "Survey of iterative methods." In *Encyclopedia of Computer Science and Technology* (J. Belzer, A. G. Holzman, and A. Kent, eds.), 354–391. New York: Dekker.

Kline, M. 1972. *Mathematical Thought from Ancient to Modern Times.* New York: Oxford University Press.

Knuth, D. E. 1969. *The Art of Computer Programming: Seminumerical Algorithms.* Vol. 2. Reading, MA: Addison-Wesley.

Knuth, D. E. 1979. "Mathematical typography." *Bulletin AMS* **2**, 337–372.

Krogh, F. 1970. "`VODQ/SVDQ/DVDQ`: Variable order integrators for the numerical solution of ordinary differential equations." Jet Propulsion Laboratory Technical Brief NPO–11643. Pasadena, CA: California Institute of Technology.

Krylov, V. I. 1962. *Approximate Calculation of Integrals.* (Transl.: A. Stroud) New York: Macmillan.

Kuang, Y. 1993. *Delay Differential Equations.* New York: Academic Press.

Kulisch, U., and W. Miranker. 1981. *Computer Arithmetic in Theory and Practice.* New York: Academic Press.

Lakshmikantham, V., and D. Trigiante. 1988. *Theory of Difference Equations, Numerical Methods and Examples.* New York: Academic Press.

Lambert, J. 1973. *Computational Methods in Ordinary Differential Equations.* New York: Wiley.

Lancaster, P. 1966. "Error analysis for the Newton-Raphson method." *NM* **9**, 55–68.

Lancaster, P., and K. Salkauskas. 1986. *Curve and Surface Fitting.* New York: Academic Press.

Lancaster, P., and M. Tismenetsky. 1985. *Theory of Matrices.* 2nd ed. New York: Academic Press.

Lanczos, C. 1966. *Discourse on Fourier Series.* Edinburgh: Oliver and Boyd.

Lapidus, L., and W. E. Schiesser. 1976. *Numerical Methods for Differential Equations.* New York: Academic Press.

Lapidus, L., and J. Seinfield. 1971. *Numerical Solution of Ordinary Differential Equations.* New York: Academic Press.

Lau, H. T. 1994. *A Numerical Library in C for Scientist and Engineers.* Boca Raton, FL: CRC Press.

Laurie, D. 1978. "Automatic numerical integration over a triangle." Technical Report. Pretoria, South Africa: National Research Center for Mathematical Sciences.

Lawson, C. L., and R. J. Hanson. 1995. *Solving Least Squares Problems.* Philadelphia: SIAM.

Lawson, C. L., R. J. Hanson, D. R. Kincaid, and F. T. Krogh. 1979. "Basic linear algebra subprograms for Fortran usage." *ACMTOMS* **5**, 308–323.

Le, D. 1985. "An efficient derivative-free method for solving nonlinear equations." *ACMTOMS* **11**, 250–262.

Lee, S. L. 1994. "A note on the total least squares problem for coplanar points." Mathematical Sciences Section Report ORNL/TM-12852. Oak Ridge, TN: Oak Ridge National Laboratory.

Levin, M. 1982. "An iterative method for the solution of systems of nonlinear equations." *Analysis* **2**, 305–313.

Li, T. -Y. 1987. "Solving polynomial systems." *MI* **9**, 33–39.

Linear, P. 1979. *Theoretical Numerical Analysis.* New York: Wiley.

Longley, J. W. 1984. *Least Squares Computations Using Orthogonalization Methods.* New York: Dekker.

Lorentz, G. G., K. Jetter, and S. D. Riemenschneider. 1983. *Birkhoff Interpolation.* Reading, MA: Addison-Wesley.

Lozier, D. W., and F. W. J. Olver. 1994. "Numerical evaluation of special functions." In *Mathematics of Computation 1943–1993: A Half-Century of Computational Mathematics* **48**, 79–125. Providence, RI : AMS.

Luenberger, D. G. 1973. *Introduction to Linear and Nonlinear Programming.* Reading, MA: Addison-Wesley.

Lyness, J. 1983. "AUG2: Integration over a triangle." Mathematics and Computer Sciences Report ANL/MCS-TM-13. Argonne, IL: Argonne National Laboratory.

Lyness, J. N., and J. J. Kaganove. 1976. "Comments on the nature of automatic quadrature routines." *ACMTOMS* **2**, 65–81.

Machura, M., and R. Sweet. 1980. "Survey of software for partial differential equations." *ACMTOMS* **6**, 461–488.

MacLeod, M. A. 1973. "Improved computation of cubic natural splines with equi-spaced knots." *MC* **27**, 107–109.

Mandelbrot, B. 1982. *The Fractal Geometry of Nature.* New York: Freeman.

Marchuk, G. I. 1994. *Numerical Methods and Applications.* Boca Raton, FL: CRC Press.

Marden, M. 1949. *The Geometry of the Zeros of a Polynomial in a Complex Variable.* Providence, RI: AMS.

Marden, M. 1966. *Geometry of Polynomials.* Providence, RI: AMS.

Maron, M. J., and R. J. Lopez. 1991. *Numerical Analysis: A Practical Approach.* 3rd ed. Belmont, CA: Wadsworth.

Marsaglia, G. 1968. "Random numbers fall mainly in the planes." *Proceedings National Academy Sciences* **61**, 25–28.

Marsden, M. J. 1970. "An identity for spline functions with applications to variation-diminishing spline approximation." *JAT* **3**, 7–49.

März, R. 1991. "Numerical methods for differential algebraic equations." *AN* 141–198.

McCormick, S. F. 1987. *Multigrid Methods.* Philadelphia: SIAM.

McKeenman, W. M. 1962. "Algorithm 145: Adaptive numerical integration by Simpson's rule." *ACMCOM* **5**, 604.

McNamee, J. M. 1985. "Numerical differentiation of tabulated functions with automatic choice of step-size." *IJNME* **21**, 1171–1185.

Meinguet, J. 1983. "Refined error analysis of Cholesky factorization." *SIAMNA* **20**, 1243–1250.

Meissner, L. P., and E. I. Organick. 1980. FORTRAN77: *Featuring Structured Programming.* Reading, MA: Addison-Wesley.

Meyer, G. H. 1968. "On solving nonlinear equations with a one-parameter operator embedding." *SIAMNA* **5**, 739–752.

Micchelli, C. A. 1986a. "Algebraic aspects of interpolation." In *Approximation Theory* (C. de Boor, ed.) Proceedings of Symposia in Applied Mathematics **36**, 81–102. Providence, RI: AMS.

Micchelli, C. A. 1986b. "Interpolation of scattered data: Distance matrices and conditionally positive definite functions." *Constructive Approximation* **2**, 11–22.

Micchelli, C. A., and T. J. Rivlin (eds.). 1977. *Optimal Estimation in Approximation Theory.* New York: Plenum.

Mickens, R. E. 1987. *Difference Equations.* New York: van Nostrand-Reinhold.

Milne, W. E. 1970. *Numerical Solution of Differential Equations.* New York: Dover.

Miranker, W. L. 1980. *Numerical Methods for Stiff Equations and Singular Perturbation Problems.* Boston: Reidel.

Mitchell, A. 1969. *Computational Methods in Partial Differential Equations.* New York: Wiley.

Mitchell, A. R., and R. Wait. 1977. *The Finite Element Method in Partial Differential Equations.* New York: Wiley.

Moler, C. B., and L. P. Solomon. 1970. "Use of splines and numerical integration in geometrical acoustics." *Journal Acoustical Society America* **48**, 739–744.

Moler, C. B., and C. F. Van Loan. 1978. "Nineteen dubious ways to compute the exponential of a matrix." *SIAMREV* **20**, 801–836.

Moore, R., 1966. *Interval Analysis.* Englewood Cliffs, NJ: Prentice Hall.

Moore, R. E. 1975. *Mathematical Elements of Scientific Computing.* New York: Holt, Reinhart & Winston.

Moore, R. E. 1979. *Methods and Applications of Interval Analysis.* Philadelphia: SIAM.

Moore, R. E. 1994. "Numerical solution of differential equations to prescribed accuracy." *Computers Math. Applic.* **28**, 253–261.

Moré, J. J., B. S. Garbow, and K. E. Hillstrom. 1980. "User guide for `MINIPACK-1`." Mathematics and Computer Sciences Report ANL-80-74. Argonne, IL: Argonne National Laboratory.

Morgan, A. 1986. "A homotopy for solving polynomial systems." *Applied Mathematical and Computation* **18**, 87–92.

Morgan, A. 1987. *Solving Polynomial Systems Using Continuation for Engineering and Scientific Problems.* Englewood Cliffs, NJ: Prentice-Hall.

Morton, K. W., and D. F. Mayers. 1995. *Numerical Solution of Partial Differential Equations.* Cambridge: Cambridge University Press.

Moses, J. 1971. "Symbolic integration: The stormy decade." *ACMCOM* **14**, 548–560.

Murota, K., and M. Iri. 1982. "Parameter tuning and repeated application of the IMT type transformation in numerical quadrature." *NM* **38**, 347–363.

Murtagh, B. A., and M. Saunders. 1978. "Large-scale linearly constrained optimization." *Mathematical Programming* **14**, 41–72.

NAG. 1995. *NAG Fortran Library Manual.* Downers Grove, IL: NAG, Inc.

Nazareth, J. L. 1986. "Homotopy techniques in linear programming." *Algorithmica* **1**, 529–535.

Nazareth, J. L. 1987. *Computer Solution of Linear Programs.* New York: Oxford University Press.

Nerinckx, D., and A. Haegemans. 1976. "A comparison of nonlinear equation solvers." *JCAM* **2**, 145–148.

Neumaier, A. 1990. *Interval Methods for Systems of Equations.* New York: Cambridge University Press.

Newman, D. J., and T. J. Rivlin. 1983. "Optimal universally stable interpolation." *Analysis* **3**, 355–367.

Niederreiter, H. 1978. "Quasi-Monte Carlo methods." *Bulletin AMS* **84**, 957–1041.

Nielson, G. M. 1974. "Some piecewise polynomial alternatives to splines under tension." In *Computer Aided Geometric Design* (R. E. Barnhill and R. F. Riesenfeld, eds.), 209–235. New York: Academic Press.

Nievergelt, Y. 1991. "Numerical linear algebra on the HP–28 or how to lie with supercalculators." *AMM* **98**, 539–544.

Nievergelt, Y. 1994. "Total least squares: State-of-the-art regression in numerical analysis." *SIAMREV* **36**, 258–264.

Nievergelt, J., J. G. Farrar, and E. M. Reingold. 1974. *Computer Approaches to Mathematical Problems.* Englewood Cliffs, NJ: Prentice-Hall.

Noble, B., and J. W. Daniel. 1988. *Applied Linear Algebra.* 3rd ed. Englewood Cliffs, NJ: Prentice-Hall.

Novak, E., K. Ritter, and H. Wozniakowski. 1995. "Average-case optimality of a hybrid secant-bisection method." *MC* **64**, 1517–1540.

Nussbaumer, H. J. 1982. *Fast Fourier Transform and Convolution Algorithms.* New York: Springer.

Oden, J. T. 1972. *Finite Elements of Nonlinear Continua.* New York: McGraw-Hill.

Oden, J. T., and J. N. Reddy. 1976. *An Introduction to the Mathematical Theory of Finite Elements.* New York: Wiley.

Oppe, T. C., W. D. Joubert, and D. R. Kincaid. 1988. "NSPCG user's guide, version 1. 0, package for solving large sparse linear systems by various iterative methods." Center for Numerical Analysis Report CNA–216. Austin, TX: University of Texas at Austin.

Oppe, T. C., and D. R. Kincaid. 1988. "Parallel *LU*-factorization algorithms for dense matrices." In *Supercomputing* (E. N. Houstis, T. S. Papatheodorou, and C. D. Polychronopoulos, eds.), *Lecture Notes in Computer Science* **297**, 576–594. New York: Springer-Verlag.

Ortega, J. M. 1972. *Numerical Analysis: A Second Course.* New York: Academic Press. (Reprinted Philadelphia: SIAM, 1990.)

Ortega, J. M. 1988. *Introduction to Parallel and Vector Solution of Linear Systems.* New York: Plenum.

Ortega, J. M., and W. C. Poole. 1986. *An Introduction to Numerical Methods for Differential Equations.* New York: Wiley.

Ortega, J., and W. C. Rheinboldt. 1970. *Iterative Solution of Nonlinear Equations in Several Variables.* New York: Academic Press.

Osborne, M. R. 1966. "On Nordsieck's method for the numerical solution of ordinary differential equations." *BIT* **6**, 51–57.

Ostrowski, A. M. 1966. *Solution of Equations and Systems of Equations.* 2nd ed. New York: Academic Press.

Paddon, D. J., and H. Holstein. 1985. *Multigrid Methods for Integral and Differential Equations.* New York: Oxford University Press.

Pan, V. 1984. "How can we speed up matrix multiplication?" *SIAMREV* **26**, 393–415.

Parlett, B. N. 1964. "Laguerre's method applied to the matrix eigenvalue problem." *MC* **18**, 464–485.

Parlett, B. N. 1981. *The Symmetric Eigenvalue Problem.* Englewood Cliffs, NJ: Prentice-Hall.

Parter, S. 1985. *Large Scale Scientific Computation.* New York: Academic Press.

PCGPAK2. 1990. "PCGPAK2 user's guide." New Haven, CT: Scientific Computing Associates, Inc.

Pearson, K. 1901. "On lines and planes of closest fit to points in space." *Phil. Mag.* **2**, 559–572.

Peitgen, H., and P. Richter. 1986. *The Beauty of Fractals.* New York: Springer-Verlag.

Peitgen, H.-O., D. Saupe, and F. V. Haeseler. 1984. "Cayley's problem and Julia sets." *MI* **6**, 11–20.

Penrose, R. 1955. "A generalized inverse for matrices." *Proceedings Cambridge Phil. Society* **51**, 406–413.

Pereyra, V. 1984. "Finite difference solution of boundary value problems in ordinary differential equations." In *Studies in Numerical Analysis* (G. H. Golub, ed.), 243–269. Washington, DC: MAA.

Perron, O. 1929. *Die Lehre von Kettenbrüchen.* Leipzig: Teubner. (Reprinted New York: Chelsea.)

Peters, G., and J. H. Wilkinson. 1971. "Practical problems arising in the solution of polynomial equations." *IMAJNA* **8**, 16–35.

Phillips, G. M., and P. J. Taylor. 1973. *Theory and Applications of Numerical Analysis.* New York: Academic Press.

Pickering, M. 1986. *An Introduction to Fast Fourier Transform Methods for Partial Differential Equations, With Applications.* New York: Wiley.

Pickover, C. A. 1988. "A note on chaos and Halley's methods." *ACMCOM* (11) **31**, 11.

Piessens, R., E. deDoncker-Kapenga, C. W. Überhuber, and D. H. Kahaner. 1983. QUADPACK: *A Subroutine Package for Automatic Integration.* New York: Springer-Verlag.

Powers, D. 1972. *Boundary Value Problems.* New York: Academic Press.

Prager, W. H. 1988. *Applied Numerical Linear Algebra.* Englewood Cliffs, NJ: Prentice-Hall.

Prenter, P. 1975. *Splines and Variational Methods.* New York: Wiley.

Press, W. H., B. P. Flannery, S. A. Teukolsky, and W. T. Vetterling. 1986. *Numerical Recipes.* New York: Cambridge University Press.

Prince, P. J., and J. R. Dormand. 1981. "High order embedded Runge-Kutta formulæ." *JCAM* **1**, 67–75.

Pritsker, A. 1986. *Introduction to Simulation and SLAM II.* New York: Wiley.

Pruess, S. 1976. "Properties of splines in tension." *JAT* **17**, 86–96.

Pruess, S. 1978. "An algorithm for computing smoothing splines in tension." *Computing* **19**, 365–373.

Rabinowitz, P. 1968. "Applications of linear programming to numerical analysis." *SIAMREV* **10**, 121–159.

Rabinowitz, P. (ed.). 1970. *Numerical Methods for Nonlinear Algebraic Equations.* London: Gordon and Breach.

Raimi, R. A. 1969. "On the distribution of first significant figures." *AMM* **76**, 342–347.

Rall, L. B. 1965. *Error in Digital Computation.* New York: Wiley.

Ralston, A., and C. L. Meek (eds.). 1976. *Encyclopedia of Computer Science.* New York: Petrocelli/Charter.

Ralston, A., and P. Rabinowitz. 1978. *A First Course in Numerical Analysis.* New York: McGraw-Hill.

Rand, R. 1984. *Computer Algebra in Applied Mathematics: An Introduction to* MACSYMA. Boston: Pitman.

Redish, K. A. 1974. "On Laguerre's method." *Int. J. Math. Educ. Sci. Technol.* **5**, 91–102.

Reid, J. K. (ed.). 1971. *Large Sparse Sets of Linear Equations.* New York: Academic Press.

Renka, R. J. 1993. "Algorithm 716: TSPACK–Tension spline curve-fitting package." *ACM-TOMS* **19**, 81–94.

Rheinboldt, W. C. 1974. *Methods for Solving Systems of Nonlinear Equations.* CBMS Series in Applied Mathematics 14. Philadelphia: SIAM.

Rheinboldt, W. C. 1980. "Solution fields of nonlinear equations and continuation methods." *SIAMNA* **17**, 221–237.

Rheinboldt, W. C. 1986. *Numerical Analysis of Parameterized Nonlinear Equations.* New York: Wiley.

Rice, J. R. 1966. "Experiments on Gram-Schmidt orthogonalization." *MC* **20**, 325–328.

Rice, J. R. 1981. *Matrix Computations and Mathematical Software.* New York: McGraw-Hill.

Rice, J. R. 1992. *Numerical Methods, Software, and Analysis.* 2nd ed. New York: Academic Press.

Rice, J. R., and R. F. Boisvert. 1985. *Elliptic Problem Solving Using* ELLPACK. New York: Springer-Verlag.

Rice, J. R., and J. S. White. 1964. "Norms for smoothing and estimation." *SIAMREV* **6**, 243–256.

Richtmeyer, R. D., and K. W. Morton. 1967. *Difference Methods for Initial Value Problems.* New York: Wiley.

Rivlin, T. J. 1990. *The Chebyshev Polynomials.* 2nd ed. New York: Wiley.

Roache, P. 1972. *Computational Fluid Dynamics.* Albuquerque, NM: Hermosa.

Roberts, S., and J. Shipman. 1972. *Two-Point Boundary Value Problems: Shooting Methods.* New York: Elsevier.

Rockafellar, R. T. 1970. *Convex Analysis.* Princeton, NJ: Princeton University Press.

Rose, D. J. 1975. "A simple proof for partial pivoting." *AMM* **82**, 919–921.

Rose, D. J., and R. A. Willoughby (eds.). 1972. *Sparse Matrices and Their Applications.* New York: Plenum.

Rosenfeld, A., and A. Kak. 1982. *Digital Picture Processing.* New York: Academic Press.

Ross, S. 1983. *Stochastic Processes.* New York: Wiley.

Rousseau, C. 1995. "The phi number system revisited." *Math. Mag.* **68**, 283–284.

Roy, M. R. 1985. *A History of Computing Technology.* Englewood Cliffs, NJ: Prentice-Hall.

Royden, H. L. 1968. *Real Analysis.* 2nd ed. New York: Macmillan.

Rozema, E. R. 1987. "Romberg integration by Taylor series." *AMM* **94**, 284–288.

Rubinstein, R. 1981. *Simulation and the Monte Carlo Method.* New York: Wiley.

Ryder, B. G. 1974. "The PFORT verifier." *Software Practice and Experience* **4**, 359–378.

Saaty, T. L. 1981. *Modern Nonlinear Equations.* New York: Dover.

Salamin, E. 1976. "Computation of π using arithmetic-geometric mean." *MC* **30**, 565–570.

Sander, L. M. 1987. "Fractal growth." *SA,* January **256**, 94–100.

Sard, A. 1963. *Linear Approximation.* Mathematical Surveys, No. 9. Providence, RI: AMS.

Schechter, M. 1984. "Summation of divergent series by computer." *AMM* **91**, 629–632.

Scheid, F. 1988. *Numerical Analysis.* New York: McGraw-Hill.

Schendel, U. 1984. *Introduction to Numerical Methods for Parallel Computers.* New York: Wiley.

Schiesser, W. E. 1994. *Computational Mathematics in Engineering and Applied Science.* Boca Raton, FL: CRC Press.

Schnabel, R. B., and P. D. Frank. 1984. "Tensor methods for nonlinear equations." *SIAMNA* **21**, 815–843.

Schnabel, R. B., J. E. Koontz, and B. E. Weiss. 1982. "A modular system of algorithms for unconstrained minimization." Computer Science Department Report CU-CS-240-82. Boulder, CO: University of Colorado.

Schoenberg, I. J. 1946. "Contributions to the problem of approximation of equidistant data by analytic functions." *Quarterly Applied Mathematics* **4**, 45–99 and 112–133.

Schoenberg, I. J. 1967. "On spline functions." In *Inequalities* (O. Shisha, ed.), 255–291. New York: Academic Press.

Schoenberg, I. J. 1982. *Mathematical Time Exposures* **5**, 132–138. Washington, DC: MAA.

Schrage, L. 1979. "A more portable Fortran random number generator." *ACMTOMS* **5**, 132–138.

Schultz, M. H. 1973. *Spline Analysis.* Englewood Cliffs, NJ: Prentice-Hall.

Schumaker, L. L. 1976. "Fitting surfaces to scattered data." In *Approximation Theory II* (G. G. Lorentz, C. K. Chui, and L. L. Schumaker, eds.), 203–268. New York: Academic Press.

Schumaker, L. L. 1981. *Spline Functions.* New York: Wiley-Interscience.

Schweikert, D. G. 1966. "An interpolation curve using splines in tension." *Journal Mathematics & Physics* **45**, 312–317.

Scott, N. R. 1985. *Computer Number Systems and Arithmetic.* Englewood Cliffs, NJ: Prentice-Hall.

Shampine, L. F. 1994. *Numerical Solution of Ordinary Differential Equations.* New York: Chapman and Hall.

Shampine, L. F., and R. C. Allen. 1973. *Numerical Computing: An Introduction.* Philadelphia: Saunders.

Shampine, L. F., and C. Baca. 1984. "Error estimators for stiff differential equations." *JCAM* **2**, 197–208.

Shampine, L. F., and P. Bogacki. 1989. "The effect of changing the stepsize in linear multistep codes." *SIAMSSC* **10**, 1010–1023.

Shampine, L. F., and C. W. Gear. 1979. "A user's view of solving stiff ordinary differential equations." *SIAMREV* **21**, 1–17.

Shampine, L. F., and M. K. Gordon. 1975. *Computer Solution of Ordinary Differential Equations: The Initial Value Problem.* San Francisco: Freeman.

Shampine, L. F., H. A. Watts, and S. M. Davenport. 1976. "Solving nonstiff ordinary differential equations—The state of the art." *SIAMREV* **18**, 376–411.

Shepard, D. 1968. "A two-dimensional interpolation function for irregularly spaced data." *Proceedings 23rd National Conference ACM,* 517–524.

Shikin, E. V. 1995. *Handbook and Atlas of Curves.* Boca Raton, FL: CRC Press.

Skeel, R. D. 1979. "Equivalent forms of multistep formulas." *MC* **33**, 1229–1250.

Skeel, R. D. 1981. "Effect of equilibration on residual size for partial pivoting." *SIAMNA* **18**, 449–454.

Sloan, I. H., and S. Joe. 1994. *Lattice Methods for Multiple Integration.* New York: Oxford University Press.

Smale, S. 1981. "The fundamental theorem of algebra and complexity theory." *Bulletin AMS* **4**, 1–36.

Smale, S. 1986. "Algorithms for solving equations." In *Proceedings International Congress of Mathematicians* (A. M. Gleason, ed.), 172–195, Providence, RI: AMS.

Smith, B. T., J. M. Boyle, B. S. Garbow, Y. Ikebe, V. C. Klema, and C. B. Moler. 1976. *Matrix Eigensystem Routines—*`EISPACK` *Guide.* 2nd ed.: *Lecture Notes in Computer Science* **6**. New York: Springer-Verlag.

Smith, G. D. 1965. *Numerical Solution of Partial Differential Equations.* New York: Oxford University Press.

Smith, K. T. 1971. *Primer of Modern Analysis.* New York: Springer-Verlag.

Sobolev, S. L. 1992. *Cubature Formulas and Modern Analysis.* Philadelphia: Gordon and Breach.

Sorenson, DC 1985. "Analysis of pairwise pivoting in Gaussian elimination." *IEEE Transactions Computing* **34**, 274–278.

Steinberg, D. I. 1975. *Computational Matrix Algebra.* New York: McGraw-Hill.

Sternbenz, P. H. 1974. *Floating-Point Computations.* Englewood Cliffs, NJ: Prentice-Hall.

Stetter, H. J. 1973. *Analysis of Discretization Methods for Ordinary Differential Equations.* New York: Springer-Verlag.

Stewart, G. W. 1973. *Introduction to Matrix Computations.* New York: Academic Press.

Stewart, G. W. 1985. "A note on complex division." *ACMTOMS* **11**, 238–341.

Stoer, J., and R. Bulirsch. 1980. *Introduction to Numerical Analysis.* New York: Springer-Verlag.

Strang, G., and G. Fix. 1973. *An Analysis of the Finite Element Method.* Englewood Cliffs, NJ: Prentice-Hall.

Street, R. L. 1973. *The Analysis and Solution of Partial Differential Equations.* Pacific Grove, CA: Brooks/Cole.

Stroud, A. H. 1965. "Error estimates for Romberg quadrature." *SIAMNA* **2**, 480–488.

Stroud, A. H. 1971. *Approximate Calculation of Multiple Integrals.* Englewood Cliffs, NJ: Prentice-Hall.

Stroud, A. H. 1974. *Numerical Quadrature and Solution of Ordinary Differential Equations.* New York: Springer-Verlag.

Stroud, A. H., and D. Secrest. 1966. *Gaussian Quadrature Formulas.* Englewood Cliffs, NJ: Prentice-Hall.

Subbotin, Y. N. 1967. "On piecewise-polynomial approximation." *Mat. Zametcki* **1**, 63–70. (Transl.: 1967. *Mathematical Notes* **1**, 41–46.)

Swarztrauber, P. N. 1975. "Efficient subprograms for the solution of elliptic partial differential equations." Report TN/LA-109. Boulder, CO: National Center for Atmospheric Research.

Swarztrauber, P. N. 1982. "Vectorizing the FFT's parallel computations." In *Parallel Computations* (G. Rodrigue, ed.). New York: Academic Press.

Swarztrauber, P. N. 1984. "Fast Poisson solvers." In *Studies in Numerical Analysis* (G. H. Golub, ed.), 319–370. Washington, DC: MAA.

Taussky, O. 1949. "A remark concerning the characteristic roots of finite segments of the Hilbert matrix." *Oxford Quarterly Journal Mathematics* **20**, 82–83.

Taussky, O. 1988. "How I became a torchbearer for matrix theory." *AMM* **95**, 801–812.

Taylor, J. R. 1982. *An Introduction to Error Analysis.* New York: University Science Books.

Tewarson, R. P. 1973. *Sparse Matrices.* New York: Academic Press.

Thomas, B. 1986. "The Runge-Kutta methods." *Byte,* April, 191–210.

Todd, J. 1961. "Computational problems concerning the Hilbert matrix." *JRNBS* **65**, 19–22.

Todd, M. J. 1982. "An introduction to piecewise linear homotopy algorithms for solving systems of equations." In *Topics in Numerical Analysis* (P. R. Turner, ed.) *Lecture Notes in Mathematics* **965**, 147–202. New York: Springer-Verlag.

Törn, A., and A. Zilinska. 1989. *Global Optimization. Lecture Notes in Computer Science* **350**. New York: Springer-Verlag.

Traub, J. F. 1964. *Iterative Methods for the Solution of Equations.* Englewood Cliffs, NJ: Prentice-Hall.

Traub, J. F. 1967. "The calculation of zeros of polynomials and analytic functions." In *Mathematical Aspects of Computer Science. Proceedings Symposium Applied Mathematics* **19**, 138–152. Providence, RI: AMS.

Trefethen, L. N. 1992. "The definition of numerical analysis." Report TR 92–1304. Ithaca, NY: Cornell University.

Trefethen, L. N., and R. S. Schreiber. 1990. "Average-case stability of Gaussian elimination." *SIAMMAA* **11**, 335–360.

Trustrum, K. 1971. *Linear Programming.* London: Routledge and Kegan Paul.

Turner, P. R. 1982. "The distribution of leading significant digits." *IMAJNA* **2**, 407–412.

Van der Corput, J. G. 1946. "Sur l'approximation de Laguerre des racines d'une équation qui a toutes ses racines réelles." *Nederl. Acad. Wetensch. Proc.* **49**, 922–929.

Van Huffel, S., and J. Vandervalle. 1991. *The Total Least Squares Problem: Computational Aspects and Analysis.* Philadelphia: SIAM.

Van Loan, C. F. 1992. *Computational Frameworks for the Fast Fourier Transform.* Philadelphia: SIAM.

Vandergraft, J. S. 1978. *Introduction to Numerical Computations.* New York: Academic Press.

Varga, R. S. 1962. *Matrix Iterative Analysis.* Englewood Cliffs, NJ: Prentice-Hall.

Vemuri, V., and W. J. Karplus. 1981. *Digital Computer Treatment of Partial Differential Equations.* Englewood Cliffs, NJ: Prentice-Hall.

Verner, J. H. 1978. "Explicit Runge-Kutta methods with estimates of the local truncation error." *SIAMNA* **15**, 772–790.

Vichnevetsky, R. 1981, 1982. *Computer Methods for Partial Differential Equations.* Vol. 1: *Elliptic Equations and the Finite Element Method,* Vol. 2: *Initial Value Problems.* Englewood Cliffs, NJ: Prentice-Hall.

Von Petersdorff, T. 1993. "A short proof for Romberg integration." *AMM* **100**, 783–785.

Von Rosenberg, D. U. 1969. *Methods for the Numerical Solution of Partial Differential Equations.* New York: American Elsevier.

Von Seggern, D. 1994. *Practical Handbook of Curve Design and Generation.* Boca Raton, FL: CRC Press.

Wachspress, E. L. 1966. *Iterative Solution of Elliptic Systems.* Englewood Cliffs, NJ: Prentice-Hall.

Wacker, H. G. (ed.). 1978. *Continuation Methods.* New York: Academic Press.

Wait, R., and A. R. Mitchell. 1986. *Finite Element Analysis and Applications.* New York: Wiley.

Walker, J. S. 1992. *Fast Fourier Transforms.* Boca Raton, FL: CRC Press.

Wall, H. S. 1948. *Analytic Theory of Continued Fractions.* Princeton: van Nostrand.

Waser, S., and M. J. Flynn. 1982. *Introduction to Arithmetic for Digital Systems Designers.* New York: Holt, Reinhart & Winston.

Wasserstrom, E. 1973. "Numerical solutions by the continuation method." *SIAMREV* **15**, 89–119.

Watkins, D. S. 1982. "Understanding the QR algorithm." *SIAMREV* **24**, 427–440.

Watson, L. T. 1979. "A globally convergent algorithm for computing fixed points of C^2 maps." *Applications in Mathematical Computing* **5**, 297–311.

Watson, L. T. 1986. "Numerical linear algebra aspects of globally convergent homotopy methods." *SIAMREV* **28**, 529–545.

Watson, L. T., S. C. Billups, and A. P. Morgan. 1987. "HOMPACK: A suite of codes for globally convergent homotopy algorithms." *ACMTOMS* **13**, 281–310.

Weaver, H. J. 1983. *Applications of Discrete and Continuous Fourier Analysis.* New York: Wiley.

Wedin, P. A. 1972a. "Perturbation bounds in connection with the singular value decomposition." *BIT* **12**, 99–111.

Wedin, P. A. 1972b. "Perturbation theory for pseudoinverses." *BIT* **13**, 217–232.
Werner, W. 1984. "Polynomial interpolation: Lagrange versus Newton." *MC* **43**, 205–217.
Wesseling, P. 1992. *An Introduction to Multigrid Methods.* New York: Wiley.
Westfall, R. S. 1980. *Never at Rest: A Biography of Isaac Newton.* New York: Cambridge University Press.
Whitehead, G. W. 1966. *Homotopy Theory.* Cambridge, MA: MIT Press.
Whittaker, E., and G. Robinson. 1924. *The Calculus of Observations.* 4th ed. London: Blackie and Son. (Reprinted New York: Dover, 1967.)
Wilkinson, J. H. 1961. "Error analysis of direct methods of matrix inversion." *ACMJ* **8**, 281–330.
Wilkinson, J. H. 1963. *Rounding Errors in Algebraic Processes.* Englewood Cliffs, NJ: Prentice-Hall.
Wilkinson, J. H. 1965. *The Algebraic Eigenvalue Problem.* New York: Oxford University Press.
Wilkinson, J. H. 1967. "Two algorithms based on successive linear interpolation." Technical Computer Science Department Report STAN-CS-67-60. Stanford, CA: Stanford University.
Wilkinson, J. H. 1984. "The perfidious polynomial." In *Studies in Numerical Analysis* (G. H. Golub, ed.), 1–28. Washington, DC: MAA.
Wilkinson, J. M., and C. Rheinsch (eds.). 1971. *Handbook for Automatic Computation II: Linear Algebra.* New York: Springer-Verlag.
Willé, D. R. 1989. "The numerical solution of delay-differential equations." Department of Mathematics, Ph.D. thesis. Manchester, England: University of Manchester.
Willé, D. R. 1994a. "New stepsize estimators for linear multistep methods." Department of Mathematics, Numerical Analysis Report No. 247. Manchester, England: University of Manchester.
Willé, D. R. 1994b. "Experiments in stepsize control for Adams linear multistep methods." Department of Mathematics, Numerical Analysis Report No. 253. Manchester, England: University of Manchester.
Willé, D. R., and C. T. H. Baker. 1992. "DELSOL – a numerical code for the solution of systems of delay-differential equations." *ANM* **9**, 223–234.
Willoughby, R. A. (ed.). 1974. *Stiff Differential Systems.* New York: Plenum.
Wimp, J. 1984. *Computation with Recurrence Relations.* Boston: Pitman.
Wouk, A. (ed.). 1986. *New Computing Environments: Parallel, Vector, and Systolic.* Philadelphia: SIAM.
Wright, M. 1991. "Interior methods for constrained optimization." *AN* 341–407.
Wright, S. J. 1993. "A collection of problems for which Gaussian elimination with partial pivoting is unstable." *SIAMSSC* **14**, 231–238.
Young, D. M. 1950. "Iterative methods for solving partial difference equations of elliptic type." Ph.D. thesis, Cambridge, MA: Harvard University.
Young, D. M. 1971. *Iterative Solution of Large Linear Systems.* New York: Academic Press.
Young, D. M., and R. T. Gregory. 1972. *A Survey of Numerical Mathematics.* Vols. 1 and 2. Reading, MA: Addison-Wesley. (Reprinted New York: Dover, 1988.)
Young, D. M., and K. C. Jea. 1980. "Generalized conjugate acceleration of nonsymmetrizable iterative methods." *Linear Algebra and Its Applications* **34**, 159–194.
Young, R. M. 1986. "A Rayleigh popular problem." *AMM* **93**, 660.
Zelkowitz, M. V., A. C. Shaw, and J. D. Gannon. 1979. *Principles of Software Engineering and Design.* Englewood Cliffs, NJ: Prentice-Hall.
Zienkiewicz, O. C., and K. Morgan. 1983. *Finite Elements and Approximation.* New York: Wiley.
Zwillinger, D. 1988. *Handbook of Differential Equations.* New York: Academic Press.

Index